CARBOHYDRATES – COMPREHENSIVE STUDIES ON GLYCOBIOLOGY AND GLYCOTECHNOLOGY

Edited by **Chuan-Fa Chang**

Carbohydrates – Comprehensive Studies on Glycobiology and Glycotechnology

http://dx.doi.org/10.5772/2702
Edited by Chuan-Fa Chang

Contributors

Chuan-Fa Chang, Aurore Richel, Michel Paquot, Brighid Pappin, Milton J. Kiefel, Todd A. Houston, Stephanie Bank, Petra Kapková, Josef Jampílek, Jiří Dohnal, Natércia F. Brás, Pedro A. Fernandes, Maria J. Ramos, Nuno M.F.S.A. Cerqueira, Marko Anderluh, Marijke Alen, Dominique Schols, Samuel J. Wadsworth, S. Jasemine Yang, Delbert R. Dorscheid, M. B. Gorishniy, S. P. Gudz, Barbara Niwińska, Katarzyna Śliżewska, Janusz Kapuśniak, Renata Barczyńska, Kamila Jochym, Suman Mishra, Allan Hardacre, John Monro, Alejandro Reyes-DelaTorre, María Teresa Peña-Rangel, Juan Rafael Riesgo-Escovar, Monia El Bour, Anna Novials, Serafín Murillo, Ali Ashraf Jafari, José Luis da Silva Nunes, Paulo Vitor Dutra de Souza, Gilmar Arduino Bettio Marodin, José Carlos Fachinello, Jorge Ernesto de Araújo Mariath, Iwona Morkunas, Sławomir Borek, Magda Formela, Lech Ratajczak, Luis Henrique Souza Guimarães, Amira El-Fallal, Mohammed Abou Dobara, Ahmed El-Sayed, Noha Omar, Stefan Kraan, Edson Holanda Teixeira, Francisco Vassiliepe Sousa Arruda, Kyria Santiago do Nascimento, Victor Alves Carneiro, Celso Shiniti Nagano, Bruno Rocha da Silva, Alexandre Holanda Sampaio, Benildo Sousa Cavada

Published by InTech

Janeza Trdine 9, 51000 Rijeka, Croatia

Copyright © 2012 InTech

Notice

Publishing Process Manager Oliver Kurelic
Typesetting InTech Prepress, Novi Sad
Cover InTech Design Team

Additional hard copies can be obtained from orders@intechopen.com

Carbohydrates – Comprehensive Studies on Glycobiology and Glycotechnology,
Edited by Chuan-Fa Chang
p. cm.
ISBN 978-953-51-0864-1

For EU product safety concerns:
IN TECH d.o.o., Prolaz Marije Krucifikse Kozulić 3, 51000 Rijeka, Croatia,
info@intechopen.com or visit our website at intechopen.com.

Meet the editor

Dr. Chuan-Fa Chang started his research work in Glycobiology for his doctorate at Institute of Biological Chemistry, Academia Sinica. He developed new methodologies for carbohydrate organic synthesis and accomplished series of carbohydrate analogs by using exo-glycals as building blocks. He then joined Dr. Chi-Huey Wong's chemical biology lab as a postdoctoral fellow and worked on discovering glycosidase inhibitors by high-throughput drug screening systems. In 2007, Dr. Chang moved to National Cheng Kung University and initiated his independent research career. He is also the member of Society of Glycobiology (since 2004) and participating investigator of Consortium for Functional Glycomics (CFG, since 2011). Recently, Dr. Chang is interested in investigating the roles of carbohydrates in infectious disease and the regulation of glycosylation in tumorigenesis and cancer metastasis.

Contents

Preface

Carbohydrates consist of carbon, oxygen, hydrogen, and other important modification atoms. They are the major component of plants and comprise more than 40% of energy in human diets. In addition to be the basic energy source and biomaterial building blocks, carbohydrates are also involved in many physiological and pathological events. Because the carbon atom has four valence electrons, it can form chemical bonds with four different atoms. As a consequence, much of constitutional isomers and stereoisomers are found in carbohydrates. In addition, the cyclization of a linear carbohydrate to form five, six or seven membered hemiacetal/hemiketal ring also increases the complexity of glycan structures. Since the late nineteenth century, many chemists devoted their life to figure out these chemical and biochemical questions. The slow development of analytical methods and instruments has prevented the biological roles of carbohydrates from being investigated and characterized until the 1970s. Due to the difficulties of getting a lot of pure and homogeneous carbohydrates from natural sources, carbohydrate organic synthesis has become an essential and flourish subject in Glycobiology.

Because carbohydrates are involved in multiple disciplines, this book consists of 23 excellent chapters, classified into five catogorics Chemistry and Biochemistry, Microbiology and Immunology, Carbohydrate metabolism, Animal and Plant, and Biotechnology. In section one, organic reactions of carbohydrates and the analysis of carbohydrate derivatives are discussed. A high-throughputable carbohydrate microarray without any wash steps is also introduced. This reliable platform not only offers consistent results with literature - it can be applied for the determination of pathogen-carbohydrate interactions. The importance of glycans in infectious disease has been well studied, as well as the immune response and signal transduction pathway. Chapters in section 2 introduce the studies of DC-SIGN antagonists and IL-13 in a viral infection, and the biosynthesis of carbohydrates in a microorganism. The metabolism of carbohydrates in bacteria and drosophila is collected in section 3. This section also discusses the effects of starvation in carbohydrate metabolism in marine bacteria. Section 4 contains some non-popular topics about the digestion of carbohydrates in ruminants, the influences of environment and fungi in plant carbohydrates, and the responses of plants under sugar starvation. The observation, discovery and results of these chapters are specialized and unique. This book is also an encouragement to people who work in the non-mainstream investigations. The final section is the technology and application of carbohydrates in microbes. They can be used as prebiotics, to form biomass, as well as healthy supplements.

I hope this comprehensive book can serve well as an uncomplicated introduction to different disciplines of carbohydrate investigators and glycobiologists. Consulting the literature cited in the chapters is suggested for readers who need further information. The corrections, improvements and suggestions from readers for all the chapters are welcome and appreciated.

I deeply appreciate the contributors to this book and the assistance from InTech in publishing this book.

Chuan-Fa Chang
Department of Medical Laboratory, Science and Biotechnology, Medical College,
National Cheng Kung University,
Tainan city,
Taiwan

Section 1

Chemistry and Biochemistry

Chapter 1

Carbohydrate Microarray

Chuan-Fa Chang

Additional information is available at the end of the chapter

http://dx.doi.org/10.5772/51971

1. Introduction

Glycosylation adorns more than one half of the proteins in eukaryotic cells [1,2]. This post-translational modification plays an indispensible role in many important biological events, especially on cell surface [1,3]. Alterations in carbohydrate structures are known to correlate with the changes in protein stability and clearance, as well as various physiological functions including cell-cell adhesion, inflammation, tumor metastasis, and infection of bacteria and viruses [4-8]. Although glycosylation is essential for the formation and progression of various diseases, study of this subject is hampered by lack of effective tools available to date, in addition to structural heterogeneity and complexity of carbohydrates. A number of techniques have been developed to analyze the binding interactions between carbohydrates and proteins [2,9]. For instances, lectin blotting/binding assay has become a routine method to determine the glycan-protein interactions [10], but the relatively low sensitivity and the necessity of multiple wash steps/time-consuming have restricted the sensitivity and application. Surface plasmon resonance is another highly sensitive method which monitors the interactions in real time and in a quantitative manner [11-16]. However, sometimes the sensitivity is relatively low toward the use of low molecular-weight carbohydrates, though the problem can be overcome by labeling sugars with heavy metal ions [17]. In addition, fluorescence polarization and two-photon fluorescence correlation have been applied to study lectin-glycan interactions [18-21]. The most applicable technique is carbohydrate microarrays which immobilize oligosaccharides to a solid supports are developed and widely used to measure the carbohydrate binding properties of proteins, cells, or viruses [22-25]. For example, a high-content glycan microarray is developed by a robotic microarray printing technology in which amine-functionalized glycans are coupled to the succinimide esters on glass slides [26,27]. These microarrays have also been subjected for profiling the carbohydrate binding specificities of lectins, antibodies, and intact viruses.

2. Carbohydrate microarray

In our recent work, we have developed two novel carbohydrate microarrays: solution microarray [28] and membrane microarray [29]. Carbohydrate solution assay is a high-throughput, homogenous and sensitive method to characterize protein-carbohydrate interactions and glycostructures by in-solution proximity binding with photosensitizers (**Figure 1**). The technology, also called AlphaScreen™, is first described by Ullman et al., and has been used to study interactions between biomolecules [30-34]. In these assays, a light signal is generated when a donor bead and an acceptor bead are brought into proximity. This method usually provides good sensitivity with *femto*-mole detection under optimized conditions, relying on the binding affinity between analytes. All the procedures are carried out in 384-welled microtiter plates, thus qualifying the protocol as high-throughput. Two particles of 200 nm are involved in this technology including streptavidin-coated particles (donor beads) and protein A-conjugated particles (acceptor beads). Biotinylated polyacrylamide (biotin-PAA)-based glycans that are immobilized on donor beads can be recognized by lectins or antibodies, and connected with acceptor beads through specific antibodies (**Figure 1**). A number of carbohydrate binding proteins, including eleven lectins and seven antibodies, are profiled for their carbohydrate binding specificity to validate the efficacy of this developed technology. This assay is performed in homogeneous solutions and does not require extra wash steps, preventing the loss of weak bindings that often occur in the repeating washes of glycan microarray. However, antigen/ligand excess effect may happen in the homogeneous solution assay if the concentrations of carbohydrate epitopes, proteins, or antibodies are too high. One mg of biotin-PAA-sugar can be applied for fifty thousand assays because minimal amount of materials are needed in this microarray system (a range of nano-gram is required per well). Although the detection limit of biotin-PAA-sugar is good (2 ng per well), the linear range is too narrow for quantitative application.

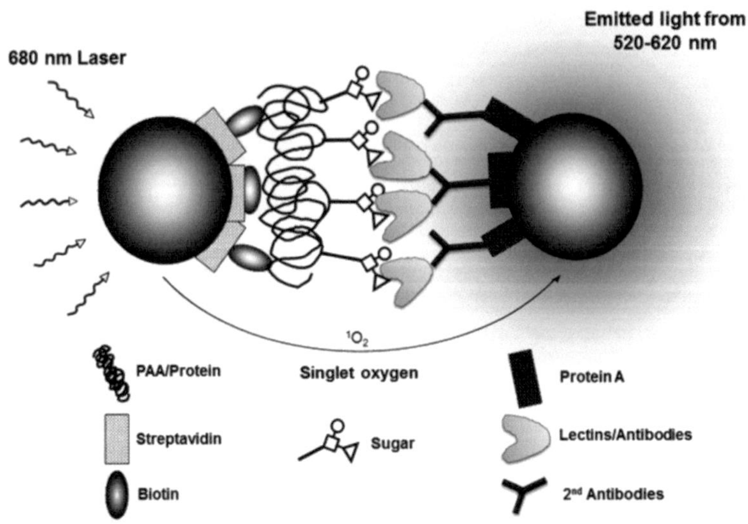

Figure 1. In-solution proximity binding with photosensitizers which was developed to characterize the protein-carbohydrate interactions [28].

Carbohydrate membrane microarray is fabricated by immobilization of the biotin-conjugated PAA-based glycans on aldehyde-functionalized UltraBind via streptavidin. Streptavidin interacted strongly with biotin and formed covalent linkage with membranes after reductive amination, which prevented the loss of glycans from membrane during repeated wash steps. The use of PAA also avoided the nonspecific interactions that take place in other studies between some lectins and non-glycosylated proteins (e.g. HAS or BSA) [35]. The operation of this carbohydrate membrane microarray is similar to that of Western blotting and can be performed easily by anyone without prior intensive training.

3. Applications

3.1. Carbohydrate binding profiles of lectins and antibodies (solution microarray)

Fifty-four biotinylated polyacrylamide backboned glycans (biotin-PAA-glycans) (**Table 1**) are collected in total to examine fifteen carbohydrate-binding proteins, including eight lectins (Con A, DBA, GS-I, PNA, SBA, UEA-1, WFA and WGA), and six antibodies (anti-Le^a, Le^b, Le^x, Le^y, sialyl Le^a and sialyl Le^x). The resulting signals are indicated with bars as relative intensities (**Figures 2 and 3**). The natural carbohydrate ligands for these lectins are listed in **Table 2**. All of the lectins showed nearly the same carbohydrate binding preferences as those in literatures. For example, concanavalin A (Con A) bound preferentially to mannose (No. 3) and biantennary *N*-glycan (No. 53), and very weakly to 3- and 6-sulfated galactosides (No. 19, 23 and 25). DBA, a GalNAc-binding lectin, recognized GalNAcα1-3Gal-containing epitopes (No. 11 and 39). ECA interacted with LacNAc disaccharide, Galβ1-4(6-sulfo)GlcNAc, and Galβ1-4(α1-2Fuc)GlcNAc (No. 17, 24, 31 and 47), and weakly bound to Le^c (Galβ1-3GlcNAc, No. 20). GS-I preferred interacting with Gal/GalNAc that contains α1-3 or 1-4 linkage (No. 11, 13, 14, 16, 40 and 42). MAA, in this study, recognized mainly to 3′-sulfated Galβ1-3GlcNAc, 3′-sulfated Galβ1-4GlcNAc and LacNAc and weakly to 3-sialylated galactosides (No. 26, 37 and 53). PNA interacted with Galβ1-3GalNAc (No. 15) and bound to some galactosides weakly (No. 12, 16, 20, 45 and 46). SBA preferentially interacted with α-linked galactosides (No. 16 and 42) and *N*-acetylgalactosaminoside (No. 11). SNA, a well-known α2-6 sialoside-binding lectin, interacted strongly to 6′-sialyl lactose and sialylated diantennary *N*-glycan (No. 36 and 53). UEA-1 specifically bound with Fucα1-2Gal-containing glycans (No. 18, 31 and 49). Due to weak interaction with PAA, WFA is the only one lectin showing higher background signals than the others. It recognized nearly half of the glycans on the glycan library, such as GlcNAc- and NeuAcα2-3-Gal/NeuAcα2-6-Gal containing saccharides. WGA also bound to terminal Gal or GalNAc epitopes (GalNAcα1-3Gal, No. 11 and Galβ1-4(6HSO_3)GlcNAc, No. 24) according to some minor signals. Interestingly, WGA showed better interactions with chitotriose than with chitobiose and GlcNAc. In addition, the binding specificities of monoclonal anti-carbohydrate antibodies also revealed some interesting features. As shown in **Figure 3**, anti-Le^a antibody bound tightly with Le^x, but less with Le^b and sialyl Le^a. Anti-Le^b antibody represented specificities for both Le^b and

Le^a, but less for Le^x and sialyl Le^x. Anti-Le^y antibody not only binding to Le^y, but also recognized lactose, Le^x, sialyl Le^x and H type 2 structures. We also compared the binding patterns of lectins with the results reported by Blixt and coworkers at CFG in which 264 different glycans are studied by using the printed microarray (Ver. 2) (http://www.functionalglycomics.org/glycomics/publicdata/ primaryscreen.jsp). There are forty-seven glyco-epitopes are found to be identical in both analyses. Even the different principles and procedures of the two systems, the binding patterns of eight lectins are nearly the same, except for a few minor differences. For example, our characterized patterns of WFA and WGA show 90% similarity to the CFG data. Nevertheless, the interactions of SBA, WFA and WGA to β-GalNAc (No. 2) in the CFG's printed microarray are not observed in our system. Both of our method and the printed microarray indicate that MAA preferentially binds to sulfated glycans [36]. Because of the observed consistency shown by the two very different methods, we conclude the protein-glycan binding interactions are not affected by the PAA linker, the assay procedure (washing vs. non-washing) and the interacting microenvironment (2D for printed microarray vs. 3D for our solution microarray).

No.	Glycan Name	No.	Glycan Name
1	PAA-biotin	28	GlcNAcβ1-4GlcNAcβ1-4GlcNAcβ, sp=-NHCOCH2NH-
2	β-GlcNAc	29	GlcNAcβ1-3Galβ1-4GlcNAcβ
3	α-Mannose	30	Fucα1-2Galβ1-3GlcNAcβ, Le^d (H type1)
4	β-GlcNAc	31	Fucα1-2Galβ1-4GlcNAcβ (H type2)
5	β-GalNAc	32	Galβ1-3(Fucα1-4)GlcNAcβ (Le^a)
6	α-Fuc	33	Galβ1-4(Fucα1-3)GlcNAcβ (Le^x)
7	α-NeuAc	34	3-HSO$_3$-Galβ1-4(Fucα1-3)GlcNAcβ (3'sulfate Le^x)
8	α-NeuGc	35	NeuAcα2-3Galβ1-3GlcNAcβ (3'Sialyl Le^c)
9	Glcα1-4Glcβ	36	NeuAcα2-6Galβ1-4Glcβ (6'Sialyl Lactose)
10	GlcNAcβ1-4GlcNAcβ	37	NeuAcα2-3Galβ1-4Glcβ (3'Sialyl Lactose)
11	GalNAcα1-3Galβ	38	NeuAcα2-3(NeuAcα2-6)GalNAcα
12	Galβ1-4Glcβ (Lactose)	39	GalNAcα1-3(Fucα1-2)Galβ (Blood Group A)
13	Galα1-3Galβ	40	Galα1-3(Fucα1-2)Galβ (Blood Group B)
14	Galα1-3GalNAcβ	41	3-HSO$_3$-Galβ1-3(Fucα1-4)GlcNAcβ (3'sulfate Le^a)
15	Galβ1-3GalNAcβ	42	Galα1-4Galβ1-4Glcβ
16	Galα1-4GlcNAcβ (αLacNAc)	43	NeuAcα2-3Galβ1-4GlcNAcβ
17	Galβ1-4GlcNAcβ (LacNAc)	44	NeuAcα2-3Galβ1-3GalNAcα
18	Fucα1-2Galβ	45	Galβ1-3(NeuAcα2-6)GalNAcα
19	3-HSO$_3$-Galβ1-4GlcNAcβ	46	Galβ1-3GlcNAcβ1-3Galβ1-4Glcβ
20	Galβ1-3GlcNAc (Le^c)	47	Galβ1-4GlcNAcβ1-3Galβ1-4Glcβ
21	NeuAcα2-6GalNAcα	48	Fucα1-2Galβ1-3(Fucα1-4)GlcNAcβ (Le^b)
22	NeuGcα2-6GalNAcα	49	Fucα1-2Galβ1-4(Fucα1-3)GlcNAcβ (Le^y)
23	3-HSO$_3$-Galβ1-3GlcNAcβ	50	NeuAcα2-3Galβ1-3(Fucα1-4)GlcNAcβ (sialyl Le^a)
24	Galβ1-4(6-HSO$_3$)GlcNAcβ	51	NeuAcα2-3Galβ1-4(Fucα1-3)GlcNAcβ (sialyl Le^x)
25	6-HSO$_3$-Galβ1-4GlcNAcβ	52	(NeuAcα2-8)$_{5-6}$
26	NeuAcα2-3Gal	53	(NeuAcα2-6Galβ1-4GlcNAcβ1-2Man)$_2$α1-3,6Manβ1-4GlcNAcβ1-4GlcNAcβ
27	NeuAcα2-3GalNAcα	54	H_2O

Table 1. List of biotin-PAA-glycans (fifty-two) used in glycan solution microarray [28].

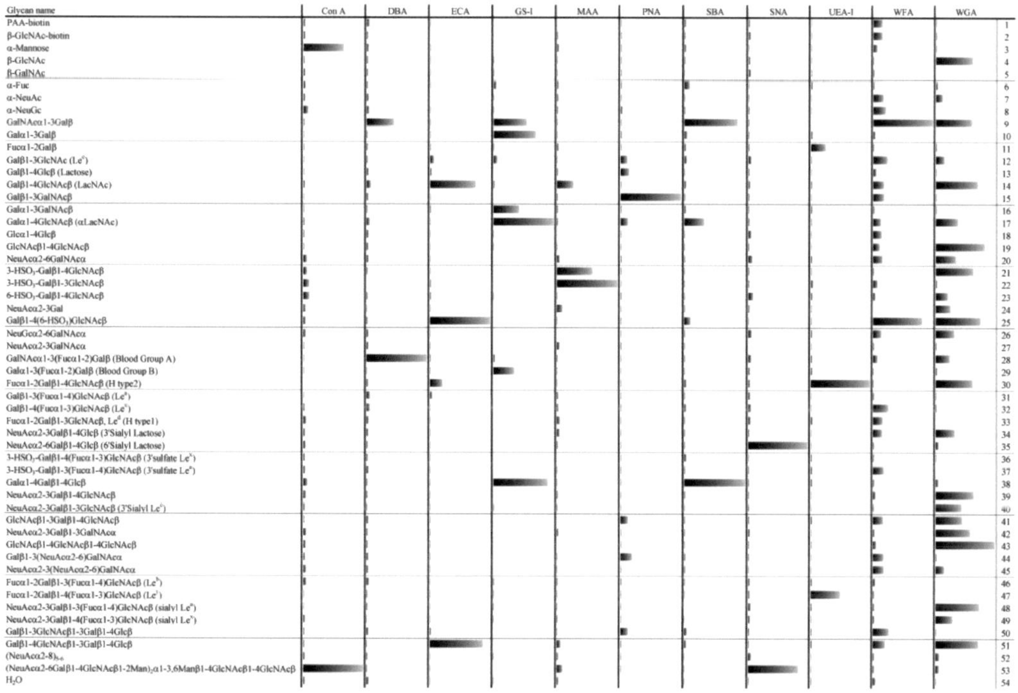

Figure 2. Carbohydrate binding specificities of eleven lectins characterized by glycan solution microarray [28].

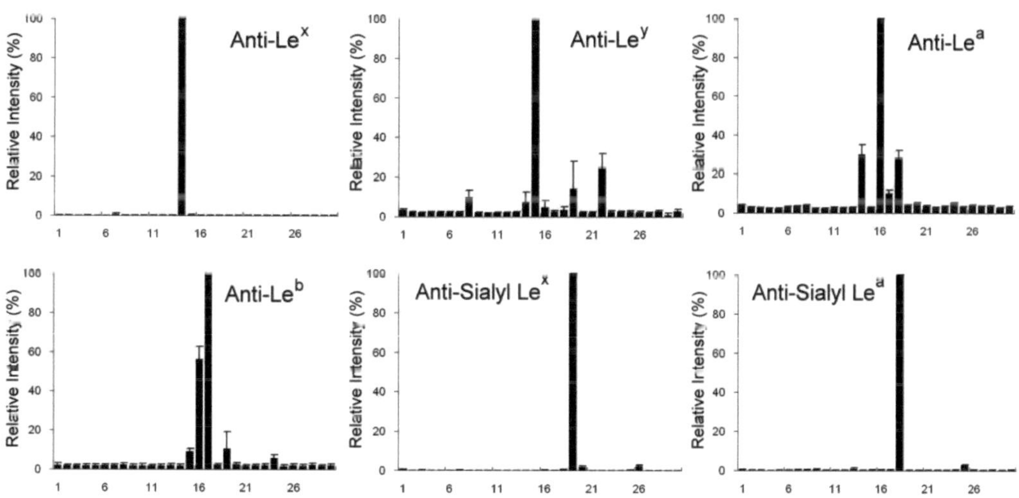

Figure 3. Carbohydrate binding specificities of six antibodies characterized by glycan solution microarray [28].

Lectins	Binding specificities	Lectins	Binding specificities
Con A	Man, Glc, GlcNAc	WGA	GlcNAcβ1-4GlcNAc, Neu5Ac
DBA	GalNAc	ECA	Galβ1-4GlcNAc
PNA	Galβ1-3GalNAc	MMA	Gal
SBA	GalNAc/Gal	GS-I	Gal
UEA-1	Fuc	SNA	Neu5Acα2-6
WFA	GalNAc		

Table 2. Carbohydrate ligands of commercial available lectins.

3.2. Binding patterns of seventeen lectins and four antibodies (membrane microarray)

The principle and procedures of carbohydrate membrane microarray are showed in **Figure 4**. The western blotting like procedures not only reduces the time and interference, but also increases the application of this platform. In order to look deep inside the carbohydrate binding preferences of proteins and microorganisms, the collections of biotin-PAA-glycans were to increased eighty-eight different structures (**Table 3**). The glycan binding specificities of sixteen lectins (six alkaline phosphatase (AP)-conjugated lectins, four FITC-conjugated lectins, six unconjugated lectins) and four Lewis blood-group antibodies are evaluated and showed in **Figures 5 and 6**. All the lectins recognized the glycans that are consistent with the literature. For instance, ECA preferentially interacted with LacNAc, lactose, GalNAc, and Gal terminal sugars; PNA specifically bound to the Galβ1-3GalNAc structure; SBA dominantly recognized α-linked GalNAc epitopes; 3-sulfate LacNAc is ligand for MAA [36]. Compare the patterns of unconjugated lectins with conjugated lectins (AP- or FITC-attached) indicated that the glycan preferences of ECA and PNA are not interfered by conjugation. More binding signals are observed in the binding profiles of AP-conjugated MAA, SBA and WGA compared with unconjugated or FITC-attached ones. Additionally, the binding patterns of DBA, ECA, GS-I, MAA, SBA and VVA are highly consistent with those reported by Consortium for Functional Glycomics (CFG, printed microarray Ver. 2,). However, few inconsistencies are also observed in the study of MPA, PNA, UEA and WGA. Furthermore, the binding patterns of four Lewis blood group antibodies represented very high specificities (**Figure 6**).

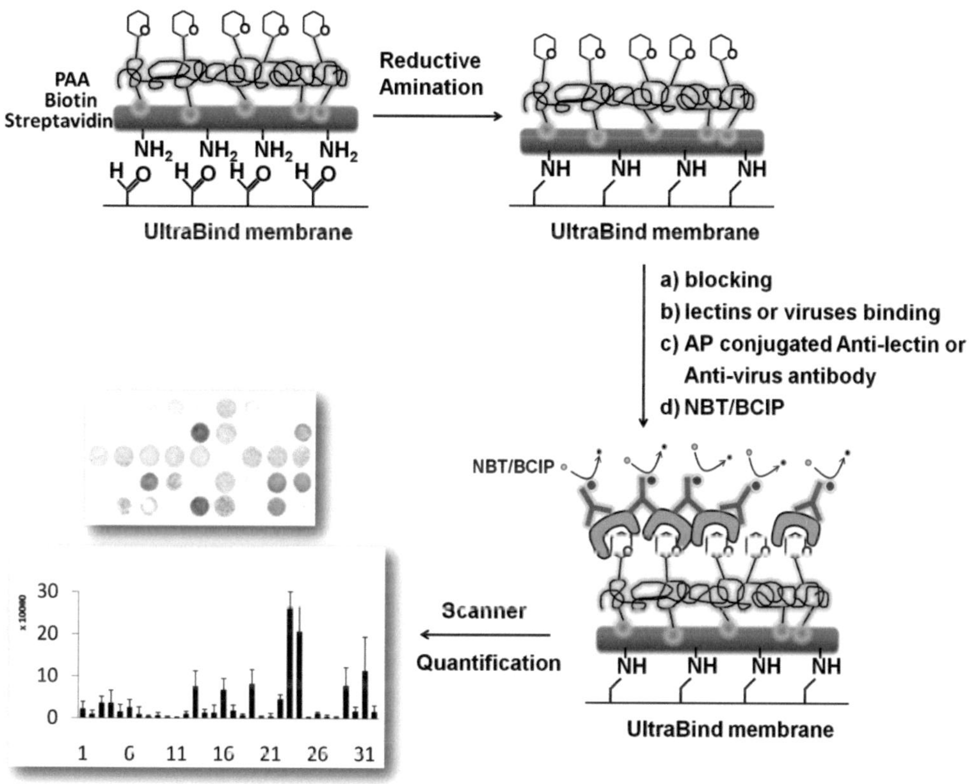

Figure 4. Fabrication, principle, and procedures of carbohydrate membrane microarray [29].

S-1	Blank-PAA-biotin	S-46	Neu5Acα2-6Galβ-PAA-biotin
S-2	β-GlcNAc-sp-biotin	S-47	Neu5Gcα2-6GalNAc-PAA-biotin
S-3	α-Mannose-PAA-biotin	S-48	Neu5Acα2-3GalNAcα-PAA-biotin
S-4	β-GlcNAc-PAA-biotin	S-49	Blood Group A-tri-PAA-biotin
S-5	β-GalNAc-PAA-biotin	S-50	Blood Group B-tri-PAA-biotin
S-6	α-L-Fuc-PAA-biotin	S-51	H(type2)-PAA-biotin
S-7	α-Neu5Ac-PAA-biotin	S-52	Le^a-PAA-biotin
S-8	α-Neu5Ac-$OCH_2C_6H_4$-p-$NHCOOCH_2$-PAA-biotin	S-53	Le^x-PAA-biotin
S-9	MDP(muramyl dipeptide)-PAA-biotin	S-54	Le^d(H type1)-PAA-biotin
S-10	α-Neu5Gc-PAA-biotin	S-55	3'Sialyl-Lactose-PAA-biotin
S-11	β-D-Gal-3-sulfate-PAA-biotin	S-56	6'Sialyl-Lactose-PAA-biotin
S-12	β-D-GlcNAc-6-sulfate-PAA-biotin	S-57	3-HSO_3-Le^x-PAA-biotin
S-13	GalNAcα1-3Galβ-PAA-biotin	S-58	3-HSO_3-Le^a-PAA-biotin
S-14	Galα1-3Galβ-PAA-biotin	S-59	Galα1-4Galβ1-4Glcβ-PAA-biotin
S-15	Fucα1-2Galβ-PAA-biotion	S-60	Galα1-3Galβ1-4Glcβ-PAA-biotin
S-16	Le^c(Galβ1-3GlcNAc)-PAA-biotin	S-61	GlcNAcβ1-2Galβ1-3GalNAcα-PAA-biotin
S-17	Galβ1-4Glcβ-PAA-biotin (Lactose)	S-62	Neu5Acα2-3Galβ1-4GlcNAcβ-PAA-Biotin
S-18	LacNAc-PAA-biotin	S-63	3'Sialyl-Le^c-PAA-biotin
S-19	Fucα1-3GlcNAcβ-PAA-biotin	S-64	Galα1-3Galβ1-4GlcNAcβ-PAA-biotin, sp=-$NHCOCH_2NH$-
S-20	Fucα1-4GlcNAcβ-PAA-biotin	S-65	GlcNAcα1-3Galβ1-3GalNAcα-PAA-biotin
S-21	GalNAcα1-3GalNAcα-PAA-biotin	S-66	GlcNAcβ1-3Galβ1-3GalNAcα-PAA-biotin
S-22	Galα1-3GalNAcα-PAA-biotin	S-67	Galβ1-3(GlcNAcβ1-6)GalNAcα-PAA-biotin
S-23	Galβ1-3GalNAcβ-PAA-biotin	S-68	Blood type A (tri)-PAA-biotin, sp=$(CH_2)_3NHCO(CH_2)_5NH$-
S-24	Galα1-3GalNAcβ-PAA-biotin	S-69	Blood type B (tri)-PAA-biotin, sp=$(CH_2)_3NHCO(CH_2)_5NH$-
S-25	Galβ1-3Galβ-PAA-biotin	S-70	GlcNAcb1-3Galβ1-4GlcNAc-PAA-biotin
S-26	GlcNAcβ1-3Galβ-PAA-biotin	S-71	Neu5Acα2-3Galβ1-3GalNAcα-PAA-Biotin
S-27	αLacNAc-PAA-biotin	S-72	GlcNAcβ1-3(GlcNAcβ1-6)GalNAcα-PAA-biotin
S-28	Glcα1-4Glcβ-PAA-biotin	S-73	Galα1-4Galβ1-4GlcNAcβ-PAA-biotin
S-29	Galβ1-3GalNAcα-PAA-biotin, sp=-p-OC_6H_4-	S-74	GlcNAcβ1-4GlcNAcβ1-4GlcNAc-PAA-biotin
S-30	Galα1-2Galβ-PAA-biotin	S-75	Galβ1-3(Neu5Acα2-6)GalNAcα-PAA-biotin
S-31	GlcNAcβ1-4GlcNAc-PAA-biotin	S-76	Neu5Acα2-3(Neu5Acα2-6)GalNAc-PAA-biotin
S-32	GlcNAcβ1-4GlcNAcβ-PAA-biotin, sp=-$NHCOCH_2NH$-	S-77	Galβ1-4GlcNAcβ1-3GalNAcα-PAA-biotin
S-33	Neu5Acα2-6GalNAc-PAA-biotin	S-78	Le^b-PAA-biotin
S-34	H(type 3)-PAA-biotin	S-79	Le^y-PAA-biotin
S-35	3-HSO_3-Galβ1-4GlcNAc-PAA-biotin	S-80	Sialyl Le^a-PAA-biotin
S-36	3-HSO_3-Galβ1-3GlcNAcβ-PAA-biotin	S-81	Sialyl Le^x-PAA-biotin
S-37	Galα1-6Glcβ-PAA-biotin (melibiose)	S-82	GlcNAcβ1-3(GlcNAcβ1-6)Galβ1-4Glcβ-PAA-biotin
S-38	Neu5Acα2-8Neu5Acα-sp**-PAA-biotin, $(Neu5Ac)_2$	S-83	Galα1-3(Fucα1-2)Galβ1-4GlcNAc-PAA-biotin
S-39	Galβ1-2Galβ-PAA-biotin	S-84	Galβ1-3GlcNAcβ1-3Galβ1-4Glcβ-PAA-biotin
S-40	6-HSO_3-Galβ1-4GlcNAc-PAA-biotin	S-85	Galβ1-4GlcNAcβ1-3Galβ1-4Glcβ-PAA-biotin
S-41	Neu5Acα2-3Gal-PAA-biotin	S-86	(NeuAcα2-8)5-6-PAA-biotin
S-42	Galβ1-4(6-HSO_3)GlcNAcb-PAA-biotin	S-87	Galβ1-4GlcNAcβ1-3(Galβ1-4GlcNAcβ1-6)GalNAcα-PAA-biotin
S-43	3-HSO_3-Galβ1-3GalNAcβ-PAA-biotin (sulfate-TF)	S-88	α2-6 sialylated diantennary N-glycans-PAA-biotin
S-44	GlcNAcβ1-3GalNAcα-PAA-biotin	S-89	GalNAc-α-Ser-PAA-biotin
S-45	GlcNAcβ1-6GalNAcα-PAA-biotin	S-90	H_2O

α2-6 sialylated diantennary N-glycans :$(NeuAcα2\text{-}6Galβ1\text{-}4GlcNAcβ1\text{-}2Man)_2$α1-3,6Manβ1-4GlcNAcβ1-4GlcNAc

Table 3. List of biotin-PAA-glycans (eighty-eight) used in carbohydrate membrane microarray [29].

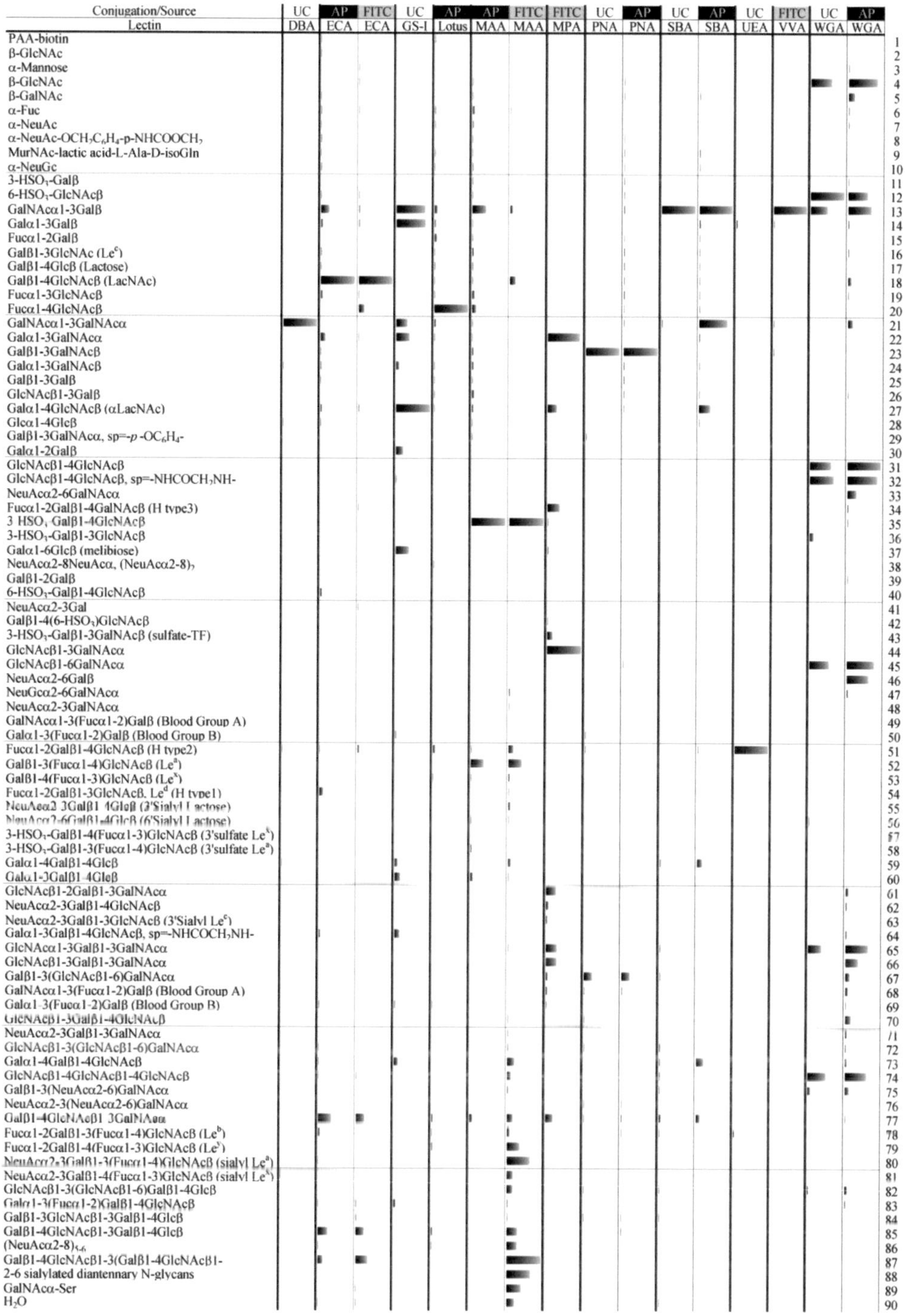

Figure 5. Carbohydrate binding specificities of sixteen lectins (six alkaline phosphatase-conjugated lectins, four FITC-conjugated lectins, and six unconjugated lectins) characterized by carbohydrate membrane microarray [29].

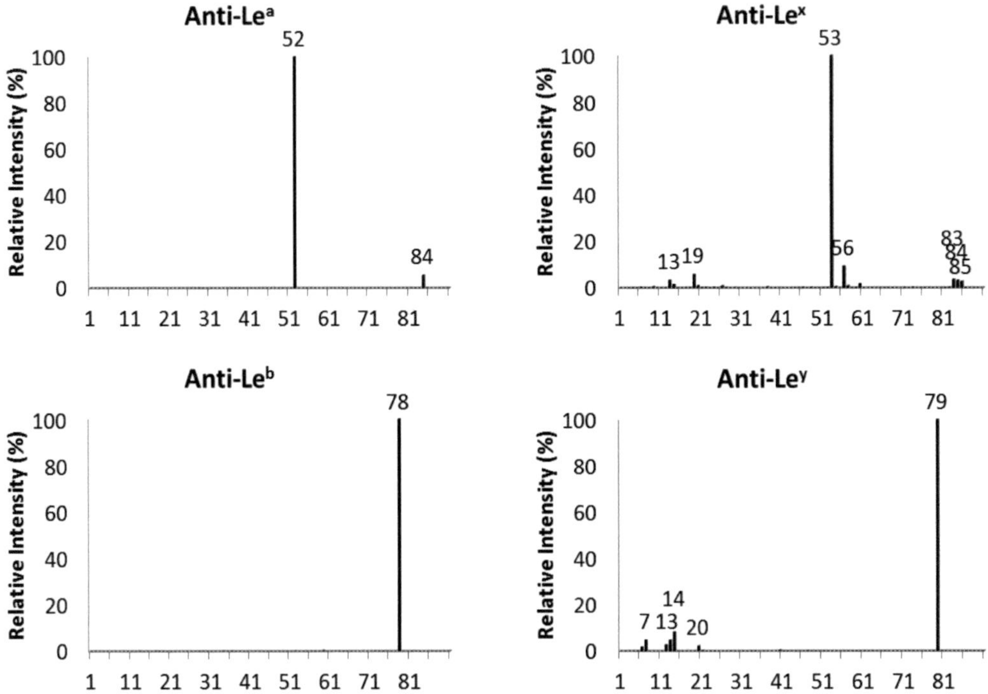

Figure 6. Carbohydrate binding specificities of four antibodies characterized by carbohydrate membrane microarray [29].

3.3. Surveillance of seasonal and pandemic H1N1 influenza A viruses (membrane microarray)

Several methods are applied for exploring the carbohydrate-binding preference of influenza viruses or recombinant HA, including cell-based assays [37-39], immobilization of virus [40-46] and glycan microarray technology [47-49]. In our study, 30 biotin-PAA-glycans (**Table 4**, most of them were sulfated and sialylated) were selected and fabricated on membrane and applied for the carbohydrate ligand surveillance of influenza clinical isolates. We subjected this influenza membrane microarray to characterize the glycan-binding features of five seasonal influenza A (H1N1) and seven A(H1N1)pdm09 clinical isolates. The binding patterns of all studied viruses are successfully profiled, with each virus exhibiting different and clear patterns, thereby enabling characterization (**Figure 7, Lane 1 to 12**). The majority of the seasonal H1N1 viruses bound strongly to 6'-sialyl lactose and sialyl biantennary *N*-glycan (**Figure 7, Lane 1 to 5**). A/Taiwan/1156/2006, A/Taiwan/510/2008, and A/Taiwan/289/2009 isolates interacted with the two glycans, and with numerous α2-3 sialylated structures, including 3'-sialyl lactose, NeuAcα2-3(NeuAcα2-6)GalNAc, sialyl Le^a, and sialyl Le^x. In addition, A/Taiwan/1156/2006 isolate also recognized NeuAc (linked to *p*-aniline), NeuGc, some of the sulfated glycans (3'-sulfated Le^a, 3'-sulfated Le^x, 6-sulfate LacNAc and 6'-sulfate LacNAc), NeuAcα2-6GalNAc, and a NeuAcα2-8 dimer. Contrary to seasonal H1N1 isolated in 2009, pandemic (H1N1) isolated in 2009 accepted more substrates

(**Figure 7, Lane 6 to 9**). California/07/2009 and A/Taiwan/2024/2009 bound with 6-sulfate GlcNAc, 3'-sulfated Le[a], 3'-sulfated Le[x], all of the 2-3 sialylated glycans with the exception of NeuAcα2-3GalNAc, 6'-sialyl lactose, sialyl biantennary *N*-glycan, and α2-8 oligomers. A/Taiwan/942/2009 favored four sulfated glycans (3-sulafe Gal, 6-sulfate GlcNAc, 3'-sulfated Le[a] and 3'-sulfated Le[x]), most of the 2-3 sialylated glycans (except NeuAcα2-3Gal and NeuAcα2-3GalNAc) and three 2-6 sialylated glycans (NeuAcα2-6Gal, 6'-sialyl lactose and sialyl biantennary *N*-glycan). A/Taiwan/987/2009 recognized most of the PAA-sugar substrates tested, with the exception of the PAA backbone, muramyl dipeptide, αNeuAc, 3-sulafe Galβ1-4GlcNAc, 3-sulafe Galβ1-3GlcNAc and Galβ1-3(NeuAcα2-6)GalNAc. In addition, we found that pandemic (H1N1) viruses isolated in 2009 represented broader substrate specificities than the pandemic viruses isolated in 2010, especially to sulfated sugars. A/Taiwan/395/2010 and A/Taiwan/1477/2010 recognized NeuAcα2-3Gal, 3'sialyl lactose, NeuAcα2-3Galβ1-3GlcNAc, sialyl-Le[a], NeuAcα2-6GalNAc, NeuAcα2-6Gal, 6'-sialyl lactose, and sialyl biantennary *N*-glycan (**Figure 7, Lane 11 to 12**). A/Taiwan/257/2010 accepted only six glycans including NeuAc (linked to *p*-aniline), 3'sialyl lactose, NeuAcα2-6GalNAc, NeuAcα2-6Gal, 6'-sialyl lactose, and sialyl biantennary *N*-glycan (**Figure 7, Lane 9**).

I-1	Blank-PAA-biotin	I-17	Neu5Gcα2-6GalNAc-PAA-biotin
I-2	α-Neu5Ac-PAA-biotin	I-18	Neu5Acα2-3GalNAcα-PAA-biotin
I-3	α-Neu5Ac-$OCH_2C_6H_4$-p-$NHCOOCH_2$-PAA-biotin	I-19	3'Sialyl-Lactose-PAA-biotin
I-4	MDP(muramyl dipeptide)-PAA-biotin	I-20	6'Sialyl-Lactose-PAA-biotin
I-5	α-Neu5Gc-PAA-biotin	I-21	3-HSO_3-Le[x]-PAA-biotin
I-6	β-D-Gal-3-sulfate-PAA-biotin	I-22	3-HSO_3-Le[a]-PAA-biotin
I-7	β-D-GlcNAc-6-sulfate-PAA-biotin	I-23	Neu5Acα2-3Galβ1-4GlcNAcβ-PAA-Biotin
I-8	Neu5Acα2-6GalNAc-PAA-biotin	I-24	3'Sialyl-Le[c]-PAA-biotin
I-9	3-HSO_3-Galβ1-4GlcNAc-PAA-biotin	I-25	Neu5Acα2-3Galβ1-3GalNAcα-PAA-Biotin
I-10	3-HSO_3-Galβ1-3GlcNAcβ-PAA-biotin	I-26	Galβ1-3(Neu5Acα2-6)GalNAcα-PAA-biotin
I-11	Neu5Acα2-8Neu5Acα-sp**-PAA-biotin, $(Neu5Ac)_2$	I-27	Neu5Acα2-3(Neu5Acα2-6)GalNAc-PAA-biotin
I-12	6-HSO_3-Galβ1-4GlcNAc PAA-biotin	I-28	Sialyl Le[a]-PAA-biotin
I-13	Neu5Acα2-3Gal-PAA-biotin	I-29	Sialyl Le[x]-PAA-biotin
I-14	Galβ1-4(6-HSO_3)GlcNAcb-PAA-biotin	I-30	(NeuAcα2-8)5-6-PAA-biotin
I-15	3-HSO_3-Galβ1-3GalNAcβ-PAA-biotin (sulfate-TF)	I-31	α2-6 sialylated diantennary N-glycans -PAA-biotin
I-16	Neu5Acα2-6Galβ-PAA-biotin	I-32	H_2O

α2-6 sialylated diantennary N-glycans :$(NeuAc\alpha2\text{-}6Gal\beta1\text{-}4GlcNAc\beta1\text{-}2Man)_2\alpha1\text{-}3,6Man\beta1\text{-}4GlcNAc\beta1\text{-}4GlcNAc$

Table 4. List of biotin-PAA-glycans (thirty) fabricated on carbohydrate membrane microarray for carbohydrate binding surveillance of influenza clinical isolates [29] [50].

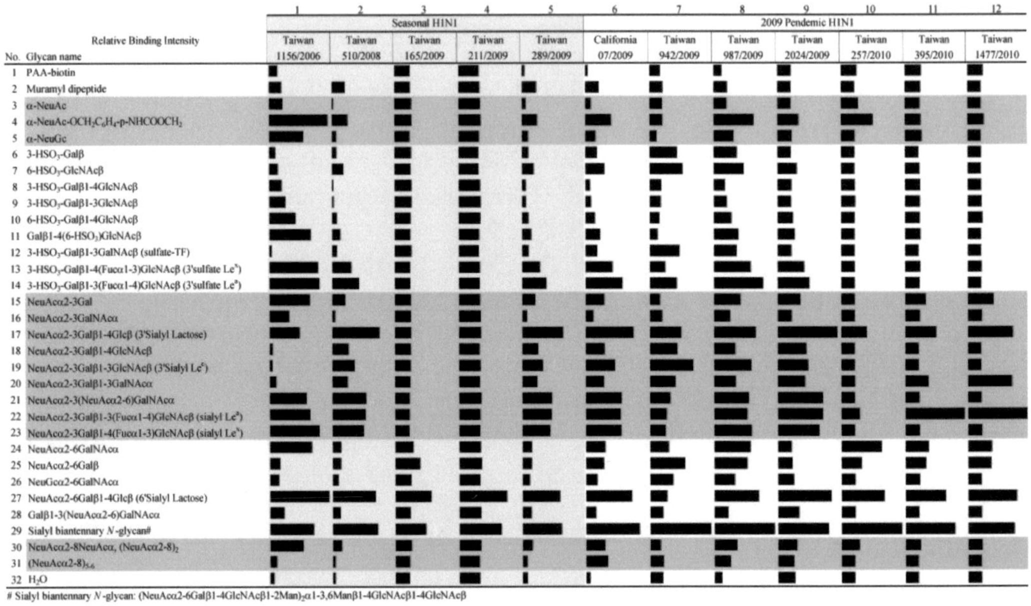

Figure 7. Carbohydrate binding profiles of twelve influenza A clinical isolates characterized by carbohydrate membrane microarray [50].

3.4. Surveillance of influenza B clinical isolates (membrane microarray)

The influenza carbohydrate membrane microarray is subsequently subjected for the surveillance of twelve influenza B (IB) clinical isolates [50]. All of the viruses are collected in the Clinical Virology Laboratory of National Cheng Kung University Hospital between 2001 and 2004. The clinical isolates are amplified in MDCK cells and purified by a sucrose gradient. The binding patterns of the influenza B viruses are investigated and are all found to exhibit clear binding profiles (**Figure 8**). Three common glycans, including α-Neu5Ac (*p*-anilinyl linked), 6'sialyl-lactose, and α2-6 sialylated biantennary *N*-glycans are bound by all of the influenza B viruses. Furthermore, some additional bindings are also observed. For instance, B/Taiwan/314/2001 and B/Taiwan/1729/2002 also recognized sulfated sugars (3-sulafe Gal, 6-sulfate GlcNAc, Galβ1-4(6-HSO_3)GlcNAc, and 3'-sulfated Le^a) and two α2-6 sialosides (NeuAcα2-6GalNAc and NeuAcα2-6Gal). B/Taiwan/288/2002 and B/Taiwan/1902/2004 portrayed minor interactions with α2-3 (3'-sialyl lactose, NeuAcα2-3Galβ1-4GlcNAc, 3'-sialyl Le^c and sialyl Le^x) and α2-6 sialylated sugars (NeuAcα2-6GalNAc and NeuAcα2-6Gal). Although the signal is relatively low, B/Taiwan/872/2004 and B/Taiwan/913/2004 also interacted with sulfated sugars. Surprisingly, seven viruses showed weak bindings to α2-8 linked sialic acids, including B/Taiwan/314/2001, B/Taiwan/174/2002, B/Taiwan/288/2002, B/Taiwan/1729/2002, B/Taiwan/872/2004, and B/Taiwan/913/2004.

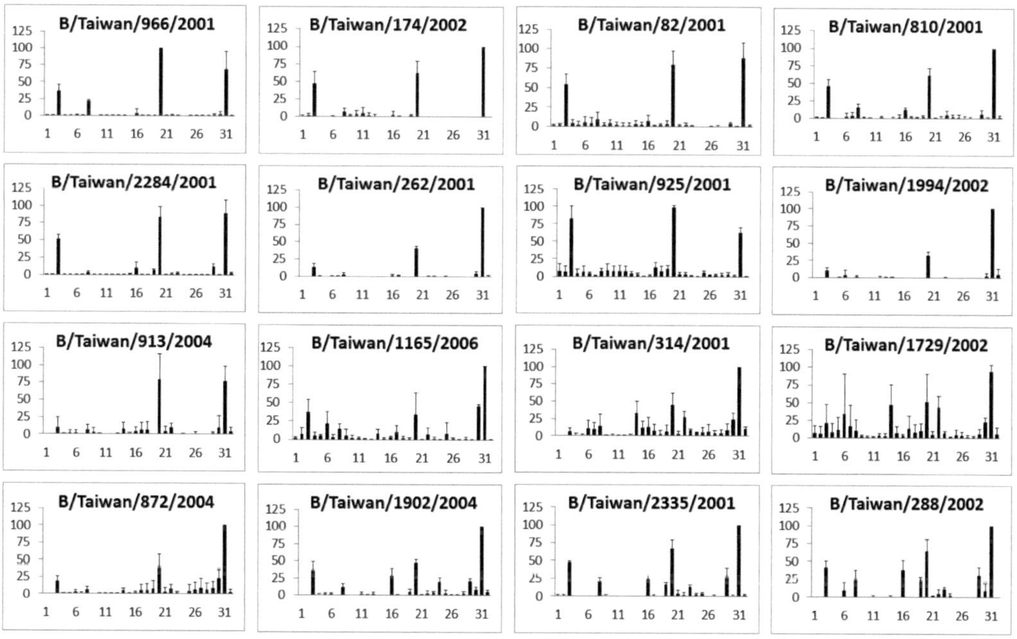

Figure 8. Carbohydrate binding profiles of sixteen influenza B clinical isolates characterized by carbohydrate membrane microarray [50].

The DNA sequences of hemagglutinin of the twelve influenza B clinical isolates were analyzed, translated into sequences of amino acids, and aligned with B/HongKong/8/73, Victoria (B/Victoria/02/87), and Yamagata (B/Yamagata/16/88) lineages using online software EBI ClustalW2. Surprisingly, the twelve influenza B clinical isolates could all be categorized into two groups depending on protein sequence alignment. Group 1 contained eight clinical isolates (B/Taiwan/314/2001, B/Taiwan/174/2002, B/Taiwan/288/2002, B/Taiwan/1729/2002, B/Taiwan/1994/2002, B/Taiwan/872/2004, B/Taiwan/913/2004 and B/Taiwan/1902/2004, **Figure 8, Table 5**) and Group 2 contained four viruses (B/Taiwan/262/2001, B/Taiwan/925/2001, B/Taiwan/966/2001, B/Taiwan/2284/2001). We found that all of the viruses in group 2 were isolated in 2001. In addition, there were twelve amino acid differences between the two groups. Most of the amino acids were located at the chain region and subunit interfaces, while four amino acids were located within the antigenic site (loop 160, amino acid number of B/HK/73) [51,52]. Additionally, four amino acids changed the charge property from group 1 to group 2 viruses (**Table 5**, No. 56, 116, 181 and 182, amino acid number of B/Yamagata/16/88). Different carbohydrate binding properties were also observed between some viruses of the two groups. Specifically, group 2 viruses interacted with three major sialylated glycans, while some of group 1 viruses bound not only strongly to the three glycans, but also weakly to sulfated and α2-3 sialylated glycans. Additionally, the number of charged amino acids was higher in group 2 than in group 1 viruses and B/HongKong/8/73 (**Table 5**).

	Amino acid number of B/HK/8/73	29	48	56	75	116	162	*	*	176	180	181	217	Number of charged amino
	B/HK/8/73	V	Q	N	T	N	K			Y	K	G	V	2
	B/Victoria/02/87	V	K	K	T	N	K	D	N	N	E	G	V	5
	B/Yamagata/16/88	V	K	N	T	N	R	D		Y	K	G	V	4
Group1	B/Taiwan/314/2001	V	R	D	A	N	K	D	N	H	E	G	V	5
	B/Taiwan/174/2002	V	R	D	T	N	K	D	N	Y	E	G	V	
	B/Taiwan/288/2002	V	R	D	T	N	K	D	N	Y	E	G	V	
	B/Taiwan/1729/2002	V	R	D	T	N	K	D	N	Y	E	G	V	
	B/Taiwan/1994/2002	V	R	D	T	N	K	D	N	Y	E	G	V	
	B/Taiwan/872/2004	V	R	D	T	N	K	D	N	Y	E	G	V	
	B/Taiwan/913/2004	V	R	D	T	N	K	D	N	Y	E	G	V	
	B/Taiwan/1902/2004	V	R	D	T	N	K	D	N	Y	E	G	V	
Group2	B/Taiwan/262/2001	A	K	T	I	K	R	E	N	H	K	E	I	7
	B/Taiwan/925/2001	A	K	T	I	K	R	D	N	H	K	E	I	
	B/Taiwan/966/2001	A	K	T	I	K	R	E	N	H	K	E	I	
	B/Taiwan/2284/2001	A	K	T	I	K	R	E	N	H	K	E	I	
	Amino acid number of B/Yamagata/16/88	29	48	56	75	116	162	163	*	177	181	182	218	

Table 5. Critical amino acid differences between the two groups of influenza B clinical isolated viruses [50].

According to the binding preference for α2-6 and α2-3 sialylated oligosaccharides displayed by these clinical isolated influenza viruses, these clinical isolates might infect not only the surface of nasal mucosa, pharynx, larynx, trachea and bronchi (that are resided in the upper respiratory tract to express α2-6 sialosides predominantly), but also alveoli, bronchi (that are located in the lower respiratory tract), eyes, and tissues in the gastrointestinal tract (that contain both α2-3 and α2-6 sialosides) [53,54]. After we surveyed more clinical isolates, factors which affect the binding preference of IB could be dissected. We believe that application of the carbohydrate microarray for detailed analysis of sugar-binding structures will eventually build the connection between clinical symptoms and the genetic specification of influenza viruses.

3.5. Characterization the glycan structure of glycoproteins (solution microarray)

The aforementioned lectins are further applied to characterize the glycan structures of six biotinylated proteins including ovalbumin, porcine mucin, human serum albumin, human transferrin, fetuin and asialofetuin. Distinctive glycopatterns are generated in accordance with the analysis of lectins (**Figure 9**). Porcine mucin is known to have *O*-linked fucosylated glycans with terminal GalNAc, GlcNAc and NeuAc residues. Our results showed positive signals in the tests with the ECA, DBA, UEA-1 and WGA lectins, indicating that the mucin contains the determinants of Galβ1-4GlcNAc, GalNAc, Fucα1-2Gal and GlcNAc/NeuAc, respectively. Manα1-3(Fucα1-6)GlcNAc and NeuAcα2-6Galβ1-4GlcNAc containing biantennary or triantennary *N*-glycans have been reported as the major glyco-structures of human transferrin [55-58]. Strong Con A and SNA signals are shown in our binding assay, but the relative intensity of GlcNAc-binding lectins is not as good as those of Con A and SNA, which is the contribution of the interference by the terminal sialic acids [59]. All of the results are similar to the analyses from lectin microarray or dot blot analysis [55-58].

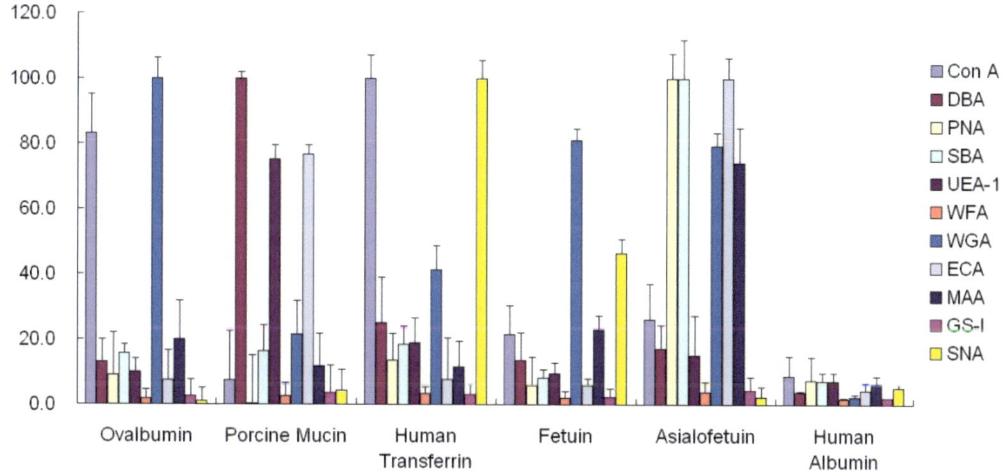

Figure 9. Glycopattern profiling of six biotinylated proteins with ten lectins by glycan solution microarray. The six proteins include ovalbumin, porcine mucin, human serum albumin, human transferrin, fetuin and asialofetuin [28].

4. Perspective

In summary, carbohydrate-protein interactions are the key steps for many physiological and pathological evens. Hence, development of new carbohydrate microarrays is important for detecting these activities. Our studies have demonstrated a rapid, highly sensitive, and reliable method to characterize carbohydrate-protein interactions with minimized materials (in the range of ng per well). Lectins and antibodies are evaluated and most of the results are coherent to previous reports, including those of CFG. Solution microarray is the first homogeneous and washless carbohydrate microarray which offers a reliable alternative for characterizing sugar-binding features of lectins and proteins, as well as antibodies. In addition, membrane microarray is also easy handled with low cost (USD$0.25 for one membrane microarray). The easy-handling feature avoids the necessity of repetitive training and accelerates the screening efficiency. Both of the microarrays represent a convenient and reliable way to examine the carbohydrate binding features of various proteins, high throughput drug screening, and the glycan binding surveillance of influenza viruses.

Author details

Chuan-Fa Chang
Department of Medical Laboratory Science and Biotechnology,
College of Medicine, National Cheng Kung University, Taiwan

5. References

[1] Varki, A. (1993) *Glycobiology* 3, 97-130

[2] Ratner, D. M., Adams, E. W., Disney, M. D., and Seeberger, P. H. (2004) *Chembiochem* 5, 1375-1383

[3] Fukuda, M. (2000) *Molecular and Cellular Glycobiology*, Oxford University Press

[4] Magnani, J. L., Brockhaus, M., Smith, D. F., Ginsburg, V., Blaszczyk, M., Mitchell, K. F., Steplewski, Z., and Koprowski, H. (1981) *Science* 212, 55-56

[5] Sacchettini, J. C., Baum, L. G., and Brewer, C. F. (2001) *Biochemistry* 40, 3009-3015

[6] Kansas, G. S. (1996) *Blood* 88, 3259-3287

[7] Geijtenbeek, T. B., Torensma, R., van Vliet, S. J., van Duijnhoven, G. C., Adema, G. J., van Kooyk, Y., and Figdor, C. G. (2000) *Cell* 100, 575-585

[8] Skehel, J. J., and Wiley, D. C. (2000) *Annu Rev Biochem* 69, 531-569

[9] Raman, R., Raguram, S., Venkataraman, G., Paulson, J. C., and Sasisekharan, R. (2005) *Nat Methods* 2, 817-824

[10] Wu, A. M. (2001) *The Molecular Immunology of Complex Carbohydreats-2*, Kluwer Academic/Plenum Publishers

[11] Ratner, D. M., Adams, E. W., Su, J., O'Keefe, B. R., Mrksich, M., and Seeberger, P. H. (2004) *Chembiochem* 5, 379-382

[12] Smith, E. A., Thomas, W. D., Kiessling, L. L., and Corn, R. M. (2003) *J Am Chem Soc* 125, 6140-6148

[13] Vila-Perello, M., Gutierrez Gallego, R., and Andreu, D. (2005) *Chembiochem* 6, 1831-1838

[14] Ofokansi, K. C., Okorie, O., and Adikwu, M. U. (2009) *Biol Pharm Bull* 32, 1754-1759

[15] Sletmoen, M., Dam, T. K., Gerken, T. A., Stokke, B. T., and Brewer, C. F. (2009) *Biopolymers* 91, 719-728

[16] Tian, P., Engelbrektson, A., and Mandrell, R. (2008) *Appl Environ Microbiol* 74, 4271-4276

[17] Beccati, D., Halkes, K. M., Batema, G. D., Guillena, G., Carvalho de Souza, A., van Koten, G., and Kamerling, J. P. (2005) *Chembiochem* 6, 1196-1203

[18] Pohl, W. H., Hellmuth, H., Hilbert, M., Seibel, J., and Walla, P. J. (2006) *Chembiochem* 7, 268-274

[19] Khan, M. I., Surolia, N., Mathew, M. K., Balaram, P., and Surolia, A. (1981) *Eur J Biochem* 115, 149-152

[20] Lee, Y. C. (2001) *Anal Biochem* 297, 123-127

[21] Sorme, P., Kahl-Knutsson, B., Huflejt, M., Nilsson, U. J., and Leffler, H. (2004) *Anal Biochem* 334, 36-47

[22] Fukui, S., Feizi, T., Galustian, C., Lawson, A. M., and Chai, W. (2002) *Nat Biotechnol* 20, 1011-1017

[23] Huang, C. Y., Thayer, D. A., Chang, A. Y., Best, M. D., Hoffmann, J., Head, S., and Wong, C. H. (2006) *Proc Natl Acad Sci U S A* 103, 15-20

[24] Feizi, T., Fazio, F., Chai, W., and Wong, C. H. (2003) *Curr Opin Struct Biol* 13, 637-645

[25] Shin, I., Park, S., and Lee, M. R. (2005) *Chemistry* 11, 2894-2901

[26] Blixt, O., Head, S., Mondala, T., Scanlan, C., Huflejt, M. E., Alvarez, R., Bryan, M. C., Fazio, F., Calarese, D., Stevens, J., Razi, N., Stevens, D. J., Skehel, J. J., van Die, I., Burton,

D. R., Wilson, I. A., Cummings, R., Bovin, N., Wong, C. H., and Paulson, J. C. (2004) *Proc Natl Acad Sci U S A* 101, 17033-17038

[27] Wong, C. H. (2005) *J Org Chem* 70, 4219-4225

[28] Chang, C. F., Pan, J. F., Lin, C. N., Wu, I. L., Wong, C. H., and Lin, C. H. (2011) *Glycobiology* 21, 895-902

[29] Lao, W. I., Wang, Y. F., Kuo, Y. D., Lin, C. H., Chang, T. C., Su, I. J., Wang, J. R., and Chang, C. F. (2011) *Future Med Chem* 3, 283-296

[30] Ullman, E. F., Kirakossian, H., Singh, S., Wu, Z. P., Irvin, B. R., Pease, J. S., Switchenko, A. C., Irvine, J. D., Dafforn, A., Skold, C. N., and et al. (1994) *Proc Natl Acad Sci U S A* 91, 5426-5430

[31] Ullman, E. F., Kirakossian, H., Switchenko, A. C., Ishkanian, J., Ericson, M., Wartchow, C. A., Pirio, M., Pease, J., Irvin, B. R., Singh, S., Singh, R., Patel, R., Dafforn, A., Davalian, D., Skold, C., Kurn, N., and Wagner, D. B. (1996) *Clin Chem* 42, 1518-1526

[32] Warner, G., Illy, C., Pedro, L., Roby, P., and Bosse, R. (2004) *Curr Med Chem* 11, 721-730

[33] Bouchecareilh, M., Caruso, M. E., Roby, P., Parent, S., Rouleau, N., Taouji, S., Pluquet, O., Bosse, R., Moenner, M., and Chevet, E. (2010) *J Biomol Screen* 15, 406-417

[34] Yi, F., Zhu, P., Southall, N., Inglese, J., Austin, C. P., Zheng, W., and Regan, L. (2009) *J Biomol Screen* 14, 273-281

[35] Manimala, J. C., Roach, T. A., Li, Z., and Gildersleeve, J. C. (2006) *Angew Chem Int Ed Engl* 45, 3607-3610

[36] Bai, X., Brown, J. R., Varki, A., and Esko, J. D. (2001) *Glycobiology* 11, 621-632

[37] Rogers, G. N., and D'Souza, B. L. (1989) *Virology* 173, 317-322

[38] Suzuki, Y., Nagao, Y., Kato, H., Suzuki, T., Matsumoto, M., and Murayama, J. (1987) *Biochim Biophys Acta* 903, 417-424

[39] Suptawiwat, O., Kongchanagul, A., Chan-It, W., Thitithanyanont, A., Wiriyarat, W., Chaichuen, K., Songserm, T., Suzuki, Y., Puthavathana, P., and Auewarakul, P. (2008) *J Clin Virol* 42, 186-189

[40] Gambaryan, A. S., Tuzikov, A. B., Piskarev, V. E., Yamnikova, S. S., Lvov, D. K., Robertson, J. S., Bovin, N. V., and Matrosovich, M. N. (1997) *Virology* 232, 345-350

[41] Gambaryan, A. S., and Matrosovich, M. N. (1992) *J Virol Methods* 39, 111-123

[42] Sauter, N. K., Hanson, J. E., Glick, G. D., Brown, J. H., Crowther, R. L., Park, S. J., Skehel, J. J., and Wiley, D. C. (1992) *Biochemistry* 31, 9609-9621

[43] Auewarakul, P., Suptawiwat, O., Kongchanagul, A., Sangma, C., Suzuki, Y., Ungchusak, K., Louisirirotchanakul, S., Lerdsamran, H., Pooruk, P., Thitithanyanont, A., Pittayawonganon, C., Guo, C. T., Hiramatsu, H., Jampangern, W., Chunsutthiwat, S., and Puthavathana, P. (2007) *J Virol* 81, 9950-9955

[44] Yamada, S., Suzuki, Y., Suzuki, T., Le, M. Q., Nidom, C. A., Sakai-Tagawa, Y., Muramoto, Y., Ito, M., Kiso, M., Horimoto, T., Shinya, K., Sawada, T., Usui, T., Murata, T., Lin, Y., Hay, A., Haire, L. F., Stevens, D. J., Russell, R. J., Gamblin, S. J., Skehel, J. J., and Kawaoka, Y. (2006) *Nature* 444, 378-382

[45] Gambaryan, A., Yamnikova, S., Lvov, D., Tuzikov, A., Chinarev, A., Pazynina, G., Webster, R., Matrosovich, M., and Bovin, N. (2005) *Virology* 334, 276-283

[46] Gambaryan, A. S., Tuzikov, A. B., Pazynina, G. V., Webster, R. G., Matrosovich, M. N., and Bovin, N. V. (2004) *Virology* 326, 310-316

[47] Gall, A., Hoffmann, B., Harder, T., Grund, C., Hoper, D., and Beer, M. (2009) *J Clin Microbiol* 47, 327-334

[48] Stevens, J., Blixt, O., Paulson, J. C., and Wilson, I. A. (2006) *Nat Rev Microbiol* 4, 857-864

[49] Stevens, J., Blixt, O., Glaser, L., Taubenberger, J. K., Palese, P., Paulson, J. C., and Wilson, I. A. (2006) *J Mol Biol* 355, 1143-1155

[50] Wang, Y. F., Lao, W. I., Kuo, Y. D., Guu, S. Y., Wang, H. C., Lin, C. H., Wang, J. R., Su, I. J., and Chang, C. F. (2012) *Fut Virol* 7, 13

[51] Wang, Q., Cheng, F., Lu, M., Tian, X., and Ma, J. (2008) *J Virol* 82, 3011-3020

[52] Wang, Q., Tian, X., Chen, X., and Ma, J. (2007) *Proc Natl Acad Sci U S A* 104, 16874-16879

[53] Kumlin, U., Olofsson, S., Dimock, K., Arnberg, N. (2008) *Influenza and Other Respiratory Viruses* 2, 8

[54] Kogure, T., Suzuki, T., Takahashi, T., Miyamoto, D., Hidari, K. I., Guo, C. T., Ito, T., Kawaoka, Y., and Suzuki, Y. (2006) *Glycoconj J* 23, 101-106

[55] Kolarich, D., and Altmann, F. (2000) *Anal Biochem* 285, 64-75

[56] Pilobello, K. T., Krishnamoorthy, L., Slawek, D., and Mahal, L. K. (2005) *Chembiochem* 6, 985-989

[57] Yang, S. N., Liu, C. A., Chung, M. Y., Huang, H. C., Yeh, G. C., Wong, C. S., Lin, W. W., Yang, C. H., and Tao, P. L. (2006) *Hippocampus* 16, 521-530

[58] Karlsson, N. G., and Packer, N. H. (2002) *Anal Biochem* 305, 173-185

[59] Angeloni, S., Ridet, J. L., Kusy, N., Gao, H., Crevoisier, F., Guinchard, S., Kochhar, S., Sigrist, H., and Sprenger, N. (2005) *Glycobiology* 15, 31-41

Chapter 2

Conversion of Carbohydrates Under Microwave Heating

Aurore Richel and Michel Paquot

Additional information is available at the end of the chapter

http://dx.doi.org/10.5772/50628

1. Introduction

The non-energetic valorisation of renewable resources using efficient and eco-friendly methodologies is the central axis of the "green chemistry" concept. In particular, the chemical and chemo-enzymatical transformation of carbohydrates arising from the hydrolysis of non-edible vegetal feedstock (i.e., lignocellulosic biomass) is a widely explored thematic for the production of new high-added value materials, synthons, and platform chemicals (Bozell, 2010). The use of mono-, di- and polysaccharides for the production of new chemicals constitutes thus a subject of special relevance from both academic and industrial points of view (Lichtenthaler, 2004). Amongst the 12 principles defining this "green chemistry" concept, the development of new effective synthetic protocols, minimising wastes and energy-consumption while enhancing purity of the final product, is the corner stone (Anastas, 1998). In this regard, the use of microwaves as a non-conventional heating method has progressively gained attention due to commonly observed acceleration in reactions rates and improved (regio- and chemo-) selectivities and yields in synthetic organic transformations (Kappe, 2004; Caddick, 2009). The claimed cleaner reaction profiles of microwave-assisted processes have thus rapidly projected this kind of heating as a popular method in chemistry, which often replaces the "classical" heating ones. Numerous organic reactions are nowadays fully depicted in the peer-reviewed literature and books. They concern typical synthetic organic approaches (substitutions, alkylations, cycloadditions, esterifications, cyclisations, etc.), organometallic reactions, oxidations and reductions, or polymerisation reactions (Bogdal, 2006).

2. Scope of this contribution

Milder reaction conditions, associated to reduced run times (from several hours to a few minutes) and improved yields and selectivities, are the key advantages usually reported for microwave-promoted reactions. Even if widely employed in common organic chemistry, microwaves find however fewer applications in carbohydrate chemistry (Corsaro, 2004; Cioffi, 2008; Richel, 2011a). This fact, usually associated to a high number of unwanted and uncontrolled typical sugars reactions such as carbonisation or browning processes, has been overcame by the design of novel sophisticated microwave reactors. Advantageously, microwaves provide nowadays, with remarkable atom efficiency and yields, novel carbohydrate-based chemicals that are not easily obtainable by any other "classical" means or only using painstaking multi-step and energy-consuming protocols (Richel, 2011b). This chapter describes thus the use of microwave processes to mediate key reactions in the field of carbohydrate chemistry. Some examples of benchmark reactions (glycosylations, hydroxyl groups' protection, etc.) under microwave conditions are displayed and highlight the benefits of this microwave approach in terms of yields, selectivities and environmental impact.

3. Fundamental of microwave technology

The pioneering works in the microwave-assisted chemistry are attributed to Gedye and Giguere using household microwave ovens (Gedye, 1986; Giguere, 1986). From a theoretical point of view, microwave heating originates from interactions between a given material (reagent, solvent or catalyst) and the electric component of the electromagnetic wave through a dipolar polarisation and/or ionic conduction mechanism. Polar substances like DMSO, ethylene glycol or ethanol are thus suited candidates for microwave applications by opposition to benzene, dioxane or carbon tetrachloride which do not possess permanent dipolar moment. It is recognised that accelerations of reactions rates under microwave conditions are to correlate with a thermal/kinetic effect. Compared to analogous heating under classical conditions (using an oil bath for instance), microwave ensures a fast, selective and homogeneous heating of the reaction vessel (Fig. 1). Convection currents and temperature gradients from vessel walls to the core of the reaction medium, commonly observed using classical heating options, are thus avoided under microwave conditions.

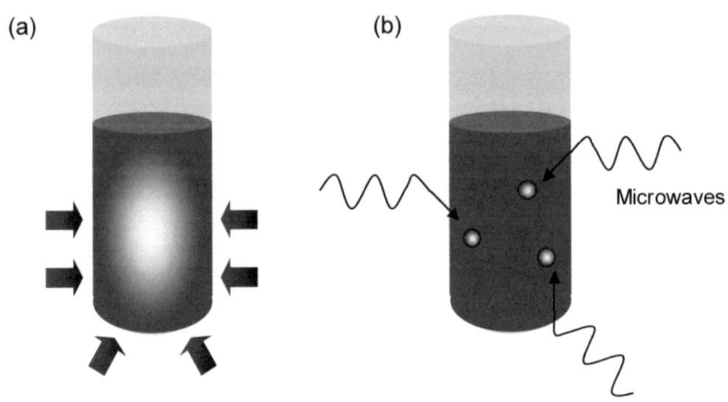

Figure 1. Difference between conventional heating (**a**) and microwave-assisted heating (**b**). In (**a**), heating is provided by thermal conduction. In (**b**), solvent and/or reagents absorb microwave heating and convert this energy into calorific components.

The ability of a given substance to convert electromagnetic energy into calorific heat, at specific frequency and temperature, is determined by the loss tangent factor ($\tan\delta$). This factor is described as the ratio between dielectric loss (ε'') and dielectric constant (ε'). Practically, substances with high $\tan\delta$ values are required to ensure a convenient and fast microwave heating. By convention, substances are thus categorised into good ($\tan\delta>0.5$), medium ($0.5>\tan\delta>0.1$) and poor ($\tan\delta<0.1$) candidates for microwave (MW) applications. At the lab-scale, MW is conveniently applied to reactions performed in the liquid state in the presence of a solvent with a high $\tan\delta$. However, in order to provide a truly eco-friendly process, syntheses without solvent or in dry-media conditions are also described with reactants and/or catalysts impregnated on a solid microwave-absorbing support such as alumina or silica (Polshettiwar, 2008). In the field of carbohydrate chemistry, reactions under microwave (MW) assistance are usually achieved with saccharides in suspension or in solution in solvent and/or liquid reagents. Highly secured MW reactors, operating in batch or continuous flow conditions, are nowadays available for chemistry. They operate typically at 2.45 GHz and allow an accurate control of the temperature within the MW reaction vessel with an adjustable power, avoiding thus browning. Reactions can also be performed under "closed vessel conditions", at temperatures exceeding the boiling point of the reaction medium.

4. Chemical transformations of carbohydrates under microwave conditions

Saccharides, which contain several hydroxyl groups, are polar molecules particularly suited for MW applications. Typical MW-assisted transformations reported in the literature include either reactions involving hydroxyl groups for the production of novel entities (category 1) or dehydration reactions leading to the formation of furfural and related platform molecules (category 2).

4.1. Microwave-assisted transformations involving hydroxyl groups (category 1)

Most common reactions of the first category include protections/deprotections of hydroxyl groups, glycosylations, glycosamines formation and halogenations. These reactions are usually catalysed by chemical entities, even if enzymes are proved to offer convenient results for specific cases under microwave conditions. The several hydroxylated stereogenic centres in carbohydrates are difficult to chemically differentiate. Thus, the regioselective derivatisation of such compounds is a challenging task and requires multi-step protocols and purifications. Microwave heating has demonstrated to be a convenient solution, allowing an accurate control of the regioselectivity and the anomeric selectivity by an appropriate tuning of operating conditions. Some relevant examples are proposed herein.

4.1.1. Hydroxyls protections/deprotections

Several methods and protocols have been reported for the MW-assisted regioselective protection of hydroxyls, especially in mono- and disaccharides (Corsaro, 2004). They include mainly acetal formation with aldehydes and acylation with acetic anhydride or an acyl chloride. All these reactions are promoted using (homogeneous or heterogeneous) acids or bases (Söderberg, 2001) or enzymes (Chen, 2001) in variable solvents, reagents concentrations and operating conditions (temperature and microwave exposure time). The total or regioselective protection (and subsequent deprotection) of specific hydroxyl groups is necessary for the synthesis of building blocks suited for the production of drugs, biologically active materials, and novel high-added value materials. Particularly, acetylation reactions are often the prerequisite step in the synthesis of such complex carbohydrates (Fig. 2). The conventional per-*O*-acetylation of saccharides is mostly achieved using anhydride acetic as both reagent and solvent. Although this step is high-yielding, long reaction times (several hours) are required and noxious pyridine is mandatory to promote the reaction. Microwave dielectric heating has thus appeared as a more eco-friendly option. Indeed, replacement of pyridine by catalytic amounts of non-toxic sodium acetate or Lewis acid such as zinc chloride is found practically quantitative after only a few minutes of MW heating (Limousin, 1997). With $ZnCl_2$, an equimolar mixture of α and β pentaacetates is obtained under MW conditions, whilst a α/β ratio of 7/3 is recovered under oil bath conventional heating (Fig. 2a). Microwave heating affords also convenient results for the complete acetylation of totally *O*-unprotected mono- and disaccharides in 90 sec at 720 Watt under closed vessel conditions using indium(III) chloride catalyst. Reactions performed in acetonitrile with stoichiometric amounts of anhydride acetic are quantitative and affords predominantly the α–peracetylated form (Das, 2005). Partial acetylation of carbohydrates is usually achieved with acetyl chloride and pyridine. With 1,2:5,6-di-*O*-isopropylidene-α-D-glucofuranose as the starting material, Söderberg demonstrates that the synergic combination of microwaves and polystyrene-supported base catalysts offers a convenient strategy to obtain in 85% yield the expected mono-acetylated analogous without formation of side-products (Fig. 2b). This MW approach provides an eco-friendly and time-saving option as the catalyst can be recovered by filtration at the end of the process, thus minimising wastes and purification (Söderberg, 2001).

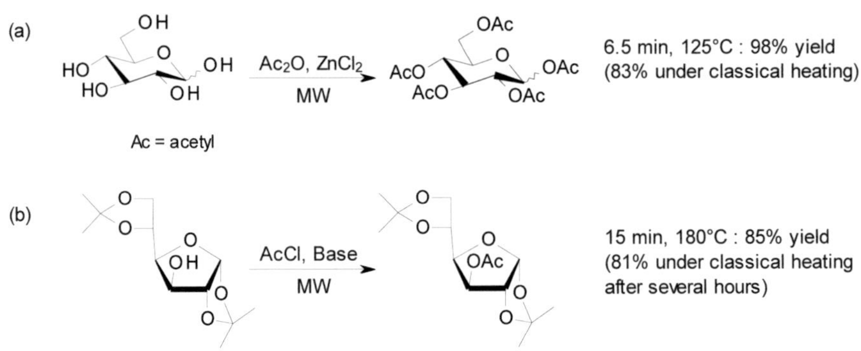

Figure 2. Peracetylation (a) and acetylation (b) of selected carbohydrates under MW conditions.

More recently, Witschi et al. describe a novel procedure for the selective acetylation of hydroxyls groups. It relies on a protecting group exchange strategy under microwave assistance starting from per-*O*-trimethylsilylated pyranosides (Fig. 3). Interestingly, whilst the 1,6-*O*-diacetate monosaccharide is formed under MW exposure, the 6-*O*-monoacetate adduct is preferentially formed under classical oil bath heating conditions. Reactions go to completion when conducted using acetic acid as a catalyst in neat anhydride acetic for 3 x 25 min under MW. For per-*O*-TMS-galactoside, a 52% yield in diacetylated adduct is recovered using 2 equivalents of acetic acid. When the galactoside is stirred for 2 days with 2 equiv of acid, the reaction shows a 50% completion with yields after column chromatography purification reaching 35% of 6-*O*-monoacetate (Witschi, 2010).

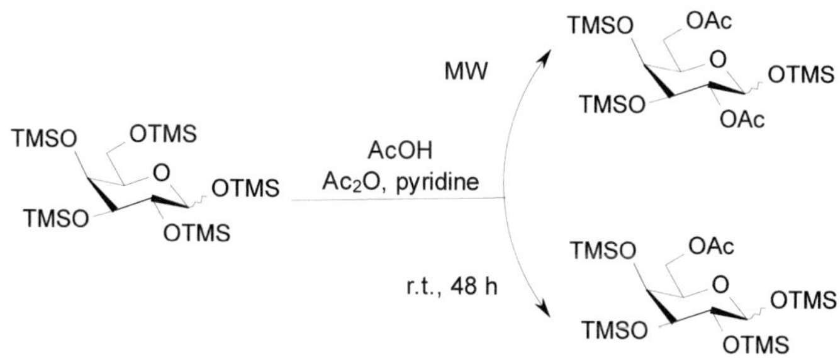

Figure 3. MW-assisted selective acetylation of per-*O*-TMS-galactoside using protecting group exchange strategy.

Regioselective benzoylation and pivaloylation of carbohydrates is another major item for free hydroxyls protection. Typically performed using benzoyl chloride and pivaloyl chloride, these reactions are only viable when using a base catalyst. Due to the steric bulk of the pivaloyl groups, the protection of hydroxyls through formation of pivaloyl esters is very slow under conventional heating method and requires several hours. Substantial reduction in reaction times (10-20 min) is however encountered under MW conditions. Pivaloylation of 1,2:5,6-di-*O*-isopropylidene-α-D-glucofuranose, catalysed by pyridine, provides thus the corresponding 3-*O*-protected target in 68% yield after 10 min at 160°C in a microwave reactor (Fig. 4a). However, presence of a side-product is recovered after MW exposure as a result of 5,6- to 3,6-acetal migration prior to acylation. The formation of this unwanted structure is repressed when using supported polystyrene-bases, i.e. *N,N*-(diisopropyl)aminoethylpolystyrene (PS-DIEA) and *N*-(methylpolystyrene)-4-(methylamino)pyridine (PS-DMAP) , as alternatives to pyridine. 79 and 88% yields are thus detected after 15-20 min at 160-180°C in the microwave cavity, whilst only 88% yield is obtained after 300 min at room temperature (Söderberg, 2001).

The regioselective benzoylation of carbohydrates affords convenient results using dibutyltin oxide mediators under microwave exposure (Herradón, 1995). Intermediate formation of dibutylstannylene acetal allows hydroxyl regioselective benzoylation after less than 10 min of microwave exposure with a 500 Watt output (Fig. 4b). This MW methodology is more advantageous than its "classical" analogous involving heating, for several hours, at reflux of a toluene suspension of saccharides and dibutyltin oxide with azeotropic removal of water. Tuning of operation conditions, including both the solvent and the MW power output, modulates the selectivity of dibutylstannylene acetal-mediated benzoylation of representative carbohydrates, such as α-D-mannopyranoside (Fig. 5).

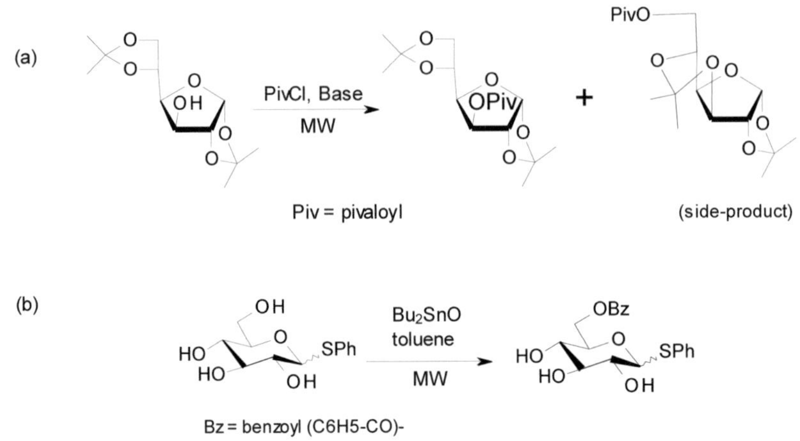

Figure 4. MW-assisted pivaloylation (a) and benzoylation (b).

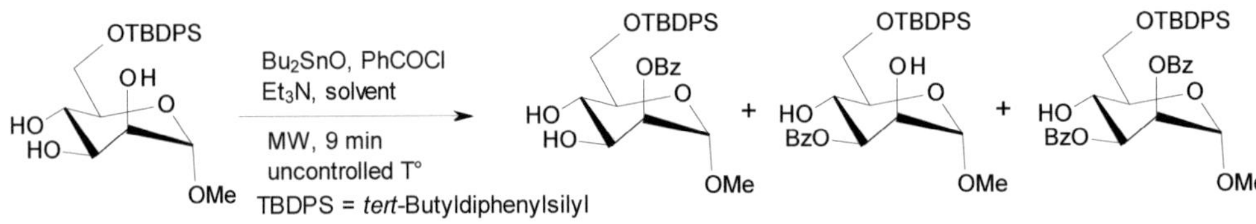

Toluene, maximum MW power output	Yield:	35%	25%	10%
Toluene, minimum MW power output		6%	33%	0%
Acetonitrile, minimum MW power output		0%	41%	0%

Figure 5. MW-promoted benzoylation of α-D-mannopyranoside with a tin catalyst.

For carbohydrates not stable in acid conditions, such as 4,6-*O*-benzylidene-α-D-glucopyranose, benzoylation can be achieved via transesterification with methyl benzoate using a week and non nucleophilic base promoter like potassium carbonate and a phase transfer agent (Limousin, 1998). MW permits again a tuning of the regioselectivity. Indeed, after 15 min in a domestic microwave oven, a 82% yield of di-benzoylated adduct is recovered, whilst classical oil bath heating affords preferentially the monobenzoylated structures (Fig. 6).

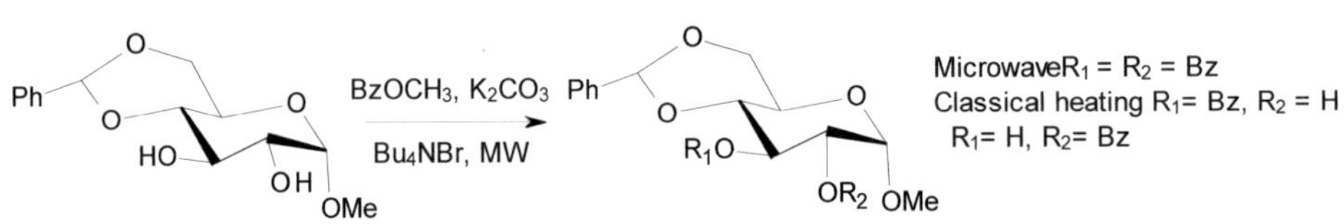

Figure 6. Competition between mono- and dibenzoylation with base catalyst.

Due to the sensitivity of some carbohydrate derivatives at elevated temperatures, microwave methodology suffers however from several limitations, notably for hydroxyls protection processes through acetal formation. As an illustration, the mono-acetalation of *O*-unprotected sucrose in the presence of *p*-toluenesulfonic acid as the catalyst is reported by Salanski and Queneau (Fig. 7). Using citral dimethylacetal, transacetalation provides, under classical heating conditions, 83% yield of a mixture of geranial and neral sucrose acetals (*E* and *Z* isomers of citral acetals) after 2 min at 100°C. When applying microwave as the heating source, yield drops to 42% after an identical runtime in open vessel conditions. Cleavage of the glycosidic linkage is denoted and leads also to the formation of unwanted side-products (17-26% yield after 2 to 10 min of MW exposure) (Salanski, 1998).

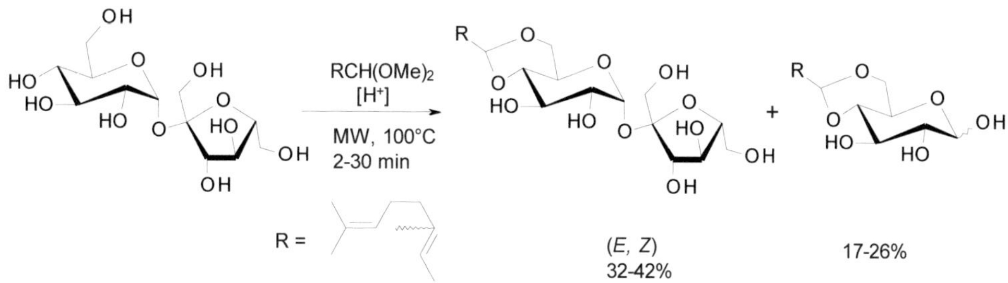

Figure 7. Protection of sucrose through acetal formation.

For deprotection processes, microwave heating is decisive and powerful, as a drastic diminution of reaction times is observed without affecting the initial anomeric configuration (Corsaro, 2004). An efficient and green protocol concerns the use of neutral alumina to ensure the regioselective depivaloylation of a set of carbohydrate derivatives (Fig 8a). Whilst the depivaloylation does not occur using a conventional heating source, the regioselective deprotection of primary hydroxyl functions is observed after a few minutes à 75°C under MW assistance (Ley, 1993). The fast, efficient and clean deacetalation of several carbohydrate 4,6-di-*O*-benzylidene acetals is also reported using silica supported reagents under microwave irradiation in solvent-free conditions (Fig. 8b). Yields are superior to 80% after 7 min in a domestic oven at 500 Watt, without anomerisation effect (Couri, 2005).

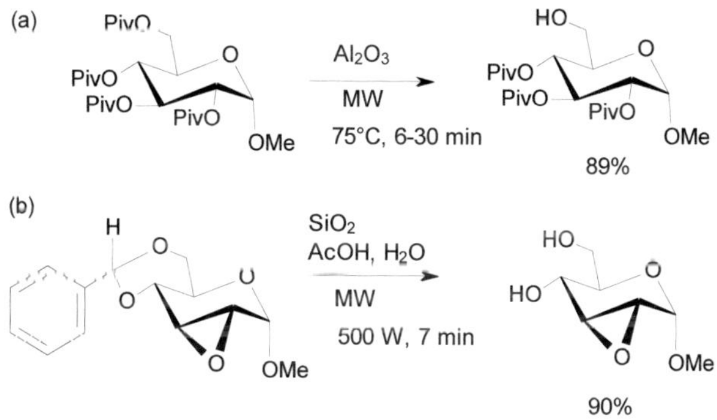

Figure 8. MW assisted deprotection initiatives: deacylation (a) and deacetalation (b).

4.1.2. *O-glycosylations*

O-glycosylation consists in the specific derivatisation of the anomeric hydroxyl position in carbohydrate structures. This reaction is one of the most emblematic in carbohydrate chemistry and is extensively reviewed (Demchenko, 2008). Several parameters are found to influence the (stereoselective) formation of glycosyl linkages such as solvent, catalyst, additives, nature of both leaving groups and surrounding protecting groups. In particular, the formation of alkyl- and aryl-glycosides has find special attention in both academic and industrial worlds, due to the importance of these molecules as non-toxic and biodegradable surfactants, liquid crystals, and pharmaceutical agents (Razafindralambo 2011 & 2012). Under microwave conditions, the introduction of alkyl substituents on the OH1 position is the most extensively developed, notably for the large-scale production of surface-active agents, and relies on the reaction of an alkyl alcohol with unprotected saccharide (Fischer-type glycosylation) or with completely *O*-protected carbohydrate compounds (Helferich-type glycosylation). These glycosylations are catalysed either by chemical promoters or by enzymes. Alkylation of D-glucose by fatty alcohol using acid catalysts is the oldest reaction reported under MW conditions (Fig. 9). Fatty alcohol advantageously plays both the role of solvent and reagent, even if glucose is insoluble in most alcohols. After 10 min of MW exposure with a 200 Watt maximum output, a 15% yield in glycosylated adduct is obtained using homogeneous Brönsted (or Lewis) acids or heterogeneous promoters (zeolites, ion exchange resins, montmorillonites). No control of the temperature is provided using this domestic MW equipment. Such a low yield is to correlate with unwanted glycoside decomposition under acidic conditions during the run and to browning reactions (Limousin, 1997). This result is not completely satisfactory but remains quite noteworthy compared to other classical protocols. The versatility of this MW Fischer glycosylation is now published for various carbohydrates, including mostly easily available monosaccharides like D-mannose, D-galactose, *N*-acetyl-D-glucosamine and *N*-acetyl-D-galactosamine. Glycosylation of representative long chain alcohols is efficient with acidic ion-exchange resins between 90 and 120°C. Typically, the reaction goes to completion after less than 2 min of MW heating, compared with 4h at reflux, and affords preferentially the thermodynamic α adduct. The anomeric distribution seems quite identical under MW and conventional oil bath heating conditions (Bornaghi, 2005). A valuable scaling-up of this Fischer-type glycosylation is proposed by Nüchter up to the kg-scale with an efficient improvement of economic efficiency, using an appropriated commercial MW batch reactor (Nüchter, 2001). An accurate control of the temperature inside the reactor cavity provides a good quality result. Practically, monosaccharide (glucose, mannose or galactose) reacts with a 3-30-fold excess of alcohol (methanol, ethanol, butanol or octanol) in the presence of sulphuric acid catalyst. Nearly quantitative yields are recovered after 20-60 min of MW heating at 60-140°C, as a function of the nature of the alcohol. Further assays, using continuous flow methods, afford quantitative yields with a MW chamber heated at 110-140°C. More recent work highlights the use of Brönsted acids immobilised on silica as convenient solid catalysts with advantages in manufacturing scale synthesis (Fig. 9). The Fischer-type glycosylation of D-glucose with *n*-decanol in solvent-less conditions leads to the expected adduct in 82% yield as an equimolar mixture of α and β anomers (Richel, 2011b). This protocol offers workup, economic and environmental advantages as catalyst can be recovered by filtration and reused.

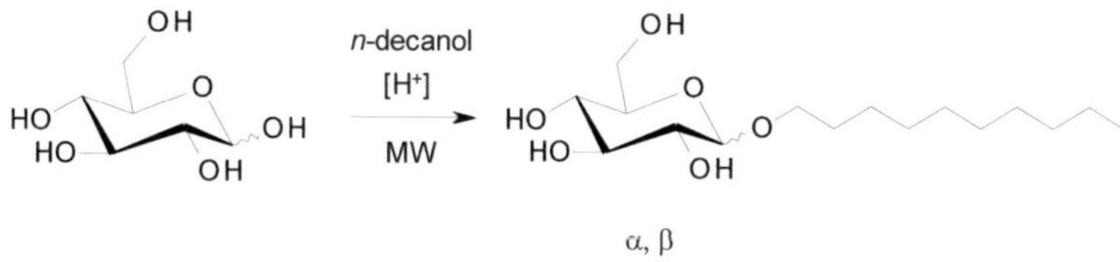

Figure 9. Fischer-type glycosylation of D-glucose with *n*-decanol under MW conditions for the production of a potent surface-active agent.

An alternative three-step protocol (complete acylation-glycosylation-deprotection) is proposed under MW. Due to the increased thermal stability of acylated compounds, MW Helferich-type reactions are expected to be more effective (Fig. 10). Indeed, addition of decanol on per-*O*-acetylated glucose offers a 74% yield (mainly the α-pyranose form) after a running time of 3 min at 113°C using zinc chloride as the catalyst. Extended MW exposure times have a detrimental effect over the yield (25% after 300 min) due to decomposition of decyl glycoside. No rationalisation between the furanose-pyranose (and anomeric) distribution and processing conditions is found at this stage (Limousin, 1997). The first elements of answer are proposed by Kovensky and Ferlin in 2008. A microwave temperature of 115°C is found as a good compromise to obtain high glycosylation rates and low decomposition products contents. Both yields and anomeric distributions are dependent on the boiling/melting points and the polarity of the selected alkyl alcohol. Polar alcohols with low boiling/melting points lead to the formation of the corresponding thermodynamic α adduct, while long chain alcohols provide preferentially the kinetic β anomer in less than 5 min. For each alcohol, an optimum reaction time, beyond which the formed alkyl glycoside is deteriorated, is estimated (Ferlin, 2008).

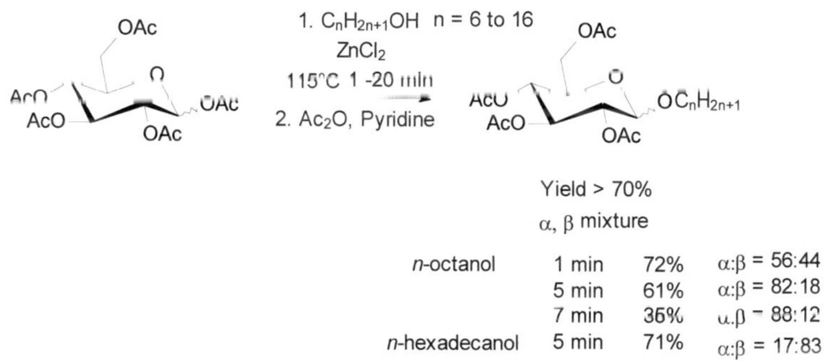

Figure 10. Helferich-type glycosylation of per-*O*-acetylated glucose with alkyl alcohols. Influence of the chain length on both yields and α/β ratio.

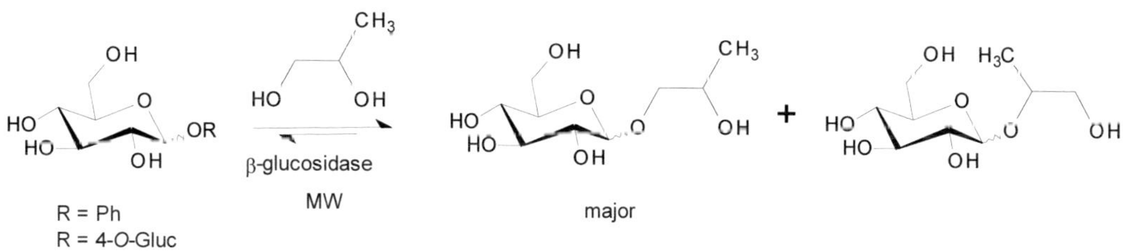

Figure 11. MW-assisted trans-glycosylation of several carbohydrates acceptors and propane-1,2-diol.

Enzymatic methods under microwave conditions are also explored as convenient option for glycosylations. Using β-glucosidase, the synthesis of several glycosides by trans-glycosylation is reported with either phenyl β-D-glucoside or cellobiose as donors and propane-1,2-diol as acceptor (Fig. 11). The process is performed in dry-media conditions, with reagents adsorbed on alumina in an open vessel system. This microwave reaction, achieved at controlled temperature (between 80 and 110°C), leads to noticeable advantages, in terms of yields and selectivity, when compared to classical heating. In particular, trans-glycosylations go to completion after less than 2h, while hydrolysis is lowered to 10% (Gelo-Pujic, 1997).

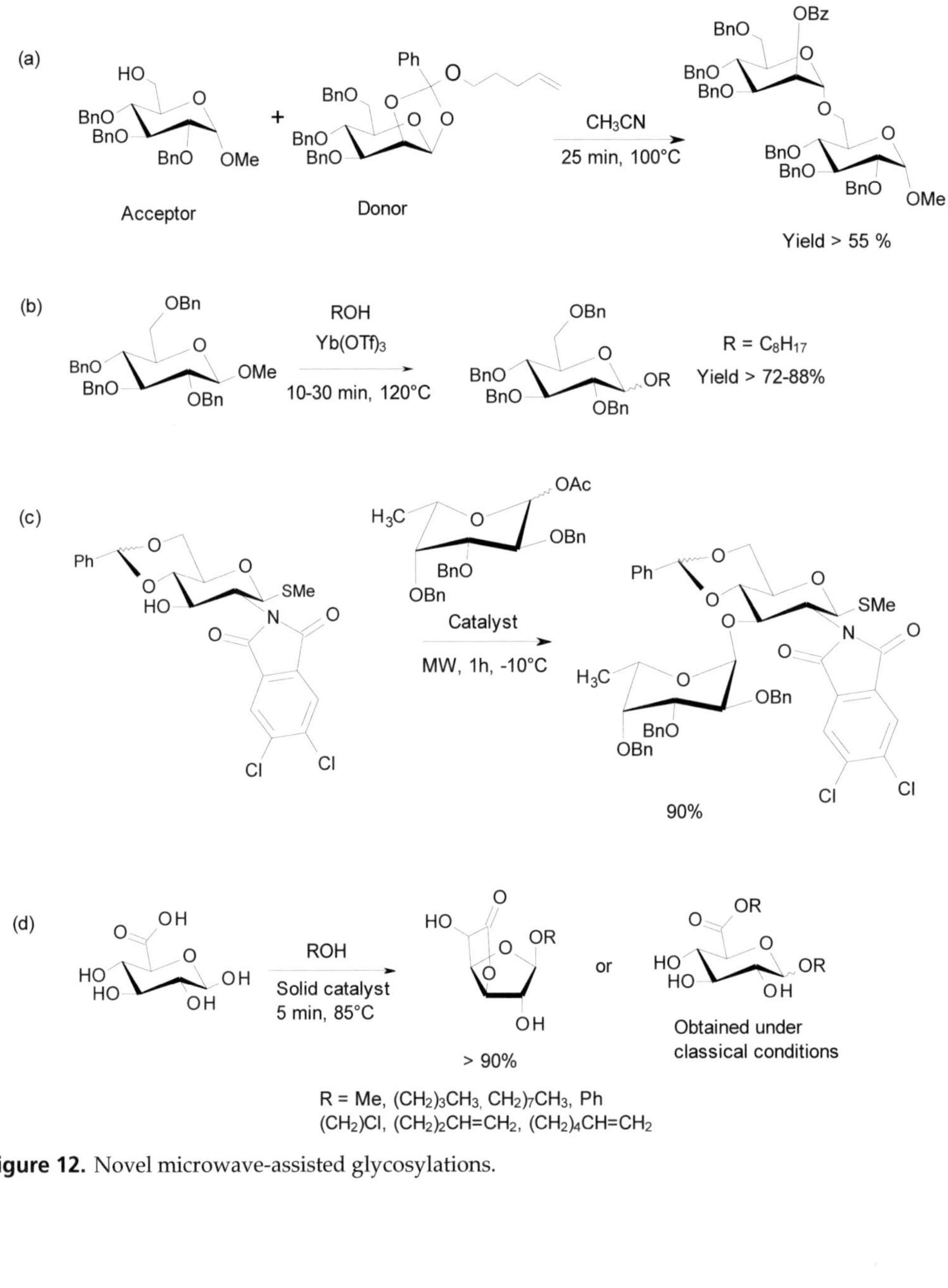

Figure 12. Novel microwave-assisted glycosylations.

Beside the benchmark Fischer- and Helferich-glycosylations, novel strategies under microwave heating are nowadays emerging in the peer-reviewed literature. Some examples are proposed in Fig. 12. Most of them concern the catalytic conjugation of alcohols (designated as acceptors) to protected 4-pentenyl glycosides or methyl glycosides (donors), notably for the production of di- and oligosaccharides. As an illustration, the synthesis of disaccharides is described in acetonitrile starting from 4-pentenyl donors and is achieved in high yields after less than 25 min at 100°C in closed vessel conditions without side-products formation (Fig. 12a) (Mathew, 2004). The use of more effective Lewis acids, like ytterbium(III) triflate, affords high yields for the anchorage of long alkyl chains on benzoylated glucose derivatives (Fig 12b). This acetal-exhange type glycosylation is performed in dichloromethane below 100°C in closed vessel conditions. This constitutes the main advantage of microwave over conventional heating techniques (Yoshimura, 2005). Application of very low temperatures (-10°C) within the MW cavity, attained by simultaneous cooling, offers a convenient approach for the synthesis of valuable saccharides and Lewis X oligosaccharides (Fig. 12c). The process involves a fucosyl acetate donor and a glucosamine acceptor in about 2h in ether. Yields culminate at about 80-90% without side-products formation, whilst the expected disaccharide is not obtained using classical protocols (Shimizu, 2008). Another strategy demonstrates the usefulness of microwave heating, in synergy with heterogeneous acid catalysts, for the regioselective functionalisation of D-glucuronic acid. Conversions are quantitative when performed in solvent-less conditions with an excess of alcohol under microwave exposure. Less than 10 min of heating at 85°C in closed vessel conditions affords selectively the corresponding β-glucurono-6,3-lactone adduct. When heating in an oil bath, only the disubstituted adducts, as a mixture of α,β-furanoses and α,β-pyranoses, are formed through competitive esterification and *O*-glycosylation processes (Fig. 12d). Comparative yields are recorded after 6 to 24 h at 85°C, evidencing that MW accelerates reactions and imparts over the selectivity. This methodology is eco-friendly as water in the only by-product and the solid catalyst can be reused for consecutive batch MW runs (Richel, 2012). These monosubstituted glucuronolactones are not easily attainable using other conventional protocols and required multi-step synthesis. These original products find special relevance in the field of surface-active agents and emulsifiers (Razafindralambo, 2011).

4.1.3. *Glycosamines synthesis*

Glycosamines are valuable compounds as glyco-amino acid building blocks for glycopeptides synthesis. Their preparation is traditionally achieved using a Kochetkov reductive amination procedure. Practically, a treatment at room temperature for 6 days of saccharides in aqueous solution with 50 equiv of ammonium bicarbonate affords the expected β-glycosaminated adducts in high yields. Another route involves the heating at 42°C for 36 h of mono- or disaccharides with an aqueous solution of ammonia in the presence of one equivalent of ammonium hydrogen carbonate. However, β-glycosamines can be deteriorated under aqueous solutions, even with accurate pH control. Nowadays, the preparation of these glycosamines is typically performed under microwave conditions, at 40°C for 90 min, by selective amination of reductive carbohydrates using reduced quantities of ammonium bicarbonate (5 equiv) (Fig. 13). Under closed vessels conditions, the reaction is achieved in methanol or in anhydrous dimethylsulfoxide to prevent degradation and side-products formation (Liu, 2010). Yields are of about 70-95%.

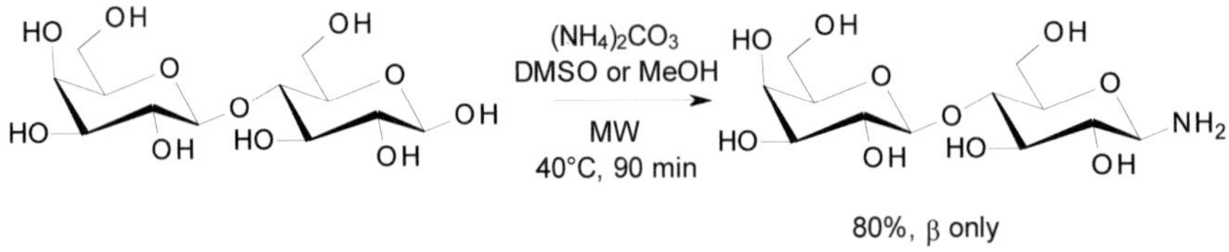

Figure 13. Synthesis of 1-amino-1-deoxy-β-D-lactoside by microwave-assisted Kochetkov amination of unprotected lactose.

In summary, reactions of the first category, involving hydroxyls manipulation appear generally faster under MW conditions than their "classical" counterparts. Expected products are indeed obtained in good yields after only a few minutes. The use of eco-friendly and non-toxic (heterogeneous) and milder catalysts in synergy with microwaves, affords ideal "green chemical" conditions and minimise workup. Microwave allows also the formation of specific protected products, not easily accessible under conventional "heating" conditions.

4.2. Microwave-assisted production of platform chemicals (category 2)

The production of 5-HMF (5-hydroxymethylfurfural), a five-membered ring compounds ranging in the US Top 10 of most valuable chemicals issued from biomass, is currently extensively investigated worldwide. Most research concerns its production via thermally-induced acid-catalysed dehydration of fructose, typically in dimethylsulfoxide, in the presence of homogeneous acid catalysts (HCl, H_2SO_4, H_3PO_4, oxalic or levulinic acids) (Fig. 14). Reactions produces selectively 40-60% yields of 5-HMF with complete conversion of starting sugar materials but suffers from drawbacks in terms of acids recovery and equipment corrosion. Heterogeneous catalysts offer workup and environmental advantages associated to high 5-HMF selectivity but provide low fructose conversions (30-60% only after up to 2h). Microwave has progressively emerged as an attractive heating source allowing energy savings and improved yields and selectivity. Recent developments in the field have been reviewed by Richel in 2011 (Richel, 2011b). They include the use of heterogeneous catalysts in various media (organic solvents, water, ionic liquids or biphasic systems).

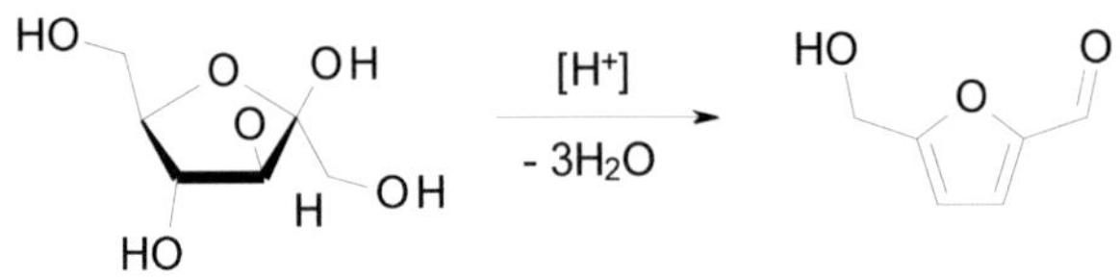

Figure 14. Catalytic dehydration of fructose for the production of 5-HMF.

Titanium and zirconium oxides, as sulphated zirconia, are effective catalysts for this dehydration, especially under microwave conditions. Yields in 5-HMF reach 35% for a fructose conversion of 78% after only 5 min heating at 200°C in closed vessels conditions ("microwave hot compressed water process"), while 5-HMF yield and fructose conversion are respectively of 12 and 27% after an identical runtime using a sand bath heating. Zirconium oxide demonstrates great performances for the catalytic dehydration of glucose, favouring isomerisation of glucose to fructose prior dehydration (Qi, 2009). When using an acid ion-exchange resin in an acetone/dimethylsulfoxide system, improvement in 5-HMF selectivity is encountered together with high yields (98% fructose conversion with 5-HMF selectivity of 92% for 10 min MW assisted reaction at 150°C in closed vessel conditions). Addition of low-boiling point acetone improves products separation and reduces adverse environmental impacts. The same results are published for strong ion-exchange resin in acetone-water systems, providing a truly eco-friendly process (Qi, 2008b). The common Lewis acid catalyst, aluminium(III) chloride, promotes also efficiently the dehydration of carbohydrate substrates, in water-organic solvents and in water. 5-HMF yield is higher when using an organic solvent than in pure water. Involvement of ionic liquids as convenient media for the conversion of D-glucose in 5-HMF is effective using chromium(III) chloride as a catalyst. With microwave heating at 140°C for 30 sec, a 5-HMF yield of 71% is obtained for 96% glucose in 1-butyl-3-methyl imidazolium chloride as the solvent (Qi, 2010). Lignin-based solid catalyst, produced by carbonisation and sulfonation of raw lignocellulosic materials, seems nowadays to be the most active catalyst for the dehydration of carbohydrates. In synergy with MW, 84% 5-HMF yield is obtained with 98% fructose conversion rate at 110°C for 10°C. A mixture of dimethylsulfoxide-ionic liquid has a beneficial effect over the selectivity of the process. Interestingly, the catalyst can be recovered after the reaction and reused for consecutive batch reactions (Guo, 2012).

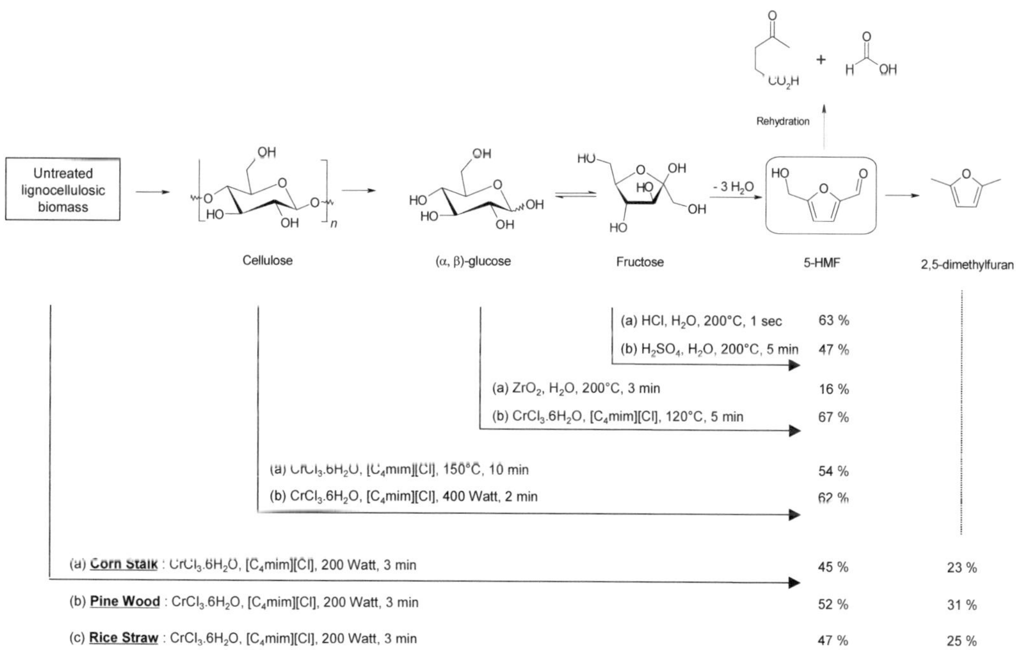

Figure 15. Overview of MW-assisted production of 5-HMF from lignocellulosic biomass materials.

Scaling up of the MW-assisted fructose dehydration process has been recently evaluated using a stop-flow MW reactor system and compared to a continuous cartridge-based reactor. Results highlight that 5-HMF yield can attain 86% yield after 2min of MW exposure (91% using a cartridge-base reactor) with a 5-HMF productivity of 0.72 g/h (versus 2.07 g/h for the continuous process). Even if quite less productive, the MW-assisted methodology is more selective and cleaner as no side-product is detected in the final mixture (Schoen, 2011).

Besides these aforementioned benchmark reactions, research has shifted toward the exploration of other carbohydrates or crude lignocellulosic feedstocks as raw materials. Some relevant illustrations are reported in Fig. 15. They concern the use of ionic liquids as solvent. For environmental concern, water is also envisioned with homogeneous Brönsted acids or metal chloride catalysts (Dutta, 2012). The production of 2-furfural, another platform chemicals, receives progressively interest. Its MW-assisted production from xylose, xylan and wheat straw is highly efficient in the presence of hydrochloric acid. A 48% furfural yield is obtained from wheat straw after 1 min of microwave heating between 140 and 190°C. (Yemis, 2011).

5. Conclusion

In the present chapter, we contribute to illustrate some relevant carbohydrate reactions under microwave heating. This heating mode appears as efficient for several key reactions like hydroxyls protection and deprotection and *O*-glycosylations. The production of 5-HMF from mono-/di-/polysaccharides is another special subject in MW science. Improvement of both yields and (regio- and anomeric) selectivities is generally reported together with a decrease in reaction times. MW offers also advantages as energy-saving and workup facilities in agreement with the green chemistry concept.

Author details

Aurore Richel and Michel Paquot
University of Liege, Belgium

Acknowledgement

This work was carried out in the framework of the "Technose" Excellence Programme (project number 716757) supported by the "Région Wallonne" (Belgium). The authors are grateful to the "Région Wallonne" for its financial support.

6. References

Anastas, P. T.; Warner, J. C. (1998). Green Chemistry: Theory and Practice; Oxford University Press: Oxford

Bogdal, D. (2006). Microwave-assisted synthesis. One hundred reaction procedures. Tetrahedron Organic Chemistry Series, Vol. 25; Elsevier.

Bornaghi, L. F.; Poulsen, S. A. (2005).

Bozell, J. J.; Petersen, G. R. (2010). Technology development for the production of biobased products from biorefinery carbohydrates—the US Department of Energy's "Top 10" revisited. *Green Chem.*, 12, 539-554

Caddick, S.; Fitzmaurice, R. (2009). Microwave enhanced synthesis. *Tetrahedron*, 65, 3325-3355

Chen, S. T.; Sookkheo, B.; Phutrahul, S.; Wang, K. T. (2001). Enzymes in nonaqueous solvents: Applications in carbohydrate and peptide preparation. *Methods in Biotechnology*, 15, 373-400.

Cioffi, E. A. (2008). High-energy glycoconjugates: synthetic transformations of carbohydrates using microwave and ultrasonic energy. Curr. Top. Med. Chem., 8, 152-158

Corsaro, A.; Chiacchio, U.; Pistarà, V.; Romeo, G. (2004). Microwave-assisted chemistry of carbohydrates. *Curr. Org. Chem.*, 8, 511-538

Das, S. K.; Reddy, K. A.; Krovvidi, V. L. N. R.; Mukkanti, K. (2005). $InCl_3$ as a powerful catalyst for the acetylation of carbohydrate alcohols under microwave irradiation. *Carbohydr. Res.*, 340, 1387-1392

Demchenko, A. V. (2008). Handbook of Chemical Glycosylation; Advances in Stereoselectivity and Therapeutic Relevance, Wiley-VCH, Weinheim

Dutta, S.; De, S.; Alam, M. I. ; Abu-Omar, M. M. ; Saha, B. (2012). Direct conversion of cellulose and lignocellulosic biomass into chemicals and biofuel with metal chloride catalysts. *J. Catal.*, 288, 8-15

Ferlin, N. ; Duchet, L. ; Kovensky, J. ; Grand, E. (2008). Microwave-assisted synthesis of long-chain alkyl glucopyranosides. *Carbohydr. Res.*, 343, 2819-2821

Gedye, R. ; Smith, F.; Westaway, K.; Ali, H.; Baldisera, L.; Laberge, L.; Roussel, J. (1986). The use of microwave ovens for rapid organic synthesis. *Tetrahedron Lett.*, 27, 279-282

Gelo-Pujic, M.; Guibé-Jampel, E.; Loupy, A.; Trincone, A. (1997). Enzymatic glycosidation in dry media under microwave irradiation. *J. Chem. Soc., Perkin Trans.* 1, 1001-1002

Giguere, R. J.; Bray, T; L.; Duncan, S. M.; Majetich, G. (1986). Application of commercial microwave ovens to organic synthesis. *Tetrahedron Lett.*, 27, 4945-4948.

Herradón, B.; Morcuende, A., Valverde, S. (1995). Microwave Accelerated Organic Transformations: Dibutylstannylene Acetal Mediated Selective Acylation of Polyols and Amino Alcohols using Catalytic Amounts of Dibutyltin Oxide. Influence of the Solvent and the Power Output on the Selectivity. *Synlett*, 455-458

Guo, F.; Fang, Z.; Zhou, T. J. (2012). Conversion of fructose and glucose with lignin-derived carbonaceous catalyst under microwave irradiation in dimethyl sulfoxide-ionic liquid mixtures. *Biores. Technol.*, 112, 313-318.

Kappe, C. O. (2004). Controlled microwave heating in modern organic synthesis. *Angew. Chem. Int. Ed.*, 43, 6250-6284

Lichtenthaler, F. W. ; Peters, S. (2004). Carbohydrates as green raw materials for the chemical industry. *C. R. Chimie*, 7, 67-90

Limousin, C.; Cléophax, J.; Petit, A.; Loupy, A. ; Lukacs, G. (1997). Solvent-Free Synthesis of Decyl D-Glycopyranosides Under Focused Microwave Irradiation. *J. Carbohydr. Chem.*, 16, 327-342

Limousin, C.; Cléophax, J.; Petit, A.; Loupy, A. (1998). Synthesis of benzoyl and dodecanoyl derivatives from protected carbohydrates under focused microwave irradiation. *Tetrahedron*, 54, 13567-13578

Liu, X.; Zhang, G.; Chan, K.; Li, J. (2010). Microwave-assisted Kochetkov amination followed by permanent charge derivatization: a facile strategy for glycomics. *Chem. Commun.*, 46, 7424-7426.

Mathew, F.; Jayaprakash, K. N.; Fraser-Reid, B.; Mathew, J.; Scicinski, J. (2004). Microwave-assisted saccharide coupling with *n*-pentenyl glycosyl donors. *Tetrahedron Lett.*, 44, 9051-9054.

Nüchter, M.; Ondruschka, B.; Lautenschlarger, W. (2001). Microwave-assisted synthesis of alkyl glycosides. *Synth. Commun.*, 31, 1277-1283

Polshettiwar, V.; Varma, R. S. (2008). Microwave-Assisted Organic Synthesis and Transformations using Benign Reaction Media. *Acc. Chem. Res.*, 41, 629–639.

Qi, X.; Watanabe, M.; Aida, T.; Smith, T. L. (2008). Catalytical conversion of fructose and glucose into 5-hydroxymethylfurfural in hot compressed water by microwave heating. *Catal. Commun.*, 9, 2244-2249.

Qi, X.; Watanabe, M.; Aida, T.; Smith, T. L. (2009). Sulfated zirconia as a solid acid catalyst for the dehydration of fructose to 5-hydroxymethylfurfural. *Catal. Commun.*, 10, 1771-1775

Qi, X.; Watanabe, M.; Aida, T.; Smith, T. L. (2010). Fast transformation of glucose and di-/polysaccharides into 5-hydroxymethylfurfural by microwave heating in an ionic liquid/catalyst system. *ChemSusChem*, 3, 1071-1077

Razafindralambo, H.; Richel, A.; Wathelet, B.; Blecker, C.; Wathelet, J. P.; Brasseur, R.; Lins, L.; Miñones, J.; Paquot, M. (2011). Monolayer Properties of Uronic Acid Bicatenary Derivatives at the Air-Water Interface: Effect of Hydroxyl Group Stereochemistry Evidenced by Experimental and Computational Approaches. *Phys. Chem. Chem. Phys.* 13, 15291–15298

Razafindralambo, H.; Richel, A.; Paquot, M. Lins, L.; Blecker, C. (2012). Liquid Crystalline Phases Induced by the Hydroxyl Group Stereochemistry of Amphiphilic Carbohydrate Bicatenary Derivatives. *J. Phys. Chem. B*, 116, 3998-4005

Richel, A.; Laurent, P.; Wathelet, B.; Wathelet, J. P.; Paquot, M. (2011a). Microwave-assisted conversion of carbohydrates. State of the art and outlook. *C. R. Chimie*, 14, 224-234.

Richel, A.; Laurent, P.; Wathelet, B.; Wathelet, J. P.; Paquot, M. (2011b). Current perspectives on microwave-enhanced reactions of monosaccharides promoted by heterogeneous catalysts. *Catal. Today*, 107, 141-147

Richel, A.; Laurent, P.; Wathelet, B.; Wathelet, J. P.; Paquot, M. (2012). Efficient microwave-promoted synthesis of glucuronic and galacturonic acid derivatives using sulfuric acid impregnated on silica. *Green Chem. Lett. Rev.*, 5, 179-186.

Salanski, P., Descotes, G., Bouchu, A., Queneau, Y. (1998). Monoacetalation of unprotected sucrose with citral and ionones. *J. Carbohydr. Chem.*, 17, 129-142

Schoen, M.; Schnuerch, M.; Mihovilovic, M. D. (2011). Application of continuous flow and alternative energy devices for 5-hydroxymethylfurfural production. *Mol. Divers.* 15, 639-643.

Shimizu, H.; Yoshimura, Y.; Hinou, H.; Nishimura, S.-I. (2008). A novel glycosylation method part 3: study of microwave effects at low temperatures to control reaction pathways and reduce byproducts. *Tetrahedron*, 2008, 10091-10096

Söderberg, E.; Westman, J.; Oscarson, S. (2001). Rapid carbohydrate protecting group manipulations assisted by microwave dielectric heating. *J. Carbohydr. Chem.*, 20, 397-410.

Witschi, M. A. & Gervay-Hague, J. G. (2010). Selective Acetylation of per-*O*-TMS-Protected Monosaccharides. *Org. Lett.*, 12, 4312-4315

Yemis, O.; Mazza, G. (2011). Acid-catalyzed conversion of xylose, xylan and straw into furfural by microwave-assisted reaction. *Biores. Technol.*, 102, 7371-7378.

Yoshimura, Y.; Shimizu, H.; Hinou, H.; Nishimura, S.-I. (2005). A novel glycosylation concept; microwave-assisted acetal-exchange type glycosylations from methyl glycosides as donors. *Tetrahedron Lett.*, 46, 4701-4705

Chapter 3

Boron-Carbohydrate Interactions

Brighid Pappin, Milton J. Kiefel and Todd A. Houston

Additional information is available at the end of the chapter

http://dx.doi.org/10.5772/50630

1. Introduction

Boron-polyol interactions are of fundamental importance to human health [1], plant growth [2] and quorum sensing among certain bacteria [3]. Such diversity is perhaps not surprising when one considers boron is one of the ten most abundant elements in sea water and carbohydrates make up the planet's most abundant class of biomass. Several boronic acids matrices are commercially available for the purification of glycoproteins by affinity chromatography [4], and boronic acids are also useful carbohydrate protecting groups.[5,6] Recently, complexes between boron and sugars have become a lynchpin for the development of synthetic carbohydrate receptors.[7] These complexes involve covalent interactions that are reversible in aqueous solution. This chapter reviews current understanding of these processes, provides a historical perspective on their discovery, identifies methods for studying these complexes and classifies these interactions by carbohydrate type. Such information is key to the design and synthesis of synthetic lectins, also termed "boronolectins" when containing boron [7].

The very nature of the reversible binding between boron acids and alcohols has been exploited in many different ways. The use of boronic acid carbohydrate recognition molecules could provide an avenue for the selective detection of specific sugars for future use in early diagnostics. By targeting cell-surface sugars, a boron-based probe could recognize particular characteristic epitopes for the identification of diseases leading to earlier treatments. In this chapter we not only review some of the fundamental aspects of boron-carbohydrate interactions but also discuss how this translates into the design of synthetic carbohydrate receptors.

2. Boron-carbohydrate interactions

2.1. Discovery of boron-sugar interactions

The first hint of the marriage between boron and polyols was detected by Biot in his seminal studies on optical rotation. In 1832 he noted that the rotation of tartaric acid changed in the presence of boric acid.[8] It would be a century later before interaction of boron acids (boric, boronic and borinic) and monosaccharides was studied in detail. In 1913, Böeseken first noted that glucose increased the acidity of boric acid solutions.[9] It was nearly another half century before Lorand and Edwards published work quantifying the affinity of boric and phenylboronic acids for simple diols (e.g.-ethylene glycol, catechol) and common monosaccharides (i.e.-glucose, fructose, mannose, galactose).[10] The covalent product between a boronic acid and a diol is termed a boronate ester, analogous to a carboxylate ester. These interactions are favoured at basic pH ranges where the tetrahedral boronate ester is formed (Figure 1). The interchange between boron acids and divalent ligands in aqueous solution can be complex and varied depending on pH.

Figure 1. Boric acid interactions with vicinal diol of sugar.

2.2. Fundamentals of boron-diol exchange

There are two general organoboron families of boric acid descent that can form esters with diols through loss of water. These are boronic acids--where one hydroxy group of the parent boric acid is substituted by carbon--and borinic acids, where two hydroxy groups are substituted by carbon-based substituents.

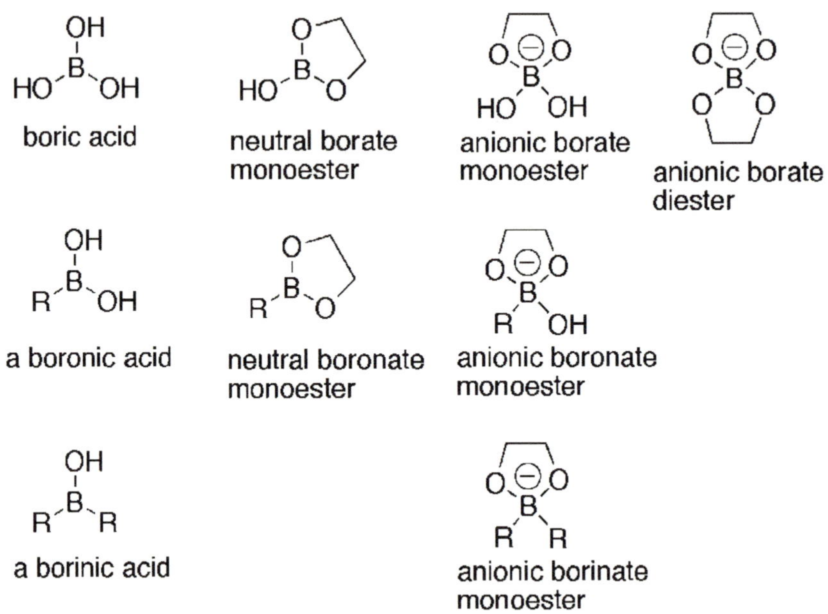

Figure 2. Boron acids and possible esters with ethylene glycol.

Boric and boronic acids can form either neutral or anionic esters depending on the pH. Diol binding by boron acids is favoured at basic pH, while esterification of boron by hydroxycarboxylic acids is favoured in acidic pH ranges. Borinic acids can only form anionic borinate esters upon dehydrative condensation with a diol or divalent ligand. Boric acid can also form an anionic, tetrahedral diester with diols and related divalent ligands (Figure 2). While boronates can form neutral esters in non-polar solvents, they tend to form anionic boronate esters in water (Figure 3). Boronate ester formation is not favoured near physiologic pH and is completely cleaved under strongly acidic conditions.

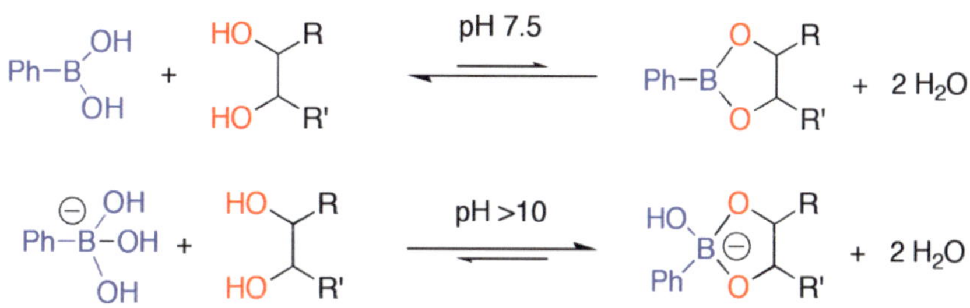

Figure 3. Diol exchange with phenylboronic acid at varied pH.

This is because the neutral boronate ester is generally more Lewis acidic than the parent boronic acid—i.e. pK_a (acid) > pK_a (ester), (Scheme 1).[11] Thus, boronate ester formation is favoured at higher pH where elevated hydroxide concentrations ensure the boronate ester is "trapped" in its more stable tetrahedral form. However, depending on the specific monosaccharide, its boronate esters are not always more Lewis acidic than the free boronic acid.[12] Rate constants for esterification of simple boronates by diols fall in the range of 10^2-10^3 $M^{-1}s^{-1}$.[13] Ishihara uncovered evidence it is the trigonal boronic acid that exchanges most rapidly with diols irregardless of pH.[14] The relative affinity of boronates for diols in most carbohydrates is of the order: *cis*-1,2-diol > 1,3-diol >> *trans*-1,2-diol. Thus, certain monosaccharides have an intrinsically higher affinity for boron acids.

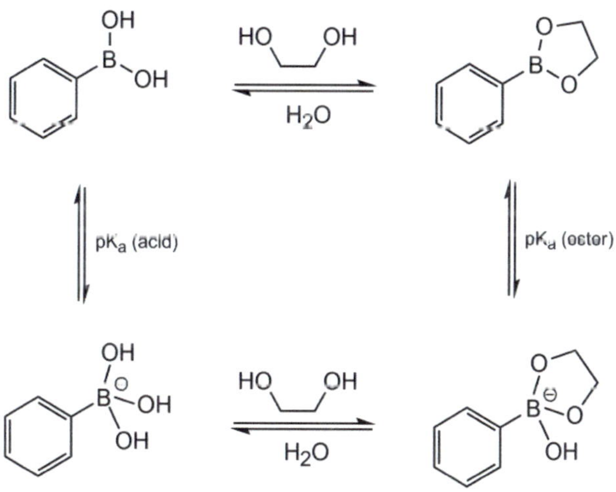

Figure 4. Multiple equilibria involved in diol exchange with phenylboronic acid.

2.3. Detection and elucidation of boron-sugar complexes

Methods for identifying boron-polyol interactions include the following, listed in roughly chronological order of the introduction of their use in this area:

a. Optical rotation/ORD: 1832 [8,15,16]
b. pH change/titration: 1913 [9]
c. Conductivity: 1928
d. Temperature jump: 1969 [17]
e. X-ray crystallography: 1973 [18,19]
f. ^{11}B-NMR: 1973 [19-22]
g. Fluorescence/CD: 1990's [23-25]
h. ESI/MALDI Mass Spectrometry: 1990's [26-28]

The second half of the list defines techniques that are most frequently used today in the study of boron-carbohydrate interactions. Obviously, X-ray crystallography provides the least ambiguous information about the structure of the boronate ester of interest. However, these boronate-sugar adducts are often amphiphilic in nature and do not lend themselves to the production of suitable crystals. The relatively slow exchange between boron acids and diols on the NMR time scale often makes it difficult to study by proton NMR. However, ^{11}B-NMR can be quite useful due to the dramatic shift of the boron resonance when it is converted from its neutral, trigonal form as a boronic acid to its anionic, tetrahedral form as a boronate ester. [20-22]

Optical methods such as fluorescence and circular dichroism (CD) are powerful tools for detecting boron-carbohydrate binding interactions. Yoon and Czarnik reported the first fluorescent boronate designed to detect binding to monosaccharides.[23] James, Shinkai and co-workers reported the first fluorescent boronates to function by photoinduced electron transfer (PET) to generate an increased fluorescence output upon carbohydrate binding.[24] This type of "turn-on" system tends to be most useful in a biological setting where background fluorescence quenching can be a problem for fluorophores that respond by fluorescence quenching ("turn-off") to ligand binding. A more complete understanding of the aminoboronate PET fluorescence mechanism has been developed by the groups of Wang [29] and Anslyn [30]. They have demonstrated that solvent insertion disrupting any dative boron-nitrogen interaction is responsible for the increased fluorescence output upon ligand binding. The Shinkai group has also designed a number of CD-active boronate receptors for oligosaccharides and have used this method to detect binding of target substrates.[25]

Advances in mass spectrometry (MS) over the past few decades, particularly electrospray ionization (ESI) and matrix-assisted laser desorption/ionization (MALDI), have revolutionized the application of this instrumental method to the study of host-guest and protein-ligand interactions. Certainly, the field of boron-based carbohydrate receptors has also benefited from the substantial improvement and refinement of these and other MS techniques. However, the tendency of boronates to dehydrate and/or oligomerize to varying degrees depending on their solvation can complicate MS analysis of boronate-carbohydrate esters. We have found that use of a glycerol matrix for fast atom bombardment (FAB) ionization is particularly useful for mass spectrometric characterization of diboronate species.[31] While other techniques are used in the study of boron-polyol complexes, those mentioned here are among the most common routinely used in the field today.

3. Boron-based carbohydrate receptors

3.1. Boron-based monosaccharide receptors

The 1992 work of Yoon and Czarnik first demonstrated the potential of boronic acids as fluorescent carbohydrate receptors for sensing applications.[23] In the past 20 years, a great deal of research has focused on the development of boron-based glucose receptors for incorporation as sensors in blood sugar monitors for diabetics.[7, 32] This has lead to the commercial development of contact lenses that can signal when circulating glucose levels drop by changing the colour of the lense to alert the wearer.[33] The affinity of mono-boronates for glucose is low at physiologic pH, but bis-boronates offer a substantial improvement in binding affinity. A landmark study from the Shinkai group involved development of a chiral glucose sensor capable of discriminating between enantiomers of glucose.[34] This utilized aminoboronates as PET sensors around a chiral binaphthol core (Figure 5). The Singaram group has developed bis-boronate bipyridinium salts (viologens) that can be tuned for selective binding of glucose (Figure 6).[35] These compounds coupled with anionic dyes are also in commercial development as blood glucose sensors.

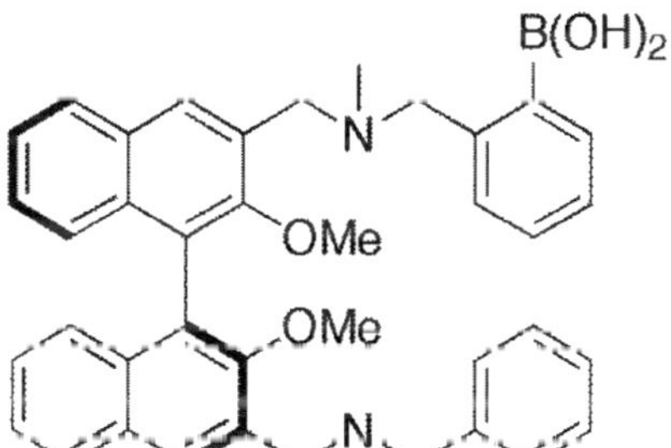

Figure 5. Shinkai's chiral binaphthol glucose sensor.

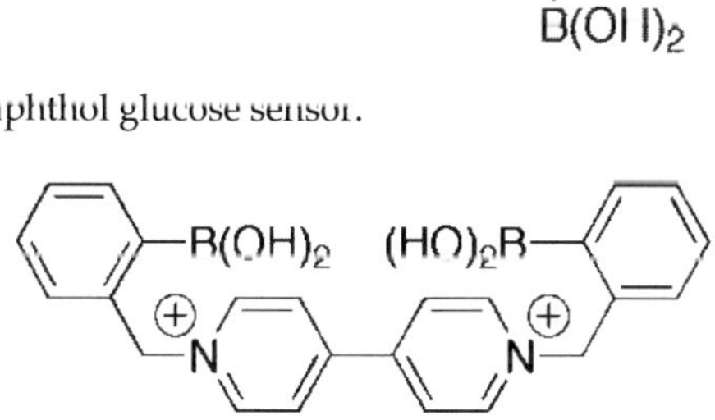

Figure 6. One isomer of Singaram's family of glucose sensors.

It was initially presumed that two boronotes could bind to the C-1/C-2 diol and the C-4/C-6 diol of glucose in its hexopyranoside form.[5] However, it has been shown that boronic acids have a much higher affinity for the furanoside form of free hexoses.[36] In fact, boronates have virtually no affinity for methyl glyocsides locked in their pyranoside form at physiologic pH. This means that boronates would not be useful components in synthetic carbohydrate receptors for many cell surface carbohydrates. Mammalian cell-surface glycoconjugates, in particular, are dominated by hexopyranoside structures. The Hall group has provided an important solution to this problem when they showed that benzoboroxoles can bind methyl hexopyranosides in water at pH 7.5.[37] For glucopyranosides, the only significant binding site is the C-4/C-6 diol as all vicinal diols in this system are of a *trans* relationship. In galactopyranosides, there is an additional possible binding site: the C-3/C-4 *cis*-diol (Figure 7):

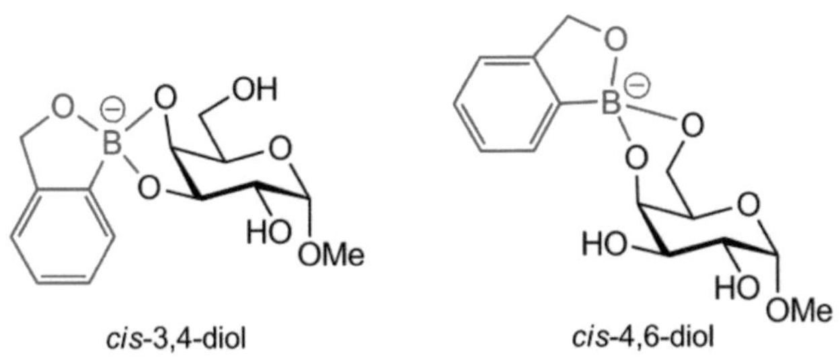

Figure 7. Potential binding modes between benzoboroxole (blue) and methyl-galactoside.

While a significant amount of research has been dedicated to the study of boronate-monosaccharide interactions, very little has been invested in borinate-monosaccharide exchange. Taylor has recently reported that borinic acids have substantial affinity for catechols and α-hydroxycarboxylates,[38] greater than that of 2-fluoro-5-nitrophenyboronic acid,[39] a boronate that is able to bind sugars at neutral pH. The affinity of this boronate for monosaccharides is greater than the affinity of a borinic acid for the same sugars, but this affinity in the latter case is still significant. Whether borinates can effectively bind to hexopyranosides under the same conditions still needs to be defined.

3.2. Boron-based sugar acid receptors

Boron-tartaric acid interactions were studied throughout the 20th century beginning with a report in 1911 on the ability of tartrate to increase the solubility of boric acid.[40] The design of sophisticated boron based receptors for tartrate did not arise until near the end of the century when Anslyn reported the first in 1999.[41] This receptor (Figure 8) also binds citrate with what is perhaps the highest association constant reported for a small molecule with a boron-based receptor ($K_a = 2x10^5$).[42] In 2002, we showed that Shinkai's binaphthol glucose receptor (Figure 5) has a high affinity for tartrate as well.[43] Bis-boronates such as this can bind simultaneously to both α-hydroxycarboxylates. James further showed that chiral discrimination between tartrate enantiomers can also be obtained with this receptor as was the case with monosaccharide enantiomers.[44] While the history of study surrounding boron-tartaric acid interactions is long and varied, study of the interaction of boron with sugar acid monosaccharides such as sialic acid and glucuronic acid has arisen much more recently. Fundamental to the understanding of these complexes is the fact that, unlike esterification with diols, boronate esterification by α-hydroxycarboxylic acids is favoured below pH 7.[44] We have recently provided a short review on the subject of boron:α-hydroxycarboxylate interactions used in sensing and catalysis.[45]

Shinkai first reported a boron-based sugar acid receptor containing a metal chelate that has significant affinity for glucuronic acid (log K_a = 3.4) and galacturonic acid (log K_a = 3.1) while the affinity for sialic acid was an order of magnitude lower (log K_a = 2.3).[46] Presumably the carboxylate of the sugar acid can coordinate to the chelated zinc while the boron binds to a vicinal diol on the monosaccharide. Smith and Taylor used a combination of electrostatic interaction and a boronate anchor within a polymeric system to bind to sialic acid selectively.[47] In 2004, Strongin identified a boronate that offered a colorimetric response to the presence of sialic acid.[48]

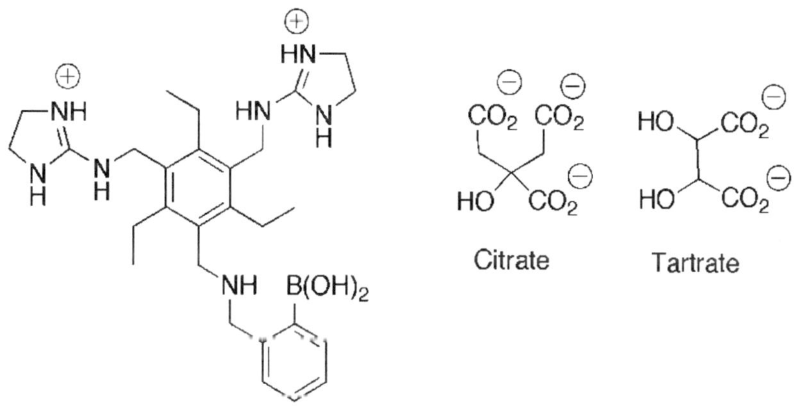

Figure 8. Anslyn's guanidino-boronate receptor and high affinity ligands.

We have recently communicated the development of a bis-boronate that can bind to sialic acid at both its α-hydroxycarboxylate-type group at the anomeric centre and its glycerol tail (Figure 9).[49] Elevated levels of free sialic acid in the blood can be indicative of the presence of certain cancers. This system uses a unique combination of boronates whose esterification has an opposing affect on the overall fluorescence output of the receptor. This diminishes signals from competing ligands such as glucose that are present at much higher concentration in the blood but cannot span both binding site to strongly quench fluorescence.

As discussed in the following sections, several groups have taken advantage of the affinity of boronates for the glycerol tail of sialic acid to target glycoconjugates on cell surfaces. Several of these synthetic compounds display lectin-like biological characteristics that offer promise of the future development of bioactive boron-based molecules.

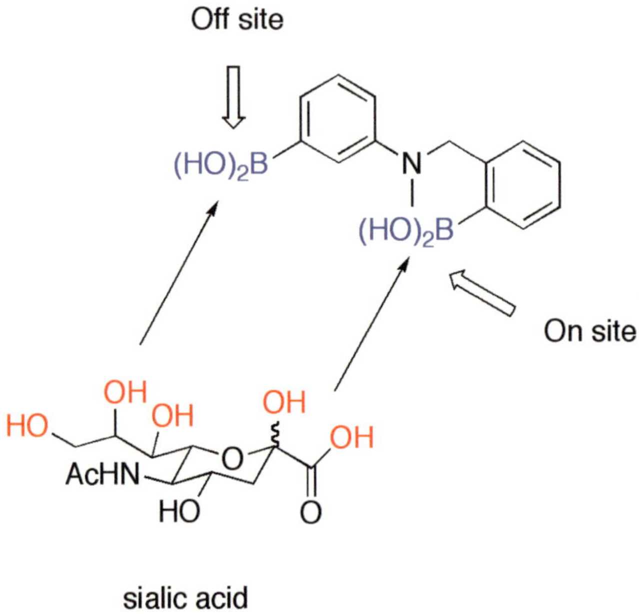

Figure 9. Our divergent response fluorescent receptor for sialic acid.

3.3. Boron-based oligosaccharide receptors

Creating receptors for oligosaccharides offers additional levels of complexity relative to monosaccharides. An obvious difference is the increased degrees of freedom available to oligosaccharides, particularly those that contain 1,6-linkages. In 2000, Shinkai reported development of a meso-meso-linked porphyrin scaffold where distance between two boronates was tuned to selectively bind to a tetrasaccharide of maltose (maltotetrose) over other oligomers containing from two to seven glucose units (Figure 10).[50] Binding of the two boronates must take place at both the reducing and non-reducing termini of the oligosaccharide. This is due to the fact that the C-1/C-2 diol at the reducing end and the C-4/C-6 at the non-reducing end are the only potential binding sites with an appreciable affinity for boronates. The ability to bind the tetramer selectively stems from the rigidly defined distance between the two boronates, i.e.-the tetrasaccharide offers the optimal fit to bridge these two boronates. The Shinkai group has also reported a similar strategy to bind to the important cell-surface trisaccharide, Lewis X.[51] In this case interaction is not with a reducing sugar but presumably with diols on both the galactose and fucose residues.

Heparin is a natural polysaccharide used clinically for its anti-coagulant properties. In 2002, Anslyn communicated a colorimetric sensing ensemble for detection of heparin.[52] As heparin has a high anionic charge density, the receptor was designed with a number of complementary cationic amino groups alongside boronic acids. This group has further reported success in sensing heparin within serum using a second generation receptor.[53] Schrader has recently developed a fluorescent polymeric heparin sensor that can quantify this polysaccharide with unprecedented sensitivity (30 nM).[54] Coupling of boron-carbohydrate interactions with electrostatic attraction in a multivalent manner is responsible for the high affinity of this receptor for its substrate. In spite of this avidity, the interaction can be controlled in a biologically relevant manner. Binding of the polymer to heparin can be reversed by the addition of protamine, similar to reversal of the complex between heparin and its natural target, anti-thrombin III. Other examples of biological mimicry by boron-based systems are delineated below and in the next section.

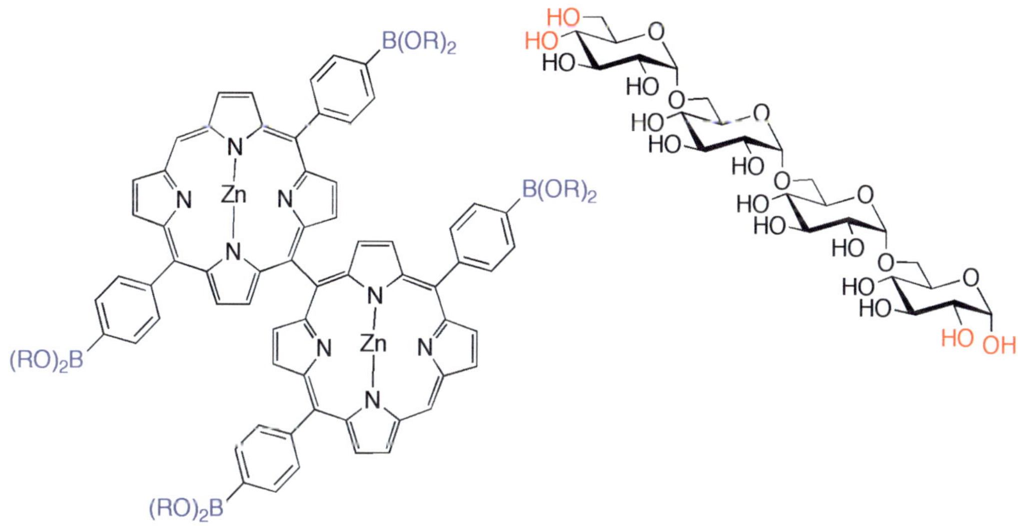

Figure 10. Shinkai's oligosaccharide receptor and maltotetrose.

The demonstrated affinity of benzoboroxoles for hexopyranosides makes these boron derivatives attractive components of receptors designed to target mammalian oligosaccharides. In 2010, Hall reported development of a bis-benzoboroxole receptor for the Thomsen-Friedenreich (TF) antigen, a tumor marker composed of consecutive galactose-based residues Figure 11.[55] This receptor was optimized within a combinatorial library constructed to add additional H-bonding and hydrophobic interactions between host and its oligosaccharide guest. A natural lectin receptor for this disaccharide, peanut agglutinin lectin (PNA), binds the TF antigen quite strongly relative to other protein-carbohydrate interactions (K_d = 10^7). However, the synthetic bis-benzoboroxole inhibits binding of PNA to TF-antigen labelled protein at low micromolar concentrations. The bis-benzoboroxole has a higher affinity for the disaccharide than the corresponding bis-boronate although the latter still has significant affinity highlighting the importance of the additional H-bonding and hydrophobic interactions in this system. This work indicates it should be possible to target multiple hexopyranoside structures with other oligomeric benzoboroxole systems.

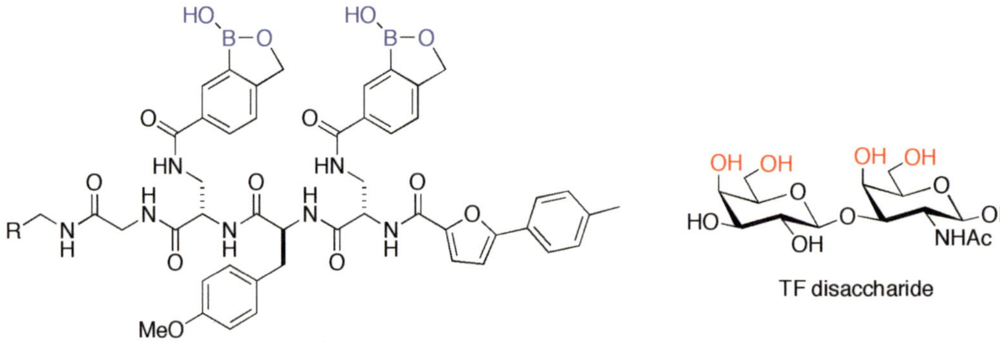

Figure 11. Hall's bis-benzoboroxole receptor for the TF antigen.

4. Boron-cell surface interactions

One ultimate goal of research into synthetic carbohydrate receptors is the development of compounds that can bind directly to cell surface glycocojugates. Such synthetic lectins may serve as diagnostics to monitor changes in cell surface structure associated with disease progression such as cancer. Additionally, they may be used as drug-targeting agents to deliver chemotherapeutic agents to specific cell types. An early and initially underappreciated demonstration of the targeting of cell-surface structures was the work of Hageman with fluorescent dansyl boronates shown to associate with *Bacillus subtilis*.[56] They were also able to show a diboronate could display other lectin like properties such as promoting the agglutination of erythrocytes. Not long after that, Gallop developed a method he defined as "boradaption" using boronates to transfer lipophilic dyes and probes into cells.[57] Although the precise mechanism was not delineated, some boronates were shown to alter the latter stages of *N*-linked glycoprotein processing.[58] A great deal of work has also gone into the development of lipophilic boronic acids as membrane transport agents for hydrophilic molecules such as sialic acid and its derivatives.[59] There is commercial interest in such artificial transporters for the extraction of monosaccharides, such as glucose and fructose, and disaccharides like lactose from natural sources.[28]

In 2002, Weston and Wang reported the ability to target a specific oligosaccharide epitope of a cell surface glycoconjugate.[60, 61] Use of a fluorescent bis-boronate to label the cancer-related antigen sialyl Lewis X on hepatocellular carcinoma cells was an important achievement. They used a combinatorial approach to optimize a bis-boronate in targeting the sialic acid and fucose residues on the tetrasaccharide. The receptor did not label cells that contained Lewis Y antigens lacking sialic acids, or were treated with fucosidase, to remove fucose from the cell surface. This indicates that both components are necessary for interaction with the synthetic receptor. The study marked the first time, as far as we are aware, that two different monosaccharide types had been targeted by design with a boronolectin on a cell surface (Figure 12).

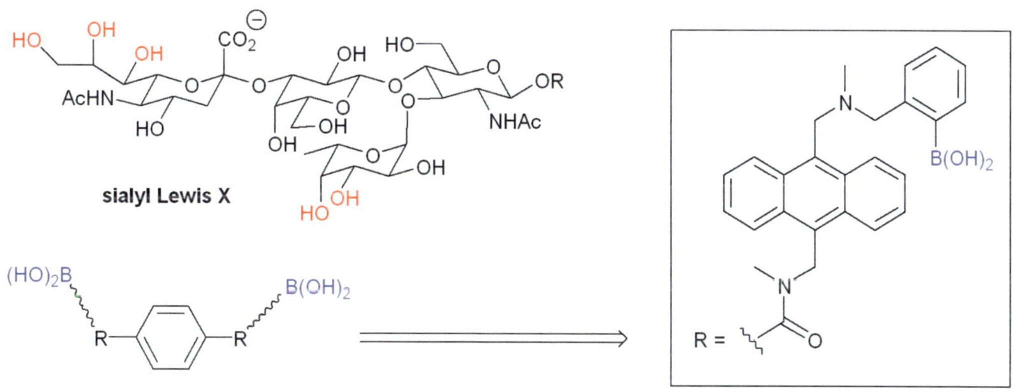

Figure 12. Wang's sialyl Lewis X receptor. Reprinted with permission. John Wiley & Sons, ©2010.

The Kataoka group has used boronic acids on a number of platforms to target cell surface sialic acids to engender a biological or analytical response. They have shown that polymeric boronates can cause the induction of lymphocytes in the same way as natural lectins do.[62] In addition these polyboronates can out-compete natural sialic acid-specific lectins for a cell surface. In collaboration with Miyahara, a powerful method for the direct determination of cell-surface sialic acid levels has been developed.[63, 64] Use of a self-assembled monolayer on a gold electrode allows a coating of boronates to be applied. Potentiometric measurements in the presence of cell suspensions containing either 0, 15, 30, or 100% metastatic cells are readily distinguishable (Figure 13).[64]

The study, application and manipulation of boron-carbohydrate interactions continues to expand into its third century. The properties of oligomeric and polymeric boronic acids in a cellular setting demonstrate that the terms "boronolectin" and/or synthetic lectin are appropriate. What remains for the field to advance are more examples targeting cell-surface carbohydrate structures beyond those containing sialic acid. The Hall group's receptor for the TF antigen marks a seminal step in this direction.[54] For a more comprehensive review of boron-based carbohydrate receptors in the context of other synthetic and biologic sugar binding systems, readers are directed to the recent publication of Wang.[7]

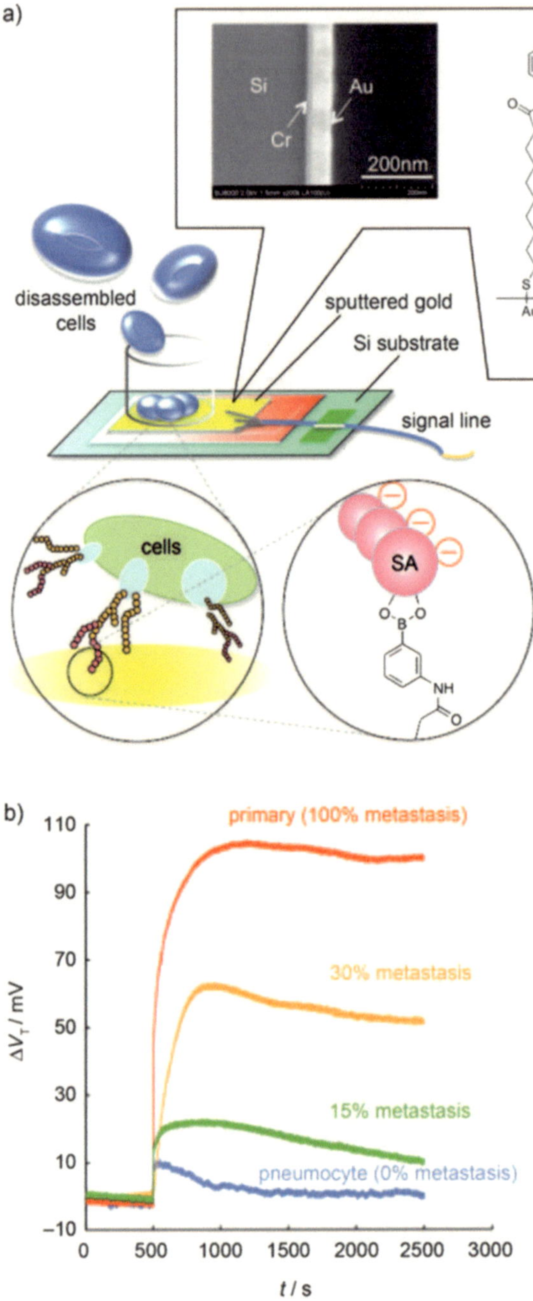

Figure 13. a) Schematic representation of potentiometric SA detection with a PBA-modified gold electrode. An SEM image of a cross-section of the electrode is shown at the top next to the chemical structure of the PBA-modified self-assembled monolayer introduced onto the electrode surface. b) Change in the threshold voltage (VT) of the PBA-modified FET as a function of time upon the addition of cell suspensions (10^6 cells/mL) with various degrees of metastasis. [64] Reprinted with permission. John Wiley & Sons, ©2010.

5. Conclusions

In spite of its long and rich history, understanding of boron acid interactions with carbohydrates continues to increase into the 21st century. In the past 20 years, much fundamental knowledge has been gained, principally from the development of boronate-based glucose receptors for application toward blood sugar monitoring in diabetics. Currently, however, significant effort is being dedicated to the development of boron-based receptors for more complex oligosaccharides. This challenge is being undertaken by an increasing number of research groups throughout the world. These designer receptors may find application in diagnostics for cancer or infectious diseases, in drug targeting, or in providing a more fundamental understanding of the biochemical roles of cell-surface carbohydrates. The ability of boron acids to distinguish between closely related polyols either stereoselectively or chemoselectively makes them an obvious choice for anchoring synthetic carbohydrate receptors.[65] Engineering these interactions to target specific oligosaccharides is currently a difficult challenge as witnessed by the limited number of boron-based oligosaccharide receptors that have been developed at this stage. However, coupling boron-carbohydrate interactions with several additional non-covalent interactions—electrostatic, H-bonding, hydrophobic—offers the best chance of success. Future endeavours will determine the scope and limitations of boron-based carbohydrate receptors and sensors.

Author details

Brighid Pappin
Institute for Glycomics, Griffith University, Gold Coast, Australia

Milton J. Kiefel
Institute for Glycomics, Griffith University, Gold Coast, Australia

Todd A. Houston
Institute for Glycomics, Griffith University, Gold Coast, Australia

School of Biomolecular and Physical Sciences, Griffith University, Nathan, Australia

6. References

[1] Scorei IR (2011) Boron Compounds in the Breast Cancer Cells Chemoprevention and Chemotherapy. In Gunduz E, Gunduz M, editors. Breast Cancer Current and Alternative Therapeutic Modalities. Rijeka: InTech. pp. 91-114.

[2] O'Neill MA, Eberhard S, Albersheim P, Darvill AG (2001) Requirement of Borate Cross-Linking of Cell Wall Rhamnogalacturonan II for Arabidopsis Growth. *Science* 294: 846-849.

[3] Chen X, Schauder S, Potier N, Dorsselaer AV, Pelczer I, Bassler BL, Hughson FM (2002) Structural identification of a bacterial quorom-sensing signal containing boron. *Nature* 415: 545-549.

[4] Hageman JH, Kuehn GD (1992) Boronic Acid Matrices for the Affinity Purification of Glycoproteins and Enzymes. In Kenney A, Fowell S, editors, Methods in Molecular Biology, Vol. 11: Practical Protein Chromatography. Totoway, NJ: The Humana Press Inc. pp. 45-71.

[5] Ferrier RJ, Prasad R (1965) Boric Acid Derivatives as Reagents in Carbohydrate Chemistry. 6. Phenylboronic Acid as a Protecting Group in Disaccharide Synthesis. *J. Chem. Soc.* 7429.

[6] Duggan PJ, Tyndall EM (2002) Boron acids as protective agents and catalysts in synthesis. *J. Chem. Soc., Perkin Trans. 1.* 1325-1339.

[7] Jin, S, Cheng Y, Reid S, Li M, Wang B, (2010) Carbohydrate recognition by boronolectins, small molecules and lectins. *Med. Res. Rev.* 30: 171-257.

[8] Lowry TM (1935) Optical Rotary Power. London: Longmans, Green, and Co. Reprinted (1964) New York: Dover Publishers, Inc.

[9] Böeseken J. (1913) On the storage of hydroxyl groups of polyoxide compounds in space. The configuration of saturated glycols and of alpha- and beta-glycose. *Ber. Dtsch. Chem. Ges.*, 46: 2612–2628.

[10] Lorand JP, Edwards JO (1959) Polyol Complexes and Structure of the Benzeneboronate Ion. *J. Org. Chem.* 24: 769-774.

[11] Yasuda M, Yshioka S, Yamasaki S, Somyo T, Chiba K, Baba A (2006) Cage-Shape Borate Esters with Enhanced Lewis Acidity and Catalytic Activity. *Org. Lett.* 8: 761-764.

[12] Springsteen G, Wang BH (2002) A detailed examination of the boronic acid-diol complexation. *Tetrahedron* 58: 5291-5300.

[13] Pizer RD, Tihal C (1992) Equilibria and Reaction Mechanism of the Complexation of Methylboronic Acid with Polyols. *Inorg. Chem.* 31: 3243-3247

[14] Iwatsuki S, Nakajima S, Inamo M, Takagi HD, Ishihara K (2007) Which is Reactive in Alkaline Solution, Boronate Ion of Boronic Acid? Kinetic Evidence for Reactive Trigonal Boronic Acid in an Alkaline Solution. *Inorg. Chem.* 46: 354-356.

[15] Katzin LI, Gulyas E (1966) Optical Rotatory Dispersion Studies on the Borotartrate Complexes and Remarks on the Aqueous Chemistry of Boric Acid. *J. Am. Chem. Soc.* 88: 5209-

[16] Jones B (1933) The rotatory dispersion of organic compounds. Part XXII Borotartrates and boromalates. *J. Chem. Soc.* 1933: 951-955.

[17] Kustin K, Pizer R (1969) Temperature-Jump Study of the Rate and Mechanism of the Boric Acid-Tartaric Acid Complexation. *J. Am. Chem. Soc.* 91: 317-322.

[18] Mariezcurrena RA, Rasmussen SE, (1973) Crystal Structure of Potassium Boromalate. *Acta Cryatallogr.* B29: 1035-1040.

[19] van Duin M, Peters JA, Kieboom APG, van Bekkum H (1987) Studies on Borate Esters. Part 5. The System Glucarate-Borate-Calcium(II)as studied by 'H, ^{11}B, and ^{13}C Nuclear Magnetic Resonance Spectroscopy. *J. Chem. Soc., Perkin Trans. 2*, 473-478.
[20] Henderson WG, How MJ, Kennedy GR, Mooney EF (1973) The Interconversion of Aqueous Boron Species and the Interaction of Borate with Diols: a ^{11}B NMR Study *Carbohydrate Res.* 28: 1-12.
[21] Collins BE, Sorey S, Hargrove AE, Shabbir SH, Lynch, VM, Anslyn, EV (2009) Probing Intramolecular B-N Interactions in Ortho-Aminomethyl Arylboronic Acids. *J. Org. Chem.* 74: 4055-4060.
[22] Johnson LL, Houston TA (2002) A Drug Targeting Motif for Glycosidase Inhibitors: An Iminosugar-Boronate Shows Unexpectedly Selective β-Galactosidase Inhibition. *Tetrahedron Lett.* 43: 8905-8908.
[23] Yoon J, Czarnik AW (1992) Fluorescent Chemosensing of Carbohydrates—A Means of Chemically Communicating the Binding of Polyols in Water by Chelation-Enhanced Quenching. *J. Am. Chem. Soc.* 114: 5874-5875.
[24] James TD, Sandanayake KRAS, Shinkai (1994) Novel Photoinduced Electron Transfer Sensor for Saccharides Based on a Boronic Acid and an Amine. *Chem. Commun.* 477-478.
[25] James TD, Shinkai S (1995) A Diboronic Acid 'Glucose Cleft' and a Biscrown Ether 'Metal Sandwich' are Allosterically Coupled. *J. Chem. Soc. Chem. Commun.* 1483-1484.
[26] Rose, M. E.; Wycherlet, D.; Preece, S. W. (1992) Negative-Ion Electrospray and Fast-Atom-Bombardment Mass Spectrometry of Esters of Boron Acids. *Org. Mass Spectrom.* 27:876-882.
[27] Penn SG, Hu H, Brown PH, Lebrilla, CB (1997) Direct Analysis of Sugar Alcohol Borate Complexes in Plant Extracts by Matrix Assisted Laser Desorption/Ionization-Fourier Transform Mass Spectrometry. *Anal. Chem.* 69: 2471-2477.
[28] Duggan PJ, Houston TA, Kiefel MJ, Levonis, SM, Smith, BD, Szydzik, ML (2008) Enhanced Fructose, Glucose and Lactose Transport Promoted by a 2-(Aminomethyl) phenylboronic Acid. *Tetrahedron.* 64: 7122-7126.
[29] Franzen S, Ni, W, Wang,B. (2003) Study of the Mechanism of Electron-Transfer Quenching by Boron-Nitrogen Adducts in Fluorescent Sensors. *J. Phys. Chem. B* 107: 12942.
[30] Zhu L, Shabbir SH, Gray M, Lynch VM, Sorey S, E. Anslyn, EV (2006) A Structural Investigation of the N-B Interaction in an *o*-(*N,N*-Dialkylaminomethyl)arylboronate System. *J. Am. Chem. Soc.* 128: 1222-1232.
[31] Kramp KL, DeWitt K, Flora J, Muddiman DC, Slunt KM, Houston, TA (2005) Derivatives of Pentamidine Designed to Target the *Leishmania* Lipophosphoglycan. *Tetrahedron Lett.* 46: 695 698.
[32] Mader HS, Wolfbeis OS (2008) Boronic acid based probes for microdetermination of saccharides and glycosylated biomolecules. Microchim. Acta 162: 1-34.

[33] Badugu R, Lakowicz JR, Geddes CD (2005) A glucose-sensing contact lens: from bench top to patient. *Curr. Opin. Biotechnol.*16: 100-107.

[34] James TD, Sandanayake KRAS, Shinkai S (1995) Chiral discrimination of monosaccharides using a fluorescent molecular sensor. *Nature* 374: 345-347.

[35] Gamsey S, Baxter NA, Sharrett Z, Cordes DB, Olmstead MM, Wessling RA, Singaram S (2006) The effect of boronic acid-positioning in an optical glucose-sensing ensemble. *Tetrahedron* 62: 6321-6331.

[36] Eggert H, Frederiksen J, Morin C, Norrild, JC (1999) A new glucose-selective fluorescent bisboronic acid. First report of strong alpha-furanose complexation in aqueous solution at physiological pH. *J. Org. Chem.* 64: 3846-3852.

[37] Dowlut M, Hall DG (2006) An Improved Class of Sugar-Binding Boronic Acids, Soluble and Capable of Complexing Glycosides in Neutral Water. *J. Am. Chem. Soc.* 130: 9809.

[38] Chudzinski MG, Chi Y, Taylor MS (2011) Borinic Acids: A Neglected Class of Organoboron Compounds for Recognition of Diols in Aqueous Solution. *Aust. J. Chem.* 64: 1466-1469.

[39] Mulla HR, Agard NJ, Basu A (2004) 3-Methoxycarbonyl-5-nitrophenyl boronic acid: h affinity diol recognition at neutral pH. *Bioorg. Med. Chem. Lett.* 14: 25-27.

[40] Werz H (1911) Solubility Studies. II. The Influence of d-Tartaric and dl-Tartaric Acids on the Solubility of Boric Acid. *Z. Anorg. Chem.* 70: 71-72.

[41] J. J. Lavigne, E. V. Anslyn,(1999) Teaching old indicators new tricks: A colorimetric chemosensing ensemble for tartrate/malate in beverages *Angew. Chem. Int. Ed.* 38: 3666-3669.

[42] Wiskur SL, Lavigne JJ, Metzger A, Tobey SL, Lynch V, Anslyn EV (2004) Thermodynamic Analysis of Receptors Based on Guanidinium/Boronic Acid Groups for the Complexation of Carboxylates, α-Hydroxycarboxylates, and Diols: Driving Force for Binding and Cooperativity *Chem. Eur. J.* 10: 3792-3804.

[43] Gray CW, Jr., Houston TA (2002) Boronic Acid Receptors for α-Hydroxycarboxylates: High Affinity of Shinkai's Glucose Receptor for Tartrate *J. Org. Chem.* 67: 5426-5428.

[44] Zhao, J, Fyles, TM, James, TD (2004) Chiral Binol–Bisboronic Acid as Fluorescence Sensor for Sugar Acids. *Angew. Chem. Int. Ed.* 43: 3461-3464.

[45] Houston TA, Levonis SM, Kiefel, MJ (2007) Tapping into Boron/α-Hydroxycarboxylic Acid Interactions in Sensing and Catalysis. *Aust. J. Chem.* 60: 811-815.

[46] Takeuchi M, Yamamoto M, Shinkai S (1997) Fluorescent sensing of uronic acids based on a cooperative action of boronic acid and metal chelate. *Chem. Commun.* 1731-1732.

[47] Patterson S, Smith BD, Taylor RE (1998) Tuning the affinity of a synthetic sialic acid receptor using combinatorial chemistry *Tetrahedron Lett.* 39: 3111-3114.

[48] Yang Y, Lewis PT, Escobedo, JO, St. Luce NN, Treleaven, WD, Cook RL, Strongin RM, (2004) Mild colorimetric detection of sialic acid. *Collect. Czech. Chem. Commun.* 69: 1282-1291.

[49] Levonis, SM, Kiefel MJ, Houston, TA (2009) Boronolectin with Divergent Fluorescent Response Specific for Free Sialic Acid *Chem. Commun.* 2278-2281.

[50] Ikeda M; Shinkai S; Osuka A (2000) Meso-meso-linked porphyrin dimmer as a novel scaffold for the selective binding of oligosaccharides. *Chem. Commun.* 1047-1048.
[51] Sugasaki A, Sugiyasu K, Ikeda M, Takeuchi M, Shinkai S (2001) First Successful Molecular Design of an Artificial Lewis Oligosaccharide Binding System Utilizing Positive Homotropic Allosterism. *J. Am. Chem. Soc.* 123: 10239-10244.
[52] Zhong Z, Anslyn EV (2002) A Colorimetric Sensing Ensemble for Heparin. *J. Am. Chem. Soc.* 124: 9014-9015.
[53] Wright AT, Zhong Z, Anslyn EV (2005) A Functional Assay for Heparin in Serum Using a Designed Synthetic Receptor. *Angew. Chem. Int. Ed.* 44: 5679-5682.
[54] Sun W, Bandmann H, Schrader T (2007) A Fluorescent Polymeric Heparin Sensor. *Chem. Eur. J.* 13: 7701-7707.
[55] Pal A, Bérubé M, Hall DG (2010) Design, Synthesis, and Screening of a Library of Peptidyl Bis(Boroxoles) as Oligosaccharide Receptors in Water: Identification of a Receptor for the Tumor Marker TF-Antigen Disaccharide. *Angew. Chem. Int. Ed.* 49: 1492-1495.
[56] Burnett TJ, Peebles HC, Hageman JH (1980) Synthesis of a Fluorescent Boronic Acid Which Reversibly Binds to Cell Walls and a Diboronic Acid Which Agglutinates Erythrocytes. *Biochem. Biophys. Res. Commun.* 96: 157-162.
[57] Gallop PM, Paz MA, E. Henson E (1982) Boradeption: A New Procedure for Transferring Water-Insoluble Agents Across Cell Membranes. *Science* 217: 166-169.
[58] Goldberger G, Paz MA, Torrelio BM, Okamoto Y, Gallop PM (1987) Effect of Hydroxyorganoboranes on Synthesis, Transport and N-Linked Glycosylation of Plasma Proteins. *Biochem. Biophys. Res. Commun.* 148: 493-499.
[59] Altamore TM, Duggan PJ, Krippner GY (2006) Improving the membrane permeability of sialic acid derivatives. *Bioorg. Med. Chem.* 14: 1126-1133.
[60] Yang W, Gao S, Gao X, Karnati VVR, Ni W, Wang B, Hooks WB, Carson J, Weston B (2002) Diboronic acids as fluorescent probes for cells expressing sialyl Lewis X. *Bioorg. Med. Chem. Lett.* 12: 2175-2177.
[61] Yang W, Fan H, Gao S, Gao X, Karnati VVR, Ni W, Hooks WB, Carson J, Weston B, B. Wang, B (2004) The first fluorescent diboronic acid sensor specific for hepatocellular carcinoma cells expressing sialyl Lewis X. *Chem. Biol.* 11: 439-448.
[62] Uchimura E, Otsuka H, Okano T, Sakurai Y, Kataoka K (2001) Totally synthetic polymer with lectin-like function: Induction of killer cells by the copolymer of 3-acrylamidophenylboronic acid with N,N-Dimethylacrylamide. *Biotechnol. Bioeng.* 72: 307-314.
[63] Matsumoto A, Sato N, Kataoka K, Miyahara Y (2009) Noninvasive Sialic Acid Detection at Cell Membrane by Using Phenylboronic Acid Modified Self-Assembled Monolayer Gold Electrode. *J. Am. Chem. Soc.*,132: 12022-12023.
[64] Matsumoto A, Cabral H, Sato N, Kataoka K, Miyahara Y (2010) Assessment of Tumor Metastasis by the Direct Determination of Cell-Membrane Sialic Acid Expression. *Angew. Chem. Int. Ed.* 49: 5494-5497.

[65] Houston TA (2010) Developing High-Affinity Boron-Based Receptors for Cell-Surface Carbohydrates. ChemBiochem. 11: 945-958.

Chapter 4

Liquid Chromatography – Mass Spectrometry of Carbohydrates Derivatized with Biotinamidocaproyl Hydrazide

Stephanie Bank and Petra Kapková

Additional information is available at the end of the chapter

http://dx.doi.org/10.5772/50634

1. Introduction

Glycans are involved in a number of cell processes. The glycocomponents of a protein encompass structural and modulatory functions, trigger receptor/ligand binding and many other processes which include mechanisms of interest to immunology, hematology, neurobiology and others (Marino *et al.*, 2010; Raman *et al.*, 2005). Because they are associated with both the physiological and the pathological processes, there is constant interest paid to analysis of protein glycosylation and its changes (North *et al.*, 2009; Leymarie & Zaia, 2012).

In order to get a full picture about the structure and function of a carbohydrate, combination of different analytical methods, detection and derivatization techniques is usually applied (Bindila & Peter-Katalinic, 2009; Geyer & Geyer, 2006; Harvey, 2011). Derivatization of sugars for the purpose of their structural or functional analysis bears various advantages (Rosenfeld, 2011). Most of the labels used for the derivatization of carbohydrates possess a chromophor or a flurophor, enabling a sensitive detection of these analytes by means of spectroscopic methods. Apart from an increase in detection sensitivity, chromatographic properties of derivatized sugars may improve in comparison with their underivatized counterparts (Melmer *et al.*, 2011; Pabst *et al.*, 2009). Furthermore, mass spectrometric signal intensity showed to profit from the derivatization step, as an effective ionization was achieved through derivatization. Moreover, depending on the type of the carbohydrate considered, mass spectrometric fragmentation showed to be influenced by the kind of label introduced, and provided additional information about the structure of sugars (Lattová *et al.*, 2005).

There are many possible labelling reactions for the generation of respective derivatives. The most applied derivatization method is the reductive amination via Schiff-base in the presence of the sodium cyanoborohydride as reductant. The greatest success and a broad area of usage over the last decades experienced the arylamine tag such as 2-aminobenzamide (2-AB) and is still extensively used in mass spectrometric studies of glycans (Bigge *et al.*, 1995). Other common tags from this group are 2-aminobenzoic acid (2-AA), 2-aminopyridine (2-AP) and 2-aminoacridone (2-AMAC). Another labelling approach which has been widely used is the formation of hydrazones, where the hydrazine label of the tag reacts with the aldehyde group of the sugar. The mentioned approaches and other derivatizations which have shown their use in analytics of glycans were reviewed recently (Harvey, 2011; Ruhaak *et al.*, 2010).

Generally, selection criteria for a certain derivatization approach depend on many different factors, including the nature of the sample, the compatibility of the label with the analytical protocol being applied and very importantly, the objective of the analysis. Here, very often, not only the structure of a glycan is a matter of a question but also its functionality, which is an essential feature, that often wants to be examined. Consequently, a derivatization method, which would allow for both the structural and functional studies of sugars can be of interest in the characterization of the glycan species. Such a possibility is given e.g. by biotinylated reagents (Grün *et al.*, 2006; Hsu *et al.*, 2006; Leteux *et al.*, 1998; Ridley *et al.*, 1997; Rothenberg *et al.*, 1993; Shinohara *et al.*, 1996; Toomre & Varki, 1994).

In our recent study, biotinamidocaproyl hydrazide (BACH) was presented as a label for carbohydrates analyzed by mass spectrometry (Kapková, 2009). This bifunctional label combines two features: the sugar reactivity and the bioaffinity. In this way, beside the structural characterization of carbohydrates, it enables also for performing interaction studies with carbohydrate-binding proteins. The derivatization was performed under non-reducing conditions via hydrazone formation and showed an increase in the ion abundance of small sugars and *N*-linked glycans.

Here, chip-based electrospray tandem mass spectrometric structural analysis of different glycan-derivatives generated via BACH-labeling using triple quadrupole instrument was performed. The methodology has been applied to glycans released from glycoproteins (ovalbumin, ribonuclease B) including different types of *N*-glycans e.g. high-mannose, complex and hybride glycans. Further, an analysis of small BACH-labeled carbohydrates was conducted as these represent important building unit of the large glycan entities. Reversed liquid chromatography and hydrophilic interaction chromatography (HILIC) coupled with ion-trap mass spectrometry (LC-ESI-IT-MS) of BACH-derivatized monosaccharides were carried out. Advantageous properties of HILIC in terms of separation of non-reducing trisaccharides and BACH-derivatized isomeric disaccharides are presented. Here, derivatization helped to obtain a favorable separation profile or a higher signal intensity. The isomers were compared in their fragmentation pattern, and a distinction of their linkage or anomeric configuration was sought.

2. Experimental

2.1. Materials

Hen ovalbumin, ribonuclease B from bovine pancreas, melezitose, raffinose and biotinamidocaproyl hydrazide were purchased from Sigma-Aldrich (Taufkirchen,

Germany). The BACH-label was used without further purification. Monosaccharides (glucose, galactose, mannose, fucose, xylose, *N*-acetylglucosamine) and disaccharides (maltose, cellobiose) were kindly provided by the Division of Chemicals (Department of Pharmacy, University Würzburg). The standard *N*-glycans MAN5, MAN6, NA2, NGA2 as well as PNGase F and the GlycoClean™ S Cartridges were from Europa Bioproducts Ltd. (Cambridge, UK). The water was from a Milli-Q system.

2.2. Enzymatic digestion of glycoproteins

About 500µg of the protein were dissolved in water (40µl) and reaction buffer (10µl) and denatured by boiling in a water-bath (100°C, 10min). After cooling, the samples were incubated with PNGase F (2µl) at 37°C for 18 hours. By adding 200µl of cold ethanol and keeping in ice for 2h, the deglycosylated protein was precipitated and centrifuged down to a pellet. The supernatant, containing the glycans, was dried in vacuum and then the residue was derivatized.

2.3. Derivatization with BACH

The released glycans from glycoproteins were mixed with 100µl of a 5mmol solution of BACH in 30% acetonitrile and subsequently evaporated. For the derivatization of standard *N*-glycans (0.5µg) the same protocol was followed. According to the evaporation, the mixture was dissolved in 30µl of methanol/water (95/5), and incubated at 90°C for 1h. Then the BACH conjugates were further purified using GlycoClean™ S cartridges according to the accompanying protocol.

2.4. Derivatization with 2-Aminobenzamid (2-AB)

50µl of the AB-derivatization solution, containing 2-AB (5mg), $NaBH_3CN$ (7.5mg), DMSO (500µl) and acetic acid (200µl), were added to the glycans or small sugars. After the incubation at 65°C for 2.5h, the AB-conjugates were purified with GlycoClean™ S cartridges according to the manufacturer's protocol.

2.5. LC/ESI-MS

2.5.1. Nano LC/ESI-MS of N-glycans

The mass spectra were measured on the triple quadrupole mass spectrometer 6460 (Agilent Technologies, Waldbronn, Germany) coupled with an Agilent 1200 liquid chromatography nanoflow system with a ChipCube interface. The HPLC-Chip used for the separation contained a 0.075 × 43mm porous graphitized carbon column and an integrated 40nl enrichment column. This column was filled with identical media, e.g. the same phase was used for both the separation and trapping. Solvents for the capillary and the nanoflow pump were A: 0.1% formic acid in water and B: 0.1% formic acid in methanol. The chromatographic conditions were as follows: capillary pump flow rate (1µl/min), nanoflow pump (0.4µl/min for standard glycans and 0.5µl/min for released glycans); 4µl of the samples were loaded. Gradient was set from 30% to 90%B in 50min. MS conditions were: positive ionization, drying gas flow 4l/min at a drying temperature of 325°C, capillary voltage 1800V, and a scan range from 300 to 2200 m/z.

The MS/MS experiments were performed based on the collision-induced dissociations (CID). In order to selectively monitor the glycans in the triple quadrupole instrument, precursor ion scanning was performed. Characteristic fragment ions used in this mode were the oxonium ions of hexose (HexNAc) at m/z 163 and of *N*-acetylhexosamine (HexNAc) at m/z 204 and the larger oxonium ion at m/z 366 (HexHexNAc). In this approach, quadrupole 1 works as a scanner across the range of interest in order to determine the mass of injected glycans, the second quadrupole acts as a collision cell and the third quadrupole serves to analyze the fragment ions generated in the collision cell. Upon the detection of precursor ions which are losing a specific fragment ion of interest (oxonium ion) the precursor ions of glycans were analyzed in the product ion scan mode which gave the fragment spectrum of the glycan.

2.5.2. LC/ESI-MS of mono-, di- and trisaccharides

Analyses of small sugars were performed on the LC-MSD ion-trap instrument (Agilent) operated with Agilent HPLC 1100 or infusion syringe pump. The separation was conducted on the C18 reversed-phase and HILIC-phase (Kinetex^{TM}, Phenomenex, 100 × 2.1 mm, 100Å, 2.6 µm). The chromatographic conditions of C18 were: A: 100mM ammonium acetate (pH 5.8), B: MeOH (0.1% formic acid), flow: 0.3 mL/min, gradient: 0 min, 11% B; 17 min, 11% B; 30 min, 50% B. The chromatographic conditions of HILIC: A: 100mM ammonium acetate (pH 5.8), B: acetonitril (0.1% formic acid), flow: 0.3 mL/min, gradient: 0 min, 99% B; 12 min, 99% B, 16min, 92% B; 30min, 60% B.

3. Results and discussion

3.1. Derivatization with BACH

Labeling reagent biotinamidocaproyl hydrazide (BACH) is a bifunctional label which possesses a hydrazide group for coupling with the reducing end of sugars and a biotin group in order to interact with a solid support which possesses affinity to biotin. This chemical process may be run under reducing or non-reducing conditions depending on the purpose of the analysis. Because the reaction with hydrazines gives much more stable products than the easily reversible Schiff-base generated by the reaction of an amine with an aldehyde, carbohydrate-hydrazone bond formed *via* hydrazide chemistry does not necessarily require a reducing agent (Hermannson, 2008).

In this way, the resultant product is a glycosylhydrazide (Fig. 1), which preserves the pyranose ring closed, i.e. it preserves its near-native conformation. This is a favorable feature in many functional studies where the closed pyranose ring is essential, in order to be recognized by a carbohydrate-binding protein. Differences in binding activity of certain sugars have been shown (Grün *et al.*, 2006), when reacting with lectins or other carbohydrate binding proteins. Hence, whether a biological interaction is affected by the reduction or not, depends on the type of the sugar and on the kind of the carbohydrate binding protein.

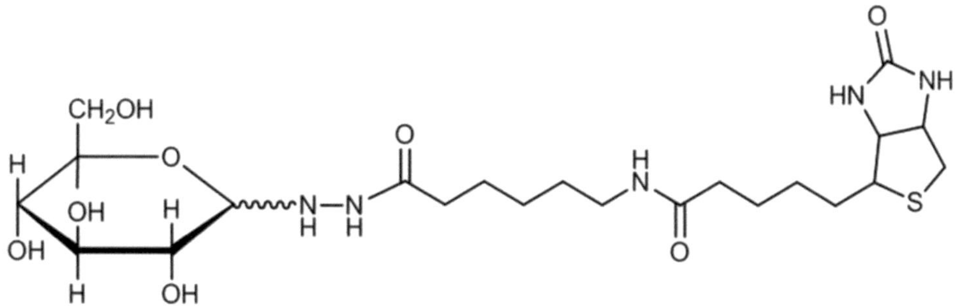

Figure 1. Scheme of glycosylhydrazide formation of carbohydrate with biotinamidocaproyl hydrazide

3.2. Structural characterization of BACH-derivatized *N*-linked glycans

In general, the structures of *N*-linked glycans fall within three main types, namely, high-mannose, complex and hybrid. These all share a common chitobiose core and differ in branching patterns mainly at the non-reducing end.

Here, different standard asparagine-linked glycans and glycans released from model glycoproteins such as ovalbumin and ribonuclease B (RNAse B) were analyzed by means of nano-LC-chip/electrospray mass spectrometry in order to cover the main types of *N*-linked glycans. The *N*-linked glycans were released by treating the glycoproteins with peptide-*N*-deglycanase F (PNGaseF). Released glycans were labeled with the biotinamidocaproyl hydrazide and analyzed by nanoLC- triple quadrupole mass spectrometry using the precursor ion scan mode. The oxonium ions chosen for the scanning mode were m/z 163 (Hex+H), m/z 325 (Hex-Hex+H) and m/z 204.1 (GlcNAc+H). Fragmentation was carried out at three different collision energies (15V, 25V and 50V). The best fragmentation spectra resulted from the application of the collision energy of 15V and this was used throughout. Tandem mass spectra of the respective glycans were recorded in the product ion scan mode by the collision induced dissociation.

RNase B has one glycosylation site occupied by five high-mannose glycans differing in one mannose residue. Because of the high-mannose nature of these glycans, precursor ion scanning with the oxonium ion of hexose (m/z 163) was performed and the resultant mass spectra showed high abundance.

Fig. 2 shows the mass spectrum of the BACH-derivatized high-mannose glycans present in RNaseB and features the characteristic spacing pattern of its glycoforms [$(Man)_nGlcNAc)_2$ with n = 5,6,7,8,9]. The signals were registered as doubly charged hydrogenated species. Hence, the spectrum exhibits the differences of 81 Da, which account for half of the mass of one mannose unit. The most abundant glycoforms were MAN6 and MAN5. This proportionality in pattern was observed also in studies on high-mannose glycopeptides (Alley *et al.*, 2009).

In Fig. 3 the tandem mass spectrum of the doubly charged BACH-labeled MAN6 glycan is shown, which is dominated by the glycosidic cleavages, which correspond with losses of mannose residues. The prominent signals were the doubly charged signals related to consecutive losses of 81Da directly from the molecular ion $[M+2H]^{2+}$. A series of weaker singly charged ions corresponded to losses of mannose residues (162Da) from the chitobiose cleavage between the two GlcNAc-moieties (B_4-ion). In this way, the collision-induced dissociation of BACH-derivatized ribonuclease B glycans provided fast and informative spectra in order to characterize the identity of the analyzed glycans. Fragments have been assigned using the systematic nomenclature for carbohydrate fragmentation proposed by Domon and Costello (Domon & Costello, 1988).

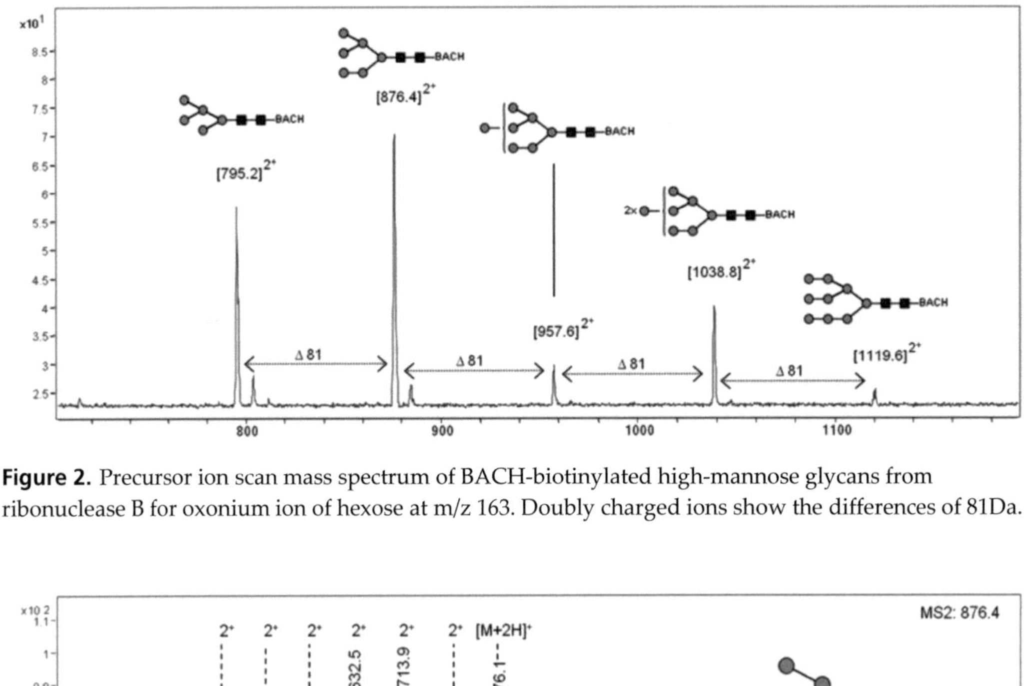

Figure 2. Precursor ion scan mass spectrum of BACH-biotinylated high-mannose glycans from ribonuclease B for oxonium ion of hexose at m/z 163. Doubly charged ions show the differences of 81Da.

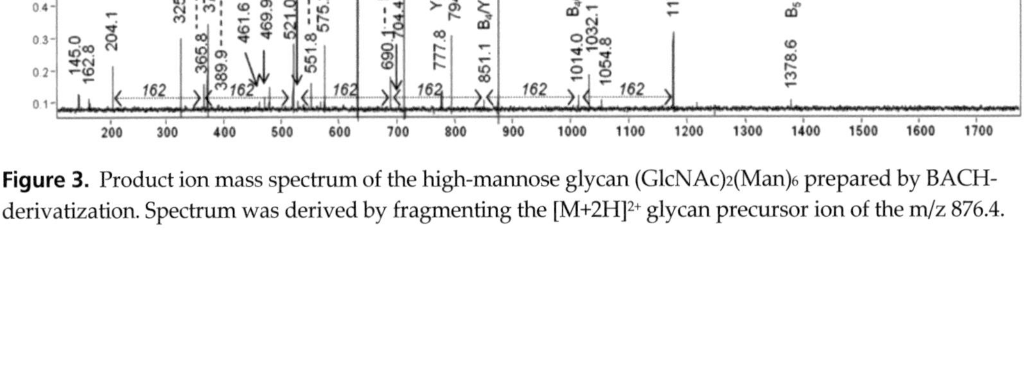

Figure 3. Product ion mass spectrum of the high-mannose glycan $(GlcNAc)_2(Man)_6$ prepared by BACH-derivatization. Spectrum was derived by fragmenting the $[M+2H]^{2+}$ glycan precursor ion of the m/z 876.4.

Ovalbumin, a major glycoprotein in hen egg white, contains one glycosylation site with a series of asparagine-linked oligosaccharides of mannose- and hybrid-type; some rare complex-type oligosaccharides are also known. Due to its structural diversity, this protein is an optimal source of a library of *N*-linked glycans and a good model glycoprotein. The mixture of glycans released from ovalbumin was analyzed by precursor ion scanning in the triple quadrupole. The oxonium ion which has been scanned for was the protonated *N*-acetylglucosamine (GlcNAc) at m/z 204. The mass spectrum of the identified glycans can be seen in the Fig. 4. Found oligosaccharides cover the glycan forms typically present in this glycoprotein (Fig. 4, Table 1). Under the conditions described in the experimental part, the glycan ions appeared as doubly or triply charged ions. Doubly charged ions were fragmented by product ion scanning and their structure was elucidated. Elucidation of the fragment spectra of triply charged ions was a cumbersome task, as the spectra contained ions with different charge states. The fragment ion with m/z 204, corresponding to the oxonium ion $[HexNAc+H]^+$, was frequently observed in the MS/MS spectra of ovalbumin glycans. Generally, this type of ion is highly diagnostic of the presence of glycans that terminate in one or more *N*-acetylglucosamine residues.

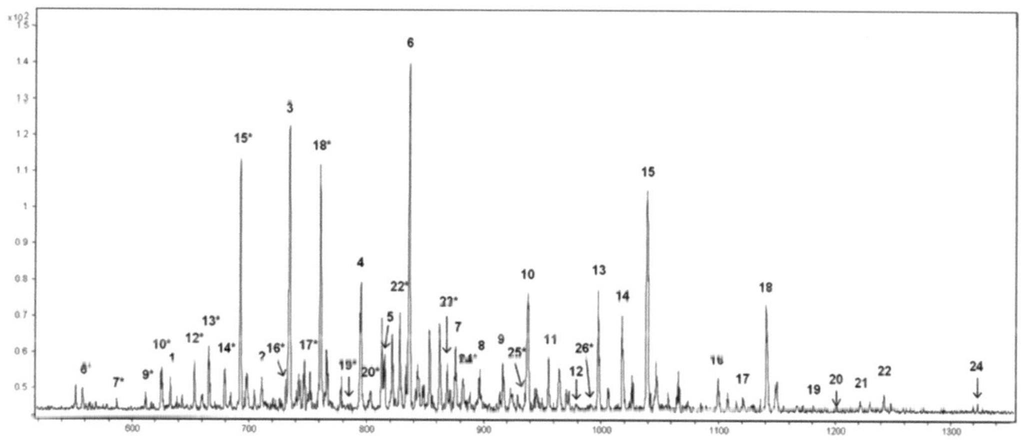

Figure 4. Precursor ion scan for oxonium ion of *N*-acetylglucosamine at m/z 204 of BACH-biotinylated glycans from ovalbumin. Doubly and triply (*) charged ions are depicted. For the structure see the Table 1.

Compd.[1]	Comp.[2]	Observed *m/z* BACH-glycans, $[M+2H]^{2+}/[M+3H]^{3+*}$	Structure
1	H_3N_2	632.9	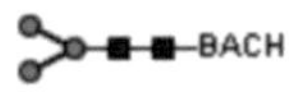
2	H_4N_2	714.0	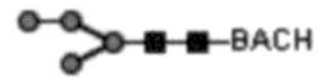
3	H_3N_3	734.5	
4	H_5N_2	795.4	
5	H_4N_3	815.4	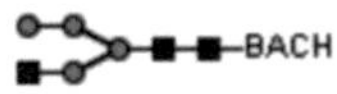
6	H_3N_4	836.4/557.4*	
7	H_6N_2	876.1/584.5*	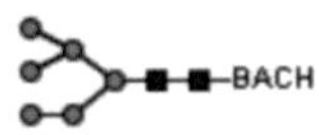
8	H_5N_3	896.7	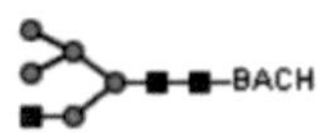
9	H_4N_4	916.5/611.9*	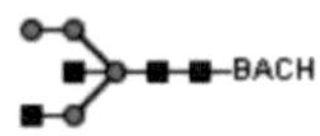
10	H_3N_5	937.8/625.6*	
11	H_7N_2	956.6	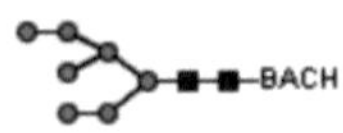
12	H_6N_3	978.0/652.3*	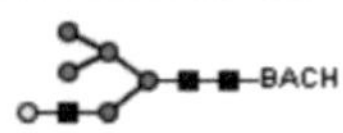
13	H_5N_4	997.6/665.4*	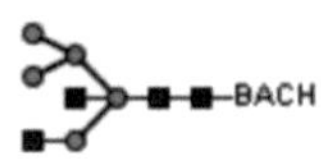
14	H_4N_5	1018.0/679.8*	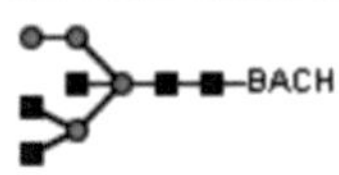
15	H_3N_6	1039.1/692.8*	
16	H_5N_5	1099.8/733.6*	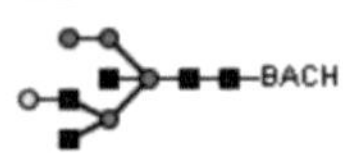

Compd.[1]	Comp.[2]	Observed m/z BACH-glycans, $[M+2H]^{2+}/[M+3H]^{3+*}$	Structure
17	H_4N_6	1120.0/747.3*	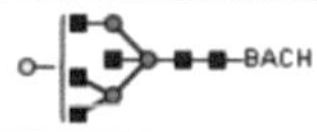
18	H_3N_7	1140.0/760.5*	
19	H_6N_5	1181.0/786.8*	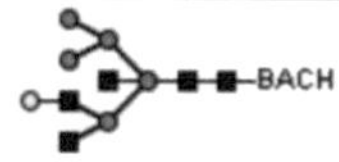
20	H_5N_6	1201.9/801.8*	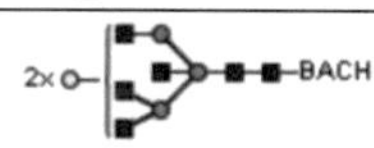
21	H_4N_7	1221.9	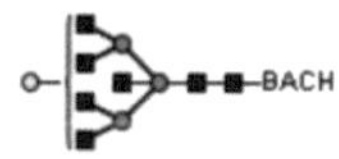
22	H_3N_8	1242.5/828.2*	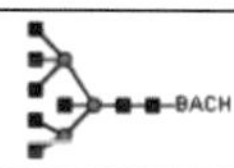
23	H_5N_7	869.2*	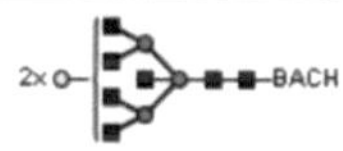
24	H_4N_8	1323.5/882.6*	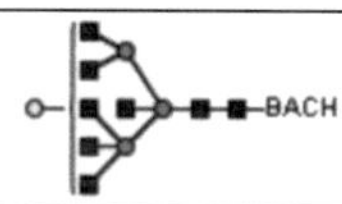
25	H_5N_8	936.8*	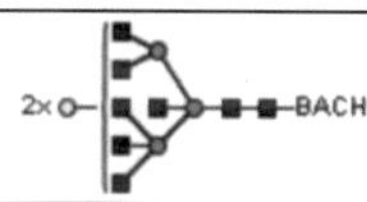
26	H_6N_8	990.9*	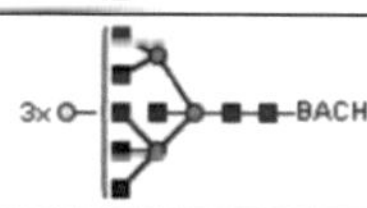

Table 1. Composition and m/z of the doubly and triply charged (*) glycans released by PNGase from the hen ovalbumin observed by precursor ion scanning experiment (m/z 204). For the spectrum see the Fig. 4. [1]Compound number, [2]Composition of *N*-glycans (H = hexose, N = GlcNAc)

Several diagnostic ions were present in the ESI-MS/MS spectra of complex and hybride glycans from ovalbumin. Fig. 5 shows the tandem mass spectrum of one of the complex glycans of ovalbumin. The MS/MS fragmentation of this BACH-derivatized glycan was characterized by sequential losses of sugar residues from its non-reducing end. Arrows with continuous lines indicate differences in the number of mannose, and the dashed lines in the number of *N*-acetylglucosamines.

The Fig. 6 represents the product ion spectrum of the galactosylated standard glycan NA2. Several diagnostic ions were present in the MS/MS spectrum. The loss of the protonated fragment with m/z 366 was the preferred route in the breaking up of this galactosylated species. The oxonium ion is diagnostic of the loss of a fragment containing a hexose residue attached to *N*-acetylhexosamine. The sequencial losses of 365 (i.a. Hex-HexNAc) from the precursor ion (m/z 1630 and m/z 1264) show that this glycan accommodates two galactosyl-*N*-acetylglucosamine residues. The appearance of the fragment at m/z 1264 distinguishes this molecule from the core-bisected species.

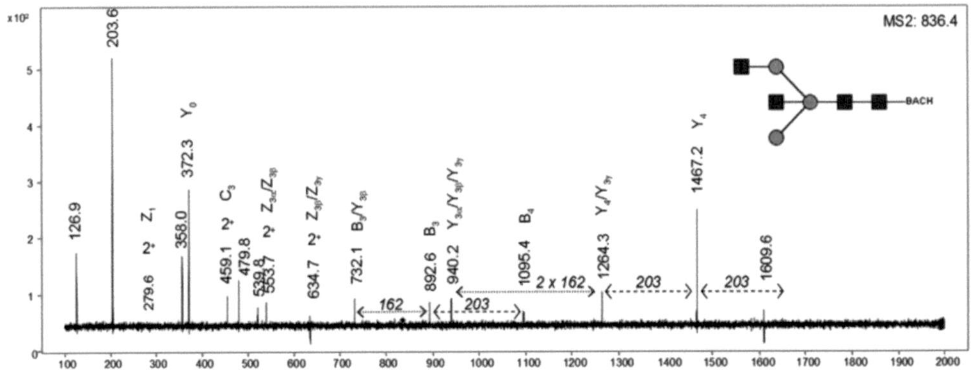

Figure 5. Product ion scan of BACH-derivatized ovalbumin glycan GlcNAc4Man3 at m/z 836.4.

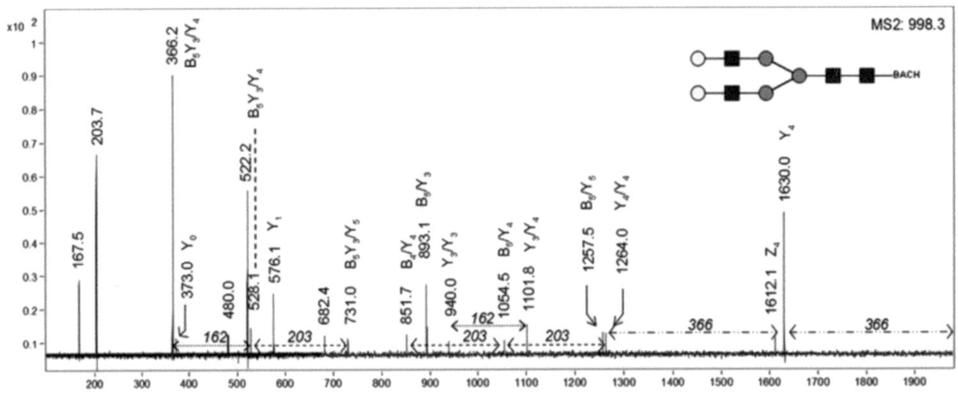

Figure 6. Product ion mass spectrum of the galactosylated NA2 glycan derivatized with BACH. The fragmentation was performed by collision-induced dissociation of the $[M+2H]^{2+}$-ion at the m/z 998.3

When compared with literature of frequently used 2-AB derivatization, tandem mass spectra of BACH-derivatives showed similar fragmentation as 2-AB-derivatives encompassing dominant glycosidic cleavages from reducing and non-reducing end (Harvey, 2000; Morelle & Michalski, 2004; Wuhrer *et al.*, 2004). These simple spectra make the MS/MS elucidation a straightforward task. Moreover, one should keep in mind that with BACH-derivatives not only the structure of the glycans could be explored, but also the function (Kapková, 2009). The bioaffinity of the BACH-label makes these derivatives suitable for the experiments exploring the interaction capabilities of a glycan (Wuhrer *et al.*, 2010).

Usually, the complete sugar analysis encompasses the characterization of the composition, sequence, branching and linkage of the monosaccharide residues constituting the oligosaccharides. The fragments arising from glycosidic cleavages are useful for the elucidation of the sugar composition, sequence and branching pattern of the glycan. Using this methodology alone, the type of linkage between the various sugar residues was not obtained. However, combining the product ion scanning with other methods e.g. enzymatic exoglycosidase digestion studies would provide some of this information.

3.3. LC-MS of BACH-derivatized monosaccharides

Mixture of carbohydrates (*N*-acetylglucosamine (GlcNAc), galactose (Gal), mannose (Man), Fucose (Fuc) and Xylose (Xyl)) which are typically present in mammalian proteins was analyzed by ESI-MS in the ion trap. Comparative examinations of pre-column labeled BACH-monosaccharides were performed by means of reversed phase liquid chromatography (RPLC) and hydrophilic interaction liquid chromatography (HILIC) as these methods shown to be highly complementary in resolution of mixtures of carbohydrates (Melmer *et al.*, 2011; Schlichtherle-Cerny *et al.*, 2003; Strege *et al.*, 2000). HILIC is based on the hydrophilic character of the analyte which accounts for the interaction of the molecule with the stationary phase (Wuhrer *et al.*, 2009). In our study, diol-modified silica phase was used. Chemically bonded diol groups were shown in the past to give similar functions as an amino-bonded silica. Moreover, they possess better stability than amino-sorbent and provide a good separation of structurally related polar molecules, e.g. sugars (Brons & Olieman, 1983; Churms, 1996; Ikegami *et al.*, 2008).

The BACH-derivatized monosaccharides, non-labeled and 2-AB-labeled counterparts (as 2-AB-labeling is one of the most widely used methods) were compared in terms of their chromatographic behavior and signal intensity of the mass spectrometric measurement.

Under the conditions used, underivatized sugars did not provide sufficient retention and separation on either of the phases (C18 and HILIC) and eluted almost at the same time (t_R = 1 min, C18; t_R = 2-5 min, HILIC) with a poor sensitivity (data not shown), especially for glucose, xylose and fucose (Intens. = 0.5 × 10^6). In order to seek an enhanced detection and better separation, derivatization of sugars was performed. After the derivatization with BACH or 2-AB, mass spectrometric signal intensity significantly enhanced about an order of a magnitude (BACH-derivatives Ints. = 0.5-4 ×10^7, 2-AB-derivatives 0.5-3 ×10^7). The extent of the intensity enhancement depended on the kind of the sugar analyzed (Fig. 7).

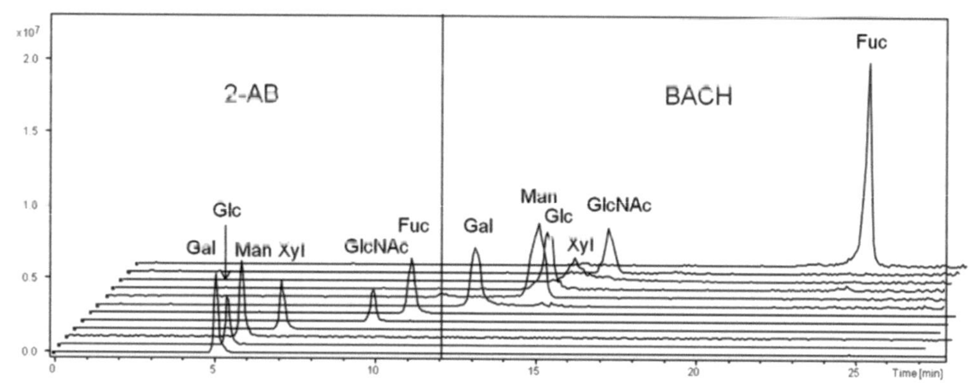

Figure 7. Total ion chromatogram (TIC) of RPLC-MS of 2-AB and BACH-derivatized monosaccharides

The BACH-derivatives interacted with the C18-phase ~8-15 min longer than the respective 2-AB-derivatives. The elution order of the sugars on the C-18-phase remained similar except for the triplet of the sugars galactose, glucose and mannose (Gal-2-AB, Glc-2AB, Man-2AB and Gal-BACH, Man-BACH, Glc-BACH). Galactose and mannose separation was achieved with both labels under the named conditions on the RP (Fig. 7), whereas the BACH-derivatives provided a clear baseline resolution.

Glucose is rarely present in glycoproteins. It can however originate from hydrolysis of contaminating polysaccharides from the environment. This problem was often encountered in reversed phase chromatography of monosaccharides (Toomre & Varki, 1994; Kwon & Kim, 1993) where glucose (at least partially) overlaps with the peak of galactose. Therefore, glucose contamination in carbohydrate samples should be addressed and examined.

In our experiments on C18 and HILIC-phase, 2-AB-derivatized glucose and galactose eluted very close to each other (Fig. 7 and Fig. 8, respectively). Through BACH-derivatization, on the other hand, glucose and galactose could have been separated on C18 (Fig. 7). However, the separation of glucose from mannose was still insufficient with BACH-labeling.

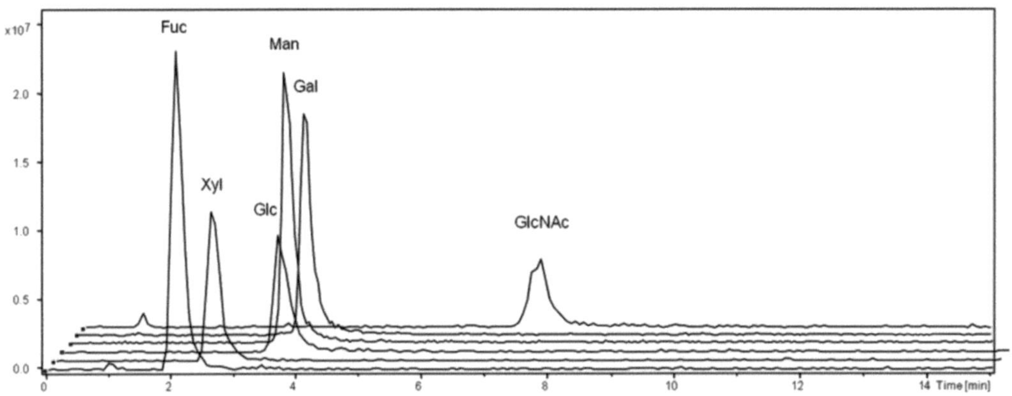

Figure 8. Hydrophilic interaction LC/MS of 2-AB-derivatized monosaccharides

A complete separation of glucose from galactose and mannose was reached only after the application of BACH-sugars to the HILIC-phase as shown in Fig. 9. However, the monosaccharides galactose and mannose could have been separated from each other only by means of RP chromatography (Fig.7). In this regard, the two phases (RP and HILIC) provided complementary features and their interchange may help to solve some analytical obstacles.

The BACH-label was used in the original form without any purification after its purchase. The presence of an isomeric byproduct was evident when the chromatography of the native label was performed. Two peaks (t_R: 16.8 min and 24 min, C18; 2 min and 6 min, HILIC) with the same m/z at 372 were present when chromatographically examined (data not shown). An example of this fact can be seen later in Fig.11 with the disaccharides, where the unreacted BACH shows two peaks. Due to this fact, we obtained after derivatization of saccharides a main product with a byproduct of the same m/z (Fig. 9). When necessary, the byproduct can be easily purified away with the excess BACH-label. Detailed structure of this byproduct is under investigation. Our further measurements concentrated only on the main product. As the BACH-tag eluted on the reversed phase between 16- 17 min with other sugars such as Man and Glc, a simultaneous clean-up of this label with the method on the C18 would not be feasible. However, when applied to HILIC-phase, the BACH-reagent eluted earlier than all the sugars and could be separated away when desired. As 2-Aminobenzamid label eluted before the mixture of monosaccharides on both stationary phases RP (6 min.) and HILIC (1.5 min.), a clean-up of the 2AB-label by this methodology would be very straightforward by both chromatographic methods.

The peaks of the investigated monosaccharides on the HILIC phase show broadening or splitting (Fig. 9). This is due to mutarotation of the sugars and consequently the presence of its anomers. The broadening is more evident with sugars, which mutarotate fast e.g. glucose and mannose. This phenomenon was observed with small reducing sugars on the HILIC-phase already in the past (Churms, 1996; Moriyasu *et al.*, 1984).

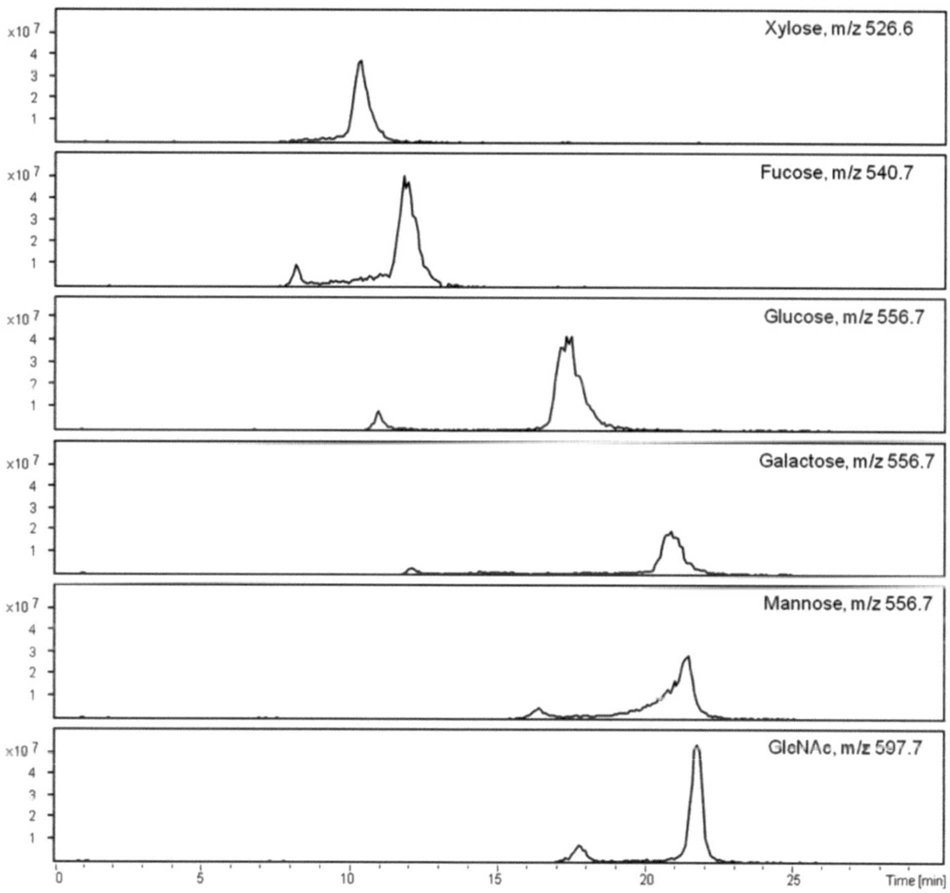

Figure 9. Extracted ion chromatograms (EIC) of BACH-derivatized monosaccharides separated by hydrophilic interaction liquid chromatography.

If the mutarotation should be suppressed, chromatography on the reversed phase can be followed and usage of a solvent with a higher content of organic phase might be preferable. A full separation of such isomeric structures (anomers) was achieved hitherto by means of chiral HPLC (Lopes & Gaspar, 2008), ion-exchange chromatography or amino-columns (Schumacher & Kroh, 1995; Nishikawa *et al.*, 1996; Moriyasu *et al.*, 1984). More recently, separation of metal adducts of anomeric methyl glycoside isomers was achieved by ion mobility mass spectrometry (Dwivedi *et al.*, 2007).

3.3.1. MS/MS of BACH-derivatized monosaccharides

The BACH-derivatized monosaccharides were subjected to tandem mass spectrometry. Analysis by ESI-ion trap showed that sodium adducts were the most abundant ions. Less abundant $[M+H]^+$ ions were detected in positive ion mode as well. Negative ion mode MS yielded $[M-H]^-$ ions of derivatized monosaccharides. All three types of ions were subjected to MS/MS in order to examine their fragmentation (Fig. 10).

MS/MS of sodium adducts $[M+Na]^+$ of neutral monosaccharides provided X-ring-cleavage in position 0-2 (m/z 436, 10% rel. intensity (RI)) and cleavage between two nitrogen atoms of the hydrazide group with m/z 379 (100% RI). With the sodiated ion of BACH-derivatized aminosugar *N*-acetylglucosamine, both of the cleavages were abundantly present in the spectrum ($^{0,2}X$ at m/z 477, 100% RI; m/z 379, ~60% RI) and were accompanied by minor water losses from the molecular ion.

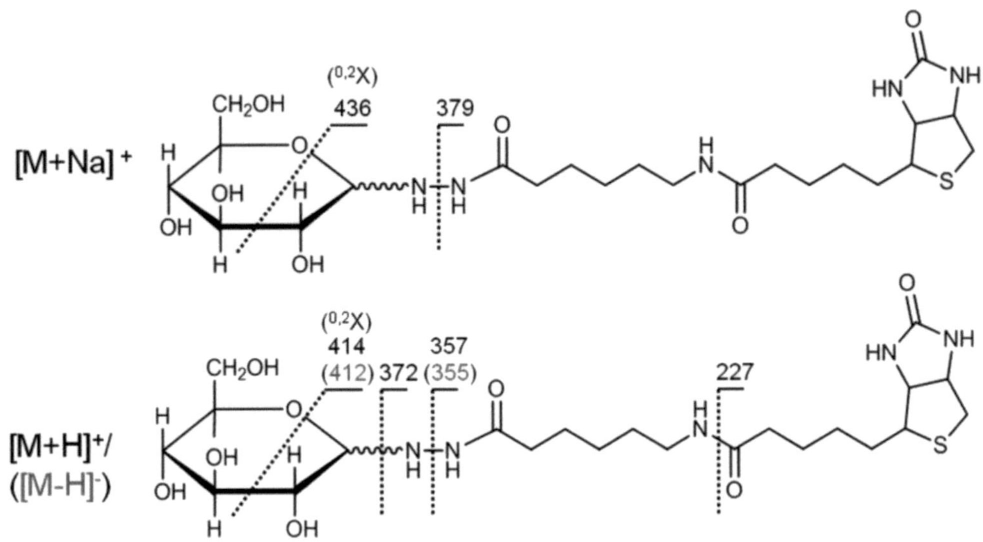

Figure 10. Scheme of fragmentation observed in MS/MS of BACH-labeled monosaccharides with different types of ions. Nominal masses are given.

The protonated ion species of the BACH-derivatized monosaccharides showed under the MS-fragmentation multiple water losses from the molecular ion (100%), ^{0,2}X-fragment at m/z 414 (~20%) and some abundant cleavages associated with the BACH label: m/z 372 (100%) – detachment of the whole label, m/z 357 (~23%) – loss of an ammonia group and m/z 227 (~43%) which presents the acylium ion of the biotin component (Fig. 10).

The deprotonated form of the molecular ion $[M-H]^-$ of BACH monosaccharides showed similarity with the fragmentation spectrum of the $[M+Na]^+$- ion and yielded weak ^{0,2}X-ring cleavage (m/z 412, ~7%) and the abundant cleavage of the hydrazine label (m/z 355, 100%). In the case of *N*-acetylglucosamine, on the other hand, the ^{0,2}X-fragment was the only prominent peak in the MS/MS spectrum with 100% of rel. intensity. Hence, variation in abundances of fragment-ions of the aminosugar GlcNAc in comparison with fragments of neutral sugars was observed when performing CID of $[M+Na]^+$ and $[M-H]^-$ ions.

3.4. Hydrophilic interaction liquid chromatography of BACH-derivatized disaccharides

A common obstacle in the mass spectrometry of saccharides is the characterization of the stereochemistry of the glycosidic bond (Zhu *et al.*, 2009; Morelle & Michalski, 2005). Isomers usually have very similar MS^2- or MS^3-fragmentation and are difficult to differentiate without a previous separation. Therefore, the MS discrimination of isomers is often aided by one or more orthogonal pre-mass separation techniques.

The formation of closed-ring derivatives (glycosylhydrazides) through non-reducing derivatization has demonstrated some advantages in the chromatographic separation and in the analysis of isomers (Li & Her, 1998, 1993). Also, differentiation of anomeric configuration has been observed in some cases following derivatization (Xue *et al.*, 2004; Ashline *et al.*, 2005). Clowers et al. achieved this with reduced form of isobaric di- and trisaccharides by ion mobility mass spectrometry (Clowers *et al.*, 2005). These research groups exploring isomeric sugars with the focus on their MS fragmentation, observed differences in the negative-ion electrospray mass spectra of structurally related compounds. The presence/absence and abundance of certain fragment ions allowed for an assembly of empirical criteria for assigning either a linkage position or the anomeric configuration of a disaccharide. For example, diagnostic ions (m/z 221, 263, 281) which were characteristic for 1-4-linkage were observed by negative ion mode mass spectrometry (Mulroney *et al.*, 1995; Garozzo *et al.*, 1991). With glucose-containing disaccharides, dissociation of the m/z 221 generated product ions that allowed for the differentiation of anomeric configuration (Fang & Bendiak, 2007; Fang *et al.*, 2007).

Here we sought to examine whether the BACH-derivatized isomeric disaccharides would be separable on HILIC and would in MS/MS dissociate to yield product ions from which stereochemical information might be obtained about the glycosidic bond. For this purpose, maltose (α-D-Glc-(1-4)-D-Glc) and cellobiose (β-D-Glc-(1-4)-D-Glc), sugars which vary solely in their anomeric configuration were derivatized by non-reducing BACH-labeling according to the protocol described in the experimental part.

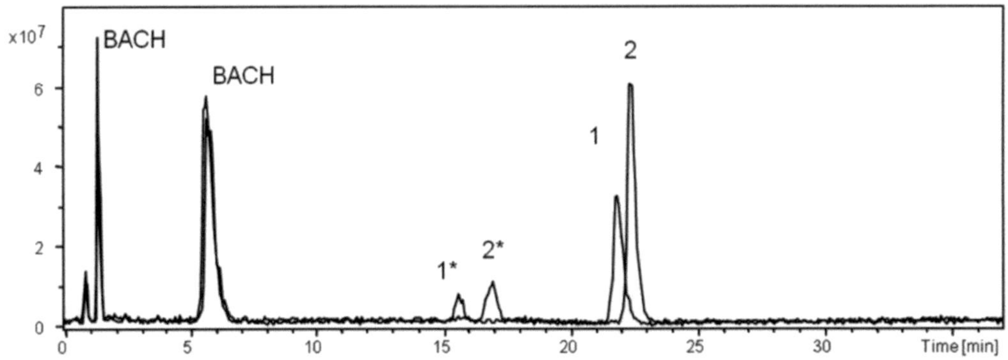

Figure 11. TIC showing separation of BACH-labeled isomeric disaccharides by means of hydrophilic interaction liquid chromatography. Peak assignements: 1 = BACH-maltose, 2 = BACH-cellobiose, * isomeric byproduct upon derivatization when BACH-label used in original unpurified form, BACH = excess of labeling reagent

With underivatized maltose and cellobiose, HPLC-separation and signal intensity were very poor (data not shown). Through BACH-derivatization, enhancement of signal intensity of two orders of magnitude (from $2x10^5$ to $6x10^7$) in comparison with native counterparts was achieved and at least a partial separation on the HILIC-phase was reached (Fig. 11). Even if they were not completely resolved, this demonstrated the feasibility of determining the presence of isomeric hits. For MS/MS investigation, the sugars were subjected separately to mass spectrometry, in order to obtain discrete fragmentation data originating only from one of the isomers. MS/MS spectra of underivatized maltose and cellobiose were without any significant difference in fragmentation pattern. Thus, no assignment could have been done with the native sugars.

In the study of BACH-labeled maltose and cellobiose, different types of molecular ions were detected: $[M+Na]^+$ (m/z 718), $[M+H]^+$ (m/z 696) and $[M-H]^-$ (m/z 694). These were subsequently subjected to tandem mass spectrometry in order to examine their fragmentation profile. Each of the precursor ions showed different fragmentation:

In positive ion mode, same fragmentation patterns for maltose and cellobiose were observed (data not shown). MS/MS of the sodium adduct contained the $^{0,2}X$-ring cleavage (m/z 436) from the reducing end of the derivatized disaccharide and the N-N-cleavage of the hydrazide group of the BACH-label (m/z 355). Fragmentation spectra of the $[M+H]^+$-ion were similar to $[M+Na]^+$. They provided week $^{0,2}X$-fragment (m/z 414) and the hydrazide group cleavage (m/z 355). Additionally to these, glycosidic cleavages (Y_0 and Y_1), which confirmed the sequence and hexose composition of the saccharide, appeared. With Y_1-ion, up to four water losses were observed. Hence, in positive ion mode, when observing either $[M+Na]^+$ or $[M+H]^+$, no difference in the fragmentation of these isomers was observed.

CID of the $[M-H]^-$ion, on the contrary (Fig. 12), yielded cleavages (m/z 412 and m/z 355) and also fragments between m/z 179 and m/z 355, having higher masses than monosaccharide and encompassing the glycosidic bond of the sugar.

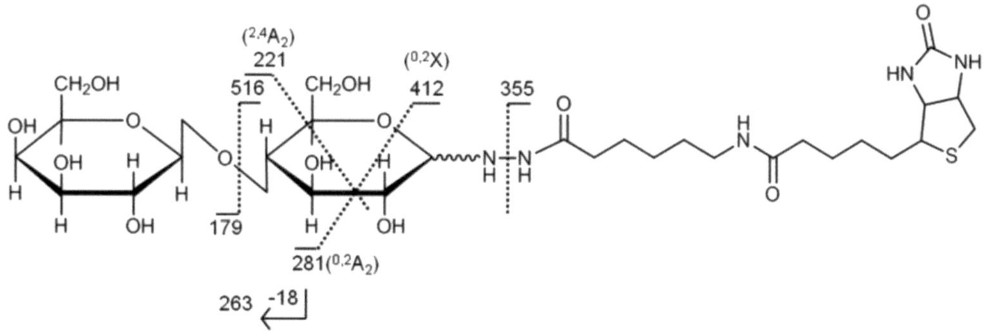

Figure 12. Fragmentation of [M-H]⁻-ion observed for BACH-labeled isomeric disaccharides maltose and cellobiose in negative ion mode

Negative MS/MS fragmentation spectra of BACH-labeled maltose and cellobiose were similar in terms of presence of diagnostic ions. However, some variation in the relative abundance of the product ions was visible. Ions observed were: m/z 179, 221, 263, 281 (Fig. 12), where m/z 179 represents the non-reducing monosaccharide containing the glycosidic oxygen; m/z 221 ($^{2,4}A_2$) comprises intact non-reducing sugar glycosidically linked to a glycoaldehyde molecule and m/z 281 corresponds to loss of one –CHOH backbone-unit and also one -CH with the BACH-label. This ion eliminated a molecule of water yielding an m/z 263 ion. This has been shown to be characteristic for 1,4-bonded underivatized glucose dimers (Dallinga & Heerma, 1991a).

In the spectrum of α-linked maltose (Fig. 13a), intensity of the m/z 221 was slightly higher than m/z 263. In the fragmentation spectra of beta-linked cellobiose, on the contrary, the ion m/z 221 appeared only in traces. With cellobiose, the intensity of m/z 263 was always higher than m/z 221. Further, the ion at m/z 281 was more abundant in the spectrum of maltose, whereas almost absent in the MS/MS of cellobiose. In order to confirm the reproducibility of the fragmentation we performed the measurements 15 times in total (5 rounds of measurements for each disaccharide on three different days) indicating the presence and intensity of respective fragments to be reproducible and constant. These observations are in accordance with the fragments reported for 4-linked disaccharides and their proportionalities correspond with dissociation studies reported for the differentiation of this type of isomeric disaccharides. In the study of Li and Her (Li & Her, 1993), where relative abundance of fragment ions of 1-4 linked disaccharides was compared, the intensity of the m/z 221 was higher than m/z 263 in alpha-linked disaccharides and smaller than m/z 263 ion in beta-linked disaccharides. As reported (Fang & Bendiak, 2007; Fang *et al.*, 2007), the cross-ring cleavage m/z 221 of the disaccharides also showed to be useful in the discrimination of anomeric configuration. Under conditions applied here, further dissociation of m/z 221 (MS^3) of the BACH- labeled maltose or cellobiose provided no additional differential information. Noneless, through the combination of BACH-labeling and tandem mass spectrometry, distinctive structural information was obtained at the MS^2-stage of fragmentation and disaccharides with the same linkage but different anomeric configurations (maltose: 1-4-α-linkage, cellobiose: 1-4-β-linkage) could have been distinguished.

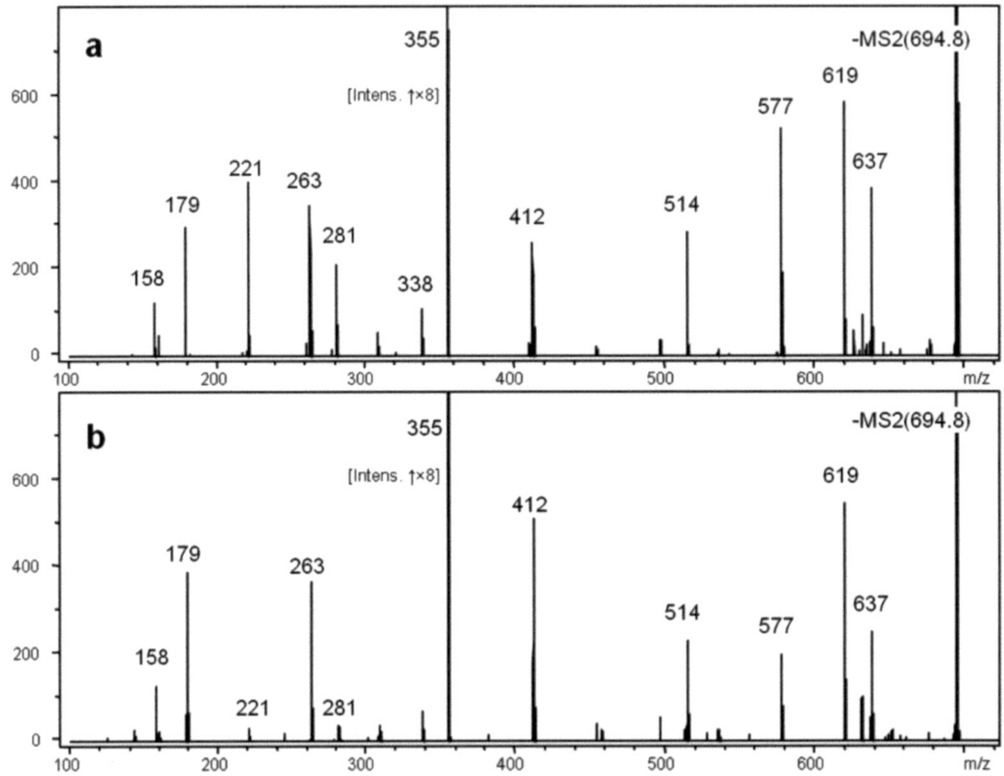

Figure 13. MS/MS of BACH-derivatized isomeric disaccharides. a) Maltose-BACH b) Cellobiose-BACH

3.5. Hydrophilic interaction liquid chromatography of isomeric non-reducing trisaccharides

In order to further examine the ability of the current analytical system to distinguish between the sugar isomers, we have analyzed two trisaccharides melezitose [α-D-Glc-(1→3)-β-D-Fru-(2→1)-α-D-Glc] and raffinose [α-D-Gal-(1→6)-α-D-Glc-(1→2)-β-D-Fru] which differ in sequence, linkage and anomeric configuration. Both of them were examined by LC-MS/MS in negative ion mode as this one showed to be the preferential mode in order to observe some distinctive fragmentation of related compounds (Section 3.4.). Figure 14a shows the total ion current of melezitose and raffinose. The chromatography of these sugars on the HILIC-phase provided a full baseline separation.

The fragmentation mass spectra of these trisaccharides were obtained from the deprotonated form of the molecular ion $[M-H]^-$ at m/z 503 (Figure 14b-e). Melezitose displayed a dominant fragment at m/z 323 which derives from cleavage of one of the two possible glycosidic bonds (Fig. 14b). This route was confirmed through the MS^3 experiment, where the dissociation of m/z 341 generated no m/z 323 but solely the ions with m/z 179 and m/z 161 (Fig. 14d). This indicates that m/z 323 occurs directly through B-fragmentation and not via formation of m/z 341 and a subsequent water loss. The peak at m/z 341 was found at

MS²-stage of both trisaccharides and is due to the elimination of a site residue from the [M-H]⁻-molecular ion. Interestingly, in positive-ion MS-studies on [M+H]⁺-ions of non-reducing oligosaccharides containing fructose (including melezitose and raffinose), the formation of m/z 323 was proposed to preferably occur via the water loss from m/z 341 (Dallinga & Heerma, 1991; Perez-Victoria *et al.*, 2008).

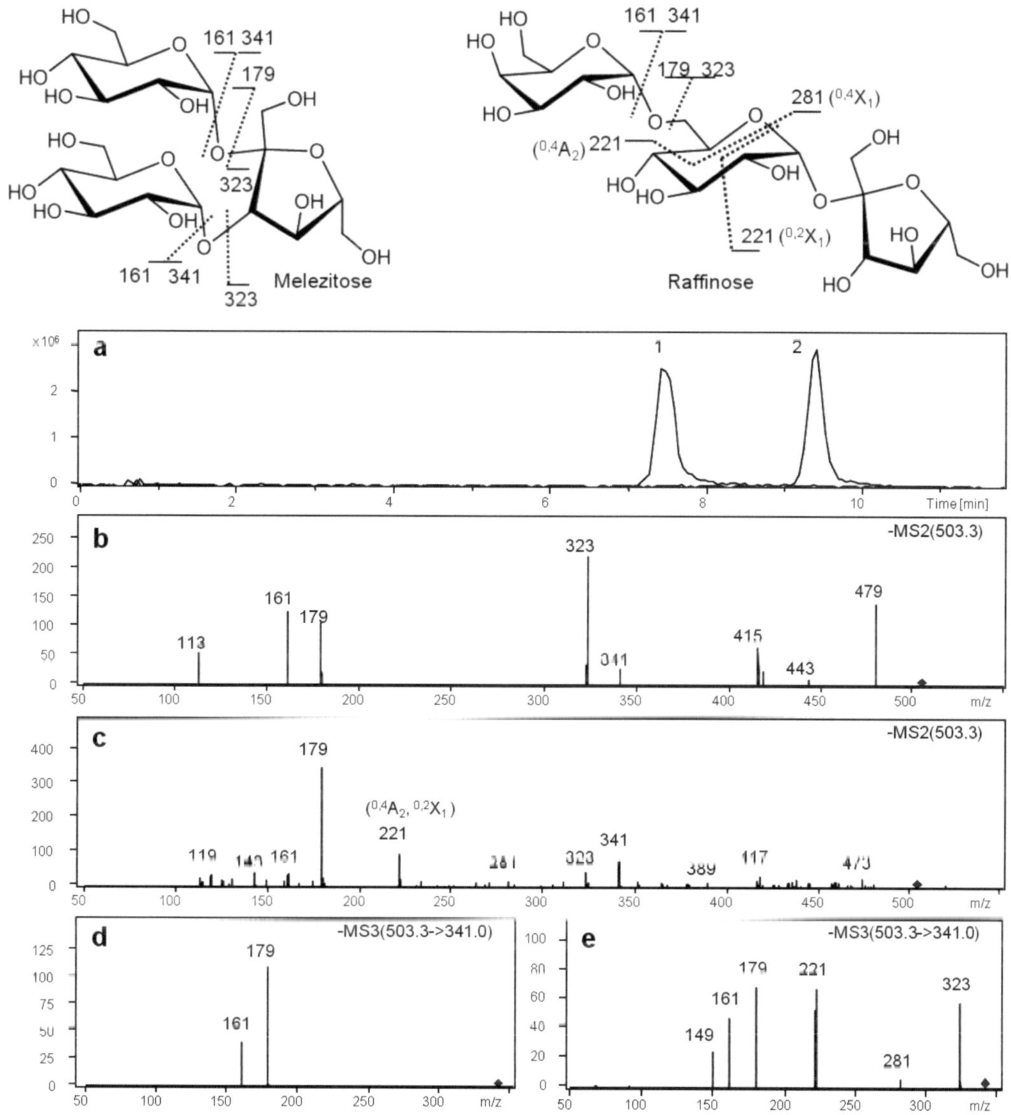

Figure 14. a) Separation of the mixture of two isomeric trisaccharides melezitose (1) and raffinose (2), demonstrating the value of coupling HILIC with ion-trap mass spectrometry performed to MS³. b) MS² of melezitose c) MS² of raffinose d) MS³ of melezitose e) MS³ of raffinose

Raffinose fragmented to form preferably m/z 179, m/z 221 and glycosidic fragment at m/z 341 (Fig. 14c). Other ions (m/z 323 and 161) were present as well. At this stage of fragmentation (MS^2), the formation of cross-ring cleavage (m/z 221) involving either the 0-2 or 0-4 bonds of the center residue appeared. This ion has not been seen in the spectrum of melezitose (compare to Fig. 14b). The same fact was reported in the studies of fragmentation of trisaccharides by negative FAB (Dallinga & Heerma, 1991b) and negative ion mobility mass spectrometry (Liu & Clemmer, 1997): raffinose yielded m/z 221 in the MS^2-spectrum, whereas melezitose showed only the product ion at m/z 323 without presence of m/z 221. The presence of m/z 221 was also evident in MS^3 of raffinose (Fig. 14e). The different fragmentation routes observed may be attributed to the difference in the steric character of these isomers: Melezitose has a branched character and raffinose represents a linear molecule.

Finally, the differences in the chromatography on the HILIC-phase and in the fragmentation pattern (presence and rel. intensity of the product ions in the MS^2 and MS^3 stage) of the melezitose and raffinose indicated the occurrence of distinct isomeric structures.

This distinguished fragmentation would not have been possible to obtain without the previous pre-mass spectrometric separation on the HILIC-phase. Considering mass spectrometric techniques, such differential identification of these isomers out of their mixture was achieved up to now, only by means of ion-mobility mass spectrometry (Zhu *et al.*, 2009; Liu & Clemmer, 1997). To our best knowledge, this is the first report on separation and identification of such isomers by coupling of hydrophilic interaction chromatography and mass spectrometry.

4. Conclusion

Different types of *N*-glycans (high-mannose, complex and hybrid) have been derivatized with biotinamidocaproyl hydrazide and analyzed by chip-based LC/ESI triple-quadrupole mass spectrometry. This approach allowed for rapid mass spectrometric selection of glycans out of glycan mixtures on the base of the scanning for a common feature ion (oxonium ion). The glycans observed upon MS were primarily in the doubly charged state. For each type of glycan, product ion spectra with the characteristic B- and Y-ions were observed which allowed for elucidation of the structure and segregation of the glycans into the respective class. This methodology provided fast structural information about the nature of the glycan pool present in a glycoprotein and may be used in the profiling of *N*-linked glycans. Beyond structural mass spectrometric studies, BACH-derivatized glycans can potentially be used also in functional studies with carbohydrate binding proteins, as recently reported.

HILIC and reversed phase liquid chromatography of BACH- and 2-AB-labeled monosaccharides enabled the investigation of different analytical objectives and showed some orthogonal features in terms of the separation of monosaccharides frequently present in glycoproteins. The HILIC-system was not sufficient for separation of anomers of BACH-derivatized monosaccharides. It was, however, capable to partially resolve the

BACH-derivatized isomeric disaccharides and to fully separate non-reducing isomeric trisaccharides.

In terms of the enhancement of mass spectrometric signal of mono- and disaccharides, BACH-labeling provided higher intense ions than 2-AB-derivatization. Analysis of the BACH-derivatives in combination with the mass spectrometry can be pursued, when investigation of isomers is desired. Unlike positive ESI-MS, negative ion MS of BACH-derivatized isomeric disaccharides provided ring-cleavage ions which were helpful in the differentiation of the anomeric configuration of the glycosidic bond. Further, separation of structurally related non-derivatized trisaccharides was achieved on HILIC and their differentiation on the basis of multiple stage mass spectrometry was possible. This differentiation would not have been possible without the suitable pre-mass spectrometric separation. In this way, coupling of the HILIC-separation with MS^n-fragmentation might be a valuable approach on the way to the distinction of certain types of isomeric structures.

Author details

Petra Kapková and Stephanie Bank
University of Würzburg/Department of Pharmacy and Food Chemistry, Germany

Acknowledgement

This work was supported by the Fond der Chemischen Industrie and the Universitätsbund of University of Würzburg.

5. References

Alley, W. R., Jr., Mechref, Y. & Novotny, M. V. (2009). Use of activated graphitized carbon chips for liquid chromatography/mass spectrometric and tandem mass spectrometric analysis of tryptic glycopeptides. *Rapid Communications in Mass Spectrometry,* Vol.23, No.4, pp. 495-505, ISSN 0951-14198

Ashline, D., Singh, S., Hanneman, A. & Reinhold, V. (2005). Congruent strategies for carbohydrate sequencing. 1. Mining structural details by MSn. *Analytical Chemistry,* Vol.77, No.19, pp. 6250-6262, ISSN 0003-2700

Bigge, J. C., Patel, T. P., Bruce, J. A., Goulding, P. N., Charles, S. M. & Parekh, R. B. (1995). Nonselective and efficient fluorescent labeling of glycans using 2-amino benzamide and anthranilic acid. *Analytical Biochemistry,* Vol.230, No.2, pp. 229-238, ISSN 0003-2697

Bindila, L. & Peter-Katalinic, J. (2009). Chip-mass spectrometry for glycomic studies. *Mass Spectrometry Reviews,* Vol.28, No.2, pp. 223-253, ISSN 1098-2787

Brons, C. & Olieman, C. (1983). Study of the high-performance liquid-chromatographic separation of reducing sugars, applied to the determination of lactose in milk. *Journal of Chromatography,* Vol.259, No.1, pp. 79-86, ISSN 0021-9673

Churms, S. C. (1996). Recent progress in carbohydrate separation by high-performance liquid chromatography based on hydrophilic interaction. *Journal of Chromatography A,* Vol.720, No.1-2, pp. 75-91, ISSN 0021-9673

Clowers, B. H., Dwivedi, P., Steiner, W. E., Hill, H. H. & Bendiak, B. (2005). Separation of sodiated isobaric disaccharides and trisaccharides using electrospray ionization-atmospheric pressure ion mobility-time of flight mass spectrometry. *Journal of the American Society for Mass Spectrometry,* Vol.16, No.5, pp. 660-669, ISSN 1044-0305

Dallinga, J. W. & Heerma, W. (1991a). Positive-Ion Fast-Atom-Bombardment Mass-Spectrometry of Some Small Oligosaccharides. *Biological Mass Spectrometry,* Vol.20, No.3, pp. 99-108, ISSN 1052-9306

Dallinga, J. W. & Heerma, W. (1991b). Reaction-Mechanism and Fragment Ion Structure Determination of Deprotonated Small Oligosaccharides, Studied by Negative-Ion Fast-Atom-Bombardment (Tandem) Mass-Spectrometry. *Biological Mass Spectrometry,* Vol.20, No.4, pp. 215-231, ISSN 1052-9306

Domon, B. & Costello, C. E. (1988). A Systematic Nomenclature for Carbohydrate Fragmentations in Fab-Ms Ms Spectra of Glycoconjugates. *Glycoconjugate Journal,* Vol.5, No.4, pp. 397-409, ISSN 0282-0080

Dwivedi, P., Bendiak, B., Clowers, B. H. & Hill, H. H. (2007). Rapid resolution of carbohydrate isomers by electrospray ionization ambient pressure ion mobility spectrometry-time-of-flight mass spectrometry (ESI-APIMS-TOFMS). *Journal of the American Society for Mass Spectrometry,* Vol.18, No.7, pp. 1163-1175, ISSN 1044-0305

Fang, T. T. & Bendiak, B. (2007). The stereochemical dependence of unimolecular dissociation of monosaccharide-glycolaldehyde anions in the gas phase: A basis for assignment of the stereochemistry and anomeric configuration of monosaccharides in oligosaccharides by mass spectrometry via a key discriminatory product ion of disaccharide fragmentation, m/z 221. *Journal of the American Chemical Society,* Vol.129, No.31, pp. 9721-9736, ISSN 0002-7863

Fang, T. T., Zirrolli, J. & Bendiak, B. (2007). Differentiation of the anomeric configuration and ring form of glucosyl-glycolaldehyde anions in the gas phase by mass spectrometry: isomeric discrimination between m/z 221 anions derived from disaccharides and chemical synthesis of m/z 221 standards. *Carbohydrate Research,* Vol.342, No.2, pp. 217-235, ISSN 0008-6215

Garozzo, D., Impallomeni, G., Spina, E., Green, B. N. & Hutton, T. (1991). Linkage analysis in disaccharides by electrospray mass-spectrometry. *Carbohydrate Research,* Vol.221, pp. 253-257, ISSN 0008-6215

Geyer, H. & Geyer, R. (2006). Strategies for analysis of glycoprotein glycosylation. *Biochimica Biophysica Acta,* Vol.1764, No. 12, pp. 1853-1869, ISSN 0006-3002

Grün, C. H., van Vliet, S. J., Schiphorst, W. E., Bank, C. M., Meyer, S., van Die, I. & van Kooyk, Y. (2006). One-step biotinylation procedure for carbohydrates to study carbohydrate-protein interactions. *Analytical Biochemistry,* Vol.354, No.1, pp. 54-63, ISSN 0003-2697

Harvey, D. J. (2000). Electrospray mass spectrometry and collision-induced fragmentation of 2-aminobenzamide-labelled neutral N-linked glycans. *Analyst*, Vol.125, No.4, pp. 609-617, ISSN 0003-2654

Harvey, D. J. (2011). Derivatization of carbohydrates for analysis by chromatography; electrophoresis and mass spectrometry. *Journal of Chromatography B*, Vol.879, No.17-18, pp. 1196-1225, ISSN 1570-0232

Hermannson, G. T. (2008). *Bioconjugate Techniques*, Academic Press, New York

Hsu, J., Chang, S. J. & Franz, A. H. (2006). MALDI-TOF and ESI-MS analysis of oligosaccharides labeled with a new multifunctional oligosacharide tag. *Journal of the American Society for Mass Spectrometry*, Vol.17, pp. 194-204, ISSN 1044-0305

Ikegami, T., Tomomatsu, K., Takubo, H., Horie, K. & Tanaka, N. (2008). Separation efficiencies in hydrophilic interaction chromatography. *Journal of Chromatography A*, Vol.1184, No.1-2, pp. 474-503, ISSN 0021-9673

Kapková, P. (2009). Mass spectrometric analysis of carbohydrates labeled with a biotinylated tag. *Rapid Communications in Mass Spectrometry*, Vol.23, No.17, pp. 2775-2784, ISSN 0951-4198

Kwon, H. & Kim, J. (1993). Determination of monosaccharides in glycoproteins by reverse-phase high-performance liquid chromatography. *Analytical Biochemistry*, Vol.215, pp. 243-252, ISSN 0003-2697

Lattová, E., Snovida, S., Perreault, H. & Krokhin, O. (2005). Influence of the labeling group on ionization and fragmentation of carbohydrates in mass spectrometry. *Journal of the American Society for Mass Spectrometry*, Vol.16, No.5, pp. 683-696, ISSN 1044-0305

Leteux, C., Childs, R. A., Chai, W., Stoll, M. S., Kogelberg, H. & Feizi, T. (1998). Biotinyl-l-3-(2-naphthyl)-alanine hydrazide derivatives of N-glycans: versatile solid-phase probes for carbohydrate recognition studies. *Glycobiology*, Vol.8, No.3, pp. 227-236, ISSN 0959 6658

Leymarie, N. & Zaia, J. (2012). Effective use of mass spectrometry for glycan and glycopeptide structural analysis. *Analytical Chemistry*, Vol.84, pp. 3040-3048, ISSN 0003-2700

Li, D. T. & Her, G. R. (1993). Linkage Analysis of Chromophore-Labeled Disaccharides and Linear Oligosaccharides by Negative-Ion Fast-Atom-Bombardment Ionization and Collisional-Induced Dissociation with B/E Scanning. *Analytical Biochemistry*, Vol.211, No.2, pp. 250-257, ISSN 0003-2697

Li, D. T. & Her, G. R. (1998). Structural analysis of chromophore-labeled disaccharides and oligosaccharides by electrospray ionization mass spectrometry and high performance liquid chromatography electrospray ionization mass spectrometry. *Journal of Mass Spectrometry*, Vol.33, No.7, pp. 644-652, ISSN 1076-5174

Liu, Y. S. & Clemmer, D. E. (1997). Characterizing oligosaccharides using injected-ion mobility mass spectrometry. *Analytical Chemistry*, Vol.69, No.13, pp. 2504-2509, ISSN 0003-2700

Lopes, J. F. & Gaspar, E. M. S. M. (2008). Simultaneous chromatographic separation of enantiomers, anomers and structural isomers of some biologically relevant monosaccharides. *Journal of Chromatography A,* Vol.1188, pp. 34-42, ISSN 0021-9673

Marino, K., Bones, J., Kattla, J. J. & Rudd, P. M. (2010). A systematic approach to protein glycosylation analysis: A path through the maze. *Nature Chemical Biology,* Vol.6, pp. 713-723, ISSN 1552-4450

Melmer, M., Stangler, T., Premstaller, A. & Lindner, W. (2011). Comparison of hydrophilic-interaction, reversed-phase and porous graphitic carbon chromatography for glycan analysis. *Journal of Chromatography A,* Vol.1218, pp. 118-123, ISSN 0021-9673

Morelle, W. & Michalski, J. C. (2004). Sequencing of oligosaccharides derivatized with benzlyamine using electrospray ionization-quadrupole time of flight-tandem mass spectrometry. *Electrophoresis,* Vol.25, No.14, pp. 2144-2155, ISSN 0173-0835

Morelle, W. & Michalski, J. C. (2005). Glycomics and mass spectrometry. *Current Pharmaceutical Design,* Vol.11, No.20, pp. 2615-2645, ISSN 1381-6128

Moriyasu, M., Kato, A., Okada, M. & Hashimoto, Y. (1984). HPLC separation of sugar anomers in a very low temperatur region. *Analytical Letters,* Vol.17, pp. 689-699, ISSN 0003-2719

Mulroney, B., Traeger, J. C. & Stone, B. A. (1995). Determination of Both Linkage Position and Anomeric Configuration in Underivatized Glucopyranosyl Disaccharides by Electrospray Mass-Spectrometry. *Journal of Mass Spectrometry,* Vol.30, No.9, pp. 1277-1283, ISSN 1076-5174

Nishikawa, T., Suzuki, S., Kubo, H. & Ohtani, H. (1996). On-column isomerization of sugars during high-performance liquid chromatography: analysis of the elution profile. *Journal of Chromatography A,* Vol.720, pp. 167-172, ISSN 0021-9673

North, S. J., Hitchen, P. G., Haslam, S. M. & Dell, A. (2009). Mass spectrometry in the analysis of N-linked and O-linked glycans. *Current Opinion in Structural Biology,* Vol.19, pp. 498-506, ISSN 0959-440X

Pabst, M., Kolarich, D., Poltl, G., Dalik, T., Lubec, G., Hofinger, A. & Altmann, F. (2009). Comparison of fluorescent labels for oligosaccharides and introduction of a new postlabeling purification method. *Analytical Biochemistry,* Vol.384, No.2, pp. 263-273, ISSN 0003-2697

Perez-Victoria, I., Zafra, A. & Morales, J. C. (2008). Positive-ion ESI mass spectrometry of regioisomeric nonreducing oligosaccharide fatty acid monoesters: In-source fragmentation of sodium adducts. *Journal of Mass Spectrometry,* Vol.43, No.5, pp. 633-638, ISSN 1076-5174

Raman, R., Raguram, S., Venkataraman, G., Paulson, J. C. & Sasisekharan, R. (2005). Glycomics: an integrated systems approach to structure-function relationships of glycans. *Nature Methods,* Vol.2, pp. 817-824, ISSN 1548-7091

Ridley, B. L., Spiro, M. D., Glushka, J., Albersheim, P., Darvill, A. & Mohnen, D. (1997). A method for biotin labeling of biologically active oligogalacturonides using a chemically stable hydrazide linkage. *Analytical Biochemistry,* Vol.249. No.1, pp. 10-19, ISSN 0003-2697

Rosenfeld, J. (2011). Enhancement of analysis by analytical derivatization. *Journal of Chromatography B,* Vol.879, pp. 1157-1158, ISSN 1570-0232

Rothenberg, B. E., Hayes, B. K., Toomre, D., Manzi, A. E. & Varki, A. (1993). Biotinylated diaminopyridine: an approach to tagging oligosaccharides and exploring their biology. *Proceedings of the National Academy of Sciences USA,* Vol.90, No.24, pp. 11939-11943, ISSN 1091-6490

Ruhaak, L. R., Zauner, G., Huhn, C., Bruggink, C., Deelder, A. & Wuhrer, M. (2010). Glycan labeling strategies and their use in identification and quantification. *Analytical and Bioanalytical Chemistry,* Vol.397, pp. 3457-3481, ISSN1618-2650

Schlichtherle-Cerny, H., Affolter, M. & Cerny, C. (2003). Hydrophilic interaction liquid chromatography coupled to electrospray mass spectrometry of small polar compounds in food analysis. *Analytical Chemistry,* Vol.75, pp. 2349-2354, ISSN 0003-2700

Schumacher, D. & Kroh, L. W. (1995). A rapid method for separation of anomeric saccharides using cyclodextrin bonded phase and for investigation of mutarotation. *Food Chemistry,* Vol.54, pp. 353-356, ISSN 0308-8146

Shinohara, Y., Sota, H., Gotoh, M., Hasebe, M., Tosu, M., Nakao, J., Hasegawa, Y. & Shiga, M. (1996). Bifunctional labeling reagent for oligosaccharides to incorporate both chromophore and biotin groups. *Analytical Chemistry,* Vol.68, No.15, pp. 2573-2579, ISSN 0003-2700

Strege, M. A., Stevenson, S. & Lawrence, S. M. (2000). Mixed mode anion-cation exchange/hydrophilic interaction liquid chromatography-electrospray mass spectrometry as an alternative to reversed phase for small molecule drug discovery. *Analytical Chemistry,* Vol.72, No.19, pp. 4629-4633, ISSN 0003-2700

Toomre, D. K. & Varki, A. (1994). Advances in the use of biotinylated diaminopyridine (BAP) as a versatile fluorescent tag for oligosaccharides. *Glycobiology,* Vol.4, No.5, pp. 653-663, ISSN 0959-6658

Wuhrer, M., de Boer, A. R. & Deelder, A. M. (2009). Structural glycomics using hydrophilic interaction chromatography (HILIC) with mass spectrometry. *Mass Spectrometry Reviews,* Vol.28, No.2, pp. 192-206, ISSN 1098-2787

Wuhrer, M., Koeleman, C. A. M., Hokke, C. H. & Deelder, A. M. (2004). Nano-scale liquid chromatography-mass spectrometry of 2-aminobenzamide-labeled oligosaccharides at low femtomole sensitivity. *International Journal of Mass Spectrometry,* Vol.232, No.1, pp. 51-57, ISSN 1387-3806

Wuhrer, M., van Remoortere, A., Balog, C. I. A., Deelder, A. M. & Hokke, C. H. (2010). Ligand identification of carbohydrate-binding proteins employing a biotinylated glycan binding assay and tandem mass spectrometry. *Analytical Biochemistry,* Vol. 406, No.2, pp. 132-140, ISSN 0003-2697

Xue, J., Song, L. G., Khaja, S. D., Locke, R. D., West, C. M., Laine, R. A. & Matta, K. L. (2004). Determination of linkage position and anomeric configuration in Hex-Fuc disaccharides using electrospray ionization tandem mass spectrometry. *Rapid Communications in Mass Spectrometry,* Vol.18, No.17, pp. 1947-1955, ISSN 0951-4198

Zhu, M. L., Bendiak, B., Clowers, B. & Hill, H. H. (2009). Ion mobility-mass spectrometry analysis of isomeric carbohydrate precursor ions. *Analytical and Bioanalytical Chemistry,* Vol.394, No.7, pp. 1853-1867, ISSN 1618-2650

Investigation of Carbohydrates and Their Derivatives as Crystallization Modifiers

Josef Jampílek and Jiří Dohnal

Additional information is available at the end of the chapter

http://dx.doi.org/10.5772/50626

1. Introduction

Development in the field of pharmaceutical administration has resulted in the discovery of highly sophisticated drug delivery systems that allow for the maintenance of a constant drug level in an organism. Contrary to these revolution biopharmaceutical results, over the last ten years, the number of poorly soluble drugs has steadily increased. Estimates state that 40% of the drugs in the pipelines have solubility problems. Progress in high throughput screening methods leads to an even greater amount of newly discovered drugs, but a lot of them have poor water solubility. Literature states that about 60% of all drugs coming directly from synthesis are nowadays poorly soluble. Meanwhile the five key physico-chemical properties, such as pK_a, solubility, permeability, stability and lipophilicity, in early compound screening should be optimized. Compounds with insufficient solubility carry a higher risk of failure during discovery and development, since insufficient solubility may compromise other property assays, mask additional undesirable properties, influence both pharmacokinetic and pharmacodynamic properties of the compound and finally may affect the developability of the compound. Poor solubility in water correlates with poor bioavailability. If there is no way to improve drug solubility, it will not be able to be absorbed from the gastrointestinal tract into the bloodstream and reach the site of action (Junghanns & Müller, 2008; Payghan et al., 2008).

Modification/optimization of unfavourable physico-chemical properties of these drugs is possible through increasing their water solubility or improving permeability. There are many ways to solubilize certain poorly soluble drugs. But these methods are limited to drugs with certain properties in regard to their chemistry or, for example, to their molecular size or conformation.

Aqueous solubility can be increased by chemical exchange: (*i*) salts, co-crystals or solvates formation (affects also chemical stability, polymorphism, technological workability);

(*ii*) substitution by hydrophilic groups (effect of drugs with small molecules can be decreased); (*iii*) prodrug preparation (hydrolyzable amides or semiesters with polybasic acids). In general, the following structural modifications are the best way to improve permeability: (*i*) replacement of ionisable groups by non-ionizable groups; (*ii*) increase of lipophilicity; (*iii*) isosteric replacement of polar groups; (*iv*) esterification of carboxylic acid; (*v*) reduction of hydrogen bonding and polarity; (*vi*) reduction of size; (*vii*) addition of a non-polar side chain; (*viii*) preparation of prodrugs. Generally, these strategies are based on a few fundamental concepts: change of ionizability, lipophilicity, polarity or change of hydrogen bond donors or acceptors. Both approaches interact logically.

Based on these facts, pre-formulation/formulation can be another and mostly successful strategy for improving aqueous solubility and/or permeability and subsequently bioavailability. For example, selection of a suitable salt, particle size reduction (till nano size) connected with an increase of the surface area, change of polymorphic forms, selection of appropriate excipients to function as solubilizers/transporters (surfactants or pharmaceutical complexing agents, permeability enhancers) can be used for the oral dosage form (Kerns & Di, 2008).

It is well-known that crystalline materials obtain their fundamental physical properties from the molecular arrangement within the solid, and altering the placement and/or interactions between these molecules can, and usually does, have a direct impact on the properties of the particular solid. Currently, solid-state chemists call upon a variety of different strategies when attempting to alter the chemical and physical solid-state properties of active pharmaceutical ingredients (APIs), namely, the formation of salts, polymorphs, hydrates, solvates and co-crystals (Seddon & Zaworotko, 1999; Datta & Grant, 2004; Grepioni & Braga, 2007; Schultheiss & Newman, 2009).

Currently, salt formation is one of the primary solid-state approaches used to modify the physical properties of APIs, and it is estimated that over half of the medicines on the market are administered as salts (Bighley et al., 1996; Stahl & Wermuth, 2002; Gu & Grant, 2003.). However, a major limitation within this approach is that the API must possess a suitable (basic or acidic) ionizable site. In comparison, co-crystals (multi-component assemblies held together by freely reversible, non-covalent interactions) offer a different pathway, where any API regardless of acidic, basic or ionizable groups, could potentially be co-crystallized. This aspect helps to complement existing methods by reintroducing molecules that had limited pharmaceutical profiles based on their non-ionizable functional groups. Since the primary structure of a drug molecule does not change in co-crystals, their development in terms of the regulation "New Chemical Entities (NCEs)" of the U.S. Food and Drug Administration (FDA) as well as European Commission Regulation (EC) No. 258/97 carries fewer risks and takes much less time; nevertheless stability and bioequivalent studies are still necessary. In addition, the number of potential non-toxic co-crystal formers (or co-formers) that can be incorporated into a co-crystalline reaction is numerous, *e.i.* pharmaceutical excipients, amino acids, food additives, nutraceuticals, see for example the GRAS list (Generally Regarded as Safe) or the EAFUS Database (Everything Added to Food in the United States), both published by the FDA.

It should be made clear that no one particular strategy offers a solution for property enhancement of all APIs. Each API must be examined and evaluated on a case-by-case basis in terms of molecular structure and desired final properties (Schultheiss & Newman, 2009).

2. Polymorphism

The term "polymorphism" (from Greek: *polys* = many, *morfé* = form) was first used by Mitscherlich in 1822. He noticed that one compound of a certain chemical composition can crystallize into several crystal forms (Mitscherlich, 1822). Polymorphism is an ability of substances to exist in two or more crystal modifications differing from each other by structure and/or molecule conformation in the crystal lattice. The concept of polymorphism is often confounded with isomorphism or pseudopolymorphism. These concepts are interconnected but there are great differences between them.

In contrast to polymorphism when one substance able to form different crystal modifications is considered, in case of isomorphism two or more different substances that have just similar structure but form the same crystal modifications are considered. Such substances can even form so-called isomorphous series. Most often they originate as a result of co-crystallization of isomorphous substances from the mixture of their saturated solutions. A typical example is sulphates: magnesium sulphate, zinc sulphate and nickel sulphate crystallizing as heptahydrates. Also pseudopolymorphism should be distinguished from polymorphism. The concept of pseudopolymorphism is used for crystalline forms which also comprise solvent molecules as an integral part of their structure. These pseudopolymorphs are sometimes called solvates or hydrates if the solvent is water (Rodríguez-Spong et al., 2004).

The fundamental forms of solid substances can be classified into polymorphs, solvates/desolvated solvates and amorphous substances. A co-crystal can be defined as a multiple-component crystal, in which two or more molecules that are solid under ambient conditions coexist through a hydrogen bond (see Fig. 1 and Fig. 2). Polymorphism can be

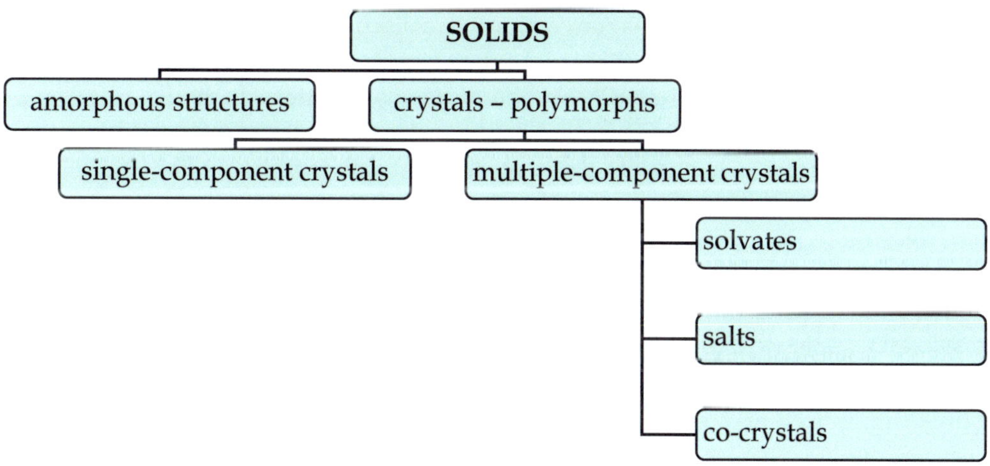

Figure 1. API solid form classification based on structure and composition. (Ref. Sekhon, 2009) (Taken and adapted with the permission of the author.)

classified depending on the fact if the molecule occurs in different conformations or is the same. If the molecule can have different conformations that crystallize differently and so the molecule is flexible, conformation polymorphism is observed. If the molecule is rigid, does not have any conformations and polymorphs differ only by their packing in the crystalline structure, this is the case of package polymorphism (Sharma et al., 2011).

Polymorphism should be distinguished from crystal morphology that represents crystallization of a substance from different solvents with a change of the form but without modification of the crystalline structure.

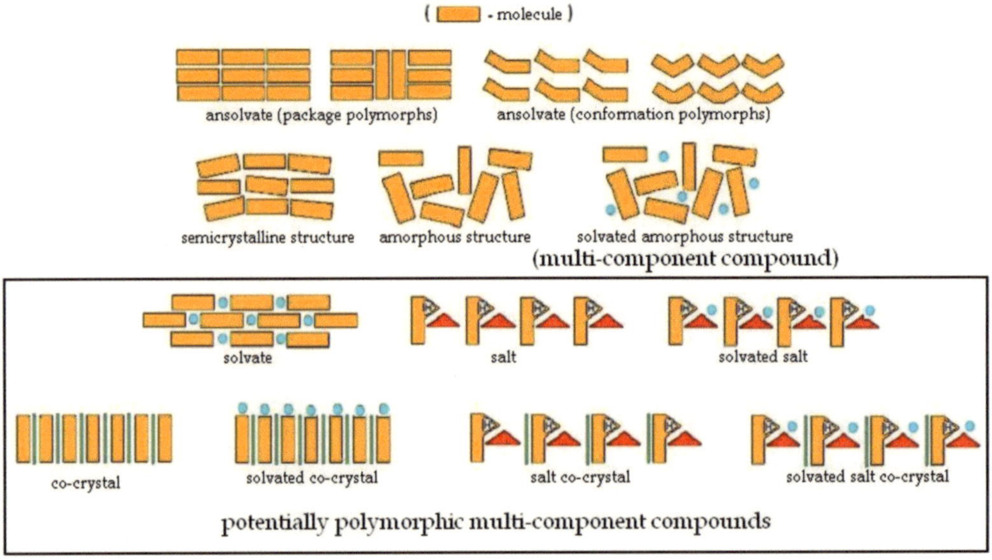

Figure 2. Schematic illustration of polymorphism of solid compounds. (Ref. Kratochvíl, 2009) (Taken and adapted from the presentation "Pharmaceutical Co-crystals" with the permission of the author.)

In the pharmaceutical practice it is conventional to collectively call all polymorphs, solvates and polyamorphs of the same API as solid forms or polymorphs in the extended meaning and designate them by, for example, Roman numerals or letters. It should be taken into consideration that designation of polymorphs is often non-uniform; it originates historically; therefore it is possible that two different polymorphs will be designated by different authors in the same way. A problem of pharmaceutical producers is polymorphic transitions in organic systems that are often hysteresis, badly defined and solvent-mediated. Polymorphic transitions can occur at all technological stages of drug production (at final API crystallization, wet granulation, micronization, tabletting or in ready tablets, for example, under the influence of auxiliaries – excipients). At drug registration, national regulation authorities demand description of all solid forms and possible phase transitions of the drug as well as a prescribed guarantee of the product polymorphic purity from pharmaceutical companies. At present approximately 85% of pharmaceutical products are solid formulations; therefore no manufacturer can afford to ignore the issue of polymorphism.

2.1. Crystallization

During crystallization a self-assembly supramolecular process takes place when randomly oriented molecules assemble into internally structured crystals (Kratochvíl, 2007). There are several ways of crystallization: (*i*) by evaporation of solvent; (*ii*) by addition of antisolvent (product precipitation); (*iii*) by cooling of solution; (*iv*) by change of pH ; (*v*) by addition of a compound that will produce the desired product by a chemical reaction; (*vi*) by lyophilisation, sublimation or cooling of melt (Kratochvíl, 2010). The first two possibilities were used in the experimental part.

The process of crystal formation is a complicated kinetic process which, if we simplify, is composed of nucleation, subsequent crystal growth around the formed crystal nuclei and growth termination. This happens at precipitation of a solid substance in the crystalline form from solution or at solidification of a substance.

The basis of nucleation is the quickest formation of the crystal nucleus. By constant sequence of molecules addition molecular aggregates, clusters, originate. The aggregates spontaneously impact and thus disintegrate but also grow. When they achieve the critical size, they become nuclei that do not disintegrate spontaneously but, to the contrary, only grow into a crystal. Nuclei can be classified into primary and secondary. Primary nucleation is further divided into homogenous and heterogeneous. Homogenous nucleation is characterized by formation of nuclei as a result of random impacts of aggregates somewhere in the solution volume without the presence of any foreign matter. By contrast, heterogeneous nucleation is assisted by foreign matters, for example, a stirrer. Secondary nucleation is also called seeded nucleation – small crystals (nuclei, seeds) of a desired crystal are added to the original solution. Seeding is successfully applied for systems where polymorphous behaviour cannot be excluded. Seeding of the system with a desired polymorph assures that an undesired polymorph will not be formed. After formation of the nucleus the system pass to the second mentioned crystallization phase. In this phase growth of crystals from the formed nucleus continues. If an API is crystallized with a co-crystallization partner, for example excipient or another API, with co-crystals formation, such process is called co-crystallization.

2.2. Properties of solid forms

For polymorphic drugs, the most thermodynamically stable polymorph is usually preferred. This polymorph normally assures reproducible bioavailability for the whole period of the pharmaceutical shelf life under different storing conditions that are possible in practice. A significant advantage is that production of such a polymorph is mostly easier controlled on an industrial scale. But in some cases a metastable crystalline form or an amorphous form is preferred due to patent and medical reasons. This happens when a higher concentration of an API in the system or faster dissolution of poorly soluble substances is required. Certain substances are used in the amorphous form also due to the fact that a crystalline form of the compound was not obtained. When an amorphous form or a metastable crystalline polymorph is used, especial attention is to be paid to safety assurance and the effectiveness of the drug for the whole shelf life period. These aspects are to be assured also for storing conditions in other climatic zones.

Though the biological effect of an API is induced by interaction of the drug molecule with a target receptor when cell natural chemistry is influenced primarily by conformation changes, it is important in what solid phase the drug is administered to the patient. That means that not only the molecular but also the crystalline structure of an API is essential. The crystalline structure influences not only chemical and physical stability but also the drug dissolution rate and so can affect markedly the drug bioavailability. On the other hand, an amorphous form may have a considerably higher dissolution rate than the most stable polymorph. Great differences in the solubility and the dissolution rate of polymorphs can be a reason of significant differences in their distribution in the organism. Low plasma concentration can theoretically cause incomplete occupation of respective membrane receptors at the site of action (they are blocked by a substrate), or biological activity can change from agonist to antagonist or vice versa.

However, the dissolution rate is not the only important parameter of the difference between polymorphs. There can be also differences in crystal size and shape that influence milling, tabletting, filtration, looseness and other important technological parameters. There are also differences in chemical reactivity, thermal stability, hygroscopicity, density, hardness, etc.

Thus it can be stated that different arrangement and conformation change of molecules in the crystal structure lead to differences in properties as follows: (*i*) mechanical (hardness, compactness, etc.); (*ii*) surface (surface energy, interfacial tension, etc.); (*iii*) kinetic (dissolution values, stability, etc.); (*iv*) spectroscopic (electronic, vibrational state transition, etc.); (*v*) thermodynamic (melting point, sublimation, enthalpy, entropy, etc.); (*vi*) packing properties (molar volume a density, hygroscopicity, etc.) (Brittain, 2009).

2.3. Multi-component solids

Usually solid crystal substances can exist as mono-component or multi-component. Mono-component systems contain, as evident from the name, only one component, and multi-component systems consist of two or more components. In pharmaceutics this is a crystal composed of an API and an additional molecule that determines the type of the multi-component system, pseudopolymorph. As illustrated in Fig. 1, the additional molecule of hydrates is water; of solvates, a solvent; of salts, an ionizable component; and of co-crystals, a co-crystallization partner (Stahly, 2007; Kratochvíl, 2010). All these systems can be polymorphic (see Fig. 1 and Fig. 2). Due to the breadth of the topic and the focus of this paper on preparation of co-crystals, only co-crystals will be described in more details.

3. Co-crystals

The term "co-crystal" (also written as cocrystal) originates from "a composite crystal" (Desiraju, 2003). The term and the definition of co-crystal is a subject of topical debate. In principle this is a multi-component system of host and guest type; both components are solid in pure state and under ambient conditions. There are a lot of definitions; for example, Stahly defines co-crystals as molecular complexes that contain two or more different molecules in the same crystal structure (Stahly, 2007). Bhogala and Nangia define co-crystals as multi-component solid-state assemblies of two or more compounds held together by any type or combination of intermolecular interactions (Bhogala & Nangia, 2008). Childs and Hardcastle define co-crystals as a crystalline material made up of two or more components, usually in a stoichiometric ratio, each component being an atom, ionic compound, or molecule (Childs & Hardcastle, 2007). The simplest definition of co-crystals was proposed by Bond: "synonym for multi-component molecular crystal" (Bond, 2007). Also the definition of Aakeröy and Salmon can be recognized. They describe co-crystal as stoichiometric structurally homogeneous multi-component crystalline material of host-guest type that contains two or more neutral building blocks (discrete neutral molecular species that are solids at ambient conditions) that are present in definite amounts whereas all solids containing ions, including complex transition-metal ions, are excluded) (Aakeröy & Salmon, 2005). Jones defines co-crystal as a crystalline complex of two or more neutral molecular constituents bound together in the crystal structure through non-covalent interactions, often including hydrogen bonding (Jones et al., 2006) and Zaworotko says that co-crystals are formed between a molecular or ionic API and a co-crystal former that is a solid under ambient conditions (Vishweshwar et al., 2006a).

The host is an API, and the guest is a co-crystallization partner (an excipient or another API). Note that according to the definition by Aakeröy and Salmon co-crystals are different from solvates by the fact that the host and the guest of co-crystals are in the solid phase, while solvates contain both a solid phase and a solvent (or its residual), *i.e.* a liquid phase. However, there is still the problem of a boundary between the salt and the co-crystal, because according to the last mentioned definition of Aakeröy and Salmon, ions as co-crystal components are excluded, so on Fig. 3 only the first example of two is a co-crystal according to the mentioned definition. The borderline between salts and co-crystals is blurred and can be distinguished by the location of the proton between an acid and a base. This state can be denoted as salt–co-crystal continuum (Childs et al., 2007).

Several types of co-crystals can be distinguished: (*i*) "simple" co-crystals; (*ii*) solvated (hydrated) co-crystals – the co-crystal contains a component that is liquid at ambient temperature; (*iii*) salt co-crystals – the host is an ionized form; (*iv*) solvated salt co-crystals; (*v*) polymorphs of all previous types of co-crystals.

3.1. Properties of co-crystals

Pharmaceutical co-crystals, that is, co-crystals that are formed between an API and pharmaceutically acceptable (GRAS) compounds (co-crystal formers, counterions) that are solid under ambient conditions, represent a new paradigm in API formulation that might address important intellectual and physical property issues in the context of drug development and delivery.

Co-crystals can be understood as "addition compounds" or "organic molecular compounds". They are attractive to the pharmaceutical industry, because they offer opportunities to modify the chemical and/or physical properties of an API without the need to make or break covalent bonds (see Fig. 4). The term "non-covalent derivatization" was coined for this approach. Co-crystals of an API with excipients become very important as a tool to tune solubility and absorption. The application of co-crystal technologies has only recently been recognised as a way to enhance solubility, stability and the intellectual property position with respect to development of APIs (Vishweshwar et al., 2006b; Sekhon, 2009; Schultheiss & Newman, 2009).

Co-crystallization with pharmaceutical excipients does not affect pharmacological activity of an API but can improve physical properties, such as solubility, hygroscopicity and compaction behaviour (Aakeröy et al., 2007; Rodríguez-Hornedo et al., 2007; Trask, 2007; Sun & Hou, 2008; Zaworotko, 2008). Co-crystals with the same active pharmaceutical ingredient will have strikingly different pharmaceutical properties (melting point, solubility, dissolution, bioavailability, moisture uptake, chemical stability, etc.), depending on the nature of the second component. Some of co-crystals formed had higher and some lower melting points as compared to their pure components, for example, succinic acid (Mp=135.3), urea (Mp=188.9), co-crystal of succinic acid-urea (Mp=149.9) (Walsh et al., 2003).

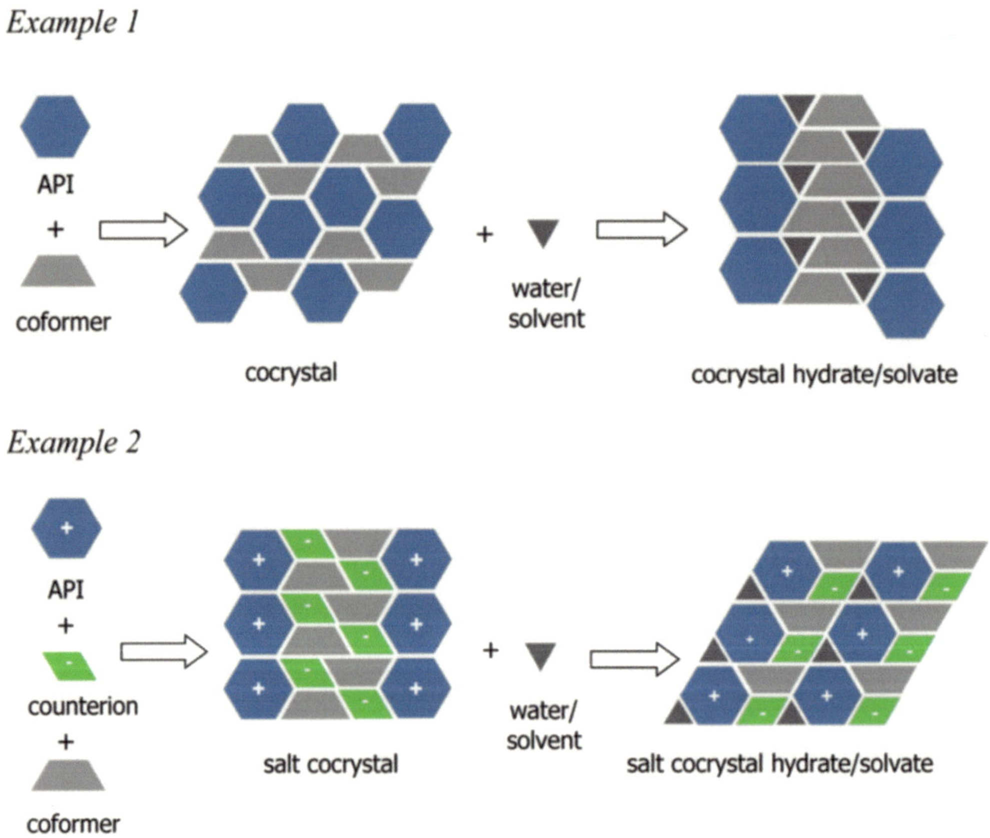

Figure 3. Possible multi-component systems: co-crystals, salt co-crystals and derived multi-component solids. (Ref. Schultheiss & Newman, 2009; reprinted and adapted with permission from Schultheiss N., Newman A. Pharmaceutical co-crystals and their physicochemical properties. Crystal Growth & Design 2009, 9(6): 2950–2967. Copyright 2009 American Chemical Society.)

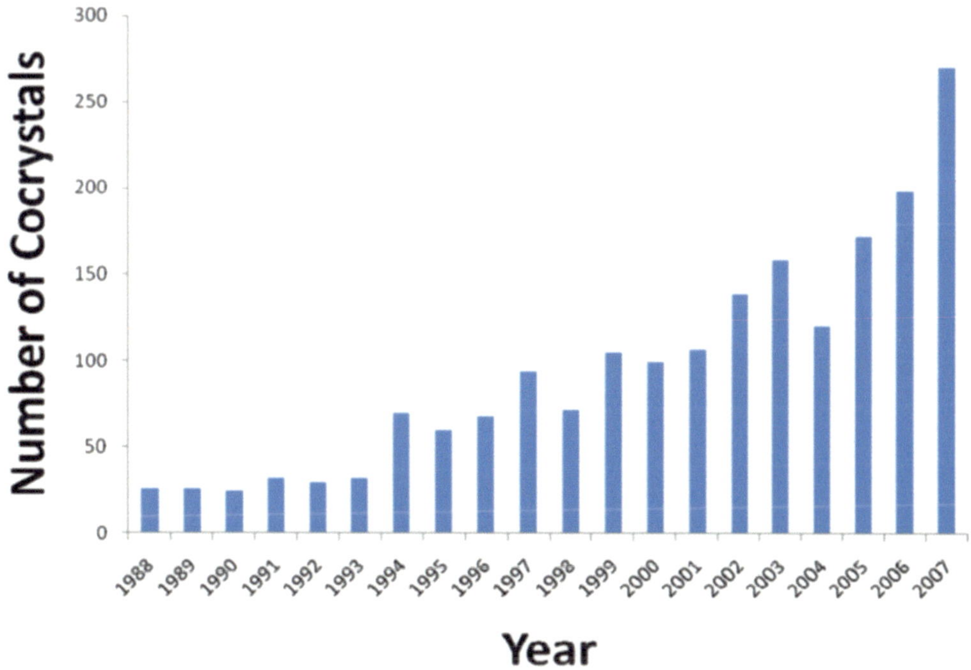

Figure 4. Frequency of occurrence of organic molecular co-crystals in the Cambridge Structural Database from 1988 to 2007. For the purposes of this graph, co-crystals are distinct from solvates, hydrates and simple salts. (Ref. Childs & Zaworotko, 2009; reprinted with permission from Childs S.L., Zaworotko M.J. The reemergence of cocrystals: The crystal clear writing is on the wall introduction to virtual special issue on pharmaceutical cocrystals. Crystal Growth&Design 2009, 9(10): 4208-4211. Copyright 2009 American Chemical Society.)

Scientists showed that modifying the physical properties of a pharmaceutical compound through pharmaceutical co-crystal formation improved the performance of a drug known to have poor solubility (Vishweshwar et al., 2006a). Pharmaceutical co-crystallization is a reliable method to modify physical and technical properties of drugs such as solubility increase/decrease, change of dissolution rate, stability, hygroscopicity and compressibility without alternating their pharmacological behaviour (Ranganathan, 1999; Fleischman et al., 2003; Almarsson & Zaworotko, 2004; Childs et al., 2004; Peterson et al.; 2006; Hickey et al., 2007; Žegarać et al., 2007; Schultheiss & Newman, 2009).

The use of co-crystals in drug design and delivery and as functional materials with potential applications as pharmaceuticals has recently attracted considerable interest (Fleischman et al., 2003; Jones et al., 2006; McNamara et al., 2006; Peterson et al., 2006; Vishweshwar et al., 2006a). Pharmaceutical co-crystals have been described for many drugs such as acetaminophen, aspirin, ibuprofen, flurbiprofen, sulfadimidine, etc. (Vishweshwar et al., 2005; Vishweshwar et al., 2006b; Caira, 2007; Rodríguez-Hornedo et al., 2007; Babu et al., 2008; Sarma et al., 2008,). Co-crystals of antitubercular drugs with dicarboxylic acids were reported using carboxylic acid-pyridine synthon as a reliable tool (Vishweshwar et al., 2003). The co-crystal of piracetam-L-tartaric acid showed improved hygroscopic properties

(Viertelhaus et al., 2009). Co-crystal forming abilities of two anti-HIV drugs (lamivudine and zidovudine) were studied to investigate the general applicability (Bhatt et al., 2009). Trimer co-crystals of *cis*-itraconazole-succinic acid were prepared and characterized by the possibility of achieving the higher oral bioavailability normally observed for amorphous forms of water-insoluble drugs (Remenar et al., 2003). The novel pharmaceutical co-crystal norfloxacin saccharinate dihydrate and its co-crystal, norfloxacin saccharinate–saccharin dihydrate were reported (Velaga et al., 2008).

3.2. Design of co-crystals

Co-crystallization is a result of competing molecular associations between similar molecules, or homomers, and different molecules, or heteromers. Instead, both components (host and guest) utilise prominent intermolecular non-covalent interactions such as hydrogen bonding, van der Waals forces and π-π stacking interactions to combine and yield a uniform crystalline material. Hydrogen bonds are the basis of molecular recognition phenomena in pharmaceutical systems and are responsible for the generation of families of molecular networks with the same molecular components (single-component crystals and their polymorphs) or with different molecular components (multiple-component crystals or co-crystals) in the crystalline state (Jayasankar et al., 2006; Sekhon, 2009). The components in a co-crystal exist in simple definite stoichiometric ratios, e.g., 1:1, 1:2, 2:1, etc. Co-crystals have different crystal structures than the pure components, contain different intermolecular packing patterns and as such they often exhibit different physical properties than the pure components. Unlike salt formation, co-crystallisation does not rely on ionisation of the API and the counterion to make a solid. Co-crystals are an alternative to salts when these do not have the appropriate solid state properties or cannot be formed due to the absence of ionization sites in the API (Aakeröy & Salmon, 2005; Miroshnyk et al., 2009).

The formation of a salt or a co-crystal can be predicted from pK_a value of an acid (A) and a base (B). Salt formation generally requires a difference of about 3 pK_a units between the conjugate base and the conjugate acid (A) *i.e.* [pK_a (B) – pK_a (A) ≥ 3] (Etter, 1990; Whitesides & Wong, 2006; Sekhon, 2009). In cases when $\Delta pK_a = pK_a (B) - pK_a (A) = 0\text{-}3$, the transfer of proton is ambiguous, and we can talk about the salt–co-crystal continuum (Childs et al., 2007).

Co-crystals can be prepared from two molecules of any shape or size having complementary hydrogen bond functionalities. The ability of an API to form a co-crystal is dependent on a range of variables, including the types of co-former, the API co-former ratio, the solvents, the temperature, the pressure, the crystallization technique, etc. Common functional groups, such as carboxylic or amino acids, amides, alcohols and carbohydrates are typically found to interact with one another in co-crystals (see Fig. 5) (Miroshnyk et al., 2009, Sarma et al., 2011; Qiao et al., 2011). Etter has studied hydrogen bonds in co-crystals and uses them as design elements. The hydrogen bond general rules are the following: (*i*) all good proton donors and acceptors are used in hydrogen bonding, (*ii*) if six-membered ring intramolecular hydrogen bonds can form, they will usually do so in preference to forming intermolecular hydrogen bonds, and (*iii*) the best proton donors and acceptors remaining after intramolecular hydrogen-bond formation form intermolecular hydrogen bonds with one another. In addition, the selectivity of hydrogen bonding in co-crystals was demonstrated by using pyridines (Etter, 1991; Sarma et al., 2011).

Based on the above mentioned facts co-crystal prediction includes the following steps: (*i*) determining whether a given set of two or more molecular components will undergo co-crystallization; (*ii*) identifying the primary intermolecular interactions, e.g., hydrogen-bond motifs that will exist within a particular co-crystal structure; and (*iii*) envisioning the overall packing arrangement in the resulting co-crystal structure (Trask, 2007).

Design and preparation of pharmaceutical co-crystals is a multi-stage process that can be schematically described in the following steps: (*i*) selection and research of APIs; (*ii*) selection of co-crystal formers; (*iii*) empirical and theoretical guidance; (*iv*) co-crystal screening; (*v*) co-crystal characterisation; (*vi*) co-crystal performance (Miroshnyk et al., 2009; Qiao et al., 2011).

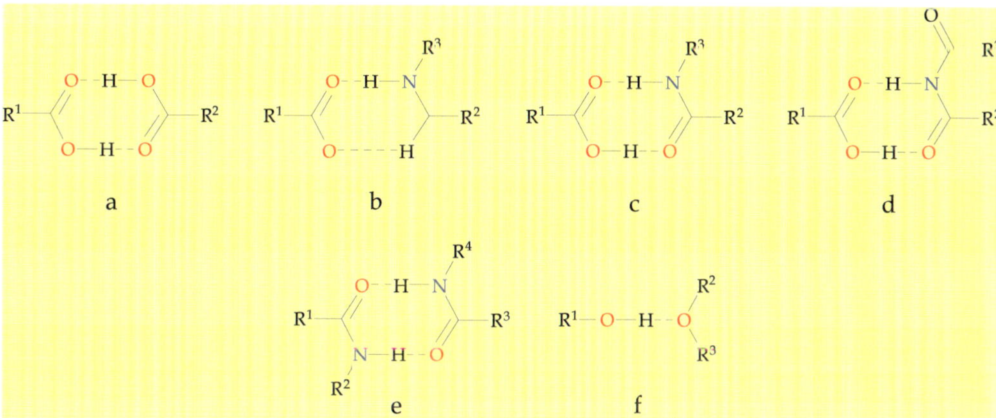

Figure 5. Possible formation of hydrogen bonds between synthons used in crystal engineering: acid-acid (a), acid-amine (b), acid-amide (c), acid-imide (d), amide-amide (e), and alcohol-ether (f).

3.3. Synthesis of co-crystals

A pharmaceutical co-crystal is a single-crystalline solid that incorporates two neutral molecules, one being an API and the other a co-crystal former (Vishweshwar et al., 2006). Co-crystal former may be an excipient or another drug (Rodríguez-Hornedo et al., 2007). Pharmaceutical co-crystal technology is used to identify and develop new proprietary forms of widely prescribed drugs and offers a chance to increase the number of forms of an API.

Crystalline forms can be generated by means of kinetically or thermodynamically controlled crystallization processes. Synthesis/processing of co-crystals can be accomplished via a number of methods, including slow solvent evaporation crystallization from solution, solvent-reduced (e.g. slurrying, solvent-drop grinding) and solvent-free (e.g. grinding, melt), high throughput crystallization and co-sublimation techniques (Shekunov & York,

2000; Morissette et al., 2004; Trask & Jones, 2005; Trask et al., 2005a; Trask et al., 2005b; Trask et al., 2006; Stahly, 2007; Berry et al., 2008; Takata et al., 2008; Friščić & Jones, 2009; Schultheiss & Newman, 2009; Qiao et al., 2011). Typically co-crystals are prepared by slow solvent evaporation that is only viable if compatible solubility in a given solvent exists between the components comprising the potential co-crystal. The potential benefits, disadvantages and methods of preparation of co-crystals were reported (Blagden et al., 2007). Solvent drop grinding has been reported to be a cost-effective, green, and reliable method for discovery of new co-crystals as well as for preparation of existing co-crystals. A slurry crystallization technique was used in co-crystal screening of two non-ionizable pharmaceutical host compounds, stanolone and mestanolone, with 11 pharmaceutically acceptable guest acids, and the results demonstrated the importance not only of hydrogen bonding but also of geometric fit in co-crystal formation (Takata et al., 2008). In addition to these classical techniques of co-crystal synthesis, also more sophisticated methods can be mentioned such as electrochemical crystallization, gel crystallization, vapour-diffusion crystallization, cryogenic grinding (cryomilling), microporous membranes crystallization, supercritical fluids crystallization and sonocrystallization (Dahlin et al., 2011; Hsu et al., 2002; Forsythe et al., 2002; He & Lavernia, 2001; Di Profio et al., 2007; Tong et al., 2001; Ruecroft et al., 2005). The combinations and variations of the above techniques may be used to cause co-crystal formation (McMahon et al., 2007). It is evident that many of the reported co-crystals appear to be the result of serendipity, although several groups have successfully exploited crystal engineering principles for design and synthesis of co-crystals.

4. Characterization of co-crystals

Co-crystal characterisation is an important constituent part of co-crystal research. The basic techniques of analysis of co-crystals involve especially solid state analysis methods (Zakrzewski & Zakrzewski, 2006; Dohnal et al., 2010), *i.e.* vibration-rotation spectroscopy, solid state NMR, thermal analysis, microscopy techniques and X-ray diffraction. The most often used solid-state analytical techniques will be discussed below.

4.1. Spectroscopy of vibration-rotation transitions

According to the used part of infrared (IR) spectra (energy), near infrared spectroscopy (NIR), middle infrared spectroscopy (MIR), Raman scattering and terahertz (THz) spectroscopy are distinguished.

4.1.1. NIR Spectroscopy

Photons in NIR region have the highest energy and can therefore vibrationally excite molecules into even higher excited vibrational states than the first level, *i.e.* the second, the third and others. These transitions are called overtons. Absorption of radiation in the NIR region is usually based on higher energy transitions between vibrational levels of molecules, namely combination transitions and overtones and not fundamental transitions, which are dominant in mid infrared region (MIR). NIR spectrometers are not so demanding of applied materials, as are the instruments working in the mid infrared region, due to different radiation frequencies used in the NIR spectrometry. Acquisition of NIR spectra requires several seconds and can be performed during manufacturing process. Therefore, the NIR spectroscopy is still more often applied as a tool of process analytical technology. In co-crystal analysis NIR spectroscopy is commonly used as a screening method of the first choice (Pekárek & Jampílek, 2010).

4.1.2. MIR spectroscopy and Raman scattering

The aim is not only to present the basic principles of both methods, but also to compare them. Raman spectroscopy and mid-infrared spectroscopy are both widely used in the pharmaceutical industry for the solid-state characterization because of their specificity. MIR region is the most important region in terms of analytical application. It is a region, where the majority of the so-called fundamental vibrations appear, *i.e.* the vibrational transitions from the basic to the first excited vibrational state. Vibrational spectroscopy provides key information about the structure of molecules. Positions and intensities of bands in a vibrational spectrum can be used to determine the structure of a molecule or to determine the chemical identity of a sample. With sufficient experience it is possible to identify chemical compounds or monitor intermolecular interactions by evaluating changes in positions and intensities of Raman bands which is extremely useful for the above mentioned purposes.

The mid-infrared and Raman spectra (with the exception of optical isomers) of a drug substance or any chemical compound are unique. Raman spectroscopy is a vibrational spectroscopy method which complements mainly the mid-IR spectroscopy. The intensity of bands in IR spectra is proportional to the dipole moment change occurring during the given type of the vibrational motion. Modes with a large change in the dipole moment having intensive bands in the IR spectra generally provide low-intensity bands in Raman spectra. Conversely, vibrations of non-polar functional groups provide intense bands in Raman spectra and weak bands in infrared spectra. (Zakrzewski & Zakrzewski, 2006, Pekárek & Jampílek, 2010). An important advantage of Raman spectroscopy over IR spectroscopy is the possibility to measure aqueous solutions. This advantage can be used in identification of APIs in aqueous solutions, emulsions or suspensions.

4.1.3. THz spectroscopy

This technique covers a wide interval, which can be roughly defined, for example, by frequencies of 100 GHz and 3 THz, which corresponds to wavelengths between 3 and 0.1 mm, *i.e.* wave numbers from 3 to 100 cm^{-1}. Hence THz region partially overlaps with far-infrared region. The characteristic frequency of 1 THz can be equivalently expressed in other spectroscopic units. The decisive factor for expansion of this method was the development of optical femtosecond lasers being an integral part of the most current laboratory and commercial THz spectrometers. Attractiveness of the THz field lies in the scale of the specific options for application of electromagnetic radiation of these frequencies. Application options in pharmaceutical analysis are mainly spectroscopy and imaging.

Solid substances often exhibit specific interactions in the THz region. While some crystals are transparent in this range, many others show low-frequency oscillations of the crystal structure (called phonon bands) with characteristic frequencies being determined by short-range inter-atom interactions and even arrangement of atoms at long distances (Kadlec & Kadlec, 2012). Similar to NIR and mid-IR spectroscopy, in THz spectroscopy reflectance (less often) and transmittance measurements are used.

It is known that vibrational modes of amorphous and crystalline substances are very different. This was experimentally proven, for example, in THz spectra of samples of crystalline and amorphous saccharides (Walther et al., 2003). The absence of sharp vibrational modes that can be observed in the crystalline form was later also confirmed in THz spectra of indomethacin (Strachan et al., 2004).

Determination of tablet coating thickness and determination of particle size belong among special applications of THz spectroscopy. Radiation in THz region has the lowest energy, which can cause changes mainly in rotational energy of molecules therefore it has the lowest (destructive) influence on measured co-crystals.

4.2. Solid state nuclear magnetic resonance

Solid state NMR (ssNMR) spectroscopy is not used for routine analyses in pharmacy because of its demands, but its role in pharmaceutical development is indispensable. It has very wide application. The most important applications can be divided into the following groups (Havlíček, 2010): (*i*) API structural analysis; (*ii*) polymorphism; (*iii*) co-crystal analysis; (*iv*) dosage form analysis; (*v*) solvate analysis; (*vi*) salt analysis.

Among the most widely used ssNMR techniques CP/MAS NMR belongs. It has three modifications: cross polarization (CP), magic angle spinning (MAS) and high-power heteronuclear decoupling. With this arrangement the sensitivity and line broadening problems were overcome, and high-resolution ssNMR was brought in practical use.

Solid-state NMR is capable of providing detailed structural information about organic and pharmaceutical co-crystals and complexes. ssNMR non-destructively analyzes small amounts of powdered material and generally yields data with higher information content than vibrational spectroscopy and powder X-ray diffraction methods. Particularly, its ability to prove or disprove molecular association and possibility to observe structural features (such as hydrogen bonding) are great advantages of this method. These advantages can be utilized in the analysis of pharmaceutical co-crystals, which are often initially produced using solvent drop grinding techniques that do not lend themselves to single-crystal growth for X-ray diffraction studies (Vogt et al., 2009).

4.3. Thermal analysis

Thermal analysis is a broad term referring to methods (see Table 1), which measure physical and chemical properties of a substance, a mixture of substances or also of a reaction mixture as a function of temperature or time during a controlled temperature programme. Most of these methods monitor corresponding system properties (mass, energy, size, conductivity, etc.) as dynamic functions of temperature.

For co-crystal analysis mainly differential scanning calorimetry (DSC) and its modifications, differential thermal analysis (DTA) and thermogravimetry analysis (TGA), are of great importance. For special cases a very useful method is thermally stimulated current (TSC). DSC and DTA are the most applied methods of thermal analysis in pharmaceutical development. In DSC, the sample is subjected to linear (or modulated) heating, and the heat flow rate in the sample is proportional to the actual specific heat and is continuously measured. Very interesting and useful modifications of DSC are hyperDSC, microDSC and modulated DSC (Krumbholcová & Dohnal, 2010).

Method name	**Tracked value**
Differential thermal analysis (DTA)	temperature difference between the studied and the standard sample
Differential scanning calorimetry (DSC)	thermal energy provided for compensation of temperature between the studied and the standard sample
Thermogravimetry analysis (TGA)	weight change
Thermomechanical analysis (TMA)	change in a mechanical property (module, hardness)
Dilatometry	change in volume
Effluent gas analysis	volume of the studied gas
Pyrolysis	pyrolysis products
Thermal luminescence analysis	emission of light
Electric conductivity analysis	change in electric conductivity

Table 1. List of the most common methods of thermal analysis. (Ref. Krumbholcová & Dohnal, 2010).

4.3.1. Thermally stimulated current

This special thermal method uses a so-called molecular mobility, which provides information about the structure of substances, about dynamic parameters ΔH and ΔS and the relaxation (release) time τ. In TSC the substance is heated to a temperature T_P in an electric field with intensity E_P for a period of time t_P, which is sufficient for differently moving particles of the studied substance to reach identical orientation in the electric field. In this state, the sample is rapidly cooled to a temperature T_0, which ensures zero motion of the particles. The effect of the electric field is then also deactivated, and the substance is then kept at temperature T_0 for time t_0. The temperature then linearly increases, and the substance returns to the previous balance, and depolarization current I_D is recorded as a function of temperature. Each depolarization peak is characterized by temperature T_{max} with intensity

I_{max}. This technique is particularly suitable for substances with polar character. It can distinguish polymorphs or co-crystals, which cannot be determined by classical DSC (Krumbholcová & Dohnal, 2010).

4.4. X-Ray diffraction

X-Ray diffraction is one of the basic solid state analytical techniques labelled as the "gold standard". It is used not only for characterization and identification of crystalline substances but also for their discrimination.

Two basic methods are discerned according to the type of the analysed sample. They are:

1. Diffraction on single crystal (single-crystal X-ray diffraction, SCXRD),
2. Powder diffraction (X-ray powder diffraction, XRPD).

Name	**single-crystal X-ray diffraction**	**X-ray powder diffraction**
Abbreviation	SCXRD	XRPD
Sample – type	Single crystal	Powder
Sample – amount	1 single crystal with size of 0.1 – 1 mm	100 – 500 mg
Sample – preparation	Selection of a suitable single crystal and its fixation onto the goniometric head.	Filling of the holder cavity or possibly milling.
Sample – consumption	Non-destructive	Non-destructive
Measurement time	Hours/days	Minutes/hours
Analysis of obtained data	Complete information about the molecule, conformation, bond lengths, chirality, spatial arrangement of molecules in the crystal, interactions between molecules, determination and location of solvents.	„Fingerprint" of the crystal structure
Principle of the method	Monochromatic light falls on a single crystal, and because only one lattice diffracts, it is necessary to rotate the crystal to obtain the desired set of reflections (the principle of four-circle diffractometer).	Monochromatic radiation impacts a polycrystalline material and all lattices that meet the diffraction condition diffract at the same time.
Practical application	Crystallographic studies: determination of the complete structure of an API, a protein, a complex, etc.	Mainly industrial, screening and control method for characterization of crystalline materials.

Table 2. Comparison of basic diffraction methods. (Ref. Brusová, 2010).

Their basic characteristics, advantages and disadvantages are summarized in Table 2. The X-ray structural analysis methods, which use single-crystal samples, often allow a complete determination of crystallographic characteristics. Diffraction on a single-crystal yields intensities of diffractions on individual configurations of the crystallography planes, individually for each of their orientation, and the number of the detected maxima is very high, in orders of $10^2 - 10^4$. In the powder diffraction, on a debyeogram or a diffractogram there at most 100 lines can be distinguished. In addition, each diffraction line is a superposition of diffractions of all inequivalent and equivalent planes (Hušák et al., 2007; Kratochvíl et al., 2008).

5. Carbohydrates and their derivatives as crystallization modifiers

As discussed above, pharmaceutical co-crystals have rapidly emerged as a new class of API solids demonstrating great promise and numerous advantages. Various co-crystals of APIs were prepared by reason of intellectual property protection, for example, imatimib (see Fig. 6, structure **I**) with co-crystal carbohydrate formers such as α-D-glucopyranose, D-fructofuranose and *N*-methyl-D-glucamine (meglumine) (Král et al., 2010); agomelatine (see Fig. 6, structure **II**) with carbohydrate counterions such as D-sorbitol, aspartame, phenyl-β-D-glucopyranoside, D-glucoheptono-1,4-lactone, D-(+)-trehalose, lactose, α-D-glucopyranose, saccharose, *N*-methyl-D-glucamine and D-(+)-glucosamine hydrochloride (Ferencova et al., 2012); or to improve permeability, for example, alendronate (see Fig. 6, structure **III**), ibandronate (see Fig. 6, structure **IV**) or risedronate (see Fig. 6, structure **V**) with a number of carbohydrates as co-crystal formers (Jampílek et al., 2009; Haroková, 2010; Havelková, 2010; Hrušková, 2010; Jampílek et al., 2010; Kos, 2010; Oktábec et. al, 2010; Kos et al., 2011; Ťažká 2011; Havelková, 2012; Oktábec, 2012).

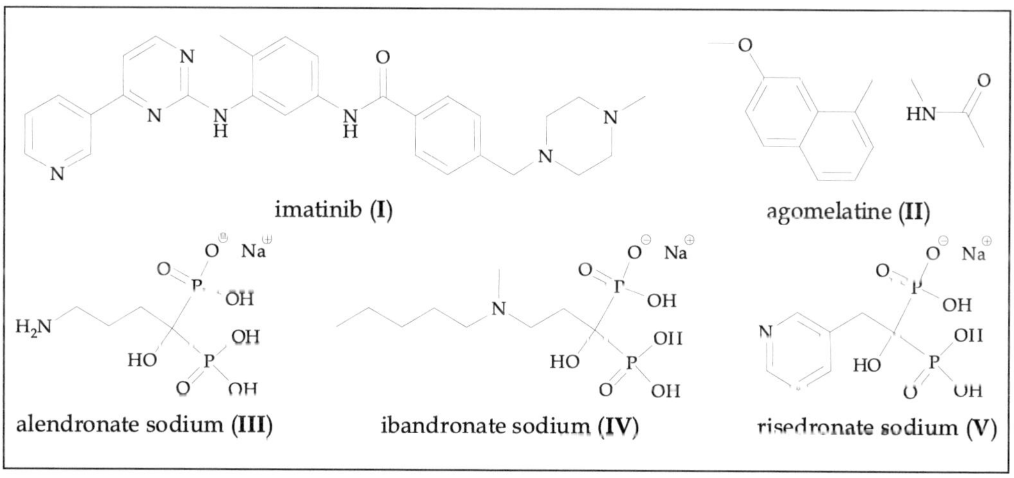

Figure 6. APIs used as host of carbohydrate formers.

The present study deals with the design and an effort to prepare co-crystals/new entities, generally new solid phases, of the above discussed bisphosphonates **III-V**. These compounds were investigated in detail, and a lot of valuable knowledge concerning generation of solids was obtained. Bisphosphonates (BPs) can be denoted as top-selling APIs. BPs are the most widely used and the most effective bone resorption inhibitors currently available for treatment of Paget's disease, tumour-associated bone disease and osteoporosis. All BPs have high affinity for bone mineral as a consequence of their P-C-P backbone structure, which allows chelation of calcium ions (Ebetino et al., 1998). Following release from bone mineral during acidification by osteoclasts, BPs appear to be internalized specifically by osteoclasts, but not other bone cells. The intracellular accumulation of BP leads to inhibition of osteoclast function due to changes in the cytoskeleton, loss of the ruffled border (Carano et al., 1990; Sato et al., 1991) and apoptosis (Hughes et al., 1995; Selander et al., 1996; Ito et al., 1999; Reszka et al., 1999). The ability of BPs to inhibit bone resorption depends on the presence of two phosphonate groups in the P-C-P structure, which appears to be required for interaction with a molecular target in the osteoclast as well as for binding bone mineral (Rogers et al., 1995; van Beek et al., 1998; Rogers et al., 2000). BPs as pyrophosphate analogues are a group of drugs that are widely used in practice. There are several injectable bisphosphonates: etidronate, pamidronate and zoledronate, which may be administered every three months or yearly. Peroral BPs alendronate and risedronate are taken daily, weekly or monthly, and ibandronate is approved to be taken monthly. Oral bioavailability of these BPs is very low (their gastrointestinal absorption is about 1%) due to their high hydrophilicity (Ezra & Golomb 2000).

A number of various patented solid forms of each API from this group can be found, which, for example, complicates their utilization for generic formulation from the intellectual property point of view. Eiermann et al. prepared crystalline forms of ibandronate (**IV**) B and A (Eiermann et al., 2006a; Eiermann et al., 2006b). Lifshitz-Liron et al. obtained forms C, D, E, F, G, H, J, K, K2, K3, Q, Q1, Q2, Q3, Q4, Q5, Q6, QQ, R, S and T (Lifshitz-Liron et al., 2006). Muddasani et al. prepared polymorphs I and II (Muddasani et al., 2007), and Devaraconda et al. generated ibandronate forms III-XXXI (Devarakonda et al., 2010). Ten different polymorphic and pseudo-polymorphic forms of sodium risedronate (**V**) identified as A, B, B1, BB, D, E, F, G and H and a semi-crystalline form were described (Cazer et al., 2001; Aronhime et al., 2003a; Aronhime et al., 2003b; Richter et al., 2007). The crystal structures of four different hydrates (monohydrate, dihydrate, hemipentahydrate and variable hydrate) and an anhydrate of sodium risedronate (**V**) have been elucidated and discussed by Redman-Furey (Redman-Furey et al., 2005) and Gossman (Gossman et al., 2003). Recently three new phases were found and named J, K and M (Bruning et al., 2011).

This paper deals with investigation of various types of carbohydrates and their derivatives as crystallization modifiers applied to crystal study concerning the BP family. Carbohydrates were used due to their hydroxyl moieties, which are able to interact with a phosphoric group and/or a nitrogen atom in the alkyl chain or heterocycle. Carbohydrates also provided a unique excellent system of hydroxyl moieties in different stereochemical modifications. These hydroxyl groups can be straightly modified, for example, by alkylation/arylation, and the structure of the carbohydrate molecule obtains absolutely different three-dimensional/space properties.

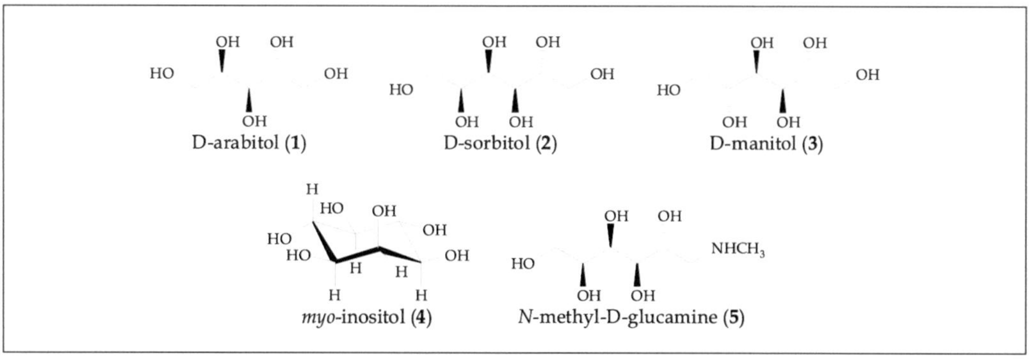

Figure 7. Structures of sugar alcohols used as potential co-crystal/crystallization formers.

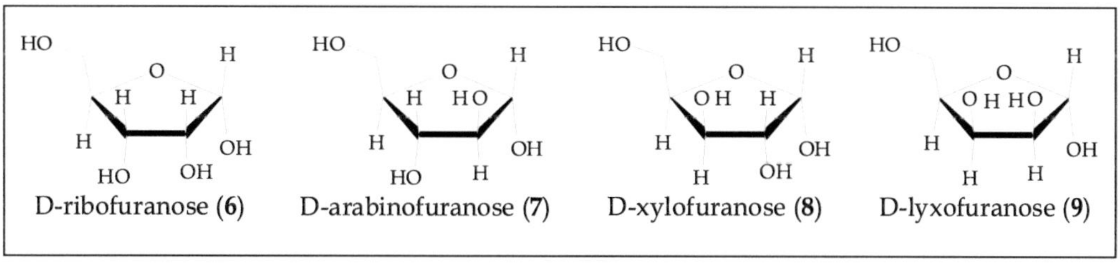

Figure 8. Structures of furanoses used as potential co-crystal/crystallization formers.

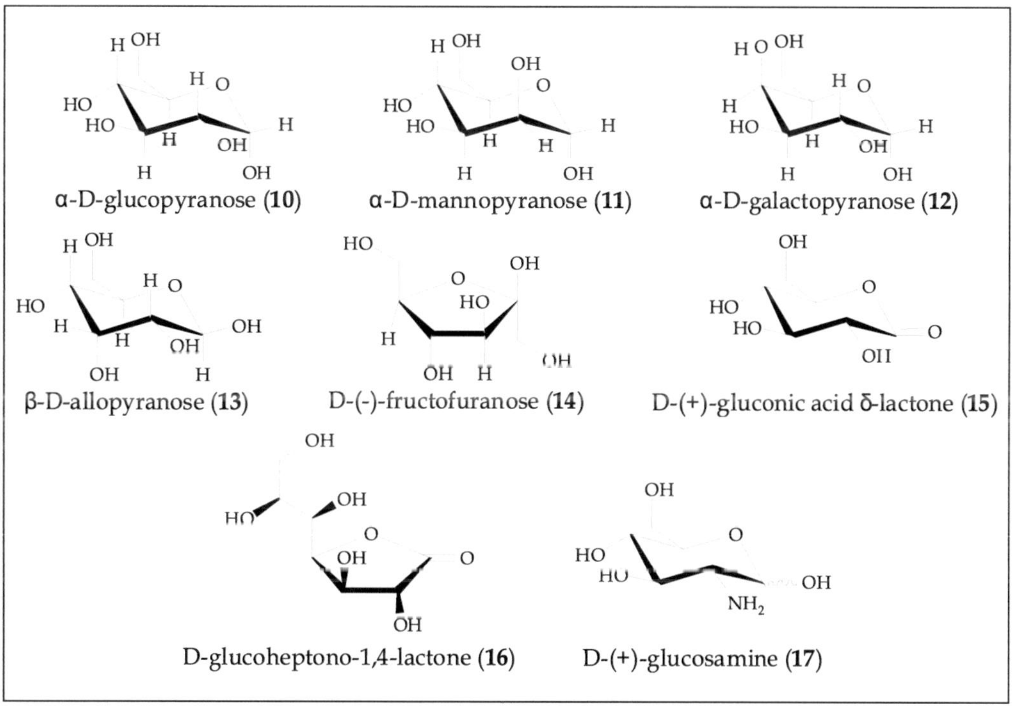

Figure 9. Structures of hexoses used as potential co-crystal/crystallization formers.

Thus mixtures of BPs with various sugar alcohols, furanoses, pyranoses and gluco-, manno- and galactopyranoside derivatives, some amino carbohydrates and disaccharides (see Figs. 7-11) as counterions were designed in an effort to prepare new crystalline forms or co-crystals/new entities. Mixtures of BPs and carbohydrates in different ratios and under various conditions were prepared. All the prepared mixtures (solid compounds) were characterized using some of the above mentioned solid state analytical techniques (Haroková, 2010; Havelková, 2010; Hrušková, 2010; Kos, 2010; Oktábec et. al, 2010; Kos et al., 2011; Ťažká 2011; Havelková, 2012; Oktábec, 2012).

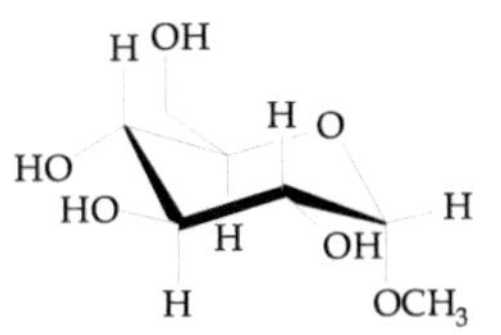

methyl-α-D-glucopyranoside (**18**)

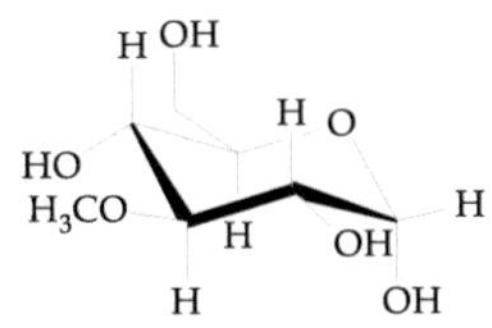

3-*O*-methyl-α-D-glucopyranoside (**19**)

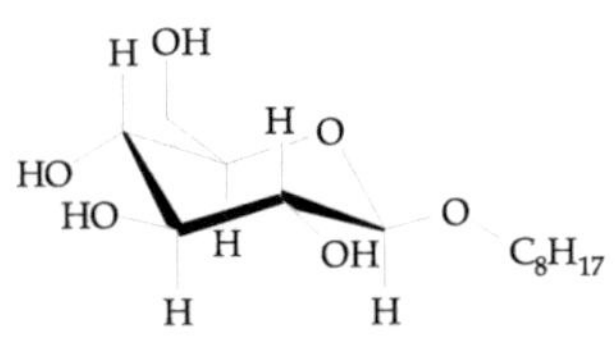

octyl-β-D-glucopyranoside (**20**)

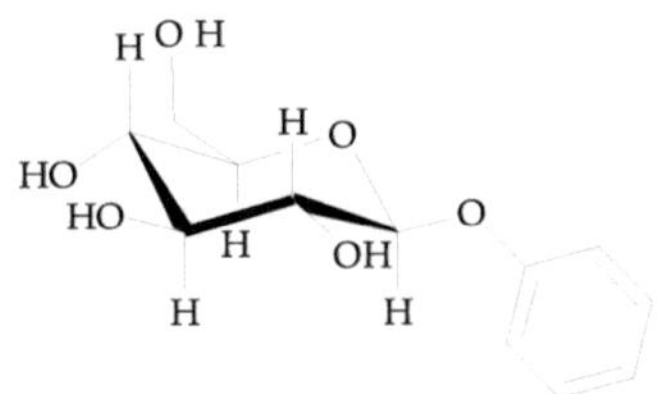

phenyl-β-D-glucopyranoside (**21**)

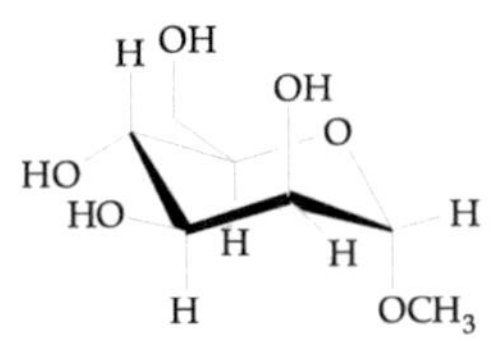

methyl-α-D-mannopyranoside (**22**)

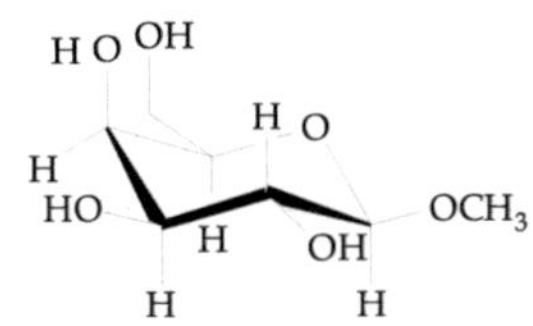

methyl-β-D-galactopyranoside (**23**)

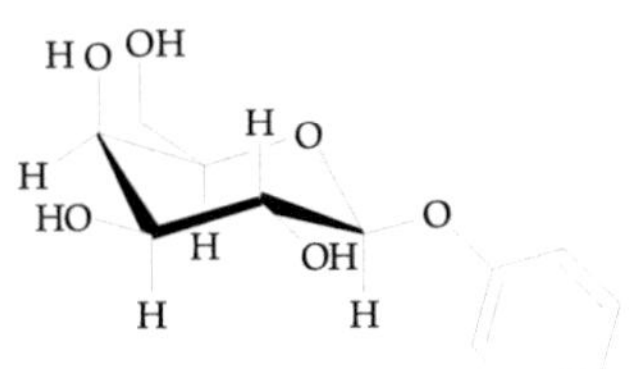

phenyl-β-D-galactopyranoside (**24**)

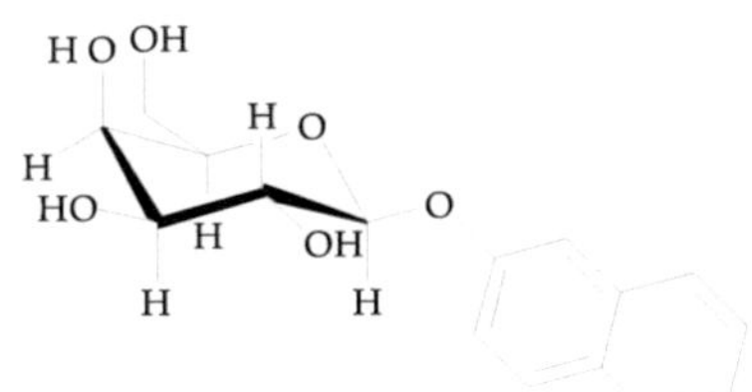

2-naphthyl-β-D-galactopyranoside (**25**)

Figure 10. Structures of pyranosides used as potential co-crystal/crystallization formers.

5.1. Generation of samples

All the evaluated samples with ratios 1:1 (**A**), 1:2 (**B**) and 1:3 (**C**) were prepared by means of dissolution of bisphosphonate monosodium salt and the excipient in water, subsequently mixed (1 h) and slowly evaporated at ambient temperature. To some samples with ratios 1:2

and 1:3 methanol was slowly added dropwise as an anti-solvent. The solid precipitated compound was filtered and dried at ambient temperature, samples 1:2 (**D**) and 1:3 (**E**), and the remaining liquid part was slowly evaporated at ambient temperature, samples 1:2 (**F**) and 1:3 (**G**). All generated solid compounds were subsequently screened by means of FT-NIR and FT-Raman spectroscopy. If a sample differing from the starting materials was found, it was additionally characterized by the below mentioned methods (Jampílek et al., 2009; Haroková, 2010; Havelková, 2010; Hrušková, 2010; Jampílek et al., 2010; Kos, 2010; Oktábec et. al, 2010; Kos et al., 2011; Ťažká 2011; Havelková, 2012; Oktábec, 2012).

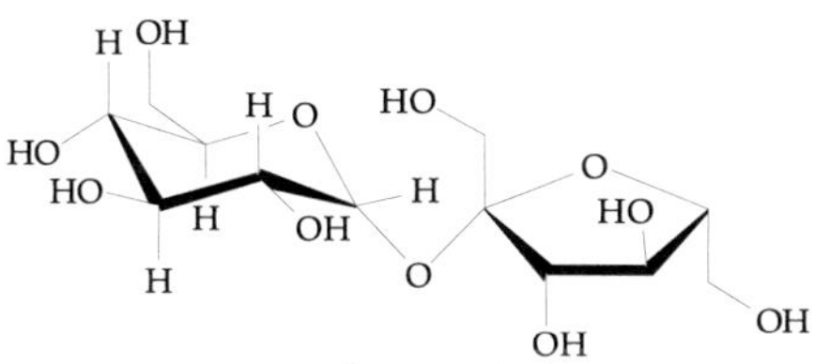

saccharose (**26**)
(4-β-D-fructofuranosyl-α-D-glukopyranoside)

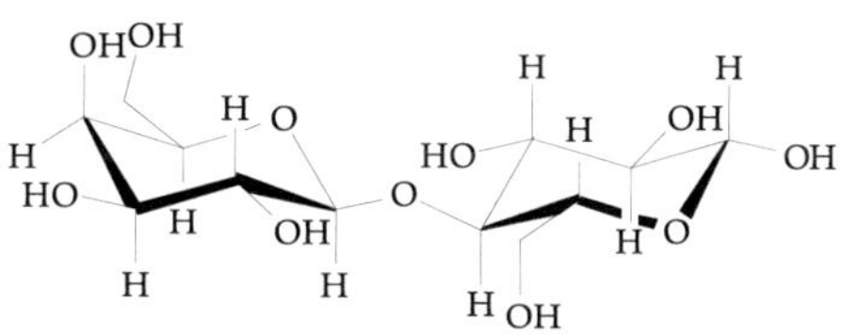

lactose (**27**)
(4-β-D-galactopyranosyl-D-glukopyranoside)

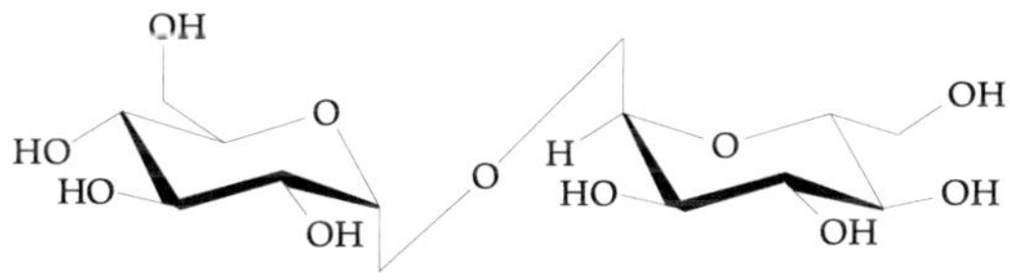

D-(+)-trehalose (**28**)
(α-D-glukopyranosyl-α-D-glukopyranoside)

Figure 11. Structures of disaccharides used as potential co-crystal/crystallization formers.

5.2. Used solid-state analytical techniques

Near infrared spectra were recorded using a Smart Near-IR UpDrift™, Nicolet™ 6700 FT-IR Spectrometer (Thermo Scientific, USA). The spectra were obtained by accumulation of 128 scans with 4 cm^{-1} resolution in the region of 12800–4000 cm^{-1}. FT-Raman spectra were accumulated by an FT-Raman spectrometer RFS 100/S (Karlsruhe, Bruker, Germany). The spectra were obtained by accumulation of 256 scans with 4 cm^{-1} resolution in the back scattering geometry with the laser wavelength of 1064 nm. ^{31}P CP/MAS NMR Spectra were recorded on a Bruker AVANCE 500 MHz spectrometer (Karlsruhe, Bruker, Germany). The ^{31}P CP/MAS spectra were measured in 4 mm rotor at 10 kHz with 2 ms contact time. ^{31}P chemical shift of $NH_4H_2PO_4$ (0 ppm) was used as an external reference for ^{31}P chemical shift. The ^{13}C CP/MAS spectra were measured in 4 mm rotor at 13 kHz with 2 ms contact time. Carbon chemical shifts were referenced to the signal for TMS via a replacement sample of glycine (176 ppm for the carbonyl group signal). The XRPD patterns were obtained on a PANalytical X'PERT PRO MPD diffractometer with Cu Kα radiation (45 kV, 40 mA). The powder samples were measured on Silica plate holder. Data were recorded in the range

2-40° 2θ, with 0.01° 2θ step size and 50 s/step scan speed. For the measurement of differential scanning calorimetry (DSC) curve an instrument DSC Pyris 1 (PerkinElmer, USA) was used. Maximum sample weight was 3.5 mg, and the standard Al sample pan was used. The record of the DSC curve was in the range of 50–300 °C at the rate of 10.0 °C/min under a nitrogen atmosphere.

5.3. Crystal products of carbohydrates with alendronate and ibandronate

Although a number of carbohydrates were tested as potential co-crystal/crystallization formers, no change in the crystal structure of the used alendronate sodium salt (**III**) was obtained. The starting material was always obtained.

Mixtures of ibandronate monosodium salt (**IV**) with twenty-eight carbohydrates were generated by means of thermodynamically and/or kinetically controlled crystallization processes. Polymorph B of ibandronate monosodium salt monohydrate (sodium hydrogen {1-hydroxy-3-[methyl(pentyl)amino]-1-phosphonopropyl}phosphonate, **IV**) was used as a starting material (Eiermann, 2006a), which is the most common in pharmaceutical formulations. It is a white powder, freely soluble in water and practically insoluble in organic solvents.

From all the tested agents, only phenyl-β-D-galactopyranoside (see Fig. 10, structure **24**) yielded noteworthy products with BP **IV**. The rest of the tested carbohydrates, except phenyl-β-D-glucopyranoside (see Fig. 10, structure **21**) and 2-naphthyl-β-D-galactopyranoside (see Fig. 10, structure **25**), generated again the starting polymorph B. The mentioned two carbohydrates **21** and **25** provided mixtures of polymorphs A+B, as shown in Table 3. Samples of **IV-24** in ratios 1:1 (**A**), 1:2 (**B**) and 1:3 (**C**) were prepared by mixing and subsequent evaporation at ambient temperature. In all three samples a change in the NIR spectra can be observed in the range of 5,300-4,800 cm^{-1}. The spectra of samples **IV-24/B** and **IV-24/C** are very similar, only slightly different from sample **IV-24/A**, probably due to the lower crystallinity of sample **IV-24/A**, which causes broader bands in the spectrum. As samples **IV-24/A-C** were prepared in the same way, it can be concluded that increasing concentration of compound **24** influences the sample crystallinity.

Comp.	A	B	C	D	E	F	G
IV-21	B	B	B	B	B	A+B	A+B
IV-24	new	new	new	B	B	new	new
IV-25	B	B	B	B	B	A+B	A+B

Table 3. Samples of ibandronate (**IV**) and used carbohydrates **21**, **24** and **25** in ratios 1:1, 1:2 and 1:3 prepared by evaporation at ambient temperature (**A, B, C**), samples in ratios 1:2 and 1:3 prepared by methanol precipitation (**D, E**) and samples in ratios 1:2 and 1:3 prepared by addition of methanol and evaporation of liquid part at ambient temperature (**F, G**). (Ref. Oktábec et al., 2010; Havelková, 2012).

Based on the NIR spectra of samples of **IV-24/D** and **IV-24/E** in ratios 1:2 and 1:3 precipitated by methanol and filtered, it can be concluded that both samples contain only form B of compound **IV**. The same characteristic bands in the range of 5,300-4,800 cm^{-1} as for samples **IV-24/A-C** can be observed for samples **IV-24/F** and **IV-24/G** in ratios 1:2 and 1:3 after addition of methanol, filtration of the obtained precipitate and evaporation at ambient temperature. Based on this fact it can be concluded that addition of methanol does not influence generation of a new unknown solid phase, because the same products were yielded with and without methanol addition. Slow evaporation seems to be important, *i.e.* thermodynamically controlled crystal modification is probable. The presence of carbohydrate **24** is fundamental for generation of a new entity. The samples **IV-24/A-C** and **IV-24/F**, **IV-24/G** were also characterized by means of the FT-Raman spectrometry and ^{31}P CP/MAS NMR spectroscopy for verification of this hypothesis. Both methods confirmed the presence of new solid phases.

From the above mentioned results (Table 3) it is evident that ibandronate (**IV**) provided a new solid phase only with phenyl-β-D-galactopyranoside (**24**). These samples, **IV-24/A-C** and **IV-24/F**, **IV-24/G**, were generated under the thermodynamic conditions (slow evaporation at ambient temperature) without or with methanol. It can be concluded that the presence of the co-crystal/crystallization former and the thermodynamic conditions were essential for generation of the new solid phase. Note that only β-D-pyranosides with substituted hydroxyl moiety in $C_{(1)}$, position 2 of the tetrahydropyran ring, *e.i.* in the equatorial position, showed interactions with BP **IV**. Although phenyl-α-D-pyranosides were not evaluated, it is possible to suppose that α-D-pyranosides possess probably a disadvantageous configuration of $C_{(1)}$ hydroxyl moiety.

Based on the above discussed results it can be also stated that substitution of $C_{(1)}$ hydroxyl with the aromatic group is necessary for interactions, because e.g. methyl-β-D-galactopyranoside (**23**) contrary to phenyl- (**24**) or naphthyl-β-D-galactopyranoside (**25**) provided no modification of the starting polymorph of compound **IV**. This hypothesis is supported by the fact that phenyl-β-D-glucopyranoside (**21**) afforded also the change of polymorph B to form A of BP **IV**. It can be assumed that naphthyl is too bulky compared to the phenyl ring. Nevertheless from all the evaluated substituted pyranosides only phenyl-β-D-galactopyranoside (**24**) yielded the new solid phase of compound **IV**. As non-covalent interactions are important for generation of crystal forms, the space configuration of all the hydroxyl moieties is probably essential for interactions between BP **IV** and carbohydrate **24**. As illustrated in Fig. 10, where basic spatial configuration of substituted pyranosides is shown, different interactions between phenyl-β-D-galactopyranoside (**24**) and phenyl-β-D-glucopyranoside (**21**) are probably caused by opposite orientation of hydroxyl moiety in $C_{(4)}$ in position 5 of the tetrahydropyran ring. In compound **21** this hydroxyl moiety is *trans*-oriented (in equatorial configuration) to the hydroxymethyl group in $C_{(5)}$ in position 6 of the tetrahydropyran ring, while in compound **24** it is *cis*-oriented (in axial configuration), which at the same time guarantees space proximity to the pyran oxygen. Pyranoside **24** possesses *cis*-orientation of hydroxyl moieties in $C_{(3)}$ and $C_{(4)}$ in positions 4 and 5 of the tetrahydropyran ring together with the phenoxy moiety in $C_{(1)}$ in position 2 of the tetrahydropyran ring. This fact together with *cis*-orientation of the hydroxymethyl group in $C_{(5)}$ in position 6 probably results in essential three-point interaction of neighbouring hydroxyl moieties between compound **24** and BP **IV** that is completed by interactions of the adjacent phenoxy moiety and pyran oxygen in comparison with carbohydrate **21**, see Fig. 10.

5.4. Crystal products of carbohydrates with risedronate

The semi-crystalline risedronate monosodium salt sodium 1-hydroxy-1-phosphono-2-(pyridin-3-yl-ethyl)phosphonate, (**V**) was used as a starting material (Richter et al., 2007). It is a white powder, freely soluble in water and practically insoluble in organic solvents. The sodium hemipentahydrate, which is the marketed form A, is the most stable of all these forms at ambient conditions (298 K, 50% room humidity) (Cazer et al., 2001).

From all the tested agents only phenyl-β-D-galactopyranoside (**24**) with risedronate (**V**) yielded noteworthy products. Other tested carbohydrates yielded either risedronate form A (in most cases), form H (in the case of the samples with *myo*-inositol (see Fig. 7, structure **4**), D-lyxofuranose (see Fig. 8, structure **9**), phenyl-β-D-glucopyranoside (**21**) and naphthyl-β-D-galactopyranoside (**25**) prepared by addition of methanol and evaporation of the liquid part at ambient temperature) or impure form B in the case of the sample with β-D-allopyranose (see Fig. 9, structure **13**) precipitated by methanol. A rapid change in solubility equilibrium and fast precipitation (kinetically controlled crystallization process) caused generation of different form B (samples **V-13/D** and **V-13/E**), while slow evaporation, *i.e.* thermodynamically controlled crystallization, led to preparation of stable polymorphs A or H. β-D-Allopyranose (**13**) modifies the environment from which compound **V** was crystallized, but this carbohydrate was not detectable in the final crystalline form. Based on this fact it can be concluded that the addition of methanol as an anti-solvent is crucial for generation of this different/uncommon solid form.

Different interactions of BP **V** with carbohydrate **13** are probably caused by the opposite orientation of hydroxyl moieties in $C_{(1)}$ and $C_{(3)}$ in positions 2 and 4 of the tetrahydropyran ring in comparison with α-D-gluco-, α-D-manno- and α-D-galactopyranose (**10-12**). The β-position of the hydroxyl moiety in $C_{(1)}$ of β-D-allopyranose (**13**) possesses also *cis*-orientation (axial configuration, see Fig. 9) with respect to the pyran oxygen in position 1 of the tetrahydropyran ring. As bonds influencing generation of crystalline forms are formed by non-binding interactions (e.g. by *H*-bonds, ionic bonds, van der Waals forces (dispersion attractions, dipole-dipole, dipole-induced dipole interactions) and hydrophobic interactions), the steric arrangement of hydroxyl moieties on pyranose skeletons is important for formation of interactions, as discussed above.

Samples of **V-24/A-C** were prepared by mixing saturated aqueous solutions and subsequent evaporation of water at ambient temperature. All three samples contained polymorph A of risedronate (**V**) (the most thermodynamically stable form). Samples **V-24/D** and **V-24/E** precipitated by methanol and filtered yielded again polymorph A of compound **V**. Samples **V-24/F** and **V-24/G** were generated by addition of methanol and filtration of the obtained precipitate with following evaporation at ambient temperature. Samples **V-24/F** and **V-24/G** were absolutely different from all the above mentioned samples. A change in the NIR spectra of samples **V-24/F** and **V-24/G** was observed in the range of 7,100–4,900 cm^{-1}. Both samples were also characterized by means of FT-Raman spectrometry and ^{31}P and ^{13}C CP/MAS NMR spectroscopy for verification of the above mentioned hypothesis. Sample **V-24/G** was a mixture of polymorph A and the amorphous form of **V**, but sample **V-24/F** was confirmed as a new crystalline form of BP **V**. Therefore solid **V-24/F** was additionally characterized by means of XRPD (Fig. 12) and also by DSC. A XRPD pattern corresponds to a crystalline sample. Visual comparison of the measured pattern with those published previously (Aronhime et al., 2003; Bruning et al., 2011) revealed that a new solid phase of BP **V** was formed. It is also supported by the absence of peaks of co-crystal former **24**. It can be concluded that the presence of compound **24** and slow evaporation, *i.e.* thermodynamically controlled crystallization process, with a small amount of methanol as anti-solvent provided risedronate (**V**) in an unknown form that was named as polymorph P.

Sugar alcohols did not provide any different forms or co-crystals with risedronate (**V**). The polyols used are acyclic compounds, or probably important heterocyclic oxygen is not present in the ring. In the case of *myo*-inositol (**4**), where only the different polymorph H of compound **V** was detected, *cis*-oriented hydroxyl moieties are in $C_{(1)}$, $C_{(2)}$, $C_{(3)}$ and $C_{(5)}$ or conversely oriented hydroxyl moieties are in $C_{(4)}$ and $C_{(6)}$, see Fig. 7. Contrary to the rest of the tested unsubstituted carbohydrates, only β-D-allopyranose (**13**) shows *cis*-orientation of hydroxyl moieties in $C_{(1)}$ and $C_{(5/6)}$ in positions 2 and 6 of the tetrahydropyran ring together with the pyran oxygen in position 1 and *cis*-orientation of hydroxyl moieties in $C_{(2)}$, $C_{(3)}$ and $C_{(4)}$ in positions 3, 4 and 5 of the tetrahydropyran ring, *i.e.* *cis*-orientation of three sequential hydroxyl moieties. These facts are probably essential for interactions between BP **V** and carbohydrate **13**. For example, α-D-galactopyranose (**12**) possesses 1, 4, 5, 6 *cis*-oriented pyran oxygen together with hydroxyl moieties; α-D-glucopyranose (**10**) possesses 1, 4, 6 *cis*-oriented pyran oxygen together with hydroxyl moiety; and α-D-mannopyranose (**11**) possesses 3, 4, 6 *cis*-oriented hydroxyl moieties together with pyran oxygen in position 1.

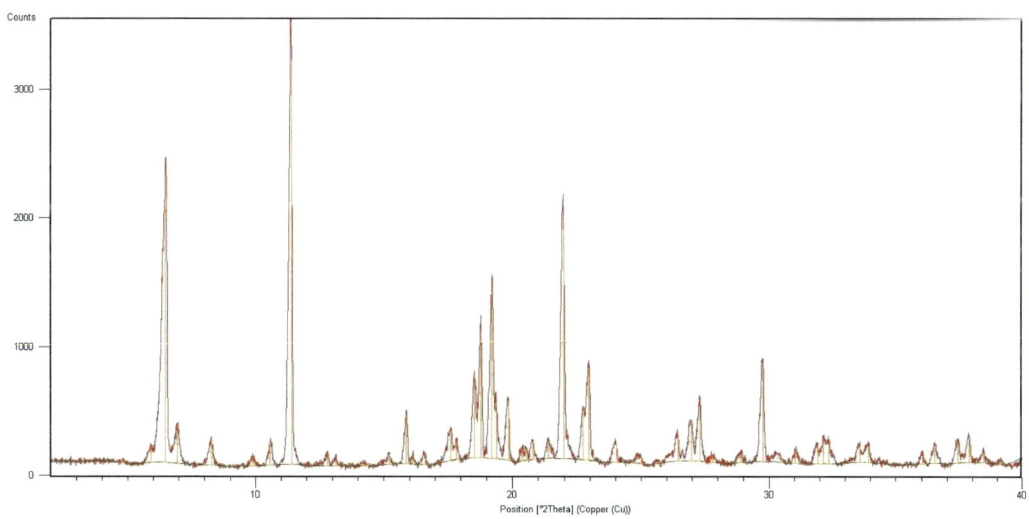

Figure 12. XRPD patterns of new solid phase **V-24/F** named as polymorph P.

According to the above mentioned hypothesis, interactions with BP **V** should be predicted only for D-lyxofuranose (Fig. 8, structure **9**) from the furanose family. D-Lyxofuranose (**9**) shows *cis*-orientation of hydroxyl moieties in $C_{(2)}$ and $C_{(3)}$ in positions 3 and 4 of the tetrahydrofuran ring together with the furan oxygen in position 1. Nevertheless, this three-point interaction of carbohydrate **9** with BP **V**, which however is not completed by other additional/secondary interactions, is not sufficient for generation of a different form or co-crystal of BP **V**, when only form H was generated using compound **9**. Probably the conformation of the tetrahydrofuran ring different from the pyranose chair conformation is also important.

Different interactions of risedronate monosodium salt (**V**) with phenyl-β-D-galactopyranoside (**24**) compared to other evaluated *O*-substituted pyranosides are probably caused by the opposite orientation of the hydroxyl moiety in $C_{(4)}$ in position 5 of the tetrahydropyran ring, as it is shown in Fig. 10 and discussed in Section 5.3. β-Position of the hydroxyl moiety in $C_{(1)}$ of β-D-gluco- and β-D-galactopyranoside (as well as in β-D-allopyranose) possessing also *cis*-orientation to the pyran oxygen together with phenyl substitution of this hydroxyl moiety seems also to be an important assumption for interactions. For example, methyl-β-D-galactopyranoside (**23**) did not show any interactions with compound **V**, whereas phenyl-β-D-glucopyranoside (**21**) and naphtyl-β-D-galactopyranoside (**25**) generated polymorph H of compound **V**, and phenyl-β-D-galactopyranoside (**24**) provided a new solid phase. Aliphatic alkoxy moieties (methoxy, octyloxy) show absolutely different physico-chemical properties, *i.e.* non-binding interactions compared with the aromatic phenyl nucleus. On the other hand, a naphthyl moiety, which is comparable with a phenyl ring, does not meet steric requirements to generate a new solid phase with compound **V**. Contrary to the rest of the tested *O*-substituted pyranosides, carbohydrate **24** shows *cis*-orientation of hydroxyl moieties in $C_{(3)}$, $C_{(4)}$ and $C_{(5\text{-}6)}$ in positions 4, 5 and 6 of the tetrahydropyran ring, *i.e.* three sequential hydroxyl moieties that possess *cis*-orientation with the phenoxy moiety in $C_{(1)}$ in position 2 of the tetrahydropyran ring together with the pyran oxygen in position 1, as was mentioned in Section 5.3. This configuration of all the hydroxyl moieties, as illustrated in Fig. 10, is probably essential for interactions between risedronate (**V**) and phenyl-β-D-galactopyranoside (**24**).

6. Future research

Samples **IV-24/A-B** and **IV-24/F**, **IV-24/G** should be characterized by XRPD. Also some of the above mentioned samples as well as sample **V-24/F** (polymorph P) should be analyzed using single-crystal X-ray diffraction (SCXRD), *i.e.* monocrystals should be prepared for absolute characterization of their structure. Based on the above described primary screening, other carbohydrate derivatives should be synthesized, especially various substituted β-D-pyranosides, e.g. *O*-arylated or *O*-alkylated. For example, based on the screening of these carbohydrates, it was confirmed that the structure derived from β-D-galactopyranose could be a successful candidate for modification of a crystalline form of BPs.

In the recent past, molecular modelling and/or molecular dynamics simulates started playing an increasingly important role in detection of molecule interactions. These computational techniques allow investigating possible binding modes of compounds. Therefore the priorities are advanced simulation of interactions between carbohydrate derivatives and various APIs and modelling/computing of physico-chemical properties of these new potential solids by means of various advanced simulating software products and subsequent transfer of the results of this systematic virtual screening to practice.

7. Conclusion

Twenty-eight carbohydrate derivatives were evaluated as formers during crystallization process of monosodium salts of alendronate (**III**), ibandronate (**IV**) and risedronate (**V**). All prepared samples were screened by FT-NIR and FT-Raman spectroscopy, and some new entities were checked by ^{31}P and ^{13}C CP/MAS NMR spectroscopy, XRPD and DSC. In the present study the relationships between the chemical structures of bisphosphonates and carbohydrates required for crystalline form change are investigated and discussed. It can be concluded that in general carbohydrates can be used as crystallization modifiers, although none of carbohydrates afforded a new solid phase with alendronate (**III**). Ibandronate (**IV**) and risedronate (**V**) generated new solid phases with phenyl-β-D-galactopyranoside (**24**). It is worth to note that both BPs **IV** and **V** contain trisubstituted nitrogen in contrast to alendronate (**III**), which can be an important factor that can cause different interactions. In case of BP **IV** it can be speculated about co-crystal generation; in case of BP **V** the new crystal form, polymorph P, is explicitly characterized. In both cases thermodynamically controlled crystallization was successful. The fundamental steric requirements to carbohydrate formers for generation of a new solid phase is β-orientation of hydroxyl moiety in $C_{(1)}$ in position 2 of the tetrahydropyran ring together with *cis*-orientation of at least three vicinal hydroxyl moieties and the pyran oxygen. β-Hydroxyl moiety in $C_{(1)}$ have to be substituted by an aromatic or heteroaromatic ring. It is also important to note that all the used carbohydrates can chelate the sodium cation in monosodium salts of BPs, and thus the sodium cation can contribute to the binding of BPs and carbohydrate with convenient conformation. The sodium cation can make the complex energetically favourable and help to retain the proper topology of the binding phosphates of BPs and the proper orientation of the hydroxyl moieties of a carbohydrate.

Author details

Josef Jampílek and Jiří Dohnal
Faculty of Pharmacy, University of Veterinary and Pharmaceutical Sciences Brno,
Research Institute for Pharmacy and Biochemistry (VUFB, s.r.o.),
Czech Republic

Acknowledgement

This study was partly supported by the Grant Agency of the Czech Republic (Czech Science Foundation), project number GACR P304/11/2246. Thanks to all our colleagues and numerous graduate and undergraduate students who participated in the synthesis and analytical evaluation of the discussed solid forms. The authors would like to thank Prof. Bohumil Kratochvíl for his constructive comments that helped us to improve the manuscript. The authors also thank Natalia Jampílková for her valuable assistance with the English proofreading of the manuscript.

8. References

Almarsson, Ö. & Zaworotko, M. J. (2004). Crystal engineering of the composition of pharmaceutical phases. Do pharmaceutical co-crystals represent a new path to improved medicines? *Chemical Communications*, Vol.2004, No.17, pp.1889-1896, ISSN 1364-548X

Aakeröy, C. B. & Salmon, D. J. (2005). Building co-crystals with molecular sense and supramolecular sensibility. *CrystEngComm*, Vol.7, No.72, pp.439-448, ISSN 1466-8033

Aakeröy, C. B.; Fasulo, M. E. & Desper, J. (2007). Cocrystal or salt: Does it really matter? *Molecular Pharmaceutics*, Vol.4, No.3, pp.317–322, ISSN 1543-8384

Aronhime, J.; Lifshitz-Liron, R.; Kovalevski-Ishai, E. & Lidor-Hadas, R. (TEVA Pharmaceutical Industries, Ltd.), (2003a). *Novel polymorphs and pseudopolymorphs of risedronate sodium*. CA 2480764

Aronhime, J.; Lifshitz-Liron, R.; Kovalevski-Ishai, E. & Lidor-Hadas, R. (TEVA Pharmaceutical Industries, Ltd.), (2003b). *Novel polymorphs and pseudopolymorphs of risedronate sodium*. WO 03086355

Babu, N. J.; Reddy, L. S.; Aitipamula, S. & Nangia, A. (2008). Polymorphs and polymorphic cocrystals of temozolomide. *Chemistry – An Asian Journal*, Vol.3, No.7, pp.1122-1133, ISSN 1861-471X

Berry, D. J.; Seaton, C. C.; Clegg, W.; Harrington, R. W.; Coles, S. J.; Horton, P. N.; Hursthouse, M. B.; Storey, R.; Jones, W.; Friščić, T. & Blagden, N. (2008). Applying hot-stage microscopy to co-crystal screening: A study of nicotinamide with seven active pharmaceutical ingredients. *Crystal Growth & Design*, Vol.8, No.5, pp.1697-1712, ISSN 1528-7483

Bhatt, P. M.; Azim, Y.; Thakur, T. S. & Desiraju, G. R. (2009). Co-crystals of the anti-HIV drugs lamivudine and zidovudine. *Crystal Growth & Design*, Vol.9, No.2, pp.951-957, ISSN 1528-7483

Bhogala, B. R. & Nangia, A. (2008). Ternary and quaternary co-crystals of 1,3-*cis*,5-*cis*-cyclohexanetricarboxylic acid and 4,4'-bipyridines. *New Journal of Chemistry*, Vol.32, No.5, pp.800-807, ISSN 1144-0546

Bighley, L. D.; Berge, S. M. & Monkhouse, D. C. (1996). Salt forms of drugs and absorption, In: *Encyclopedia of Pharmaceutical Technology*, Vol.13, Swarbick, J. & Boylan, J. C. (Eds.), pp.453-499, Marcel Dekker, ISBN 978-0-8247-2814-9, New York, NY, USA

Blagden, N.; de Matas, M.; Gavan, P. T. & York, P. (2007). Crystal engineering of active pharmaceutical ingredients to improve solubility and dissolution rates. *Advanced Drug Delivery Reviews*, Vol.59, No.7, pp.617-630, ISSN 0169-409X

Bond, A. B. (2007). What is a co-crystal? *CrystEngComm*, Vol.9, No.9, pp.833–834, ISSN 1466-8033

Brittain, H. (Ed.). (2009). *Polymorphism in Pharmaceutical Solids*. Informa Healthcare, ISBN 978-1-4200-7321-8, New York, NY, USA

Bruning, J.; Petereit, A. C.; Alig, E.; Bolte, M.; Dressman, J. B. & Schmidt, M. U. (2011). Characterization of a new solvate of risedronate. *Journal of Pharmaceutical Sciences*, Vol.100, No.3, pp.863–873, ISSN 1520-6017

Brusová, H. (2010). X-Ray diffraction, In: *Modern Approaches to Pharmaceutical Analysis*, Dohnal, J.; Jampílek, J.; Král, V. & Řezáčová, A. (Eds.). pp.433-477, Faculty of Pharmacy, University of Veterinary and Pharmaceutical Sciences Brno, ISBN 978-80-7305-085-6, Prague, Czech Republic

Caira, M. R. (2007). Sulfa drugs as model co-crystal formers. *Molecular Pharmaceutics*, Vol.4, No.3, pp.310–316, ISSN 1543-8384

Carano, A.; Teitelbaum, S. L.; Konsek, J. D.; Schlesinger, P. H. & Blair, H. C. (1990). Bisphosphonates directly inhibit the bone resorption activity of isolated avian osteoclasts *in vitro*. *Journal of Clinical Investigation*, Vol.85, No.2, pp.456–461, ISSN 0021-9738

Cazer, F. D.; Perry, G. E.; Billings, D. M. & Redman-Furey, N. L. (Warner Chilcott Company, LLC.), (2001). *Selective crystallization of 3-pyridyl-1-hydroxy-ethylidene-1,1-bisphosphonic acid sodium as the hemipentahydrate or monohydrate*. EP 1252170

Childs, S. L.; Chyall, L. J.; Dunlap, J. T.; Smolenskaya, V. N.; Stahly, B. C. & Stahly, G. P. (2004). Crystal engineering approach to forming cocrystals of amine hydrochlorides with organic acids. Molecular complexes of fluoxetine hydrochloride with benzoic, succinic, and fumaric acids. *Journal of American Chemical Society*, Vol.126, No.41, pp.13335–13342, ISSN 0002-7863

Childs, S. L. & Hardcastle, K. I. (2007). Cocrystals of piroxicam with carboxylic acids. *Crystal Growth & Design*, Vol.7, No.7, pp.1291–1304, ISSN 1528-7483

Childs, S. L.; Stahly, G. P. & Park, A. (2007). The salt–cocrystal continuum: The influence of crystal structure on ionization state. *Molecular Pharmaceutics*, Vol.4, No.3, pp.323-338, ISSN 1543-8384

Childs, S. L. & Zaworotko, M. J. (2009). The reemergence of cocrystals: The crystal clear writing is on the wall introduction to virtual special issue on pharmaceutical cocrystals. *Crystal Growth & Design*, Vol.9, No.10, pp.4208-4211, ISSN 1528-7483

Dahlin, A. B.; Sannomiya, T.; Zahn, R.; Sotiriou, G. A. & Vőrős, J. (2011). Electrochemical crystallization of plasmonic nanostructures. *Nano Letters*, Vol.11, No.3, pp.1337-1343, ISSN 1530-6984

Datta, S. & Grant, D. J. W. (2004). Crystal structures of drugs: Advances in determination, prediction and engineering. *Nature Reviews Drug Discovery*, Vol.3, No.1, pp.42–57, ISSN 1474-1776.

Desiraju, G. R. (2003). Crystal and co-crystal. *CrystEngComm*, Vol.5, No.82, pp.466–467, ISSN 1466-8033

Devarakonda, S. N.; Thaimattam, R.; Raghupati, B.; Asnani, M.; Vasamsetti, S. K. Rangineni, S. & Muppidi, V. K. (Dr. Reddy's Laboratories, Ltd.), (2010). *Ibandronate sodium polymorphs*. US 20100125149

Di Profio, G.; Tucci, S.; Curcio, E. & Drioli, E. (2007). Selective glycine polymorph crystallization by using microporous membranes. *Crystal Growth & Design*, Vol.7, No.3, pp.526-530, ISSN 1528-7483

Dohnal, J.; Jampílek, J.; Král, V. & Řezáčová, A. (Eds.). (2010). *Modern Approaches to Pharmaceutical Analysis*. Faculty of Pharmacy, University of Veterinary and Pharmaceutical Sciences Brno, ISBN 978-80-7305-085-6, Prague, Czech Republic

Ebetino, F. H.; Francis, M. D.; Rogers, M. J.; Russell, R. G. G. (1998). Mechanisms of action of etidronate and other bisphosphonates. *Reviews in Contemporary Pharmacotherapy*, Vol.9, No.4, pp.233–243, ISSN 0954-8602

Eiermann, U.; Junghans, B.; Knipp, B. & Sattelkau, T. (Hoffmann-La Roche, Inc.), (2006a). *Ibandronate polymorph B*. WO 2006081962

Eiermann, U.; Junghans, B.; Knipp, B. & Sattelkau, T. (Hoffmann-La Roche, Inc.), (2006b). *Ibandronate polymorph A*. WO 2006081963

Etter, M. C. (1990) Encoding and decoding hydrogen bonds patterns of organic compounds. *Accounts of Chemical Research*, Vol.23, No.4, pp.120–126, ISSN 0001-4842

Etter, M. C. (1991). Hydrogen bonds as design elements in organic chemistry. *Journal of Physical Chemistry*, Vol.95, No.12, pp.4601-4610, ISSN 0022-365

Ezra, A. & Golomb, G. (2000). Administration routes and delivery systems of bisphosphonates for the treatment of bone resorption. *Advanced Drug Delivery Reviews*, Vol.42, No.3, pp.175–195, ISSN 0169-409X

Ferencová, Z. (2012). *Physico-Chemical Properties Modification of Biologically Active Compounds I*, Diploma Thesis, Faculty of Pharmacy, University of Veterinary and Pharmaceutical Sciences Brno

Fleischman, S. G.; Kuduva, S. S.; McMahon, J. A.; Moulton, B.; Walsh, R. B.; Rodríguez-Hornedo, N. & Zaworotko, M. J. (2003). Crystal engineering of the composition of pharmaceutical phases. Multiple component crystalline solids involving carbamazepine. *Crystal Growth & Design*, Vol.3, No.6, pp.909-919, ISSN 1528-7483

Forsythe, E.L.; Maxwell, D.L. & Pusey, M. (2002). Vapor diffusion, nucleation rates and the reservoir to crystallization volume ratio. *Acta Crystallographica Section D*, Vol.58, No.10-1, pp.1601-1605, ISSN 0907-4449

Friščić, T. & Jones, W. (2009). Recent advances in understanding the mechanism of cocrystal formation via grinding. *Crystal Growth & Design*, Vol.9, No.3, pp.1621-1637, ISSN 1528-7483

Gossman, W. L.; Wilson, S. R. & Oldfield, E. (2003). Three hydrates of the bisphosphonate risedronate, consisting of one molecular and two ionic structures. *Acta Crystallographica Section C*, Vol.C59, No.2, pp.m33–m36, ISSN 0108-2701

Grepioni, F. & Braga, D. (Eds.). (2007). *Making Crystals by Design - from Molecules to Molecular Materials, Methods, Techniques, Applications*, Wiley-VCH, ISBN 978-3-527-31506-2, Weinheim, Germany.

Gu, C. H. & Grant, D. J. W. (2003). In: *Handbook of Experimental Pharmacology: Stereochemical Aspects of Drug Action and Disposition*; Eichelbaum, M.; Testa, B. & Somogyi, A. (Eds.), pp.113–137, Springer, ISBN 978-3-5404-1593-0, Berlin, Germany

Haroková, P. (2010). *Study of Polymorphism of Biologically Active Compounds I*, Diploma Thesis, Faculty of Pharmacy, University of Veterinary and Pharmaceutical Sciences Brno

Havelková, L. (2010). *Study of Drug Polymorphism I*, Diploma Thesis, Faculty of Pharmacy, University of Veterinary and Pharmaceutical Sciences Brno

Havelková, L. (2012). *Crystallization Products of Ibandronate with Gluco- and Galactopyranoside Derivatives*, Rigorous Thesis, Faculty of Pharmacy, University of Veterinary and Pharmaceutical Sciences Brno

Havlíček, J. (2010). NMR Spectroscopy, In: *Modern Approaches to Pharmaceutical Analysis*, Dohnal, J.; Jampílek, J.; Král, V. & Řezáčová, A. (Eds.). pp.261-313, Faculty of Pharmacy, University of Veterinary and Pharmaceutical Sciences Brno, ISBN 978-80-7305-085-6, Prague, Czech Republic

He, J. H. & Lavernia, E. J. (2001). Development of nanocrystalline structure during cryomilling of Inconel 625. *Journal of Materials Research*, Vol.16, No.9, pp.2724-2732, ISSN 0884-2914

Hickey, M. B.; Peterson, M. L.; Scoppettuolo, L. A.; Morrisette, S. L.; Vetter, A.; Guzmán, H.; Remenar, J. F.; Zhang, Z.; Tawa, M. D.; Haley, S.; Zaworotko, M. J. & Almarsson, Ö. (2007). Performance comparison of a co-crystal of carbamazepine with marketed product. *European Journal of Pharmaceutics and Biopharmaceutics*, Vol.67, No.1, pp.112–119, ISSN 0939-6411

Hilfiker, R. (Ed.). (2006). *Polymorphism in the Pharmaceutical Industry*. Wiley-VCH, ISBN 978-3-527-31146-0, Weinheim, Germany

Hrušková, J. (2010). *Study of Polymorphism of Biologically Active Compounds II*, Diploma Thesis, Faculty of Pharmacy, University of Veterinary and Pharmaceutical Sciences Brno

Hsu, W. P.; Koo, K. K. & Myerson, A. S. (2002). The gel-crystallization of 1-phenylalanine and aspartame from aqueous solutions. *Chemical Engineering Communications*, Vol.189, No.8, pp.1079-1090, ISSN 0098-6445

Hughes, D. E.; Wright, K. R.; Uy, H. L.; Sasaki, A.; Yoneda, T.; Roodman, G. D.; Mundy, G. R. & Boyce, B.F. (1995). Bisphosphonates promote apoptosis in murine osteoclasts *in vitro* and *in vivo*. *Journal of Bone and Mineral Research*, Vol.10, No.10, pp.1478–1487, ISSN 0884-0431

Hušák, M.; Rohlíček, J.; Čejka, J. & Kratochvíl, B. (2007). Structure determination from powder diffraction data - it is unrealizable dream or daily use? *Chemické Listy*, Vol.101, No.9, pp. 697–705, ISSN 0009-2770

Ito, M.; Amizuka, N.; Nakajima, T. & Ozawa, H. Ultrastructural and cytochemical studies on cell death of osteoclasts induced by bisphosphonate treatment. *Bone*, Vol.25, No.4, pp.447–452, ISSN 8756-3282

Jampílek, J.; Oktábec, Z.; Řezáčová, A.; Plaček, L.; Kos, J.; Havelková, L.; Dohnal, J. & Král V. (2009). Preparation and Properties of New Co-crystals of Ibandronate. *Proceedings of the 13th International Electronic Conference on Synthetic Organic Chemistry (ECSOC-13)*, ISBN 3-906980-23-5, Basel, Switzerland, November, 2009, Available from: <http://www.sciforum.net/presentation/201>

Jampílek, J.; Kos, J.; Oktábec, Z.; Mandelová, Z.; Pekárek, T.; Tkadlecová, M.; Havlíček, J.; Dohnal, J. & Král V. (2010). Co-crystal Screening Study of Risedronate and Unsubstituted Hexoses. *Proceedings of the 14th International Electronic Conference on Synthetic Organic Chemistry (ECSOC-14)*, ISBN 3-906980-21-3, Basel, Switzerland, November, 2010, Available from: <http://www.sciforum.net/presentation/421>

Jayasankar, A.; Somwangthanaroj, A.; Shao, Z. J. & Rodríguez-Hornedo, N. (2006). Co-crystal formation during cogrinding and storage is mediated by amorphous phase. *Pharmaceutical Research*, Vol.23, No.10, pp.2381-2392, ISSN 0724-8741

Jones, W.; Motherwell, W. D. S. & Trask, A. V. (2006). Pharmaceutical co-crystals: An emerging approach to physical property enhancement. *Materials Research Society Bulletin*, Vol.31, No.11, pp.875-879, ISSN 0883-7694

Junghanns, J. U. A. H. & Müller, R. H (2008). Nanocrystal technology, drug delivery and clinical applications. *International Journal of Nanomedicine*, Vol.3, No.3, pp.295–309, ISSN 1178-2013.

Kadlec, F. & Kadlec, Ch. (2012). Terahertz spectroscopy in time-domain, In: *Modern Pharmaceutical Analysis of Solid State*. Dohnal, J.; Jampílek, J. & Čulen, M. (Eds.), in press, Research Institute for Pharmacy and Biochemistry (VUFB), ISBN 978-80-905220-0-8, Brno, Czech Republic

Kerns, E. H. & Di, L. (2008). *Drug-like Properties: Concepts, Structure, Design and Methods: From ADME to Toxicity Optimization*. Elsevier/Academic Press, ISBN 978-0-1236-9520-8, San Diego, CA, USA.

Kos, J. (2010). *Study of Drug Polymorphism II*, Diploma Thesis, Faculty of Pharmacy, University of Veterinary and Pharmaceutical Sciences Brno

Kos, J.; Pěntáková, M.; Oktábec, Z.; Krejčík, L.; Mandelová, Z.; Haroková, P.; Hrušková, J.; Pekárek, T.; Dammer, O.; Tkadlecová, M.; Havlíček, J.; Král, V.; Vinšová, J.; Dohnal, J. & Jampílek, J. (2011). Crystallization products of risedronate with carbohydrates and their substituted derivatives. *Molecules*, Vol.16, No.5, pp.3740-3760, ISSN 1420-3049

Král, V.; Jampílek, J.; Havlíček, J.; Brusová, H. & Pekárek, T. (Zentiva, k.s.), (2010). *Dosage forms of tyrosine kinase inhibitors*. WO 2010081443

Kratochvíl, B. (2007). Crystallization of pharmaceutical substances. *Chemické Listy*, Vol.101, No.1, pp. 3–12, ISSN 0009-2770

Kratochvíl, B.; Hušák, M.; Brynda, J. & Sedláček, J. (2008). What can the current x-ray structure analysis offer? *Chemické Listy*, Vol.102, No.10, pp. 889–901, ISSN 0009-2770

Kratochvíl, B. (2009). Pharmaceutical cocrystals. *Proceedings of the XXXVIII Conference Drug Synthesis and Analysis*, ISBN 978-80-7305-078-8, Hradec Králové, Czech Republic, September, 2009

Kratochvíl, B. (2010). Cocrystals and their expected pharmaceutical applications. *Chemické Listy*, Vol.104, No.9, pp.823-830, ISSN 0009-2770

Krumbholcová, L. & Dohnal, J. (2010). Thermal analysis, In: *Modern Approaches to Pharmaceutical Analysis*, Dohnal, J.; Jampílek, J.; Král, V. & Řezáčová, A. (Eds.). pp.347-381, Faculty of Pharmacy, University of Veterinary and Pharmaceutical Sciences Brno, ISBN 978-80-7305-085-6, Prague, Czech Republic

Lifshitz-Liron, R.; Bayer, T.; Aronhime, J. (TEVA Pharmaceutical), (2006). *Solid and crystalline ibandronate sodium and processes for preparation thereof*. WO 2006024024

McMahon, J.; Peterson, M.; Zaworotko, M. J.; Shattock, T. & Magali, B. H. (Trans Form Pharmaceuticals, Inc.), (2007). *Pharmaceutical co-crystal compositions and related methods of use*. US 20070299033

McNamara, D. P.; Childs, S. L.; Giordano, J.; Iarriccio, A.; Cassidy, J.; Shet, M. S.; Mannion, R.; O'Donnell, E & Park, A. (2006). Use of a glutaric acid cocrystal to improve oral bioavailability of a low solubility API. *Pharmaceutical Research*, Vol.23, No.8, pp.1888-1897, ISSN 0724-8741

Miroshnyk, I.; Mirza, S. & Sandler, N. (2009). Pharmaceutical co-crystals – An opportunity for drug product enhancement. *Expert Opinion Drug Delivery*, Vol.6, No.4, pp.333-341, ISSN 1742-5247

Mitscherlich, E. (1822). Über das Verhältnis der Krystalform zu den chemischen Proportionen. *Annales de Chimie et de Physique*, Vol.19, pp.350–419, ISSN 0003-4169

Morissette, S.L.; Almarsson, Ö.; Peterson, M. L.; Remenar, J. F.; Read, M. J.; Lemmo, A. V.; Ellis, S.; Cima, M. J. & Gardner, C. R. (2004). High-throughput crystallization: polymorphs, salts, co-crystals and solvates of pharmaceutical solids. *Advanced Drug Delivery Reviews*, Vol.56, No.3, pp.275-300, ISSN 0169-409X

Muddasani, P. R.; Vattikuti, U. & Nannapaneni, V. C. (NATCO Pharma, Ltd.), (2007). *Novel polymorphic forms of ibandronate*. WO 2007074475

Oktábec, Z.; Kos, J.; Mandelová, Z.; Havelková, L.; Pekárek, T.; Řezáčová, A.; Plaček, L.; Tkadlecová, M.; Havlíček, J.; Dohnal, J. & Jampílek, J. (2010). Preparation and properties of new co-crystals of ibandronate with gluco- or galactopyranoside derivatives. *Molecules*, Vol.15, No.12, pp.8973-8987, ISSN 1420-3049

Oktábec, Z. (2012). *Modification of Physico-Chemical Properties of Biologically Active Compounds*, Ph.D. Thesis, Faculty of Pharmacy in Hradec Králové, Charles University in Prague

Payghan, S. A.; Bhat, M.; Savla, A.; Toppo, E. & Purohit, S. (2008). Solubility of active pharmaceutical ingredients (API) has always been a concern for formulators, since inadequate aqueous solubility may hamper development of parenteral products and limit bioavailability of oral products. [cited 2012 February 28], Retrieved from <http://www.pharmainfo.net/reviews/potential-solubility-drug-discovery-and-development>

Pekárek, T. & Jampílek, J. (2010). Infrared and Raman spectroscopy, In: *Modern Approaches to Pharmaceutical Analysis*, Dohnal, J.; Jampílek, J.; Král, V. & Řezáčová, A. (Eds.). pp.222-260, Faculty of Pharmacy, University of Veterinary and Pharmaceutical Sciences Brno, ISBN 978-80-7305-085-6, Prague, Czech Republic

Peterson, M. L.; Hickey, M. B.; Zaworotko, M. J. & Almarsson Ö. (2006). Expanding the scope of crystal form evaluation in pharmaceutical science. *Journal of Pharmacy & Pharmaceutical Sciences*, Vol.9, No.3, pp.317-326, ISSN 1482-1826

Qiao, N.; Li, M. Z.; Schlindwein, W.; Malek, N.; Davies, A. & Trappitt, G. (2011). Pharmaceutical cocrystals: An overview. *International Journal of Pharmaceutics*, Vol.419, No.1-2, pp.1-11, ISSN 0378-5173

Ranganathan, A. (1999). Hydrothermal synthesis of organic channel structures: 1:1 Hydrogen-bonded adducts of melanine with cyanuric amd trithiocyanuric acids. *Journal of American Chemical Society*, Vol.121, No.8, pp.1752–1753, ISSN 0002-7863

Redman-Furey, N. L.; Ficka, M.; Bigalow-Kern, A.; Cambron, R. T.; Lubey, G.; Lester, C. & Vaughn, D. (2005). Structural and analytical characterization of three hydrates and an anhydrate form. *Journal of Pharmaceutical Sciences*, Vol.94, No.4, pp.893–911, ISSN 1520-6017

Remenar, J. F.; Morissette, S. L.; Peterson, M. L.; Moulton, B.; MacPhee, J. M.; Guzmán, H. & Almarsson, Ö. (2003). Crystal engineering of novel co-crystals of a triazole drug with

1,4-dicarboxylic acids. *Journal of American Chemical Society*, Vol.125, No.28, pp.8456-8457, ISSN 0002-7863

Reszka, A. A.; Halasy-Nagy, J. M.; Masarachia, P. J. & Rodan, G. A. (1999). Bisphosphonates act directly on the osteoclast to induce caspase cleavage of Mst1 kinase during apoptosis. A link between inhibition of the mevalonate pathway and regulation of an apoptosis-promoting kinase. *Journal of Biological Chemistry*, Vol.274, No.49, pp.34967–34973, ISSN 0021-9258

Richter, J. & Jirman, J. (Zentiva, a.s.), (2007). *Crystalline form of the sodium salt of 3-pyridyl-1-hydroxyethylidene-1,1-bisphosphonic acid*. US 7276604

Rodríguez-Hornedo, N.; Nehm, S. J. & Jayasankar, A. (2007). Cocrystals: Design, properties and formation mechanisms, In: *Encyclopedia of Pharmaceutical Technology*, 3rd ed.; Swarbrick, J. (Ed.), pp 615–635, Taylor & Francis, ISBN 978-1841848198, London, UK

Rodríguez-Spong, B.; Price, C. P.; Jayasankar, A.; Matzger, A. J. & Rodríguez-Hornedo, N. (2004). General principles of pharmaceutical solid polymorphism: A supramolecular perspective. *Advanced Drug Delivery Reviews*, Vol.56, No.3, pp.241-274, ISSN 0169-409X

Rogers, M. J.; Xiong, X.; Brown, R. J.; Watts, D. J.; Russell, R. G.; Bayless, A. V. & Ebetino, F. H. (1995). Structure-activity relationships of new heterocycle-containing bisphosphonates as inhibitors of bone resorption and as inhibitors of growth of *Dictyostelium discoideum* amoebae. *Molecular Pharmacology*, Vol.47, No.2, pp.398–402, ISSN 0026-895X

Rogers, M. J.; Gordon, S.; Benford, H. L.; Coxon, F. P.; Luckman, S. P.; Monkkonen, J. & Frith, J. C. (2000). Cellular and molecular mechanisms of action of bisphosphonates. *Cancer*, Vol.88, No.12, pp.2961–2978, ISSN 1097-0142

Ruecroft, G.; Hipkiss, D.; Ly, T.; Maxted, N. & Cains, P. W. (2005). Sonocrystallization: The use of ultrasound for improved industrial crystallization. *Organic Process Research & Development*, Vol.9, No.6, pp.923–932, ISSN 1083-6160

Sarma, B.; Reddy, L. S. & Nangia, A. (2008). The role of π-stacking in the composition of phloroglucinol and phenazine cocrystals. *Crystal Growth & Design*, Vol.8, No.12, pp.4546-4552, ISSN 1528-7483

Sarma, B.; Chen, J.; Hsi H.-Y. & Myerson, A. S. Solid forms of pharmaceuticals: Polymophs, salts and cocrystals. (2011). *Korean Journal of Chemical Engineering*, Vol.28, No.2, pp.315-322, ISSN 0256-1115

Sato, M.; Grasser, W.; Endo, N.; Akins, R.; Simmons, H.; Thompson, D. D.; Golub, E. & Rodan, G. A. (1991). Bisphosphonate action. Alendronate localization in rat bone and effects on osteoclast ultrastructure. *Journal of Clinical Investigation*, Vol.88, No.6, pp.2095–2105, ISSN 0021-9738

Schultheiss, N. & Newman, A. (2009). Pharmaceutical co-crystals and their physicochemical properties. *Crystal Growth & Design*, Vol.9, No.6, pp.2950–2967, ISSN 1528-7483

Seddon, K. R. & Zaworotko, M. J. (Eds.). (1999). *Crystal Engineering: The Design and Application of Functional Solids*, NATO-ASI Series, Vol. 539, Kluwer Academic Publisher, ISBN 978-0-7923-5905-4, Dordrecht, the Netherlands

Sekhon, B. S. Pharmaceutical co-crystals - A review. (2009). *Ars Pharmaceutica*, Vol.50, No.3, pp.99-117, ISSN 0004-2927

Selander, K. S.; Monkkonen, J.; Karhukorpi, E. K.; Harkonen, P.; Hannuniemi, R. & Vaananen, H. K. (1996). Characteristics of clodronate-induced apoptosis in osteoclasts and macrophages. *Molecular Pharmacology*, Vol.50, No.5, pp.1127–1138, ISSN 0026-895X

Shan, N. & Zaworotko, M. J. (2008). The role of cocrystals in pharmaceutical science. *Drug Discovery Today*, Vol.13, No.9-10, pp.440-446, ISSN 1359-6446

Shekunov, B. Y. & York, P. (2000). Crystallization processes in pharmaceutical technology and drug delivery design. *Journal of Crystal Growth*, Vol.211, No.1-4, pp.122-136.

Stahl, P. H. & Wermuth, C. G. (Eds.). (2008). *Handbook of Pharmaceutical Salts: Properties, Selection and Use*, Verlag Helvetica Chimica Acta, ISBN 978-3-906390-58-1, Zürich, Switzerland

Stahly, G. P. (2007). Diversity in single-and multiple-component crystals. The search for and prevalence of polymorphs and co-crystals. *Crystal Growth & Design*, Vol.7, No.6, pp.1007-1026, ISSN 1528-7483

Strachan, C. J.; Rades, T.; Newnham, D. A.; Gordon, K. C.; Proper, M. & Taday P. F. (2004). Using terahertz pulsed spectroscopy to study crystallinity of pharmaceutical materials. *Chemical Physics Letters*, Vol.390, No.1-3, pp. 20–24, ISSN 0009-2614

Sun, C. C. & Hou, H. (2008). Improving mechanical properties of caffeine and methyl gallate crystals by co-crystallization. *Crystal Growth & Design*, Vol.8, No.5, pp.1575-1570, ISSN 1528-7483

Takata, N.; Shiraki, K.; Takano, R.; Hayashi, Y. & Terada, Y. (2008). Co-crystal screening of stanolone and mestanolone using slurry crystallization. *Crystal Growth & Design*, Vol.8, No.9, pp.3032-3037, ISSN 1528-7483

Ťažká, A. (2011). *Study of Polymorphism of Biologically Active Compounds II*, Diploma Thesis, Faculty of Pharmacy, University of Veterinary and Pharmaceutical Sciences Brno

Tong, H. H.; Shekunov, B.Y.; York, P. & Chow, A. H. (2001). Characterization of two polymorphs of salmeterol xinafoate crystallized from supercritical fluids. *Pharmaceutical Research*, Vol.18, No.6, pp.852-858, ISSN 0724-8741

Trask, A. V. & Jones, W. (2005). Crystal engineering of organic co-crystals by the solid-state grinding approach. *Topics in Current Chemistry*, Vol.254, pp.41-70, ISSN 0340-1022

Trask, A. V.; van de Streek, J.; Motherwell, W. D. S. & Jones, W. (2005a). Achieving polymorphic and stoichiometric diversity in co-crystal formation: Importance of solid-state grinding, powder X-ray structure determination and seeding. *Crystal Growth & Design*, Vol.5, No.6, pp.2233-2241, ISSN 1528-7483

Trask, A. V.; Motherwell, W. D. S. & Jones, W. Pharmaceutical co-crystallization: Engineering a remedy for caffeine hydration. (2005b). *Crystal Growth & Design*, Vol.5, No.3, pp.1013-1021, ISSN 1528-7483

Trask, A. V.; Haynes, D. A.; Motherwell, W. D. S. & Jones, W. (2006). Screening for crystalline salts via mechanochemistry. *Chemical Communications*, Vol.2006, No.1, pp.51-53, ISSN 1364-548X

Trask, A. V. (2007). An overview of pharmaceutical co-crystals as intellectual property. *Molecular Pharmaceutics* Vol.4, No.3, pp.301-309, ISSN 1543-8384 van Beek, E. R.; Lowik, C. W.; Ebetino, F. H. & Papapoulos, S. E. (1998). Binding and antiresorptive properties of heterocycle-containing bisphosphonate analogs: Structure-activity relationships. *Bone*, Vol.23, No.5, pp.437–442, ISSN 8756-3282

Velaga, S. P.; Basavoju, S. & Boström D. (2008). Norfloxacin saccharinate–saccharin dihydrate cocrystal – A new pharmaceutical co-crystal with an organic counter ion. *Journal of Molecular Structure*, Vol.889, No.1-3, pp.150-153, ISSN 0022-2860

Viertelhaus, M.; Hilfiker, R. & Blatter, F. (2009). Piracetam co-crystals with OH-group functionalized carboxylic acids. *Crystal Growth & Design*, Vol.9, No.7, pp.2220–2228, ISSN 1528-7483

Vishweshwar, P.; Nangia, A. & Lynch, V. M. (2003). Molecular complexes of homologous alkanedicarboxylic acids with isonicotinamide: X-Ray crystal structures, hydrogen bond synthons, and melting point alternation. *Crystal Growth & Design*, Vol.3, No.5, pp.783-790, ISSN 1528-7483

Vishweshwar, P.; McMahon, J. A.; Peterson, M. L.; Hickey, M. B.; Shattock, T. R. & Zaworotko, M. J. (2005). Crystal engineering of pharmaceutical co-crystals from polymorphic active pharmaceutical ingredients. *Chemical Communications*, Vol.2005, No.36, pp.4601-4603, ISSN 1364-548X

Vishweshwar, P.; McMahon, J. A., Bis, J. A. & Zaworotko, M. J. (2006a). Pharmaceutical co-crystals. *Journal of Pharmaceutical Sciences*, Vol.95, No.3 , pp.449-516, ISSN 1520-6017

Vishweshwar, P.; McMahon, J. A. & Zaworotko, M. J. (2006b). Crystal engineering of pharmaceutical co-crystals, In: *Frontiers in Crystal Engineering*, Tiekink, E. R. T. & Vittal, J. J. (Eds.)., pp.25-50, John Wiley & Sons, ISBN 978-0-470-02258-0, Chichester, UK

Vogt, F. G; Clawson, J. S.; Strohmeier, M.; Edwards, A. J.; Pham, T. N. & Watson, S. A. (2009). Solid-state NMR analysis of organic cocrystals and complexes. *Crystal Growth & Design*, Vol.9, No.2, pp. 921–937, ISSN 1528-7483

Walsh, R. D. B.; Bradner, M. W.; Fleischman, S.; Morales, L. A.; Moulton, B.; Rodríguez-Hornedo, N. & Zaworotko, M. J. (2003). Crystal engineering of the composition of pharmaceutical phases. *Chemical Communications*, Vol.2003, No.2, pp.186-187, ISSN 1364-548X

Walther, M.; Fischer, B. M. & Jepsen, P. U. (2003). Noncovalent intermolecular forces in polycrystalline and amorphous saccharides in the far infrared. *Chemical Physics*, Vol.288, No.2-3, pp. 261–288, ISSN 0301-0104

Whitesides, G. M. & Wong, A. P. (2006). The intersection of biology and materials science. *Materials Research Society Bulletin*, Vol. 31, No.1, pp.19-27, ISSN 0883-7694

Zakrzewski, A. & Zakrzewski, M. (Eds.) (2006). *Solid State Characterization of Pharmaceuticals*. Assa, ISBN 83-920584-5-3, Danbury, CT, USA

Zaworotko, M. (2008). Crystal engineering of co-crystals and their relevance to pharmaceuticals and solid-state chemistry. *Acta Crystallographica Section A*, Vol.A64, No.Supplement, pp.C11-C12, ISSN 0108-7673

Žegarać, M.; Meštrović, E.; Dumbović, A.; Devčić, M.; Tudja, P. (Pliva Hrvatska D.o.o.), (2007). *Pharmaceutically acceptable co-crystalline forms of sildenafil*. WO 2007080362

Chapter 6

Glycosidases – A Mechanistic Overview

Natércia F. Brás , Pedro A. Fernandes,
Maria J. Ramos and Nuno M.F.S.A. Cerqueira

Additional information is available at the end of the chapter

http://dx.doi.org/10.5772/52019

1. Introduction

Carbohydrates are the most abundant and structurally diverse class of biological compounds in nature. However, our current understanding regarding the relationship between carbohydrate structure and its biological function is still far from what is known regarding proteins and nucleic acids. Initially, carbohydrates were only recognized as structural and energy storage molecules (e.g. cellulose, chitin and glycogen), but recent developments in the field have shown that carbohydrates are also involved in numerous biological events, such as cancer, inflammations, pathogen infections, cell-to-cell communication, etc. In addition, carbohydrate-processing enzymes have become the choice in many industrial applications due to their stereo selectivity and efficiency[1].

Carbohydrates can be found in nature in many forms, from simple monomers to more complex oligomers, polymers or glicoconjugates. The complexity of these structures can be reasonably high since each carbohydrate monomer can accommodate multiple linkages and/or branches in its structure. Moreover, as the glycosidic linkage between each monosaccharide can have two anomeric configurations (α or β), even in small oligosaccharides, the potential number of structures that can exist is huge.

In the last decades, there has been a great effort to synthesize oligo- and poly-saccharides in the laboratory, mainly due to their key role in many biological events but also to the interest expressed by the food and technical industries. The chemical approaches to carbohydrate synthesis have been known since Arthur Michael first reported the synthesis of a natural glycoside in 1879[2]. However, the construction of complex carbohydrates and glycoconjugates in the laboratory remains a challenging endeavor. The causes for these difficulties are several but they mainly rely on the exceptional complexity and diversity that some compounds may show. Indeed, unlike the systematic processes of proteins and nucleic acids synthesis, in which the order of attachment of amino acids and nucleotides is read from a nucleic acid matrix, the synthesis of carbohydrates is a non-template-directed process that is controlled by a complex stereo- and regio-specific process. It requires a special regioselective reaction at a particular position of the sugar unit, in which the hydroxyl group that is available in such position must be distinguished from all the other hydroxyl groups in the structure that have similar properties. Additionally, the linkage between sugars must proceed through a stereoselective manner, since the linkage can produce two stereoisomers and one of them must be preferred to the other. Many carbohydrates are also found linked to protein and lipids. The synthesis of glycoconjugates has also proven to be a difficult task because it generally involves the participation of multiple transporters and enzymes. The mechanisms governing the regulation of these pathways are still being elucidated, but so far it has been found that the assembly of carbohydrates to proteins and lipids requires a specific chemistry that is far from being universal.

The production of oligosaccharides and polysaccharides has been deeply studied in the past decades and revealed to be, as expected, a challenging task[3]. In spite of the advances observed in organic chemistry, the chemical synthetic routes addressed to synthesize these compounds have proven to be inefficient in the majority of the cases. This happens because the preparation of complex oligosaccharides and polysaccharides require multiple protection/deprotection and purification steps, which often lead to a tedious and time-consuming process and normally result in poor yields. To overcome these limitations, the enzymatic synthesis rapidly gained more prominence. The attractiveness of enzymatic synthesis is that protecting groups are not required and the stereo- and regio-selectivity chemistry is always followed in the formation of the glycosidic linkages, in the majority of the cases.

Enzymatic formation and cleavage of the bond between two sugars or between a sugar and another group can occur by hydrolysis to give the free sugar (glycosidases), by transglycosylation to give a new glycoside (glycosyltransferases), by phosphorolysis to give the sugar-1-phosphate (phosphorylases) or by elimination to give unsaturated sugar products (lyases). Currently, glycosidases and glycosyltransferases are the major classes of biocatalysts that are available for the enzymatic synthesis of polysaccharides and oligosaccharides.[1] As the structure of lysozyme was first solved in 1965[4], glycosidases have long been the subject of structural biology studies in order to understand the molecular details of substrate recognition and of catalysis. As a result, about three quarters of the 113 known families of glycosidases have at least a structural representative. In contrast, progress in the structural biology of glycosyltransferases has been slower[1]. Part of the success in characterizing glycosidases is due to the high stability of these enzymes when compared with the glycosyltransferases and because they are very easy to isolate, being generally available from natural sources like seeds, micro-organisms or fungal cultures, as well as in higher organisms (typically plant seed, mollusks, etc)[5]. These facts have turned glycosidases into an attractive target for many industries involved in the food, the paper and pulp industry, as well as in organic chemistry, where glycosidases have proven to be extremely efficient catalysts, being capable of hydrolyzing the very stable glycosidic bonds in glycoconjugates, oligo- and poly-saccharides[6]. The importance of glycosidases has also attracted the attention of many pharmaceutical industries since they are involved in many biological processes such as cell-cell or cell-virus recognition, immune responses, cell growth, and viral and parasitic infections. Currently, they have been associated with many diseases, which result from the lack or dysfunction of a glycosidase and are used in the treatment of metabolic disorders, viral infections and even cancer.

Despite the current advances in the field and the exponential interest in glycosidases, many aspects of the mechanism of action of these enzymes remain hidden in the available experimental data, in particularly in the X-ray structures that figure in the protein databank. Taking this into account, we focus in this review in the current literature regarding the catalytic mechanisms of glycosidases.

2. Catalytic mechanism of glycosidases

Glycosidases (GH) are present in almost all living organisms (exceptions are some Archaeans and a few unicellular parasitic eukaryotes)[7,8] where they play diverse and different roles. Taking into account the diversity of reactions that they catalyze as well as amino acid sequence and folding, glycosidases have been classified in many different ways. According to The IUBMB (International Union of Biochemistry and Molecular Biology) glycosidases are classified based on their substrate specificity and/or their molecular mechanism.[9,10] However, this classification is far from gaining consensus. A necessary consequence of the EC classification scheme is that codes can be applied only to enzymes for which a function has been biochemically identified. Additionally, certain enzymes can catalyse reactions that fall in more than one class, which makes them bear more than one EC number. Furthermore, this classification does not reflect the structural features and evolutionary relations of enzymes. In order to overcome these limitations, a new type of classification was proposed based on the amino acid similarity within the protein. This new classification is available at the Carbohydrate-Active Enzymes database (CAZy - http://www.cazy.org/)[7,10] and provides a direct relationship between sequence and folding similarities, that can be found in 130 amino acid sequence-based families. Some families with apparently unrelated sequence similarities show some uniformity in their three-dimensional structures. In those cases, these structures have been assigned to the so-called "Clans", that have been numbered from A to N.[7] In general, GHs belonging to the families of the same clan have common ancestry, similar 3D structure and are characterized by an identical catalytic mechanism of action.[7,10-16]

The two most commonly employed mechanisms used by glycosidases to achieve glycosidic bond cleavage with overall inversion or retention of anomeric stereochemistry are shown schematically at Figure 1. These mechanisms can be generally divided in two main groups: the retaining GHs and the inverting GHs. [17,18] Generally, enzymes of the same family have the same mechanism (but not specificity) [41, 43, 63], and the only exception are the GH23 and GH97 families that combine retaining and inverting GHs.[19] Table 1 summarizes the information about all GHs discovered until now.

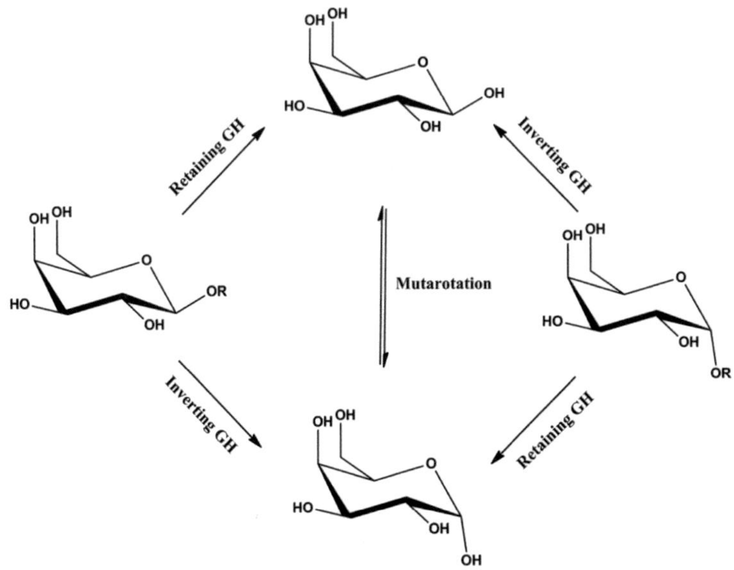

Figure 1. Mechanistic overview of retaining and inverting GHs activities.

Clan	Families (GHs)	Type of Mechanism	Three-dimensional structure
A [7,10,16,20-33]	1, 2, 5, 10, 17, 26, 30, 35, 39, 42, 50, 51, 53, 59, 72, 79, 86, 113, 128	Retaining	$(\beta/\alpha)_8$ barrel
B [34]	7, 16	Retaining	β-sandwich
C [35,36]	11, 12	Retaining	β-sandwich
D [37-43]	27, 31, 36	Retaining	$(\beta/\alpha)_8$ barrel
E [7,38]	33, 34, 83, 93	Retaining	6-bladed β-propeller
F [44,45]	43, 62	Inverting	5-bladed β-propeller
G [30]	37, 63	Inverting	$(\alpha/\alpha)_6$ barrel
H [46-52]	13, 70, 77	Retaining	$(\beta/\alpha)_8$ barrel
I [7,53-55]	24, 46, 80	Inverting	α+β lysozyme
J [26,44,45,56,57]	32, 68	Retaining	5-bladed β-propeller
K [7,25,28,30]	18, 20, 85	Retaining	$(\beta/\alpha)_8$ barrel
L [7,18,32]	15, 65, 125	Inverting	$(\alpha/\alpha)_6$ barrel
M [30]	8, 48	Inverting	$(\alpha/\alpha)_6$ barrel
N [30,32,58,59]	28, 49	Inverting	$(\beta)_3$ solenoid

Table 1. Organization of glycosidases families in clans and their correlation with the type of mechanism that they catalyze, and their protein folding.

2.1. Retaining GHs

The catalytic mechanism of retaining glycosidases was proposed about 58 years ago by Koshland *et al* [60] (Figure 2). According to this proposal, the mechanism occurs as a double displacement involving two steps: a glycosylation and a deglycosylation step. In the first step, the enzyme is glycosylated by the concerted action of the carboxylates of two residues, either Asp or Glu, or both that are found on opposite sides of the enzyme active site and are normally close to each other (around 5.5 Å). One of these residues functions as a general acid in the first step of the mechanism where the glycosidic bond starts to break. The acid residue donates a proton to the dissociated sugar. During the same step, the second deprotonated carboxylate acts as a nucleophile, attacking the anomeric carbon at the oxocarbenium ion-like transition state. This step, referred to as the glycosylation step, leads to the formation of a covalently linked glycosyl-enzyme intermediate that has an anomeric configuration opposite to that of the starting material. The second step of this reaction, the deglycosylation step, involves the hydrolytic breakdown of the glycosyl-enzyme intermediate[61]. The carboxylate that first functioned as an acid catalyst now acts as a base by abstracting a proton from the incoming nucleophile, usually a water molecule. Simultaneously, the water molecule attacks the carbohydrate-enzyme linkage in a reverse mode of the first step. At the end of the reaction, the enzymatic turnover is obtained and a hemi-acetal is formed with the same anomeric configuration as the starting material. Recent studies have shown that the transition states (TS1 and TS2) of both glycosylation and deglycosylation steps have a dissociative nature. Both reactions are favored by the distortion of the substrate during catalysis, but this effect is more evident in the first step of the reaction[62-65]. The glycosylation process is also favored by the hydrogen bond between the nucleophilic carboxylate and the hydroxyl group of position 2 in the substrate. It behaves almost as an anchor that aligns the substrate in the active site and facilitates the glycosylation process.

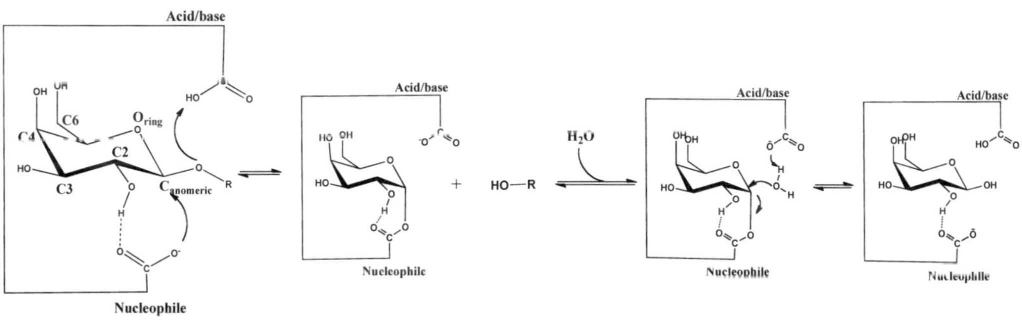

Figure 2. Catalytic mechanism of retaining GHs.

A variation of the general mechanism for retaining enzymes has been demonstrated for the N-acetyl-β-hexosaminidases, belonging to families 18 and 20[66,67]. Unlike the most retaining glycosidases, these enzymes lack a catalytic nucleophile, e.g. the water molecule. Instead, it is the acetamido substituent of the substrate that acts as an intramolecular catalytic nucleophile.[68-71] As it is shown in Figure 3, the general acid/base residue protonates the oxygen of the scissile glycosidic bond. The other charged carboxylate residue stabilizes the positive charge developed on the nitrogen of the oxazolinium ion that is formed after the intramolecular attack of the N-acetamido oxygen to the anomeric carbon.[68,72] To complete the double displacement mechanism, in the second step, an incoming water molecule attacks the anomeric carbon, resulting in a product with retention of the initial configuration.[73] In this reaction, several aromatic residues available in the active site have a key role to endorse the correct orientation of the nucleophilic carbonyl oxygen of the substrate and in this way promote and stabilize the formation of the oxazolinium ion.

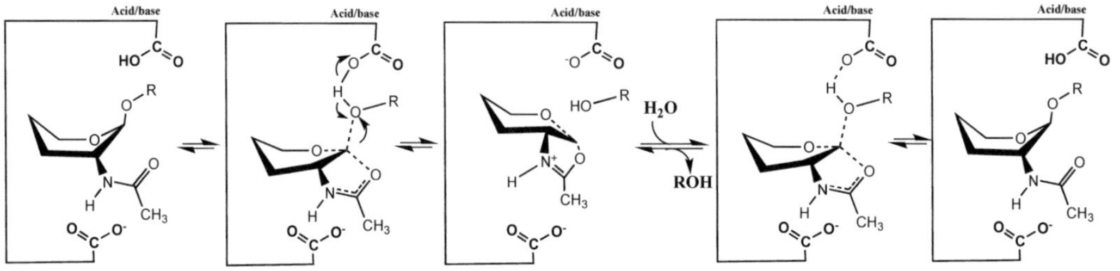

Figure 3. Catalytic mechanism of retaining β-hexosaminidase from families GH18 and GH20.

2.2. Inverting GHs

In inverting GHs, there is an inversion of the anomeric configuration of the starting material. Here, the two crucial carboxylic residues act as general acid and base catalysts and these groups are *circa* 10.5 Å apart from each other. It this specific case, this distance is larger than in retaining GHs because the substrate and the water molecule must be present simultaneously in the active site of the enzyme during the hydrolytic process.[18,66,74-78] Figure 4 shows the proposed mechanism of action for inverting GHs, which occurs via a single-displacement type of mechanism. In this case, one of the carboxylate residues protonates the scissile glycosidic oxygen atom while the other coordinates the nucleophile (i.e. the water molecule) to assist its deprotonation and in this way complete the hydrolysis reaction.[79]

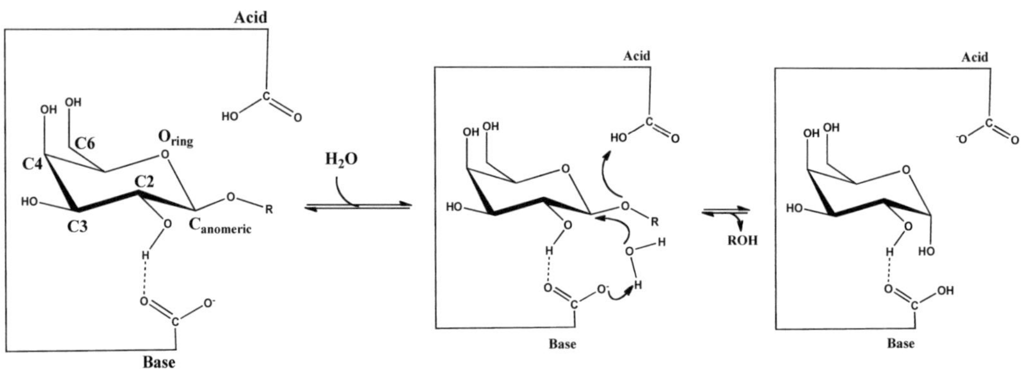

Figure 4. Catalytic mechanism of inverting GHs.

In contrast with the retaining mechanism, this reaction is completed in a single step and it is supposed that it requires the formation of a single transition state structure. Moreover, it does not involve the formation of any covalent enzyme intermediate during the course of catalysis and induces the inversion of the anomeric configuration of the starting material.

2.3. Cofactor dependent GHs

There are other GHs whose catalytic mechanism is substantially different from the mechanisms described above. One of the most interesting ones requires the presence of an NADH cofactor. The retaining 6-phospho-α-GH enzymes from family 4 are among these enzymes (Figure 5) in which the cofactor is perfectly positioned to remove the hydride from carbon C3 of the substrate[80]. Consequently, the acidity of the hydrogen atom that is attached to carbon C2 of the substrate increases and helps its abstraction by one of the tyrosine residues that is available in the active site and acts as a base. The hydroxyl group that is initially attached to carbon C3 of the substrate is subsequently oxidized to a ketone forming the 1,2-unsaturated reactive intermediate. Simultaneously, one of the carboxylate residues assists the cleavage of the sugar bound and the proton that is attached to carbon C2 of the substrate is abstracted by the base. This step is favoured by the presence of a metal ion in the active site of the enzyme that polarizes the carbonyl at carbon C3 and stabilizes the enolate species. The last step of this mechanism involves the nucleophilic attack of one water molecule to the double bond of the ketone. Simultaneously, the re-protonation of the enolate is catalysed by the close presence of the tyrosine residue, and the reduction of the ketone located at carbon C3 is accomplished by the NADH, favouring the enzymatic turnover.

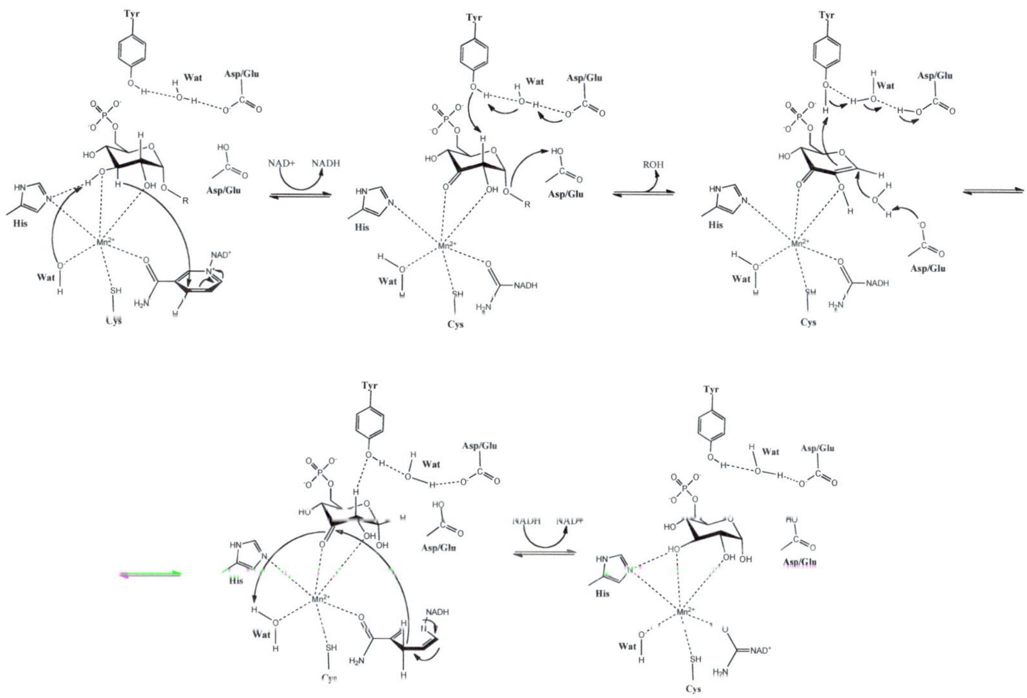

Figure 5. Catalytic mechanism for NADH dependent GHs.

2.4. Transglycosylation activity of GHs

In addition to the hydrolytic ability, GHs can also be used under appropriate conditions for the reverse reaction, thus promoting the formation of glycosidic linkages. This type of reactions are called transglycosylation[62] and generally require high concentration of substrate. The proposed catalytic mechanism is depicted in Figure 6. Similarly to the previous described mechanisms, the first step leads to the departure of the aglycon group and the formation of the covalent intermediate. The second step of the reaction involves the attack of the carbohydrate-enzyme linkage by another sugar molecule, and the proton transfer from the sugar to the active site acid/base carboxylate.

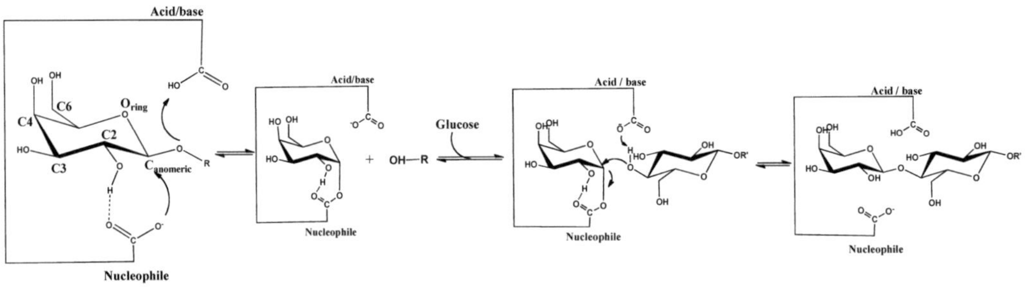

Figure 6. Transglycosylation reaction catalyzed by retaining GHs.

Usually, the synthesis of glycosidic linkages in nature is carried out natively by glycosyltransferases (that use activated glycosides as the glycosyl donors). Typical glycoside donors are expensive nucleotide sugars such as ADP-glucose, UDP-glucose, and UDP-galactose. In contrast, the transglycosylation activity of GHs employs a considerably inexpensive substrate (such as simple sugars) as a glycoside donor molecule leading to large industrial interest in employing these enzymes for biotechnological synthesis. However, the yields for these transglycosylation reactions are typically low because the product itself is a substrate for the enzyme and undergoes hydrolysis. As their hydrolytic activities compete with this mechanistic pathway, it is necessary to displace the equilibrium of the glycosidic bond formation using excessive substrate concentration (thermodynamic control) or using good activated glycosyl donors, such as an aryl glycoside (kinetic control). Another disadvantage of the transglycosylation reactions catalysed by GHs is their limited efficiency for the glycosides synthesis in disaccharides or trisaccharides.[3,81] This happens because these reactions require high degrees of chemo, regio and stereo-selectivity.[82] The same is also true for oligosaccharides, but in this case the problems arise from their complex structure turning their chemical synthesis difficult to achieve, namely the production of glycosides with a mixture of various linkages (i.e., formation of 1-2, 1-3, 1-4 and 1-6 glycosidic bonds) and both anomers (α and β).[83] In this regard, the control of the stereospecificity and the regiospecificity of bond formation remains a challenging problem in the chemical synthesis of oligosaccharides [84,85]. A solution for this unsolved problem would be very important, as industrially there is only interest in the oligosaccharide target.

In order to overcome most of the limitations of the transglycosylation reaction in glycosidases, many enzymes have been mutated in the region of the active site in order to enhance the rate of this reaction. A classical example of mutated glycosidases are the glycosynthases, in which the hydrolytic activity has been inactivated through the mutation of their catalytic nucleophile residues by small non-nucleophilic residues, such as alanine, glycine or serine These enzymes possess a high activity because they are able to accept an activated glycosyl donor group (generally glycosyl fluorides or nitrophenyl glycosides) and catalyse transglycosylation reactions to an acceptor molecule.[81,85] In opposition to the native GHs, these engineered enzymes produce carbohydrates with elevated molecular weight and with higher product yields.[81] The first glycosynthase enzyme was reported in 1998 by Withers and colleagues[86] but, currently many other glycosynthases have been developed that posses specific substrate specificity. Figure 7 shows the reaction mechanisms of several types of glycosynthases. As native glycosidases, the glycosynthases can also show retaining and inverting mechanism. In the inverting α-glycosynthases, the donor group is transferred

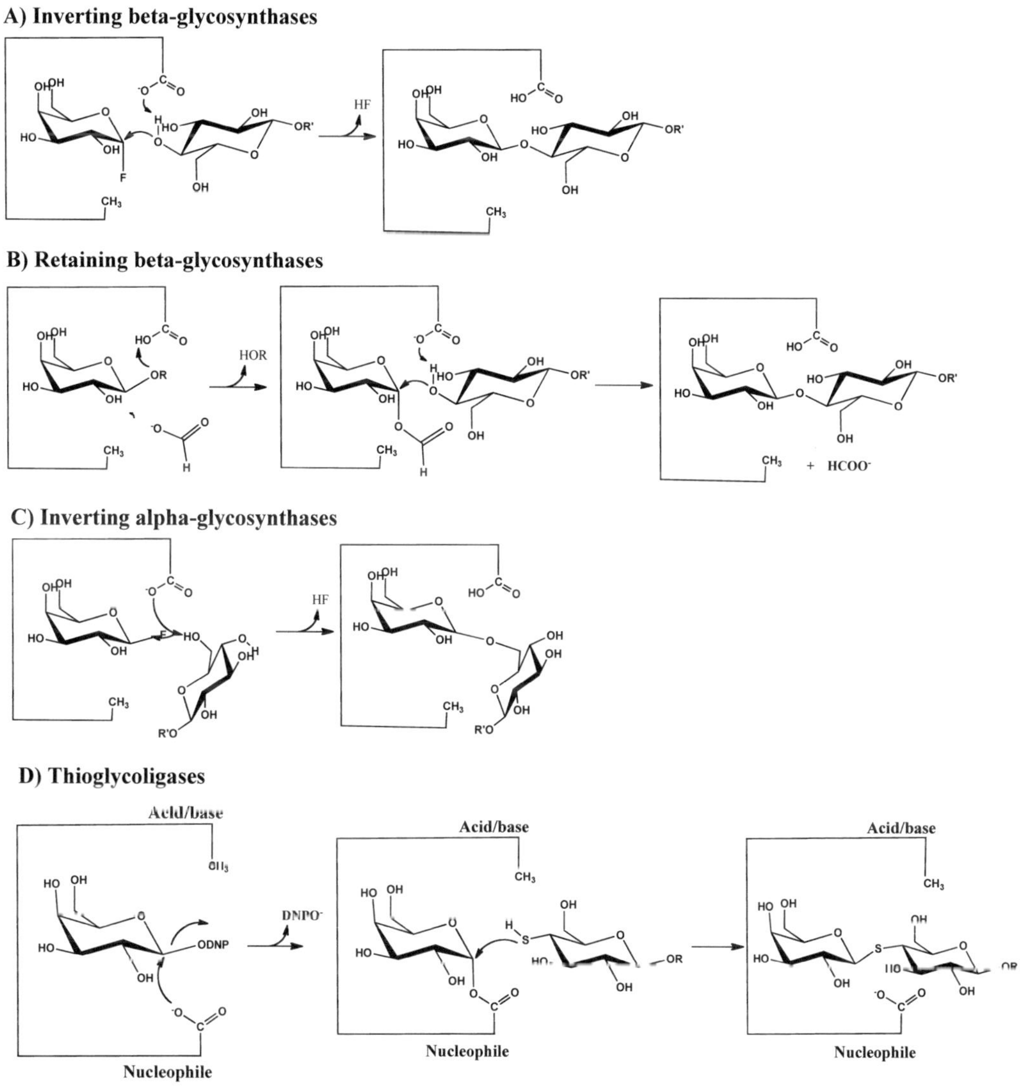

Figure 7. Catalytic mechanism of glycosynthases: a) inverting β-glycosynthase, b) retaining β-glycosynthase, c) inverting α-glycosynthase and d) thioglycoligase.

to the 4-nitrophenyl-α-glucoside acceptor group and the deglycosylation step proceeds similarly to what is observed with the retaining GHs (Figure 7a). The retaining glycosynthases act within the presence of one external nucleophile, such as sodium formate, and an activated donor group with the anomeric configuration of the native substrate (commonly 2-nitrophenyl- or 2,4-dinitrophenyl-β-glucoside) (Figure 7b). Therefore, the nucleophile mimics the catalytic active-site carboxylate of the enzyme and builds the formyl-glycoside intermediate. Subsequently, the donor carbohydrate is transferred to an acceptor sugar, promoting the polysaccharide synthesis.[85,87-90]

Some retaining glycosynthases can also have inverting mechanisms. This occurs when the donor sugar has a glucosyl fluoride in an opposite anomeric configuration relatively to the native substrate, thus mimicking the intermediate of the reaction (Figures 7a and 7c).[86]

Another type of glycosynthases are the thioglycoligase engineering enzymes, in which the mutated residue is the acid/base carboxylate instead of the nucleophile residue as in the previously described glycosynthases (Figure 7d). In these cases, a good leaving group such as dinitrophenyl, is placed in the substrate, which allows the formation of the glycosyl-enzyme intermediates that favors the catalytic process.[91]

3. Structural aspects that influence the GHs catalytic mechanism

The structural studies addressed at GHs have also provided important clues about how these enzymes enhance the catalytic process. As mentioned before, the distortion of the substrate along the full catalytic process is one of these mechanisms and this effect is found in many studies.[92-99] The available data reveals that GHs are able to selectively bind and stabilize high energy substrate conformations before the hydrolysis takes place. Such distortion of the substrate favours, in the Michaelis complex, the attack of the catalytic acid/base carboxylates to the glycosidic oxygen of the substrate. At the same time, it helps to guide the leaving group in a pseudoaxial position in relation to the substrate, facilitating the nucleophilic attack on carbon C1 and the subsequent cleavage of the glycosidic bond. It has been proposed also that the conformation changes of the ring along the catalytic process might determine the efficiency of the polysaccharides degradation. Taking into account these results and the available X-ray structures of several GHs that contain the substrates with different conformational distortions, Stoddart [100-102] proposed a diagram to classify the conformation of a α-glucopyranose molecule ring along the reaction pathway (Figure 8). In this diagram C, B, S corresponds to the chair, boat, and skew conformations, respectively. This diagram includes the most energetically stable $^4C_1$ chair, six boat-type and six skew-type conformations, as well as several transient structures (e.g. half-chair and envelope conformations) between the transition of $^4C_1$ chair to the boat/skew conformations.

The itinerary map of Stoddart gives therefore all the possible conformational pathways that a hypothetical substrate may follow as it moves from one conformation to another[94]. However, no energetically information can be extracted regarding the relative stability of different conformations, nor can it be assumed that all conformations on this map correspond to stationary points in the free energy landscape with respect to ring distortion.

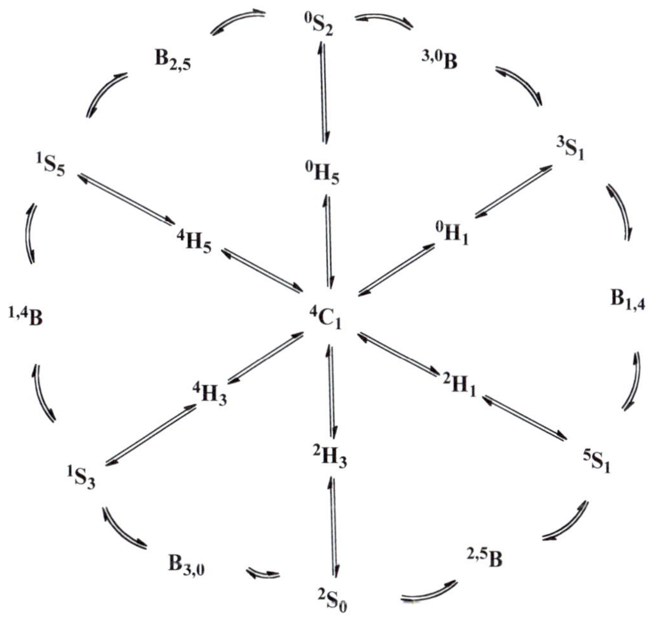

Figure 8. Stoddart's diagram.

Nevertheless, over the years, this diagram has been actively used as an "itinerary map" to design new enzyme inhibitors for therapeutic activities. In this regard, the conformational itinerary pathway of several GHs families has been studied, such as GH29 enzymes and α-xylosidases from the GH31 family that catalyze the hydrolysis using the $^4C_1 \leftrightarrow {}^3H_4 \leftrightarrow {}^3S_1$ glycosylation itinerary [93,103]; inverting endoglucanases from the GH8 family that use the β-$^2S_0/^{2,5}B \leftrightarrow {}^{2,5}B \leftrightarrow$ α-$^5S_1$ glycosylation itinerary of the glycon ring [104]; the glycosylation reaction of golgi α-mannosidase II from the GH38 family following an $^0S_2/B_{2,5} \leftrightarrow B_{2,5} \leftrightarrow {}^1S_5$ itinerary [105]; the catalytic itinerary of 1,3-1,4-β-glucanase from the GH16 family 16 pursue the $^{1,4}B/^1S_3 \leftrightarrow {}^4E/^4H_3 \leftrightarrow {}^4C_1$ [95,106] itinerary, and the inverting α-mannosidases from the GH47 family that follow the $^3S_1 \leftrightarrow {}^3H_4 \leftrightarrow {}^4C_1$ glycosylation itinerary.[107]

Many GHs also contain cations in the region of the active site. The presence of these species in the structure of GHs appears to be more common than it was initially imagined, and are believed to be very important for the stabilization of the transition states during catalysis. For instance, the Golgi α-mannosidase II from the GH38 family has a zinc ion in its 3D structure. Studies on the catalytic mechanism of this enzyme revealed that the Zn^{2+} ion is fundamental to coordinate the hydroxyl groups attached to carbons C2 and C3 of the mannosyl group, which stabilize the transition states, and thus reduces the overall activation energy required for the reaction. Furthermore, QM/MM metadynamics simulations also demonstrate that the zinc ion helps to lengthen the C2 hydroxyl bond when the substrate acquires the oxocarbenium character, facilitating the electron reduction of this species.[105] A similar role has been proposed for the calcium ion present in the structure of the endoplasmic reticulum α-mannosidase I from the GH47 family. The crystallographic structure shows that the cation coordinates with the hydroxyl groups that are attached to carbons C2 and C3 of 1-deoxymannojirimycin or kifunensin inhibitors.[108] A magnesium ion is also found close to the region of the active site of β-galactosidase from *Escherichia coli*. In these case, theoretical calculations have clearly demonstrated that the presence of the cation has a key role in lowering down the activation barrier by 14.9 kcal/mol, emphasizing its importance during the catalytic process.[63]

4. Conclusions

GHs are impressive nano-molecular machines that are present in almost all living organisms (exceptions are some Archaeans and a few unicellular parasitic eukaryotes). These enzymes catalyze the hydrolysis of the glycosidic linkage in a myriad of biological reactions and under specific conditions can also catalyse the reverse reaction promoting the formation of glycoside linkages.

The interest of GHs has started long ago but the catalytic power behind glycosidases is only now being established. Structural analyses of various enzyme complexes representing stable intermediates along the reaction coordinate together with detailed mechanistic and spectroscopic studies on wild type and mutant enzymes, have revealed that the source for their catalytic power is based on nucleophilic and general acid/base catalysis. These enzymes develop finely tuned active sites that contain two carboxylates residues (Asp and Glu) carefully aligned and positioned on opposite sides of the enzyme active site that embrace the substrate upon substrate binding. The active site also provides an extensive network of hydrogen bonds that endorse a conformational distortion of the substrate. This induces the substrate to adopt a higher energetic conformation before the hydrolysis takes place, and such configuration is maintained during the full catalytic process. This effect is very important for the stabilization of the transition-state structure and therefore to lower the activation barrier of the full process. Some GHs also possess positive ions (Zn^{2+}, Mn^{2+}, and Mg^{2+}) into their structure and these cations have been found also to be essential for the stabilization of the transition states during catalysis. The two most commonly employed mechanisms used to hydrolyze the glycoside linkage of substrates by glycosidases involve the retention or the inversion mechanisms. These mechanisms are conserved within the majority of GHs families. The only exception are the glycosidases from family 4 and 109, in which the hydrolytic process occurs via an elimination type of mechanism, and requires the presence of the NADH cofactor.

The structural and the mechanistic studies addressed to glycosidases provided and continue to provide important clues about the catalytic power of these enzymes. This knowledge is very important as it offers new ways to improve, modify or even inhibit the activity of GHs. These developments are particularly important for the biotechnology industries that have been increasing the commercial uses of glycosidases in several areas. Indeed, specific glycosidases are increasingly used for food processing, for bio-bleaching in the pulp and paper industry, as well as for biomass degradation with the potential to convert solid biomass into liquid fuels.

In the last two decades, it has also been watched an increasing interest of glycosidases for therapeutic uses. Glycosidases are involved in the biosynthesis of the oligosaccharide chains and quality control mechanisms in the endoplasmic reticulum of the N-linked glycoproteins. Inhibition of these glycosidases can have profound effects on quality control, maturation, transport, and secretion of glycoproteins and can alter cell–cell or cell–virus recognition processes. This principle is the basis for the potential use of glycosidase inhibitors in viral infection, diabetes and genetic disorders. [109,110]. Most of these drugs are glycosidase inhibitors that can bind and block the active sites of these enzymes. Some successful examples are the α-amylase inhibitor Acarbose and Miglitol that were approved by the FDA in 1990 and 1996, respectively, and are used to inhibit some of the intestinal glycosidases and pancreatic α-amylase in order to regulate the absorption of carbohydrates. These inhibitors are currently used therapeutically in the oral treatment of the non-insulin-dependent diabetes mellitus (typr II diabetes). Other glycosidases inhibitors are used as anti-viral agents[111]. Here the inhibitors were developed to inhibit the formation of glycoproteins of the viral envelopes, which are essential for virion assembly and secretion and/or infectivity. A successful example was the development of Zanamivir(Relenza) and Oseltamivir (Tamiglu), approved by FDA in 1999, that are used in the treatment and prophylaxis of influenza caused by influenza A virus and influenza B virus[112]. These compounds efficiently inhibit a glycosidases called neuraminidase. Glycosidases are also used in the therapy of human genetic disorders. The glycosphingolipid storage diseases (GSD, also glycogenosis and dextrinosis) are a rare hereditary disorders that are severe in nature and frequently fatal. These diseases result from defects in the processing of glycogen synthesis or breakdown within muscles, liver, and other cell types[113]. An example of such disorders is the Fabry disease that is caused by the deficiency of the essential enzyme α-glycosidase A, resulting in renal failure along with premature myocardial infarction and strokes. The only successful treatment is, to date, the enzyme replacement therapy. Fabrazyme was approved by FDA in 2003 and is intended to replace the missing enzyme in patients with this progressive disease.

Taking into account that almost two-thirds of all carbon that exist in the biosphere is carbohydrate, we believe that the current applications of GHs are only a small group of the many very important applications that these enzymes may find in the future. It is therefore expected that a wide variety of relevant and new applications will arise in the near future involving glycosidases. To stimulate these developments, the continuous study regarding glycosidases will be very important as they will provided crucial knowledge to turn their use more efficient and effective.

Author details

Natércia F. Brás, Pedro A. Fernandes, Maria J. Ramos and Nuno M.F.S.A. Cerqueira*
REQUIMTE, Departamento de Química e Bioquímica da Faculdade de Ciências,
Universidade do Porto, Porto, Portugal

* Corresponding Author

5. References

[1] Henrissat, B.; Sulzenbacher, G.; Bourne, Y. *Current Opinion in Structural Biology* 2008, *18*, 527.

[2] Michael, A. *Am. Chem. J.* 1879, *1879*, 305.

[3] Bucke, C. *J Chem Technol Biot* 1996, *67*, 217.

[4] Blake, C. C. F.; Koenig, D. F.; Mair, G. A.; North, A. C. T.; Phillips, D. C.; Sarma, V. R. *Nature* 1965, *206*, 757.

[5] Flitsch, S. L. Current Opinion in Chemical Biology 2000, 4, 619.

[6] Bojarova, P.; Kren, V. *Trends Biotechnol* 2009, *27*, 199.

[7] *Carbohydrate Active Enzymes server*, http://www.cazy.org/ 2012.

[8] Cantarel, B. L.; Coutinho, P. M.; Rancurel, C.; Bernard, T.; Lombard, V.; Henrissat, B. *Nucleic Acids Research* 2009, *37*, D233.

[9] IUBMB: Enzyme Nomenclature. Recommendations.; Academic Press: San Diego, 1992.

[10] Henrissat, B.; Davies, G. *Current Opinion in Structural Biology* 1997, *7*, 637.

[11] Davies, G. J. Biochemical Society Transactions 1998, 26, 167.

[12] Coutinho, P. M.; Henrissat, B. In *Recent Advances in Carbohydrate Bioengineering*; Gilbert, H. J., Davies, G. J., Henrissat, B., Svensson, B., Eds. 1999, p 3.

[13] Henrissat, B. Biochemical Society Transactions 1998, 26, 153.

[14] Gebler, J.; Gilkes, N. R.; Claeyssens, M.; Wilson, D. B.; Beguin, P.; Wakarchuk, W. W.; Kilburn, D. G.; Miller, R. C.; Warren, R. A. J.; Withers, S. G. *Journal of Biological Chemistry* 1992, *267*, 12559.

[15] Henrissat, B.; Bairoch, A. *Biochemical Journal* 1996, *316*, 695.

[16] Henrissat, B.; Callebaut, I.; Fabrega, S.; Lehn, P.; Mornon, J. P.; Davies, G. *Proceedings of the National Academy of Sciences of the United States of America* 1995, *92*, 7090.

[17] Davies, G. J., Sinnott, M.L., Withers, S.G. *Glycosyl transfer. In Comprehensive Biological Catalysis. Edited by Sinnott ML.*; Academic Press: London, 1997; Vol. 1.

[18] Vasella, A.; Davies, G. J.; Bohm, M. *Current Opinion in Chemical Biology* 2002, *6*, 619.

[19] Gloster, T. M.; Turkenburg, J. P.; Potts, J. R.; Henrissat, B.; Davies, G. J. *Chemistry & Biology* 2008, *15*, 1058.

[20] Rabinovich, M. L.; Melnick, M. S.; Bolobova, A. V. *Biochemistry-Moscow* 2002, *67*, 850.

[21] Moller, P. L.; Jorgensen, F.; Hansen, O. C.; Madsen, S. M.; Stougaard, P. *Applied and Environmental Microbiology* 2001, *67*, 2276.

[22] Henrissat, B.; Teeri, T. T.; Warren, R. A. J. *Febs Letters* 1998, *425*, 352.

[23] Mian, I. S. Blood Cells Molecules and Diseases 1998, 24, 83.

[24] Juers, D. H.; Huber, R. E.; Matthews, B. W. *Protein Science* 1999, *8*, 122.

[25] Nagano, N.; Porter, C. T.; Thornton, J. M. *Protein Engineering* 2001, *14*, 845.

[26] Henrissat, B.; Bairoch, A. *Biochemical Journal* 1993, *293*, 781.

[27] Himmel, M. E.; Karplus, P. A.; Sakon, J.; Adney, W. S.; Baker, J. O.; Thomas, S. R. *Applied Biochemistry and Biotechnology* 1997, *63-5*, 315.

[28] Rigden, D. J.; Jedrzejas, M. J.; de Mello, L. V. *Febs Letters* 2003, *544*, 103.

[29] Jenkins, J.; Leggio, L. L.; Harris, G.; Pickersgill, R. *Febs Letters* 1995, *362*, 281.

[30] Pickersgill, R.; Harris, G.; Lo Leggio, L.; Mayans, O.; Jenkins, J. *Biochemical Society Transactions* 1998, *26*, 190.

[31] St John, F. J.; Gonzalez, J. M.; Pozharski, E. *Febs Letters* 2010, *584*, 4435.

[32] Stam, M. R.; Blanc, E.; Coutinho, P. M.; Henrissat, B. *Carbohydrate Research* 2005, *340*, 2728.

[33] Sakamoto, Y.; Nakade, K.; Konno, N. *Applied and Environmental Microbiology* 2011, *77*, 8350.

[34] Divne, C.; Stahlberg, J.; Reinikainen, T.; Ruohonen, L.; Pettersson, G.; Knowles, J. K. C.; Teeri, T. T.; Jones, T. A. *Science* 1994, *265*, 524.

[35] Tomme, P.; Warren, R. A. J.; Miller, R. C.; Kilburn, D. G.; Gilkes, N. R. In *Enzymatic Degradation of Insoluble Carbohydrates*; Saddler, J. N., Penner, M. H., Eds. 1995; Vol. 618, p 142.

[36] Torronen, A.; Kubicek, C. P.; Henrissat, B. *Febs Letters* 1993, *321*, 135.

[37] Naumoff, D. G. *Molecular Biology* 2004, *38*, 388.

[38] Henrissat, B. *Biochemical Journal* 1991, *280*, 309.

[39] Henrissat, B.; Romeu, A. *Biochemical Journal* 1995, *311*, 350.

[40] Naumoff, D. G.; Carreras, M. *Molecular Biology* 2009, *43*, 652.

[41] Dagnall, B. H.; Paulsen, I. T.; Saier, M. H. *Biochemical Journal* 1995, *311*, 349.

[42] MargollesClark, E.; Tenkanen, M.; Luonteri, E.; Penttila, M. *European Journal of Biochemistry* 1996, *240*, 104.

[43] Liebl, W.; Wagner, B.; Schellhase, J. *Systematic and Applied Microbiology* 1998, *21*, 1.

[44] Naumoff, D. G. Proteins-Structure Function and Bioinformatics 2001, 42, 66.

[45] Pons, T.; Naumoff, D. G.; Martinez-Fleites, C.; Hernandez, L. *Proteins-Structure Function and Bioinformatics* 2004, *54*, 424.

[46] Naumoff, D. G.; Rassb The alpha-galactosidase superfamily: Sequence based classification of alpha-galactosidases and related glycosidases, 2001.

[47] Ferretti, J. J.; Gilpin, M. L.; Russell, R. R. B. *Journal of Bacteriology* 1987, *169*, 4271.

[48] Rojas, A.; Garcia-Vallve, S.; Palau, J.; Romeu, A. *Biologia* 1999, *54*, 255.

[49] MacGregor, E. A.; Jespersen, H. M.; Svensson, B. *Febs Letters* 1996, *378*, 263.

[50] MacGregor, E. A.; Janecek, S.; Svensson, B. Biochimica Et Biophysica Acta-Protein Structure and Molecular Enzymology 2001, 1546, 1.

[51] Mooser, G.; Hefta, S. A.; Paxton, R. J.; Shively, J. E.; Lee, T. D. *Journal of Biological Chemistry* 1991, *266*, 8916.

[52] Janecek, S. *Biologia* 2005, *60*, 177.

[53] Monzingo, A. F.; Marcotte, E. M.; Hart, P. J.; Robertus, J. D. *Nature Structural Biology* 1996, *3*, 133.

[54] Wohlkonig, A.; Huet, J.; Looze, Y.; Wintjens, R. *Plos One* 2010, *5*.

[55] Tremblay, H.; Blanchard, J.; Brzezinski, R. *Canadian Journal of Microbiology* 2000, *46*, 952.

[56] Naumov, D. G.; Doroshenko, V. G. *Molecular Biology* 1998, *32*, 761.

[57] Pons, T.; Olmea, O.; Chinea, G.; Beldarrain, A.; Marquez, G.; Acosta, N.; Rodriguez, L.; Valencia, A. *Proteins-Structure Function and Genetics* 1998, *33*, 383.

[58] Jenkins, J.; Mayans, O.; Pickersgill, R. *Journal of Structural Biology* 1998, *122*, 236.

[59] Rigden, D. J.; Franco, O. L. *Febs Letters* 2002, *530*, 225.

[60] Koshland, D. E.; Stein, S. S. *Journal of Biological Chemistry* 1954, *208*, 139.
[61] Rempel, B. P.; Withers, S. G. *Glycobiology* 2008, *18*, 570.
[62] Bras, N. F.; Fernandes, P. A.; Ramos, M. J. *Theoretical Chemistry Accounts* 2009, *122*, 283.
[63] Bras, N. F.; Fernandes, P. A.; Ramos, M. J. *Journal of Chemical Theory and Computation* 2010, *6*, 421.
[64] Bras, N. F.; Moura-Tamames, S. A.; Fernandes, P. A.; Ramos, M. J. *Journal of Computational Chemistry* 2008, *29*, 2565.
[65] Bras, N. F.; Ramos, M. J.; Fernandes, P. A. *Journal of Molecular Structure-Theochem* 2010, *946*, 125.
[66] Davies, G.; Henrissat, B. *Structure* 1995, *3*, 853.
[67] Jefferson, T.; Jones, M. A.; Doshi, P.; Del Mar, C. B.; Heneghan, C. J.; Hama, R.; Thompson, M. J. *Cochrane Db Syst Rev* 2012.
[68] Passos, O.; Fernandes, P. A.; Ramos, M. J. *Journal of Physical Chemistry B* 2011, *115*, 14751.
[69] Passos, O.; Fernandes, P. A.; Ramos, M. J. *Theoretical Chemistry Accounts* 2011, *129*, 119.
[70] Bottoni, A.; Miscione, G. P.; Calvaresi, M. *Physical Chemistry Chemical Physics* 2011, *13*, 9568.
[71] He, Y.; Macauley, M. S.; Stubbs, K. A.; Vocadlo, D. J.; Davies, G. J. *Journal of the American Chemical Society* 2010, *132*, 1807.
[72] Mark, B. L.; Mahuran, D. J.; Cherney, M. M.; Zhao, D. L.; Knapp, S.; James, M. N. G. *Journal of Molecular Biology* 2003, *327*, 1093.
[73] Mark, B. L.; Vocadlo, D. J.; Knapp, S.; Triggs-Raine, B. L.; Withers, S. G.; James, M. N. G. *Journal of Biological Chemistry* 2001, *276*, 10330.
[74] Krasikov, V. V.; Karelov, D. V.; Firsov, L. M. *Biochemistry-Moscow* 2001, *66*, 267.
[75] Sinnott, M. L. *Chemical Reviews* 1990, *90*, 1171.
[76] Wang, Q. P.; Graham, R. W.; Trimbur, D.; Warren, R. A. J.; Withers, S. G. *Journal of the American Chemical Society* 1994, *116*, 11594.
[77] Rye, C. S.; Withers, S. G. *Current Opinion in Chemical Biology* 2000, *4*, 573.
[78] Zechel, D. L.; Withers, S. G. *Accounts of Chemical Research* 2000, *33*, 11.
[79] Koshland, D. E. Biological Reviews of the Cambridge Philosophical Society 1953, 28, 416.
[80] Rajan, S. S.; Yang, X. J.; Collart, F.; Yip, V. L. Y.; Withers, S. G.; Varrot, A.; Thompson, J.; Davies, G. J.; Anderson, W. F. *Structure* 2004, *12*, 1619.
[81] Faijes, M.; Planas, A. *Carbohydrate Research* 2007, *342*, 1581.
[82] Crout, D. H. G.; Vic, G. *Current Opinion in Chemical Biology* 1998, *2*, 98.
[83] Ajisaka, K.; Yamamoto, Y. Trends in Glycoscience and Glycotechnology 2002, 14, 1.
[84] Mayer, C.; Jakeman, D. L.; Mah, M.; Karjala, G.; Gal, L.; Warren, R. A. J.; Withers, S. G. *Chemistry & Biology* 2001, *8*, 437.
[85] Perugino, G.; Trincone, A.; Rossi, M.; Moracci, M. *Trends Biotechnol* 2004, *22*, 31.
[86] Mackenzie, L. F.; Wang, Q. P.; Warren, R. A. J.; Withers, S. G. *Journal of the American Chemical Society* 1998, *120*, 5583.
[87] Viladot, J. L.; de Ramon, E.; Durany, O.; Planas, A. *Biochemistry* 1998, *37*, 11332.

[88] Moracci, M.; Trincone, A.; Perugino, G.; Ciaramella, M.; Rossi, M. *Biochemistry* 1998, *37*, 17262.
[89] Trincone, A.; Perugino, G.; Rossi, M.; Moracci, M. *Bioorganic & Medicinal Chemistry Letters* 2000, *10*, 365.
[90] Perugino, G.; Trincone, A.; Giordano, A.; van der Oost, J.; Kaper, T.; Rossi, M.; Moracci, M. *Biochemistry* 2003, *42*, 8484.
[91] Witczak, Z. J. *Current Medicinal Chemistry* 1999, *6*, 165.
[92] Davies, G. J.; Planas, A.; Rovira, C. *Accounts of Chemical Research* 2012, *45*, 308.
[93] Ducros, V. M. A.; Zechel, D. L.; Murshudov, G. N.; Gilbert, H. J.; Szabo, L.; Stoll, D.; Withers, S. G.; Davies, G. J. *Angewandte Chemie-International Edition* 2002, *41*, 2824.
[94] Biarnes, X.; Ardevol, A.; Planas, A.; Rovira, C.; Laio, A.; Parrinello, M. *Journal of the American Chemical Society* 2007, *129*, 10686.
[95] Biarnes, X.; Nieto, J.; Planas, A.; Rovira, C. *Journal of Biological Chemistry* 2006, *281*, 1432.
[96] Mulakala, C.; Nerinckx, W.; Reilly, P. J. *Carbohydrate Research* 2006, *341*, 2233.
[97] Tailford, L. E.; Offen, W. A.; Smith, N. L.; Dumon, C.; Morland, C.; Gratien, J.; Heck, M. P.; Stick, R. V.; Bleriot, Y.; Vasella, A.; Gilbert, H. J.; Davies, G. J. *Nature Chemical Biology* 2008, *4*, 306.
[98] Greig, I. R.; Zahariev, F.; Withers, S. G. *Journal of the American Chemical Society* 2008, *130*, 17620.
[99] Soliman, M. E. S.; Ruggiero, G. D.; Pernia, J. J. R.; Greig, I. R.; Williams, I. H. *Organic & Biomolecular Chemistry* 2009, *7*, 460.
[100] Stoddart, J. F. *In Stereochemistry of Carbohydrates.*; Wiley-Interscience: Toronto, Canada, 1971.
[101] Davies, G. J.; Ducros, V. M. A.; Varrot, A.; Zechel, D. L. *Biochemical Society Transactions* 2003, *31*, 523.
[102] Taylor, E. J.; Goyal, A.; Guerreiro, C.; Prates, J. A. M.; Money, V. A.; Ferry, N.; Morland, C.; Planas, A.; Macdonald, J. A.; Stick, R. V.; Gilbert, H. J.; Fontes, C.; Davies, G. J. *Journal of Biological Chemistry* 2005, *280*, 32761.
[103] van Bueren, A. L.; Ardevol, A.; Fayers-Kerr, J.; Luo, B.; Zhang, Y. M.; Sollogoub, M.; Bleriot, Y.; Rovira, C.; Davies, G. J. *Journal of the American Chemical Society* 2010, *132*, 1804.
[104] Petersen, L.; Ardevol, A.; Rovira, C.; Reilly, P. J. *Journal of Physical Chemistry B* 2009, *113*, 7331.
[105] Petersen, L.; Ardevol, A.; Rovira, C.; Reilly, P. J. *Journal of the American Chemical Society* 2010, *132*, 8291.
[106] Biarnes, X.; Ardevol, A.; Iglesias-Fernandez, J.; Planas, A.; Rovira, C. *Journal of the American Chemical Society* 2011, *133*, 20301.
[107] Karaveg, K.; Siriwardena, A.; Tempel, W.; Liu, Z. J.; Glushka, J.; Wang, B. C.; Moremen, K. W. *Journal of Biological Chemistry* 2005, *280*, 16197.
[108] Vallee, F.; Karaveg, K.; Herscovics, A.; Moremen, K. W.; Howell, P. L. *Journal of Biological Chemistry* 2000, *275*, 41287.
[109] vonItzstein, M.; Colman, P. *Current Opinion in Structural Biology* 1996, *6*, 703.
[110] Taylor, G. Current Opinion in Structural Biology 1996, 6, 830.

[111] Asano, N. *Glycobiology* 2003, *13*, 93R.
[112] von Itzstein, M. *Nat Rev Drug Discov* 2007, *6*, 967.
[113] Aerts, J. M.; Hollak, C.; Boot, R.; Groener, A. *Philos T Roy Soc B* 2003, *358*, 905.

Section 2

Microbiology and Immunology

Chapter 7

DC-SIGN Antagonists – A Paradigm of C-Type Lectin Binding Inhibition

Marko Anderluh

Additional information is available at the end of the chapter

http://dx.doi.org/10.5772/50627

1. Introduction

Carbohydrates have rarely been a matter of research by medicinal chemistry-oriented scientists in the past, but has recently gained substantially more publicity [1]. Although equally or even more abundant in nature than peptides/proteins or nucleic acids, they were often simply neglected as potential drug targets and/or drug leads or even drug candidates. There are a number of reasons for this. Glycans – complex carbohydrates linked to proteins or lipids are essential components of every cell surface as all cells are coated with a so-called glycocalyx layer. These cell surface carbohydrates allow for intercellular communication by binding to the carbohydrate-binding proteins (CBPs). Many physiological and even pathophysiological processes like pathogen-cell contact rely upon these interactions. Currently, more than 80 human CBPs have been identified [2,3], so one might immediately recognise CBPs as promising targets for anti-infective therapy. However, only a few of them have been thoroughly studied and as a result, few CBPs have been recognised and validated as drug targets. Furthermore, CBPs' affinity is generally weak per monosaccharide unit, so CBPs usually form strong interactions by binding massive glycans that bear several hundred terminal carbohydrate units, many of which form contact with a single CBP or even a cluster of CBPs on the cellular surface. To inhibit glycan-protein interaction efficiently, one would have to consider designing and synthesizing multivalent carbohydrate ligands. This is by no means an easy task, and makes CBPs relatively unattractive targets in terms of druggability. Even if one would consider producing multivalent carbohydrates with optimal pharmacodynamic properties, they possess unattractive pharmacokinetic properties. Carbohydrates, especially large multivalent ones, share some intrinsic pharmacokinetic shortcomings as drugs; they are rapidly digested in the gut in most cases, and even if they survive the gut metabolism obstacles, they are practically unable to passively diffuse through the enterocyte layer in the small intestine. This inevitably means that no oral application can be guaranteed for a carbohydrate drug, which is the preferred application route. Once in the body, carbohydrates are mostly a source of energy, but are also excreted rapidly with glomerular filtration. Due to these drawbacks, carbohydrates or carbohydrate-derived drugs were often considered unappealing even before their design would take place.

To tackle the first drawback, advances in CBP biochemistry has led to a progressive gain of knowledge on CBPs. A vast international research network named *Consortium for Functional Glycomics* (CFG, http://www.functionalglycomics.org/static/consortium/consortium.shtml) was founded in 2001 with the ultimate goal of enlightening the role of glycans and glycan-binding proteins. To date, hundreds of members of this network as well as outside researchers have revealed the secrets of CBPs and many are now considered potential drug targets: their specific ligands were discovered with screening on large glycoarrays and their binding sites were mostly elucidated [3]. Understanding the 3D structures of CBPs and their binding modes is indispensable for drug design. In order to offer a central repository of knowledge available about CBPs, PROCARB has been constructed [4]. PROCARB is a database of known and modelled carbohydrate-binding protein structures with sequence-based prediction tools, and is a single resource where all the relevant information about a pair of interacting proteins and carbohydrates is available. A similar, although a bit less appealing and informative database on lectin 3D structure is the database of lectins of *Centre de Recherches sur les Macromolécules Végétales* (CERMAV), a French fundamental research centre devoted to glycosciences (http://www.cermav.cnrs.fr/lectines/). Glycoscience.de (http://www.glycosciences.de/) is a German internet portal to support glycomics and glycobiology research. These web sites offer proof that the scientific community has recognised the potential of CBPs. Accordingly, many national and international associations have been involved in deciphering CBPs fundamental roles in human pathophysiological processes, which inevitably leads to on-going cognizance of how CBPs might be exploited to help develop novel therapeutic approaches for treatment of human diseases.

CBPs may be classified primarily into two broad categories: lectins (which are further subdivided into specific types, like C-, H-, I-, L-, P-, R-, and S-type lectins) and sulphated glycosaminoglycan-binding proteins [3]. Extracellular lectins are the most promising drug targets among CBPs and CFG divides them into 4 specific groups: C-type lectins, galectins (or S-type lectins), siglecs (a subclass of I-type lectins that bind sialic acid) and other. Out of 174 total CBPs records on the CFG webpage, 120 of them are classified as C-type lectins, which in itself highlights the importance of this class of CBPs. Like all extracellular lectins, C-type lectins bind terminal carbohydrate epitopes of glycans that originate from either pathogens or other cells. DC-SIGN, (Dendritic Cell-Specific Intercellular adhesion molecule-3-Grabbing Non-integrin) is a type II C-type lectin that functions as an adhesion molecule located exclusively on dendritic cells (DCs). Originally discovered in 1992 [5], it was defined as a C-type lectin capable of binding the HIV-1 envelope glycoprotein gp120, but it took time until Steinman [6] and van Kooyk [7] unravelled its specific role in immunology and pathology and revealed the information to the broader scientific community. Since then, it has been shown to have a major role in primary immune response [8], but it also enhances several pathogens (like HIV, Ebola) infection of T cells and other cells of the immune system

[7,9]. This makes DC-SIGN a very interesting CBP and a target of interest towards novel immunomodulatory and anti-infective therapy. DC-SIGN function can be modulated by small molecules termed DC-SIGN antagonists that bind to DC-SIGN and prevent binding of native DC-SIGN ligands. These molecules could act as novel anti-infectives, and are currently in the early phase of drug development. Latest *in vitro* studies demonstrate that DC-SIGN antagonists effectively block the transmission of pathogens like HIV-1 and Ebola to CD4+ T cells [10]. Although DC-SIGN has not been validated *in vivo* as a drug target yet, DC-SIGN antagonists are a fruitful example of how inhibition of a C-type lectin function might be achieved by small synthetic molecules. As exposed before, CBPs' affinity is generally weak per carbohydrate unit, so the design of molecules that bind to DC-SIGN with high affinity is a demanding assignment. In general, the affinity issue and pharmacokinetic drawbacks of carbohydrates may be overcome by two predominant strategies implied in the design of DC-SIGN antagonists: screening of non-carbohydrate compounds to obtain ligands that are completely devoid of carbohydrate nature [11] and the design of glycomimetics, the compounds designed based on carbohydrate leads which usually still retain some or even a significant degree of carbohydrate nature [3,12]. DC-SIGN binds mannose- and fucose-based oligo- and polysaccharides, so their glycomimetics have been designed and proved to inhibit pathogen-DC-SIGN interaction potently. The author foresees that the approach used to design glycomimetic DC-SIGN antagonists is of general applicability when designing lectin antagonists and will be the thoroughly presented in the present chapter.

2.1. C-type lectins

Probably the largest type or family of lectins is the C-type found on animal (and human) cells [13]. This family includes several endocytic receptors and proteoglycans, and all collectins and selectins identified to date. They are of paramount importance for normal function of the immune system, as they mediate innate immunity, inflammation and immunity to tumour and virally infected cells. Although these CBPs vary greatly in structure among themselves, they have in common a domain, named C-type lectin-like domain (CTLD) or carbohydrate recognition domain (CRD). It is a compact globular structure responsible for selective binding of terminal units found in large carbohydrates [14]. The unique structural hallmark of such a domain is that it binds a carbohydrate by Ca^{2+} ion, which acts as a bridge between the protein and the "core monosaccharide" unit through complex interactions with sugar hydroxyl groups [15]. Namely, the binding with Ca^{2+} ion involves just one terminal saccharide unit – the "core monosaccharide" (as illustrated in Figure 1 for two distinct C-type lectins, the mannose-binding protein and DC-SIGN), while other ligand carbohydrate units (if present) form structural and bonding complementarity with the CRD. Several amino acid residues of the CRD offer 6 coordinate bonds for a Ca^{2+} ion and the carbohydrate donates 2 coordinate bonds with its hydroxyls, so that the Ca^{2+} ion is octacoordinated [16]. Distinct CRD aminoacid residues form hydrogen bonds with other hydroxyl groups on the carbohydrate directly or through water bridges. Changes in the amino acid residues that interact with the "core monosaccharide" modify the carbohydrate-binding specificity of the lectin, so that specific carbohydrate is recognised.

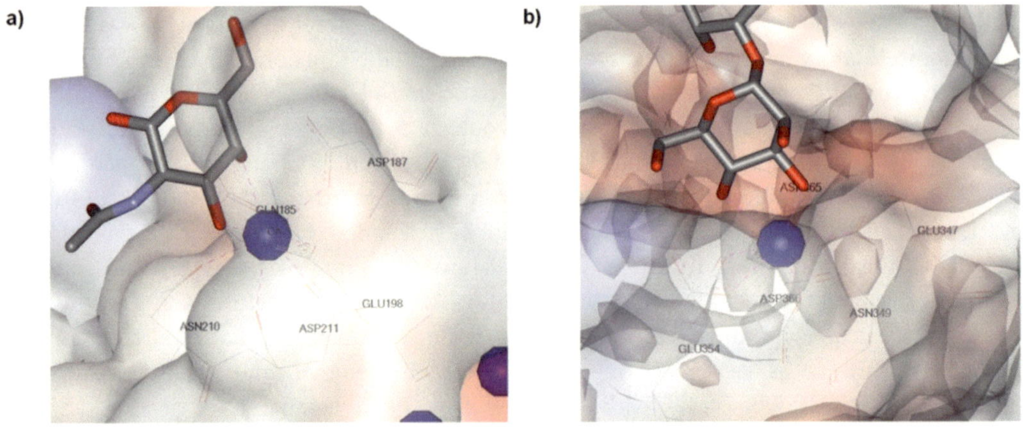

Figure 1. Binding mode of 2 C-type lectin CRDs with the "core monosaccharide". a) Crystal structure of mannose-binding protein with *N*-acetylgalactosamine (GalNAc) in the binding site (PDB code: 1AFB) [17]. b) Crystal structure of DC-SIGN in complex with Man_4 tetrasaccharide (PDB code: 1SL4) [18]. The 3- and 4-OH groups of the "core monosaccharide" (sticks rendering, without hydrogens) directly coordinate Ca^{2+} (blue sphere) and form hydrogen bonds with amino acids that also serve as Ca^{2+} ligands. Ca^{2+} is octacoordinated (violet broken line). For the sake of clarity, proteins are presented as transparent surfaces with amino acid residues (thin sticks) that coordinate Ca^{2+}.

The free energy of such interactions is relatively weak per carbohydrate unit, as we have to take into account high desolvatation penalties of numerous hydroxyls upon carbohydrate binding. The C-type lectin family however has means to obtain high binding affinity, or better, avidity; some C-type lectins oligomerize in order to promote high avidity for specific glycans [19]. The clustering of the CRDs influences not only avidity, but also the lectin selectivity, since each individual CRD can act independently to bind end mono- or oligosaccharide moiety [20]. Although the majority of lectins contain a single C-type CRD, the macrophage mannose receptor has multiple independent CRDs in a single polypeptide. The adjacent CRDs in the mannose receptor may help to direct its binding to specific multivalent, mannose-containing glycans [21].

2.2. DC-SIGN function and structure

DC-SIGN (Dendritic Cell-Specific Intercellular adhesion molecule-3-Grabbing Non-integrin) is a C-type lectin that functions as an adhesion molecule expressed specifically by dendritic cells (DCs), a class of professional antigen presenting cells (APCs). The intrinsic role of DCs is guidance of adaptive immune responses, since they are the major APCs that capture, process and present antigens [22]. DC-SIGN is a specific pattern recognition receptor (PRR) that recognizes distinct molecular patterns (PAMPs – Pathogen-Associated Molecular Patterns) of a number of pathogens [23]. It induces intracellular signalling pathways and triggers DC maturation upon pathogen binding [8,9]. To promote efficient transport of DCs towards effector cells – T cells, DC-SIGN binds human adhesion molecules ICAM-2 on vascular and lymphoid endothelium and enables cell interactions during DC migration [24]. Furthermore, DC-SIGN binding to ICAM-3 allows early antigen-nonspecific contact between DC and T cells in the lymph nodes, enabling T cell receptor engagement by stabilizing the DC-T-cell interaction [25]. Thus, we may conclude that DC-SIGN enables some of the normal DC functions by binding endogenous ligands, but also modulates immune responses to diverse pathogens via its ability to induce antigen-specific intracellular signalling.

DC-SIGN is a transmembrane C-type lectin that consists of one CRD, which defines the ligand specificity of the receptor, a neck region composed of seven complete and one incomplete tandem repeats, and a transmembrane region followed by a cytoplasmic tail containing recycling, internalization and intracellular signalling motifs (Figure 2) [9].

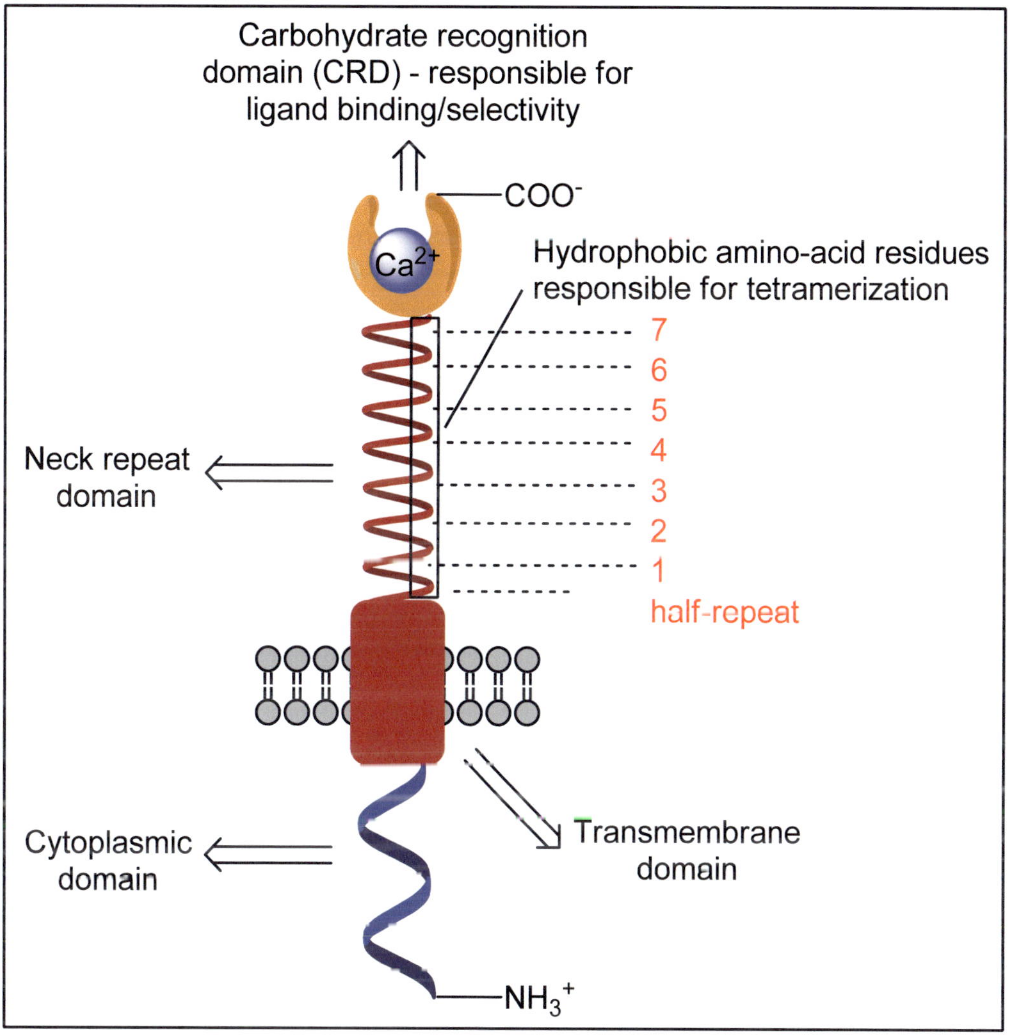

Figure 2. DC-SIGN schematic structure: Ca^{2+} CRD responsible for ligand binding/selectivity, a neck region composed of seven complete and one incomplete repeats, and a transmembrane region followed by a cytoplasmic tail (modified from Švajger et al.) [9].

Like many other C-type lectins, DC-SIGN tetramerizes to increase binding affinity and oligomerization occurs through association of the DC-SIGN neck domain (Figure 3) [26]. The oligomerization status of the DC-SIGN and related C-type lectins depends on the number of helical repeats of the neck region; at least 6 repeats are needed for tetramerization [27]. The stacked CRDs of tetramerized DC-SIGN provides the means of increasing the specificity for multiple repeating (oligo)saccharide units on host molecules, thereby defining the set of pathogens or endogenous molecules that are recognized by DC-SIGN. DC-SIGN neck domain, while allowing stacking and tetramerization, plays a central role as a pH-sensor that balances the equilibrium between the monomeric and tetrameric states of DC-SIGN [28]. In this sense, affinity for carbohydrates may be changed markedly by changing the pH, which helps DC-SIGN to realize its native function. Namely, upon binding, pathogen particles are internalized and further degraded into smaller particles and conjugated to MHC class-II proteins. As degradation proceeds in an acidic endosomal environment, the acid-triggered pathogen release from DC-SIGN is needed for successful degradation. Apart from tetramerization, DC-SIGN forms clusters that organize in membrane microdomains [29]. This organization on the plasma membrane is important for the binding and internalization, suggesting that clustered assemblies act as functional docking sites for pathogens (Figure 3).

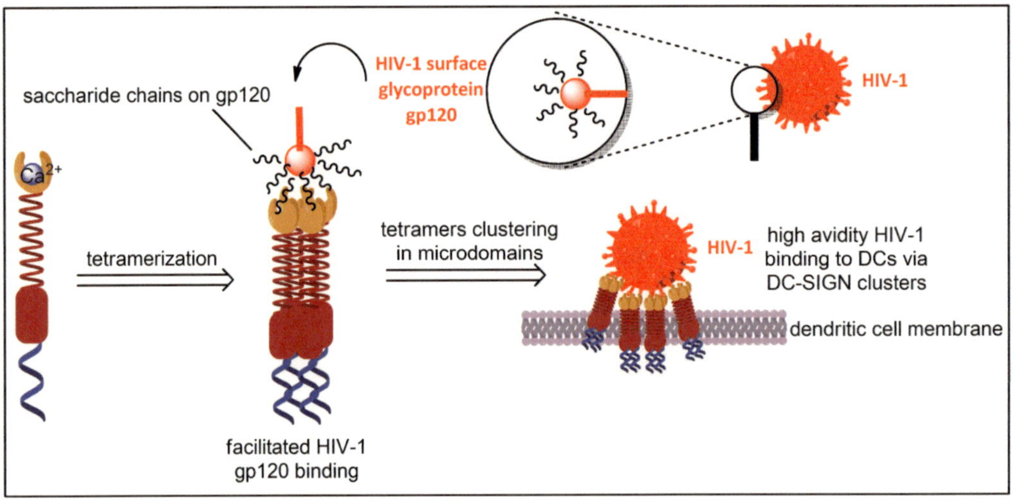

Figure 3. Tetramerization of DC-SIGN and further clustering allows high binding avidity and influences pathogen binding, as depicted with the example of HIV-1 binding to DCs: tetramerization increases avidity to PAMPs like gp120 on HIV-1 surface, while DC-SIGN clusters act as functional docking site for HIV-1 (modified from Švajger et al.) [9].

Apart from allowing oligomerization, DC-SIGN neck repeat domains separate CRDs from the cell surface to enable multivalent interaction with glycans. DC-SIGN binds particularly well to viral and bacterial glycans with closely spaced terminal oligo- or monosaccharides with roughly 5 nm between the units [18,30,31]. DC-SIGN is not a totally rigid macromolecule and shows a degree of flexibility upon ligand binding [31]. Namely, DC-

SIGN adapts to an arrangement of terminal oligo- and monosaccharides, so all CDRs are allowed to interact with their ligands. Taken together, the tetrameric form of DC-SIGN and its conformational flexibility enable effective and selective binding of various glycans. Furthermore, due to the nature of the receptor-mediated cellular signalling, tetramerization of DC-SIGN could contribute to signal transduction after ligand binding.

2.3. DC-SIGN, a target for anti-infective therapy

Probably the most important feature of DC-SIGN is its ability to bind a great number of highly virulent pathogens [32]. Accordingly, it has been recognised as a potential new target for anti-infective therapy that perpetuates basic research of considerable importance. Apart from the already mentioned CFG, the importance of DC-SIGN as a new drug target is reflected in several other projects aimed to thoroughly clarify its function in pathogen infection. For example, several prominent European research teams have joined to form CARMUSYS, a collaborative training project aimed at designing and synthesizing carbohydrate multivalent systems to be used as inhibitors for pathogen attachment and penetration into target cells (CARMUSYS - *Carbohydrate Multivalent System as Tools to Study Pathogen Interactions with DC-SIGN*).

DC-SIGN modulates the outcome of the immune response of DCs by binding and recognizing a variety of microorganisms, including viruses (HIV-1, HCV, CMV, Dengue, Ebola, SARS-CoV, HSV, coronaviruses, H5N1, West Nile virus, measles virus), bacteria (*H. pylori, M. tuberculosis, L. interrogans*), fungi (*C. albicans, A. fumigatus*) and parasites (Leishmania, *S. mansoni*) [9]. On mucosal surfaces of the body where immature dendritic cells sample pathogens, DC-SIGN serves as one of the very first pathogen attachment points and the usual result of this interaction is pathogen internalization, degradation and exposure of the pathogen PAMPs to recruit CD4+ T-cells and to start the humoral immune response [6,22]. HIV-1 exploits native DC-SIGN functions to enslave DCs as carriers to boost T cell infection with or without becoming infected themselves [33]. The very first interaction between HIV-1 and DCs occurs via HIV-1 envelope glycoprotein gp120 with DC-SIGN on immature DCs (Figure 3). The HIV-1-DC-SIGN complex is rapidly endocyted into early endosomes, where the acidic medium causes ligands to dissociate from DC-SIGN [34]. Upon dissociation, most DC-SIGN-bound ligands are lyzed and processed as a normal degradation pathway. Some HIV-1 particles however remain bound to DC-SIGN and thus protected from the host immune system, so a fraction of HIV-1 retains its infectiveness [35]. HIV-1 may rest in DCs in an infectious state for days, hidden in undefined bodies that differ from both endosomes and lysosomes [36]. N-glycan composition of surface proteins governs the viral faith upon interaction of viral envelope with DC-SIGN [37]. By altering the N-linked glycan composition from mixed to oligomannose-enriched, one increases the affinity of HIV-1 for DC-SIGN, which enhances the viral degradation and reduces virus transfer to target cells. On the contrary, HIV-1 with oligomannose-enriched N-glycans is presented to viral envelope-specific CD4+ T cells more efficiently. In the alternative scenario, HIV-1 binds to DC-SIGN and facilitates lateral binding of HIV-1 to CD4 and CCR5 receptors expressed on the same immature DCs [33]. The direct consequence of this interaction is HIV-1 infection of DCs [38]. To conclude, HIV-1 uses DC-SIGN as an entry mode to DCs, and DCs may be regarded as a Trojan horse that takes HIV-1 to CD4+ T-cells while protecting it from the host immune system [39]. A similar entry mode has been observed for some other pathogens like viruses (Ebola) and bacteria (*M. tuberculosis, H. pylori*) [32]. Apart from this straightforward infection pathway, HIV-1-infected DCs are able to mediate transmission of HIV-1 to CD4+ cells by means of immune response modulation and infectious synapse formation. Namely, HIV-1 causes both inhibition of DC maturation while inducing formation of viral synapse, a process previously attributed to mature DCs only [40]. As only several pathogens are able to modulate DCs maturation process, it may be concluded that the DC maturation depends upon selective ligand recognition, possibly also by DC-SIGN.

The fact that DC-SIGN acts as an entry point and a mediator of pathogen infections points out the possible therapeutic usefulness of DC-SIGN. DC-SIGN antagonists work by inhibiting pathogen interaction with DC-SIGN, so the very first phase of pathogen infection can be inhibited, as proven in *in vitro* experiments [10,41,42]. DC-SIGN antagonists may be designed as monovalent glycomimetics based on the DC-SIGN-binding oligosaccharides and their multimeric presentation [43]. Alternatively, screening of compound libraries has been successful in obtaining non-carbohydrate DC-SIGN antagonists [44]. As mentioned, potential carbohydrate-derived drugs have poor pharmacokinetic properties in general and this might raise some concern as to whether there is potential for therapeutically useful glycomimetic DC-SIGN antagonists. In the case of HIV-1, the major initial contact site between HIV-1 and DC-SIGN is in the vaginal mucosa, so a DC-SIGN antagonist could be administered topically to prevent HIV-1 transmission without systemic application [45]. It has to be stressed that no clinically proven therapy based on inhibition of DC-SIGN-mediated pathogen infection has been presented [43].

2.4. How does DC-SIGN choose among terminal monosaccharides?

DC-SIGN has a highly regulated recognition of its ligands as it selectively binds glycans with terminal D-mannose- and L-fucose expressed on a number of bacteria, parasites, fungi and viruses [46]. However, the mere presence of D-mannose and L-fucose does not assure binding selectivity itself. Namely, monosaccharide binding to DC-SIGN CRD is generally very weak, with Ki(D-mannose) = 13.1 mM and Ki(L-fucose) = 6,7 mM being the strongest binders among monosaccharides [26]. DC-SIGN CRD forms a 1-to-1 complex with terminal mono- or oligosaccharides, which relies upon already mentioned octacoordination of Ca^{2+} ion in the binding site by the "core monosaccharide"; the equatorial 3- and 4-hydroxyls each form coordination bonds with the Ca^{2+} in the binding site common to all C-type lectins, but also offer hydrogen bonds with amino acids that also serve as Ca^{2+} ligands [47]. A crucial structural feature of a mannose residue is an axial position of the 2-hydroxyl group; this allows tight surface complementarity of the core mannose with the binding site. Equatorial position of the 2-hydroxyl group would probably prevent this tight binding due to steric clash, so hexopyranoses with equatorial 2-hydroxyl groups do not form strong interactions. To increase binding affinity, other saccharide units form additional interactions with the binding site, while binding specificity is based on spatial constraints. An excellent description of selectivity/specificity mechanism may be found in the work of Guo et al., who demonstrated that DC-SIGN selectively binds high-mannose and fucosylated oligosaccharides [18]. The difference in the affinity of each oligosaccharide results from a different spatial arrangement of the mannose- and fucose-based ligands demonstrated in two crystal structures of fucose-based tetrasaccharide LNFP III and Man_4 tetrasaccharide bound to DC-SIGN CRD; the Man3 mannose moiety of Man_4 (the "core monosaccharide") inclines the rest of the molecule towards Phe313 with high surface complementarity, while the fucose moiety (the "core monosaccharide" of LNFP III) makes hydrophobic contact with Val351 and positions the second saccharide in a vertical manner, away from the protein surface (Figure 4). The binding mode presented for Man_4 tetrasaccharide (PDB code: 1SL4) in which Phe313 forms a steric hindrance is a prevalent one for a mannose-containing oligosaccharides. However several other PDB structures of DC-SIGN CRD (PDB code: 1K9I, 1K9J, 2IT5 and 2IT6) show that Phe313 residue is rather flexible and allows 2 distinct binding modes, both including coordination of the Ca2+ by one mannose residue. One mode is shown in Figure 4, while in the second, the binding orientation of the mannose at the principal Ca^{2+} site is reversed, and creates different interactions between the terminal mannose and the region around Phe313 [47,48]. Thus we conclude that the binding mode of a specific oligosaccharide does not depend solely on the "core monosaccharide" involved, but is highly sensitive to the substitution pattern and 3D structure of adjacent monosacccharides. Furthermore, the mannose- and fucose-based ligands have overlapping, but different binding modes, which might offer a rational explanation for the different biological effects of mannose- and fucose-based ligands [49]. Consequently, the two observed binding modes offer a solid basis upon which DC-SIGN antagonists with either fucose or mannose anchors can be designed [50].

On the supramolecular level, DC-SIGN tetramerization has a major impact on binding affinity. By forming tetramers, binding affinity for glycans with repetitive sugar motifs with high-mannose or fucose N-linked oligosaccharides increases markedly [46]. This simple observation drives us to the conclusion that only polyvalent ligands could efficiently bind to DC-SIGN and offer a rationale for design of multivalent carbohydrate systems as DC-SIGN antagonists [51].

DC-SIGN was first regarded as the major binding lectin for various mannose-glycosylated PAMPS including HIV-1 gp120. However, other C-type lectins bind both D-mannose and L-fucose-containing glycans, which brings under question DC-SIGN's exclusive role in pathogen binding, but above all, raises the concern of binding selectivity when designing DC-SIGN antagonists. In particular, two C-type lectins - Langerin on epithelial Langerhans cells and a mannose receptor on dermal DCs - bind high-mannose oligosaccharides and HIV-1 gp120 with high affinity [52]. Their function in HIV-1 (and possibly other pathogen) infection differs: interaction of HIV-1 with DC-SIGN enables HIV-1 to survive a host immune system, while Langerin helps Langerhans cells to eradicate HIV-1 [53,54]. On the other hand, mannose receptor facilitates HIV-1 infection of DCs by the CD4/CCR5 entry pathway [53]. Therefore, blockade of both DC-SIGN and mannose receptors seems the right strategy for prevention of HIV-1 entry into DCs while Langerin inhibition suppresses viral

a)

b)

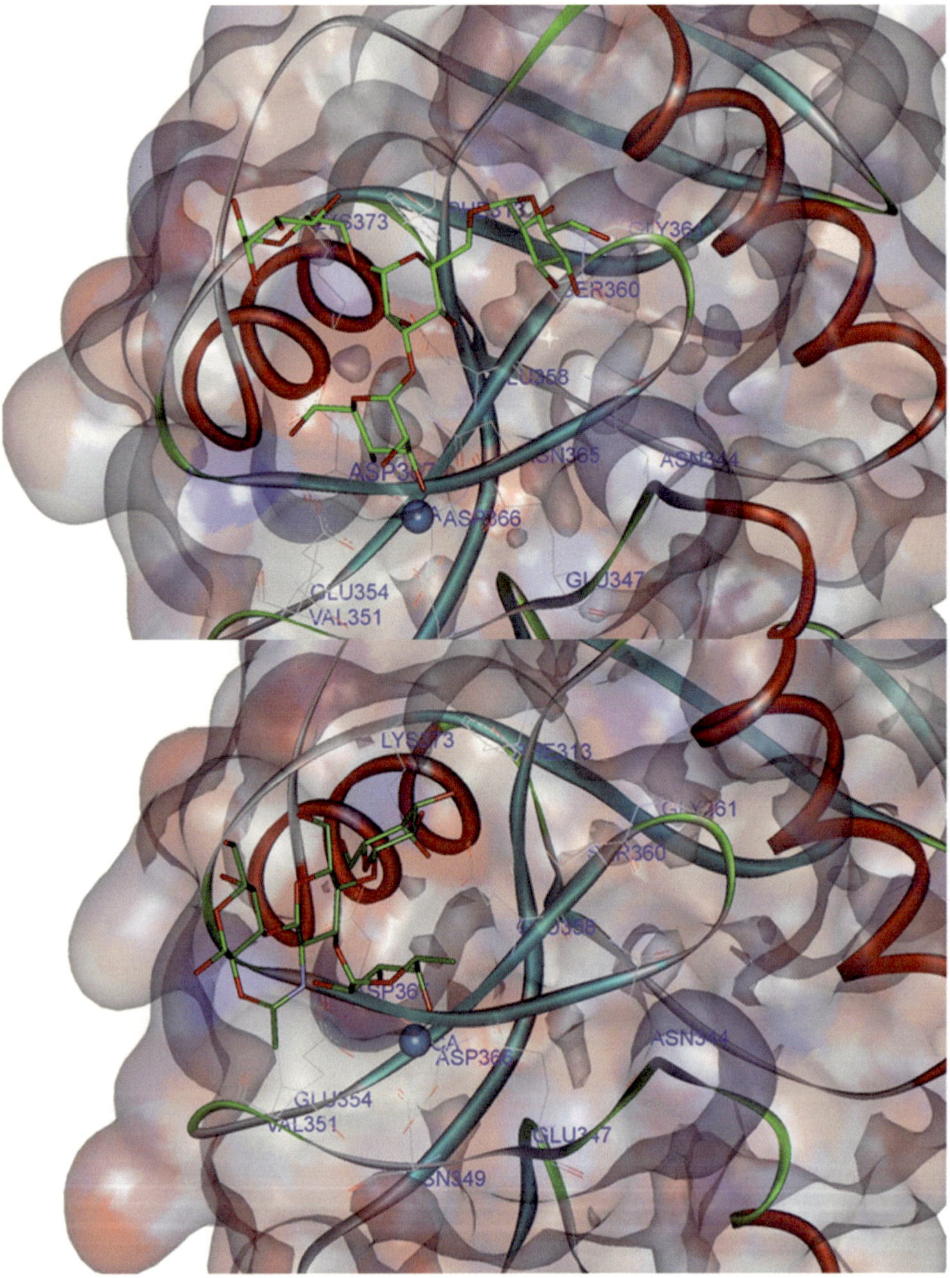

Figure 4. Selected amino acid residues (thin sticks) of DC-SIGN CRD (transparent surface) in complex with: a) Man_4 tetrasaccharide (solid sticks, PDB code: 1SL4), and b) fucose-based tetrasaccharide LNFP III (solid sticks, PDB code: 1SL5) [18]. The Man3 mannose moiety of Man_4 or the "core monosaccharide" inclines the rest of the molecule towards Phe313 with high surface complementarity, while the fucose moiety or the "core monosaccharide" of LNFP III makes hydrophobic contact with Val351 and positions the second saccharide in a vertical manner, away from the protein surface, so only one galactose moiety makes additional contacts with DC-SIGN CRD.

clearance thus allowing a boost of HIV-1 infection. When designing an efficient HIV-1 entry inhibitor based on C-type lectin antagonism, an agent of choice should bind to both DC-SIGN and mannose receptors, while having no or marginal affinity to Langerin. Although all three receptors bind to virtually the same ligands, selectivity against Langerin is an achievable objective since Langerin binding sites differ from that of DC-SIGN [55]. As expected, selectivity versus Langerin does not necessarily rely on the "core monosaccharide" involved in the binding process, but on spatial constraints formed by adjacent glycan monosaccharide units [56].

2.5. Binding of monovalent DC-SIGN antagonists

As mentioned earlier, binding of DC-SIGN natural ligands depends upon the presence of an L-fucose or D-mannose hexopyranose unit. Their inherent disadvantages in terms of low activity and/or insufficient drug-like properties can be modified by the design of glycomimetics – compounds that mimic the bioactive function of carbohydrates but have far better drug-like properties [3]. This concept has been successfully used in the design of monovalent DC-SIGN antagonists. The term "monovalent" is used here to describe low-molecular weight molecules that can occupy only one DC-SIGN CRD at a time, so it incorporates mono- and oligosaccharide structures and their mimetics. The design of these monovalent glycomimetics can be structurally divided into the following sections, as depicted in figure 5:

- the choice of a monosaccharide unit,
- the choice of glycosidic bond surrogate,
- the choice of adjacent saccharides or structures that contribute to overall binding affinity.

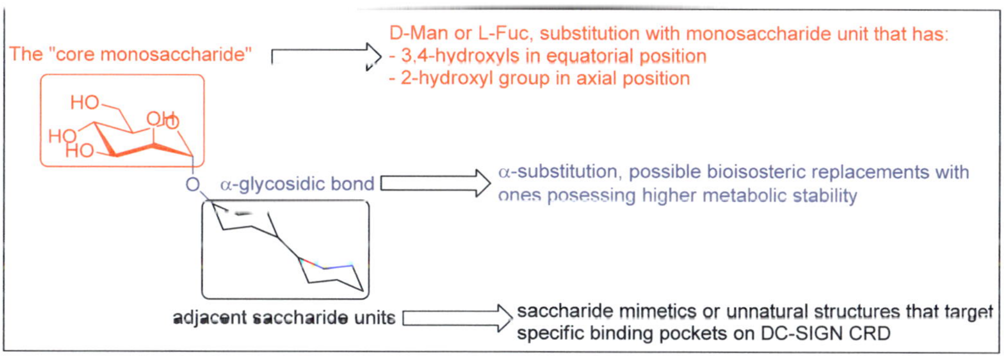

Figure 5. Schematic presentation of the design of monovalent glycomimetics that act as DC-SIGN antagonists: systematic replacements in structure that lead to efficient DC-SIGN ligands.

2.5.1. *The choice of a monosaccharide unit*

The most extensive work regarding the choice of the monosaccharide unit comes from the work of Bernardi and Rojo's groups from CARMUSYS [41,50,57]. In particular, they have shown that L-fucose can be incorporated in glycomimetic surrogates of Lewis-X trisaccharide to obtain even better affinity than the native trisaccharide. The full Lewis-X mimetic **1** was shown to inhibit DC-SIGN binding of mannosylated BSA in the upper micromolar range (IC_{50}=350 μM), but the second generation of analogous compounds failed to give any significant improvement over **1** (Figure 6) [50,58]. STD-NMR studies of **1** with DC-SIGN ECD have shown that only fucose residue makes strong contact with the DC-SIGN CRD. A reasonable explanation for this observation might be the before mentioned binding mode of the L-fucose moiety: it positions the second saccharide in a vertical manner away from the protein surface, and thus the rest of the molecule fails to form tight interactions with protein. Just a glance at figure 4 reveals that DC-SIGN CRD is quite flat and high structural complementarity is one of the requirements for strong binding. In the absence of functional groups that would allow ionic interactions which are not highly dependent on the distance, the rest of the Lewis-X mimetic **1** and its analogues probably form the majority of interactions with the solvent. Lewis-X mimetics however share one important figure; their affinity for Langerin is insignificant, so they are selective DC-SIGN antagonists [58].

The L-fucose monosaccharide has the highest affinity for DC-SIGN among monosaccharides, and L-fucose should be the logical choice as the "core monosaccharide" when designing DC-SIGN antagonists. On the contrary, D-mannose has received most of the attention as the majority of mono- and polyvalent DC-SIGN antagonists incorporate D-mannose as the "core monosaccharide".

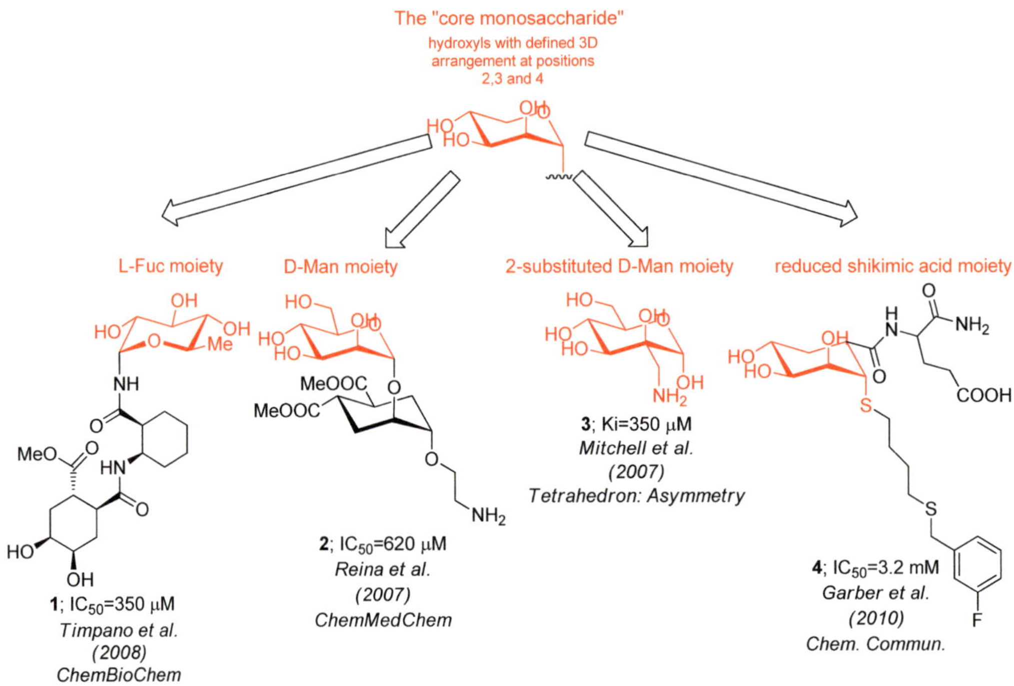

Figure 6. The choice of "core monosaccharide" of DC-SIGN antagonists; L-fucose (L-Fuc), D-mannose (D-Man), 2-substituted D-mannose and reduced shikimic acid are useful "core monosaccharides".

Pseudo-1,2-mannobioside **2** and its analogues of Reina et al. (Figure 6) contain a D-mannose unit substituted at the anomeric position with conformationally constrained cyclohexanediol [41]. The STD-NMR of DC-SIGN ECD with an azido derivative of **2** shows that the compound makes close contact with the protein, which is in agreement with the binding mode of Man_4 tetrasaccharide. The inhibitory activity of **2** on Ebola virus entry into DC-SIGN expressing Jurkat cells was quite high (IC_{50}=0.62 mM) and this was the first functional assay showing that DC-SIGN antagonism with small molecules might be used to inhibit viral transfection mediated by DC-SIGN. Another example of substituted D-mannose as the "core monosaccharide" might be found in the work of Mitchell et al.: they have synthesized a small library of 2-C-substituted branched D-mannose analogues of which compound **3** exhibited a 48-fold stronger binding to DC-SIGN (Ki=0.35 mM, Ki(mannose)=17.1 mM) [59].

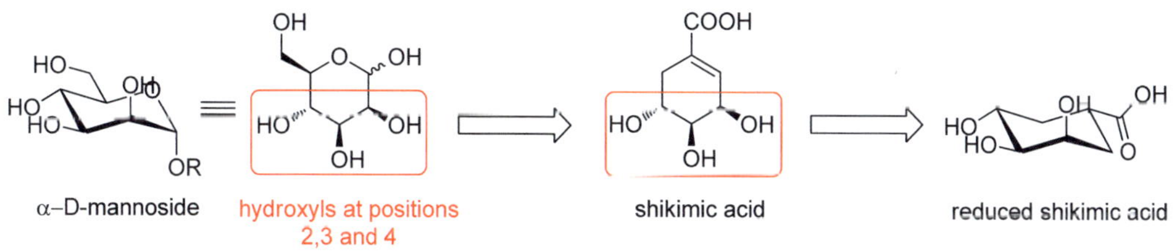

Figure 7. Design of reduced shikimic acid "core monosaccharide" as D-mannose mimetic [60].

An innovative approach was used by Garber et al.: taking D-mannose as the lead structure, they have concentrated on the spatial relationship of hydroxyls at positions 2,3 and 4 and concluded that reduced shikimic acid should enable the same spatial relationship of hydroxyls (Figure 7) [60]. They have synthesized 192 derivatives of reduced shikimic acid and compound **4** (Figure 6) was found to be the most potent hit of this focused library.

2.5.2. *The choice of glycosidic bond surrogate*

The metabolic instability of glycosidic bonds renders it inappropriate for the design of drugs, and hence requires an appropriate surrogate when designing mimetics of oligosaccharides. Numerous changes in carbohydrate structures have been successful in the design of glycosidase inhibitors, and exactly these structures could be used to modify glycosidic bond instability [61]. However, unwanted inhibition of glycosidases may as well provoke side effects of potential drugs, so careful choice of glycosidic bond surrogates has to be made if only its stability is the ultimate goal. The α-glycosidic bond found in both L-fucose and D-mannose containing oligosaccharides that bind to DC-SIGN was successfully replaced by Timpano et al. by a stable α fucosylamide structure (compound **1**, Figure 6) and shown not to affect the binding affinity in a detrimental sense [50]. The derivatives of reduced shikimic acid (compound **4,** Figure 6) have 2 features that influence glycosidic bond stability: first, they have a thioglycosidic bond, which is proven to be metabolically more stable towards glycosidases [62], and second, they are a constitute of reduced shikimic acid which is a carbasugar. Carbasugars lack anomeric reactivity, which implies their metabolic stability towards glycosidases and glycosyltransferases. Although glycosidic bond surrogates were often not used in the design of DC-SIGN antagonists, the stability of glycosidic bonds has to be challenged when designing stable glycomimetics. According to the latest literature, a large number of alternatives already exist [63].

2.5.3. The choice of adjacent saccharides or structures that contribute to overall binding affinity

Monosaccharide moieties other than "core monosaccharide" in the structure of DC-SIGN oligosaccharide ligands form not only additional interactions with the binding site, but also influence the binding specificity with spatial constraints and point other monosaccharide units towards or away from the protein surface, as seen in the Figure 4. Notable quality of the adjacent monosaccharide units is that they do not form the same interactions like the "core monosaccharide", but instead form a network of H-bonds, possibly through water molecules. For example, Man_4 tetrasaccharide makes contact with DC-SIGN through at least two water molecules, while one stabilizes its binding conformation [18,64]. Alternatively, both "core" and adjacent monosaccharides make surface complementarity and hydrophobic interactions. In particular, Van Liempt et al. demonstrated that Val351 in DC-SIGN creates a hydrophobic pocket that strongly interacts with the Fucalpha1,3/4-GlcNAc moiety of the Lewis antigens [65]. So, although highly polar in nature, the adjacent monosaccharide units contribute significantly to overall binding free energy also by hydrophobic interactions apart from being mere "linker" to other structures. This implies that altering hydrophilic character of adjacent monosaccharide units to more hydrophobic surrogates should increase free binding energy by increasing hydrophobic interactions. With this in mind, Timpano et al. have used the (1*S*,2*R*)-2-aminocyclohexanecarboxylic acid as a scaffold/linker to attach D-galactose mimetic into the structure of compound **1** and its derivatives (Figure 8) [50]. The molecule was carefully chosen to mimic Lewis-X structure while minimizing the hydrophilic ballast of the Lewis-X, and galactose mimetic was incorporated based on the observation, that galactose residue makes contact with the DC-SIGN CRD surface and is thus important in binding [18].

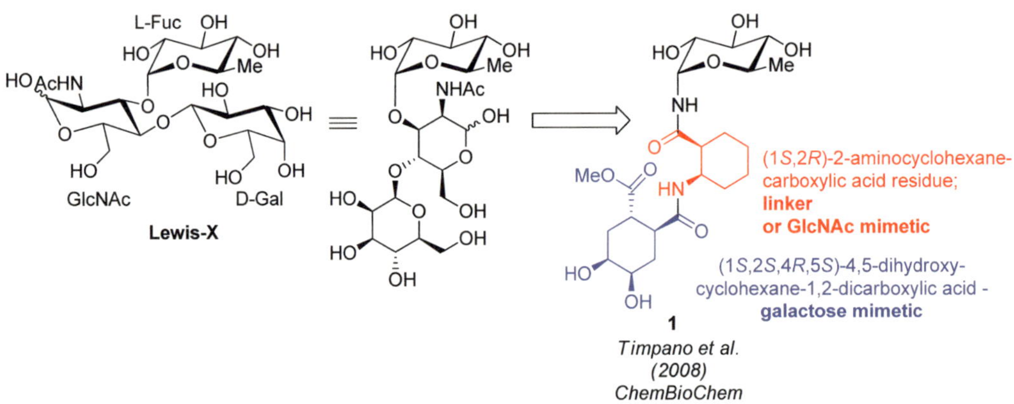

Figure 8. (1*S*,2*R*)-2-aminocyclohexanecarboxylic acid as a scaffold/linker to attach D-galactose mimetic into the structure of compound **1** and its derivatives, all mimic Lewis-X [50].

In the series of mannose-based DC-SIGN antagonists, Reina et al. and Sattin et al. (groups of Rojo and Bernardi) have incorporated (1*S*,2*S*,4*S*,5*S*)-dimethyl 4,5-dihydroxycyclohexane-1,2-dicarboxylate as adjacent monosaccharide mimicking "trans" conformation of 1,2-hydroxyls in D-mannose (Figure 9, compounds **2** and **5**) [10,41]. Again, the rational for this change was to imitate the 3D relationship of key hydroxyls in the D-mannose moiety, while lowering the overall hydrophilicity. Furthermore, the cyclohexane saccharide mimic lacks glycosidic bond and is thus metabolically stable.

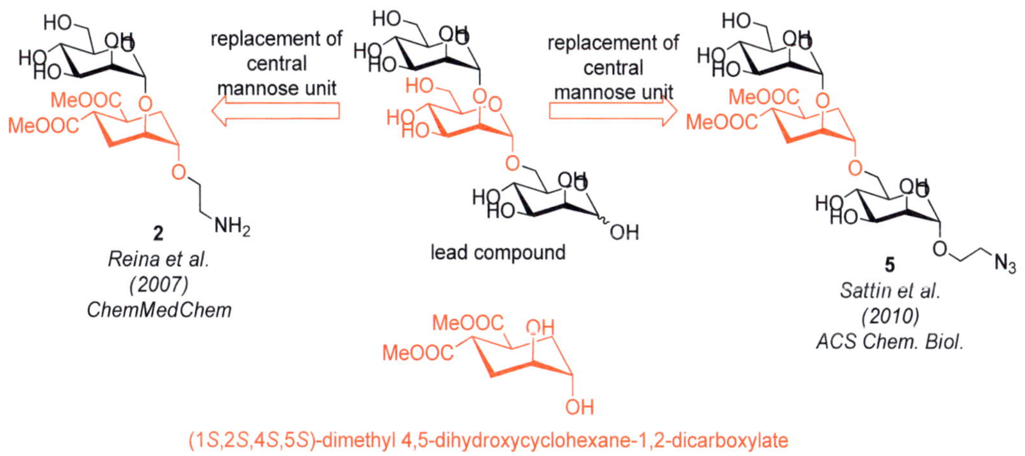

Figure 9. (1*S*,2*S*,4*S*,5*S*)-dimethyl 4,5-dihydroxycyclohexane-1,2-dicarboxylate as central monosaccharide surrogate that mimics "trans" conformation of 1,2-hydroxyls in D-mannose.

Starting from compound **2** or its azido derivative, the group of Bernardi (Obermajer et al.) continued to pursue the idea of increasing the binding affinity by identifying two binding areas around Phe313 in the DC-SIGN binding site that were only partially occupied by co-crystallized tetramannoside Man4 [64]. These hydrophobic areas were targeted by attaching different hydrophobic moieties to deprotected carboxylates of pseudo-1,2-mannobioside **2** (Figure 10). A number of mannose-based DC-SIGN antagonists were synthesized (an illustrative example is compound **6**), and the majority of them inhibited DC adhesion at low micromolar concentrations improving the potency of the starting compound **2** by two orders of magnitude. Probably the same hydrophobic areas have contributed to the affinity of **4**.

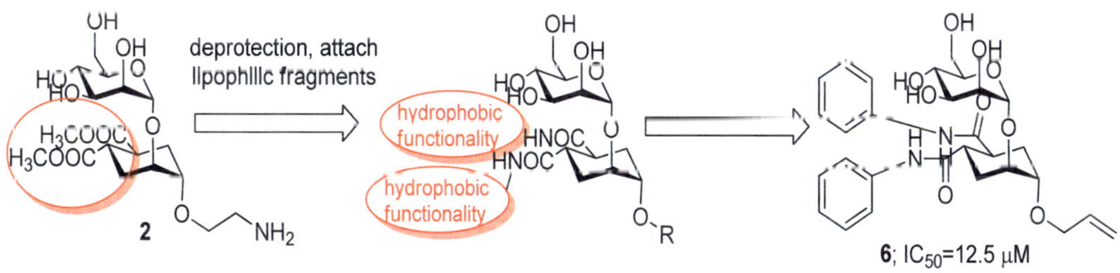

Figure 10. Increasing potency of DC-SIGN antagonists by attaching hydrophobic moieties to deprotected carboxylates of **2** to afford **6** and its derivatives [64].

2.6. Increasing affinity and/or avidity?

The evident disadvantage of monovalent DC-SIGN antagonists – their low affinity for DC-SIGN and thus low potency – can be overcome by conjugating monovalent units to various scaffolds for polyvalent presentation (Figure 11) [51]. Namely, even the most potent monovalent ligands have inhibitory constants in low micromolar concentrations, while therapeutically useful compounds have the same effect in nanomolar or even picomolar concentrations. The polyvalent antagonists are believed to interact with possibly all DC-SIGN CRDs on tetramers, or perhaps, they might interact and collate DC-SIGN clusters on the cell surface in the same manner as HIV-1 increases its avidity to DC surface (depicted in Figure 3). The avidity observed for polyvalent ligands in general originates from multiple binding: the polyvalent molecule with reversible mechanism of binding has higher possibility for being bound to at least one of receptor binding sites (i.e. one of CRDs), so that dissociation rate constant significantly decreases [66]. The other rationale may be derived from the observation of Andrews et al., who concluded that the average loss of overall rotational and translational entropy accompanying drug-receptor interaction is a constant for relatively small molecules and was estimated to be approx. 14 kcal/mol at 310 K [67]. In other words, more favourable free binding energy is obtained by combining binding epitopes/structures into one larger molecule, and this is also true for uniting same monovalent structures into a large polyvalent one because all rotational and translational entropic losses that occur during binding of individual monovalent molecules are reduced to one entropic loss for a larger molecule. According to both theories, when larger polyvalent molecules are employed, binding avidity is observed instead of just a linear increase in binding affinity. The main concern when choosing appropriate polyvalent support is the spatial relationship between individual monovalent ligands: monovalent ligands should be appropriately spaced to allow binding to at least two binding sites (or CRDs in case of C-type lectins), otherwise avidity cannot be achieved.

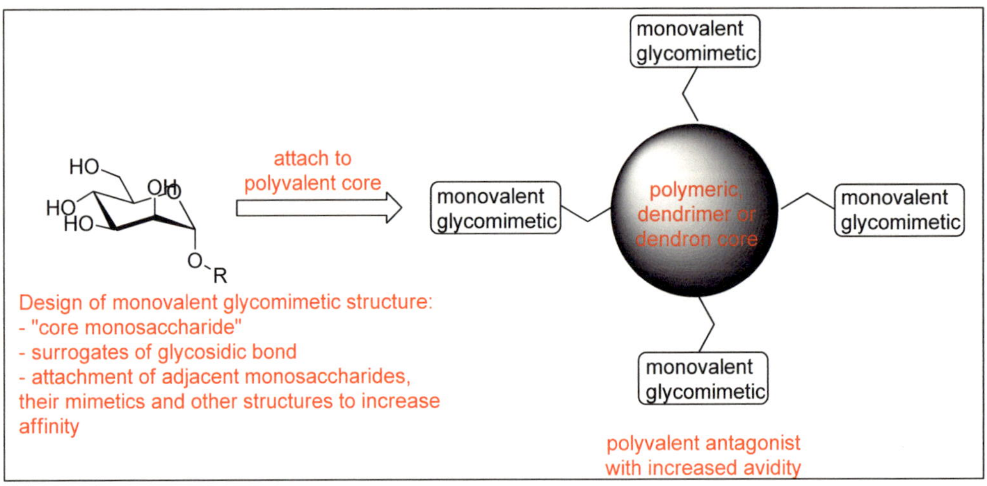

Figure 11. Strategy for increasing binding affinity/avidity of DC-SIGN antagonists: monosaccharides or monovalent glycomimetics are attached to polyvalent dendrimer or dendron core (modified from Anderluh et al.) [43].

The first polyvalent DC-SIGN antagonists were designed, synthesized and assayed in the laboratories of Rojo and Delgado [68]. A simple glycodendritic core bearing 32 mannose residues conjugated to BoltornH30 dendrimer through a succinic acid spacer (**7**, Figure 12) hindered DC-SIGN mediated Ebola virus infection at nanomolar concentrations (IC_{50}=337 nM). In comparison, the same inhibition was achieved with only millimolar concentrations of α-methyl-D-mannopyranoside (IC_{50}=1.27 mM); the core bears only 32 mannose residues and the difference in binding affinity is more than 3000-fold, which clearly indicates high binding avidity.

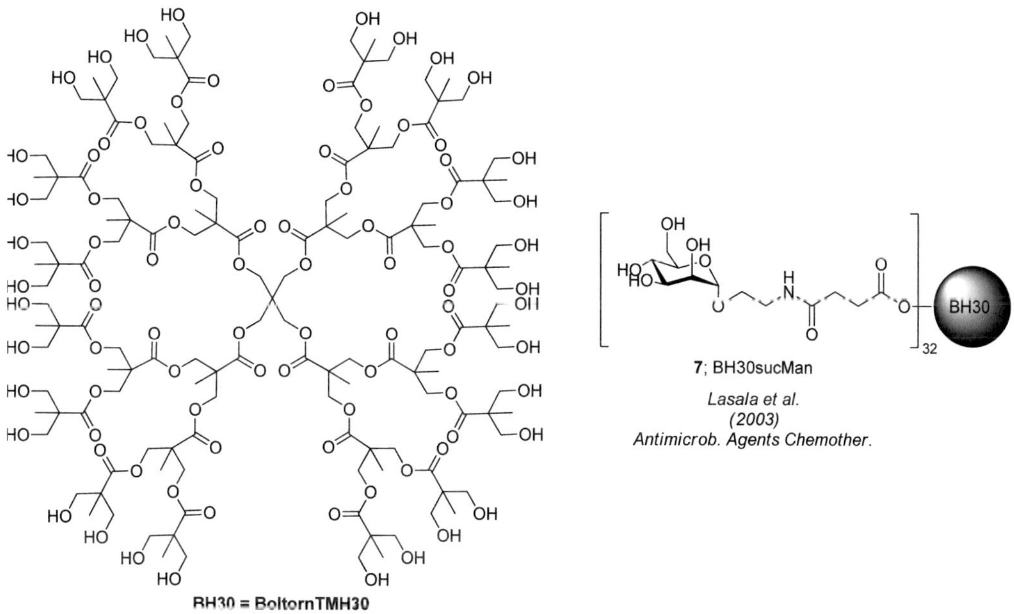

Figure 12. Glycodendritic structure bearing 32 mannose residues conjugated to BoltornH30 dendrimer as DC SIGN antagonist [68].

Relatively simple glycopolymers with a different load of combined α-mannose and β-galactose in ratios were synthesized by Becer et al. and evaluated in inhibition of DC-SIGN-gp120 binding (**8**, Figure 13) [69]. Glycopolymer with 25% mannose failed to inhibit gp120 binding potently (IC_{50} of 1.45 μM), while glycopolymer with 100% mannose was far more efficient (IC_{50} of 37 nM) even when calculated per mannose unit, as it had approx. 40 times higher affinity compared with 4-times higher load of the mannose unit.

Polyvalent DC-SIGN antagonists described before relied on the use of "core monosaccharide" only. The design and choice of a potent monovalent DC-SIGN ligand might reduce the requirement for huge dendrimeric presentation while retaining desired effect. Designed monovalent glycomimetic **4** (Figure 6) of Garber et al. was loaded onto a carefully selected multivalent core designed to link at least 2 CRDs of DC-SIGN tetramer as monovalent units were roughly 40Å apart, which is exactly the width of DC-SIGN CRD (**8**, Figure 13) [60]. Glycopolymer **8** with 29 units of **4** was found to have 1000-fold higher affinity for DC-SIGN (IC_{50}=2.9 μM) than **4**, showing that high avidity binding was obtained.

It is hard though to assess the relevance of designed monovalent ligand versus D-mannose, as the analogous polyvalent DC-SIGN antagonist bearing D-mannose or reduced shikimic acid was not synthesized. The direct evidence however came from the work of Sattin et al [10]. They have attached four copies of monovalent trimannoside mimetic **5** (Figure 9) to a tetravalent dendron **10** (Figure 14), which potently inhibited HIV-1 transfection to CD4+ T lymphocytes (>94 % inhibition at 100 μM). For comparison only, a tetrameric dendron bearing 4 copies of simple D-mannose (**11**, Figure 14) failed to inhibit HIV-1 transfection with even comparable potency (65 % inhibition at 100 μM). Furthermore, Dendron **10** inhibited Ebola *cis* infection of Jurkat T cells one order of magnitude more potently than dendron **11** [70]. This data clearly demonstrates that potent monovalent DC-SIGN antagonists reduce the need for high polyvalency number and influences binding affinity markedly.

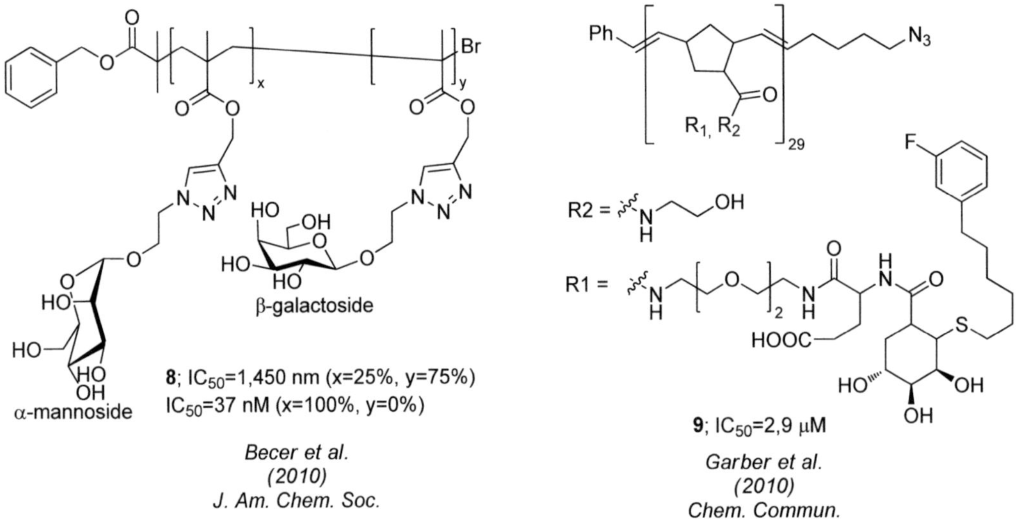

Figure 13. Glycopolymers of Becer et al. and Garber et al. with high binding avidity [60,69].

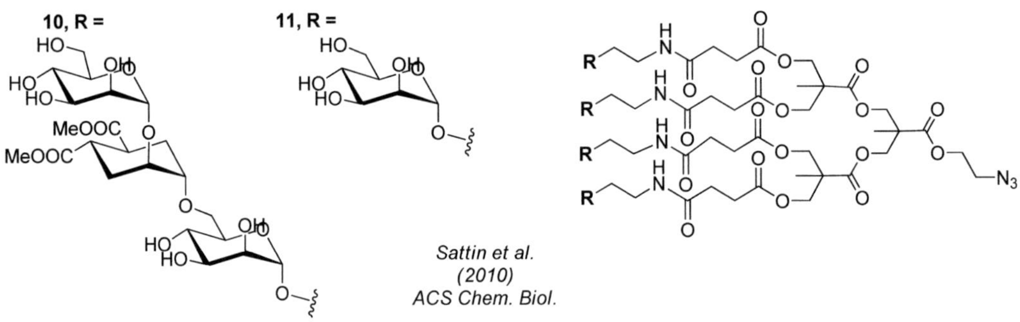

Figure 14. Tetravalent dendrons of Sattin et al. bearing four copies of monovalent trimannoside mimetic **5** (**10**) or D-mannose residue (**11**) [10].

3. Conclusion; could it work for all C-type lectins?

As mentioned earlier, all C-type lectins share a structural feature, namely C-type lectin-like domain (CTLD) or carbohydrate recognition domain (CRD) responsible for Ca^{2+}-dependent selective binding of terminal mono- or oligosaccharide units of large carbohydrates. This implies that only the "core monosaccharide" makes contact with Ca^{2+} ion while other ligand carbohydrate units (if present) form structural and bonding complementarity with the CRD. From this point of view, the systematic approach presented herein could be of general applicability when designing glycomimetic C-type lectin antagonists. It consists of designing the monovalent ligand based on three distinct steps: the choice/design of a "core" monosaccharide unit, the choice/design of glycosidic bond surrogate and the choice/design of adjacent saccharides or structures that contribute to overall binding affinity. Still, the evident drawback of monovalent glycomimetics is their low affinity not only to DC-SIGN, but to lectins in general [3]. The other characteristic of C-type lectins is their ability to oligomerize, and further make clusters of functional oligomers. The tactics of polyvalent presentation targets exactly the oligomerized or even clustered structures and is a prerequisite of polyvalent structures as C-type lectin antagonists. Accordingly, high avidity for DC-SIGN and other C-type lectins can be achieved with high loading of monovalent ligands to various polyvalent systems. Taken together, the results of several groups presented in this chapter clearly demonstrate that this general procedure for designing glycomimetic DC-SIGN antagonists gives notable results, and according to the structural resemblance of diverse C-type lectin CRDs, there is a high probability of its applicability to distinct C-type lectins.

Author details

Marko Anderluh
Department of Medicinal Chemistry, University of Ljubljana,
Faculty of Pharmacy, Ljubljana, Slovenia

Acknowledgement

This work is dedicated to Prof. Dr. Slavko Pečar, a great mentor and teacher, who taught me not only pure science, but above all how to nourish my enthusiasm for science.

4. References

[1] Finkelstein J (2007) Insight - Glycochemistry & glycobiology. Nature 446: 999-999.
[2] Ernst JD (1998) Macrophage receptors for Mycobacterium tuberculosis. Infect. Immun. 66: 1277-1281.
[3] Ernst B, Magnani JL (2009) From carbohydrate leads to glycomimetic drugs. Nat. Rev. Drug Discover. 8: 661-677.

[4] Malik A, Firoz A, Jha V, Ahmad S (2010) PROCARB: A Database of Known and Modelled Carbohydrate-Binding Protein Structures with Sequence-Based Prediction Tools. Adv. Bioinformatics. 436036.

[5] Curtis BM, Scharnowske S, Watson AJ (1992) Sequence And Expression Of A Membrane-Associated C-Type Lectin That Exhibits CD4-Independent Binding Of Human-Immunodeficiency-Virus Envelope Glycoprotein-Gp120. Proc. Natl. Acad. Sci. USA 89: 8356-8360.

[6] Steinman RM (2000) DC-SIGN: A guide to some mysteries of dendritic cells. Cell. 100: 491-494.

[7] Geijtenbeek TBH, Kwon DS, Torensma R, van Vliet SJ, van Duijnhoven GCF, Middel J, Cornelissen I, Nottet H, KewalRamani VN, Littman DR (2000) DC-SIGN, a dendritic cell-specific HIV-1-binding protein that enhances trans-infection of T cells. Cell.100: 587-597.

[8] Geijtenbeek TBH, Torensma R, van Vliet SJ, van Duijnhoven GCF, Adema GJ, van Kooyk Y, Figdor CG (2000) Identification of DC-SIGN, a novel dendritic cell-specific ICAM-3 receptor that supports primary immune responses. Cell 100: 575-585.

[9] Svajger U, Anderluh M, Jeras M, Obermajer N (2010) C-type lectin DC-SIGN: An adhesion, signalling and antigen-uptake molecule that guides dendritic cells in immunity. Cell. Signall. 22: 1397-1405.

[10] Sattin S, Daghetti A, Thepaut M, Berzi A, Sanchez-Navarro M, Tabarani G, Rojo J, Fieschi F, Clerici M, Bernardi A (2010) Inhibition of DC-SIGN-Mediated HIV Infection by a Linear Trimannoside Mimic in a Tetravalent Presentation. ACS Chem. Biol. 5: 301-312.

[11] Komath SS, Kavitha M, Swamy MJ (2006) Beyond carbohydrate binding: new directions in plant lectin research. Org. Biomol. Chem. 4: 973-988.

[12] Becker B, Condie GC, Le GT, Meutermans W (2006) Carbohydrate-based scaffolds in drug discovery. Mini-Rev. Med. Chem. 6: 1299-1309.

[13] Varki A (1999) Essentials of glycobiology. Cold Spring Harbor, NY: Cold Spring Harbor Laboratory Press; 1999. xvii, 653 p.

[14] Zelensky AN, Gready JE (2005) The C-type lectin-like domain superfamily. FEBS J. 272: 6179-6217.

[15] Drickamer K (1993) Ca2+-Dependent Carbohydrate-Recognition Domains In Animal Proteins. Curr. Opin. Struct. Biol. 3: 393-400.

[16] Harding MM (2006) Small revisions to predicted distances around metal sites in proteins. Acta Crystallogr. Sect. D-Biol. Crystallogr. 62: 678-682.

[17] Kolatkar AR, Weis WI (1996) Structural basis of galactose recognition by C-type animal lectins. J. Biol. Chem. 271: 6679-6685.

[18] Guo Y, Feinberg H, Conroy E, Mitchell DA, Alvarez R, Blixt O, Taylor ME, Weis WI, Drickamer K (2004) Structural basis for distinct ligand-binding and targeting properties of the receptors DC-SIGN and DC-SIGNR. Nat. Struct. Mol. Biol. 11: 591-598.

[19] Teillet F, Dublet B, Andrieu JP, Gaboriaud C, Arland GJ, Thielens NM (2005) The two major oligomeric forms of human mannan-binding lectin: Chemical characterization,

carbohydrate-binding properties, and interaction with MBL-associated serine proteases. J. Immunol. 174: 2870-2877.

[20] Rini JM (1995) Lectin Structure. Annu. Rev. Biophys.Biomol. Struct. 24: 551-577.

[21] Martinez-Pomares L, Linehan SA, Taylor PR, Gordon S (2001) Binding properties of the mannose receptor. Immunobiology 204: 527-535.

[22] Banchereau J, Steinman RM (1998) Dendritic cells and the control of immunity. Nature 392: 245-252.

[23] Janeway CA, Medzhitov R (2002) Innate immune recognition. Annu. Rev. Immunol. 20: 197-216.

[24] Geijtenbeek TB, Krooshoop DJ, Bleijs DA, van Vliet SJ, van Duijnhoven GC, Grabovsky V, Alon R, Figdor CG, van Kooyk Y (2000) DC-SIGN-ICAM-2 interaction mediates dendritic cell trafficking. Nat. Immunol. 1: 353-357.

[25] Geijtenbeek TBH, Torensma R, van Vliet SJ, van Duijnhoven GCF, Middel J, Cornelissen ILMHA, Adema GJ, Nottet HSLM, Figdor CG, van Kooyk Y (1999) DC-SIGN, a novel dendritic cell-specific adhesion receptor for ICAM-3 mediates DC-T cell interactions and HIV-1 infection of DC. Blood 94: 434a-434a.

[26] Mitchell DA, Fadden AJ, Drickamer K (2001) A novel mechanism of carbohydrate recognition by the C-type lectins DC-SIGN and DC-SIGNR. Subunit organization and binding to multivalent ligands. J. Biol. Chem. 276: 28939-28945.

[27] Feinberg H, Guo Y, Mitchell DA, Drickamer K, Weis WI (2005) Extended neck regions stabilize tetramers of the receptors DC-SIGN and DC-SIGNR. J. Biol. Chem. 280: 1327-1335.

[28] Tabarani G, Thepaut M, Stroebel D, Ebel C, Vives C, Vachette P, Durand D, Fieschi F (2009) DC-SIGN Neck Domain Is a pH-sensor Controlling Oligomerization: Saxs and Hydrodynamic Studies of Extracellular Domain. J. Biol. Chem. 284: 21229-21240.

[29] Cambi A, de Lange F, van Maarseveen NM, Nijhuis M, Joosten B, van Dijk EM, de Bakker BI, Fransen JA, Bovee-Geurts PH, van Leeuwen FN and others (2004) Microdomains of the C-type lectin DC-SIGN are portals for virus entry into dendritic cells. J. Cell Biol. 164: 145-155.

[30] Cambi A, Koopman M, Figdor CG (2005) How C-type lectins detect pathogens. Cell. Microbiol. 7: 481-488.

[31] Menon S, Rosenberg K, Graham SA, Ward EM, Taylor ME, Drickamer K, Leckband DE (2009) Binding-site geometry and flexibility in DC-SIGN demonstrated with surface force measurements. Proc. Natl. Acad. Sci. USA 106: 11524-11529.

[32] Khoo US, Chan KY, Chan VS, Lin CL. DC-SIGN and L-SIGN: the SIGNs for infection (2008) J. Mol. Med. 86: 861-874.

[33] Tsegaye TS, Pohlmann S (2010) The multiple facets of HIV attachment to dendritic cell lectins. Cell. Microbiol. 12: 1553-1561.

[34] Cambi A, Beeren I, Joosten B, Fransen JA, Figdor CG (2009) The C-type lectin DC-SIGN internalizes soluble antigens and HIV-1 virions via a clathrin-dependent mechanism. Eur. J. Immunol. 39: 1923-1928.

[35] Kwon DS, Gregorio G, Bitton N, Hendrickson WA, Littman DR (2002) DC-SIGN-mediated internalization of HIV is required for trans-enhancement of T cell infection. Immunity 16: 135-144.

[36] Trumpfheller C, Park CG, Finke J, Steinman RM, Granelli-Piperno A (2003) Cell type-dependent retention and transmission of HIV-1 by DC-SIGN. Int. Immunol. 15: 289-298.

[37] van Montfort T, Eggink D, Boot M, Tuen M, Hioe CE, Berkhout B, Sanders RW (2011) HIV-1 N-Glycan Composition Governs a Balance between Dendritic Cell-Mediated Viral Transmission and Antigen Presentation. J. Immunol. 187: 4676-4685.

[38] Sodhi A, Montaner S, Gutkind JS (2004) Viral hijacking of G-protein-coupled-receptor signalling networks. Nat. Rev. Mol. Cell. Biol. 5: 998-1012.

[39] van Kooyk Y, Appelmelk B, Geijtenbeek TB (2003) A fatal attraction: Mycobacterium tuberculosis and HIV-1 target DC-SIGN to escape immune surveillance. Trends Mol. Med. 9: 153-159.

[40] Hodges A, Sharrocks K, Edelmann M, Baban D, Moris A, Schwartz O, Drakesmith H, Davies K, Kessler B, McMichael A, Simmons A (2007). Activation of the lectin DC-SIGN induces an immature dendritic cell phenotype triggering Rho-GTPase activity required for HIV-1 replication. Nat Immunol 2007;8:569-77.

[41] Reina JJ, Sattin S, Invernizzi D, Mari S, Martinez-Prats L, Tabarani G, Fieschi F, Delgado R, Nieto PM, Rojo J, Bernardi A (2007) 1,2-Mannobioside mimic: synthesis, DC-SIGN interaction by NMR and docking, and antiviral activity. ChemMedChem 2: 1030-1036.

[42] Alen MMF, Kaptein SJF, De Burghgraeve T, Balzarini J, Neyts J, Schols D (2009) Antiviral activity of carbohydrate-binding agents and the role of DC-SIGN in dengue virus infection. Virology 387: 67-75.

[43] Anderluh M, Jug G, Svajger U, Obermajer N (2012) DC-SIGN Antagonists, a Potential New Class of Anti-Infectives. Curr. Med. Chem. 19: 992-1007.

[44] Borrok MJ, Kiessling LL (2007) Non-carbohydrate inhibitors of the lectin DC-SIGN. J. Am. Chem. Soc. 129: 12780-12785.

[45] Lederman MM, Offord RE, Hartley O (2006) Microbicides and other topical strategies to prevent vaginal transmission of HIV. Nat. Rev. Immunol. 6: 371-382.

[46] Appelmelk BJ, van Die I, van Vliet SJ, Vandenbroucke-Grauls CM, Geijtenbeek TB, van Kooyk Y (2003) Cutting edge: carbohydrate profiling identifies new pathogens that interact with dendritic cell-specific ICAM-3-grabbing nonintegrin on dendritic cells. J. Immunol. 170: 1635-1639.

[47] Feinberg H, Mitchell DA, Drickamer K, Weis WI (2001) Structural basis for selective recognition of oligosaccharides by DC-SIGN and DC-SIGNR. Science 294: 2163-2166.

[48] Feinberg H, Castelli R, Drickamer K, Seeberger PH, Weis WI (2007) Multiple modes of binding enhance the affinity of DC-SIGN for high mannose N-linked glycans found on viral glycoproteins. J. Biol. Chem. 282: 4202-4209.

[49] Gringhuis SI, den Dunnen J, Litjens M, van der Vlist M, Geijtenbeek TB (2009) Carbohydrate-specific signaling through the DC-SIGN signalosome tailors immunity to Mycobacterium tuberculosis, HIV-1 and Helicobacter pylori. Nat. Immunol. 10: 1081-1088.

[50] Timpano G, Tabarani G, Anderluh M, Invernizzi D, Vasile F, Potenza D, Nieto PM, Rojo J, Fieschi F, Bernardi A (2008) Synthesis of novel DC-SIGN ligands with an alpha-fucosylamide anchor. Chembiochem 9: 1921-1930.
[51] Rojo J, Delgado R (2004) Glycodendritic structures: promising new antiviral drugs. J. Antimicrob. Chemother. 54: 579-581.
[52] Cunningham AL, Turville S, Wilkinson J, Cameron P, Dable J (2003) The role of dendritic cell C-type lectin receptors in HIV pathogenesis. J. Leukoc. Biol. 74: 710-718.
[53] Turville SG, Cameron PU, Handley A, Lin G, Pohlmann S, Doms RW, Cunningham AL (2002) Diversity of receptors binding HIV on dendritic cell subsets. Nat. Immunol. 3: 975-983.
[54] de Witte L, Nabatov A, Pion M, Fluitsma D, de Jong M, de Gruijl T, Piguet V, van Kooyk Y, Geijtenbeek TBH (2007) Langerin is a natural barrier to HIV-1 transmission by Langerhans cells. Nat. Med. 13: 367-371.
[55] Chatwell L, Holla A, Kaufer BB, Skerra A (2008) The caxbohydrate recognition domain of Langerin reveals high structural similarity with the one of DC-SIGN but an additional, calcium-independent sugar-binding site. Mol. Immunol. 45: 1981-1994.
[56] Holla A, Skerra A (2011) Comparative analysis reveals selective recognition of glycans by the dendritic cell receptors DC-SIGN and Langerin. Prot. Eng. Des. & Sel. 24: 659-669.
[57] Reina JJ, Diaz I, Nieto PM, Campillo NE, Paez JA, Tabarani G, Fieschi F, Rojo J (2008) Docking, synthesis, and NMR studies of mannosyl trisaccharide ligands for DC-SIGN lectin. Org. Biomol. Chem. 6: 2743-2754.
[58] Andreini M, Doknic D, Sutkeviciute I, Reina JJ, Duan JX, Chabrol E, Thepaut M, Moroni E, Doro F, Belvisi L, Weiser J, Rojo J, Fieschi F, Bernardi A (2011) Second generation of fucose-based DC-SIGN ligands: affinity improvement and specificity versus Langerin. Org. Biomol. Chem. 9: 5778-5786.
[59] Mitchell DA, Jones NA, Hunter SJ, Cook JMD, Jenkinson SF, Wormald MR, Dwek RA, Fleet GWJ (2007) Synthesis of 2-C-branched derivatives of D-mannose: 2-C-aminomethyl-D-mannose binds to the human C-type lectin DC-SIGN with affinity greater than an order of magnitude compared to that of D-mannose. Tetrahedron-Asymmetry 18: 1502-1510.
[60] Garber KCA, Wangkanont K, Carlson EE, Kiessling LL (2010) A general glycomimetic strategy yields non-carbohydrate inhibitors of DC-SIGN. Chem. Commun. 46: 6747-6749.
[61] Gloster TM, Davies GJ (2010) Glycosidase inhibition: assessing mimicry of the transition state. Org. Biomol. Chem. 8: 305-320.
[62] Almendros M, Danalev D, Francois-Heude M, Loyer P, Legentil L, Nugier-Chauvin C, Daniellou R, Ferrieres V (2011) Exploring the synthetic potency of the first furanothioglycoligase through original remote activation. Org. Biomol. Chem. 9: 8371-8378.
[63] Cipolla L, La Ferla B, Airoldi C, Zona C, Orsato A, Shaikh N, Russo L, Nicotra F (2010) Carbohydrate mimetics and scaffolds: sweet spots in medicinal chemistry. Future Med. Chem. 2:587-599.

[64] Obermajer N, Sattin S, Colombo C, Bruno M, Svajger U, Anderluh M, Bernardi A (2011) Design, synthesis and activity evaluation of mannose-based DC-SIGN antagonists. Mol. Divers. 15: 347-360.

[65] Van Liempt E, Imberty A, Bank CMC, Van Vliet SJ, Van Kooyk Y, Geijtenbeek TBH, Van Die I (2004) Molecular basis of the differences in binding properties of the highly related C-type lectins DC-SIGN and L-SIGN to Lewis x trisaccharide and Schistosoma mansoni egg antigens. J. Biol. Chem. 279: 33161-33167.

[66] Shinohara Y, Hasegawa Y, Kaku H, Shibuya N (1997) Elucidation of the mechanism enhancing the avidity of lectin with oligosaccharides on the solid phase surface. Glycobiology 7: 1201-1208.

[67] Andrews PR, Craik DJ, Martin JL (1984) Functional-Group Contributions to Drug Receptor Interactions. J. Med. Chem. 27: 1648-1657.

[68] Lasala F, Arce E, Otero JR, Rojo J, Delgado R (2003) Mannosyl glycodendritic structure inhibits DC-SIGN-mediated Ebola virus infection in cis and in trans. Antimicrob. Agents Chemother. 47: 3970-3972.

[69] Becer CR, Gibson MI, Geng J, Ilyas R, Wallis R, Mitchell DA, Haddleton DM (2010) High-Affinity Glycopolymer Binding to Human DC-SIGN and Disruption of DC-SIGN Interactions with HIV Envelope Glycoprotein. J. Am. Chem. Soc. 132: 15130-15132.

[70] Luczkowiak J, Sattin S, Sutkeviciute I, Reina JJ, Sanchez-Navarro M, Thepaut M, Martinez-Prats L, Daghetti A, Fieschi F, Delgado R Bernardi A, Rojo J (2011) Pseudosaccharide Functionalized Dendrimers as Potent Inhibitors of DC-SIGN Dependent Ebola Pseudotyped Viral Infection. Bioconjug. Chem. 22: 1354-1365.

Chapter 8

Broad Antiviral Activity of Carbohydrate-Binding Agents Against Dengue Virus Infection

Marijke Alen and Dominique Schols

Additional information is available at the end of the chapter

http://dx.doi.org/10.5772/50631

1. Introduction

1.1. Origin and epidemiology

Dengue virus (DENV) is a member of the *Flavivirus* genus within the family of the *Flaviviridae* and is the most common mosquito-borne viral disease. *Flaviviruses* derived from a common viral ancestor 10,000 years ago. DENV has a relative recent revolutionary history originating 1000 years ago and establishes transmission in humans since a few hundred years. There is strong evidence that DENV was originally a monkey virus in non-human primates in Africa and Asia. Cross-species transmission to humans has occurred independently for all four DENV serotypes [1,2]. Each serotype shares around 65% of the genome and despite of the differences, each serotype causes nearly identical syndromes in humans and circulates in the same ecological niche [3]. First clinical symptoms of dengue infections date from the 10th century but it is not for sure that this was a dengue epidemic. The first large dengue epidemics were in 1779 in Asia, Africa and North America. The first reported epidemic of dengue hemorrhagic fever (DHF) was in Manilla, Philippines, in 1953 after World War II. It was suggested that he movement of the troops during World War II has led to the spread of the virus. It has been shown that in the 19th and 20th century, the virus was widespread in the tropics and subtropics where nowadays 3.6 billion people are at risk of getting infected with DENV (Figure 1). Every year, 50 million infections occur, including 500,000 hospitalizations for DHF, mainly among children, with a case fatality rate exceeding 15% in some areas [4,5]. In 40 years of time, DENV became endemic in more than 100 countries because of the increase in human population, international transport and the lack of vector control.

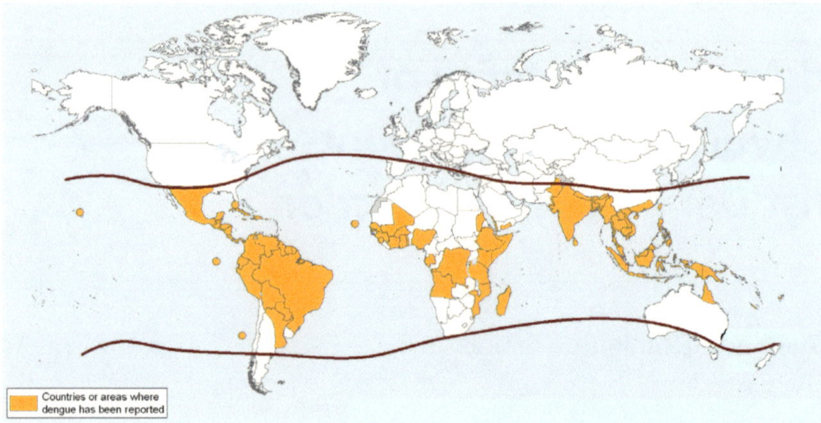

Figure 1. Global distribution of dengue virus infections in 2011. Contour lines represent the areas at risk (Source: WHO, 2012).

1.2. Transmission

The transmission of DENV can only occur by the bite of an infected female mosquito, the *Aedes aegypti* and the *Aedes albopictus*. *Aedes aegypti* originated, and is still present, in the rainforests of Africa feeding on non-human primates (sylvatic cycle, Figure 2). DENV infection in non-human primates occurs asymptomatically. However, the mosquito became domesticated due to massive deforestation and breeds in artificial water holdings, like automobile tires, discarded bottles and buckets that collect rainwater [4]. On one hand *Aedes aegypti* is not an efficient vector because it has a low susceptibility for oral infection with virus in human blood. Since mosquitoes ingest 1 µl of blood, the virus titer in human blood has to be 10^5-10^7 per ml for transmission to be sustained. After 7-14 days the virus has passed the intestinal tract to the salivary glands and can be transmitted by the infected mosquito to a new host. On the other hand, *Aedes aegypti* is an efficient vector because it has adapted to humans and they repeatedly feed themselves in daylight on different hosts. After a blood meal, the oviposition can be stimulated and the virus can be passed transovarially to the next generation of mosquitoes (vertical transmission, Figure 2) [6].

The tiger mosquito *Aedes albopictus* is spreading his region from Asia to Europe and the United States of America (USA). In the 1980s, infected *Aedes Albopictus* larvae were transported in truck tires from Asia to the United States. Dengue viruses were introduced into port cities, resulting in major epidemics [6].

Because there is no vaccine available, the only efficient way to prevent DENV infection is eradication of the mosquito. In the 1950s and 1960s there was a successful vector control program in the USA organized by the Pan American Health Organization. They eradicated the mosquito from 19 countries. Unfortunately, the vector control program was stopped in 1972 because the government thought that DENV was not important anymore [2,4]. This resulted in a re-emergence of the mosquito and DENV infections in the USA. Both demographic and ecological changes contributed to the world wide spread of DENV infections.

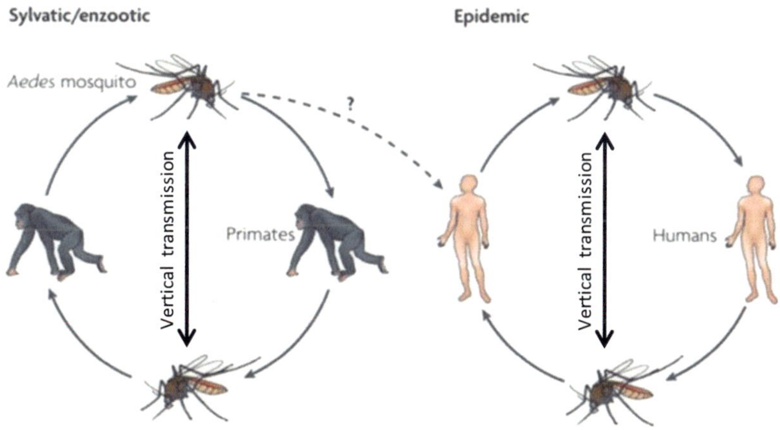

Figure 2. Transmission of DENV by Aedes mosquitoes. Modified from Whitehead et al. [7].

Very recently, another approach to attack the vector has been documented [8,9]. There was a mosquito made resistant to DENV infection after trans infection with the endosymbiont *Wolbachia* bacterium which can infect a lot of insects. A certain strain of the *Wolbachia* bacteria was trans infected in *Aedes* mosquitoes and was reported to inhibit the replication and dissemination of several RNA viruses, such as DENV. Embryos of a *Wolbachia* uninfected female die if the female has bred with a *Wolbachia* infected male. This means that *Wolbachia* infected mosquitoes can take over the natural population. This was recently tested in Australia, where dengue is endemic, and after 2 months the *Wolbachia* infected mosquitoes resistant to DENV had taken over the natural mosquito population. Thus, this indicates the beginning of a new area in vector control efforts with a high potential to succeed.

1.3. Pathogenesis

Although DENV infections have a high prevalence, the pathogenesis of the disease is not well understood. The disease spectrum can range from an asymptomatic or flu like illness to a lethal disease. After a bite of an infected mosquito, there is an incubation period of 3 to 8 days. Then there is on acute onset of fever (≥ 39°C) accompanied by nonspecific symptoms like severe headache, nausea, vomiting, muscle and joint pain (dengue fever). Clinical findings alone are not helpful to distinguish dengue fever from other febrile illnesses such as malaria or measles. Half of the infected patients report a rash and is most commonly seen on the trunk and the insides of arms and thighs. Skin hemorrhages, including petechiae and purpura, are very common. Liver enzyme levels of alanine aminotransferase and aspartate aminotransferase can be elevated. Dengue fever is generally self-limiting and is rarely fatal [5,10,11].

The disease can escalate into dengue hemorrhagic fever (DHF) or dengue shock syndrome (DSS). DHF is primarily a children's disease and is characterized as an acute febrile illness with thrombocytopenia (≤ 100,000 cells/mm^3). This results in an increased vascular permeability and plasma leakage from the blood vessels into the tissue. Plasma leakage has been documented by an increased hematocrit and a progressive decrease in platelet count. Petechiae and subcutaneous bleedings are very common [12].

DSS is defined when the plasma leakage becomes critical resulting in circulatory failure, weak pulse and hypotension. Plasma volume studies have shown a reduction of more than 20% in severe cases. A progressively decreasing platelet count, a rising hematocrit, sustained abdominal pain, persistent vomiting, restlessness and lethargy may be all signs for DSS. Prevention of shock can only be established after volume replacement with intravenous fluids [5,11]. When experienced clinicians and nursing staff are available in endemic areas, the case fatality rate is < 1%.

DHF and DSS occur during a secondary infection with a heterologous serotype. The first infection with one of the four serotypes provides lifelong immunity to the homologous virus. During a second infection with a heterologous serotype, non-neutralizing IgG antibodies can enhance disease severity. This phenomenon is called antibody-dependent enhancement (ADE). The pre-existing non-neutralizing heterotypic antibodies can form a complex with DENV and enhance the access to Fc-receptor bearing cells such as monocytes and macrophages [13,14] (Figure 3).

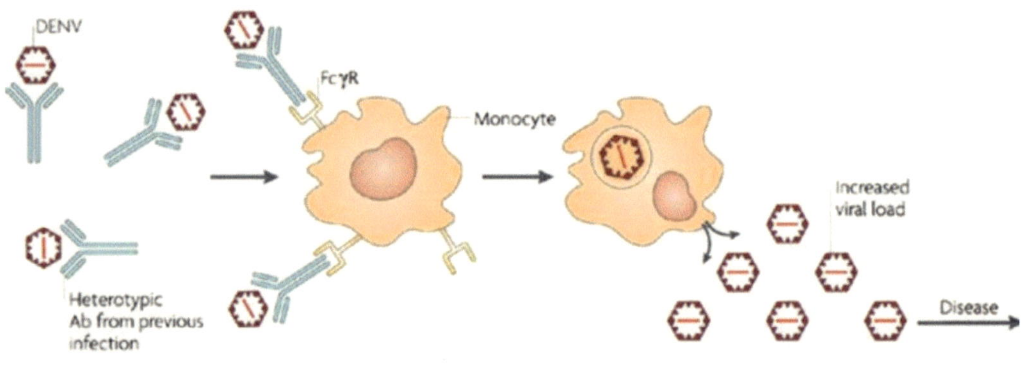

Figure 3. Mechanism of antibody-dependent enhancement (ADE). During a secondary infection caused by a heterologous virus, the pre-existing heterotypic antibodies can cross-react with the other DENV serotypes. The non-neutralizing antibody-virus complex can interact with the Fc-receptor on monocytes or macrophages. This will lead to an increased viral load and a more severe disease. Figure derived from Whitehead et al. [7].

This will lead to an increase in viral load and a more severe disease. These non-neutralizing antibodies can cross-react with all four virus serotypes, as well as with other flaviviruses. This phenomenon explains why young infants born to dengue immune mothers often experience a more severe disease due to transplacental transfer of DENV-specific antibodies [15]. Another approach to assist this phenomenon is the observation of increased viremia in non-human primates which received passive immunization with antibodies against DENV [16].

A second mechanism to explain ADE of flaviviruses is the involvement of the complement system. It has been shown that monoclonal antibodies against complement receptor 3 inhibit ADE of West Nile virus *in vitro* [14]. But Fc-receptor-dependent ADE is believed to be the most common mechanism of ADE.

2. DENV entry

2.1. Entry process

The infectious entry of DENV into its target cells, mainly dendritic cells [17], monocytes and macrophages, is mediated by the viral envelope glycoprotein E via receptor-mediated endocytosis [18]. The E-glycoprotein is the major component (53 kDa) of the virion surface and is arranged as 90 homodimers in mature virions [19]. Recent reports demonstrated also that DENV enters its host cell via clathrin-mediated endocytosis [20,21], as observed with other types of flaviviruses [22,23]. Evidence for flavivirus entry via this pathway is based on the use of inhibitors of clathrin-mediated uptake, such as chlorpromazine. However, DENV entry via a non-classical endocytic pathway independent from clathrin has also been described [24]. It seems that the entry pathway chosen by DENV is highly dependent on the cell type and viral strain. In case of the classical endocytic pathway, there is an uptake of the receptor-bound virus by clathrin coated vesicles. These vesicles fuse with early endosomes to deliver the viral RNA into the cytoplasm. The E-protein responds to the reduced pH of the endosome with a large conformational rearrangement [25,26]. The low pH triggers dissociation of the E-homodimer, which then leads to the insertion of the fusion peptide into the target cell membrane forming a bridge between the virus and the host. Next, a stable trimer of the E-protein is folded into a hairpin-like structure and forces the target membrane to bend towards the viral membrane and eventually fusion takes place [25,27,28]. The fusion results in the release of viral RNA into the cytoplasm for initiation of replication and translation (Figure 4).

2.2. The DENV envelope

The DENV E-glycoprotein induces protective immunity and flavivirus serological classification is based on its antigenic variation. During replication the virion assumes three conformational states: the immature, mature and fusion-activated form. In the immature state, the E-protein is arranged as a heterodimer and generates a "spiky" surface because the premembrane protein (prM) covers the fusion peptide. In the Golgi apparatus, the virion maturates after a rearrangement of the E-protein. The E-heterodimer transforms to an E-homodimer and results in a "smoothy" virion surface. After a furin cleavage of the prM to pr and M, the virion is fully maturated and can be released from the host cell. Upon fusion, the low endosomal pH triggers the rearrangement of the E-homodimer into a trimer [29].

The E-protein monomer is composed out of β-barrels organized in three structural domains (Figure 5).

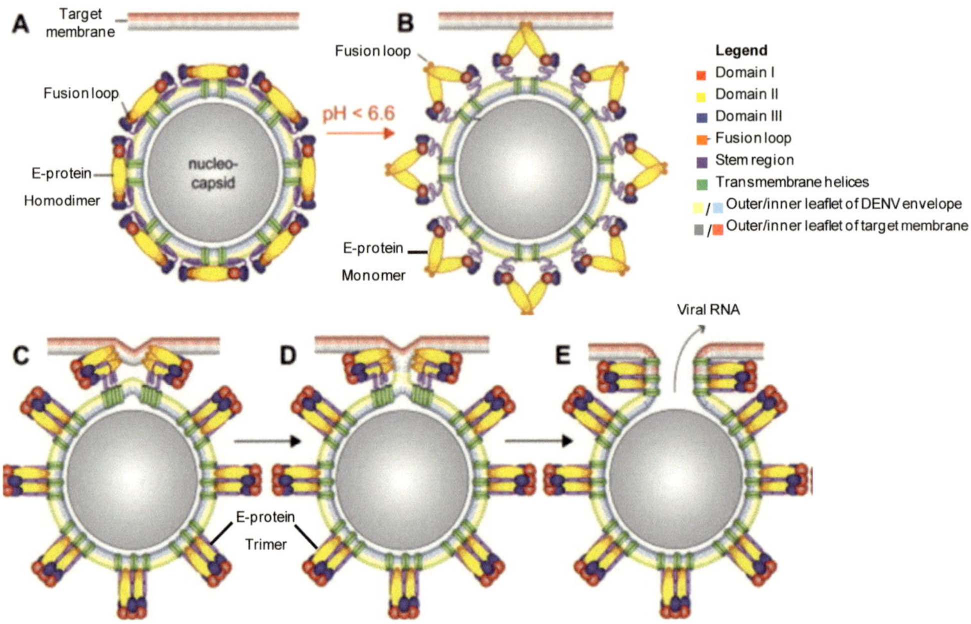

Figure 4. Schematic overview of the DENV membrane fusion process. (A) Pre-fusion conformation of the E-protein consists of homodimers on the virus surface. (B) Low endosomal pH triggers dissociation of the E-dimers into monomers which leads to the insertion of the fusion peptide with the endosomal target membrane. (C) A stable E-protein trimer is folded in a hairpin-like structure. (D) Hemifusion intermediate in which only the outer leaflets of viral and target cellular membranes have fused. (E) Formation of the post-fusion E-protein trimer and opening of the fusion pore allows the release of the viral RNA into the cytoplasm. Modified from Stiasny et al. [26].

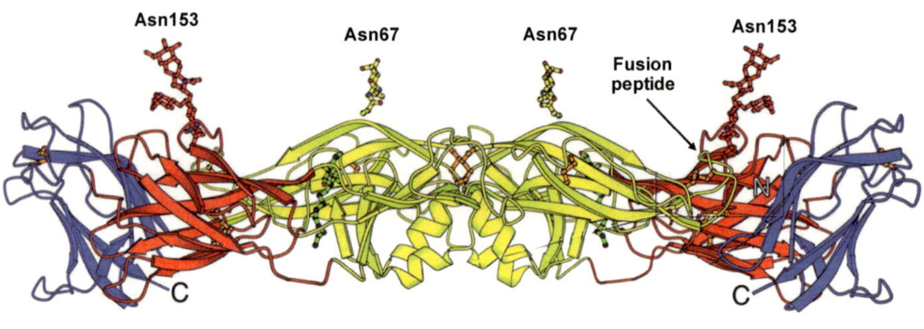

Figure 5. Location of the 2 N-glycans on the envelope protein of DENV. The DENV E-protein dimer carries 2 N-glycans on each monomer at Asn67 and Asn153. β-strands are shown as ribbons with arrows, α-helices are shown as coiled ribbons. Thin tubes represent connecting loops. Domain I is shown in red, domain II is shown in yellow and contains the fusion peptide near Asn153. Domain III is shown in blue. Disulfide bridges are shown in orange. In green, the ligand N-octyl-D-glucoside is shown, which interacts with the hydrophobic pocket between domain I and II. Modified from Modis et al. [30].

The central domain I contains the aminoterminus and contains two disulfide bridges. Domain II is an extended finger-like domain that bears the fusion peptide and stabilizes the dimer. This sequence contains three disulfide bridges and is rich in glycine. Between domain I and domain II is a binding pocket that can interact with a hydrophobic ligand, the detergent β-N-octyl-glucoside. This pocket is an important target for antiviral therapy because mutations in this region can alter virulence and the pH necessary for the induction of conformational changes. The immunoglobulin-like domain III contains the receptor binding motif, the C-terminal domain and one disulfide bond [30,31]. Monoclonal antibodies recognizing domain III are the most efficient of blocking DENV [32,33] and this domain is therefore an interesting target for antiviral therapy.

Because dendritic cell-specific intercellular adhesion molecule 3-grabbing non-integrin (DC-SIGN) (See 1.3.1) is identified as an important receptor for DENV in primary DC in the skin and DC-SIGN recognizes high-mannose sugars, carbohydrates present on the E-protein of DENV could be important for viral attachment. The E-protein has two potential glycosylation sites: asparagines 67 (Asn67) and Asn153. Glycosylation at Asn153 is conserved in flaviviruses, with the exception of Kunjin virus, a subtype of West Nile virus [34] and is located near the fusion peptide in domain II [30,31] (Figure 5). Glycosylation at Asn67 is unique for DENV [31].

3. Role of DC-SIGN in DENV infection

Prior to fusion, DENV needs to attach to specific cellular receptors. Because DENV can infect a variety of different cell types isolated from different hosts (human, insect, monkey and even hamster), the virus must interact with a wide variety of cellular receptors. In the last decade, several candidate attachment factor/receptors are identified. DC-SIGN is described as the most important human cellular receptor for DENV.

Since 1977, monocytes are considered to be permissive for DENV infection [35]. More recent, phenotyping of peripheral blood mononuclear cells (PBMCs) from pediatric DF and DHF cases resulted in the identification of monocytes as DENV target cells [36].

First, it was believed that monocytes are important during secondary DENV infections during the ADE process, because of their Fc-receptor expression. The complex formed between the non-neutralizing antibody and the virus can bind to Fc-receptors and enhance infection in neighboring susceptible cells [14,18,37]. However, *in vitro*, monocytes isolated from PBMCs, apparently have a very low susceptibility for DENV infection for reasons that remain to be elucidated.

More detailed observation of the natural DENV infection, changes the idea of monocytes being the first target cells. Following intradermal injection of DENV-2 in mice, representing the bite of an infected mosquito, DENV occurs to replicate in the skin [38]. The primary DENV target cells in the skin are believed to be immature dendritic cells (DC) or Langerhans cells [17,39-41]. Immature DC are very efficient in capturing pathogens whereas mature DC are relatively resistant to infection. The search for cellular receptors responsible for DENV capture leads to the identification of cell-surface C-type lectin DC-specific intercellular adhesion molecule 3-grabbing non-integrin (DC-SIGN; CD209) [42-45]. DC-SIGN is mainly expressed by immature DC, but also alveolar macrophages and interstitial DC in the lungs, intestine, placenta and in lymph nodes express DC-SIGN [46]. DC-SIGN is a tetrameric transmembrane receptor and is a member of the calcium-dependent C-type lectin family. The receptor is composed of four domains: a cytoplasmic domain responsible for signaling and internalization due to the presence of a dileucine motif, a transmembrane domain, seven to eight extracellular neck repeats implicated in the tetramerization of DC-SIGN and a carbohydrate recognition domain (CRD) (Figure 6) [47].

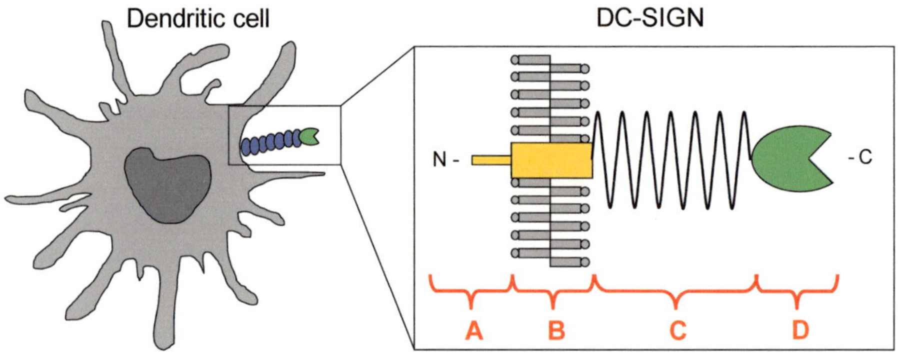

Figure 6. Structure of DC-SIGN. DC-SIGN, mainly expressed by human dendritic cells in the skin, is composed out of four domains: (A) cytoplasmic domain containing internalization signals, (B) transmembrane domain, (C) 7 or 8 extracellular neck repeats implicated in the oligomerization of DC-SIGN and (D) carbohydrate recognition domain which can interact calcium-dependent with a variety of pathogens.

Alen *et al.* [42] investigated the importance of DC-SIGN receptor in DENV infections using DC-SIGN transfected Raji cells versus Raji/0 cells. A strong contrast in DENV susceptibility was observed between Raji/DC-SIGN$^+$ cells and Raji/0 cells. DC-SIGN expression renders cells susceptible for DENV infection. Also in other cell lines, the T-cell line CEM and the astroglioma cell line U87, expression of DC-SIGN renders the cells permissive for DENV infection. To evaluate the importance of DC-SIGN, Raji/DC-SIGN$^+$ cells were incubated with a specific anti-DC-SIGN antibody prior to DENV infection. This resulted in an inhibition of the DENV replication by ~90%, indicating that DC-SIGN is indeed an important receptor for DENV. Also 2 mg/ml of mannan inhibited the DENV infection in Raji/DC-SIGN$^+$ cells by more than 80%. This data indicate that the interaction between DC-SIGN and DENV is dependent on mannose-containing N-glycans present on the DENV envelope [42].

Thus, the CRD of DC-SIGN recognizes high-mannose N-glycans and also fucose-containing blood group antigens [48,49]. Importantly, DC-SIGN can bind a variety of pathogens like human immunodeficiency virus (HIV) [50], hepatitis C virus (HCV) [51], ebola virus [52] and several bacteria, parasites and yeasts [46]. Many of these pathogens have developed strategies to manipulate DC-SIGN signaling to escape from an immune response [46]. Following antigen capture in the periphery, DC maturate by up regulation of the costimulatory molecules and down regulation of DC-SIGN. By the interaction with ICAM-2 on the vascular endothelial cells, DC can migrate to secondary lymphoid organs [53]. Next, the activated DC interact with ICAM-3 on naïve T-cells. This results in the stimulation of the T-cells and subsequently in the production of cytokines and chemokines [54]. Inhibition of the initial interaction between DENV and DC could prevent an immune response. DC-SIGN could be considered as a target for antiviral therapy by interrupting the viral entry process. But caution must be taken into account as the DC-SIGN receptor has also an important role in the activation of protective immune responses instead of promoting the viral dissemination. However, several DC-SIGN antagonists have been developed such as small interfering RNAs (siRNA) silencing DC-SIGN expression [55], specific anti-DC-SIGN antibodies [56] and glycomimetics interacting with DC-SIGN [57]. The *in vivo* effects of DC-SIGN antagonists remain to be elucidated.

Besides DC, macrophages play a key role in the immune pathogenesis of DENV infection as a source of immune modulatory cytokines [58]. Recently, Miller *et al.* showed that the mannose receptor (MR; CD206) mediates DENV infection in macrophages by recognition of the glycoproteins on the viral envelope [59]. Monocyte-derived DC (MDDC) can be generated out of monocytes isolated from fresh donor blood incubated IL-4 and GM-CSF. After a differentiation process MDDC were generated highly expressing DC-SIGN (Figure 7A, B) and showing a significantly decrease in CD14 expression in contrast to monocytes [59,60]. Again, DC-SIGN expression on MDDC renders cells susceptible for DENV in contrast to monocytes (Figure 7A, B). MR is also present on monocyte-derived DC (MDDC) and anti-MR antibodies can inhibit DENV infection, although to a lesser extent than anti-DC-SIGN antibodies do (Figure 7C) [61]. Furthermore, the combination of anti-DC-SIGN and anti-MR antibodies was even more effective in inhibiting DENV infection. Yet, complete inhibition of DENV infection was not achieved, indicating that other entry pathways are potentially involved. Two other receptors on DC reported to be responsible for HIV attachment are syndecan-3 (a member of the heparan sulfate proteoglycan family) [62] and the DC immune receptor [63]. Since DENV interacts with heparan sulfate, syndecan-3 may be a possible (co)-receptor on DC. It has been hypothesized that DENV needs DC-SIGN for attachment and enhancing infection of DC *in cis* and needs MR for internalization [59]. In fact, cells expressing mutant DC-SIGN, lacking the internalization domain, are still susceptible for DENV infection because DC-SIGN can capture the pathogen [43].

Another C-type lectin, CLEC5A (C-type lectin domain family 5, member A) expressed by human macrophages can also interact with DENV and acts as a signaling receptor for the release of proinflammatory cytokines [64]. However, whereas the DC-SIGN-DENV interaction is calcium-dependent, CLEC5A binding to its ligand is not dependent on calcium. Mannan and fucose can inhibit the interaction between CLEC5A and DENV, indicating that the interaction is carbohydrate-dependent [64]. However, a glycan array demonstrated no binding signal between CLEC5A and N-glycans of mammals or insects [65]. The molecular interaction between CLEC5A and DENV remains to be elucidated.

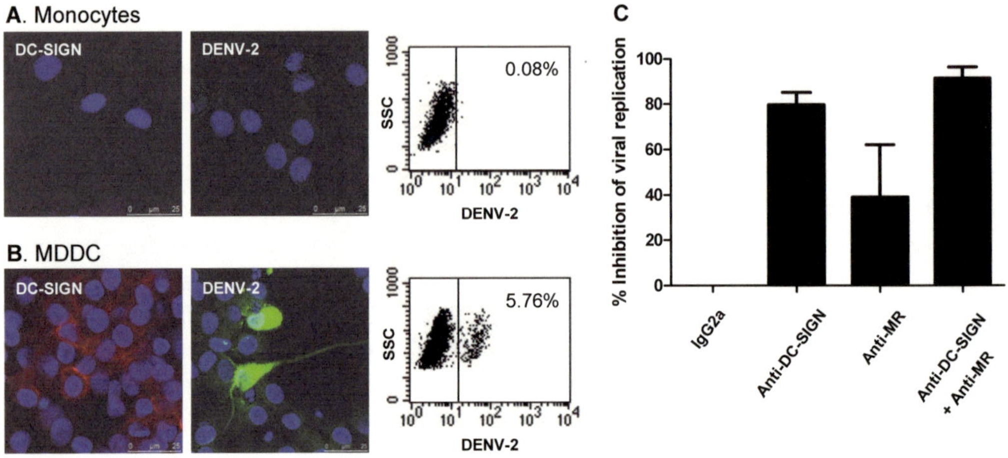

Figure 7. Infection of MDDC by DENV. Monocytes isolated from PBMCs were untreated (A) or treated with 25 ng/ml IL-4 and 50 ng/ml GM-CSF (B) for 5 days prior to DENV-2 infection. Two days after infection the cells were permeabilized and analyzed for DC-SIGN expression and DENV infection by confocal microscopy and flow cytometry. Uninfected cells were stained with a PE-labeled monoclonal DC-SIGN-antibody (red). Infected cells were stained with a mixture of antibodies recognizing DENV-2 E-protein and PrM protein (green). Nuclei were stained with DAPI (blue). Infected monocytes (A) and MDDC (B) were analyzed by flow cytometry to detect DENV-2 positive cells. The values indicated in each dot plot represent the % of DENV-2 positive cells. (C) MDDC were preincubated with 10 μg/ml of isotype control IgG_{2a}, anti-DC-SIGN or anti-MR antibody for 30 minutes before DENV-2 infection. Viral replication was analyzed by flow cytometry. % Inhibition of viral replication ± SEM of 4 different blood donors is shown.

Immune cells, in particular dendritic cells, are the most relevant cells in the discovery of specific antiviral drugs against dengue virus, but the isolation of these cells and the characterization is unfortunately labour intensive and time consuming.

Liver/lymph node-specific ICAM-3 grabbing non-integrin (L-SIGN) is a DC-SIGN related transmembrane C-type lectin expressed on endothelial cells in liver, lymph nodes and placenta [66,67]. Similar to DC-SIGN, L-SIGN is a calcium-dependent carbohydrate-binding protein and can interact with HIV [67], HCV [51], Ebola virus [52], West Nile virus [68] and DENV [45]. Zellweger *et al.* observed that during antibody-dependent enhancement in a mouse model that liver sinusoidal endothelial cells (LSEC) are highly permissive for antibody-dependent DENV infection [69]. Given the fact that LSEC express L-SIGN, it is interesting to focus on the role of L-SIGN in DENV infection. L-SIGN expression on LSEC has probably an important role in ADE *in vivo* and therefore it is interesting to find antiviral agents interrupting the DENV-L-SIGN interaction and subsequently prevent the progression to the more severe and lethal disease DHF/DSS. Although endothelial cells [70] and liver endothelial cells [71] are permissive for DENV and L-SIGN-expression renders cells susceptible for DENV infection, the *in vivo* role for L-SIGN in DENV entry remains to be established.

4. Antiviral therapy

At present, diagnosis of dengue virus infection is largely clinical, treatment is supportive through hydration and disease control is limited by eradication of the mosquito. Many efforts have been made in the search for an effective vaccine, but the lack of a suitable animal model, the need for a high immunogenicity vaccine and a low reactogenicity are posing huge challenges in the dengue vaccine development [7,72]. There are five conditions for a dengue vaccine to be effective: (i) the vaccine needs to be protective against all four serotypes without reactogenicity, because of the risk of ADE, (ii) it has to be safe for children, because severe dengue virus infections often affects young children, (iii) the vaccine has to be economical with minimal or no repeat immunizations, because dengue is endemic in many developing countries, (iv) the induction of a long-lasting protective immune response is necessary and finally (v) the vaccine may not infect mosquitoes by the oral route [7,73].

As there is no vaccine available until now, the search for antiviral products is imperative. The traditional antiviral approach often attacks viral enzymes, such as proteases and polymerases [74,75]. Because human cells lack RNA-dependent polymerase, this enzyme is very attractive as antiviral target without cytotoxicity issues. Nucleoside analogues and non-nucleoside compounds have previously been shown to be very effective in anti-HIV therapy and anti-hepatitis B virus therapy. The protease activity is required for polyprotein processing which is necessary for the assembly of the viral replication complex. Thereby, the protease is an interesting target for antiviral therapy. However, the host cellular system has similar protease activities thus cytotoxic effects form a major recurrent problem. Very recently, many efforts have been made in the development of polymerase and protease inhibitors of DENV, but until today, any antiviral product has reached clinical trials. This chapter is focusing on a different step in the virus replication cycle, namely, the viral entry process. In the past few years, progression has been made in unraveling the host cell pathways upon DENV infection. It is proposed that viral epitopes on the surface of DENV can trigger cellular immune responses and subsequently the development of a severe disease. Therefore, these epitopes are potential targets for the development of a new class of antiviral products, DENV entry inhibitors. Inhibition of virus attachment is a valuable antiviral strategy because it forms the first barrier to block infection. Several fusion inhibitors, glycosidase inhibitors and heparin mimetics have been described to inhibit DENV entry in the host cell. Here, specific molecules, the carbohydrate-binding agents (CBAs), preventing the interaction between the host and the N-glycans present on the DENV envelope are discussed.

4.1. Carbohydrate-binding agents (CBAs)

The CBAs form a large group of natural proteins, peptides and even synthetic agents that can interact with glycosylated proteins. CBAs can be isolated from different organisms: algae, prokaryotes, fungi, plants, invertebrates and vertebrates (such as DC-SIGN and L-SIGN) [76,77]. Each CBA will interact in a specific way with monosaccharides, such as mannose, fucose, glucose, N-acetylglucosamine, galactose, N-acetylgalactosamine or sialic acid residues present in the backbone of N-glycan structures. Because a lot of enveloped viruses are glycosylated at the viral surface, such as HIV, HCV and DENV (Figure 5), CBAs could interact with the glycosylated envelope of the virus and subsequently prevent viral entry into the host cell [78,79]. Previously, antiviral activity against HIV and HCV [78-80] was demonstrated of several CBAs isolated from plants (plant lectins) and algae specifically binding mannose and N-acetylglucosamine residues.

Here, we focus on the antiviral activity of three plant lectins, *Hippeastrum* hybrid agglutinin (HHA), *Galanthus nivalis* agglutinin (GNA) and *Urtica dioica* (UDA) isolated from the amaryllis, the snow drop and the stinging nettle, respectively. In general, plant lectins form a large diverse group of proteins, exhibiting a wide variety of monosaccharide-binding properties which can be isolated from different sites within the plant, such as the bulbs, leaves or roots. HHA (50 kDa) and GNA (50 kDa) isolated from the bulbs are tetrameric proteins. For GNA, each monomer contains two carbohydrate-binding sites and a third site is created once if tetramerization had occurred, resulting in a total of 12 carbohydrate-binding sites (Figure 8). HHA specifically interacts with α1-3 and α1-6 mannose residues and GNA only recognizes α1-3 mannose residues. UDA, isolated from the rhizomes of the nettle, is active as a monomer containing 2 carbohydrate-binding sites recognizing N-acetylglucosamine residues (Figure 8). In 1984, UDA was isolated for the first time and with its molecular weight of 8.7 kDa, UDA is the smallest plant lectin ever reported [81]. The plant lectins have been shown to possess both antifungal and insecticidal activities playing a role in plant defense mechanisms. Here, the antiviral activity of the plant lectins against DENV will be further highlighted and discussed in detail.

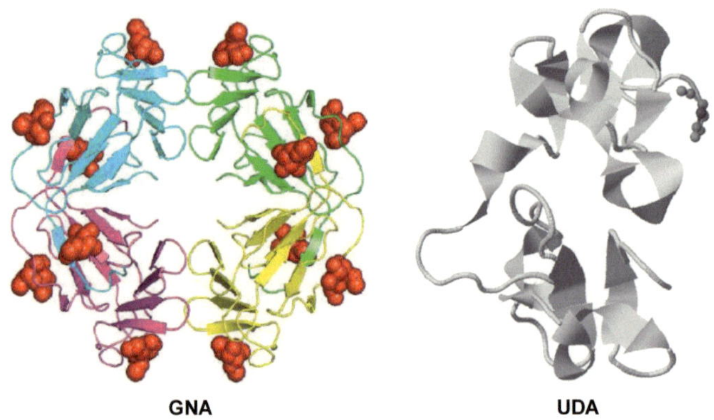

Figure 8. Structure of GNA and UDA. GNA is isolated from the snow drop and is a tetrameric protein. UDA is isolated from the stinging nettle and is a monomeric protein composed out of hevein domains.

Previously, concanavalin A (Con A), isolated from the Jack bean, binding to mannose residues and wheat germ agglutinin binding to N-acetylglucosamine (Glc-NAc) residues, were shown to reduce DENV infection *in vitro*. A competition assay, using mannose, proved that the inhibitory effect of Con A was due to binding α-mannose residues on the viral protein, because mannose successfully competed with Con A [82]. Together with the fact that HHA, GNA and UDA act inhibitory against HIV and HCV, we hypothesized that these plant lectins had antiviral activity against DENV, because DENV has two N-glycosylation sites on the viral envelope and uses DC-SIGN as a cellular receptor to enter DC. The antiviral activity of the three plant lectins was investigated in DC-SIGN+ and L-SIGN+ cells and the infection was analyzed by flow cytometry, RT-qPCR and confocal microscopy.

4.2. Broad antiviral activity of CBAs against DENV

Because DC-SIGN interacts carbohydrate-dependent with DENV, the antiviral activity of the three plant lectins, HHA, GNA and UDA, recognizing monosaccharides present in the backbone of N-glycans on the DENV E-protein, was evaluated. A consistent dose-dependent antiviral activity was observed in DC-SIGN transfected Raji cells against DENV-2 analyzed by flow cytometry (detecting the presence of DENV Ag) and RT-qPCR (detecting viral RNA in the supernatants) [42].

Next, the antiviral potency of the three plant lectins was determined against all four serotypes of DENV, of which DENV-1 and DENV-4 are low-passage clinical virus isolates, in both Raji/DC-SIGN+ cells and in primary immature MDDC. The use of MDDC has much more clinical relevance than using a transfected cell line. MDDC resemble DC in the skin [83] and mimic an *in vivo* DENV infection after a mosquito bite. Moreover, cells of the hematopoietic origin, such as DC, have been shown to play a key role for DENV pathogenesis in a mouse model [84]. A dose-dependent and a DENV serotype-independent antiviral activity of HHA, GNA and UDA in MDDC was demonstrated as analyzed by flow cytometry (Figure 9). These CBAs proved about 100-fold more effective in inhibiting DENV infection in primary MDDC compared to the transfected Raji/DC-SIGN+ cell line.

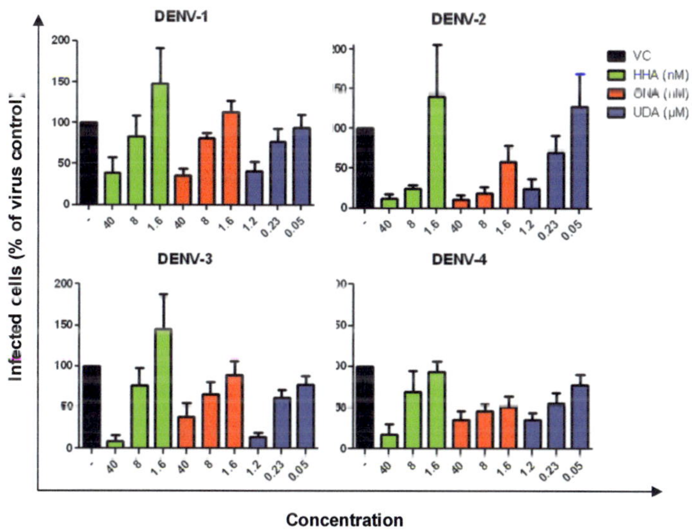

Figure 9. Dose-dependent antiviral activity of HHA, GNA and UDA in DENV-infected MDDC. MDDC were infected with the four serotypes of DENV in the presence or absence of various concentrations of HHA, GNA and UDA. DENV infection was analyzed by flow cytometry using an anti-PrM antibody recognizing all four DENV serotypes (clone 2H2). % of infected cells compared to the positive virus control (VC) ± SEM of 4 to 12 different blood donors is shown. (Adapted from Alen et al. [61]).

When DENV is captured by DC, a maturation and activation process occurs. DC require downregulation of C-type lectin receptors [85], upregulation of costimulatory molecules, chemokine receptors and enhancement of their APC function to migrate to the nodal T-cell areas and to activate the immune system [86]. Cytokines implicated in vascular leakage are produced, the complement system becomes activated and virus-induced antibodies can cause DHF via binding to Fc-receptors. Several research groups demonstrated maturation of DC induced by DENV infection [87,88]. Some groups made segregation in the DC population after DENV infection, the infected DC and the uninfected bystander cells. They found that bystander DC, in contrast to infected DC, upregulate the cell surface expression of costimulatory molecules, HLA and maturation molecules. This activation is induced by TNF-α and IFN-α secreted by DENV-infected DC [40,89,90]. Instead, Alen *et al.* observed an upregulation of the costimulatory molecules CD80 and CD86 and a downregulation of DC-SIGN and MR on the total (uninfected and infected) DC population following DENV infection [61]. This could indicate that the DC are activated and can interact with naive T-cells and subsequently activate the immune system resulting in increased vascular permeability and fever. When the effect of the CBAs was examined on the expression level of the cell surface markers of the total DC population, it was shown that the CBAs are able to inhibit the activation of all DC caused by DENV and can keep the DC in an immature state. Furthermore, DC do not express costimulatory molecules and thus do not interact or significantly activate T-cells. An approach to inhibit DENV-induced activation of DC may prevent the immunopathological component of DENV disease.

However, since plant lectins are expensive to produce and not orally bioavailable, the search for non-peptidic small molecules is necessary. PRM-S, a highly soluble non-peptidic small-size carbohydrate-binding antibiotic is a potential new lead compound in HIV therapy, since PRM-S efficiently inhibits HIV replication and prevents capture of HIV to DC-SIGN$^+$ cells [91]. PRM-S also inhibited dose-dependently DENV-2 replication in MDDC but had only a weak antiviral activity in Raji/DC-SIGN$^+$ cells [61]. Actinohivin (AH), a small prokaryotic peptidic lectin containing 114 amino acids, exhibits also anti-HIV-1 activity by recognizing high-mannose-type glycans on the viral envelope [92]. Although DENV has high mannose-type glycans on the E-protein, there was no antiviral activity of AH against DENV infection. Other CBAs such as microvirin, griffithsin and Banlec have been shown to exhibit potent activity against HIV replication [93-95], but these CBAs did not show antiviral activity against DENV. Previously, it has been shown that the CBAs HHA, GNA and UDA also target the N-glycans of other viruses, such as HIV, HCV [79] and HCMV [80]. This indicates that the CBAs can be used as broad-spectrum antiviral agents against various classes of glycosylated enveloped viruses. Although, the three plant lectins did not act inhibitory against parainfluenza-3, vesicular stomatitis virus, respiratory syncytial virus or herpes simplex virus [79]. Together, these data indicate a unique carbohydrate-specificity, and thus also a specific profile of antiviral activity of the CBAs.

4.3. Molecular target of the CBAs on DENV

It was demonstrated in time of drug addition assays that the mannose binding lectin HHA prevents DENV-2 binding to the host cell and acts less efficiently when the virus had already attached to the host cell. It was shown that HHA interacts with DENV and not with cellular membrane proteins such as DC-SIGN on the target cell. The potency of HHA to inhibit attachment of DENV to Raji/DC-SIGN$^+$ cells is comparable to its inhibitory activity of the capture of HIV and HCV to Raji/DC-SIGN$^+$ cells [79]. CBAs could thus be considered as unique prophylactic agents of DENV infection.

To identify the molecular target of the CBAs on DENV, a resistant DENV to HHA was generated in the mosquito cell line C6/36 by Alen *et al.* (HHA^{res} DENV). Compared with the WT DENV, two highly prevalent mutations were found, namely N67D and T155I, present in 80% of all clones sequenced. Similar mutational patterns destroying both glycosylation motifs (T69I or T69A each in combination with T155I) were present in another 10% of the clones analyzed. The N-glycosylation motif $_{153}$N-D-T$_{155}$ is conserved among the majority of all flaviviruses, while a second N-glycosylation motif $_{67}$N-T-T$_{69}$ is unique among DENV [96]. In the HHA^{res} virus both N-glycosylation motifs were mutated either directly at the actual N-glycan accepting a residue of the first site (Asn67) or at the C-proximal Thr155 being an essential part of the second N-glycosylation site [97], thus both N-glycosylation sites on the viral envelope protein can be considered to be deleted. This indicates that HHA directly targets the N-glycans on the viral E-protein. In fact, all clones sequenced showed the deletion of the N-glycan at Asn153. However, 10% of the clones sequenced had no mutation at the glycosylation motif $_{67}$N-T-T$_{69}$, indicating that this glycosylation motif [96] has a higher genetic barrier compared to $_{153}$N-D-T$_{155}$. Though there are multiple escape pathways to become resistant to HHA, it seems not to be possible to fully escape the selective pressure of favoring a deglycosylation of the viral E-protein. In addition, there were no mutations found either apart from the N-glycosylation sites of the E-protein or in any of the five WT DENV-2 clones passaged in parallel. This is not fully unexpected as flaviviruses replicate with reasonable fidelity and DENV does not necessarily exist as a highly diverse quasispecies neither *in vitro* nor *in vivo* [98,99].

There are some contradictions in terms of necessity of glycosylation of Asn67 and Asn153 during DENV viral progeny. Johnson *et al.* postulated that DENV-1 and DENV-3 have both sites glycosylated and that DENV-2 and DENV-4 have only one N-glycan at Asn-67 [100]. In contrast, a study comparing the number of glycans in multiple isolates of DENV belonging to all four serotypes led to the consensus that all DENV strains have two N-glycans on the E-protein [101]. However, mutant DENV lacking the glycosylation at Asn153 can replicate in mammalian and insect cells, indicating that this glycosylation is not essential for viral replication [96,102]. There is a change in phenotype because ablation of glycosylation at Asn153 in DENV is associated with the induction of smaller plaques in comparison to the wild type virus [96]. Asn153 is proximal to the fusion peptide and therefore deglycosylation at Asn153 showed also an altered pH-dependent fusion activity and displays a lower stability [103,104]. In contrast, Alen *et al.* showed that the mutant virus, HHA^{res} lacking both

N-glycosylation sites, had a similar plaque phenotype in BHK cells (manuscript submitted). It has been shown that DENV lacking the glycosylation at Asn67 resulted in a replication-defective phenotype, because this virus infects mammalian cells weakly and there is a reduced secretion of DENV E-protein. Replication in mosquito cells was not affected, because the mosquito cells restore the N-glycosylation at Asn67 with a compensatory site-mutation (K64N) generating a new glycosylation site [96,105]. These data are in contrast with other published results, where was demonstrated that DENV lacking the Asn67-linked glycosylation can grow efficiently in mammalian cells, depending on the viral strain and the amino acid substitution abolishing the glycosylation process [102]. A compensatory mutation was detected (N124S) to repair the growth defect without creating a new glycosylation site. Thus, the glycan at Asn67 is not necessary for virus growth, but a critical role for this glycan in virion release from mosquito cells was demonstrated [102]. However, HHA resistant virus was found to replicate efficiently in mosquito and insect cells indicating an efficient carbohydrate-independent viral replication in these cell lines. A possible explanation for the differences between our data and data from previous studies could be that the mutant virus has been generated in mosquito C6/36 cells (during replication under antiviral drug pressure) and not in mammalian cells (after introducing the mutations by site-directed mutagenesis). In addition, in previous studies, other amino acid substitutions were generated, resulting in different virus genotypes and subsequently resulting in poorly to predict virus phenotypes.

The glycosylation at Asn67 is demonstrated to be essential for infection of monocyte-derived dendritic cells (MDDC), indicating an interaction between DC-SIGN and the glycan at Asn67 [96,106]. Also the HHA^{res} DENV was not able to infect efficiently DC-SIGN^{+} cells or cells that express the DC-SIGN-related liver-specific receptor L-SIGN. Interestingly, MDDC are also not susceptible for HHA^{res} DENV infection, indicating the importance of the DC-SIGN-mediated DENV infection in MDDC. Moreover, cells of the hematopoietic origin, such as DC, are described to be necessary for DENV pathogenesis [84]. If the CBA resistant DENV in not able to infect DC anymore, it can be stated that the CBAs interfere with a physiologically highly relevant target. DC-SIGN is postulated as the most important DENV entry receptor until now. The entry process of DENV in Vero, Huh-7, BHK-21 and C6/36 cell lines is DC-SIGN-independent and also carbohydrate-independent. Indeed, HHA^{res} DENV can efficiently enter and replicate in these cell lines. HHA^{res} DENV lacking both N-glycans on the envelope E-glycoprotein is able to replicate efficiently in mammalian cells, with the exception of DC-SIGN^{+} cells.

The HHA^{res} virus was used as a tool to identify the antiviral target of other classes of compounds as it could replicate in human liver Huh-7 cells. The use of Huh-7 cells has much more clinical relevance than using monkey (Vero) or hamster (BHK) kidney cells. The HHA^{res} DENV was found cross-resistant to GNA, that recognizes like HHA, α-1,3 mannose residues. UDA, which recognizes mainly the N-acetylglucosamine residues of the N-glycans, also lacked antiviral activity against HHA^{res} DENV in Huh-7 cells. This indicates that the entire backbone of the N-glycan is deleted. Likewise, pradimicin-S (PRM-S), a small-size α-1,2-mannose-specific CBA, was also unable to inhibit HHA^{res} DENV. This demonstrates that PRM-S targets also the N-glycans on the DENV envelope. In contrast, ribavirin (RBV), a nucleoside analogue and inhibitor of cellular purine synthesis [74], retained as expected wild-type antiviral activity against HHA^{res} DENV. This argues against that there would be compensatory mutations in the non-structural proteins of DENV which are responsible for an overall enhanced replication of the viral genome [107,108]. SA-17, a novel doxorubicine analogue that inhibits the DENV entry process [109], was equipotent against WT and HHA^{res} DENV. The SA-17 compound is predicted to interact with the hydrophobic binding pocket of the E-glycoprotein which is independent from the N-glycosylation state of the E-glycoprotein [109]. These data confirm the molecular docking experiments of SA-17.

Generally, the function of glycosylation on surface proteins is proper folding of the protein, trafficking in the endoplasmic reticulum, interaction with receptors and influencing virus immunogenicity [110]. Virions produced in the mosquito vector and human host may have structurally different N-linked glycans, because the glycosylation patterns are fundamentally different [101,111]. N-glycosylation in mammalian cells is often of the complex-type because a lot of different processing enzymes could add a diversity of monosaccharides. Glycans produced in insect cells are far less complex, because of less diversity in processing enzymes and usually contain more high-mannose and pauci-mannose-type glycans. DC-SIGN can distinguish between mosquito- and mammalian cell-derived alphaviruses [112] and West Nile virus [68], resulting in a more efficient infection by a mosquito-derived virus, but this was not the case for DENV [101].

Although the CBAs HHA and GNA are not mitogenic and not toxic to mice when administered intravenously [113], caution must be taken in the development of the CBAs to use as antiviral drug in the clinic. First, the natural plant lectins are expensive to produce and hard to scale-up, but efforts have been made to express CBAs in commensal bacteria which provide an easy production process of this class of agents. Second, there can be a systemic reaction against the lectins such as in food allergies against peanut lectin or banana lectin [114,115]. Third, the CBAs can recognize aspecifically cellular glycans and could interfere with host cellular processes. But, DENV glycosylation is of the high-mannose or pauci-mannose type, which is only rare on mammalian proteins. The synthetic production of small non peptidic molecules, such as PRM-S, with CBA-like activity, could overcome the pharmacological problems of the plant lectins. Therefore, PRM-S forms a potential lead candidate in the development of more potent and specific DENV entry inhibitors.

5. Conclusion

In conclusion, besides active vector control in tropical and subtropical regions, there is an urgent need for antiviral treatment to protect half of the world's population against severe DENV infections. DC-SIGN is thought to be the most important DENV receptor and that the DC-SIGN-DENV envelope protein interaction is an excellent target for viral entry inhibitors such as the CBAs. Resistance against HHA forces the virus to delete its N-glycans and subsequently this mutant virus is not able anymore to infect its most important target cells.

Thus the CBAs act in two different ways: prevention of viral entry by directly binding N-glycans on the viral envelope and indirectly forcing the virus to delete its N-glycans and loose the capability to infect DC. The plant lectins provided more insight into the entry pathway of the virus into the host cell. Hopefully some of these future derivatives with a comparable mode of action will reach clinical trials in the near future.

Author details

Dominique Schols* and Marijke Alen
Department of Microbiology and Immunology, Rega Institute for Medical Research, University of Leuven, Leuven, Belgium

Acknowledgement

We are grateful to S. Claes, B. Provinciael, E. Van Kerckhove and E. Fonteyn for excellent technical assistance. This work was supported by the FWO (G-485-08N and G.0528.12N), the KU Leuven (PF/10/018 and GOA/10/014) and the Dormeur Investment Service Ltd.

6. References

[1] Holmes EC, Twiddy SS (2003) The origin, emergence and evolutionary genetics of dengue virus. Infect Genet Evol 3: 19-28.
[2] Mackenzie JS, Gubler DJ, Petersen LR (2004) Emerging flaviviruses: the spread and resurgence of Japanese encephalitis, West Nile and dengue viruses. Nat Med 10: S98-S109.
[3] Halstead SB (2008) Dengue virus - Mosquito interactions. Annu Rev Entomol 53: 273-291.
[4] Gubler DJ (1998) Dengue and dengue hemorrhagic fever. Clin Microbiol Rev 11: 480-496.
[5] Halstead SB (2007) Dengue. Lancet 370: 1644-1652.
[6] Monath TP (1994) Dengue - the risk to developed and developing countries. Proc Natl Acad Sci U S A 91: 2395-2400.
[7] Whitehead SS, Blaney JE, Durbin AP, Murphy BR (2007) Prospects for a dengue virus vaccine. Nat Rev Microbiol 5: 518-528.
[8] Walker T, Johnson PH, Moreira LA, Iturbe-Ormaetxe I, Frentiu FD, McMeniman CJ et al. (2011) The wMel Wolbachia strain blocks dengue and invades caged Aedes aegypti populations. Nature 476: 450-453.
[9] Hoffmann AA, Montgomery BL, Popovici J, Iturbe-Ormaetxe I, Johnson PH, Muzzi F et al. (2011) Successful establishment of Wolbachia in Aedes populations to suppress dengue transmission. Nature 476: 454-457.

* Corresponding Author

[10] Mairuhu ATA, Wagenaar J, Brandjes DPM, van Gorp ECM (2004) Dengue: an arthropod-borne disease of global importance. Eur J Clin Microbiol Infect Dis 23: 425-433.

[11] Rigau-Pérez JG, Clark GG, Gubler DJ, Reiter P, Sanders RJ, Vorndam AV (1998) Dengue and dengue haemorrhagic fever. Lancet 352: 971-977.

[12] Solomon T, Mallewa M (2001) Dengue and other emerging flaviviruses. J Infect 42: 104-115.

[13] Halstead SB (1988) Pathogenesis of dengue: challenges to molecular biology. Science 239: 476-481.

[14] Takada A, Kawaoka Y (2003) Antibody-dependent enhancement of viral infection: molecular mechanisms and in vivo implications. Rev Med Virol 13: 387-398.

[15] Kliks SC, Nimmanitya S, Nisalak A, Burke DS (1988) Evidence that maternal dengue antibodies are important in the development of dengue hemorrhagic fever in infants. Am J Trop Med Hyg 38: 411-419.

[16] Halstead SB (1979) In vivo enhancement of dengue virus infection in rhesus monkeys by passively transferred antibody. J Infect Dis 140: 527-533.

[17] Wu SJL, Grouard-Vogel G, Sun W, Mascola JR, Brachtel E, Putvatana R et al. (2000) Human skin Langerhans cells are targets of dengue virus infection. Nat Med 6: 816-820.

[18] Kou Z, Quinn M, Chen HY, Rodrigo WW, Rose RC, Schlesinger JJ et al. (2008) Monocytes, but not T or B cells, are the principal target cells for dengue virus (DV) infection among human peripheral blood mononuclear cells. J Med Virol 80: 134-146.

[19] Kuhn RJ, Zhang W, Rossmann MG, Pletnev SV, Corver J, Lenches E et al. (2002) Structure of dengue virus: Implications for flavivirus organization, maturation, and fusion. Cell 108: 717-725.

[20] Acosta EG, Castilla V, Damonte EB (2008) Functional entry of dengue virus into Aedes albopictus mosquito cells is dependent on clathrin-mediated endocytosis. J Gen Virol 89: 474-484.

[21] van der Schaar HM, Rust MJ, Chen C, van der Ende-Metselaar H, Wilschut J, Zhuang XW et al. (2008) Dissecting the cell entry pathway of dengue virus by single-particle tracking in living cells. Plos Pathog 4: e1000244.

[22] Chu JJH, Ng ML (2004) Infectious entry of West Nile virus occurs through a clathrin-mediated endocytic pathway. J Virol 78: 10543-10555.

[23] Nawa M, Takasaki T, Yamada K, Kurane I, Akatsuka T (2003) Interference in Japanese encephalitis virus infection of vero cells by a cationic amphiphilic drug, chlorpromazine. J Gen Virol 84: 1737-1741.

[24] Acosta EG, Castilla V, Damonte EB (2009) Alternative infectious entry pathways for dengue virus serotypes into mammalian cells. Cell Microbiol 11: 1533-1549.

[25] Modis Y, Ogata S, Clements D, Harrison SC (2004) Structure of the dengue virus envelope protein after membrane fusion. Nature 427: 313-319.

[26] Stiasny K, Fritz R, Pangerl K, Heinz FX (2009) Molecular mechanisms of flavivirus membrane fusion. Amino Acids 41: 1159-1163.

[27] Martin CSS, Liu CY, Kielian M (2009) Dealing with low pH: entry and exit of alphaviruses and flaviviruses. Trends Microbiol 17: 514-521.

[28] Zhang Y, Zhang W, Ogata S, Clements D, Strauss JH, Baker TS et al. (2004) Conformational changes of the flavivirus E glycoprotein. Structure 12: 1607-1618.

[29] Perera R, Khaliq M, Kuhn RJ (2008) Closing the door on flaviviruses: Entry as a target for antiviral drug design. Antiviral Res 80: 11-22.

[30] Modis Y, Ogata S, Clements D, Harrison SC (2003) A ligand-binding pocket in the dengue virus envelope glycoprotein. Proc Natl Acad Sci U S A 100: 6986-6991.

[31] Rey FA, Heinz FX, Mandl C, Kunz C, Harrison SC (1995) The envelope glycoprotein from tick-borne encephalitis virus at 2 angstrom resolution. Nature 375: 291-298.

[32] Crill WD, Roehrig JT (2001) Monoclonal antibodies that bind to domain III of dengue virus E glycoprotein are the most efficient blockers of virus adsorption to Vero cells. J Virol 75: 7769-7773.

[33] Rajamanonmani R, Nkenfou C, Clancy P, Yau YH, Shochat SG, Sukupolvi-Petty S et al. (2009) On a mouse monoclonal antibody that neutralizes all four dengue virus serotypes. J Gen Virol 90: 799-809.

[34] Scherret JH, Mackenzie JS, Khromykh AA, Hall RA (2001) Biological significance of glycosylation of the envelope protein of Kunjin virus. Ann N Y Acad Sci 951: 361-363.

[35] Halstead SB, O'rourke EJ, Allison AC (1977) Dengue viruses and mononuclear phagocytes .II. Identity of blood and tissue leukocytes supporting in vitro infection. J Exp Med 146: 218-229.

[36] Durbin AP, Vargas MJ, Wanionek K, Hammond SN, Gordon A, Rocha C et al. (2008) Phenotyping of peripheral blood mononuclear cells during acute dengue illness demonstrates infection and increased activation of monocytes in severe cases compared to classic dengue fever. Virology 376: 429-435.

[37] Kliks SC, Nisalak A, Brandt WE, Wahl L, Burke DS (1989) Antibody-dependent enhancement of dengue virus growth in human-monocytes as a risk factor for dengue hemorrhagic-fever. Am J Trop Med Hyg 40: 444-451.

[38] Taweechaisupapong S, Sriurairatana S, Angsubhakorn S, Yoksan S, Khin MM, Sahaphong S et al. (1996) Langerhans cell density and serological changes following intradermal immunisation of mice with dengue 2 virus. J Med Microbiol 45: 138-145.

[39] Ho LJ, Wang JJ, Shaio MF, Kao CL, Chang DM, Han SW et al. (2001) Infection of human dendritic cells by Dengue virus causes cell maturation and cytokine production. J Immunol 166: 1499-1506.

[40] Libraty DH, Pichyangkul S, Ajariyakhajorn C, Endy TP, Ennis FA (2001) Human dendritic cells are activated by dengue virus infection: Enhancement by gamma interferon and implications for disease pathogenesis. J Virol 75: 3501-3508.

[41] Marovich M, Grouard-Vogel G, Louder M, Eller M, Sun W, Wu SJ et al. (2001) Human dendritic cells as targets of dengue virus infection. J Invest Dermatol 6: 219-224.

[42] Alen MMF, Kaptein SJF, De Burghgraeve T, Balzarini J, Neyts J, Schols D (2009) Antiviral activity of carbohydrate-binding agents and the role of DC-SIGN in dengue virus infection. Virology 387: 67-75.

[43] Lozach PY, Burleigh L, Staropoli I, Navarro-Sanchez E, Harriague J, Virelizier JL et al. (2005) Dendritic cell-specific intercellular adhesion molecule 3-grabbing non-integrin (DC-SIGN)-mediated enhancement of dengue virus infection is independent of DC-SIGN internalization signals. J Biol Chem 280: 23698-23708.

[44] Navarro-Sanchez E, Altmeyer R, Amara A, Schwartz O, Fieschi F, Virelizier JL et al. (2003) Dendritic-cell-specific ICAM3-grabbing non-integrin is essential for the productive infection of human dendritic cells by mosquito-cell-derived dengue viruses. Embo Rep 4: 723-728.

[45] Tassaneetrithep B, Burgess TH, Granelli-Piperno A, Trumpfherer C, Finke J, Sun W et al. (2003) DC-SIGN (CD209) mediates dengue virus infection of human dendritic cells. J Exp Med 197: 823-829.

[46] van Kooyk Y, Geijtenbeek TBH (2003) DC-SIGN: Escape mechanism for pathogens. Nat Rev Immunol 3: 697-709.

[47] Mitchell DA, Fadden AJ, Drickamer K (2001) A novel mechanism of carbohydrate recognition by the C-type lectins DC-SIGN and DC-SIGNR - Subunit organization and binding to multivalent ligands. J Biol Chem 276: 28939-28945.

[48] Appelmelk BJ, van Die I, van Vliet SJ, Vandenbroucke-Grauls CMJE, Geijtenbeek TBH, van Kooyk Y (2003) Cutting edge: Carbohydrate profiling identifies new pathogens that interact with dendritic cell-specific ICAM-3-grabbing nonintegrin on dendritic cells. J Immunol 170: 1635-1639.

[49] Feinberg H, Mitchell DA, Drickamer K, Weis WI (2001) Structural basis for selective recognition of oligosaccharides by DC-SIGN and DC-SIGNR. Science 294: 2163-2166.

[50] Geijtenbeek TBH, Kwon DS, Torensma R, van Vliet SJ, van Duijnhoven GCF, Middel J et al. (2000) DC-SIGN, a dendritic cell-specific HIV-1-binding protein that enhances trans-infection of T cells. Cell 100: 587-597.

[51] Pöhlmann S, Zhang J, Baribaud F, Chen ZW, Leslie G, Lin G et al. (2003) Hepatitis C virus glycoproteins interact with DC-SIGN and DC-SIGNR. J Virol 77: 4070-4080.

[52] Marzi A, Möller P, Hanna SL, Harrer T, Eisemann J, Steinkasserer A et al. (2007) Analysis of the interaction of Ebola virus glycoprotein with DC-SIGN (dendritic cell-specific intercellular adhesion molecule 3 grabbing nonintegrin) and its homologue DC-SIGNR. J Infect Dis 196 (Suppl 2): S237-S246.

[53] Geijtenbeek TBH, Krooshoop DJEB, Bleijs DA, van Vliet SJ, van Duijnhoven GCF, Grabovsky V et al. (2000) DC-SIGN-ICAM-2 interaction mediates dendritic cell trafficking. Nat Immunol 1: 353-357.

[54] Banchereau J, Steinman RM (1998) Dendritic cells and the control of immunity. Nature 392: 245-252.

[55] Nair MPN, Reynolds JL, Mahajan SD, Schwartz SA, Aalinkeel R, Bindukumar B et al. (2005) RNAi-directed inhibition of DC-SIGN by dendritic cells: Prospects for HIV-1 therapy. AAPS J 7: E572-E578.

[56] Dakappagari N, Maruyama T, Renshaw M, Tacken P, Figdor C, Torensma R et al. (2006) Internalizing antibodies to the C-type lectins, L-SIGN and DC-SIGN, inhibit viral glycoprotein binding and deliver antigen to human dendritic cells for the induction of T cell responses. J Immunol 176: 426-440.

[57] Anderluh M, Jug G, Svajger U, Obermajer N (2012) DC-SIGN antagonists, a potential new class of anti-infectives. Curr Med Chem 19: 992-1007.

[58] Chen YC, Wang SY (2002) Activation of terminally differentiated human monocytes/macrophages by dengue virus: Productive infection, hierarchical production of innate cytokines and chemokines, and the synergistic effect of lipopolysaccharide. J Virol 76: 9877-9887.

[59] Miller JL, Dewet BJM, Martinez-Pomares L, Radcliffe CM, Dwek RA, Rudd PM et al. (2008) The mannose receptor mediates dengue virus infection of macrophages. Plos Pathog 4: e17.

[60] Kwan WH, Helt AM, Maranon C, Barbaroux JB, Hosmalin A, Harris E et al. (2005) Dendritic cell precursors are permissive to dengue virus and human immunodeficiency virus infection. J Virol 79: 7291-7299.

[61] Alen MMF, De Burghgraeve T, Kaptein SJF, Balzarini J, Neyts J, Schols D (2011) Broad antiviral activity of carbohydrate-binding agents against the four serotypes of dengue virus in monocyte-derived dendritic cells. Plos One 6: e21658.

[62] de Witte L, Bobardt M, Chatterji U, Degeest G, David G, Geijtenbeek TB et al. (2007) Syndecan-3 is a dendritic cell-specific attachment receptor for HIV-1. Proc Natl Acad Sci U S A 104: 19464-19469.

[63] Lambert AA, Gilbert C, Richard M, Beaulieu AD, Tremblay MJ (2008) The C-type lectin surface receptor DICIR acts as a new attachment factor for HIV-1 in dendritic cells and contributes to trans- and cis-infection pathways. Blood 112: 1299-1307.

[64] Chen ST, Lin YL, Huang MT, Wu MF, Cheng SC, Lei HY et al. (2008) CLEC5A is critical for dengue-virus-induced lethal disease. Nature 453: 672-676.

[65] Watson AA, Lebedev AA, Hall BA, Fenton-May AE, Vagin AA, Dejnirattisai W et al. (2011) Structural flexibility of the macrophage dengue virus receptor CLEC5A implications for ligand binding and signaling. J Biol Chem 286: 24208-24218.

[66] Bashirova AA, Geijtenbeek TBH, van Duijnhoven GCF, van Vliet SJ, Eilering JBG, Martin MP et al. (2001) A dendritic cell-specific intercellular adhesion molecule 3-grabbing nonintegrin (DC-SIGN)-related protein is highly expressed on human liver sinusoidal endothelial cells and promotes HIV-1 infection. J Exp Med 193: 671-678.

[67] Pöhlmann S, Soilleux EJ, Baribaud F, Leslie GJ, Morris LS, Trowsdale J et al. (2001) DC-SIGNR, a DC-SIGN homologue expressed in endothelial cells, binds to human and simian immunodeficiency viruses and activates infection in trans. Proc Natl Acad Sci U S A 98: 2670-2675.

[68] Davis CW, Nguyen HY, Hanna SL, Sanchez MD, Doms RW, Pierson TC (2006) West Nile virus discriminates between DC-SIGN and DC-SIGNR for cellular attachment and infection. J Virol 80: 1290-1301.

[69] Zellweger RM, Prestwood TR, Shresta S (2010) Enhanced infection of liver sinusoidal endothelial cells in a mouse model of antibody-induced severe dengue disease. Cell Host Microbe 7: 128-139.

[70] Avirutnan P, Malasit P, Seliger B, Bhakdi S, Husmann M (1998) Dengue virus infection of human endothelial cells leads to chemokine production, complement activation, and apoptosis. J Immunol 161: 6338-6346.

[71] Seneviratne SL, Malavige GN, de Silva HJ (2006) Pathogenesis of liver involvement during dengue viral infections. Trans R Soc Trop Med Hyg 100: 608-614.

[72] Murrell S, Wu SC, Butler M (2011) Review of dengue virus and the development of a vaccine. Biotechnol Adv 29: 239-247.

[73] Murphy BR, Whitehead SS (2011) Immune response to dengue virus and prospects for a vaccine. Annu Rev Immunol 29: 587-619.

[74] Leyssen P, De Clercq E, Neyts J (2008) Molecular strategies to inhibit the replication of RNA viruses. Antiviral Res 78: 9-25.

[75] Sampath A, Padmanabhan R (2009) Molecular targets for flavivirus drug discovery. Antiviral Res 81: 6-15.

[76] Balzarini J (2006) Inhibition of HIV entry by carbohydrate-binding proteins. Antiviral Res 71: 237-247.

[77] François KO, Balzarini J (2010) Potential of carbohydrate-binding agents as therapeutics against enveloped viruses. Med Res Rev .

[78] Balzarini J, Van Herrewege Y, Vermeire K, Vanham G, Schols D (2007) Carbohydrate-binding agents efficiently prevent dendritic cell-specific intercellular adhesion molecule 3 grabbing nonintegrin (DC-SIGN)-directed HIV-1 transmission to T lymphocytes. Mol Pharmacol 71: 3-11.

[79] Bertaux C, Daelemans D, Meertens L, Cormier EG, Reinus JF, Peumans WJ et al. (2007) Entry of hepatitis C virus and human immunodeficiency virus is selectively inhibited by carbohydrate-binding agents but not by polyanions. Virology 366: 40-50.

[80] Balzarini J, Schols D, Neyts J, Van Damme E, Peumans W, De Clercq E (1991) Alpha-(1-3)-D-mannose-specific and alpha-(1-6)-D-mannose-specific plant lectins are markedly inhibitory to human immunodeficiency virus and cytomegalovirus infections in vitro. Antimicrob Agents Chemother 35: 410-416.

[81] Harata K, Muraki M (2000) Crystal structures of Urtica dioica agglutinin and its complex with tri-N-acetylchitotriose. Journal of Molecular Biology 297: 673-681.

[82] Hung SL, Lee PL, Chen HW, Chen LK, Kao CL, King CC (1999) Analysis of the steps involved in dengue virus entry into host cells. Virology 257: 156-167.

[83] Grassi F, Dezutter-Dambuyant C, McIlroy D, Jacquet C, Yoneda K, Imamura S et al. (1998) Monocyte-derived dendritic cells have a phenotype comparable to that of dermal

dendritic cells and display ultrastructural granules distinct from Birbeck granules. J Leukoc Biol 64: 484-493.

[84] Pham AM, Langlois RA, TenOever BR (2012) Replication in cells of hematopoietic origin is necessary for Dengue virus dissemination. Plos Pathog 8: e1002465.

[85] Relloso M, Puig-Kröger A, Pello OM, Rodriguez-Fernandez JL, de la Rosa G, Longo N et al. (2002) DC-SIGN (CD209) expression is IL-4 dependent and is negatively regulated by IFN, TGF-beta, and anti-inflammatory agents. J Immunol 168: 2634-2643.

[86] Tacken PJ, de Vries IJM, Torensma R, Figdor CG (2007) Dendritic-cell immunotherapy: from ex vivo loading to in vivo targeting. Nat Rev Immunol 7: 790-802.

[87] Ho LJ, Wang JJ, Shaio MF, Kao CL, Chang DM, Han SW et al. (2001) Infection of human dendritic cells by Dengue virus causes cell maturation and cytokine production. J Immunol 166: 1499-1506.

[88] Sun P, Fernandez S, Marovich MA, Palmer DR, Celluzzi CM, Boonnak K et al. (2009) Functional characterization of ex vivo blood myeloid and plasmacytoid dendritic cells after infection with dengue virus. Virology 383: 207-215.

[89] Dejnirattisai W, Duangchinda T, Lin CLS, Vasanawathana S, Jones M, Jacobs M et al. (2008) A complex interplay among virus, dendritic cells, T cells, and cytokines in Dengue virus infections. J Immunol 181: 5865-5874.

[90] Palmer DR, Sun PF, Celluzzi C, Bisbing J, Pang SN, Sun W et al. (2005) Differential effects of dengue virus on infected and bystander dendritic cells. J Virol 79: 2432-2439.

[91] Balzarini J, François KO, Van Laethem K, Hoorelbeke B, Renders M, Auwerx J et al. (2010) Pradimicin S, a highly soluble nonpeptidic small-size carbohydrate-binding antibiotic, is an anti-HIV drug lead for both microbicidal and systemic use. Antimicrob Agents Chemother 54: 1425-1435.

[92] Hoorelbeke B, Huskens D, Férir G, François KO, Takahashi A, Van Laethem K et al. (2010) Actinohivin, a broadly neutralizing prokaryotic lectin, inhibits HIV-1 infection by specifically targeting high-mannose-type glycans on the gp120 envelope. Antimicrob Agents Chemother 54: 3287-3301.

[93] Alexandre KB, Gray ES, Mufhandu H, McMahon JB, Chakauya E, O'Keefe BR et al. (2012) The lectins griffithsin, cyanovirin-N and scytovirin inhibit HIV-1 binding to the DC-SIGN receptor and transfer to CD4(+) cells. Virology 423: 175-186.

[94] Huskens D, Férir G, Vermeire K, Kehr JC, Balzarini J, Dittmann E et al. (2010) Microvirin, a novel alpha(1,2)-mannose-specific lectin isolated from microcystis aeruginosa, has anti-HIV-1 activity comparable with that of cyanovirin-N but a much higher safety profile. J Biol Chem 285: 24845-24854.

[95] Swanson MD, Winter HC, Goldstein IJ, Markovitz DM (2010) A lectin isolated from bananas is a potent inhibitor of HIV replication. J Biol Chem 285: 8646-8655.

[96] Mondotte JA, Lozach PY, Amara A, Gamarnik AV (2007) Essential role of dengue virus envelope protein N-glycosylation at asparagine-67 during viral propagation. J Virol 81: 7136-7148.

[97] Ploegh H, Neefjes JJ (1990) Protein glycosylation. Curr Opin Cell Biol 2: 1125-1130.

[98] Pugachev KV, Guirakhoo F, Ocran SW, Mitchell F, Parsons M, Penal C et al. (2004) High fidelity of yellow fever virus RNA polymerase. J Virol 78: 1032-1038.

[99] Thai KTD, Henn MR, Zody MC, Tricou V, Nguyet NM, Charlebois P et al. (2012) High-resolution analysis of intrahost genetic diversity in dengue virus serotype 1 infection identifies mixed infections. J Virol 86: 835-843.

[100] Johnson AJ, Guirakhoo F, Roehrig JT (1994) The envelope glycoproteins of dengue-1 and dengue-2 viruses grown in mosquito cells differ in their utilization of potential glycosylation sites. Virology 203: 241-249.

[101] Hacker K, White L, de Silva AM (2009) N-Linked glycans on dengue viruses grown in mammalian and insect cells. J Gen Virol 90: 2097-2106.

[102] Lee E, Leang SK, Davidson A, Lobigs M (2010) Both E protein glycans adversely affect dengue virus infectivity but are beneficial for virion release. J Virol 84: 5171-5180.

[103] Guirakhoo F, Hunt AR, Lewis JG, Roehrig JT (1993) Selection and partial characterization of dengue-2 virus mutants that induce fusion at elevated pH. Virology 194: 219-223.

[104] Lee E, Weir RC, Dalgarno L (1997) Changes in the dengue virus major envelope protein on passaging and their localization on the three-dimensional structure of the protein. Virology 232: 281-290.

[105] Bryant JE, Calvert AE, Mesesan K, Crabtree MB, Volpe KE, Silengo S et al. (2007) Glycosylation of the dengue 2 virus E protein at N67 is critical for virus growth in vitro but not for growth in intrathoracically inoculated Aedes aegypti mosquitoes. Virology 366: 415-423.

[106] Pokidysheva E, Zhang Y, Battisti AJ, Bator-Kelly CM, Chipman PR, Xiao CA et al. (2006) Cryo-EM reconstruction of dengue virus in complex with the carbohydrate recognition domain of DC-SIGN. Cell 124: 485-493.

[107] Wang QY, Bushell S, Qing M, Xu HY, Bonavia A, Nunes S et al. (2011) Inhibition of dengue virus through suppression of host pyrimidine biosynthesis. J Virol 85: 6548-6556.

[108] Xie X, Wang QY, Xu HY, Qing M, Kramer L, Yuan Z et al. (2011) Inhibition of dengue virus by targeting viral NS4B protein. J Virol 85: 11183-11195.

[109] Kaptein SJF, De Burghgraeve T, Froeyen M, Pastorino B, Alen MMF, Mondotte JA et al. (2010) A derivate of the antibiotic doxorubicin is a selective inhibitor of dengue and yellow fever virus replication in vitro. Antimicrob Agents Chemother 54: 5269-5280.

[110] Vigerust DJ, Shepherd VL (2007) Virus glycosylation: role in virulence and immune interactions. Trends Microbiol 15: 211-218.

[111] Rogers KM, Heise M (2009) Modulation of cellular tropism and innate antiviral response by viral glycans. J Innate Immun 1: 405-412.

[112] Klimstra WB, Nangle EM, Smith MS, Yurochko AD, Ryman KD (2003) DC-SIGN and L-SIGN can act as attachment receptors for alphaviruses and distinguish between mosquito cell- and mammalian cell-derived viruses. J Virol 77: 12022-12032.

[113] Balzarini J, Hatse S, Vermeire K, Princen K, Aquaro S, Perno CF et al. (2004) Mannose-specific plant lectins from the Amaryllidaceae family qualify as efficient microbicides for prevention of human immunodeficiency virus infection. Antimicrob Agents Chemother 48: 3858-3870.

[114] Rougé P, Culerrier R, Granier C, Rancé F, Barre A (2010) Characterization of IgE-binding epitopes of peanut (Arachis hypogaea) PNA lectin allergen cross-reacting with other structurally related legume lectins. Molecular Immunology 47: 2359-2366.

[115] Sanchez-Monge R, Blanco C, Diaz-Perales A, Collada C, Carrillo T, Aragoncillo C et al. (1999) Isolation and characterization of major banana allergens: identification as fruit class I chitinases. Clin Exp Allergy 29: 673-680.

Chapter 9

IL-13, Asthma and Glycosylation in Airway Epithelial Repair

Samuel J. Wadsworth, S. Jasemine Yang and Delbert R. Dorscheid

Additional information is available at the end of the chapter

http://dx.doi.org/10.5772/51970

1. Introduction

1.1. Clinical assessment of asthma

Pulmonary function tests (PFTs/spirometry) are routinely used clinically to diagnose asthma. Forced expiratory volume in one second (FEV_1) is a measure of airflow and is influenced largely by the resistance in the airways. During an "asthma attack", the airways narrow owing to mucus hypersecretion and smooth muscle contraction, which results in a sharp increase in airway resistance and reduction in FEV_1. In the clinics, this can be reproduced by exposing patients with asthma to non-specific smooth muscle contractile agonists such as methacholine. A fall in FEV_1 at a relatively low dose of methacholine indicates airway hyperresponsiveness (AHR). In asthmatic patients this drop in FEV_1 can be reversed by inhaled bronchodilators such as β2-agonists (e.g. salbutamol), which relaxes airway smooth muscles. AHR is often used as a diagnostic criterion for asthma.

Asthma is increasingly considered a syndrome, with diverse overlapping pathologies and phenotypes contributing to significant heterogeneity in clinical manifestation, disease progression, and treatment response [1]. Severe uncontrolled asthmatics make up approximately 5-10% of the population of asthmatics, yet they consume around 50% of the treatment resources [2]. Spirometric tests are incapable of exposing the underlying pathologies or phenotypes which combine within an individual to produce the asthmatic disease. In contrast specific biomarkers may enable more accurate sub-phenotyping of disease by indicating the pathology within an individual, and assist clinicians to better tailor the type and/or dose of therapy. In this way biomarkers have the potential to guide more effective personalized treatment regimes in asthma.

1.2. The causes of asthma

Several studies have used lung biopsies taken from non-asthmatic patients and those who died from fatal asthma to show the airways of asthmatics possess characteristic structural differences, shown in Figure 1. These include but are not limited to; increased eosinophil numbers in asthmatics, and areas of focal bronchial epithelial denudation [3]. The contents of the bronchial lumen can also be examined using the bronchoalveolar lavage (BAL) technique, in which warm saline is instilled into a segmental bronchus of a patient and removed. BAL fluid-based studies confirmed the previous research that inflammatory cell influx and epithelial damage occur in the airways of asthmatics [3] [4]. It is a point of some debate as to whether the epithelial damage is caused by inflammation, is a cause of inflammation, or whether epithelial damage and airway inflammation are concurrent but unrelated processes. We shall discuss the potential role of the epithelium and epithelial repair in asthma in more detail later.

1.3. Inflammation in asthma

The rationale for treating severe asthmatics with steroids is centred on the assumption that the disease is driven by uncontrolled airway inflammation. The classical paradigm of the acute inflammatory response in allergic asthma is briefly; inhaled allergenic antigens are captured by professional antigen-presenting dendritic cells, macrophages, or epithelial cells. Antigens presented by these cells are recognized by T-cells which proliferate and differentiate. Primed Th_2 cells bind to B cells and release cytokines including triggering maturation of antigen-specific B cell populations into plasma cells. Plasma B cells release antigen-specific IgE that binds to IgE-receptors on mast cells in the airways, causing release of histamine-containing granules. Extracellular histamine released from mast cells binds to membrane receptors on airway smooth muscle cells triggering a rise in intracellular calcium, muscle contraction and airway narrowing. Although this paradigm has gained acceptance as an important mechanism in asthma, there is significant variation in specific cell types involved between individuals. In the setting of chronic asthma, eosinophils are thought to play a major role in maintaining airway inflammation in the long-term. Neutrophils are occasionally the predominate inflammatory cell present in the airways of chronic asthmatics, suggesting multiple underlying pathologies of disease. Neutrophilic vs. eosinophilic asthma are clinically indistinguishable by pulmonary function testing, but such pathological variation may have important implications regarding treatment as patients with "neutrophilic asthma" tend to be relatively unresponsive to steroid treatment [5]. A rapid and definitive method for determining the nature of airways inflammation in a diagnosed asthmatic is therefore needed to guide the clinician's selection of therapy.

1.4. IL-13 and asthma

Interleukin-13 (IL-13) is a pleiotropic Th_2 cytokine secreted from a variety of inflammatory cells and structural cells, including the airway epithelium. IL-13 is upregulated in the airways of patients with allergic asthma exposed to allergen; segmental allergen challenge of atopic asthmatics triggered an increase of IL-13 protein and transcript in bronchoalveolar lavage, and the source was identified as mononuclear inflammatory cells [6]. Experiments in small mammals and *in vitro* studies using human cells, have demonstrated IL-13 triggers many of the pathophysiological characteristics of asthma, independent of the IgE and eosinophil inflammatory mechanisms classically associated with allergic disease [7] [8]. Genetic studies in humans have demonstrated the gene loci for IL-13 and the IL-4 receptor portion of the IL-13Rα1 receptor have a strong genetic association to asthma susceptibility and severity [9]. IL-13 can act directly on the airway epithelium in various ways; previous studies have demonstrated IL-13 exposure of cultured human airway epithelial cells triggers an matrix-metalloproteinase-7 (MMP-7)-dependent release of the pro-inflammatory chemokine, soluble Fas-ligand, potentially leading to local inflammation and epithelial damage via removal of the protective Fas-ligand-mediated epithelial immune barrier [10]. Mucous hypersecretion is a common observation in asthmatic airways; acute treatment of differentiated cultures of primary human airway epithelial cells with IL-13 induces a hypersecretory phenotype [11], which develops into an asthma-like mucous hyperplasia with a 5-10-fold increase in goblet cell density and MUC5AC (the predominant airway mucin) detection after 14 day treatment [12]. Murine studies confirm IL-13 is required and sufficient to directly induce epithelial mucous hyperplasia *in vivo* in the absence of intermediate inflammatory cells [13]. Mucins including MUC5AC are highly glycosylated proteins, it thus follows that increased mucous production will necessitate an increase in the activity of associated transferase enzymes. In the H292 airway epithelial cell line, IL-13 treatment has been shown to increase the activity of the M and L isoforms of core 2 beta1,6 N-acetylglucosaminyltransferase (C2GnT) via activation of a JAK/STAT signaling pathway [14]. Thus, IL-13 acts directly on the airway epithelium to trigger development of an asthma-like phenotype, at least in part by increasing protein glycosylation via upregulation of glucosyltransferase activity.

1.5. Current treatments for asthma

Asthma is rarely fatal but often leads to significant patient morbidity. For over thirty years the mainstay therapy for asthma has been inhaled β2-agonists (βA), which relax the airway smooth muscles, causing bronchodilation, and inhaled glucocorticosteroids (i.e. steroids), which ameliorate airway inflammation and reduce the risk of asthma exacerbations. Steroids are anti-inflammatory drugs but also have many unwanted side-effects, particularly at high doses, including; weight-gain due to changes in body metabolism, and a reduction in bone mineral density, potentially leading to brittle bones. Therefore the decision to treat a patient with steroids should not be taken lightly. Inhaled β_2-agonists (s) are the most effective bronchodilators available and their main mechanism of action is via the relaxation of airway smooth muscle [15]. βAs interact with G-protein-coupled cell surface β_2-adrenoceptors (β_2AR) on bronchial smooth muscle cells and activate adenyl cyclase, causing a rise in intracellular cAMP, this results in relaxation of the smooth muscle and bronchodilation. Long term βA treatment can result in a loss of response which shows variation between individuals. Desensitisation due to chronic βA treatment is caused by a down-regulation of receptor transcription, secondary to a reduced activity of the CREB (cyclic AMP response element binding protein) transcription factor [15]. The β2-adrenoceptor gene possesses at least three possible GRE binding sites for the activated GCR, and GCs have been shown to increase β2AR transcription two-fold in human lung preparations *in vitro* and also in rat lungs *in vivo*. The downregulation of β2AR transcription resulting from long-term βA treatment can therefore be reversed by a GC, so there is no net change in the level of β2ARs expressed in the bronchial epithelium or airway smooth muscle of animals treated with a combination of GCs and βAs [15]. The very large β2AR reservoir in airway smooth muscle cells may mean that this upregulation by GCs is functionally unimportant *in vivo* [15]. Many asthmatics who are unresponsive to high doses of steroids show a marked improvement in symptoms when given steroids with βA together. Thus there may be a true synergy between the two classes of drugs as β2-adrenoceptor agonists enhance the function of GCs both *in vitro* and *in vivo*, possibly by increasing the nuclear localisation of activated GCRs. Low-dose theophylline has been prescribed as an asthma treatment for many years, however it has now been relegated to a third-line therapy according to the global asthma guidelines (reviewed in [16]). Theophylline is a weak, non-specific cyclic nucleotide phosphodiesterase (PDE) inhibitor, and recently these PDE inhibitory characteristics have re-ignited interest in this drug as a possible asthma treatment.

Monoclonal antibodies have gained recent credence as a potential means to target therapy against specific receptors and signalling pathways. Omalizumab, a biologic agent that binds to IgE, has shown some promise as a means of controlling some difficult to treat asthma patients who are poorly controlled with standard inhaled steroid and β-agonist combination therapy (reviewed in [17]). Interleukin-5 (IL-5) is an important cytokine for eosinophil differentiation, maturation, migration into the circulation and survival. Various studies have implicated the eosinophil as being the primary cell responsible for airway hyperresponsiveness in asthma, thus anti-IL-5 antibodies have been developed (such as mepolizumab) with the intention of reducing airways hyperresponsiveness by preventing eosinophil recruitment and survival in the airway. Clinical trials have demonstrated such inhaled anti-IL-5 therapy does indeed reduce airway eosinophilia, but does not have a clinically-beneficial effect on lung function [18, 19]. Considering the highly important role IL-13 is thought to play in asthma, it is unsurprising that therapies are being developed to specifically target IL-13. Lebrikizumab, an anti-IL-13 monoclonal antibody, has just completed phase II clinical trials in which patients with poorly-controlled asthma were given the drug subcutaneously at 4 weekly intervals for 6 months [20]. This study found anti-IL-13 therapy did significantly improve lung function, but only in a sub-group of asthma patients who exhibited high serum periostin levels, a surrogate marker of high airway IL-13 [20]. These studies demonstrate highly-targeted therapies may be useful in asthma, but each individual drug will probably only prove effective in a specific sub-group of asthmatic patients. The potential for such personalised medicine highlights the importance for accurate phenotyping of asthmatic individuals, discussed in the biomarkers section of this chapter (Section 5).

1.6. Carbohydrates and asthma

Carbohydrate decoration of proteins has been associated with asthmatic disease. Two case-control studies have investigated the role of the histoblood group antigens; H(O), A, B, or AB in asthma susceptibility. They found in human subjects the O-secretor mucin glycan (H-antigen) phenotype is associated with an increased susceptibility to recurrent asthma exacerbations [21].

2. Normal airway structure

From the level of the bronchioles upwards, the stratified epithelium lining the airway lumen rests on a basal lamina or 'true' basement membrane, a specialised 'mat' of extracellular matrix proteins. Below the basal lamina is a layer of collagenous matrix, termed the reticular basement membrane, in which fibroblasts are embedded in a sporadic arrangement. Farther below the fibroblastic layer are arranged bundles of smooth muscle myocytes, blood vessels and afferent nerve endings.

2.1. Structural changes to the airway in asthma

Although asthma is considered an inflammatory disease, there are many structural changes in the airways. Figure 1. shows cross-sections through the large airways of two patients, one normal and one severe asthmatic. The Movat's pentachrome stain clearly highlights the various architectural remodeling events occurring in the asthmatic airway. Obstruction of the airways from excessive mucus production is a common finding in severe asthmatics, blue staining in the epithelium and the lumen demonstrates mucous cell hyperplasia, with excessive mucus deposition into the airway. Under the asthmatic epithelium a thicker basement membrane is present which contains several different extracellular matrix (ECM) factors (including tenascin-C) compared to normals [22]. Deeper into the airway, red-stained muscle mass is increased in the asthmatic owing to a combination of smooth muscle hypertrophy and hyperplasia (reviewed in [23]). The airways of severe asthmatics also demonstrate a fibrotic response with increased connective tissue deposition and fibroblast and myofibroblast proliferation. Opinion is divided as to whether inflammation precedes airway remodeling, or whether the two occur in parallel (reviewed in [15, 24, 25]). Evidence tends to favour the latter because; firstly, remodeling occurs very early on in the disease, and in some cases in the absence of inflammation [26], secondly, there is only a weak link between airway inflammation and symptoms [27], and thirdly, epidemiological data demonstrate steroids do not work in all asthmatics [5]. In reality it is likely effective therapies will need to target both airway inflammation and remodeling.

2.2. Airway epithelial structure and function

The bronchial epithelium lines the inner wall of the respiratory tract in a continuous layer; it forms the interface between inspired air and the internal milieu as well as being the primary target for inhaled respiratory drugs. In the larger airways and down the respiratory tree to the level of the bronchioles the epithelium is at least two cells thick (reviewed. in [28]). Although there are at least eight morphologically and functionally distinct epithelial cell types present in the respiratory tract, they can be classified into three main groups; basal, ciliated columnar, and secretory columnar (reviewed in [29]).

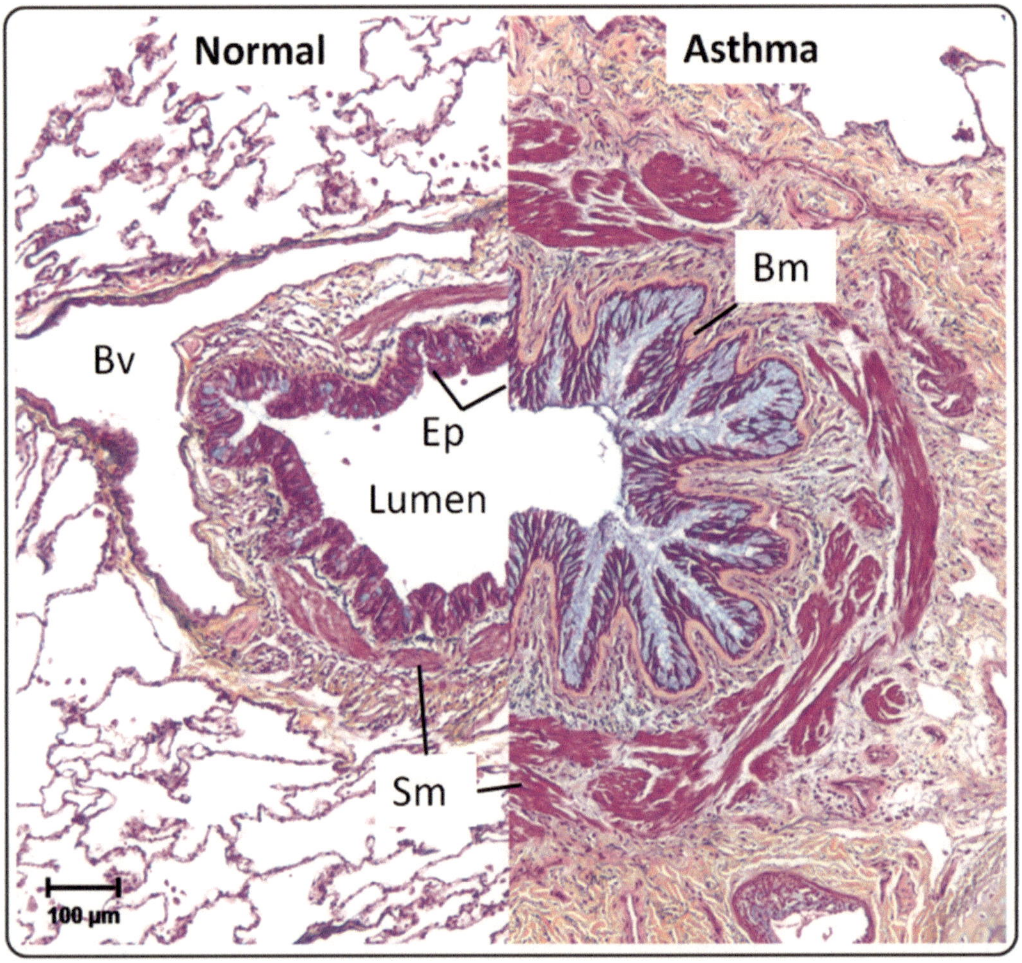

Figure 1. The airways in asthma undergo significant structural remodeling. Medium-sized airways from a normal and severe asthmatic patient were sectioned and stained using Movat's pentachrome stain. The epithelium (Ep) in asthma shows mucous hyperplasia and hyper secretion (blue), and significant basement membrane (Bm) thickening. Smooth muscle (Sm) volume is also increased in asthma. Bv = blood vessel. Scale bar = 100µm.

Basal cells

Basal epithelial cells are pyramidal-shaped cells with a low cytoplasm to nucleus ratio, although found throughout the airway, their contribution to epithelial volume decreases with airway size (reviewed in [29]). Basal cells are anchored to the underlying basement membrane and to other cells via specialised adhesion structures discussed below (Figure 2.)

[4, 30]. Historically the basal cell has been considered to be the stem cell of the bronchial epithelium (reviewed in [31]), however animal studies suggest there may be alternative stem cells located in the airway epithelium, including secretory cells, and cells located at specialised niches including the broncho-alveolar junction and the neuroepithelial bodies.

Columnar ciliated cells

Columnar epithelial cells lie above the basal cell layer and line the airway lumen, in direct contact with the inspired air. Ciliated cells are terminally differentiated columnar cells and are the most common cell-type in the bronchial epithelium, accounting for around 50% of all epithelial cells and 80% of terminally-differentiated apical cells in the human trachea [32], reviewed in [29]. Their main function is the removal of particulate matter by means of the mucociliary pathway. Ciliated cells arise from either a basal or secretory cell pre-cursor and possess around 300 cilia per cell, directly beneath the cilia are observed numerous mitochondria and a dense microtubule system, reflecting the high metabolic demands of particle clearance [29] [33].

Secretory cells

Secretory cells are also present at the apical surface, comprising 15 to 20% of the normal tracheobronchial epithelium [32]. In the larger airways the mucus-secreting goblet cell represents the predominant secretory cell and is the main source of airway mucus. These cells are characterised by membrane-bound electron-lucent acidic-mucin granules which are released into the airway lumen to trap inhaled particles and pathogens, prior to removal from the respiratory tract by coordinated cilia beating. Serous cells and Clara cells are relatively rare secretory cells in the larger airways present in large airways, but Clara cells are the main secretory epithelial cell type of the bronchioles, particularly the most distal bronchioles (reviewed in [34]). Several studies in rodents indicate the progenitor cell of the bronchial epithelium is not the basal cell but it is in fact the non-ciliated, secretory columnar cell. Both goblet and clara cells have been observed to undergo de-differentiation and proliferation during *in vivo* wound repair [4] [35-38].

2.3. Airway epithelial carbohydrate expression

The airway epithelial surface is covered by a layer of airway surface liquid (ASL), mainly of epithelial origin. A layer of mucins form a mucous layer that overlies the thin watery periciliary layer (PCL). The predominant mucins in the airways are MUC5AC and MUC5B secreted by goblet cells and submucosal glands respectively. Mucins are highly glycosylated proteins; 70-80% of their molecular weight are carbohydrate, and the structural characteristics of the mucous layer depends on interaction between the carbohydrate side-chains of mucin proteins. Studies of airway mucins in disease, demonstrate mucins are both oversulphated and hyper-sialylated in patients with cystic fibrosis and in chronic bronchitis patients with significant infections [39]. Pseudomonas aeruginosa, the pathogen responsible for the majority of morbidity and mortality in cystic fibrosis patients, uses the sialylated and sulphated Lewis-x determinants as attachment sites. Thus mucin glycosylation patterns may significantly affect infection in the airway. Specific lectin-binding assays have been used to characterise carbohydrate expression patterns in differentiated human airway epithelium [40]. Out of 38 lectin probes tested, seven bound specifically to basal cells, seven to columnar cells, and three specifically labelled secretory cells. The 1HAEo- and 16HBE14o- airway epithelial cell lines were also probed with the same lectin panel, revealing identical carbohydrate expression to the basal epithelial cells in the human tissue [40]. Thus the different cell types in the pseudostratified epithelium of the airways, express specific patterns of carbohydrates, possibly reflecting their different functions.

2.4. Airway epithelial functions

The primary function of the bronchial epithelium is to serve as a continuous physical barrier and as such it functions as part of the non-specific immune system, preventing pollutants, bacteria, viruses, allergens, and other potentially noxious substances transferring from inspired air into the underlying mesenchyme. This defence is mediated by several adhesive mechanisms which have been elucidated through transmission electron microscopy and immunohistochemical studies of bronchial epithelial biopsies [4] [30] (Figure 2.). Belt-like tight junctions (zonula occludens) seal the lateral apices of columnar cells, regulating para-cellular transport, whilst columnar cells adhere to each other via classical E-cadherin mediated adherens junctions [30]. Desmosomal cadherins mediate strong cell-cell adhesion and desmosomes are present between basal cells and particularly at the junction of basal and columnar cell layers [30]. Basal cells are in turn anchored to the basement membrane via specialised integrin-mediated cell-ECM junctions. Integrin heterodimers function by dynamically linking the contractile machinery of the cell's cytoskeletal network to the external matrix at specialised sites termed focal adhesions or focal contacts. The airway epithelium expresses several integrin heterodimers, including the $\alpha6\beta4$ integrin which binds to laminin-5 in the basement membrane forming a hemidesmosome anchor with the ECM.

The bronchial epithelium is more than a passive barrier; it performs a variety of roles allowing it to function as a dynamic regulator of the innate immune system [39]. The mucociliary pathway represents the principle mechanism of particle clearance in the airway. Inhaled particles and pathogens (bacteria, viruses) are trapped in mucus prior to removal from the respiratory tract by coordinated cilia beating. In addition to producing mucus, secretory cells are the source of a variety of mediators capable of having direct effects on inhaled noxious agents; including anti-oxidants and a variety of anti-bacterial agents; lactoferrin, lysozyme, β-defensins and also opsonins, components of the complement system that coat bacteria to facilitate phagocytosis by macrophages (reviewed in [39]). Various immunoglobin isotypes are secreted onto mucosal surfaces where they act in a protective manner by opsonising bacteria and other pathogens, rendering them relatively harmless and activating phagocytosis and digestion by tissue macrophages. Allergen-specific IgA is secreted by activated B lymphocytes (plasma cells) in the sub-mucosa, dimeric IgA (dIgA) selectively binds to the polymeric Ig receptor (pIgR) expressed on the basal surface of basal airway epithelial cells. The pIgR-IgA complex is subsequently internalised and is transported to the apical surface via the endosomal pathway ready for secretion into the airway lumen. Dimeric secretory IgA (sIgA) is released into the airway lumen as a 1:1 combination of dIgA and a portion of the pIgR known as secretory component (SC), which protects the IgA from digestion by proteases on the mucosal surface [40, 41]. SC is a highly glycosylated protein containing 15% N-linked carbohydrate. Glycosylated SC binds IL-8, inhibiting IL-8-mediated neutrophil chemotaxis and transendothelial migration. This interaction is dependent on SC glycosylation state as de-glycosylation with peptide N-glycosidase F abolishes the SC-IL-8 complex [42].

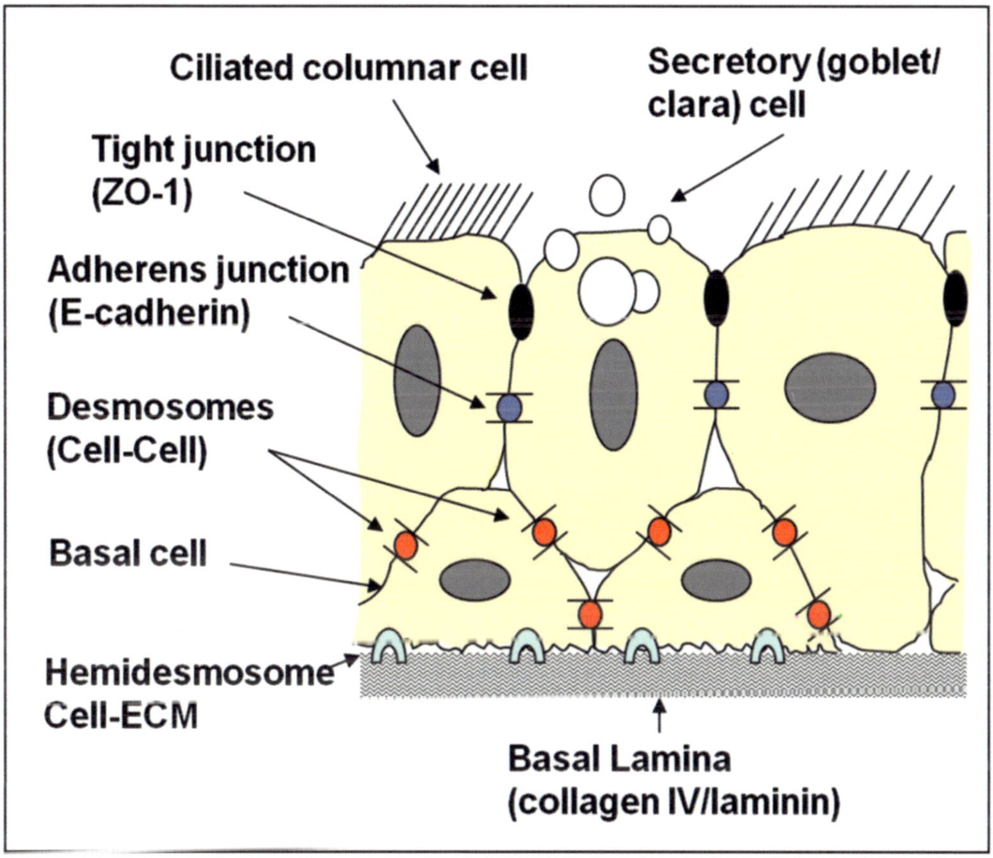

Figure 2. Adhesion mechanisms in the human bronchial epithelium. Tight junctions seal the lateral apices of columnar cells which adhere to each other via adherens junctions. Desmosomes mediate basal-basal and columnar-basal cell attachments, whilst hemidesmosomes anchor basal cells to the basal lamina. (Adapted from Roche et al. [30]).

3. Epithelial injury and repair

All epithelial tissues are regularly damaged as a consequence of exposure to environmental insults. Rapid repair following injury is crucial for restoring adequate barrier function, with subsequent cellular differentiation required to regenerate normal epithelial structure and function. The repair processes in all epithelial types have common elements producing an orderly progression of events including; cell spreading at the margins of the wound, cell migration into the wound, cell proliferation and finally re-differentiation (Figure 3). *In vivo* investigations of wound repair have demonstrated that migration and proliferation are both critical for the rapid restitution of the bronchial epithelium after injury [35-38]. Hamster tracheal epithelia were mechanically denuded to leave behind a bare (and sometimes damaged) basal lamina and wound repair was monitored until a morphologically normal epithelium was restored. Following epithelial removal, plasma promptly exudes into the injured site from the underlying vasculature to cover the denuded basement membrane, facilitating the binding of serum proteins to cellular receptors that stimulate repair of the damaged epithelium. At 12h post-wounding, viable secretory and basal cells at the wound margins de-differentiate, flatten into a squamous morphology, and migrate at about 0.5μm/min to cover the wound site.

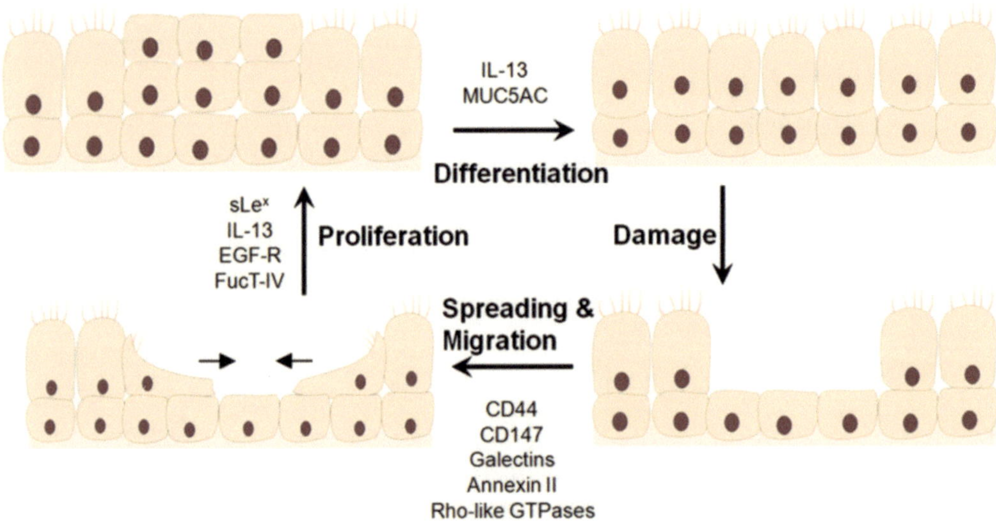

Figure 3. Airway epithelial repair involves a series of events including, cell spreading, migration, proliferation and differentiation. Following epithelial injury, neighboring cells flatten, spread and migrate to cover the site of injury. Subsequent proliferation and differentiation restores the epithelial structure and barrier function. The different stages of repair are mediated and facilitated by a diverse array of growth factors, cytokines and cell surface receptors, adhesion molecules, and intracellular enzymes, many of which are glycosylated. These glycans can modulate the function of proteins and lipids thereby regulating the repair process.

Proliferation is also an intrinsic component of epithelial repair. In the aforementioned hamster tracheal injury studies, mitosis rates at 12h post-injury are low (0.4%) but by 24h a wave of proliferation occurs primarily in secretory cells, resulting in a multi-layered epidermoid metaplasia [35]. At 48h post-wounding proliferation rates decrease (although mitosis is still significantly above basal levels), from this time point onwards the upper-most cells of the metaplastic epidermis start to slough off, resulting in a gradual loss of the metaplastic phenotype, with subsequent regeneration of a functional mucociliated phenotype achieved by apical cell re-differentiation. In small wounds normal mucociliary structure is restored by 120h but some persistent metaplasia persists in larger wounds through 168h post-wounding [38]. Although both secretory and basal cells are involved in epithelial repair, secretory cells play a dominant role in these experiments, consistently demonstrating higher mitosis rates than any other epithelial cell type [35-38]. *In vitro* studies using muco-ciliated human bronchial epithelial cells grown in air-liquid interface (ALI) culture, demonstrate that scrape wound repair is achieved by rapid spreading and migration of cells at the wound edge, followed by proliferation of basal cells outside of the wound area [43]. Acute wound repair speeds were inhibited slightly by the steroid dexamethasone, but the same drug potentiated the ability of long-term ALI cultures to repair repeated wounds [43].

In vivo studies demonstrate epithelial repair is accompanied by a corresponding proliferation of the underlying mesenchyme. Hamster tracheal epithelial damage is followed at 24h by capillary endothelial cell division, and at 36h many fibroblasts were observed to be in mitosis [35-37]. Mesenchymal proliferation persisted for a longer period than the epithelial response and supports several *in vitro* studies demonstrating epithelial damage can influence the phenotype and proliferation of underlying fibroblasts [44, 45].

3.1. Airway epithelial wound repair is compromised in asthmatics

The patho-physiological changes observed in the asthmatic airway may be due to reactivation of the epithelial-mesenchymal trophic unit (EMTU), triggered by an abnormal epithelium that is held in a repair phenotype [24]. *In vitro* studies support the hypothesis that epithelial-derived factors are capable of altering the phenotype of adjacent mesenchymal cells [43, 44]. The mechanisms involved in initiating and maintaining the epithelial repair response in asthma are largely unknown, however recent studies have demonstrated the airway epithelium in asthmatics is intrinsically altered compared to normals, and that these differences persist for extended periods in culture [45, 46]. There is considerable evidence to suggest the airway epithelium is capable of orchestrating the inflammatory immune response, as following injury, epithelial cells release a variety of pro-inflammatory cytokines, chemokines and growth factors. Acute inflammation is a necessary aspect of wound repair, however if epithelial damage persists, or the epithelial repair response is not shut-off, then potentially deleterious chronic inflammation may result (Figure 4.).

3.2. Mechanisms of cellular migration

Many disorders characterized by impaired re-epithelialization, are not a result of inadequate proliferation, but are due to impaired cell migration over the denuded site [47-49]. Cell migration requires a coordinated, highly complex series of events including; changes to the cell cytoskeleton, modification of the surrounding ECM and modulation of adhesions to ECM substrate and to other cells. The majority of these processes are regulated by the Rho-like family of small GTPases including, RhoA, B, C, E, Rac1, 2, 3 and Cdc42 (Figure 3.) (reviewed in [50, 51]) that function by acting as "molecular switches", cycling between an

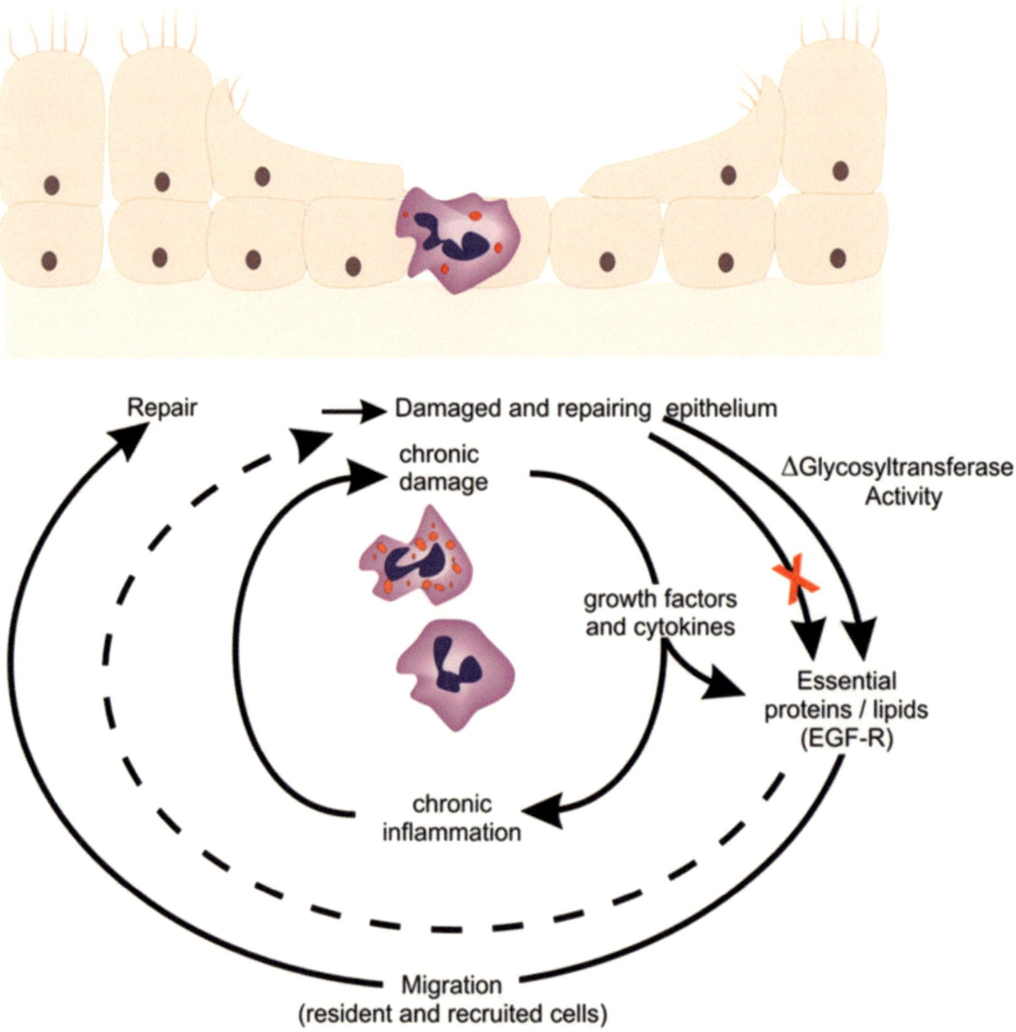

Figure 4. Injury and repair cycle of the airway epithelium in asthma. Complete repair requires glycosylation as a means of regulation of essential elements. Aberrant glycosylation would result in a defect in the mechanisms of repair, the accumulation of epithelial damage and persistent airway inflammation. Modified from Davies [46].

inactive GDP-bound state and an active GTP-bound form at the plasma membrane [52]. GTPases are able to regulate the transmission of signals from cell surface receptors, such as integrin-mediated cell-ECM adhesion, or growth factor ligation, to downstream intracellular signalling pathways, influencing the cell cytoskeleton, gene transcription, and the cell cycle (reviewed in [53]). Rho induces formation of stress fibres, integrin-mediated focal adhesions, and actomyosin-mediated cell body contraction [54], Rac-dependent actin polymerisation leads to lamellipodium extension and membrane ruffles at the leading edge of the cell [55], and Cdc42 induces formation of microspikes/filopodia [56].

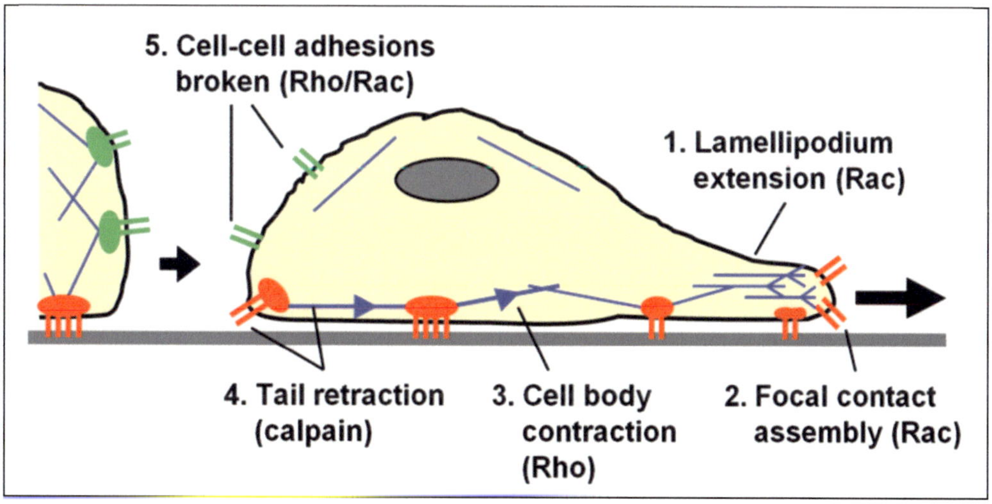

Figure 5. Cellular processes occurring during migration. The transition from a stationary state to a migratory phenotype involves several distinct mechanisms, Rho-like GTPases regulate the majority of these processes and the individual GTPases most directly linked to each are: (1.) Rac-dependent actin (blue) polymerisation leads to lamellipodium extension at the leading edge of the cell, (2.) new ECM adhesions are formed at the lamellipodium via focal contacts (red), also regulated by Rac, (3.) cell body contraction is dependent on actomyosin contraction (blue arrows) and is dependent on Rho activity, (4.) tail detachment is possibly affected by GTPase activity but the protease calpain is important for degradation of focal adhesions at the rear of migrating cells, (5.) cadherin-mediated cell-cell adhesions (green) are dependent on Rho and Rac activity and are often downregulated during migration. MMP-mediated ECM degradation is necessary for migration, even in cells migrating over a basement membrane.

Several *in vitro* studies have demonstrated activity of the RhoGTPase family is essential for airway epithelial wound repair; activation of protein kinase A (PKA) facilitates scrape wound closure in cultured bovine airway epithelial cells via inhibition of Rho activity and decreased focal contact attachments [57]. The 16HBE human airway epithelial cell line has been used to demonstrate coordination of RhoA and Rac1 activities are vital for efficient wound closure [58]. Detachment of the trailing edge is often a limiting factor of cellular migration speed, detachment is possibly affected by GTPase activity but the protease calpain is important for degradation of focal adhesions at the rear of migrating cells. Rho-GTPases are also involved in cell proliferation during airway epithelial repair; the scaffold IQ domain GTPase-activating protein (IQGAP1), an effector of Rho GTPases, is upregulated during 16HBE scrape wound repair and facilitates closure by increasing via a ß-catenin-mediated transcription of genes involved in cell cycle regulation such as CyclinD1 [59, 60].

3.3. The effect of intracellular protein glycosylation on epithelial repair

The activity of the Rho-GTPase family is highly regulated by glycosylation. *Clostridium difficile* toxins A and B (TcdA and TcdB) are major virulence factors for the induction of antibiotic-associated diarrhea and pseudomembranous colitis. These toxins possess potent cytotoxicity against cultured cell lines by inducing disaggregation of the cytoskeleton. TcdA and TcdB cellular toxicity is dependent on their intrinsic glucosyltransferase activity, they modify Rho-like GTPases by using UDP-glucose as a co-substrate [61]. RhoA undergoes mono-O-glucosylation at threonine-37, and Rac1 and Cdc42 are glucosylated at the homologous amino acid threonine-35. The threonine residue at position 37 of Rho (position 35 of Rac or Cdc42) is highly conserved and is located in the so-called effector region, it participates in co-ordination of the bound magnesium ion and thereby plays a pivotal role in nucleotide binding [62]. Glucose binding to residues within the effector region inhibits Rho-GTPase activity leading to drastic alteration of the cell cytoskeleton via several potential mechanisms including; inhibition of effector coupling, inhibition of nucleotide exchange (thereby directly blocking GTPase activity), by altering the sub-cellular localization of the GTPase, and finally by altering the stability of the enzyme within the cell (Reviewed in [52], [63]). TcdA has also been shown to induce apoptosis of intestinal epithelial cells via activation of caspases-3, -8 and -9 [64]. To date there is no direct experimental evidence linking Rho-GTPase glucosylation to the inhibition of wound repair in airway epithelial cells. However as this family of enzymes play such a fundamental role in repair of the airway epithelium, and the structure and function of the Rho-family of GTPases is very highly conserved, it is logical to presume that glycosylation would also play a pivotal role in regulating GTPase activity and thus wound repair.

3.4. Role of cell surface glycoconjugates in epithelial repair

Many proteins essential for normal cell physiology including adhesion molecules and cell surface receptors are glycosylated [65,66]. Alteration in the glycosylation pattern of many glycoproteins leads to changes in their function. It has been shown that impaired glycosylation of receptors often leads to abnormal intracellular trafficking, ligand binding and downstream signal transduction ability [67-70]. Tsuda *et al.* demonstrated that removal of sialic acid from erythropoietin leads to loss of its *in vivo* activity [71]. Complex carbohydrate structures attached to cell surface proteins and lipids have functional roles in cell motility [72], adhesion [73], proliferation [74], and growth potential [75] in several cell types. It also has been demonstrated that certain apical cell-surface carbohydrates are altered during cellular differentiation [76]. Thus many cellular functions are regulated and dependent upon glycoproteins. Because of the role of glycosylated structures in cell-cell and cell-matrix interaction, it is not surprising that there is a growing interest to explore the role for and regulation of these cell-surface carbohydrates in epithelial repair.

Lectins are naturally occurring proteins that can be isolated from a variety of plants and animals. Each lectin binds to a specific sugar moiety. As such, lectins are exquisitely selective tools to identify or block specific glycoconjugate motifs and have been extensively employed to study the role of cell-surface sugars and complex carbohydrates in cellular function [77, 78]. A variety of approaches have been employed to unravel the role(s) of carbohydrates in the multiple steps of the repair process. Some studies have investigated the expression pattern of different carbohydrates after injury using lectins as probes. Using three methods for localizing or quantifying lectin-binding sites Gipson *et al.* compared cell-surface of normal and migrating corneal epithelium of the rat. In their study they found that cell-surface of migrating epithelia express different sugar moieties relative to cell membranes of stratified stationary epithelia. There was a dramatic increase in concanavalin A (ConA) and wheat germ agglutinin (WGA) binding on migrating cells relative to stationary cells. The authors also showed that migrating cells have an increase in glycoprotein production determined by an increase in the incorporation of radiolabelled leucine and glucosamine. In addition they found that N-glycosylation of epithelial cells is necessary for epithelial cell migration [79, 80]. Sweatt *et al.* demonstrated an increase in N-acetylgalactosamine in cell-surface glycoconjugates at the site of epithelial injury in pig cornea [81]. Dorscheid *et al.* has previously characterized cell-surface glycosylation in non-secretory cells of central human airway epithelium and airway epithelial cell lines utilizing lectin-binding patterns [82]. In this study it was shown that galactose- or galactosamine-specific lectins labeled basal epithelial cells and cell lines derived from basal cells. Lectins specific for several different carbohydrate structures bound columnar epithelial cells, and certain fucose-specific lectins labeled subsets of the airway epithelial cells. The cellular specificity of these differences suggests they may be relevant in various cellular functions. It has also been demonstrated that following mechanical injury of guinea pig tracheal epithelium, glycosylation profiles in the repairing epithelium change over time [83]. These changes may represent either the expression of one or more new glycoproteins, or changes in glycosylation of constitutive proteins, required for activation or a change in cell function needed for repair to proceed. Studies have shown that injury of the respiratory epithelium enhances *P. aeruginosa* adhesion and it has been speculated that changes of cell surface glycoconjugates related to wound repair, cell migration and/or spreading may favor *P. aeruginosa* adhesion [84].

In a previous study, Dorscheid *et al.* examined the functional role of cell surface carbohydrates in an *in vitro* model of wound repair after mechanical injury of human airway epithelial cells. The results demonstrated that N-glycosylated glycoproteins, particularly those with a terminal fucose residue, are essential in the adhesion and migration of airway epithelial cells and facilitate closure of epithelial wounds in monolayer culture [85]. Recently, Allahverdian *et al.* studied the role of a fucose containing tetrasaccharide, sialyl-Lewis X (sLex) [NeuAcα2-3Galβ1-4(Fucα1-3)GlcNAc], in airway epithelial repair. Increased presentation of sLex was observed in areas of epithelial damage relative to areas of intact epithelium. In an *in vitro* model of bronchial epithelial repair, cell surface expression of sLex was shown to be significantly increased following mechanical injury of airway epithelial cell monolayers and inhibition of sLex completely prevented repair [86]. Further studies demonstrated that sLex decoration of epidermal growth factor receptor (EGFR) plays an important role in mediating airway epithelial wound repair [87]. Allahverdian *et al.* demonstrated that sLex co-localizes with EGFR and blocking of sLex with a neutralizing antibody resulted in reduced phosphorylation of EGFR and prevented repair. The final step in the synthesis of sLex is catalysed by a specific α-1,3-fucosyltransferase, fucosyltransferase-IV (FucT-IV). Reduction in EGFR phosphorylation and repair were similarly observed when FucT-IV gene expression was knocked down using small interfering RNA (siRNA). These studies demonstrate the importance of cell surface carbohydrates in mediating airway epithelial repair.

It has been shown previously by Donaldson *et al.* that the plant lectin concanavalin A (ConA), which binds glucose and mannose moieties, can inhibit migration of newt epidermal cells [88]. Gipson *et al.* cultured rat corneas with 3 mm central epithelial abrasions in the presence of four plant lectins. In this study the authors demonstrated that blocking glucose, mannose, and glucosamine sites on corneal epithelial cell surfaces and/ or the epithelial basement membrane reversibly slows or inhibits epithelial migration [78]. Using the same culture model of bronchial epithelial repair, Patchell *et al.* has demonstrated that following mechanical wounding of intact monolayers, the lectins *Allomyrina dichotoma* (AlloA) and chick pea agglutinin (CPA) differentially stain human airway epithelial cells in damaged areas relative to the staining of intact epithelial monolayers [89]. While AlloA positive staining cells are those that appear to be migrating from areas distant to the wound and accumulating in the wound, CPA positive staining cells are restricted to the leading edge of the wound. Moreover, the addition of the above lectins following mechanical wounding inhibited the repair. These results suggest that AlloA and CPA bind specific carbohydrate structures involved in normal epithelial repair. Further work is being carried out to determine the identity of the relevant proteins associated with these carbohydrate ligands.

In a study by Trinkaus-Randall *et al.*, the effect of specific carbohydrate moieties of the basal lamina on the attachment and spreading of rabbit corneal epithelial cells was studied. Corneal epithelial basal cells were plated onto freshly denuded basal lamina and three lectins WGA, ConA and RCA were used to block specific sugar moieties in the basal lamina. This study showed that lectin binding of glucose, mannose, and galactose moieties on the basal lamina significantly altered the extent of cellular spreading, while the binding of glucosamine inhibited attachment. This study demonstrated that alteration of specific sugar moieties on the native basal lamina dramatically affects the ability of basal cells of corneal epithelium to attach or spread [90]. Using an *in vitro* model of airway epithelial repair Adam *et al.* demonstrate that lectin WGA which binds to N-acetyl glucosamine residues inhibits the repair of epithelial damage without altering cell viability, while other N-acetyl glucosamine binding lectins do not affect the repair process [81].

These studies clearly demonstrate the critical role of carbohydrates in the process of epithelial repair. However, carbohydrates must either modify a protein or lipid to regulate its function or participate in binding to a specific receptor to effect the desired action. As such, the biological role of glycans can be broadly divided into two groups. One group depends on the structural and modulatory properties of glycans and the other relies on specific recognition of glycan structures (generally receptor proteins or lectins).

3.5. Galectins, annexins and epithelial injury – repair

Galectins and annexins are two of possibly many families of proteins with carbohydrate binding capabilities. Although their role in epithelial repair remains unclear, recent studies highlight their importance in these processes. Galectins are a specific family of lectins with an affinity for β-galactose-containing oligosaccharides that have no enzymatic activity. To date, as many as 14 galectins have been identified, and they have been found in non-mammalian species such as birds, amphibians, fish, worms, sponges and fungi. The binding affinities of galectins are often greater for the oligosaccharides than the monosaccharide galactose. The expression of galectins is conserved however; their expression is often tissue specific and developmentally regulated. Different cells express a unique complement of galectins. Of the 14 galectins, eight have been identified in the nucleus and cytoplasm and participate in specific cellular functions. Nine galectins have been found extracellularly. Their presentation on the cell surface remains a mystery since they lack a signal sequence for secretion via the classical secretory pathway for membrane proteins.

Studies have shown that galectin-1 and galectin-3 have the potential to mediate cell-matrix interactions [92-95]. Their ubiquitous expression makes it difficult to understand the mechanism of their involvement in this process. In corneal epithelial cells, galectin-3 and galectin-7 have been characterized as mediators of epithelial cell migration, an essential component of epithelial wound repair [96, 97]. Galectin immunostaining in healing wounds is more intense relative to normal epithelium, primarily at the leading edge and at areas of cell-matrix interactions [97]. Furthermore, wound repair was impaired in gal-3$^{-/-}$ mice in the wound model systems of either alkali burn wounds that leave the corneal stroma intact or laser ablation wounds that result in damage to the underlying basement membrane. This impaired repair was not seen in gal-1$^{-/-}$ mice. The authors propose that this effect is mediated by the lectin domain binding complementary glycoconjugates in the ECM and cell surface molecules resulting in enhanced cell-cell and cell-matrix interactions [97]. Similar work has demonstrated that galectin-7 also plays a role in corneal epithelial repair [96]. Although obvious differences exist between tissues, galectins may play a similar role in other epithelial organs and epithelial repair such as the lung and intestinal tract.

Annexins are a family of proteins that bind acidic phospholipids in a calcium dependant manner. Similar to galectins, annexins are largely cytoplasmic, however; several annexins have been detected on the cell surface of a variety of cells. Like galectins, annexins also lack a signal sequence for the cell surface presentation [98]. Annexins are proteins that have been associated with many cellular functions however their role remains poorly understood. Recent work has suggested that annexins possess carbohydrate binding abilities. This was first observed with annexin A4 when it was shown that it binds a variety of glycans [99]. Recent studies have suggested that annexin II (AII) is also capable of binding carbohydrates however this work is still in preliminary stages. Previously, the binding of tissue-type plasminogen activator (t-PA) was shown to bind HepG2 cells, however, following enzymatic removal of α-fucose residues on t-PA, this binding was dramatically decreased [100]. Subsequently, AII has been shown to be a cell surface receptor for t-PA [101]. The importance of α-fucose on t-PA may be a result of the carbohydrate binding of AII. Patchell *et al.* has shown that AII is presented on the cell surface and acts as a mediator of epithelial repair. AII was shown to be actively involved in wound repair in cells at the leading edge of the wound. Interestingly, the initial isolation purification of AII was achieved using a lectin, chick pea agglutinin, suggesting that itself, or an associated protein is glycosylated. Their tight association with the membrane and potential carbohydrate binding may influence the several functions associated with these relatively poorly understood proteins. On the surface of cells, AII has previously been shown to be involved in the migration of prostate cancer and lung carcinoma cells [102, 103]. In the case of metastatic cells, the regulation of cell migration and cell death is lost, however, under the appropriate levels of control, similar mechanisms of migration may be involved in the normal processes of wound repair. Following mechanical injury to rat cornea, at the leading edge of the wound AII translocates to the cell surface and interacts with the matrix protein tenascin-C [104]. The interaction of AII with tenascin-C results in the loss of focal adhesions and cell migration in endothelial cells and has been shown to initiate downstream signaling pathways. These findings combined with previous results suggest that AII is involved in cell migration and repair. However, the mechanism of AII involvement in these cellular events is unknown but likely involves a fucose binding lectin activity to coordinate the needed response.

As a cell surface molecule, it has been suggested that AII is the human cytomegalovirus receptor [105]. Following the initial interaction with heparan sulfate proteoglycan, the virus particle is primed for membrane fusion and infection [106, 107]. As a membrane fusion protein, AII could potentially act as a receptor to other enveloped viruses such as respiratory syncytial virus (RSV). Recently, AII has been shown to bind RSV and was characterized as a potential cell surface protein that can promote RSV infection [108]. These findings along with data that has identified AII on the surface of airway epithelial cells provide strong evidence AII may be a cell fusion receptor allowing RSV infection of the respiratory tract. Viruses are a common source of lung injury, specifically the epithelium. Epithelial cells, as a protective mechanism, become highly apoptotic following virus infection to prevent further infection and viral persistence [109]. Viral infectivity of several enveloped viruses is dependent upon carbohydrate interactions between the viral proteins and their receptors. The end result is injury to the epithelium that requires repair. Altered glycosylation then may either increase the resulting injury from RSV infection or impair the AII coordinated repair.

3.6. MMPs in asthma, glycosylation and repair

Matrix-metalloproteinases (MMPs) are a family of zinc-dependant endopeptidases that digest all components of the extracellular matrix (ECM) and many non-ECM substrates including growth factors, cytokines, and their receptors [110, 111]. Several MMPs are increased in airway tissue and broncho-alveolar lavage (BAL) fluid of asthmatics including MMP-1, -2, -3, -7, -8 and -9 [112, 113], suggesting a link between increased airway proteolytic activity and the asthma phenotype. *In vivo* studies have shown MMPs play a direct role in the development of allergic airway inflammation [114] and hyperresponsiveness [115-117]. Inflammatory cells, particularly neutrophils and alveolar macrophages, were thought to be the major source of increased proteinases in asthma and other inflammatory lung diseases [118, 119], however several groups have demonstrated structural cells of the lung including the epithelium, also synthesize MMPs and their endogenous tissue inhibitors (TIMPs) [10, 120-125]. MMPs perform many biological functions, they are integral to connective tissue homeostasis and their activities are strictly coordinated during epithelial wound repair [124-

126]. Many growth factors such as TGF-β and VEGF bind to sites in the ECM forming "reservoirs", and are released during ECM digestion [112]. Recent studies have provided strong evidence that dysregulated airway MMP activity in the lung epithelium may contribute to airway inflammation and the progression of tissue remodeling in asthma. MMP-7 expression is altered in the epithelium of asthmatics which may contribute to airway inflammation via cleavage and release of cytokines and chemokines including IL-25 and soluble FasL [10, 127]. It is possible that aberrant MMP expression in the airway epithelium contributes to inflammation and lung remodeling in asthma via proteolytic alterations in cytokine and chemokine signaling.

3.7. IgCAMs in asthma, repair and the effects of IL-13

Immunoglobulin-cell adhesion molecules (Ig-CAMs) are a diverse family of cell adhesion molecules characterised by the presence of one or more copies of a structure known as the Ig fold [128]. These receptors are expressed on a variety of cell types and are vital for many different biological processes, including neural development and immune system function [129]. In the adult, Ig-CAMs regulate the recruitment and subsequent activation of circulating lymphocytes, studies show they are also expressed in the epithelium and vasculature of the bronchial mucosa and levels of two Ig-CAMs, ICAM-1 and CD44 (hyaluronic acid receptor), are increased in the epithelium of asthmatics [130, 131]. Recent data also suggests a third Ig-CAM, CD147 (EMMPRIN – Extracellular matrix Metalloprotease Inducer) is increased in the epithelium of patients with chronic obstructive pulmonary disease (COPD) [132]. CD147 is over-expressed by metastatic cells of the skin, bladder, breast, and lung [92]. In cancer CD147 stimulates synthesis of various MMPs via homophilic interaction of CD147 oligomers on adjacent cells [134]. CD147 is potentially involved in various signaling pathways relevant to remodeling in airways disease, for an in depth review see [135]. Studies have shown CD147 is endogenously expressed at low levels in the basal cells of various stratified epithelia including the airway. Airway epithelial CD147 expression is increased in murine models of lung injury [136] and in several inflammatory airway diseases including pulmonary fibrosis [137] and interstitial pneumonias, where increased CD147 is associated with MMP-2, -7 (matrilysin) and -9 levels in the lungs of patients [138]. Recent studies have demonstrated CD147 is increased in the airway epithelium and BAL of patients with COPD, and signaling via CD147 regulates MMP-9 expression in cultured bronchial epithelial cells [132]. CD147 is potentially important in asthma; in a mouse model of allergic asthma, inhalation of a CD147-neutralizing monoclonal antibody reduced ovalbumin (OVA)-induced airway mucin production and airway hyperreactivity, whilst partially inhibiting the recruitment of eosinophils and effector $CD4^+$ T cells into lung tissue [139]. The authors attributed the reduced airways inflammation to an inhibition of the interaction between CD147 on inflammatory cells with cyclophilin chemokines. The reduction in OVA-induced mucous production suggests CD147 inhibition directly affects the epithelium.

MMP activity is dependent upon the close spatial juxtaposition of enzyme and substrate, therefore targeted MMP localization is a further mechanism of regulation. Specific MMP-substrate interactions can be facilitated and regulated at the cell surface via association of MMPs with chaperone molecules. CD44 is strongly upregulated in injured asthmatic airways and at the leading edge of repairing bronchial epithelial cells [43, 140], yet its function during airway epithelial repair is unknown. Studies suggest CD44 may co-ordinate MMP-substrate interactions at the plasma membrane by functioning as a docking molecule. In repairing reproductive epithelium, the v3 isoform of CD44 (CD44v3) forms a signaling complex with MMP-7 which is presented at the cell surface in close proximity to a potential substrate, pro-Heparin-Binding Epidermal-like Growth Factor (pro-HB-EGF). MMP-7 cleaves pro-HB-EGF into an active soluble form which triggers recruitment and activation of the Erb-B4 receptor (a member of the EGFR family) [141]. The EGF pathway is essential for efficient repair of the bronchial epithelium [43], and a similar mechanism of targeted MMP-7 presentation may occur during lung epithelial repair. Carbohydrate modification of proteins is essential for the creation of such micro-environments at the cell surface; in this case the interaction of MMP-7 (an non-glycosylated protease) with CD44 is dependent on the presence of branched glycosylated side-chains on the extracellular portion of the CD44 molecule. It is logical to extrapolate that a deficiency in the post-translational modification of CD44 could lead to a lack of MMP-7 binding, a reduction in HB-EGF activation, and a consequent inhibition of epithelial repair.

3.8. How carbohydrates mediate cell-cell interaction and migration

The glycans attached to matrix molecules such as collagens and proteoglycans are important for the maintenance of tissue structure, porosity, and integrity. Such molecules also contain binding sites for specific types of glycans, which in turn help with the overall organization of the matrix. Glycans are also involved in the proper folding of newly synthesized polypeptides in the endoplasmic recticulum and /or in the subsequent maintenance of protein solubility and conformation [66]. In this manner, altered glycosylation of matrix proteins may change adhesion properties and the potential of cell to migrate. This is due to either an altered conformation of the extracellular matrix of a change in the carbohydrate ligands available to bind.

Glycosylation of growth factor receptors has tremendous effects on the receptor function. Several studies have focused on the role of EGF and its family of receptors on epithelial repair [24, 142]. EGF, a well known mitogen for epithelial cells, has been used in to stimulate epithelial wound healing in guinea pigs as well as human airway cell monolayers and differentiated cells *in vitro* [43, 143]. EGFR is known to be up-regulated upon the creation of a wound on airway epithelial monolayers in culture and correlated to the damaged areas of epithelium [24]. Increased tyrosine phosphorylation of the EGFR has also been observed after mechanical injury even in the absence of exogenous ligand [24]. While a critical role of signaling mediated by EGFR in repairing damaged epithelium has been well demonstrated in many epithelial systems, an important role for core glycosylation of extracellular domain of EGFR in ligand binding and tyrosine kinase activity has also been documented [105-107]. The extracellular domain of EGFR has 10 to 11 potential site for N-glycosylation. It has been shown that glycosylation is necessary for ligand binding and tyrosine kinase activity of

EGFR [144-146]. Carbohydrate moieties of EGFR are critical for the direct interaction of the receptor with other structures [147]. Moreover, glycosylation defines localization of EGFR to specific domain of plasma membrane which could facilitate association of the receptor with other molecules and its transactivation [148]. Modification of receptor N-glycans can also regulate receptor trafficking and duration of cell surface residency [149]. A direct association between changes in glycosylation of EGFR and cellular events involved in epithelial repair such as migration and proliferation remains to be identified. It could be inferred that role of carbohydrates in epithelial repair is partly through their remarkable role in modulation of EGFR and other growth factor receptors.

The specificity of carbohydrate interactions allows for the high degree of selectivity found within the cell. Whether it is in regulating receptor activation, cell-cell or cell-matrix interactions or cell migration, defects in these processes may result in abnormalities and altered phenotypes. Their understanding will provide new avenues for research and therapies.

4. IL-13 structure

Interleukin-13 (IL-13) is a type I cytokine, comprised of four short α-helical hydrophobic bundles joined by two disulfide bonds [111]. The cDNA of IL-13 has a single open reading frame that encodes 132 amino acids, including a 20-amino acid signal sequence that is cleaved from the mature secreted protein [112]. IL-13 has four predicted N-linked glycosylation sites and can either exist as a 17kDa glycoprotein or 10-12kDa unglycosylated protein [113]. IL-13 has been demonstrated to be secreted predominantly as the 10-12kDa unglycosylated protein [113]. In a study conducted by Liang *et al.*, the distribution and expression levels of unglycosylated and glycosylated IL-13 was demonstrated to differ between human bronchial smooth muscle cells (BSMC) and lymphocytes [114]. Intracellularly, BSMC were shown to express higher levels of 12kDa unglycosylated IL-13, while lymphocytes expressed more of the 17kDa glycosylated form. However, lymphocytes secreted significantly higher levels of 12kDa unglycosylated IL-13 compared to BSMC. These findings suggest that cellular distribution of unglycosylated and glycosylated forms of IL-13 is tissue and cell type-dependent and may reflect the role of IL-13 in each cell type. The study also demonstrated that IL-13 secretion is regulated by human sulfatase modifying factor 2 (SUMF2), a protein predominantly found in the endoplasmic reticulum and highly expressed in BSMC [114]. SUMF2 was shown to physically interact with IL-13 and regulate its secretion irrespective of the glycosylation status of IL-13 in BSMC. Lymphocytes were shown to express lower levels of SUMF2 and to secrete higher levels of IL-13 compared to BSMC, further supporting the role of SUMF2 in regulating IL-13 secretion.

The glycosylation state of IL-13 can also affect its activity and potency to induce certain effector functions such as the up-regulation of low affinity IgE receptor, CD23 on the surface of monocytes. A functional study conducted by Vladich *et al.* [115] demonstrated that eukaryotic COS-7 cell-derived IL-13 was significantly more active in inducing the surface expression of CD23 on peripheral blood mononuclear cells (PBMC) compared to two different commercially available prokaryotic *E. Coli*-derived IL-13. The difference in bio-activity between the prokaryotic and eukaryotic-derived IL-13 molecules can be attributed to the fact that prokaryotic systems often lack post-translational modification steps such as glycosylation that take place in eukaryotic cells. These findings suggest the importance of glycosylation for IL-13 bio-activity.

4.1. IL-13 function

IL-13 is a pleiotropic cytokine that exerts its effect on a wide variety of cell types. IL-13 has a variety of functions which are critical to immune homeostasis and repair, but when uncontrolled contribute to the asthma phenotype. IL-13 is predominantly produced by T helper type 2 (Th2) cells [116] and to a lesser extent by mast cells, basophils, and eosinophils [117, 118]. IL-13 is also produced by a variety of non-hematopoietic cells including airway epithelial cells [119] and bronchial smooth muscle cells [114]. As a Th2 cytokine, IL-13 shares numerous overlapping characteristics with IL-4, where both cytokines promote B cell proliferation, IgE class switching and synthesis in B cells, and induce surface expression of antigens such as CD23 and major histocompatibility complex (MHC) class II [120]. These similarities between IL-13 and IL-4 function can be explained by the fact that the cytokines share a common receptor subunit, IL-4 receptor α subunit (IL-4Rα). However unlike IL-4, IL-13 does not exert direct effects on T cells and cannot elicit Th2 differentiation in naïve T cells; IL-13 is instead involved in mediating downstream Th2-effector functions [121]. In monocytes and macrophages, IL-13 can inhibit the production of various pro-inflammatory mediators such as prostaglandins, reactive oxygen species, IL-1, and TNF-α [113, 122]. IL-13 can also enhance the expression of several members of the integrin family that play important roles in adhesion, including CD11b, CD11c, CD18, and CD29 as well as induce surface expression of CD23 and MHC class II in monocytes and macrophages [121]. Furthermore, IL-13 has been reported to have direct effects on eosinophils, including promoting eosinophil survival, activation, and recruitment [123, 124].

IL-13 has important functions on non-hematopoietic cells, including endothelial cells, smooth muscle cells, fibroblasts, and epithelial cells. In endothelial cells IL-13 is a potent inducer of vascular cell adhesion molecule 1 (VCAM-1), which plays a role in eosinophil recruitment [125]. IL-13 has been shown to enhance proliferation and cholinergic-induced contractions of smooth muscle cells *in vitro* [126, 127]. In epithelial cells, IL-13 is a potent inducer of growth factors [128, 129] and chemokine expression [130]. It also induces epithelial cell proliferation [128], alters mucociliary differentiation [131], resulting in mucin production and goblet cell metaplasia [7, 132, 133].

4.2. IL-13Rα1 structure and function

IL-13 signals via a receptor system that consists of IL-13 receptor α1 (IL-13Rα1)/IL-4 receptor α subunit (IL-4Rα) and IL-13 receptor α2 (IL-13Rα2). Both IL-13Rα1 and IL-13Rα2 belong to the type I cytokine receptor family, which possess several definitive features, including a W-S-X-W-S motif, four conserved cysteine residues, fibronectin type II modules in the extracellular domain and proline-rich box regions in the intracellular domain [134]. The two receptors share 33% homology and 21% identity at the amino acid level [135] and their respective genes have both been mapped to the X chromosome [136]. Kinetic analysis have revealed that IL-13Rα1 and IL-13Rα2 both bind specifically only to IL-13 and not IL-4, while IL-13 alone has no measurable affinity for IL-4Rα [137].

IL-13Rα1 is widely expressed on both hematopoietic and nonhematopoietic cells except human and mouse T cells and mouse B cells [112]. It is a 65-70kDa glycosylated trans-membrane receptor protein with 10 predicted N-glycosylation sites in the extracellular domain [138]. The cDNA for human IL-13Rα1 encodes a 427-amino acid sequence, including a 60-amino acid intracellular domain and 21-amino acid signal sequence [139]. A mutational analysis demonstrated that Leu319 and Tyr321 in the cytokine receptor homology module (CRH) of human IL-13Rα1 are critical residues for binding to IL-13 [135]. Interestingly, these residues are in close proximity to the predicted glycosylation sites of IL-13Rα1, suggesting that the glycosylation state of IL-13Rα1 could potentially influence receptor-ligand interactions.

IL-13Rα1 binds IL-13 with low affinity by itself but binds IL-13 with high affinity when heterodimerized with IL-4Rα to form a functional signalling receptor [138]. This signalling complex is known as the type II IL-4/IL-13 receptor and it also serves as an alternate receptor for IL-4. IL-13 is known to primarily signal through IL-13Rα1/IL-4Rα by first binding to IL-13Rα1 chain with low affinity and then heterodimerizing with IL-4Rα to become a stable, high affinity signalling complex [137]. The formation of IL-13Rα1/IL-4Rα results in the activation of the Jak kinases, Jak1 and Tyk2, followed by the recruitment of the transcription factor, signal transducer and activator of transcription 6 (STAT6) to the receptor. STAT6 is then phosphorylated and forms functional dimers that translocate to the nucleus to bind specific canonic DNA elements and initiate transcription of downstream genes [140]. Some examples of downstream targets of IL-13Rα1 activation include eotaxin [141], MUC5AC [142] , and, arginase I which is an enzyme important in the development of airway hyperreactivity [143].

4.3. IL-13Rα2 structure and function

IL-13Rα2 is a 56kDa glycoprotein that is expressed by a variety of cell types including monocytes, airway epithelial cells, fibroblasts, and keratinocytes [144, 145]. The cDNA of human IL-13Rα2 encodes a 380-amino acid protein with a 26-amino acid signal sequence and a 17-amino acid intracellular domain [146]. IL-13Rα2 contains four predicted glycosylation sites [147] and the glycosylation state of IL-13Rα2 has been demonstrated to be important in its interactions with IL-13 [148]. IL-13Rα2 exists in three cellular compartments: on the cell surface as a single trans-membrane receptor, in the cytosol within large intracellular pools, and in the extracellular space as a soluble form [149]. In fibroblasts and airway epithelial cells, IL-13Rα2 has been demonstrated to be predominantly localized to intracellular pools that can be rapidly mobilized to the membrane upon stimulation by IL-13 and IL-4 [150].

IL-13Rα2 has been considered for a long time as a decoy receptor which does not directly contribute to IL-13 signaling, but serves as a negative regulator to terminate IL-13 responses by directly binding to IL-13. The notion that this receptor had no signaling function arose from the fact that it has a short cytoplasmic tail that is missing two known signalling motifs [151] and does not bind JAKs or STATs [152]. IL-13Rα2 has also been shown to be internalized quickly upon IL-13 binding [152]. Unlike IL-13Rα1, IL-13Rα2 alone has very high affinity for IL-13 [146] and this is further increased when IL-13Rα2 exists as a soluble form [153]. N-linked glycosylation in the extracellular domain of IL-13Rα2 has shown to be necessary for optimal IL-13 binding, demonstrating the importance of glycosylation in IL-13 signalling [148].

Recent investigations have suggested that IL-13Rα2 acts as a signaling receptor not merely as a decoy receptor. Dienger *et al.* showed that IL-13Rα2 knockout mice have attenuated rather than enhanced allergic airway responses, suggesting that under some circumstances, IL-13Rα2 may contribute to IL-13 signaling [118] and reduce inflammation. Further evidence of anti-inflammatory properties comes from models of allergic asthma and chronic helminth infection where IL-10 and IL-13Rα2 coordinately suppress Th_2-mediated inflammation and pathology [154]. Fichtner-Feigl *et al.* has demonstrated that IL-13 signals through the IL-13Rα2 to mediate TGF-β production via the transcription factor, AP-1 in macrophages. [129]. These studies provide evidence that IL-13Rα2 can serve as a functional signalling receptor but the role of glycosylation in regulating its signalling function has yet to be elucidated.

4.4. IL-13 genetic and clinical linkage to asthma

The gene encoding IL-13 consists of four exons and three introns and is located on chromosome 5q31 [155]. The chromosomal region 5q31 also contains genes for other molecules associated with asthma such as IL-4, IL-3, IL-9 and GM-CSF [156]. Previous studies have demonstrated strong associations between asthma and several single nucleotide polymorphisms (SNP) in the IL-13 gene, including +2043G>A and -1111C>T [156, 157]. +2043G>A is within the IL-13 coding sequence and causes a positively charged arginine (R) to be substituted by a neutral glutamine (Q) at position 130, forming the IL-13 R130Q variant [158]. Position 130 is in the α-D segment of the IL-13 molecule, where it has been demonstrated to interact with IL-13Rα1 [159]. Compared to wildtype IL-13, the R130Q variant has previously been reported to have lower affinity for soluble IL-13Rα2 due to slower association rates with the receptor [176]. IL-13R130Q is also more active compared to wildtype IL-13 in inducing STAT-6 activation and CD23 expression in monocytes [115]. The enhanced activity of IL-13R130Q may be explained by its decreased affinity to IL-13Rα2.

4.5. Roles of IL-13 in airway epithelial repair

IL-13 has been demonstrated to be critical in mediating normal airway epithelial wound repair. In an *in vitro* model, Allahverdian *et al.* demonstrated that normal bronchial epithelial cells produce and secrete IL-13 in response to mechanical injury. IL-13 then mediates repair by inducing the production and autocrine/ paracrine release of HB-EGF, which subsequently activates EGFR and downstream signalling pathways required for repair [119]. In another *in vitro* study, thymic stromal lymphopoietin (TSLP) was shown to promote bronchial epithelial proliferation and repair via the upregulation of IL-13 production, providing further evidence for the critical role of IL-13 in repair [160]. IL-13 also plays a role in promoting normal airway epithelial health. Treatment of normal bronchial epithelial cells with IL-13 and IL-9 alone and in combination was shown to significantly reduce spontaneous cell apoptosis and to be protective against dexamethasone-induced cell apoptosis [161].

5. The use of biomarkers in asthma diagnosis and therapeutic intervention

Asthma is increasingly recognised as a heterogeneous syndrome with multiple patient phenotypes reflecting the varied underlying pathologies present in each individual. Pulmonary function testing has severe limitations; spirometry can identify a broad spectrum of asthmatics, but it is incapable of discerning sub-types of disease and therefore which individuals will respond to specific treatment regimes. Tissue biopsies and inflammatory cell counting in induced sputum, are accepted measures of determining airway inflammation, however both techniques are invasive, expensive and difficult to standardize, making them unsuitable for routine clinical use. Biomarkers hold the promise of being able to rapidly and specifically diagnose and monitor various sub-types of asthma in a non-invasive manner. The advantages and disadvantages of current and potential future biomarkers of airway inflammation for the diagnosis and monitoring of asthma in the clinic are reviewed in [162].

5.1. "Invasive" airway biomarkers

Tissue biopsies: To date the most accurate method to assess lung inflammation (and remodeling) is by histological examination of lung tissue. Sub-types of inflammatory cells can be identified in tissue sections using specific stains. Unfortunately the process of taking bronchial biopsies via bronchoscopy is invasive and requires experienced, skilled pathologists for tissue examination. In addition, clinical studies demonstrate a disconnect between the numbers of inflammatory cells counted in airway biopsies, and lung function in asthmatics [27].

Induced sputum: One alternative method of directly assessing airway inflammation is by sputum induction. A patient inhales nebulized hypertonic saline to trigger sputum production in the airways. The sputum is then coughed out along with any inflammatory cells present in the airway lumen. The main assumption is that the inflammatory infiltrate in the airway lumen reflects that in the tissue. Cytospins of the resulting sputum samples are then stained using similar techniques to tissue biopsies to examine the specific type of cellular infiltrate in the sputum. The principal output is the differential inflammatory cell count, expressed as a percentage, based on the manual counting of cells (eosinophils, neutrophils, lymphocytes, macrophages, and epithelial cells). The protein content of induced sputum may also have some value in diagnosing asthma. High mobility group box-1 (HMGB-1), a ligand of the receptor for advanced glycation end products (RAGE), is a mediator in many inflammatory disorders. HMGB-1 is increased in the sputum of asthmatic patients, along with endogenous secretory RAGE, a soluble receptor that inhibits RAGE signalling [163]. Sputum induction is considerably less invasive than a tissue biopsy, but this technique is nonetheless uncomfortable for the patient. As many children are unwilling to undergo sputum collection in follow-up visits this technique tends to be limited to patients aged 8 years and above (although sputum induction has been performed successfully on younger children)[164]. Although the techniques of sputum induction and processing are well validated, they are time-consuming, require skilled people, and results are often difficult to reproduce and vary across centers.

5.2. Exhaled biomarkers

Less invasive means of obtaining airway biomarkers are desirable, particularly as diagnostic or disease monitoring tools in pediatric patients. The composition of exhaled breath is correlated to various disease states. This has been exploited by cancer researchers who have pioneered the use of gas sensor arrays, or "electronic noses", to aid the diagnosis of lung cancer [165]. As asthma is also a disease of the airways, it follows that the composition of exhaled breath will also be altered in the disease. The advent of so-called "breathomics" promises to revolutionize the way clinicians diagnose and treat asthmatics.

Fractional exhaled nitric oxide (FeNO): To date FeNO is the most widely used exhaled biomarker of airway inflammation in asthma. Levels of NO in exhaled breath can be measured relatively quickly in the clinic although the gas analyzers required are expensive [164]. FeNO is often increased in steroid naïve asthmatics and severe asthmatics, and is correlated with airway eosinophilia [166, 167]. FeNO is derived from the action of inducible NO synthase (iNOS) expressed by the airway epithelium [168] however the precise mechanism of how eosinophilia triggers iNOS activity in epithelial cells is undefined. It is unlikely that FeNO measurements per se would give any indication of the carbohydrate modifications of proteins in a patient, therefore this method will not be discussed further herein. For a more in-depth discussion of FeNO as a biomarker in asthma readers are referred to a recent review by Wadsworth *et al.* [162].

Exhaled Breath Condensate: The collection of exhaled breath condensate (EBC) and subsequent measure of inflammatory biomarkers is a relatively recent development. Exhaled breath condenses when it comes into contact with a cooled collector, allowing the collection of respiratory particles, droplets and water vapour. The pH of the EBC has been shown to relate to airway inflammation, low EBC pH indicates poorly controlled eosinophilic asthma in a similar manner to high FeNO [169]. It is unlikely that a single biomarker will be able to reflect the various pathologies which are present in a heterogeneous disease such as asthma. Therefore additional markers of airway inflammation will be needed to provide information complementary to that gained from FeNO or pH measurements. Many proteins are present in EBC and this method has been postulated as a means to allow the objective proteomic analysis of exhaled breath. As with pH, several of these proteins are markers of oxidative stress including, cysteinyl leukotrienes, leukotriene B4, 8-isoprostane and hydrogen peroxide, although inflammatory proteins such as IL-6, IL-8, TNF-α may also be useful markers [170] (reviewed in [164]). Other proteins which do not obviously fit into any inflammatory pathway including; actin, cytokeratins, albumin, and hemoglobin have also been shown to be increased in the EBC of asthmatic patients [171].

5.3. Non-exhaled biomarkers

Serum proteins: As described above, biomarker studies in asthma have tended to concentrate on changes to the composition of exhaled breath. Although primarily a disease of the airways, there is mounting evidence to suggest there is also a systemic component to asthma [172]. If this is the case then circulating metabolites may be able to act as biomarkers of disease. Blood collection, serum isolation and analysis are highly standardized techniques of a minimally invasive nature and are therefore an ideal source of reproducible data. Serum proteins are already gaining credence as biomarkers in other inflammatory lung diseases. In COPD for example, not all smokers develop the disease, so being able to identify those who are at risk would be useful. Studies have shown circulating levels of pulmonary and activation-regulated chemokine (PARC)/CCL-18 are increased in COPD patients [173] and COPD patients who exhibit a rapid fall in FEV_1 tend to have high circulating levels of fibrinogen [174]. Even accepted markers of cardiac dysfunction, cardiac troponin-T and N-terminal pro-brain natriuretic peptide (NT-proBNP) have been associated with increased mortality in patients with COPD, suggesting a cardiac component to the disease [175]. Several serum biomarkers have been demonstrated to be associated with asthma, including eosinophil cationic protein (ECP). ECP levels increase in response to allergen challenge, and decrease after allergen avoidance or inhaled corticosteroids (ICS) therapy, albeit in a less responsive manner than sputum eosinophils or FeNO (reviewed in [164]). In the clinic however, serum ECP levels do not reflect treatment-induced functional changes in chronic asthmatics, and serum ECP is unable to predict steroid responsiveness (reviewed in [164]). One randomized trial demonstrated that patients whose asthma management was based on serum ECP levels, experienced no improvement in symptoms compared to those treated using traditional monitoring techniques [176]. In addition, raised serum ECP levels may not be a specific marker of asthma; studies in pediatric patients have demonstrated serum ECP is also raised in cystic fibrosis and viral bronchiolitis [177]. Studies using purified ECP demonstrate it undergoes N-linked glycosylation and its cytotoxic activity against a cultured human small cell lung cancer cell line is dependent on both gene polymorphisms and glycosylation state [178]. Thus, rather than measuring total ECP, it is likely that analysing the proportion of total ECP which is glycosylated would prove to be a more specific biomarker for monitoring asthma progression. Other novel biomarkers are also altered in the circulation of patients with asthma. CCL-17 is a chemokine released from dendritic cells and epithelial cells after allergen contact, and is involved in the recruitment of Th_2 cells into the lungs. Studies have shown sputum levels of CCL-17 is increased in the lungs of asthmatic adults,[179] whilst serum levels of CCL-17 is increased in children with asthma and is lowered in steroid-treated children with asthma [180]. In asthma, structural cells of the airway including the epithelium, release various pro-inflammatory chemokines such as IL-6 and IL-8 which trigger subsequent infiltration by immune cells. The epithelium also releases proteinases which are capable and necessary for the cleavage and activation of chemokinetic activity of molecules such as Fas-ligand and IL-25 [10] [127]. It is possible circulating levels of these chemokines and proteinases may provide an early warning for an imminent asthma exacerbation, or even indicate the particular sub-type of inflammation occurring in the patient.

Urinary metabolites: Data mentioned above suggests that the metabolism of patients with asthma is altered compared to normals. Urine is possibly the least invasive biofluid for biomarker measurements and is therefore highly suitable for the study and assessment of asthma in young children. Clinical studies have shown urinary biomarkers are potentially useful in asthma; levels of the downstream histamine metabolite, N-methylhistamine, is increased in the urine of patients with asthma, is increased after allergen challenge or exacerbation, and is reduced in asthmatic children taking anti-allergy medication [183, 184]. The human histamine receptor (hH1R) is a G-protein-coupled receptor, GPCRs are a large family of transmembrane receptors that are dependent on glycosylation for correct expression, trafficking and signaling. Studies using hH1R-expressing insect cells have shown that the receptor is N-glycosylated at a specific asparagene residue, Asn_5 [185], whether glycosylation affects receptor activity, and how hH1R glycosylation could be measured in patients is unknown.

One recent study used nuclear magnetic resonance spectroscopy (^{1}H-NMR) to measure levels of 70 metabolites in the urine of children with and without asthma. NMR is an attractive method for urinary biomarker examination as it is able to provide qualitative and quantitative data on multiple compounds in a complex biofluid, without requiring significant pre-treatment of the sample. Urine was collected from control children without asthma (C), with stable asthma in the outpatient department (AO), and in children with unstable disease hospitalized for an asthma exacerbation (AED). NMR examination of urinary metabolites showed a 94% success rate in identifying AO children versus C, and a similar success rate was seen when diagnosing AED versus AO asthma [186]. This study suggests the measurement of urinary metabolites is a potentially valuable technique to help clinicians diagnose and monitor asthma in children.

6. Closing remarks

In addition to forming a barrier against inhaled toxins, pathogens and allergens, the airway epithelium performs numerous innate immune and transport functions necessary for airway health and homeostasis. Damage to the epithelial layer is a common occurrence, thus complex repair mechanisms have evolved to rapidly restore the epithelial barrier, with subsequent regeneration of the fully functional differentiated tissue. In asthmatic airways, these epithelial repair mechanisms are compromised.

The post-translational modification of proteins by glycosylation is capable of drastically affecting function via changes to; sub-cellular localisation and secretion, enzyme substrate binding, receptor-ligand binding, protein stability and degradation amongst others. Many of the proteins involved in the complex repair mechanisms defined in the airway epithelium are dependent on correct glycosylation in order to function correctly. There is accumulating evidence to suggest glycosylation is altered in airway diseases such as asthma. In addition, an important cytokine in asthma, IL-13, has a significant effect on the glycosylation state of the epithelial lining of the airway, and IL-13 signalling itself is affected by glycosylation of its receptors. Thus altered glycosylation of the airway epithelium, will have significant effects on the repair and regeneration of a functional airway epithelial barrier.

Asthma is a complex syndrome, individuals with asthma can be classified into a number of sub-phenotypes who will respond to different types of therapies. In order to facilitate the accurate diagnosis of asthmatics the development of specific bio-markers is required. Considerable research is being conducted into the use of non-invasive exhaled, urinary, or blood-borne markers of asthmatic disease to aid diagnosis in adults and children. To date the majority of biomarker research has concentrated on the identification of proteins linked to disease, rather than identifying differences in post-translational modifications. As glycosylation plays a fundamental role in regulating protein function, it is highly likely that future biomarkers for asthma will need focus on identifying alterations to the carbohydrate structures of proteins, particularly those involved in airway epithelial repair.

Author details

S. Jasemine Yang and Delbert R. Dorscheid and Samuel J. Wadsworth*
James Hogg Research Centre, Institute for Heart + Lung Health, St. Paul's Hospital, University of British Columbia, Vancouver, Canada

7. References

[1] Wenzel, S.E., *Asthma: defining of the persistent adult phenotypes.* Lancet, 2006. 368(9537): p. 804-13.

[2] Adcock, I.M. and K. Ito, *Steroid resistance in asthma: a major problem requiring novel solutions or a non-issue?* Curr Opin Pharmacol, 2004. 4(3): p. 257-62.

[3] Djukanovic, R., et al., *Quantitation of mast cells and eosinophils in the bronchial mucosa of symptomatic atopic asthmatics and healthy control subjects using immunohistochemistry.* Am Rev Respir Dis, 1990. 142(4): p. 863-71.

[4] Montefort, S., et al., *The site of disruption of the bronchial epithelium in asthmatic and non-asthmatic subjects.* Thorax, 1992. 47(7): p. 499-503.

[5] Green, R.H., et al., *Analysis of induced sputum in adults with asthma: identification of subgroup with isolated sputum neutrophilia and poor response to inhaled corticosteroids.* Thorax, 2002. 57(10): p. 875-9.

* Correspong Author

[6] Huang, S.K., et al., *IL-13 expression at the sites of allergen challenge in patients with asthma.* J Immunol, 1995. 155(5): p. 2688-94.

[7] Wills-Karp, M., et al., *Interleukin-13: central mediator of allergic asthma.* Science, 1998. 282(5397): p. 2258-61.

[8] Cohn, L., et al., *Th2-induced airway mucus production is dependent on IL-4Ralpha, but not on eosinophils.* J Immunol, 1999. 162(10): p. 6178-83.

[9] Beghe, B., et al., *Polymorphisms in IL13 pathway genes in asthma and chronic obstructive pulmonary disease.* Allergy, 2010. 65(4): p. 474-81.

[10] Wadsworth, S.J., et al., *IL-13 and TH2 cytokine exposure triggers matrix metalloproteinase 7-mediated Fas ligand cleavage from bronchial epithelial cells.* J Allergy Clin Immunol, 2010. 126(2): p. 366-74, 374 e1-8.

[11] Danahay, H., et al., *Interleukin-13 induces a hypersecretory ion transport phenotype in human bronchial epithelial cells.* Am J Physiol Lung Cell Mol Physiol, 2002. 282(2): p. L226-36.

[12] Atherton, H.C., G. Jones, and H. Danahay, *IL-13-induced changes in the goblet cell density of human bronchial epithelial cell cultures: MAP kinase and phosphatidylinositol 3-kinase regulation.* Am J Physiol Lung Cell Mol Physiol, 2003. 285(3): p. L730-9.

[13] Whittaker, L., et al., *Interleukin-13 mediates a fundamental pathway for airway epithelial mucus induced by CD4 T cells and interleukin-9.* Am J Respir Cell Mol Biol, 2002. 27(5): p. 593-602.

[14] Beum, P.V., et al., *Mucin biosynthesis: upregulation of core 2 beta 1,6 N-acetylglucosaminyltransferase by retinoic acid and Th2 cytokines in a human airway epithelial cell line.* Am J Physiol Lung Cell Mol Physiol, 2005. 288(1): p. L116-24.

[15] Barnes, P.J., *The role of inflammation and anti-inflammatory medication in asthma.* Respir Med, 2002. 96 Suppl A: p. S9-15.

[16] Barnes, P.J., *Update on asthma.* Isr Med Assoc J, 2003. 5(1): p. 68-72.

[17] Kraft, M., *Asthma phenotypes and interleukin-13--moving closer to personalized medicine.* N Engl J Med, 2011. 365(12): p. 1141-4.

[18] Flood-Page, P., et al., *A study to evaluate safety and efficacy of mepolizumab in patients with moderate persistent asthma.* Am J Respir Crit Care Med, 2007. 176(11): p. 1062-71.

[19] O'Byrne, P.M., *The demise of anti IL-5 for asthma, or not.* Am J Respir Crit Care Med, 2007. 176(11): p. 1059-60.

[20] Corren, J., et al., *Lebrikizumab treatment in adults with asthma.* N Engl J Med, 2011. 365(12): p. 1088-98.

[21] Innes, A.L., et al., *The H antigen at epithelial surfaces is associated with susceptibility to asthma exacerbation.* Am J Respir Crit Care Med, 2011. 183(2): p. 189-94.

[22] Laitinen, A., et al., *Tenascin is increased in airway basement membrane of asthmatics and decreased by an inhaled steroid.* Am J Respir Crit Care Med, 1997. 156(3 Pt 1): p. 951-8.

[23] Hirst, S.J., *Airway smooth muscle as a target in asthma.* Clin Exp Allergy, 2000. 30 Suppl 1: p. 54-9.

[24] Holgate, S.T., et al., *Epithelial-mesenchymal interactions in the pathogenesis of asthma.* J Allergy Clin Immunol, 2000. 105(2 Pt 1): p. 193-204.

[25] Hackett, T.L. and D.A. Knight, *The role of epithelial injury and repair in the origins of asthma.* Curr Opin Allergy Clin Immunol, 2007. 7(1): p. 63-8.

[26] Malmstrom, K., et al., *Lung function, airway remodelling and inflammation in symptomatic infants: outcome at 3 years.* Thorax, 2011. 66(2): p. 157-62.

[27] Sont, J.K., et al., *Relationship between the inflammatory infiltrate in bronchial biopsy specimens and clinical severity of asthma in patients treated with inhaled steroids.* Thorax, 1996. 51(5): p. 496-502.

[28] Jeffery, P.K., *Remodeling in asthma and chronic obstructive lung disease.* Am J Respir Crit Care Med, 2001. 164(10 Pt 2): p. S28-38.

[29] Knight, D.A. and S.T. Holgate, *The airway epithelium: structural and functional properties in health and disease.* Respirology, 2003. 8(4): p. 432-46.

[30] Roche, W.R., et al., *Cell adhesion molecules and the bronchial epithelium.* Am Rev Respir Dis, 1993. 148(6 Pt 2): p. S79-82.

[31] Evans, M.J., et al., *Cellular and molecular characteristics of basal cells in airway epithelium.* Exp Lung Res, 2001. 27(5): p. 401-15.

[32] Rhodin, J.A., *The ciliated cell. Ultrastructure and function of the human tracheal mucosa.* Am Rev Respir Dis, 1966. 93(3): p. Suppl:1-15.

[33] Philippou, S., et al., *The morphological substrate of autonomic regulation of the bronchial epithelium.* Virchows Arch A Pathol Anat Histopathol, 1993. 423(6): p. 469-76.

[34] Nettesheim, P., J.S. Koo, and T. Gray, *Regulation of differentiation of the tracheobronchial epithelium.* J Aerosol Med, 2000. 13(3): p. 207-18.

[35] Keenan, K.P., J.W. Combs, and E.M. McDowell, *Regeneration of hamster tracheal epithelium after mechanical injury. I. Focal lesions: quantitative morphologic study of cell proliferation.* Virchows Arch B Cell Pathol Incl Mol Pathol, 1982. 41(3): p. 193-214.

[36] Keenan, K.P., J.W. Combs, and E.M. McDowell, *Regeneration of hamster tracheal epithelium after mechanical injury. III. Large and small lesions: comparative stathmokinetic and single pulse and continuous thymidine labeling autoradiographic studies.* Virchows Arch B Cell Pathol Incl Mol Pathol, 1982. 41(3): p. 231-52.

[37] Keenan, K.P., J.W. Combs, and E.M. McDowell, *Regeneration of hamster tracheal epithelium after mechanical injury. II. Multifocal lesions: stathmokinetic and autoradiographic studies of cell proliferation.* Virchows Arch B Cell Pathol Incl Mol Pathol, 1982. 41(3): p. 215-29.

[38] Keenan, K.P., T.S. Wilson, and E.M. McDowell, *Regeneration of hamster tracheal epithelium after mechanical injury. IV. Histochemical, immunocytochemical and ultrastructural studies.* Virchows Arch B Cell Pathol Incl Mol Pathol, 1983. 43(3): p. 213-40.

[39] Tam, A., et al., *The airway epithelium: more than just a structural barrier.* Ther Adv Respir Dis, 2011. 5(4): p. 255-73.

[40] Johansen, F.E. and C.S. Kaetzel, *Regulation of the polymeric immunoglobulin receptor and IgA transport: new advances in environmental factors that stimulate pIgR expression and its role in mucosal immunity.* Mucosal Immunol, 2011. 4(6): p. 598-602.

[41] Woof, J.M. and M.W. Russell, *Structure and function relationships in IgA.* Mucosal Immunol, 2011. 4(6): p. 590-7.

[42] Martin, N. and I.D. Pavord, *Bronchial thermoplasty for the treatment of asthma.* Curr Allergy Asthma Rep, 2009. 9(1): p. 88-95.

[43] Wadsworth, S.J., H.S. Nijmeh, and I.P. Hall, *Glucocorticoids increase repair potential in a novel in vitro human airway epithelial wounding model.* J Clin Immunol, 2006. 26(4): p. 376-87.

[44] Zhang, S., et al., *Growth factors secreted by bronchial epithelial cells control myofibroblast proliferation: an in vitro co-culture model of airway remodeling in asthma.* Lab Invest, 1999. 79(4): p. 395-405.

[45] Morishima, Y., et al., *Triggering the induction of myofibroblast and fibrogenesis by airway epithelial shedding.* Am J Respir Cell Mol Biol, 2001. 24(1): p. 1-11.

[46] Davies, D.E., *The bronchial epithelium: translating gene and environment interactions in asthma.* Curr Opin Allergy Clin Immunol, 2001. 1(1): p. 67-71.

[47] Woodley, D., *The Molecular and Cellular Biology of Wound Repair.* 2nd ed, ed. R.A.F. Clark1996, New York: Plenum Publishing Corp. 339-354.

[48] Seiler, W.O., et al., *Impaired migration of epidermal cells from decubitus ulcers in cell cultures. A cause of protracted wound healing?* Am J Clin Pathol, 1989. 92(4): p. 430-4.

[49] Hanna, C., *Proliferation and migration of epithelial cells during corneal wound repair in the rabbit and the rat.* Am J Ophthalmol, 1966. 61(1): p. 55-63.

[50] Ridley, A.J., *Rho family proteins: coordinating cell responses.* Trends Cell Biol, 2001. 11(12): p. 471-7.

[51] Ridley, A.J., *Rho GTPases and cell migration.* J Cell Sci, 2001. 114(Pt 15): p. 2713-22.

[52] Schirmer, J. and K. Aktories, *Large clostridial cytotoxins: cellular biology of Rho/Ras-glucosylating toxins.* Biochim Biophys Acta, 2004. 1673(1-2): p. 66-74.

[53] Sander, E.E. and J.G. Collard, *Rho-like GTPases: their role in epithelial cell-cell adhesion and invasion.* Eur J Cancer, 1999. 35(9): p. 1302-8.

[54] Ridley, A.J. and A. Hall, *The small GTP-binding protein rho regulates the assembly of focal adhesions and actin stress fibers in response to growth factors.* Cell, 1992. 70(3): p. 389-99.

[55] Ridley, A.J., et al., *The small GTP-binding protein rac regulates growth factor-induced membrane ruffling.* Cell, 1992. 70(3): p. 401-10.

[56] Kozma, R., et al., *The Ras-related protein Cdc42Hs and bradykinin promote formation of peripheral actin microspikes and filopodia in Swiss 3T3 fibroblasts.* Mol Cell Biol, 1995. 15(4): p. 1942-52.

[57] Spurzem, J.R., et al., *Activation of protein kinase A accelerates bovine bronchial epithelial cell migration.* Am J Physiol Lung Cell Mol Physiol, 2002. 282(5): p. L1108-16.

[58] Desai, L.P., et al., *RhoA and Rac1 are both required for efficient wound closure of airway epithelial cells.* Am J Physiol Lung Cell Mol Physiol, 2004. 287(6): p. L1134-44.

[59] Wang, Y.P., et al., *IQ domain GTPase-activating protein 1 mediates the process of injury and repair in bronchial epithelial cells.* Sheng Li Xue Bao, 2008. 60(3): p. 409-18.

[60] Wang, Y., et al., *IQGAP1 promotes cell proliferation and is involved in a phosphorylation-dependent manner in wound closure of bronchial epithelial cells.* Int J Mol Med, 2008. 22(1): p. 79-87.

[61] Just, I., et al., *Glucosylation of Rho proteins by Clostridium difficile toxin B.* Nature, 1995. 375(6531): p. 500-3.

[62] Herrmann, C., et al., *Functional consequences of monoglucosylation of Ha-Ras at effector domain amino acid threonine 35.* J Biol Chem, 1998. 273(26): p. 16134-9.

[63] Sehr, P., et al., *Glucosylation and ADP ribosylation of rho proteins: effects on nucleotide binding, GTPase activity, and effector coupling.* Biochemistry, 1998. 37(15): p. 5296-304.

[64] Gerhard, R., et al., *Glucosylation of Rho GTPases by Clostridium difficile toxin A triggers apoptosis in intestinal epithelial cells.* J Med Microbiol, 2008. 57(Pt 6): p. 765-70.

[65] Zanetta, J.P., et al., *Glycoproteins and lectins in cell adhesion and cell recognition processes.* Histochem J, 1992. 24(11): p. 791-804.

[66] Lis, H. and N. Sharon, *Protein glycosylation. Structural and functional aspects.* Eur J Biochem, 1993. 218(1): p. 1-27.

[67] Hoe, M.H. and R.C. Hunt, *Loss of one asparagine-linked oligosaccharide from human transferrin receptors results in specific cleavage and association with the endoplasmic reticulum.* J Biol Chem, 1992. 267(7): p. 4916-23.

[68] Rands, E., et al., *Mutational analysis of beta-adrenergic receptor glycosylation.* J Biol Chem, 1990. 265(18): p. 10759-64.

[69] Feige, J.J. and A. Baird, *Glycosylation of the basic fibroblast growth factor receptor. The contribution of carbohydrate to receptor function.* J Biol Chem, 1988. 263(28): p. 14023-9.

[70] Leconte, I., et al., *N-linked oligosaccharide chains of the insulin receptor beta subunit are essential for transmembrane signaling.* J Biol Chem, 1992. 267(24): p. 17415-23.

[71] Tsuda, E., et al., *The role of carbohydrate in recombinant human erythropoietin.* Eur J Biochem, 1990. 188(2): p. 405-11.

[72] Schnaar, R.L., *Glycosphingolipids in cell surface recognition.* Glycobiology, 1991. 1(5): p. 477-85.

[73] Phillips, M.L., et al., *ELAM-1 mediates cell adhesion by recognition of a carbohydrate ligand, sialyl-Lex.* Science, 1990. 250(4984): p. 1130-2.

[74] Chammas, R., et al., *Functional hypotheses for aberrant glycosylation in tumor cells.* Braz J Med Biol Res, 1994. 27(2): p. 505-7.

[75] Lowe, J.B., et al., *ELAM-1--dependent cell adhesion to vascular endothelium determined by a transfected human fucosyltransferase cDNA.* Cell, 1990. 63(3): p. 475-84.

[76] Mann, P.L., I. Lopez-Colberg, and R.O. Kelley, *Cell surface oligosaccharide modulation during differentiation. I. Modulation of lectin binding.* Mech Ageing Dev, 1987. 38(3): p. 207-17.

[77] Lis, H. and N. Sharon, *The biochemistry of plant lectins (phytohemagglutinins).* Annu Rev Biochem, 1973. 42(0): p. 541-74.

[78] Gipson, I.K. and R.A. Anderson, *Effect of lectins on migration of the corneal epithelium.* Invest Ophthalmol Vis Sci, 1980. 19(4): p. 341-9.

[79] Gipson, I.K. and T.C. Kiorpes, *Epithelial sheet movement: protein and glycoprotein synthesis.* Dev Biol, 1982. 92(1): p. 259-62.

[80] Gipson, I.K., et al., *Lectin binding to cell surfaces: comparisons between normal and migrating corneal epithelium.* Dev Biol, 1983. 96(2): p. 337-45.

[81] Sweatt, A.J., R.M. Degi, and R.M. Davis, *Corneal wound-associated glycoconjugates analyzed by lectin histochemistry.* Curr Eye Res, 1999. 19(3): p. 212-8.

[82] Dorscheid, D.R., et al., *Characterization of cell surface lectin-binding patterns of human airway epithelium.* Histochem J, 1999. 31(3): p. 145-51.

[83] Xiantang, L., et al., *Glycosylation profiles of airway epithelium after repair of mechanical injury in guinea pigs.* Histochem J, 2000. 32(4): p. 207-16.
[84] Plotkowski, M.C., et al., *Differential adhesion of Pseudomonas aeruginosa to human respiratory epithelial cells in primary culture.* J Clin Invest, 1991. 87(6): p. 2018-28.
[85] Dorscheid, D.R., et al., *Role of cell surface glycosylation in mediating repair of human airway epithelial cell monolayers.* Am J Physiol Lung Cell Mol Physiol, 2001. 281(4): p. L982-92.
[86] Allahverdian, S., B.J. Patchell, and D.R. Dorscheid, *Carbohydrates and epithelial repair - more than just post-translational modification.* Current Drug Targets, 2006. 7(5): p. 597-606.
[87] Allahverdian, S., et al., *Sialyl Lewis X modification of the epidermal growth factor receptor regulates receptor function during airway epithelial wound repair.* Clin Exp Allergy, 2010. 40(4): p. 607-18.
[88] Donaldson, D.J. and J.M. Mason, *Inhibition of epidermal cell migration by concanavalin A in skin wounds of the adult newt.* J Exp Zool, 1977. 200(1): p. 55-64.
[89] Patchell, B.J. and D.R. Dorscheid, *Repair of the injury to respiratory epithelial cells characteristic of asthma is stimulated by Allomyrina dichotoma agglutinin specific serum glycoproteins.* Clinical and experimental allergy : journal of the British Society for Allergy and Clinical Immunology, 2006. 36(5): p. 585-593.
[90] Trinkaus-Randall, V., et al., *Carbohydrate moieties of the basal lamina: their role in attachment and spreading of basal corneal epithelial cells.* Cell Tissue Res, 1988. 251(2): p. 315-23.
[91] Adam, E.C., et al., *Role of carbohydrates in repair of human respiratory epithelium using an in vitro model.* Clin Exp Allergy, 2003. 33(10): p. 1398-404.
[92] Sato, S. and R.C. Hughes, *Binding specificity of a baby hamster kidney lectin for H type I and II chains, polylactosamine glycans, and appropriately glycosylated forms of laminin and fibronectin.* J Biol Chem, 1992. 267(10): p. 6983-90.
[93] Woo, H.J., et al., *The major non-integrin laminin binding protein of macrophages is identical to carbohydrate binding protein 35 (Mac-2).* J Biol Chem, 1990. 265(13): p. 7097-9.
[94] Warfield, P.R., et al., *Adhesion of human breast carcinoma to extracellular matrix proteins is modulated by galectin-3.* Invasion Metastasis, 1997. 17(2): p. 101-12.
[95] Matarrese, P., et al., *Galectin-3 overexpression protects from apoptosis by improving cell adhesion properties.* Int J Cancer, 2000. 85(4): p. 545-54.
[96] Cao, Z., et al., *Galectin-7 as a potential mediator of corneal epithelial cell migration.* Arch Ophthalmol, 2003. 121(1): p. 82-6.
[97] Cao, Z., et al., *Galectins-3 and -7, but not galectin-1, play a role in re-epithelialization of wounds.* J Biol Chem, 2002. 277(44): p. 42299-305.
[98] Hughes, R.C., *Secretion of the galectin family of mammalian carbohydrate-binding proteins.* Biochim Biophys Acta, 1999. 1473(1): p. 172-85.
[99] Kojima, K., et al., *Affinity purification and affinity characterization of carbohydrate-binding proteins in bovine kidney.* J Chromatogr, 1992. 597(1-2): p. 323-30.
[100] Hajjar, K.A. and C.M. Reynolds, *alpha-Fucose-mediated binding and degradation of tissue-type plasminogen activator by HepG2 cells.* J Clin Invest, 1994. 93(2): p. 703-10.
[101] Kim, J. and K.A. Hajjar, *Annexin II: a plasminogen-plasminogen activator co-receptor.* Front Biosci, 2002. 7: p. d341-8.

[102] Liu, J.W., et al., *Annexin II expression is reduced or lost in prostate cancer cells and its re-expression inhibits prostate cancer cell migration.* Oncogene, 2003. 22(10): p. 1475-85.

[103] Balch, C. and J.R. Dedman, *Annexins II and V inhibit cell migration.* Exp Cell Res, 1997. 237(2): p. 259-63.

[104] Matsuda, A., et al., *Identification and immunohistochemical localization of annexin II in rat cornea.* Curr Eye Res, 1999. 19(4): p. 368-75.

[105] Raynor, C.M., et al., *Annexin II enhances cytomegalovirus binding and fusion to phospholipid membranes.* Biochemistry, 1999. 38(16): p. 5089-95.

[106] Compton, T., R.R. Nepomuceno, and D.M. Nowlin, *Human cytomegalovirus penetrates host cells by pH-independent fusion at the cell surface.* Virology, 1992. 191(1): p. 387-95.

[107] Compton, T., D.M. Nowlin, and N.R. Cooper, *Initiation of human cytomegalovirus infection requires initial interaction with cell surface heparan sulfate.* Virology, 1993. 193(2): p. 834-41.

[108] Malhotra, R., et al., *Isolation and characterisation of potential respiratory syncytial virus receptor(s) on epithelial cells.* Microbes Infect, 2003. 5(2): p. 123-33.

[109] Kotelkin, A., et al., *Respiratory syncytial virus infection sensitizes cells to apoptosis mediated by tumor necrosis factor-related apoptosis-inducing ligand.* J Virol, 2003. 77(17): p. 9156-72.

[110] Chakraborti, S., et al., *Regulation of matrix metalloproteinases: an overview.* Mol Cell Biochem, 2003. 253(1-2): p. 269-85.

[111] Van Lint, P. and C. Libert, *Chemokine and cytokine processing by matrix metalloproteinases and its effect on leukocyte migration and inflammation.* J Leukoc Biol, 2007. 82(6): p. 1375-81.

[112] Gueders, M.M., et al., *Matrix metalloproteinases (MMPs) and tissue inhibitors of MMPs in the respiratory tract: potential implications in asthma and other lung diseases.* Eur J Pharmacol, 2006. 533(1-3): p. 133-44.

[113] Greenlee, K.J., Z. Werb, and F. Kheradmand, *Matrix metalloproteinases in lung: multiple, multifarious, and multifaceted.* Physiol Rev, 2007. 87(1): p. 69-98.

[114] Trifilieff, A., et al., *Pharmacological profile of PKF242-484 and PKF241-466, novel dual inhibitors of TNF-alpha converting enzyme and matrix metalloproteinases, in models of airway inflammation.* Br J Pharmacol, 2002. 135(7): p. 1655-64.

[115] Lee, K.S., et al., *Doxycycline reduces airway inflammation and hyperresponsiveness in a murine model of toluene diisocyanate-induced asthma.* J Allergy Clin Immunol, 2004. 113(5): p. 902-9.

[116] Kumagai, K., et al., *Inhibition of matrix metalloproteinases prevents allergen-induced airway inflammation in a murine model of asthma.* J Immunol, 1999. 162(7): p. 4212-9.

[117] Demedts, I.K., et al., *Matrix metalloproteinases in asthma and COPD.* Curr Opin Pharmacol, 2005. 5(3): p. 257-63.

[118] Russell, R.E., et al., *Alveolar macrophage-mediated elastolysis: roles of matrix metalloproteinases, cysteine, and serine proteases.* Am J Physiol Lung Cell Mol Physiol, 2002. 283(4): p. L867-73.

[119] Bracke, K., et al., *Matrix metalloproteinase-12 and cathepsin D expression in pulmonary macrophages and dendritic cells of cigarette smoke-exposed mice.* Int Arch Allergy Immunol, 2005. 138(2): p. 169-79.

[120] Miller, T.L., et al., *Expression of matrix metalloproteinases 2, 7 and 9, and their tissue inhibitors 1 and 2, in developing rabbit tracheae.* Biol Neonate, 2006. 89(4): p. 236-43.
[121] Prikk, K., et al., *In vivo collagenase-2 (MMP-8) expression by human bronchial epithelial cells and monocytes/macrophages in bronchiectasis.* J Pathol, 2001. 194(2): p. 232-8.
[122] Lopez-Boado, Y.S., C.L. Wilson, and W.C. Parks, *Regulation of matrilysin expression in airway epithelial cells by Pseudomonas aeruginosa flagellin.* J Biol Chem, 2001. 276(44): p. 41417-23.
[123] Puchelle, E., et al., *[Regeneration of injured airway epithelium].* Ann Pharm Fr, 2006. 64(2): p. 107-13.
[124] Parks, W.C., Y.S. Lopez-Boado, and C.L. Wilson, *Matrilysin in epithelial repair and defense.* Chest, 2001. 120(1 Suppl): p. 36S-41S.
[125] Coraux, C., et al., *Differential expression of matrix metalloproteinases and interleukin-8 during regeneration of human airway epithelium in vivo.* J Pathol, 2005. 206(2): p. 160-9.
[126] Adiseshaiah, P., et al., *A Fra-1-dependent, matrix metalloproteinase driven EGFR activation promotes human lung epithelial cell motility and invasion.* J Cell Physiol, 2008. 216(2): p. 405-12.
[127] Goswami, S., et al., *Divergent functions for airway epithelial matrix metalloproteinase 7 and retinoic acid in experimental asthma.* Nat Immunol, 2009. 10(5): p. 496-503.
[128] Nabeshima, K., et al., *Emmprin (basigin/CD147): matrix metalloproteinase modulator and multifunctional cell recognition molecule that plays a critical role in cancer progression.* Pathol Int, 2006. 56(7): p. 359-67.
[129] Kadomatsu, K. and T. Muramatsu, *[Role of basigin, a glycoprotein belonging to the immunoglobulin superfamily, in the nervous system].* Tanpakushitsu Kakusan Koso, 2004. 49(15 Suppl): p. 2417-24.
[130] Vignola, A.M., et al., *HLA-DR and ICAM-1 expression on bronchial epithelial cells in asthma and chronic bronchitis.* Am Rev Respir Dis, 1993. 148(3): p. 689-94.
[131] Lackie, P.M., et al., *Expression of CD44 isoforms is increased in the airway epithelium of asthmatic subjects.* Am J Respir Cell Mol Biol, 1997. 16(1): p. 14-22.
[132] Jouneau, S., et al., *EMMPRIN (CD147) regulation of MMP-9 in bronchial epithelial cells in COPD.* Respirology, 2011. 16(4): p. 705-12.
[133] Caudroy, S., et al., *Expression of the extracellular matrix metalloproteinase inducer (EMMPRIN) and the matrix metalloproteinase-2 in bronchopulmonary and breast lesions.* J Histochem Cytochem, 1999. 47(12): p. 1575-80.
[134] Caudroy, S., et al., *EMMPRIN-mediated MMP regulation in tumor and endothelial cells.* Clin Exp Metastasis, 2002. 19(8): p. 697-702.
[135] Huet, E., et al., *Role of emmprin/CD147 in tissue remodeling.* Connect Tissue Res, 2008. 49(3): p. 175-9.
[136] Betsuyaku, T., et al., *Increased basigin in bleomycin-induced lung injury.* Am J Respir Cell Mol Biol, 2003. 28(5): p. 600-6.
[137] Guillot, S., et al., *Increased extracellular matrix metalloproteinase inducer (EMMPRIN) expression in pulmonary fibrosis.* Exp Lung Res, 2006. 32(3-4): p. 81-97.
[138] Odajima, N., et al., *Extracellular matrix metalloproteinase inducer in interstitial pneumonias.* Hum Pathol, 2006. 37(8): p. 1058-65.

[139] Gwinn, W.M., et al., *Novel approach to inhibit asthma-mediated lung inflammation using anti-CD147 intervention.* J Immunol, 2006. 177(7): p. 4870-9.

[140] Leir, S.H., et al., *Increased CD44 expression in human bronchial epithelial repair after damage or plating at low cell densities.* Am J Physiol Lung Cell Mol Physiol, 2000. 278(6): p. L1129-37.

[141] Yu, W.H., et al., *CD44 anchors the assembly of matrilysin/MMP-7 with heparin-binding epidermal growth factor precursor and ErbB4 and regulates female reproductive organ remodeling.* Genes Dev, 2002. 16(3): p. 307-23.

[142] Polosa, R., et al., *Expression of c-erbB receptors and ligands in the bronchial epithelium of asthmatic subjects.* J Allergy Clin Immunol, 2002. 109(1): p. 75-81.

[143] White, S.R., et al., *Role of very late adhesion integrins in mediating repair of human airway epithelial cell monolayers after mechanical injury.* Am J Respir Cell Mol Biol, 1999. 20(4): p. 787-96.

[144] Bishayee, S., *Role of conformational alteration in the epidermal growth factor receptor (EGFR) function.* Biochem Pharmacol, 2000. 60(8): p. 1217-23.

[145] Stroop, C.J., et al., *Characterization of the carbohydrate chains of the secreted form of the human epidermal growth factor receptor.* Glycobiology, 2000. 10(9): p. 901-17.

[146] Soderquist, A.M., G. Todderud, and G. Carpenter, *The role of carbohydrate as a post-translational modification of the receptor for epidermal growth factor.* Adv Exp Med Biol, 1988. 231: p. 569-82.

[147] Wang, X.Q., et al., *Epidermal growth factor receptor glycosylation is required for ganglioside GM3 binding and GM3-mediated suppression [correction of suppresion] of activation.* Glycobiology, 2001. 11(7): p. 515-22.

[148] Konishi, A. and B.C. Berk, *Epidermal growth factor receptor transactivation is regulated by glucose in vascular smooth muscle cells.* J Biol Chem, 2003. 278(37): p. 35049-56. Epub 2003 Jun 26.

[149] Partridge, E.A., et al., *Regulation of cytokine receptors by Golgi N-glycan processing and endocytosis.* Science, 2004. 306(5693): p. 120-4.

[150] Eisenmesser, E.Z., et al., *Solution structure of interleukin-13 and insights into receptor engagement.* Journal of Molecular Biology, 2001. 310(1): p. 231-241.

[151] Hershey, G.K.K., *IL-13 receptors and signaling pathways: An evolving web.* Journal of Allergy and Clinical Immunology, 2003. 111(4): p. 677-690.

[152] Minty, A., et al., *Interleukin-13 is a new human lymphokine regulating inflammatory and immune responses.* Nature, 1993. 362(6417): p. 248-250.

[153] Liang, H., et al., *SUMF2 interacts with interleukin-13 and inhibits interleukin-13 secretion in bronchial smooth muscle cells.* Journal of cellular biochemistry, 2009. 108(5): p. 1076-1083.

[154] Vladich, F.D., et al., *IL-13 R130Q, a common variant associated with allergy and asthma, enhances effector mechanisms essential for human allergic inflammation.* The Journal of clinical investigation, 2005. 115(3): p. 747-754.

[155] McKenzie, A.N., et al., *Interleukin 13, a T-cell-derived cytokine that regulates human monocyte and B-cell function.* Proceedings of the National Academy of Sciences of the United States of America, 1993. 90(8): p. 3735-3739.

[156] Schmid-Grendelmeier, P., et al., *Eosinophils express functional IL-13 in eosinophilic inflammatory diseases.* Journal of immunology (Baltimore, Md.: 1950), 2002. 169(2): p. 1021-1027.

[157] Wills-Karp, M., *Interleukin-13 in asthma pathogenesis.* Immunological reviews, 2004. 202(Journal Article): p. 175-190.

[158] Allahverdian, S., et al., *Secretion of IL-13 by airway epithelial cells enhances epithelial repair via HB-EGF.* Am J Respir Cell Mol Biol, 2008. 38(2): p. 153-60.

[159] Zurawski, G. and J.E. de Vries, *Interleukin 13 elicits a subset of the activities of its close relative interleukin 4.* Stem cells (Dayton, Ohio), 1994. 12(2): p. 169-174.

[160] Zurawski, G. and J.E. de Vries, *Interleukin 13, an interleukin 4-like cytokine that acts on monocytes and B cells, but not on T cells.* Immunology today, 1994. 15(1): p. 19-26.

[161] Hart, P.H., et al., *Potential antiinflammatory effects of interleukin 4: suppression of human monocyte tumor necrosis factor alpha, interleukin 1, and prostaglandin E2.* Proceedings of the National Academy of Sciences of the United States of America, 1989. 86(10): p. 3803-3807.

[162] Pope, S.M., et al., *IL-13 induces eosinophil recruitment into the lung by an IL-5- and eotaxin-dependent mechanism.* The Journal of allergy and clinical immunology, 2001. 108(4): p. 594-601.

[163] Luttmann, W., et al., *Activation of human eosinophils by IL-13. Induction of CD69 surface antigen, its relationship to messenger RNA expression, and promotion of cellular viability.* Journal of immunology (Baltimore, Md.: 1950), 1996. 157(4): p. 1678-1683.

[164] Bochner, B.S., et al., *IL-13 selectively induces vascular cell adhesion molecule-1 expression in human endothelial cells.* Journal of immunology (Baltimore, Md.: 1950), 1995. 154(2): p. 799-803.

[165] Grunstein, M.M., et al., *IL-13-dependent autocrine signaling mediates altered responsiveness of IgE-sensitized airway smooth muscle.* American journal of physiology.Lung cellular and molecular physiology, 2002. 282(3): p. L520-8.

[166] Akiho, H., et al., *Role of IL-4, IL-13, and STAT6 in inflammation-induced hypercontractility of murine smooth muscle cells.* American journal of physiology.Gastrointestinal and liver physiology, 2002. 282(2): p. G226-32.

[167] Booth, B.W., et al., *Interleukin-13 induces proliferation of human airway epithelial cells in vitro via a mechanism mediated by transforming growth factor-alpha.* American journal of respiratory cell and molecular biology, 2001. 25(6): p. 739-743.

[168] Fichtner-Feigl, S., et al., *IL-13 signaling through the IL-13alpha2 receptor is involved in induction of TGF-beta1 production and fibrosis.* Nature medicine, 2006. 12(1): p. 99-106.

[169] Li, L., et al., *Effects of Th2 cytokines on chemokine expression in the lung: IL-13 potently induces eotaxin expression by airway epithelial cells.* Journal of immunology (Baltimore, Md.: 1950), 1999. 162(5): p. 2477-2487.

[170] Laoukili, J., et al., *IL-13 alters mucociliary differentiation and ciliary beating of human respiratory epithelial cells.* The Journal of clinical investigation, 2001. 108(12): p. 1817-1824.

[171] Zhu, Z., et al., *Pulmonary expression of interleukin-13 causes inflammation, mucus hypersecretion, subepithelial fibrosis, physiologic abnormalities, and eotaxin production.* The Journal of clinical investigation, 1999. 103(6): p. 779-788.

[172] Shim, J.J., et al., *IL-13 induces mucin production by stimulating epidermal growth factor receptors and by activating neutrophils.* American journal of physiology.Lung cellular and molecular physiology, 2001. 280(1): p. L134-40.
[173] Leonard, W.J. and J.X. Lin, *Cytokine receptor signaling pathways.* The Journal of allergy and clinical immunology, 2000. 105(5): p. 877-888.
[174] Arima, K., et al., *Characterization of the interaction between interleukin-13 and interleukin-13 receptors.* The Journal of biological chemistry, 2005. 280(26): p. 24915-24922.
[175] Guo, J., et al., *Chromosome mapping and expression of the human interleukin-13 receptor.* Genomics, 1997. 42(1): p. 141-145.
[176] Andrews, A.L., et al., *Kinetic analysis of the interleukin-13 receptor complex.* The Journal of biological chemistry, 2002. 277(48): p. 46073-46078.
[177] Miloux, B., et al., *Cloning of the human IL-13Rα1 chain and reconstitution with the IL-4Rα of a functional IL-4/IL-13 receptor complex.* FEBS letters, 1997. 401(2-3): p. 163-166.
[178] Aman, M.J., et al., *cDNA cloning and characterization of the human interleukin 13 receptor alpha chain.* The Journal of biological chemistry, 1996. 271(46): p. 29265-29270.
[179] Wang, I.M., et al., *STAT-1 is activated by IL-4 and IL-13 in multiple cell types.* Molecular immunology, 2004. 41(9): p. 873-884.
[180] Hirst, S.J., et al., *Selective induction of eotaxin release by interleukin-13 or interleukin-4 in human airway smooth muscle cells is synergistic with interleukin-1beta and is mediated by the interleukin-4 receptor alpha-chain.* American journal of respiratory and critical care medicine, 2002. 165(8): p. 1161-1171.
[181] Yasuo, M., et al., *Relationship between calcium-activated chloride channel 1 and MUC5AC in goblet cell hyperplasia induced by interleukin-13 in human bronchial epithelial cells.* Respiration; international review of thoracic diseases, 2006. 73(3): p. 347-359.
[182] Wei, L.H., et al., *IL-4 and IL-13 upregulate arginase I expression by cAMP and JAK/STAT6 pathways in vascular smooth muscle cells.* American journal of physiology.Cell physiology, 2000. 279(1): p. C248-56.
[183] Daines, M.O. and G.K. Hershey, *A novel mechanism by which interferon-gamma can regulate interleukin (IL)-13 responses. Evidence for intracellular stores of IL-13 receptor alpha - 2 and their rapid mobilization by interferon-gamma.* The Journal of biological chemistry, 2002. 277(12): p. 10387-10393.
[184] David, M., et al., *Induction of the IL-13 receptor alpha2-chain by IL-4 and IL-13 in human keratinocytes: involvement of STAT6, ERK and p38 MAPK pathways.* Oncogene, 2001. 20(46): p. 6660-6668.
[185] Caput, D., et al., *Cloning and characterization of a specific interleukin (IL)-13 binding protein structurally related to the IL-5 receptor alpha chain.* The Journal of biological chemistry, 1996. 271(28): p. 16921-16926.
[186] Lupardus, P.J., M.E. Birnbaum, and K.C. Garcia, *Molecular basis for shared cytokine recognition revealed in the structure of an unusually high affinity complex between IL-13 and IL-13Ralpha2.* Structure (London, England : 1993). 18(3): p. 332-342.
[187] Kioi, M., S. Seetharam, and R.K. Puri, *N-linked glycosylation of IL-13R alpha2 is essential for optimal IL-13 inhibitory activity.* FASEB journal : official publication of the Federation of American Societies for Experimental Biology, 2006. 20(13): p. 2378-2380.

[188] Konstantinidis, A.K., et al., *Cellular localization of interleukin 13 receptor alpha2 in human primary bronchial epithelial cells and fibroblasts.* Journal of investigational allergology & clinical immunology : official organ of the International Association of Asthmology (INTERASMA) and Sociedad Latinoamericana de Alergia e Inmunologia, 2008. 18(3): p. 174-180.

[189] Andrews, A.L., et al., *IL-13 receptor alpha 2: a regulator of IL-13 and IL-4 signal transduction in primary human fibroblasts.* The Journal of allergy and clinical immunology, 2006. 118(4): p. 858-865.

[190] Donaldson, D.D., et al., *The Murine IL-13 Receptor {alpha}2: Molecular Cloning, Characterization, and Comparison with Murine IL-13 Receptor {alpha}1.* The Journal of Immunology, 1998. 161(5): p. 2317-2324.

[191] Kawakami, K., et al., *The interleukin-13 receptor alpha2 chain: an essential component for binding and internalization but not for interleukin-13-induced signal transduction through the STAT6 pathway.* Blood, 2001. 97(9): p. 2673-2679.

[192] Zhang, J.G., et al., *Identification, purification, and characterization of a soluble interleukin (IL)-13-binding protein. Evidence that it is distinct from the cloned Il-13 receptor and Il-4 receptor alpha-chains.* The Journal of biological chemistry, 1997. 272(14): p. 9474-9480.

[193] Wilson, M.S., et al., *IL-13Ralpha2 and IL-10 coordinately suppress airway inflammation, airway-hyperreactivity, and fibrosis in mice.* The Journal of clinical investigation, 2007. 117(10): p. 2941-2951.

[194] Smirnov, D.V., et al., *Tandem arrangement of human genes for interleukin-4 and interleukin-13: resemblance in their organization.* Gene, 1995. 155(2): p. 277-281.

[195] Heinzmann, A., et al., *Genetic variants of IL-13 signalling and human asthma and atopy.* Human molecular genetics, 2000. 9(4): p. 549-559.

[196] Howard, T.D., et al., *Gene-gene interaction in asthma: IL4RA and IL13 in a Dutch population with asthma.* American Journal of Human Genetics, 2002. 70(1): p. 230-236.

[197] Tarazona-Santos, E. and S.A. Tishkoff, *Divergent patterns of linkage disequilibrium and haplotype structure across global populations at the interleukin-13 (IL13) locus.* Genes and immunity, 2005. 6(1): p. 53-65.

[198] Madhankumar, A.B., A. Mintz, and W. Debinski, *Alanine-scanning mutagenesis of alpha-helix D segment of interleukin-13 reveals new functionally important residues of the cytokine.* The Journal of biological chemistry, 2002. 277(45): p. 43194-43205.

[199] Semlali, A., et al., *Thymic stromal lymphopoietin-induced human asthmatic airway epithelial cell proliferation through an IL-13-dependent pathway.* The Journal of allergy and clinical immunology, 2010. 125(4): p. 844-850.

[200] Singhera, G.K., R. MacRedmond, and D.R. Dorscheid, *Interleukin-9 and -13 inhibit spontaneous and corticosteroid induced apoptosis of normal airway epithelial cells.* Experimental lung research, 2008. 34(9): p. 579-598.

[201] Wadsworth, S., D. Sin, and D. Dorscheid, *Clinical update on the use of biomarkers of airway inflammation in the management of asthma.* J Asthma Allergy, 2011. 4: p. 77-86.

[202] Watanabe, T., et al., *Increased levels of HMGB-1 and endogenous secretory RAGE in induced sputum from asthmatic patients.* Respir Med, 2011. 105(4): p. 519-25.

[203] Reddel, H.K., et al., *An official American Thoracic Society/European Respiratory Society statement: asthma control and exacerbations: standardizing endpoints for clinical asthma trials and clinical practice.* Am J Respir Crit Care Med, 2009. 180(1): p. 59-99.
[204] D'Amico, A., et al., *An investigation on electronic nose diagnosis of lung cancer.* Lung Cancer, 2010. 68(2): p. 170-6.
[205] Payne, D.N., et al., *Relationship between exhaled nitric oxide and mucosal eosinophilic inflammation in children with difficult asthma, after treatment with oral prednisolone.* Am J Respir Crit Care Med, 2001. 164(8 Pt 1): p. 1376-81.
[206] Schleich, F.N., et al., *Exhaled nitric oxide thresholds associated with a sputum eosinophil count >/=3% in a cohort of unselected patients with asthma.* Thorax, 2010. 65(12): p. 1039-44.
[207] Jiang, J., et al., *Nitric oxide gas phase release in human small airway epithelial cells.* Respir Res, 2009. 10: p. 3.
[208] Kostikas, K., et al., *Exhaled NO and exhaled breath condensate pH in the evaluation of asthma control.* Respir Med, 2011. 105(4): p. 526-32.
[209] Gessner, C., et al., *Angiogenic markers in breath condensate identify non-small cell lung cancer.* Lung Cancer, 2010. 68(2): p. 177-84.
[210] Bloemen, K., et al., *A new approach to study exhaled proteins as potential biomarkers for asthma.* Clin Exp Allergy, 2011. 41(3): p. 346-56.
[211] Lessard, A., et al., *Obesity and asthma: a specific phenotype?* Chest, 2008. 134(2): p. 317-23.
[212] Sin, D.D., et al., *Serum PARC/CCL-18 Concentrations and Health Outcomes in Chronic Obstructive Pulmonary Disease.* Am J Respir Crit Care Med, 2011. 183(9): p. 1187-1192.
[213] Wedzicha, J.A., et al., *Acute exacerbations of chronic obstructive pulmonary disease are accompanied by elevations of plasma fibrinogen and serum IL-6 levels.* Thromb Haemost, 2000. 84(2): p. 210-5.
[214] Chang, C.L., et al., *Biochemical markers of cardiac dysfunction predict mortality in acute exacerbations of COPD.* Thorax, 2011.
[215] Lowhagen, O., et al., *The inflammatory marker serum eosinophil cationic protein (ECP) compared with PEF as a tool to decide inhaled corticosteroid dose in asthmatic patients.* Respir Med, 2002. 96(2): p. 95-101.
[216] Dosanjh, A., et al., *Elevated Serum Eosinophil Cationic Protein Levels in Cystic Fibrosis, Pediatric Asthma, and Bronchiolitis.* Pediatric Asthma, Allergy & Immunology, 1996. 10(4): p. 169-173.
[217] Trulson, A., et al., *The functional heterogeneity of eosinophil cationic protein is determined by a gene polymorphism and post-translational modifications.* Clin Exp Allergy, 2007. 37(2): p. 208-18.
[218] Sekiya, T., et al., *Increased levels of a TH2-type CC chemokine thymus and activation-regulated chemokine (TARC) in serum and induced sputum of asthmatics.* Allergy, 2002. 57(2): p. 173-7.
[219] Leung, T.F., et al., *Plasma concentration of thymus and activation-regulated chemokine is elevated in childhood asthma.* J Allergy Clin Immunol, 2002. 110(3): p. 404-9.
[220] Stephan, V., et al., *Determination of N-methylhistamine in urine as an indicator of histamine release in immediate allergic reactions.* J Allergy Clin Immunol, 1990. 86(6 Pt 1): p. 862-8.

[221] Takei, S., et al., *Urinary N-methylhistamine in asthmatic children receiving azelastine hydrochloride.* Ann Allergy Asthma Immunol, 1997. 78(5): p. 492-6.

[222] Sansuk, K., et al., *GPCR proteomics: mass spectrometric and functional analysis of histamine H1 receptor after baculovirus-driven and in vitro cell free expression.* J Proteome Res, 2008. 7(2): p. 621-9.

[223] Saude, E.J., et al., *Metabolomic profiling of asthma: diagnostic utility of urine nuclear magnetic resonance spectroscopy.* J Allergy Clin Immunol, 2011. 127(3): p. 757-64 e1-6.

Chapter 10

Metabolism of Carbochidrates in the Cell of Green Photosintesis Sulfur Bacteria

M. B. Gorishniy and S. P. Gudz

Additional information is available at the end of the chapter

http://dx.doi.org/10.5772/50629

1. Introduction

Green bacteria - are phylogenetic isolated group photosyntetic microorganisms. The peculiarity of the structure of their cells is the presence of special vesicles - so-called chlorosom containing bacteriochlorofils and carotenoids. These microorganisms can not use water as a donor of electrons to form molecular oxygen during photosynthesis. Electrons required for reduction of assimilation CO_2, green bacteria are recovered from the sulfur compounds with low redox potential.

Ecological niche of green bacteria is low. Well known types of green bacteria - a common aquatic organisms that occur in anoxic, was lit areas of lakes or coastal sediments. In some ecosystems, these organisms play a key role in the transformation of sulfur compounds and carbon. They are adapted to low light intensity. Compared with other phototrophic bacteria, green bacteria can lives in the lowest layers of water in oxygen-anoxic ecosystems.

Representatives of various genera and species of green bacteria differ in morphology of cells, method of movement, ability to form gas vacuoles and pigment structure of the complexes. For most other signs, including metabolism, structure photosyntetic apparats and phylogeny, these families differ significantly. Each of the two most studied families of green bacteria (*Chlorobium* families and *Chloroflexus*) has a unique way of assimilation of carbon dioxide reduction. For species of the genus *Chlorobium* typical revers tricarboxylic acids cycle, and for members of the genus *Chloroflexus* - recently described 3-hidrocsypropionatn cycle. Metabolism of organic compounds, including carbohydrates in the cells of representatives of genera *Chlorobium* and *Chloroflexus* remains poorly understood. Anabolism and catabolism of monomeric and polymeric forms of carbohydrates in the cells of green bacteria is discussed.

Representatives of the green sulfur bacteria family *Chlorobiaceae* and green nonsulfur bacteria family *Chloroflexaceae* grow with CO_2 as the sole carbon source. In addition, the growth process, they can use some organic carbon sources, particularly carbohydrates. Species of the genus *Chlorobium* able to assimilate organic compounds is limited only in the presence of CO_2 and inorganic electron donors. Instead, representatives of the genus *Chloroflexus* grow on different carbon sources anaerobically and aerobically under illumination in the dark. Significant differences between species and genera *Chlorobium, Chloroflexus* due to the nature of their photosyntetic apparat. Despite the fact that members of both families contain the same types of chlorosoms, species of the genus *Chlorobium* reaction centers have photosystemy I (PS I), while both species of the genus *Chloroflexus* contain reaction centers photosystemy II (PS II). Because of the low redox potential (~ -0,9 V) of the primary electron acceptor in green sulfur bacteria, reaction center capable reduce ferredocsyn. In green bacteria nonsulfur bacteria redox potential of the primary acceptor is less negative (~ -0,5 V), resulting in these organisms synthesize reduction equivalents by reverse electron transport, like the purple bacteria. Thus, differences in the structure of the apparatus of green sulfur and green nonsulfur bacteria are reflected in their exchange carbohidrats compounds, and there fore also in their evolution and ecology.

In the evolution of autotrophic organisms formed several ways to assimilate CO_2, each of which is characterized by biochemical reactions that require the appropriate enzymes and reduction equivalents [9, 11]. The most common mechanism for CO_2 assimilation is Calvin cycle, which was found in most plants, algae and most famous groups of autotrophic prokaryotes. In green bacteria described two alternative ways of assimilation of CO_2. Revers cycle of tricarboxylic acids (RTAC) in green sulfur bacteria, first proposed by Evans in 1966. In 1989, Holo described 3-hidroksypropionat way that is characteristic of green non sulfur bacteria.

Larsen, using washed cells of *C. thiosulfatophilum,* for the first time found that they can absorb light in only small amounts of CO_2. They found that most carbon dioxide *C. thiosulfatophilum* records in an atmosphere containing H_2S, H_2 and CO_2. Most data on how carbon dioxide conversion and other compounds related to green sulfur bacteria genus *Chlorobium,* including *C. thiosulfatophilum, C. phaeobacteroides* and *C. limicola.*

Green sulfur bacteria can use some organic compounds (sugars, amino acids and organic acids). However, adding these compounds to the environment leads only to a slight stimulation of growth of culture in the presence of CO_2 and is to ensure that they are used only as additional sources of carbon [13]. In any case they are electron donors or major source of carbon. The use of these substances only if there among CO_2 and H_2S.

In the cells of *C. thiosulfatophilum* not identified Calvin cycle enzyme activity. The main role in the transformation of CO_2 is open in this group of bacteria (RTAC). Here in green sulfur bacteria is reduction of CO_2 assimilation. The cycle was proposed by the opening of phototrophic bacteria and other anaerobs two new ferredocsyn - dependent carboxylation reactions [13]:

$$\text{Acetyl} - \text{CoA} + CO_2 \rightarrow \text{pyruvate} + \text{CoA}$$
$$\text{Succinyl} - \text{CoA} + \text{CO2} \rightarrow \alpha - \text{ketoglutarate} + \text{CoA}$$

They make possible (RTAC), in which two molecules of CO_2 formed a molecule of acetyl - CoA (Fig. 1).

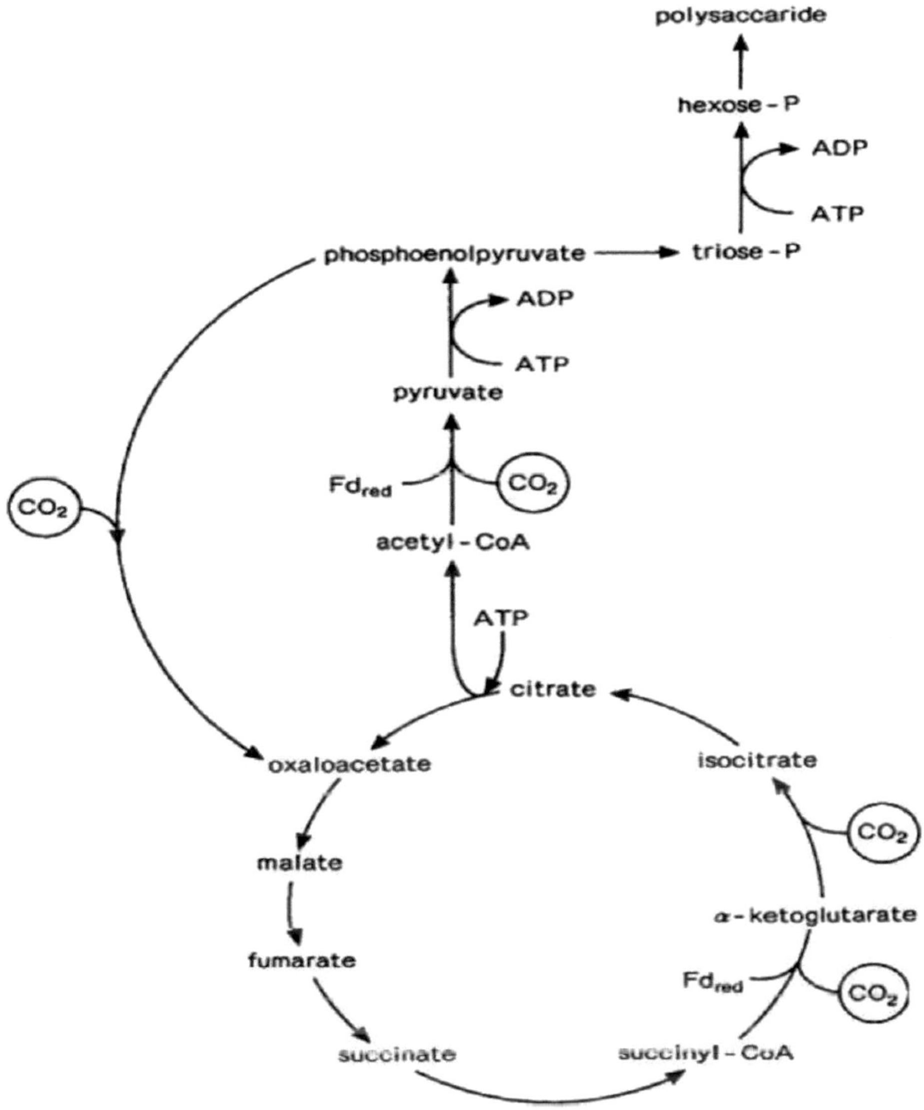

Figure 1. The revers tricarboxylic acid cycle (RTAC) in *Chlorobium* .

First, revers tricarboxylic acid cycle (RTAC) considered an additional mechanism for better functioning of rehabilitation Calvin cycle of the genus *Chlorobium.* Assumed that its main function is the formation of precursors for the synthesis of amino acids, lipids and porphyrins, while restorative Calvin cycle given the main role in the synthesis of carbohydrates. However, the lack of activity in cells rubisco put the availability of restorative Calvin cycle in cells of *Chlorobium limicola* doubt. Confirmation of operation of restorative tricarboxylic acid cycle in cells of green sulfur bacteria was discovered in them a key enzyme of this cycle. Using tracer and fractionation of isotopes of carbon have shown that (RTAC) is the only recovery mechanism for fixation of CO_2 in green sulfur bacteria, and the product cycle acetyl - CoA directly used for the synthesis of carbohydrates. It was also found that the genes of cells rubisco *Rhodospirillum rubrum*, not related of DNA isolated from cells of bacteria genus *Chlorobium*. Similar negative results were obtained using genes to cells rubisco *Anacystis nidulans*.

The study of restorative (RTAC) can explain the inability of green sulfur bacteria photoheterotroph. Simultaneously with the operation of the mechanism fixation of CO_2 cycle intermediates also provide cells needed organic matter for the synthesis of fatty acids (from acetyl-CoA), amino acids (from pyruvate, α-ketoglutarat acid) and carbohydrates (with pyruvate). However, since the activity of α-ketoglutaratdehidrohenaz not found in species of the genus *Chlorobium*, this cycle can operate only in recreation and hence organic compounds can not oxidate with the formation of reduction equivalents .

Recovery (RTAC) provides fixation of CO_2, to be based on restorative carboxylation reaction of organic acids. Fixation of carbon dioxide occurs in three enzymatic reactions, two of which occur with photochemically reduced ferredoksyn, and one - the same way formed provided with (H^+). As a result of a turnover cycle of four molecules of CO_2 and 10 [H^+] using the energy of three molecules of ATP synthesized molecule oxaloacetat acid is the end product cycle.

Described as "short" version of the cycle, in which 2 molecules of CO_2 are fixed using for their restoration 8 [H^+] and the energy of ATP. The final product in this case is acetyl-CoA, which is used to build components of cells. Addition of acetate in the culture medium promotes the accumulation of biomass and stimulates the formation of reserve polysacharides in the cells of green sulfur bacteria. Representatives of the family *Chlorobiaceae*, including *Chlorobium limicola* and *C. thiosulfatophilum*, often accumulate in cells poliglycose and / or glycogen. Accumulation of polysacharides increases in carbon dioxide assimilation by cells under conditions of deficiency of nitrogen and phosphorus. Under certain conditions of cultivation the level of glycogen in the cells can exceed 12% of dry weight of cells. Formed spare polysaccharides play an important role in changing the conditions of cultivation of green sulfur bacteria, especially when ingested bacteria in the extreme conditions of growth.

Larsen and collaborators found that the bacteria *C. thiosulfatophilum* not grow on media containing traces of hydrogen sulfide (0.01%) and various organic compounds: alcohols, sugars, organic acids. Only media with acetic, lactic or pyruvatic acid was seen a slight increase in biomass under conditions of hydrogen sulfide and carbon dioxide in the environment. Regarding the nature of organic compounds green sulfur bacteria similar to purple sulfur bacteria.

Larsen found that washed suspensions of cells *C. thiosulfatophilum* the light can absorb only small amounts of CO_2. The greatest amount of carbon dioxide *C. thiosulfatophilum* assimilates in an atmosphere containing 86% N_2, 9,2% CO_2, 3,9% H_2S and 0,5% H_2. Most data on how metabolic carbon dioxide and other carbon compounds obtained in experiments using *C. thiosulfatophilum* and *C. phaeobacteroides*.

Found that in cells *C thiosulfatophilum* not Calvin cycle enzyme activity. Important role in the assimilation of CO_2 is open in this group of microorganisms revers Crebs cycle, which was named Arnon cycle. This series provides a record of CO_2 through renewable carboxylation of organic acids. As a result of the work cycle in cells of green sulfur bacteria in the process of photosynthesis, glucose is formed, which is the first product of photosynthesis, carbohydrate nature. Ways to transform cells in green sulfur bacteria studied not enough. According to it becomes poliglucose, other authors believe that the glucose immediately polymerizes to form glycogen.

To detect sugars that accumulate in cells of *C. limicola* IMB- K-8 in the process of photosynthesis they were grown under illumination and in the presence of electron donor, which served as H_2S. After 10 days culturing cells destroy bacteria and cell less extract analyzed for the presence in it is reduced sugars. The total number of cell less extracts was determined by Shomodi-Nelson and for determination of glucose using enzymatic set "Diaglyc- 2". It turned out that the total number is reduced sugars determined by the method of Shomodi-Nelson did not differ from the rate obtained specifically for glucose.

It follows that the sugar is reduced *C. limicola* IMB- K-8 represented only glucose, which is the first carbohydrate, which is formed during photosynthesis.To test whether glucose in cells is in free or bound state spent acid hydrolysis cell less extract. It found that the content is reduced sugars increased approximately two fold. It follows that the glucose in the cells located in the free and in a bound state. In these experiments investigated the dynamics of accumulation of intracellular glucose bacterium *C. limicola* IMB- K-8 in the process of growth. It was found that the formation of glucose in the cells is observed throughout the period of growth and completed the transition culture stationary phase.

Growth of *C. limicola* IMB K 8 and glucose in cells growing in culture in the light in a mineral medium with $NaHCO_3$ and Na_2S. We investigated the growth and accumulation of glucose in the cells of *C. limicola* IMB- K-8 for varying light intensity.

It was found that light intensity plays an important role in CO_2 assimilation in *C. limicola* IMB- K-8. More intensive process proceeded in low light, which does not exceed 40 lux. Reduction or increase in illumination intensity was accompanied by reduced productivity of photosynthesis.

On the intensity of photosynthesis reveals a significant influence of mineral nutrition of bacteria We shows the influence of different sources nitrogen and phosphorus supply of glucose in the cells *C. limicola* IMB- K-8.

Simultaneous limitation of growth of culture nitrogen and phosphorus accompanied by increase in glucose in the cells. Her level of these compounds for the deficit grew by about 60%. Separately salts of nitrogen and phosphorus showed much less effect In these experiments investigated how bacteria use glucose under various conditions of cultivation. This used washed cells were incubated under light and dark. When incubation of cells at the light in the presence of CO_2 and H_2S levels of glucose in the cells practically did not change while under these conditions in the dark glucose concetration in the cells decreased about 2.5 times.

Obviously, in the dark using glucose as an energy source, turning towards Embdena-Meyerhof-Parnas. The level of intracellular glucose is reduced and the conditions of incubation of cells at the light in the environment without hydrogen sulfide, indicating that the use of glucose under these conditions as the sole source of renewable equivalents. In the dark, without glucose hydrogen sulfide is the only source of energy. Thus, the glucose formed by cells plays an important role in the life of cell *C. limicola* IMB - K-8. When staying in the light cells in the process of photosynthesis observed formation of glucose in the cells, and in darkness it is used to maintain cell viability.

2. Isolation, identification and patterns of accumulation poliglucose *C. limicola* IMB- K-8

Nature poliglucose formed in the cells of *C. limicola* IMB- K-8. As already mentioned above green sulfur bacteria in the process of growth can form glucose, a small part of which the cells are in a free state, and the part becomes glycogen. To test the ability of *C. limicola* IMB-K-8 form glycogen was held their extraction from the cells by the method. Grown under light conditions cells in acetic acid. Polisacharide precipitate obtained by adding to the supernatant concentrated ethanol. The obtained precipitate distroy 10M H_2SO_4 obtained hydrolyzate were separated by chromatography. As witnesses used the glucose and galactose, and atzer. After manifestation chromatography revealed only one spot, which is slowly moving in the system butanol - water and Rf value was identical to glucose. It follows that polisaharide that piled up in the cells *C. limicola* IMB- K-8, is polisaharide. As in the literature found allegations that members of *Chlorobium* form glycogen, we extracted polisaharide by Zacharova-Kosenko, which is specific for bacterial glycogen deposition. The formation of glycogen in the cells is only the lighting conditions and the presence of carbon dioxide and hydrogen sulfide in the culture medium. In the absence of H_2S and CO_2 accumulation of glycogen in the cells was observed. In microsections of cells grown under different light, the presence of carbon dioxide and hydrogen sulfide, clearly visible rozet not surrounded by a membrane, glycogen granules (Fig. 2).

Comparative analysis of selected polisaharide and glycogen company "Sigma" showed that the resulting sample shows identical chemical and physical properties: white crystalline powder soluble in water, not soluble alcohol, hydrolyzed in acidic medium to form glucose. Infrared spectroscopy etylaceton extract the studied sample and glycogen "Sigma" has shown that these substances are characterized by the presence of identical functional groups, O-H bonds (the interval 3608 - 3056 cm^{-1}), revealed specific absorption in the carbonyl group (1656 cm^{-1}),- CH2-group (2932 cm^{-1}), and -C-O-H groups (1048 cm^{-1}) and others, indicating the identity of the investigated sample of bovine liver glycogen (the drug company "Sigma") (Fig. 3.) Therefore, we first selected polisacharide of cells *C. limicola* IMB-K-8, which by the nature of the infrared spectrum identical with glycogen "bovine liver".

Accumulation of glycogen in the cells may be an indicator of flow speed. Therefore, we first selected polisacharide of cells *C. limicola* IMB- K-8, which by the nature of the infrared spectrum identical with glycogen "bovine liver". Accumulation of glycogen in the cells may be an indicator of flow speed.

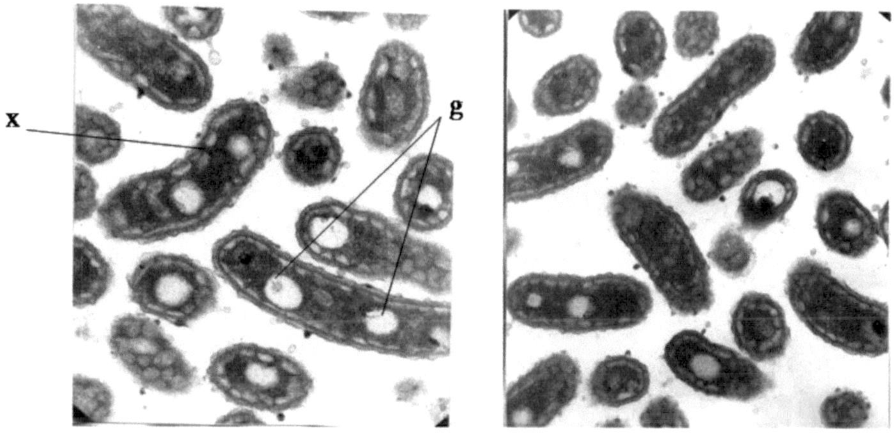

A B

Figure 2. Microsections of cells *C. limicola* IMB- K-8, grown under different light intensity (A - 40lk, B - 100lk): g – glycogen granules, x – chlorosomu.

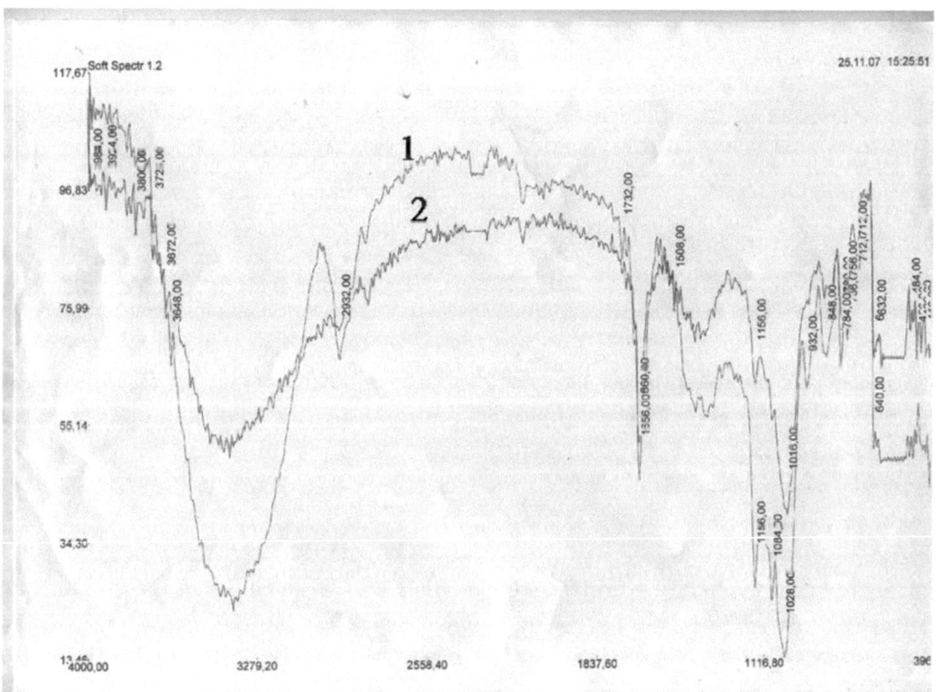

Figure 3. Infrared spectrum of glycogen company "Sigma" (1) and glycogen cells of *C. limicola* IMB- K-8 (2)

The laws of accumulation and utilization of glycogen *C. limicola*. In the presence of light *C. limicola* IMB- K-8 can use organic compounds only, subject to the availability of hydrogen sulfide as an additional source of carbon and continue in their presence actively assimilate carbon dioxide. Assimilation green sulfur bacteria carbon dioxide and organic compounds leads not only to form cells of substances necessary for their growth, but can also affect the synthesis of glycogen. Assuming in these experiments, we investigated the influence of organic carbon sources of power in the process of accumulation of this compound. It turned out that adding to the medium glucose, sucrose, maltose, lactate, not accompanied by changes in intracellular glucose and glycogen.

Only adding to the environment pyruvate and acetate stimulated the growth of glycogen content in cells of *C. limicola* IMB- K-8 which clearly explains the functioning of the studied bacteria cycle Arnon, in the process which produced acetate. Notably, cells with elevated levels of glycogen synthesis that is caused by the addition of pyruvate and acetate, in contrast to cells grown in the presence of other sources of carbon, used almost entirely endogenous glucose. Only adding to the environment pyruvate and acetate stimulated the growth of glycogen content in cells of *C. limicola* IMB- K-8 which clearly explains the functioning of the studied bacteria cycle Arnon, in the process which produced acetate. Notably, cells with elevated levels of glycogen synthesis that is caused by the addition of pyruvate and acetate, in contrast to cells grown in the presence of other sources of carbon, used almost entirely endogenous glucose.

The results obtained give grounds to assert that *C. limicola* IMB- K-8 the most effective use as an additional source of carbon nutrition acetate. It is used only in the presence of hydrogen sulfide and carbon dioxide in the environment and occurs through the inclusion of this compound in the Arnon cycle with the formation of cell components and glycogen. In the presence of pyruvate and acetate in the environment there are some differences in photoreduction CO_2 cells.

So when the concentration of CO_2 in the atmosphere 60mM observed maximum cell growth and increased by 50% the level of glycogen. A slight reduction of carbon dioxide in the environment (20%) accompanied by a reduction in biomass, while increasing the level of glycogen in the cells by about 30%. Further reduction of CO_2 was accompanied by decrease in the intensity of photosynthesis. Increase in glycogen levels in cells with the shortage of carbon dioxide in the atmosphere, apparently, can be explained by inhibition of pyruvate carboxylation reaction and its conversion in to oksaloatsetat and then using it in a constructive metabolism. Note that formed in the process of photosynthesis annoxy carbohydrates not allocated to the environment and stockpiled exclusively in the cell. As evidenced by a negative test for glucose and other sugars is reduced before and after hydrolysis of culture broth. To find ways of further use of glycogen in these experiments, free cells of *C. limicola* IMB K-8 with a high content of polisacharide, incubated in light and in darkness, and then determined the level of glycogen in the cells and analyzed the nature of the organic matter accumulating in the environment. It turned out that the absence of light and presence of CO_2 and H_2S in the medium, cells of *C. limicola* IMB K-8 used a significant amount of glycogen, which testified to a significant reduction of its level in cells.

Analysis of the products of glucose catabolism, obtained after deposition of the mixture (acetone - petroleum ether) from the environment showed that the cells incubated in the dark in the environment accumulate organic compounds as evidenced by their elemental analysis (C-40.25%, H-4.5 %, N-0%). Infrared spectrometry etylatseton hoods showed that these substances are characterized by the presence of O-H bonds (the interval 3608 - 3056cm^{-1}), CH_3-CH_2 bonds (1456sm^{-1}) and specific absorption in the carbonyl (1656 cm^{-1}) and methyl group (2920 cm^{-1}) and R-COOH groups (2700 cm^{-1}) and others that indicate the presence in culture fluid of carboxylic acids (Fig. 4).

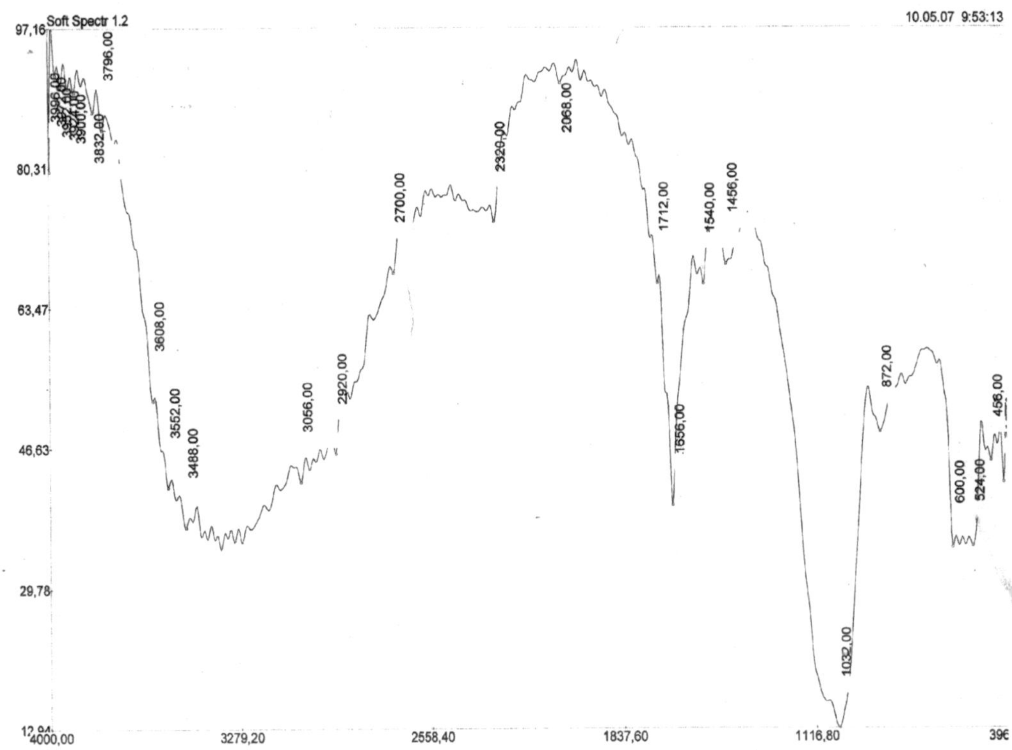

Figure 4. Infrared spectrum of culture fluid components *C. limicola* IMB K-8 cells under incubation in the dark

These results are consistent with data Sirevag, under which the cells incubated with *C. thiosylfatophillum* in the dark in culture fluid accumulated carboxylic acids: acetate (the main component that makes up 70%), propionate and succinate. They are the authors produced by reactions of glycolysis, pyruvate decarboxylation and other reactions.

Under the conditions of incubation, washed cells *C. limicola* IMB K-8 in light of the formation of organic compounds in the culture fluid was observed, and the total content of glucose after hydrolysis of glycogen, not significantly different from control. We shows the dynamic changes in the concentration of glucose (after hydrolysis of glycogen) and the accumulation of carboxylic acids during incubation of cells *C. limicola* IMB K-8 in the dark.

As seen from for 40h incubation, the contents of glycogen (for glucose) in the incubation mixture decreased almost three times while there was accumulation of carboxylic acids in the environment. Thus, synthesized by cells during photosynthesis *C. limicola* IMB K-8 - glucose and glycogen play an important role in the life of these bacteria during their stay on the light and in darkness. In the first case in the photoreduction CO_2 observed accumulation of glucose and its conversion into glycogen. In the darkness degradation glycogen to glucose, catabolism of which provides energy and constructive metabolism of green sulfur bacteria

In addition to the family *Chlorobiaceae* green bacteria carry the family *Chloroflexaceae*, which is called green nonsulfur bacteria. Green nonsulfur bacteria form filaments capable of sliding movement, optional anaerobes that can use organic compounds as sources of carbon and energy. By type of metabolism they phototropy under anaerobic conditions and under aerobic heterotrophs. Their cells contain bakteriohlorophily and carotenoids. Some molecules of green pigments nonsulfur bacteria contained directly in the cytoplasmic membrane, and part of chlorosom. Protein membrane chlorosom similar representatives of families *Chlorobiaceae* and *Chloroflexaceae*. Slow growth of photoavtotroph on the environment of sulphide was first described Median of employes in 1974. The representatives of green nonsulfur bacteria detected actively functioning oxidative tricarboxylic acid cycle. Like most photobacter bacteria genus *Chloroflexus* can grow using CO_2 as the sole carbon source. Found that green nonsulfur bacteria can use hydrogen sulfide as electron donor for photosynthesis and *Chloroflexus aurantiacus* may molecular hydrogen in the process reduce CO_2.

Found that one of the key enzymes - piruvatsyntaza that catalyzes the formation of pyruvate from acetyl-CoA and CO_2 detects activity in *Ch. aurantiacus*. The activity of other specific enzymes that are restorative (RTAC) was absent. On this basis it was concluded that cells of these bacteria, acetyl-CoA is synthesized from CO_2. The mechanism of this synthesis is different from what is in *C. limicola*.

Holo and Grace in 1987 found that in autotrophic conditions is inhibiting the tricarboxylic acid cycle and gliocsilate shunt, and in the cells is a new metabolic pathways in which acetyl-CoA is an intermediate product. Later Holo found that in autotrophic conditions *Ch. aurantiacus* converts acetyl-CoA in 3-hidroksypropionat, which is an intermediate product in the fixation of CO_2. Further to its transformation leads to the formation of malate and succinate. The results were confirmed by Strauss in 1992, which showed that the autotrophic cell growth *Ch. aurantiacus* isolated succinate and many 3-hidroksypropionat in the period from the late exponential phase to early stationary.

When culture *Ch. aurantiacus* was placed in an environment of ^{13}C labeled succinate and analyzed the different components of cells using ^{13}C spectroscopy, which determines the distribution of ^{13}C isotope in various compounds of cells, the results confirmed the role as an intermediate metabolite 3-hidroksypropionatu in CO2 fixation.

Hidrokspropionat role as intermediate in the fixation of CO_2 was investigated Fuchs and Staff in experiments using ^{13}C. The relative amount of ^{13}C after growth *Ch. aurantiacus* in the presence of ^{13}C and ^{13}C 3-hidroksypropionatu acetate. From the samples were labeled ^{13}C marker central intermediate metabolite as trioz and dicarboxylic acids. These experiments showed that cell growth *Ch. aurantiacus* was determined by adding ^{13}C 3-hidroksypropionat for several generations of cells, where it was concluded that this substance is a precursor of all cellular compounds *Ch. aurantiacus*. Thus, 3-hidroksypropionat functions in the body of *Ch. aurantiacus* as a central intermediate metabolite. The data obtained with labeled acetate, also confirmed the key role 3-hidroksypropionat as intermediate in the cyclic mechanism of CO_2 fixation (Fig. 5).

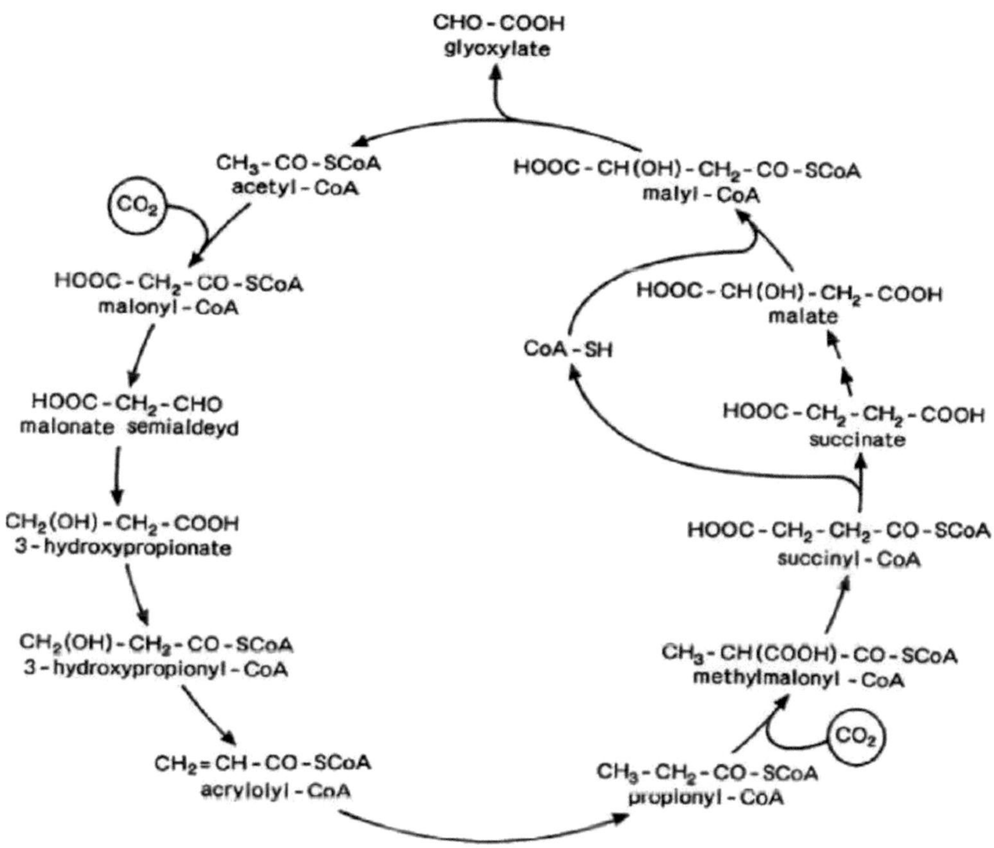

Figure 5. 3-Hidroksypropionatnyy cycle CO_2 fixation (Holo 1989 p.)

For a final check of the cycle Strauss and Fuchs had enzymatic studies and showed that the cells of green bacteria is nonsulfur activity of all enzymes required for assimilation cycle 3-hidroksypropionat reduction of carbon dioxide. In this cycle acetyl - CoA in malonil - CoA and then, reducing turns through 3-hidroksypropionat to propionil - CoA.

Thus, in green bacteria nonsulfur *Ch. aurantiacus* operating mechanism of autotrophic fixation of CO_2, the key intermediates which are 3-hidroksypropionat.. The final product of this cycle is glyoxylate, who fotoheterotrofiv becomes a backup compound poli -*β*-hidrocsybutyrat.

Thus, green bacteria families *Chlorobiaceae* and *Chloroflexaceae,* despite the similarity of their photosintetic system, assimilation of CO_2 reduction carried out in different ways. In the family *Chlorobiaceae* CO_2 fixation reactions proceeding with revers tricarboxylic acid cycle. Carbohydrates - products of photosynthesis, they lay in store as glycogen, which is used in extreme conditions for energy and carbon. Green nonsulfur bacteria family *Chloroflexaceae* used for CO_2 fixation reaction 3-hidroksypropionat way. Under these conditions produced a poli-β-hidroksybutyrat, which, like glycogen in the family *Chlorobiaceae,* is used in the energy and constructive exchanges.

Author details

M. B. Gorishniy and S. P. Gudz
National University of Lviv Ivan Franko, Lviv, Ukraine

3. References

[1] Bergstein T., Henis Y., Cavari B. Investigation of the photosynthetic bacterium *Chlorobium phaeobacteroides* causing seasonal blooms in Lake Kinneret // Canada J. Microbiol. – 1979. – Vol. 25. – P. 999–1007.

[2] Bergstein T., Henis Y., Cavari B. Uptake and metabolism of organic compounds by Chlorobium isolated from Lake Kinneret // Microbiol. – 1981. – Vol. 27. – P. 1087–1091.

[3] Castenholz R. W., Pierson B. K. The prokaryotes. – New York: Springer, 1978. – P. 290–298.

[4] Cork D., CarunasR., Sajjad A. *Chlorobium limicola* forma *thiosulfatophilum* biocatalyst in the production of sulfur and organic carbon from gas stream containing H_2S and CO_2 // Appl. And Envir. Microbiol. – 1983. – Vol. 45. – P. 913–918.

[5] Gemerden H. Physiological ecology of green sulfur bacteria // Ann. Microbiol. – 1983. – Vol. 134. – P. 73–92.

[6] Gorishniy M. B. Ecological significance of green sulfur bacteria in the utilization of hydrogen sulfide / / Thesis for PhD degree in specialty 03.00.16. - Ecology - Institute of Agroecology biochemistry of Ukraine, Kyiv, 2008.

[7] Gorishni y M. B, Gudz S. P, Hnatush S. O The metabolism of glucose and glycogen in the cells of green sulfur bacteria fotosyntezuvalnyh / / Herald of Lviv. Univ. Biological Series. - 2008. - Vol. 46. S. 129-136.

[8] Gorishniy M. B., Gudz. S. P., Hnatush S. O. Metabolism of carbohydrates in the cells of green sulfur bacteria *Chlorobium limicola* Ya-2002 "Ukrainian Biochemical Journal" - 2009. - N. 5. T-81. - S. 26-33.

[9] Gudz S. P, Gorishniy M. B., Hnatush S. O. Bacterial photosynthesis - Lions: Ivan Franko Lviv National University, 2011. – 180 p.

[10] Herter S., Farfsing J., Gad'On N. et al. Autotrophic CO_2 fixation by *Chloroflexus aurantiacus*: study of glyoxylate formation and assimilation via the 3-hydroxypropionate cycle // J. Biol. Chem. – 2001. – Vol. 267. – P. 20256–20273.

[11] Mas J., Gemerden H. Storage products in purple and green sulfur bacteria. In: Anoxygenic photosynthetic – D.: KmwerAcad. Pub. (Netherlands), 1995. – P. 973–990.
[12] Overmann J. Green Sulfur Bacteria. In: Bergey's Manual Systematic Bacteriology. – 2002. – 2nd ed. – Vol. 1. – P. 601–605.
[13] Pfennig N. Photoprophic green bacteria: a comparative, systematic survey // Ann. Rev. Microbiol. – 1977. – Vol. 31. – P. 275–290.
[14] Pfennig N. The phototrophic bacteria and their role in the sulfur cycle // Plant and Soil. – 1975. – Vol. 43. – P. 1–16.
[15] Pfennig N., Widdel F. The bacteria of sulfur cycle // Phil. Trans. R. Soc. Lond. – 1982. – Vol. 298. – P. 433–441.
[16] Repeta D. J., Simpson D.J., Jorgensen B. B., Janasch H. W. Evidence for anoxygenic photosynthesis from the distribution of bacteriochlorophylls in the Black Sea // Nature. – 1989. – Vol. 12. – P. 69–72.
[17] Ugolkova N. V., Ivanovsky R.N. On the mechanism of autotrophic fixation of CO_2 by Chloroflexus aurantiacus // Microbiology. – 2000. – Vol. 5. – P. 139–142.
[18] Van Niel C. B. Natural selection in the microbial world // Gen. Microbiol. – 1955. – Vol. 13, N 1. – P. 201.
[19] Overmann J., Garcia-Pichel F. The Phototrophic Way of Life. The Prokaryotes. – New York: Springer, 2000. – 887 p.
[20] Blankenship M. T. Madigan C. E. Anoxygenic Photosynthetic Bacteria. – Boston: Kluwer Academic Publishers Dordrecht, 1997. – P. 49–89.107.
[21] Casamayor E., Mas J., Pedro-Alio C. In situ assessment on the physiological state of Purple and Green Sulfur Bacteria through the analyses of pigment and 5S rRNA content // Microbiol. Ecol. – 2001. – Vol. 42. – P. 427–437.
[22] Gemerden H. Physiological ecology of green sulfur bacteria // Ann. Microbiol. – 1983. – Vol. 134. – P. 73–92.
[23] Gest H., Favinger J. *Heliobacterium chlorum,* an anoxygenie brownish-green photosyn thetic bacterium, containing a new form of bacteriochlorophill // Arch. Microbiol. – 2002. – Vol. 136 – P. 11-16.
[24] Gorlenko V. M., Pivovarova T. A. On the belonging of blue-green alga *Oscillatoria coerulescens,* to a new genus of chlorobacteria *Oscillochloris* nov. gen. // Izd. Acad. Nauk SSSR. – Ser. Biol. – 1977. – P. 396–409.
[25] Guerrero R., Pedros-Alio C, Esteve I., Mas J. Communities of phototrophic sulfur bacteria in lakes of the Spanish Mediterranean region // Acta Acad. Abo. – 1987. – Vol. 47. – P. 125–151.
[26] Gusev M. V., Shenderova L. V., Kondratieva E. N. The relation to molecular oxygen in different species of photoprophic bacteria // Microbiol. – 1969. – Vol. 38. – P. 787–792.
[27] Hanson T. E., Tabita F. R. A ribulose-1, 5-bisphosphate carboxylase / oxygenase (Rubisco)-like protein from *Chlorobium tepidum*that is involved with sulfur metabolism and the response to oxidative stress. Proc. Natl. Acad. Sci. – 2001. – P. 4397–4402
[28] Holf G., Brattar I. Taxonomic diversity and metabolic activity of microbial communities in the water column of the central Baltic Sea // Limnol. Oceanogr. – 1995. – Vol. 40. – P. 868–874.

[29] IKerryon C. N. Comlex lipids and fatty acids of photosynthetic bacteria. In: The photosynthetic bacteria – New York: Plenum publ. Co, 1978. – P. 281–313.

[30] Kuster E., Dorusch F., Vogt C., Weiss H., Altenburger R. Online biomonitors used as a tool for toxicity reduction evaluation of in situ ground water remediation techniques // Biosens Bioelectron. – 2004. – Vol. 19. – P. 1711–1722.

[31] Larsen H. On the culture and general physiology of the green sulfur bacteria // J. Bacteriol. – 1952. – Vol. 64. – P. 187–196.

[32] Miller M., Liu X., Snyder S. et al. Photosynthetic electron-transferreactions in the green sulphur bacterium *Chlorobium vibrioforme.* Evidence forthe functional involvement of iron-sulfur redox centers on the acceptorside of the reaction center // Biochemistry. – 1992. – Vol. 31. – P. 4354–4363.

[33] Okkels J., Kjaer B., Hansson O., Svendsen I., Moller B., Scheller H. A membrane bound monogeme *cyt. c*551 of a nowel type is the immediate electron donor to P_{840} of the *Chlorobium vibrioforme* photosynthetic reaction centre complex // J. Biol. Chem. – 1992. – Vol. 267. – P. 21139–21145.

[34] Otte S. C., van de Meent E. J., van Veelen P. A. et al. Identification of the major chromosomal bacteriochlorophills of the green sulfur bacteria *Chlorobium vibrioforme* and *Chlorobium phaeobacteroides;* their function in lateral energy transfer // Photosynt. Res. – 1993. – Vol. 35. – P. 159–169.

[35] Overmann J., Cypionka H., Pfennig N. An extremely low-light-adapted prototrophic sulphur bacterium from the Black Sea // Limnol. Oceanogr. – 1992. – Vol. 37. – P. 150–155.

[36] Parkin T. B., Brock T. D. The effects of light quality on the growth of phototrophic in lakes // Arch. Microbiol. – 1980. – Vol. 125. – P. 19–27.

[37] Parkin T. B., Brock T. D. The role of phototrophic in the sulfur cycle of a meromictic lake // Limnol. Oceanogr. – 1981. – Vol. 26. – P. 880–890.

[38] Peschek G., Loffelhardt W. The phototrophic prokaryotes. – New York: Plenum, 1999. – P. 763–774.

[39] Pfennig N. Syntrophic mixed cultures and symbiotic consortia with phototrophic bacteria. In: Anaerobes and Anaerobic Infections. – Studgard: Fischer, 1980. – P. 127-131.

[40] Pfennig N., Widdel F. The bacteria of sulfur cycle // Phil. Trans. R. Soc. Lond. – 1982. – Vol. 298. – P. 433–441.

[41] Pierson B. K., Kenn L. M., Leovy J. G. Isolation of Pigmentation Mutants of the green Filamentous Photosynthetic Bacterium *Chloroflexus aurantiacus* of Bacteriology // Arch. Microbiol. – 1984. – Vol. 159. – P. 222–227.

[42] Powell E.O. The growth rate of microorganism as a function of substrate concentration. In: Microbial physiology and continuous culture, N. Y.: Porton P, 1967. – P. 365–369.

[43] Pringault O., Kühl R. Growth of green sulphur bacteria in experimental benthic oxygen, sulphide, pH and light gradients // Microbiol. – 1998. – Vol. 144. – P. 1051–1061.

[44] Puchkova N. N. Green sulphur bacteria as a component of the sulfureta of shallow saline waters of the Crimea and northern Caucasus // Microbiol. – 1984. – Vol. 53. – P. 324–328.

[45] Puchkova N. N., Gorlenko V. M. A new green sulfur bacterium *Chlorobium chlorovibrioides* spec. // Microbiol. – 1984. – Vol. 51. – P. 118–124.

[46] Rocap G. F., Larimer J. F. Genome divergence in two Prochlorococcus ecotypes reflects oceanic niche differentiation // Nature. – 2003. – Vol. 424. – P. 1042–1047.

[47] Savikhin S., Zhou W., Blankenship R., Struve W. Femtosecond energy transferand spectral equilibration in bacteriochlorophyll α-protein antenna trimers from the green bacterium *Chlorobium tepidum* // Biophys. J. – 1994. – Vol. 66. – P. 110–114.

[48] Scheer H. Structure and occurrence of chlorophylls – Boca Raton.: CRC Press, 1991. – P. 3–30.

[49] Shill D. A., Wood P. M. Light-driven reduction of oxygen as a method forstuduing electron transport in the green photosynrhetic *Chlorobium limicola* // Arch. Microbiol. – 1985. – Vol. 143. – P. 82–87.

[50] Sirevåg R., Buchanan B. Mechanisms of CO_2 fixation in bacterial photosynthesis studied by carbon isotope fractionation technique // Arch. Microbiol. – 1977. – Vol. 16, N 112. – P. 35–38.

[51] Smith H. et al. Nomenclature of the bacteriochlorophills *c*, *d* and *e* II Photosynth. Res. – 1994. – Vol. 41. – P. 23-26.

[52] Stackebrandt E., Embly M., Weckesser J. Phylogenetic evolutionary, and taxonomic aspects of Phototrophic bacteria // New York: Plenum Press, 1988. – P. 201–215.

[53] Stackebrandt E., Woese C. The evolution of prokaryotes. In: Molecularand cellular aspects of microbioal evolution // Symp. Soc. Gen. Microbiol. – 1981. – Vol. 32. – P. 1–31.

[54] Takaichi S., Wang Z-Y., Umetsu M. et al. New carotenoids from the thermophilic green sulfurbacteria *Chlorobium tepidum*: 1,2-dihydro-carotene, 1,2-dihydrochlorobactene, OH-chlorobactene glucoside ester, and the carotenoid composition of different strains // Arch. Microbiol. – 1997. – Vol. 168. – P. – 270–276.

[55] Tamiaki H. Supramolecuiar structure in extramembraneous antennae of green photosynthetic bacteria // Coord. Chem. Rev. – 1996. – Vol. 148. – P. 183–197.

[56] Thompson R. W., Valentine H. L., Valentine W. N. Cytotoxic mechanisms of hydrosulfide anion and cyanide anion in primaryrat hepatocyte cultures // Toxicology. – 2003. – Vol. 12. – P. 149–159.

[57] Truper H.G. Culture and isolation of phototrophic sulfur bacteria from the marine environment // Helgol. Wiss. Meeresunters. – 1970. – Vol. 20. – P. 6–16.

[58] Truper H., Fisher U. Anaerobic oxidation of sulphur compounds as electron donors for bacterial photosynthesis // Phil. Trans. R. Soc. Lond. – 1982. – Vol. 298. – P. 254–258.

[59] Tuschak C., Glaeser J., Overmann J. Specific detection of green sulfur bacteria by in situ-hybridization with a fluorescently labeled oligonucleotide probe // Arch. Microbiol. – 2003. – Vol. 22. – P. 34–39.

[60] Van Niel C. B. On the nuwphology and physiology of the green sulfur bacteria // Arch. Microbiol. – 1932. – Vol. 3. – P. 11–112.

[61] Van Noort P. I., Zhu Y. R., LoBrutto R. E. Redox effects on the excited-state lifetime in chlorosomes and bacteriochlorophyll c oligomers // Biophys. J. – 1997. – Vol. 72. – P. 316–325

[62] Veldhuis M. J., van Gemerden W. H. Competition between purple and brown phototrophic in stratified lakes: sulfide, acetate, and light as limiting factors // FEMS Microbiol. Ecol. – 1998. – Vol. 38. – P. 31–38.
[63] Vignais P. M., Colbeau J. C. et al. Hydrogenase, nitrogenase, and hydrogen metabolism in photosynthetic bacteria // Adv. Microb. Physiol. – 1985. – Vol. 26. – P. 155–234.
[64] Wahlund T. M., Tabita R. F. The reductive tricarboxylic acid cycle of carbon dioxide assimilation: initial studies and purification of ATP citratlyase from the green sulfur bacteria *Chlorobium* // Bacteriology. – 1997. – Vol. 179. – P. 48–59.
[65] Welsh D. T., Herbert R. A. Identification of organic solutes accumulated by purple and green sulfur bacteria during osmotic stress using natural abundance nuclearmagnetic resonance spectroscopy // FEMS Microbiol. EcoL. – 1993. – Vol. 13. – P. 145–150.
[66] Wu H., RatsepM., Young C. et al. High pressure and stark hole burning studies of chlorosome antennas from green sulfur bacterium *Chlorobium tepidum* // Biophys. J. – 2000. – Vol. 79. – P. 1561–1572.
[67] Xiong J., Inoue K., Nakahara M., Bauer E. Molecularevidence forthe early evolution of photosynthesis // Science. – 2000. – Vol. 289. – P. 1724–1730.
[68] Overmann J., Tuschak C. Phylogeny and molecular fingerprinting of green sulfur bacteria // Arch. Microbiol. – 1997. – Vol. 167. – P. 302–309.

Chapter 11

Digestion in Ruminants

Barbara Niwińska

Additional information is available at the end of the chapter

http://dx.doi.org/10.5772/51574

1. Introduction

Ruminants, cloven-hoofed mammals of the order *Artiodactyla*, obtain their food by browsing or grazing, subsisting on plant material (Hungate, 1966). Today, 193 species of living ruminants exist in 6 families: *Antilocapridae, Bovidae, Cervidae, Giraffidae, Moschidae* and *Tragulidae* (Nowak, 1999). The number of wild ruminants is about 75 million and of domesticated about 3.6 billion (Hackmann and Spain, 2010). Approximately 95% of the population of domesticated ruminants constitute species: cattle, sheep and goats, all of them belong to the *Bovidae* family. Cattle and sheep are the two most numerous species and cattle is of the most economic importance. The economic value of milk and beef production in the EU is almost 125 billion Euro per year and accounts for 40% of total agricultural production (FAIP, 2003). The dairy cows is unique among all other mammalian species because of the intense artificial transgenerational genetic selection for milk production during the last 50 yr, so that annual averages of more than 12,500 kg/cow of milk per lactation are not uncommon (Eastridge, 2006). The selection has increased their peak energy yield by about 250% (20 Mcal × d^{-1} observed vs. 7.76 Mcal × d^{-1} expected) (Hackmann and Spain, 2010). Genetic improvement is accompanied by increasing metabolic demands for energy. The efficient use of energy of the feed resources is the main reason for the numerous and multilateral studies on carbohydrates digestion processes in cattle.

2. Digestive tract

Ruminants digestive system is characterized by functional and anatomical adaptations that allowed them to unlock otherwise unavailable food energy in fibrous plant material, mainly in cellulose and others recalcitrant carbohydrates (Van Soest, 1994). This property gives them an advantage over nonruminants. An important characteristic of ruminants digestive system is the occurrence of the microbial fermentation prior to the gastric and intestinal digestion activity. Their unique digestive system integrates a large microbial population with the animal's own system in the symbiotic relationship. The microbial fermentation occurs mainly in the rumen, the first chamber of the four-compartment stomach, which consists also of the reticulum and omasum (act as filters), and the abomasum (the true enzymatic stomach).

3. Rumen function

The feedstuffs consumed by ruminants are all initially exposed to the fermentative activity in the rumen, the place of more or less complete microbial fermentation of dietary components. Ruminal fermentation initially results in the degradation of carbohydrates and protein to short-term intermediates such as sugars and amino acids. The products of this initial degradation are readily metabolized to microbial mass and carbon dioxide, methane, ammonia and volatile fatty acids (VFA): primarily acetate, propionate and butyrate and to a lesser degree branched chain VFA and occasionally lactate. The rate and extent of fermentation are important parameters that determine protein, vitamins, and short-chain organic acids supply to the animal (Koenig et al., 2003; Hall, 2003). The host ruminant animal absorbs VFA (mostly through the rumen wall) and digests proteins, lipids, and carbohydrate constituents of microbes and feed residues entering the small intestine to supply its maintenance needs and for the production of meat and milk. Ruminant animals derive about 70% of their metabolic energy from microbial fermentation of feed particles and microbial protein accounts for as much as 90% of the amino acids reaching the small intestine (Nocek and Russell, 1988; Bergman, 1990).

The rumen has a complex environment composed of microbes, feed at various stages of digestion, gases, and rumen fluid. Rumen microorganisms usually adhere to feed particles and form biofilms to degrades plant material. The efficiency of ruminants to utilize of feeds is due to highly diversified rumen microbial ecosystem consisting of bacteria (10^{10}–10^{11} cells/ml, more than 50 genera), ciliate protozoa (10^4–10^6/ml, 25 genera), anaerobic fungi (10^3–10^5 zoospores/ml, 5 genera) and bacteriophages (10^8–10^9/ml) (Hobson, 1989). The synergism and antagonism among the different groups of microbes is so diverse and complicated that it is difficult to quantify the role played by any particular group of microbes present in the rumen (Kamra, 2005). Bacterial numbers in the rumen are the highest and bacteria play a dominant role in all facets of ruminal fermentation. They are adopted to live at acidities between pH 5.5 and 7.0, in the absence of oxygen, at the temperature of 39-40°C, in the presence of moderate concentration of fermentation products, and at the expense of the ingesta provided by ruminant (Hungate, 1966). Rumen digesta volume accounts for 8-14% of body weight of cows and is characterized by dry matter content about 15% (Dado and Allen, 1995; Reynolds et al., 2004; Kamra, 2005).

4. Techniques for estimating rumen digestibility

The rumen digestibility of feeds can be estimated by biological methods. The "basic model" which gives the value utilized for defining the nutritive value of a feed is the *in vivo* digestibility, which represents the entire process occurring in the gastro-intestinal tract. *In vitro* methods which simulate the digestion process, have being less expensive and less time-consuming, and they allow to maintain experimental conditions more precisely than do *in vivo* trials. Three major *in vitro* digestion techniques currently available to determine the nutritive value of ruminant feeds are: digestion with rumen microorganisms (Tilley and Terry, 1963; Menke et al., 1979), digestion with enzymes (De Boever et al., 1986), and *in situ* the nylon bag technique (Mehrez and Ørskov, 1977). The nylon bag technique *(in sacco)* has been used for many years to provide estimates of both the rate and the extent of disappearance of feed constituents. Those characteristics are measured by placing feedstuffs in fabric bag and then incubating the bag by certain time intervals in the rumen of animal. However, the single technique does not provide accurate estimation of *in vivo* digestion. Judkins et al. (1990) compared 11 techniques for estimating diet dry matter digestibility across six different diets in experiment with rams. Authors found, that the rumen digestibility of feeds nutrients was influenced by diets composition, feeding conditions and physiological status of animals. It therefore seems appropriate that the developments and use of various modification of mentioned experimental techniques have enabled much progress in rumen studies.

5. Carbohydrates classification in ruminants feeds

Carbohydrates constitute the highest proportion of diets and are important for meeting the energy needs of animals and of rumen microbes, and are important for maintaining the health of the gastrointestinal tract. Typically, carbohydrates make up 70 to 80% of the diets fed to dairy cattle and are composed of mixture of numerous monomers and polymers (Nocek and Russell, 1988). The carbohydrates fraction of feeds are defined according the chemical or enzymatic methods used for their analysis and availability to the ruminants. Broadly, carbohydrates are classified as nonstructural that are found inside the cells of plants or structural that are found in plant cell walls, but these fractions are not chemically uniform (Van Soest et al., 1991).

Fraction of nonstructural carbohydrates (NSC) includes organic acid, mono- di- and oligosaccharides, starches, and other reserve carbohydrates. Total NSC includes pectin is referred as nonfibrous carbohydrates (NFC), calculated as 100–(CP+ether extract+ash+NDF) (Mertens, 1992). NFC are the highly digestible and are the major source of energy for high producing cattle. Fraction of structural carbohydrates is characterized by neutral detergent fiber (NDF) and acid detergent fiber (ADF) contents. NDF includes the crosslinked matrix of the plant cell wall with cellulose, hemicellulose, and lignin as the major components and ADF does not include hemicelluloses (Van Soest, 1963). NDF, ADF, and cellulose content are measured according to methods described by Van Soest et al. (1991). The content of hemicellulose was calculated as NDF – ADF (Mertens, 1992).

Fractions of carbohydrates described above are subdivided by chemical composition, physical characteristics, ruminal degradation, and postruminal digestibility characteristics, because of these factors, various modifications of the analytical methods have been proposed (Hall et al., 1999; Nie et al., 2009).

6. Degradation and utilization of carbohydrates by rumen microbial ecosystem

Dietary carbohydrates are the main rumen microbial fermentation substrates. Microbial yields are related primarily to the growth rate that carbohydrate permits. The individual carbohydrates characterized by faster rumen degradation rates result in greater microbial yield (Hall and Herejk, 2001). The enzyme systems produced by microorganisms for carbohydrates hydrolysis are complex; they usually comprise hydrolases from several families, and there may be multiple enzymes hydrolysing each polysaccharide. Nearly all carbohydrate digestion occurs (>90%) within the rumen, but under certain circumstances (e.g., high rate of passage), a significant amount of carbohydrate digestion can occur in the small and large intestine.

7. Nonfibrous carbohydrates

Nonfiber carbohydrates may provide 30 to 45% of the diet on a dry matter basis (Hall et al., 2010). The NFC fraction is considered a source of readily available energy for microbial growth (Ariza et al., 2001).

7.1. Mono- di- and oligosaccharides

The concentration of monosaccharides, glucose and fructose was estimated from 1% to 3% (in grasses and herbage) and of sucrose from 2% to 8% (Smith, 1973). Sucrose formed from α-D-glucose and β-d-fructose linked by 1, 2 glycosidic linkage is digested by enzyme sucrose phosphorylase (EC 2.4.1.7, according to the IUB-MB enzyme nomenclature; Stan-Glasek et al., 2010). Maltose formed from two units of glucose joined with an α(1–4) bond is digested by enzyme α–glucosidase (EC 3.2.1.20). Oligosaccharides concentration in the different plants ranges between 0.3% and 6% and represent a wide diversity of biomolecules (including stachyose and raffinose), they are chains of monosaccharides that are two to approximately 20 units long. The enzymes belonging to the group of polysaccharide hydrolases (EC 3.2.1.-) which hydrolyse the glycosidic bond between two or more carbohydrates utilize oligosaccharides (Courtois, 2009).

Mono-and disaccharidesare rapidly fermented within the rumen to yield VFA. The rate of glucose fermentation after glucose dosing varied from 422 to 738% h^{-1} and the rate of fermentation of monosaccharides originating from disaccharide hydrolysis was 300 to 700% h^{-1}(Wejsberg et al., 1998).

Ruminal bacteria that ferment sucrose include *Streptococcus bovis, Lachnospira multiparus, Lactobacillus ruminis, Lactobacillus vitulinis, Clostridium longisporum, Eubacterium cellulosolvens,* and some strains of *Eubacterium ruminantium, Butyrivibrio fibrisolvens, Ruminococcus albus, Ruminococcus flavefaciens, Megaspaera elsdenii, Prevotella spp., Selenomonas ruminantium, Pseudobutyrivibrio ruminis* strain A and *Succinivibrio dextrinosolvens* (Stewart et al., 1997, Martin and Russell, 1987, Stan-Glasek et al., 2010). Maltose utilize *Ruminobacter amylophilus*) and oligosaccharides *Actinomyces ruminicola* as a sources of energy (Anderson, 1995; An et al., 2006).

7.2. Pectic substances

Pectic substances are a group of galacturonan polymers with neutral sugars (largely arabinose and galactose) substitutions (Jung, 1997). Pectic substances are found in the middle lamella and other cell wall layers (Van Soest, 1994). The most important pectinolytic activity represents pectin lyase (EC 4.2.2.10) (Wojciechowicz, 1982).

Grasses contain from 3 to 4% of pectin in the dry matter, leguminous plants from 5 to 12%, and sugar beet pulp 25% (Aspinall, 1970; Van Soest, 1983; Cassida et al., 2007). The utilization of pectin varied from 79.4 to 95.9% (Marounek and Duškovă, 1999). Most of pectin degrades at a rate of 13% h^{-1} (Hall et al., 1998).

Pectin-utilizing bacteria include *Butyrivibrio fibrisolvens* and *Prevotella ssp.*(the principal rumen pectin-utilizing bacteria) and *Fibrobacter succinogenes, S. bovis* and *Lachnospira multiparus* (Czerkawski and Breckenridge, 1969; Gradel and Dehority, 1972; Baldwin and Allison, 1983). Pectin is reported to ferment primarily to acetate (Czerkawski and Breckenridge, 1969; Marounek and Duškovă, 1999).

7.3. Starch

Starch is a complex of two structurally distinct polymers: amylose and amylopectin (Chesson, 1997). Amylose is chemically composed of α-*1,4*- linked polymers of glucose. It is degraded by α-amylases (EC 3.2.1.1), which releases oligosaccharides maltodextrins, and β-amylases (EC 3.2.1.2), which remove maltose units. Amylopectin is a highly branched molecule of α-*1,4*- linked polymers of glucose joined *1,6* at intervals along the backbone moleculr. Amylopectin is degraded to maltose by β-amylases action (in 50%), glucanohydrolases (EC 3.2.1.3 and EC 3.2.1.41) and isoamylase (EC 3.2.1.68). Maltose and maltodextrins are degraded to glucose by α-glucosidase (EC 3.2.1.20) (Hobson, 1989). Starch can be degraded by ruminal microbial enzymes as well as enzymes in the small intestine of the ruminats.

Of the nonfibrous fraction, starches are the highest proportion in the diet, and cereal grains is the major source of starch in ruminants diet. The cereal grains differ in their starch content, with wheat containing (on dry matter basis) 77% starch, corn 72%, and barley and oats 57 to 58% (Huntington, 1997). Differences exist among cereal grains in their extents and rates of ruminal starch degradation. In the rumen has been digested from 55 to 70% of corn starch, 80 to 90% of barley and wheat starch, and 92 to 94% of oats starch (Huntington, 1997). The degradation rates is estimated from 4.0 to 6.4% h^{-1} for corn starch and from 14.7 to 24.5% h^{-1} for barley starch (Herrera-Saldana et al., 1990; Tamminga et al., 1990). On average, 5 to 20% of starch consumed is digested postruminally, mainly in the small intestine (from 45 to 85% of starch entering the duodenum), this capacity is limited by the supply of pancreatic amylase (Hunhington, 1997).

The bacteria *Ruminobacter amylophilus, Prevotela ruminicola, Streptococcus bovis, Succinimonas amylolytica* and many strains of *Selenomonas ruminantium, Butyrivibrio fibrisolvens, Eubacterium ruminantium* and *Clostridium* ssp., all of the entodiniomorph protozoa and the chytrid fungi are amylolitic (Chesson, 1997). The high-starch concentrate diets favor the development of propionate producing bacteria species (Ørskov, 1986; France and Dijkstra, 2005). The fermentation of starch in the rumen depending on factors such as structure (amylose/amylopectin ratio), plant source, mechanical alterations (grain processing, chewing), diet composition, amount of feed consumed per unit time and degree of adaptation of ruminal microbiota to the diet (Piva and Masoero, 1996, Huntington, 1997; Eastridge, 2006).

8. Structural carbohydrates

Structural carbohydrates is less digestible than NFC and is negatively correlated with energy concentration in the diet for ruminants, but is important for rumination, saliva flow, ruminal buffering, and health of the rumen wall, however, high dietary concentrations can limit dry matter intake by increased rumen fill. The retention time of plant fiber in the rumen is sufficiently long (48 h or more in some species) to allow the growth of a fibrolytic microbial population whose extensive fiber utilization contributes a major portion of the energy for the animal (Van Soest, 1994). The cellulose fibers are embedded in a matrix of other structural biopolymers, primarily hemicelluloses and lignin (Lynd, 1999; Marchessault and Sundararajan, 1993; Van Soest, 1994). The high-fibre forage diets encourage the grown of acetate producing bacterial species, the acetate : propionate : butyrate molar proportion would typically be in region 70:20:10 (France and Dijkstra, 2005). Fiber digestion may be reduced due to decreased rumen pH (the fiber digesters are most active at a pH of 6.2 to 6.8), and the availability of surface area for colonization (Sutton et al., 1987; Chesson and Forsberg, 1997). Fungi plays an active and positive role in fiber degradation (Williams and Orpin, 1987).

8.1. Cellulose

Cellulose content is in the range from 35 to 50% of plant dry weight (Lynd et al., 1999). It is chemically composed of a homogenous polymers of β-1,4-D-glucose linked through β-1,4-glycosidic bonds. Native cellulose exists as fibrils which are composed of amorphous and crystalline regions formed from cellulose chains, each of them contains between 500 and 14,000 b-1,4D-glucose units (Bazooyar et al., 1012). The digestion of cellulose necessitates a combination of many classes of cellulases. The digestion process including activity of endoglucanases (EC 3.2.1.4), that cut randomly at internal amorphous sites in the cellulose chain; exoglucanases (cellodextrinases EC 3.2.1.74 and cellobiohydrolases EC 3.2.1.91), that act processively on the reducing or non-reducing ends of cellulose chains, releasing either cellobiose or glucose as major products; and glucosidases (EC 3.2.1.21) that hydrolyze soluble cellodextrins and cellobiose to glucose (Lynd et al., 2002).

Fibrobacter succinogenes, Ruminococcus flavefaciens and *R. albus* are considered to be the predominant cellulolytic bacteria present in the rumens, these species gain selective advantage in the rumen is by optimizing two catabolic activities: cellulose hydrolysis (depolymerization) and efficient utilization of the hydrolytic products (cellodextrins) (Weimer, 1996; Koike and Kobayashi, 2001). *F. succinogenes* has a potent ability to solubilize crystalline chains of cellulose (Halliwell and Bryant, 1963; Shinkai and Kobayashi, 2007). *F. succinogenes* produces primarily succinate (a propionate precursor), and lesser amounts of acetate, *R. flavefaciens* produces primarily acetate and lesser amounts of succinate converts to propionate by *Selenomonas ruminantium* (Weimer at al., 1999).

The predominant ruminal cellulolytic species digest cellulose at rate approximately from 5 to 10% h^{-1}, however, the extent to which native cellulose is utilizes by ruminal microorganisms is limiting by the cellulose association with lignin (Weimer, 1996; Chesson, 1993). Cellulose degradability of forages varies from 25 to 90% (Pigden and Heaney, 1969).

8.2. Hemicelluloses

Hemicellulose concentration varies from 6 to 22% (on dry matter basis) in leaves of grasses and herbs (Schädel at al., 2010). Hemicelluloses are composed of complex heteropolymers that vary considerably in primary composition, substitution and degree of branching, and can be grouped into four classes: xylans, xyloglucans, mannans and mixedlinkage β-glucans (Ebringerova et al., 2005).

Bacteroides (Fibrobacter) succinogenes, Ruminococcus albus, and *Ruminococcus flavefaciens* and same strains of *Butyrivibrio fibrisolvens* and *Bacteroides ruminicola* are considered to be the organisms responsible for most of the degradation of hemicelluloses (Hespell, 1988). Rumen degradation of hemicelluloses varies from 16 to 90%, depending on their composition (Pigden and Heaney, 1969; Coen and Dehority, 1970).

8.3. Lignin

Lignin, a complex phenolic polymer, is indigestible by rumen microbes, but their concentration limits digestibility of structural carbohydrates (Van Soest, 1994). The main reason for reduction of accessibility for the hydrolases secreted by ruminal microbes is the presence of strong covalent bonds between lignin and the cell wall polysaccharides (Chesson, 1993).

9. Interactions between energy and protein metabolism in rumen

Manipulation of rumen fermentation through proper diet formulation changes microbial population in a way that improved efficiency of microbial protein synthesis. A major factor in maximizing microbial protein synthesis is the ruminally available energy and N in the diet. There are many interactions of dietary conditions on bacterial populations and on protein and carbohydrates digestion in rumen. For instance: microbial fermentation releases organic acids that readily dissociate to decrease pH that influence on the microbial ecosystem and determining the selective growth of certain microbial species, and the types and quantities of fermentation products (Russell and Rychlik, 2001).

10. "Synchrony" hypothesis

The purpose of proper nutrition is "nutritional synchrony" refers to provision of dietary protein (N sources, true protein) and energy (ruminally fermented carbohydrates) to the rumen in such a manner that they are available simultaneously in proportions needed by the ruminal microorganisms (Hall and Weimer, 2007). Synchronous nutrient availability should allow more efficient use of nutrients, thus enhancing production of microbial products, increasing nutrient supply to the animal, and potentially improving animal production performance (Sinclair et al., 1993; Hall and Huntington, 2008).

10.1. Production efficiency

A number of studies have been conducted to evaluate the effects of "nutritional synchrony" conception on production efficiency, but the results are not consistently. There are many results which confirm increase in the yield of microbial protein when highly degradable carbohydrates were synchronized with rapidly degraded protein (Kovler et al., 1998; Charbonneau et al., 2006). Result from the *in sacco* study confirms that the better synchronization also affects degradation rate of diet components (Niwinska, 2009; Niwińska and Andrzejewski, 2011). However, nutrient synchrony has generally not resulted in improved animal performance (Yang et al., 2010). The fundamental reason is the following: ruminants animal have the ability to recycle N from blood and saliva to the gastrointestinal tract during periods of dietary protein deficiency or during periods of asynchronous carbohydrate and protein supply. Hall and Huntington (2008) suggested that for the optimal use of "nutritional synchrony" conception we may need to look at the whole animal, not just the rumen, and that a term such as the optimal balance may be more appropriate when considering the complexity of the ruminant animal.

10.2. Product composition

Volatile fatty acids, produced in the rumen, can have a major effect on fat composition of ruminant products. Results of studies with animal models, in tissue culture systems and in clinical research indicate that the functional health-related properties of milk and beef fat appear to be linked to the presence of rumenic acid and vaccenic acid (Parodi, 2005; Lee, 2008; Field et al., 2009). Milk fat provides 30% of fat consumed by humans and is the richest natural dietary source of those valuable fatty acids (Ritzenthaler et al., 2001). Research undertaken over the past decade has indicated, that concentration of those favorable fatty acids in milk fat may be controlled by the starch/fibre ratio in the diet of dairy cows (Niwińska et al., 2011).

10.3. Pollution emissions

An excessive supply of feed nutrients results in an increase in waste excreted to the environment. The pollutants produced by ruminants are nitrogen and methane, their production is dependent on carbohydrate composition of diet. Improved efficiency of microbial protein synthesis is considered as the most important target in reduction emissions of N, while synchronization of carbohydrate and protein supply in the rumen has been suggested as one possible solution to achieve this aim (Kaswari et al., 2007; Reynolds and Kristensen, 2008; Yang et al., 2010). The rumen microbial ecosystem produces methane as a result of anaerobic fermentation. Methane production results in losses of 5 to 12% of gross energy of diet and is estimated to be about 15% of total atmospheric methane emissions (Reid et al., 1980; Moss et al., 2000). The proper selection of carbohydrates in the ration, taking into account the structural and nonfibrous carbohydrates content, can reduce the formation of carbon dioxide, hydrogen and formate the major precursors of methane production in the rumen (Mitsumori and Sun, 2008).

11. Conclusion

The current data indicated, that the world production of dairy products is expected to grow 26% by 2020 and beef production worldwide grew by 30 million tons during 1965-2005 (OECD, 2011; FAO; 2006). The projected increase in cattle production directs our attention to the better utilization of feed resources. Understanding the effects of carbohydrates types and rumen microbial population shifts in response to nutrients contained in different feeds may be valuable to improve production efficiency, to modify the composition of the product and to minimize pollution emissions.

Author details

Barbara Niwińska
Department of Animal Nutrition and Feed Science,
National Research Institute of Animal Production, Kraków, Poland

12. References

An D., Cai S., Dong X. 2006. Actinomyces ruminicola sp. nov., isolated from cattle rumen. Int. J. Sys. Evol. Microbiol. 56:2043–2048.

Anderson K.L. 1995. Biochemical analysis of starch degradation by *Ruminobacter amylophilus* 70. Appl. Environ. Microbiol., 61:1488–1491.

Ariza P., Bach A., Stern M.D., Hall M.B. 2001. Effects of carbohydrates from citrus pulp and hominy feed on microbial fermentation in continuous culture. J. Anim. Sci. 79:2713-2718.

Aspinall G.O. 1970. Polysaccharides. Oxford: Pergamon Press.

Baldwin R.L., Allison M.J. 1983. Rumen metabolism. J. Anim. Sci. 57 (Suppl. 2):461-477.

Bazooyar G., Momany F.A., Bolton K. 2012. Validating empirical force fields for molecular-level simulation of cellulose dissolution. Comp. Theor. Chem. 984:119–127.

Bergman E.N. 1990. Energy contribution of VFA from the gastrointestinal tract in various species. Physiol. Rev. 70:567–590.

Cassida K.A., Turner K.E., Foster J.G., Hesterman O.B. 2007. Comparison of detergent fiber analysis methods for forages high in pectin. Anim. Feed Sci. Tech. 135:283–295.

Charbonneau E., Chouinard P.Y., Allard G., Lapierre H., Pellerin D. 2006. Milk from forage as affected by carbohydrate source and degradability with alfalfa silage based diets. J. Dairy Sci. 89:283–293.

Chesson A. 1993. Mechanistic model of forage cell wall degradation. In: Rorage cell wall structure and digestibility. Ed: H.G.Jung, D.R.Buxton, R.D. Hatfield, J. Ralph. ASA-CSSA-SSSA, Madison, WI, pp. 347-376.

Chesson A., Forsberg C.W. 1997. Polysaccharide degradation by rumen microorganism. In: Hobson PN, Stewart CS (eds). The rumen microbial ecosystem. Blackie, London. 329-381.

Coen JA. Dehority BA. 1970. Degradation and utilisationof hemicelluloses from intact forage by pure culture of rumen bacteria. Appl. Microbiol. 20:362-368.

Courtois J. 2009. Oligosaccharides from land plants and algae: production and applications in therapeutics and biotechnology. Curr. Opinion Microbiol. 12:261–273.

Czerkawski J.W., Breckenridge G. 1969. Fermentation of various soluble carbohydrates by rumen micro-organisms with particular reference to methane production. Br. J. Nutr. 23:925-937.

Dado R.G., Allen M.S. 1995. Intake limitations, feeding behavior, and rumen function of cows challenged with rumen fill from dietary fiber or inert bulk. J. Dairy Sci. 78:118–133.

De Boever J.L., Cottyn B.G., Buysse F.X., Wainman F.W., Vanacker J.M. 1986. The use of an enzymatic technique to predict digestibility, metabolisable and net energy of compound feedstuffs for ruminants. Anim. Feed Sci. Technol. 14:203–214.

Eastridge M.L. 2006. Major advances in applied dairy cattle nutrition. J. Dairy Sci. 89:1311–1323.

Ebringerova A., Hromadkova Z., Heinze T. 2005. Hemicellulose. Polysaccharides 1: Structure, Characterization and Use. Springer-Verlag, Berlin, pp 1–67.

FAIP. 2003. The economic value of livestock production in the EU 2003. Farm Animal Industrial Platform. AnNe Publishers, ISBN 90-76642-19-2.

FAO 2006. *Food and Agriculture Organization.* Livestock's Long Shadow. Environmental Issues and Options.

Field C.J., Blewett H.H., Proctor S., Vine D. 2009. Human health benefits of vaccenic acid. Appl. Physiol. Nutr. Metab. 34:979–991.

France J., Dijkstra J. 2005. Volatile fatty acid production. In: J. Dijkstra, J. Forbes, J.M. France (Ed) Quantitative aspects of ruminant digestion and metabolism. 2nd Ed. CAB International, Wallingford, UK, pp.

Gradel C.M., Dehority B.A. 1972. Fermentation of isolated pectin and pectin from intact forages by pure cultures of rumen bacteria. Appl. Microbiol. 23:332-340.

Hackmann T.J., Spain J.N. 2010. *Invited review:* Ruminant ecology and evolution: Perspectives useful to ruminant livestock research and production. J. Dairy Sci. 93:1320–1334.

Hall M.B. 2003. Challenges with nonfiber carbohydrate methods. J. Anim. Sci. 81:3226–3232.

Hall M.B., Herejk C. 2001. Differences in yields of microbial crude protein from in vitro fermentation of carbohydrates. J. Dairy Sci. 84:2486-2493.

Hall M.B., Hoover W.H., Jennings J.P., Webster T.K.M. 1999. A method for partitioning neutral detergent-soluble carbohydrates. J. Sci. Food Agric. 79:2079–2086.

Hall M.B., Huntington G.B. 2008. Nutrient synchrony: Sound in theory, elusive in practice. J. Anim. Sci. 86(E. Suppl.):E287–E292.

Hall M.B., Larson C.C., Wilcox C.J. 2010. Carbohydrate source and protein degradability alter lactation, ruminal, and blood measures. J. Dairy Sci. 93:311–322.

Hall M.B., Pell A.N., Chase L.E. 1998. Characteristics of neutral detergent soluble fiber fermentation by mixed ruminal microbes. Anim. Feed Sci. Technol. 70:23–29.

Hall M.B., Weimer P.J. 2007 Sucrose concentration alters fermentation kinetics, products, and carbon fates during in vitro fermentation with mixed ruminal microbes. J. Animal Sci. 8:1467-1478.

Halliwell G., Bryant M.P. 1963. The cellulolytic activity of pure strains of bacteria from the rumen of cattle. J. Gen. Microbiol. 32:441–448.

Herrera-Saldana R.E., Huber J.T., Poore M.H. 1990. Dry matter, crude protein, and starch degradability of five cereal grains. J. Dairy Sci. 73:2386–2393.

Hespell R.B. 1988. Microbial digestion of hemicelluloses in the rumen. Microbiol. Sci. 5:362-365.

Hobson P.N. 1989. The rumen microbial eco-system. Elsevier Applied Science, London, 1989.

Hungate R.E. 1966. The rumen and its microbes. Academic Press Inc. (London) LTD.

Huntington G.B. 1997. Starch utilization by ruminants: From basics to the bunk. J. Anim. Sci. 75:852–867.

IUB-M. Enzyme nomenclature database. International Union of Biochemistry and Molecular Biology, http://enzyme.expasy.org/

Judkins M.B., Krysl L.J., Barton R.K. 1990. Estimating diet digestibility: a comparison of 11 techniques across six different diets fed to rams. J. Anim. Sci. 68:1405-1415.

Jung H.J.G. 1997. Analysis of forage fiber and cell walls in ruminant nutrition. J. Nutr. 127: 810S–813S.

Kamra D.N. 2005. Rumen microbial ekosystem. Curr. Sci. 89:124-135.

Kaswari T., Lebzien P., Flachowsky G., ter Meulen U. 2007. Studies on the relationship between the synchronization index and the microbial protein synthesis in the rumen of dairy cows. Anim. Feed Sci. Technol. 139:1-22.

Koenig K.M., Beauchemin K.A., Rode L.M. 2003. Effect of grain processing and silage on microbial protein synthesis and nutrient digestibility in beef cattle fed barley-based diets. J. Anim. Sci. 81:1057–1067.

Koike S., Kobayashi Y. 2001. Development and use of competitive PCR assays for the rumen cellulolytic bacteria: *Fibrobacter succinogenes, Ruminococcus albus* and *Ruminococcus flavefaciens*. FEMS Microbiol. Lett. 204:361–366.

Kolver E., Muller L.D., Varga G.A., Cassidy T.J. 1998. Synchronization of ruminal degradation of supplemental carbohydrate with pasture nitrogen in lactating dairy cows. J. Dairy Sci. 81:2017–2028.

Lee Y. 2008. Isomer specificity of conjugated linoleic acid (CLA): 9*E*,11*E*-CLA. Nutr. Res. Pract. 2:326–330.

Lynd L.R., Weimer P.J., van Zyl W.H., Pretorius I.S. 2002. Microbial cellulose utilization: fundamentals and biotechnology. Microbiol. Mol. Biol. Rev. 66:506–577.

Lynd L.R., Wyman C.E., Gerngross T.U. 1999. Biocommodity engineering. Biotechnol. Prog. 15:777–793.

Marchessault R.H., Sundararajan P.R. 1993. Cellulose, p. 11–95 *In* G. O. Aspinall (ed.), The polysaccharides, vol. 2. Academic Press, Inc., NewYork, N.Y.

Marounek M., Dušková D. 1999. Metabolism of pectin in rumen bacteria *Butyrivibrio fibrisolvens* and *Prevotella ruminicola.* Lett. Appl. Microbiol. 29:429–433.

Martin S.A., Russel J.B. 1987. Transport and phosphorylation of disaccharides by the ruminal bacterium *Streptococcus bovis*. Appl. Environ. Microbiol. 53:2388–2393.

Mehrez A.Z., Ørskov E.R. 1977. A study of artificial fibre bag technique for determining the digestibility of feeds in the rumen. J. Agric. Sci. 88:645–650.

Menke K.H., Raab L., Salewski A., Steingass H., Fritz D., Schneider W. 1979. The estimation of the digestibility and metabolisable energy content of ruminant feedingstuffs from the gas production when they are incubated with rumen liquor. J. Agric. Sci. 93:217–222.

Mertens D.R. 1992. Nonstructural and structural carbohydrates. In: Van Horn H.H., Wilcox, C.J. (Eds.), Am. Dairy Sci. Assoc., Champaign, IL, USA. p. 219.

Mitsumori M., Sun W. 2008. Control of Rumen Microbial Fermentation for mitigating Methane Emissions from the Rumen. Asian-Aust. J. Anim. Sci. 21:144-154.

Moss A.R., Jouany J.P.,, Newbold J. 2000. Methane production by ruminants: its contribution to global warming. Ann. Zootech. 49:231–253.

Nie Z., Tremblay G. F., Bélanger G., Berthiaume R., Castonguay Y., Bertrand A., Michaud R., Allard G., Han J. 2009. Near-infrared reflectance spectroscopy prediction of neutral detergent-soluble carbohydrates in timothy and alfalfa. J. Dairy Sci. 92:1702–1711.

Niwińska B. 2009: Effect of carbohydrates in grass silage- based diets on *in sacco* ruminal degradability of barley grain (*Hordeum vulgare* L. cv. Lomerit) ground to different particle sizes. Czech J. Anim. Sci. 54: 260–269.

Niwińska B., Andrzejewski M. 2011. Effect of carbohydrates in grass silage-based diets on *in sacco* ruminal degradability and nutritional value of lupin seeds (*Lupinus angustifolius* L *.cv.* Sonet) ground to different particle sizes. Czech J. Anim. Sci. 56: 231–241.

Niwińska B., Bilik K., Andrzejewski M. 2011.Factors influencing rumenic acid and vaccenic acid content in cow's milk fat. Ann. Anim. Sci. 11:3–16.

Nocek J.E., Russell J.B. 1988. Protein and energy as an integrated system: Relationship of ruminal protein and carbohydrate availability to microbial synthesis and milk production. J. Dairy Sci. 71:2070–2107.

Nowak R.M. 1999. Walker's Mammals of the World. Vol 2. 6th ed. John Hopkins Univ. Press, Baltimore, MD.

OECD-FAO 2011. Organisation for Economic Co-operation and Development-*Food and Agriculture Organization.* Agricultural Outlook 2011-2020; *http://dx.doi.org/10.1787/888932428329.*

Ørskov E.R. 1986. Starch digestion and utilization in ruminants. J. Anim. Sci. 63:1624-1633.

Parodi P.W. 2005. Dairy product consumption and the risk of breast cancer. J. Am. Coll. Nutr. 24:S556-S568.

Pigden W.J., Heaney D.P. 1969. Lignocellulose in ruminat nutrition. Adv. Chem. 95:245-261.

Piva G., Masoero F. 1996. Carbohydrates with different ruminal fermentability: starch and fibrous fractions. Zoot. Nutr. Anim. 22:215-229.

Reid J.T., White O.D., Anrique R., Forstin A. 1980. Nutritional energetic of livestock: Some present boundaries of knowledge and future research needs. J. Animal Sci. 51:1393-1415.

Reynolds C.K., Dürst B., Lupoli B., Humphries D.J., Beever D.E. 2004. Visceral tissue mass and rumen volume in dairy cows during the transition from late gestation to early lactation. J. Dairy Sci. 87:961–971.

Reynolds C.K., Kristensen N.B. 2008. Nitrogen recycling through the gut and the nitrogen economy of ruminants: An asynchronous symbiosis. J. Animal Sci. 86(Suppl.):E293–E305.

Ritzenthaler K.L., McGuire M,K., Falen R., Shultz T.D., Dasgupta N., McGuire M.A. 2001. Estimation of conjugated linoleic acid intake by written dietary assessment methodologies underestimates actual intake evaluated by food duplicate methodology. J. Nutr. 131:1548–1554.

Russell J.B., Rychlik J.L. 2001. Factors that alter rumen microbial ecology. Science 292:1119-1222.

Schädel C., Richter A., Blöchl A., Hoch G. 2010. Hemicellulose concentration and composition in plant cell walls under extreme carbon source–sink imbalances. Physiol. Plantarum 139:241–255.

Shinkai T., Kobayashi Y. 2007. Localization of ruminal cellulolytic bacteria on plant fibrous materials as determined by fluorescence in situ hybridization and Real-Time PCR. Appl. Environ. Microbiol. 73:1646 –1652.

Sinclair L.A., Garnsworthy P.C., Newbold J.R., Buttery P.J. 1993. Effect of synchronizing the rate of dietary energy and nitrogen release on rumen fermentation and microbial protein synthesis in sheep. J. Agric. Sci. 120:251–263.

Smith D. 1973. The nonstructural carbohydrates. Chemistry and Biochemistry of Herbage. Ed: G.W. Butler and R. W. Bailey, Academic Press Inc. (London) LTD.

Stan-Glasek K., Kasperowicz A., Guczyńska W., Piknováb M., Pristaš P., Nigutová K., Javorský P., Michałowski T. 2010. Phosphorolytic Cleavage of Sucrose by sucrose-grown ruminal bacterium *Pseudobutyrivibrio ruminis* strain k3. Folia Microbiol. 55:383–385.

Stewart C.S., Flint H.J., Bryant M.P. 1997. The rumen bacteria. The Rumen Microbial Ecosystem. P. N. Hobson and C. S. Stewart, eds. Chapman & Hall, London, UK.

Sutton J.D., Bines J.A., Morant S.V., Napper D.J., Givens D.I. 1987. A comparison of starchy and fibrous concentrates for milk production, energy utilization and hay intake by Friesian cows. J. Agric. Sci. (Camb.) 109:375–386.

Tamminga S., Van Vuuren A.M., Van Der Koelen C.J., Ketelaar R.S., Van Der Togt P.L. 1990. Ruminal behavior of structural carbohydrates, non-structural carbohydrates and crude protein from concentrate ingredients in dairy cows. Neth. J. Agric. Sci. 38:513–526.

Tilley J.M.A., Terry R.A. 1963. A two-stage technique for the digestion of forage crops. J. Br. Grassl. Soc. 18:104–111.

Van Soest P.J. 1963. Use of detergents in the analysis of fibrous feeds. I. Preparation of fiber residues of low nitrogen content. J. Assoc. Off. Anal. Chem. 46:825–829.

Van Soest P.J. 1983. Nutritional Ecology of the Ruminant. 2nd edn.Orvallis, OR: O & B Books.

Van Soest P.J. 1994. Nutritional ecology of the ruminant. 2nd. ed. Cornell University Press.

Van Soest P.J., Robertson J.B., Lewis B.A. 1991. Methods for dietary fiber, neutral detergent fiber, and nonstarch polysaccharides in relation to animal nutrition. J. Dairy Sci. 74:3583-3597.

Weimer P.J. 1996. Why don't ruminal bacteria digest cellulose faster? J. Dairy Sci. 79:1496-1502.

Weimer P.J., Waghorn G.C., Odt C.L., Mertens D.R. 1999. Effect of diet on populations of three species of ruminal cellulolytic bacteria in lactating dairy cows. J. Dairy Sci. 82:122–134.

Wejsberg M.R., Hvelplund T., Bibby B.M. 1998. Hydrolysis and fermentation rate of glucose, sucrose and lactose in the rumen. Acta Agric. Scand. A. 48:12-18.

Williams A.G. Orpin C.G. 1987. Polysaccharide degrading enzymes formed by three species of anaerobic fungi grown on a range of carbohydrate substrates. Can. J. Bot. 33:418–426.

Wojciechowicz M., Heinrichova K., Ziolecki A. 1982. An exopectate lyase of *Butyrivibrio fibrisolvens* from the bovine rumen. J. Gen. Microbiol. 128:2661–2665.

Yang J.Y., Seo J., Kim H.J., Seo S., Ha J.K. 2010. Nutrient Synchrony: Is it a suitable strategy to improve nitrogen utilization and animal performance? Asian-Aust. J. Anim. Sci. 23:972-979.

Section 3

Carbohydeate Metabolism

Chapter 12

Resistant Dextrins as Prebiotic

Katarzyna Śliżewska, Janusz Kapuśniak, Renata Barczyńska and Kamila Jochym

Additional information is available at the end of the chapter

http://dx.doi.org/10.5772/51573

1. Introduction

In current days, the way each of human beings live indicates his or her health in the future time. Many factors determine the risk of illnesses, or reversibly, the possibility of being healthy. Being physically active and consumption of appropriate diet are examples of daily routines that may influence the condition of an organism. Lack of physical activity, particularly if associated with over consumption, increases the risk of development of nutrition related chronic diseases, such as obesity, hypertension, cardiovascular diseases, osteoporosis, type II diabetes, and several cancers. Over the last decade, drastic changes have taken place in the image and assessment of the importance of the daily diet. Foods are no longer judged in terms of taste and immediate nutritional needs, but also in terms of their ability to improve the health and well-being of consumers. The role of diet in human health has led to the recent development of the so-called functional food concept. A functional food is dietary ingredient, that has cellular or physiological effects above the normal nutritional value. Functional food can contain probiotics and/or prebiotics.

2. The prebiotic concept

A number of different strategies can be applied to modify microbial intestinal populations. Antibiotics can be effective in eliminating pathogenic organisms within the intestinal microbiota. However, they carry the risk of side effects and cannot be routinely used for longer periods or prophylactically [17, 33].

The consumption of probiotics aims to directly supplement the intestinal microbiota with live beneficial organisms. Lactobacilli and bifidobacteria are numerically common members of the human intestinal microbiota, and are nonpathogenic, nonputrefactive, nontoxigenic, saccharolytic organisms that appear from available knowledge to provide little opportunity for deleterious activity in the intestinal tract. As such, they are reasonable candidates to target in terms of restoring a favorable balance of intestinal species [18, 85].

Prebiotics represent a third strategy to manipulate the intestinal microbiota. Rather than supplying an exogenous source of live bacteria, prebiotics are nondigestible food ingredients that selectively stimulate the proliferation and/or activity of desirable bacterial populations already resident in the consumer's intestinal tract. Most prebiotics identified so far are nondigestible, fermentable carbohydrates. Intestinal populations of bifidobacteria, in particular, are stimulated to proliferate upon consumption of a range of prebiotics, increasing in numbers by as much as 10–100-fold in faeces [9, 17].

3. Advantages and disadvantages of the prebiotic

The prebiotic strategy offers a number of advantages over modifying the intestinal microbiota using probiotics or antibiotics.

Advantages over probiotics [17]:

- Stable in long shelf life foods and beverages;
- Heat and pH stable and can be used in a wide range of processed foods and beverages;
- Have physicochemical properties useful to food taste and texture;
- Resistant to acid, protease, and bile during intestinal passage;
- Stimulate organisms already resident in the host, and so avoid host/strain compatibilities, and the need to compete with an already established microbiota;
- Stimulate fermentative activity of the microbiota and health benefits from SCFA (short chain fatty acids);
- Lower intestinal pH and provide osmotic water retention in the gut.

Advantages over antibiotics [18]:

- Safe for long-term consumption and prophylactic approaches;
- Do not stimulate side effects such as antibiotic-associated diarrhea, sensitivity to UV radiation, or liver damage;
- Do not stimulate antimicrobial resistance genes;
- Not allergenic;

Disadvantages of prebiotics [17]:

- Unlike probiotics, overdose can cause intestinal bloating, pain, flatulence, or diarrhea.
- Not as potent as antibiotics in eliminating specific pathogens.
- May exacerbate side effects of simple sugar absorption during active diarrhea.

A consumed probiotic strain must compete with an already established microbiota, and in most cases they persist only transiently in the intestine. Individuals also harbor their own specific combination of species and unique strains within their intestinal bacteria suggesting that certain host–microbiota compatibilities exist. By targeting those strains that are already resident in the intestinal tract of an individual, the prebiotic strategy overcomes the need for probiotic bacteria to compete with intestinal bacteria that are well established in their niche [12, 17, 89, 103].

4. Definition of term "prebiotic"

The term prebiotics was first introduced in 1995 by Gibson and Roberfroid, defining "non-digestible food ingredient that beneficially affects the host by selectively stimulating the growth and/or activity of one or a limited number of bacteria in the colon, and thus improves host health" [33]. Definition brought up to date by Gibson specified prebiotic as "selectively fermented ingredient that allows specific changes, both in the composition and/or activity in the gastrointestinal microflora that confers benefits upon host well-being and health" [32]. Current definition of prebiotics was suggested during ISAPP experts' meeting in 2008 and it states that prebiotic is "dietary prebiotic is a selectively fermented ingredient that results in specific changes, in the composition and/or activity of the gastrointestinal microbiota, thus conferring benefit(s) upon host health" [24, 44].

Substances with prebiotic properties have to possess following properties [42, 74, 114, 116]:

- selectively stimulate growth and activity of chosen bacterial strains that have positive influence on health,
- lower pH of the bowel content,
- show positive for human spot action in the intestinal tract,
- be resistant to hydrolysis, action of intestinal tract enzymes and gastric acids,
- should not get soaked up in the upper part of the intestinal tract,
- should act as a selective substrate for one or for determined amount of beneficial species of microorganisms in the colon,
- should be stable in the process of food processing.

In order to evaluate and reason, if the given product is a prebiotic, the source of the substance should be given, as well as its purity, chemical composition and structure. It is very important to specify the carrier, concentration and amount in which it should be given to the host. Relating to the newest definition of the prebiotic, it was decided to type out three main criteria that have to be fulfilled by the substance in order to include it to the group of prebiotics [24].

1. Substance (component) – it is neither an organism, nor a medicine; substance that may be characterized chemically; in most cases this is a nutrient component.
2. Health benefits – calculable, exceeding any adverse effects.
3. Modulation – represents, that the presence of the substance and the preparatory, in which it is handled, changes the composition or activity of host microflora.

Prebiotics, similarly to other nutrient elements, have to fulfill certain safety parameters established in a given county. In the assessment of the final product following points should be taken into account [24, 39], (Figure 1):

1. If according to the legislation in the country, the history of safe use of the product in host is known (GRAS or its equivalent). If yes, the conductance of the following toxicological tests on animals and humans may not be necessary.
2. Safe, allowable norms for the consumption with minimal symptoms and adverse effects.

3. Product must not be infected and it should not contain any impurities.
4. Prebiotic cannot change the microflora in such a way, to cause a long-lasting harmful effect on host.

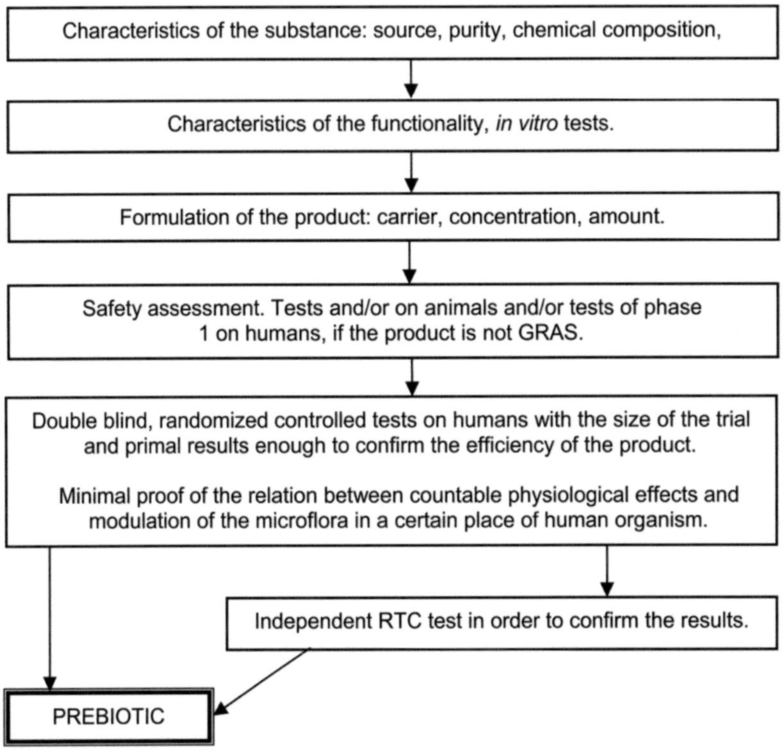

Figure 1. Guidelines to the assessment and proof of the action of prebiotics

5. Criteria of prebiotics classification

According to Wang [114] there are 5 most important basic criteria for classification of prebiotics (Figure 2). The first one assumes, that prebiotics are undigested in upper parts of intestinal tract and thus they are able to get to the large bowel where they can be fermented by potentially beneficial bacteria, which in the meantime meets the second criterion [61]. This fermentation may lead to increase in expression of short chain fatty acids, enlargement of faecal mass, some small reduction of large bowel pH, reduction of end nitrogen compounds and decrease of faecal enzymes, as well as general improvement of immunological system of host organism [17]. All these features contribute to improvement of consumer's health, which is the third criterion for prebiotics. The following one, that has to be fulfilled for the product to be recognized as prebiotic, is selective stimulation of growth of bacteria potentially thought to be connected with health improvement [32]. In order to assess the ability of prebiotic to selectively stimulate positive bacteria of species *Bifidobacterium* and *Lactobacillus*, so-called Prebiotic Index (PI) was introduced, and it can be calculated from the following formula [71]:

$$PI = (Bif/Total) - (Bac/Total) + (Lac/Total) - (Clos/Total)$$

Where:
Bif – *Bifidobacterium*
Bac – *Bacteroides*
Lac – *Lactobacillus*
Clos – *Clostridium*
Total – total bacteria

This PI allows to track the changes in the population in the given time in vitro conditions. At last, but not least, prebiotic should be able to survive the conditions in which the food would be stored, remain unchanged chemically and be accessible for bacteria metabolism [71].

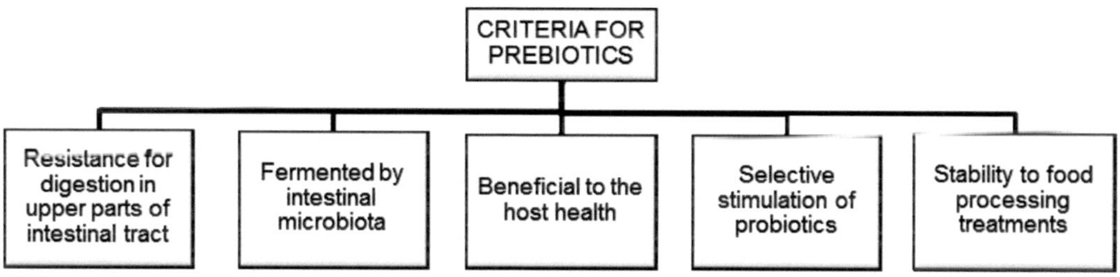

Figure 2. Criteria for prebiotics classification

6. Prebiotic mechanisms

The impact of prebiotics on the organism is indirect, because prebiotics do not do anything healthy for it, but they improve microorganisms that are beneficial [43]. The mechanism of how prebiotics influence the human health is presented on the Figure 3.

It is thought, that molecular structure of prebiotics is important taking into consideration the physiological effects, and that it determines which microorganisms are actually able to use that prebiotics. However, the way and progress of the stimulation of bacterial growth still remains unknown.

The most important function of prebiotic action is its influence on the microorganisms' growth and number in the large bowel [55, 90]. Going further, the tests have been conducted in order to investigate the potential ant pathogenic and anticancer action of prebiotics, their ability to decrease the presence of large bowel diseases [57]. A lot of different potential beneficial influences on human organisms are being sought, and those are, among the others: increase of the volume and improvement of stool moisture, lowering of the cholesterol level, decrease of the amount long chain fatty acids in bowels, decrease of pH in bowels, increase of mineral compounds absorption and raised short chain fatty acids production [1, 21, 90, 114], and the mechanism can be observed on the Figure 4.

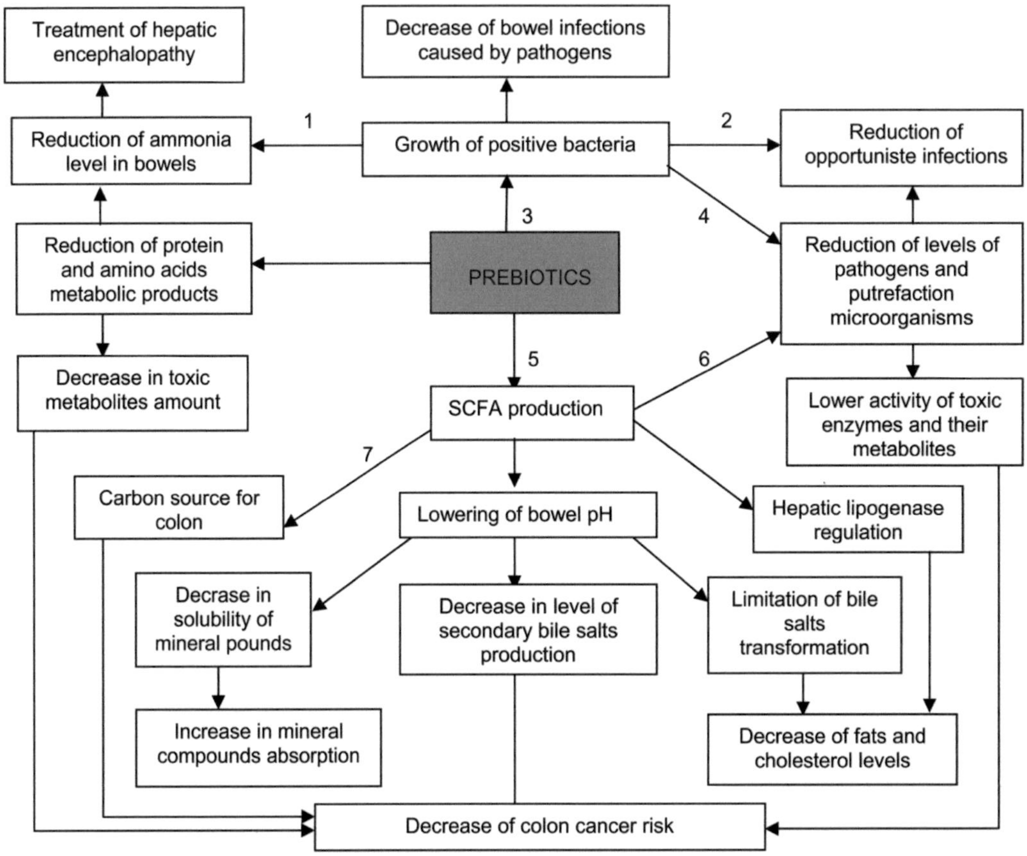

1 - bacterial proteins synthesis
2 - colonization counteraction
3 - selective carbon and energy source
4 - contention
5 - fermentation
6 - inhibition
7 - butyl acid

Figure 3. Proposed mechanism of prebiotic action

7. Production of prebiotics

Some prebiotics can be extracted from plant sources, but most are synthesized commercially using enzymatic or chemical methods. Overall, prebiotics are manufactured by four major routes (Table 1). Food-grade oligosaccharides are not pure products, but are mixtures containing oligosaccharides of different degrees of polymerization (DP), the parent polysaccharide or disaccharide, and monomer sugars. Oligosaccharide products are sold at this level of purity, often as syrups. Chromatographic purification processes are used to remove contaminating mono- and disaccharides to produce higher purity oligosaccharide products containing between 85 and 99% oligosaccharides, which are often dried to powders [17].

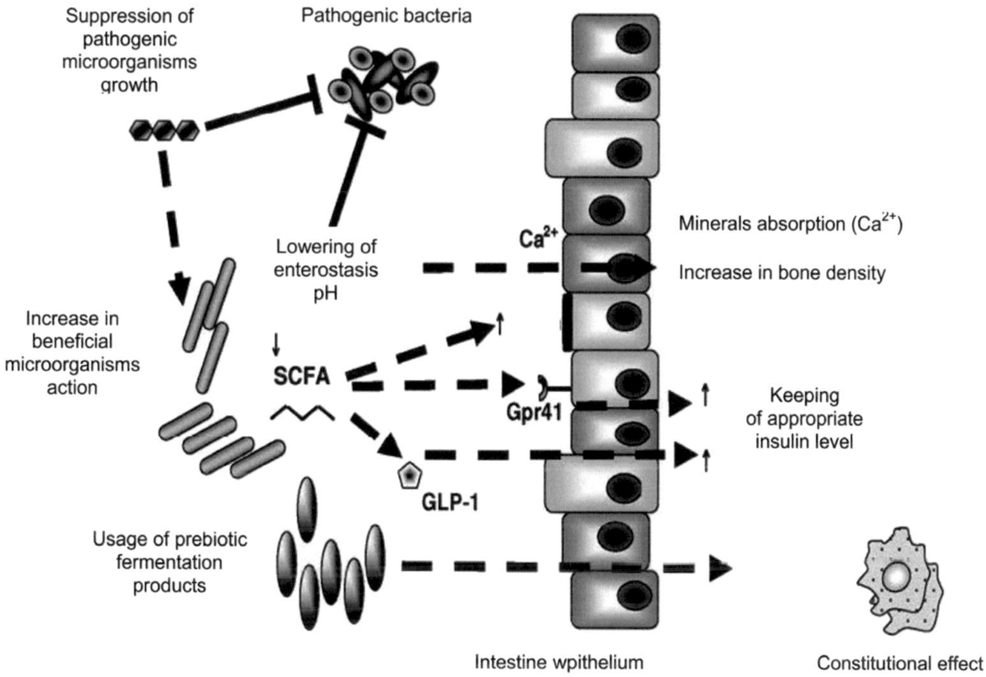

Figure 4. Mechanism of prebiotic action

Approach	Process	Prebiotic Examples
Direct extraction	Extraction from raw plant materials	Resistant starch from maize Inulin from chicory Soybean oligosaccharides from soybean whey
Controlled hydrolysis	Controlled enzymatic hydrolysis of polysaccharides; may be followed chromatography to purify the prebiotics	Fructooligosaccharides from inulin Xylooligosaccharides from arabinoxylan
Transglycosylation	Enzymatic process to build up oligosaccharides from disaccharides; may be followed by chromatography to purify the prebiotics	Fructooligosaccharides from sucrose Galactooligosaccharides from lactose Lactosucrose from lactose + sucrose
Chemical processes	Catalytic conversion of carbohydrates	Lactitol from hydrogenation of lactose Lactulose from alkaline isomerization of lactose

Table 1. Production of prebiotic carbohydrates

Carbohydrate	Chemical structure	Degree of polymerisation	Method of manufacture
Inulin	β(2-1)-Fructans	2 - 65	Extraction from chicory root and *Agave tequilana*.
Fructooligosaccharides (FOS)	β(2-1)-Fructans	2 - 9	Transfructosylation from sucrose or hydrolysis of chicory inulin.
Galactooligosaccharides (GOS)	Galactose oligomers and some glucose/ lactose/ galactose units	2 - 5	Produced from lactose by β-galactosidase.
Soya-oligosaccharides	Mixture of raffinose and stachyose	3 - 4	Extracted from soya bean whey.
Xylooligosaccharides (XOS)	β(1–4)-Linked xylose	2 - 4	Enzymatic hydrolysis of xylan. Enzyme treatments of native lignocellulosic materials. Hydrolytic degradation of xylan by steam, water or dilute solutions of mineral acids.
Isomaltooligosaccharides (IMO)	α(1–4)-glucose and branched α(1–6)-glucose	2 - 8	Microbial or enzymatic transgalactosylation of maltose. Enzymatic synthesis from sucrose.
Dextrins	Mixture of glucose-containing oligosaccharides	Various	Chemical modification of starch.

Table 2. Main candidates for prebiotic status

Different manufacturing processes also produce slightly different oligosaccharide mixtures. For example, FOS mixtures produced by transfructosylation of sucrose contain oligosaccharides between three and five monomer units, with the proportion of each oligosaccharide decreasing with increasing molecular size. These oligosaccharides contain a terminal glucose with β-1→2 linked fructose moieties. FOS produced by the controlled hydrolysis of inulin contain a wider range of β-1→2 fructooligosaccharide sizes (DP 2–9), relatively few of which possess a terminal glucose residue. Even different b-galactosidases used in the production of GOS will produce oligosaccharide mixtures with different proportions of β-1→4 and β-1→6 linkages. Hence, there can be some diversity between the structures of oligosaccharides produced by different manufacturers. The precise impact of these differences in their health effects remains to be determined [17, 100].

There are many oligosaccharides under investigation for their prebiotic potential. Fermentation of some oligosaccharides is not as selective as that of FOS, and their prebiotic status therefore remains in doubt. The main candidates for prebiotic status is provided in Table 2 [26]. There is, therefore, a need for new prebiotic substances of distinct, selective stimulation of growth of lactic acid bacteria, and non-fermented or slightly fermented by other, sometimes pathogenic intestinal bacteria. The search for functional food or functional food ingredients is beyond any doubt one of the leading trends in today's food industry.

8. Resistant dextrins as prebiotics

8.1. Resistant starch

Resistant starch (RS) includes the portion of starch that can resist digestion by human pancreatic amylase in the small intestine and thus, reach the colon. The general behaviour of RS is physiologically similar to that of soluble, fermentable fibre, like guar gum. The most common results include increased faecal bulk and lower colonic pH and improvements in glycaemic control, bowel health, and cardiovascular disease risk factors, so it has shown to behave more like compounds traditionally referred to as dietary fibre [31, 60, 97, 126].

Resistant starch is found in many common foods, including grains, cereals, vegetables (especially potatoes), legumes, seeds, and some nuts [31, 35].

Resistant starch may not be digested for four reasons [30, 38, 60]:

- this compact molecular structure limits the accessibility of digestive enzymes, various amylases, and explains the resistant nature of raw starch granules. The starch may not be physically bio accessible to the digestive enzymes such as in grains, seeds or tubers,
- the starch granules themselves are structured in a way which prevents the digestive enzymes from breaking them down (e.g. raw potatoes, unripe bananas and high-amylose maize starch),
- starch granules are disrupted by heating in an excess of water in a process commonly known as gelatinization, which renders the molecules fully accessible to digestive enzymes. Some sort of hydrated cooking operation is typical in the preparation of starchy foods for consumption, rendering the starch rapidly digestible. However, if these
- starch gels are then cooled, they form starch crystals that are resistant to enzymes digestion. This form of "retrograded" starch is found in small quantities (approximately 5%) in foods such as "corn-flakes" or cooked and cooled potatoes, as used in a potato salad.
- selected starches that have been chemically modified by etherisation, esterisation or cross-bonding, cannot be etherisation, esterisation or cross-bonding, cannot be broken down by digestive enzymes.

Resistance starch is the sum of starch itself and products of her decomposition, that are neither being digested nor absorbed in the small bowel of healthy human [23]. Resistant starch is the difference between amount of the starch exposed to the action of amylolytic enzyme complex and the amount of starch decomposed to glucose during hydrolysis performed by those enzymes [83, 123].

$$RS = TS - (RDS + SDS)$$

Where:
RS – resistant starch
TS – total starch
RDS – rapidly digestible starch
SDS – slowly digestible starch
Few types of digestible starch are recognized nowadays [36, 52, 113].

Resistant starch of type 1 – RS1 covers the starch in plant cells with undestroyed cell walls. This starch is unavailable for digestive enzymes present in human intestinal tract, and thus together with fragments of plant tissues passes through the small bowel getting to the large bowel untouched, and there it can undergo fermentation [70, 104]. RS1 is heat stable in most normal cooking operations, which enables its use as an ingredient in a wide variety of conventional foods [31].

Resistant starch type 2 – RS2 is composed of native starch granules from certain plants containing uncooked starch or starch that was gelatinized poorly and hydrolyzed slowly by R-amylases (e.g., high-AM corn starches). RS2 covers scoops of raw starch of some plant species, especially high-amylase corn, potato and banana [70, 104]. Huge size of raw potato flour scoops and hence combined with it limited area of the access for enzymes was considered as the cause of its resistance [73]. But the main reason for the resistance of the raw starch of some plant species on amylolytic enzymes is the structure of its scoops and crystallization type B that us present within them (in the scoop of potato and corn starch). Also other elements of the scoop structure have an impact on the resistance of the starch – such as the shape of the area, size of pores or susceptibility of the starch to germinate. A particular type of RS2 is unique as it retains its structure and resistance even during the processing and preparation of many foods; this RS2 is called high-AM maize starch [31, 117].

Resistant starch type 3 – RS3 covers the substance precipitated from pap or starch gel during the process of retrograding. During the germination of the starch in the lowered temperature and with the proper concentration (1.5% amylose, 10% amylopectin) colloidal solution is formed. Stable starch phase existing as double helix forms reticular structure binding water phase in its 'eyes'. During the storage of the gel (few hours in lowered temperature) helixes undergo aggregation forming thermally stable crystal structures. Such structures show the resistance to amylolytic enzymes [66, 104]. RS3 is of particular interest, because of its thermal stability. This allows it to be stable in most normal cooking operations, and enables its use as an ingredient in a wide variety of conventional foods [15].

During food processing, in most cases in which heat and moisture are involved, RS1 and RS2 can be destroyed, but RS3 can be formed. Storey et al. [99], classified a soluble polysaccharide called 'retrograded resistant maltodextrins' as type 3 RS. They are derived from starch that is processed to purposefully rearrange or hydrolyze starch molecules, and subsequent retrogradation, to render them soluble and resistant to digestion. This process results in the formation of indigestible crystallites that have a molecular similarity to type 3 RS but with a smaller degree of polymerization as well as a lower MW, converting a portion of the normal α- 1,4-glucose linkages to random 1,2-, 1,3-, and 1,4-α or β linkages [25, 31,64].

The definition of presented forms of resistant starch may be presented according to the formulas [83]:

$$RS1 = TS - (RDS + SDS) - RS2 - RS3$$

$$RS2 = TS - (RDS + SDS) - RS1 - RS3$$

$$RS3 = TS - (RDS + SDS) - RS2 - RS1$$

Where:
RS1 – resistant starch type 1
RS2 – resistant starch type 2
RS3 – resistant starch type 3
TS – total starch
RDS – rapidly digestible starch
SDS – slowly digestible starch

Resistant starch type 4 – RS4 covers the starch chemically or physically modified and achieved by combination of these two processes. During chemical modification new functional groups are brought into the starch chain, and they bind to glucose residues. Presence of the substituents and spatial changes in the chain prevent proper functioning of human digestive enzymes. In physical method, during warming of starch in high temperature process of dextrinization occurs, and it may also occur in the presence of acid as a catalyst. One of the products of dextrinization is free glucose, which binds to the chains randomly. As a result of such process between glucose residues, bonds typical for starch and those normally not existing in its chains, arise [13, 70, 121].

Resistant starch type 5 – RS5 is an AM-lipid complexed starch [31, 46], which is formed from high AM starches that require higher temperatures for gelatinization and are more susceptible to retrograde [20, 31]. In general, the structure and amount of starch-lipid in foods depend on their botanical sources. Also, Frohberg and Quanz [29] defined as RS5 a polysaccharide that consists of water-insoluble linear polyα- 1,4-glucan that is not susceptible to degradation by alpha-amylases. They also found that the poly-α-1,4-D-glucans promote the formation of short-chain fatty acids (SCFA), particularly butyrate, in the colon and are thus suitable for use as nutritional supplements for the prevention of colorectal diseases [31].

RS is the fraction of starch which is not hydrolyzed to D-glucose in the small intestine within 120 min of being consumed, but which is fermented in the colon. Many studies have shown that RS is a linear molecule of a-1,4-D-glucan, essentially derived from the retrograded AM fraction, and has a relatively low MW (1.2 x 105 Da) [31].

Resistant starch obtained during chemical or physical modification is being investigated nowadays, due to the fact that it possesses some specific physical properties, as well as because of its health benefits [49, 86, 87, 96]. During chemical modification functional groups are introduced to the starch molecule, which then leads to the changes of physical and chemical properties of obtained product, and also it lowers the availability of the starch to amylolytic enzymes, because new functional groups prevent occurring of the enzyme-substrate complex [7]. Chemical modification was found to be advantageous method of decrease of starch digestion, and therefore starch modified chemically may be the source of resistance starch RS4 [15, 36, 70].

8.2. Resistant dextrin

Resistant dextrins are defined as short chain glucose polymers, without sweet taste and performing strong resistance to hydrolytic action of human digestive enzymes [68]. In accessible throughout the whole products (resistant dextrin Nutriose, Fibersol) bigger percentage presence of (1→2)-, (1→3)-, (1→6)- α and β-glycoside bonds than in native starch which is the source of getting them [62, 115].

During warming of starch in high temperature, with or without addition of catalyst (usually acidic) dextrinization of starch is observed. Dextrinization is a complex process taking chemical side of it into account. It covers depolymerization, transglucolyzation and repolymerization [122].

During warming of wet starch random bonds (1→4) and rarely (1→6) hydrolytically break. Intermediate form in this reaction is either oxycarbonic ion, or free radicals [105].

Most probably, dextrinization undergoes the mixed mechanism. In case of warming of dry starch (or with a low moisture content), bonds (1→6) are made between two starch chains and intramolecular dehydration coincides, which results in development of 1,6-anhydro-β-D-glucose. In such a way only extreme glucose units with free hydroxyl group within anomeric carbon atom may react. In both cases exuded water has hydrolytic character. With the temperature equal to about 290°C α-(1→4)-glycosidic bonds begin to break. However during dextrinization not only bond breaking is observed, but also the isomerization (e.g. through mutarotaion) or formation of new bonds. Hydroxyl groups at C-2, C-3 or C-6 glucose unit act on oxycarbonic ions or free radicals and transglycolization, which is based on formation of (1→2), (1→3) and (1→6) bonds, undergoes. This process leads to formation of branched dextrins. Because of spherical considerations, and maybe also thermodynamic ones, formation of (1→6) bonds is privileged. 1,6-anhydro-β-D-glucose is formed, which easily forms polymers, leads to formation of 4-O-α-D-, 4-O-β-D-, 2-O-α-D- and 2-O-β-D-glukopiranozylo-1,6-dihydro-β-D-glucopyranose [105, 122].

In the next stage following reactions have to be taken into account:

- reversion, that is the reaction between glucose units leading to formation of (1→6)-glycosidic bonds,
- reaction between (1→6)-anhydro-β-D-glucopyranose and free radicals with formation of 1,6-glycosidic bonds,
- recombination.

From all presented reactions the most characteristic and dominating one is transglycolisation. Formation of bonds other than typical for starch (1→4) and (1→6) causes that the received product becomes unavailable for human digestive enzymes and shows properties of resistant starch.

In the presence of acidic catalyst dextrinization process progresses a bit differently. First the hydrolysis undergoes, and in the result of it (1→6) bonds break, but (1→4) bonds stay untouched. In this way, white dextrins are formed. Because (1→6) bonds are more resistant to hydrolysis than (1→4) bonds, these last ones undergo transformation into (1→6) bonds [105].

The role of basic catalyst in dextrinization process is not well known. The only thing know is that in this process deprotonating of hydroxyl groups at C-2 and C-3 is the first stage. In the presence of oxidating agents atom C-1 of terminal glucose units are being oxidized to carboxyl group [105]. Summing up, as a result of hydrolysis the reductive ends of the starch become glucose cations, which undergo intramolecular dehydration forming (1→6)-anhydro-β-D-glucopyranose unit or they take part in formation of intermolecular bonds (transglycolization). As a result of this process random glycosidic (1→2), (1→3) bonds are formed [68]. Formation of bonds other than typical for starch, i.e. (1→4) and (1→6) causes that end product doesn't undergo hydrolysis through the human digestive enzymes [115].

In the process of formation of resistant dextrin, piroconversion is the first stage, and it covers following steps: thermolysis, transglucolysis, regrouping and repolymerisation. Starch thermolysis leads to breaking of α-D-(1→4) and α-D-(1→6) glycosidic bonds, which then leads to the formation of of products with lower molecular mass and higher viscosity and reducing sugars content. After transglucolysis recombination of hydrolyzed starch fragments with free hydroxyl groups happens, and formation of strongly branched structures. Repolymerisation of glucose and oligosaccharides with formation of high molecular compounds is done in high temperature and presence of acidic catalyst (e.g. hydrochloric acid) [68]. Achieved pirodextrins are the mixture of poli- and oligosaccharides with a different degree of polimerisation (DP), and simultaneously with different molecular mass. Pirodextrins are subjected to enzymatic hydrolysis or chromatography – stages, which aim is to reduce the fractions other than typical for the starch (i.e. containing bonds other than α-D-(1→4) and α-D-(1→6) glycosidic ones [6, 81, 118].

Chemical modification has long been known to inhibit in vitro digestibility of starch, the extent of which is related to the type and degree of modification, the extent of gelatinization, and the choice of enzyme [117, 119]. Starch phosphates [45], hydroxypropyl starches [51,

121], starch acetates [120], phosphorylated starch [86, 94, 119], and citrate starches [117, 124] have been tested for enzymatic degradation previously. In previous studies have been suggested that the substituted groups hindered enzymatic attack and thus also made neighboring bonds resistant to degradation. Chemical substitution of starch reduces its enzyme digestibility, probably because the bulky derivatizing groups sterically hinder formation of the enzyme-substrate complex [7]. The largest change in digestibility has been achieved through cross-linking of starch [119]. The application of organic acids, as citric acid and tartaric acid as derivatizing agents seemed to be profoundly safe. These acids are nutritionally harmless compared to other substances used for chemical modification [123]. When citrate starches were fed to rats no pathological changes could be found in comparison to native wheat and corn starches [117].

Potato starch in a natural form has limited possibilities to be used, and its chemical structure and physical properties give possibilities to many modifications, also those leading to formation of resistant substances to amylolytic enzymes. Big hopes are laid on usage of products with modified starch, especially resistant starch and resistant dextrin as substances with prebiotic properties.

Kapusniak et al. [47] resistant dextrin was receive by simultaneous pyroconversion and chemical modification (esterification/ cross-linking) of potato starch in the presence of hydrochloric acid as catalyst of dextrinization process, and citric acid as derivatizing agent. Potato starch was modified by thermolysis in the presence of acid catalyst in a sealed container at 130°C for 180 min. The effect of addition of multifunctional polycarboxylic acids (citric and tartaric) on the progress of dextrinization process, structure and properties of resulting products was investigated [47]. It seems likely that probiotic activity will be exhibited by dextrin obtained by simultaneous thermolysis and chemical modification of potato starch in the presence of a volatile inorganic acid (hydrochloric acid) as a catalyst of the dextrinization process and an excess amount of an organic acid (tartaric acid) as a modifying factor. Kapuśniak et al. [47, 48] analyzed this dextrin in terms of the solubility and pH of its 1% aqueous solution, the content of reducing sugars, molecular mass distribution, weight average molecular mass using high performance size-exclusion chromatography (HPSEC), average chain length using high performance anion exchange chromatography with pulsed amperometric detection (HPAEC-PAD), and the content of the resistant fraction using the enzymatic-gravimetric method AOAC 991.43, the enzymatic-gravimetric-chromatographic method AOAC 2001.03 [69], the enzymatic-spectrophotometric method [23] and the pancreatin-gravimetric method [94]. It was shown that the use of tartaric acid in the process of starch thermolysis yielded acidic dextrin characterized by high water solubility (about 68%) and a high content of reducing sugars (about 29%). The studies showed that dextrin modified with tartaric acid did not contain any traces of unreacted starch, and the percentage share of the main fraction (having a weight average molecular mass of about 1.800 g/mol) was 80%. The average length of the carbohydrate chain in dextrin obtained with tartaric acid was 8.2 as determined by means of HPAEC. A study by Kapuśniak et al. [47] revealed that the content of the resistant fraction in dextrin modified with tartaric acid, determined by means of the AOAC 991.43, amounted to 44.5%. However, results obtained by the Englyst [43] enzymatic-spectrophotometric method showed that the actual content of the resistant fraction was above 68%. Kapuśniak et al. [47, 48] used the official AOAC 2001.03 method to determine the content of the resistant fraction in dextrin modified with tartaric acid. This method is the latest approved method for determining total content of dietary fiber in foods containing resistant maltodextrins. Apart from measuring the content of insoluble dietary fiber and the high molecular weight fractions of soluble fiber, this method makes it possible to determine resistant oligosaccharides (by using high-performance liquid chromatography, HPLC). The total content of dietary fiber in dextrin modified with tartaric acid was about 50% [47, 48]. In the Englyst method, fractions undigested after 120 min are considered resistant. In the pancreatin-gravimetric method, similarly as in the Englyst method, samples are digested with pancreatin, but resistant fractions are determined gravimetrically only after 16 h. In the case of dextrin modified with tartaric acid, the results of determination by the pancreatin-gravimetric method (67%) were similar to those obtained in the previous studies using the Englyst method (68%), but much higher than those obtained using the AOAC 2001.03 method (50%) [43, 47]. The observed differences among the various methods in terms of the measured content of the resistant fraction in dextrin modified with tartaric acid was caused by the fact that, according to the latest reports, enzymatic-gravimetric methods (including AOAC 2001.03) using thermostable α-amylase can determine only part of resistant starch type 4 [124]. Based on the enzymatic tests, it can be argued that dextrin obtained using an excessive amount of tartaric acid may be classified as resistant starch type 4.

8.3. Prebiotic effects of resistant starch and resistant dextrin

Resistant starch has a long history of safe consumption by humans and is a natural component of some foods. Intakes vary but are generally low, particularly in Western diets. Similar to soluble fibre, a minimum intake of resistant starch (5 - 6 g) appears to be needed in order for beneficial reductions in insulin response to be observed. Estimates of daily intake of resistant starch range from 3 to 6 g/day (averaging 4.1 g/day) [4, 31]. As a food ingredient, resistant starch has a lower calorific (8 kJ/g) value compared with fully digestible starch (15 kJ/g) [31, 78], therefore it can be a substitutive of digestible carbohydrates, lowering the energy content of the final formulation. Some resistant starch products are also measured as total dietary fibre in standard assays, potentially allowing high-fibre claims [31, 35].

Resistant starch can be fermented by human gut microbiota, providing a source of carbon and energy for the 400 - 500 bacteria species present in this anaerobic environment and thus potentially altering the composition of the microbiota and its metabolic activities. The fermentation of carbohydrates by anaerobic bacteria yields SCFA, primarily composed of acetic, propionic, and butyric acids, which can lower the lumen pH, creating an environment less prone to the formation of cancerous tumours [31, 126].

RS consumption has also been related to reduced post-prandial glycemic and insulinemic responses, which may have beneficial implications in the management of diabetes, and is associated with a decrease in the levels of cholesterol and triglycerides. Other effects of resistant starch consumption are increased excretion frequency and faecal bulk, prevention of constipation and hemorrhoids, decreased production of toxic and mutagenic compounds, lower colonic pH, and ammonia levels. Considering that nowadays several diseases result from inadequate feeding, and that some may be related to insufficient fibre intake, it is reasonable to assume that an increased consumption of indigestible components would be important [31].

Resistant starch enhance the ileal absorption of a number of minerals in rats and humans. Lopez et al. [59] and Younes et al. [128] reported an increased absorption of calcium, magnesium, zinc, iron and copper in rats fed RS-rich diets. In humans, these effects appear to be limited to calcium [16, 30, 107]. Resistant starch could have a positive effect on intestinal calcium and iron absorption. A study to compare the apparent intestinal absorption of calcium, phosphorus, iron, and zinc in the presence of either resistant or digestible starch showed that a meal containing 16.4% RS resulted in a greater apparent absorption of calcium and iron compared with completely digestible starch [30, 65].

Liu and Xu [58] showed that resistant starch dose-dependently suppressed the formation of colonic aberrant crypt foci only when it was present during the promotion phase to a genotoxic carcinogen in the middle and distal colon, suggesting that administration of resistant starch may retard growth and/or the development of neoplastic lesions in the colon. Therefore, colon tumorigenesis may be highly sensitive to dietary intervention. Adults with preneoplastic lesions in their colon may therefore benefit from dietary resistant starch. This suggests the usefulness of resistant starch as a preventive agent for individuals at high risk for colon cancer development [30, 58].

Short chain fructooligosaccharides (FOS) and resistant starch (RS) may act synergistically (by combining, and thus increasing, their prebiotic effects) [31, 79], the administration of the combination of FOS and RS induced changes in the intestinal microbiota, by increasing lactobacilli and bifidobacteria in caecum and colonic contents. Several types of prebiotic fibres can be distinguished considering their rate of fermentability. Such role depends on the carbohydrate chain length as it has been demonstrated *in vitro* in a fermentation system, showing that FOS are rapidly fermented whereas long chain prebiotic, like inulin, are steadily fermented. These observations have been confirmed *in vivo* once the different prebiotics reach the large intestine: FOS are rapidly fermented, whereas RS is slowly degraded. In consequence, the particular kinetics would determine the region of the intestine where the effects will be clearer. Thus, FOS would be more active in the first parts of the large bowel whereas RS would reach the distal part of the colon. In fact, Le Blay et al. [31, 50] have reported that administration of FOS or raw potato starch induces different changes in bacterial populations and metabolites in the caecum, proximal, and distal colon, as well as in faeces. As compared with RS FOS doubled the pool of faecal fermentation products, like lactate, while the situation was just the opposite distally. These observations confirm that each prebiotic shows particular properties, which should be considered before their application for intestinal diseases; thus, rapidly fermentable prebiotics are particularly useful in those affecting the proximal part of the large intestine, while slowly fermentable prebiotics should be chosen for more distal intestinal conditions. Moreover, an association with different prebiotics with complementary kinetics should be considered when a health-promoting effect throughout the entire colon is required. So, functional foods based on the combination of two different dietary fibres, with different rate of fermentability along the large intestine, may result in a synergistic effect, and thus, in a more evident prebiotic effect that may confer a greater health benefit to the host [31, 127].

Younes et al. [127] was to examine the potential synergistic effect of a combination of these two fermentable carbohydrates (inulin and resistant starch). For this purpose, thirty-two adult male Wistar rats weighing 200 g were used in the present study. The rats were distributed into four groups, and fed for 21 d a fibre-free basal purified diet or diet containing 100 g inulin, or 150 g resistant starch (raw potato starch)/kg diet or a blend of 50 g inulin and 75 g resistant starch/kg diet. After an adaptation period of 14 d, the rats were then transferred to metabolic cages and dietary intake, faeces and urine were monitored for 5 d. The animals were then anaesthetized and faecal Ca and Mg absorption were measured. Finally, the rats were killed and blood, caecum and tissues were sampled. Ca and Mg levels were assessed in diets, faeces, urine, caecum and plasma by atomic absorption spectrometry. The inulin and resistant starch ingestion led to considerable faecal fermentation in the three experimental groups compared with the control group diet. Moreover, both carbohydrates significantly increased the intestinal absorption and balance of Ca and Mg, without altering the plasma level of these two minerals. Interestingly, the combination of the studied carbohydrates increased significantly the faecal soluble Ca and Mg concentrations, the apparent intestinal absorption and balance of Ca, and non-significantly the plasma Mg level. The combination of different carbohydrates showed synergistic effects on intestinal Ca absorption and balance in rats [127].

The example of commercially available resistant dextrin is Nutriose. It is a non-viscous soluble fiber made from starch using a highly controlled process of dextrinization. It is mostly resistant to digestion in the small intestine and largely fermented in the colon. A process of dextrinization includes a degree of hydrolysis followed by repolymerization that converts the starch into fiber by forming no digestible glycosidic bonds. Nutriose is totally soluble in cold water without inducing viscosity [37, 53]. It is produced from wheat or maize starch using a highly controlled process of dextrinization followed by chromatographic fractionation step [28]. Nutriose FB®06, produced from wheat starch, contains approx. 13% of 1,2- and 14% of 1,3- glycosidic linkages [81]. The weight average molecular weight (M_w) and the number average molecular weight (M_n) for that dextrin were nearly 5000 and 2800 g/mole respectively. The residual content of sugars (DP1-2) of Nutriose FB®06 was below 0.5% and it could be considered as sugar free [80]. The enzyme-resistant fraction content, determined according to AOAC official method 2001.03 for total dietary fiber in foods containing resistant maltodextrins, was nearly 85% for Nutriose®06 and nearly 70% for Nutriose®10 [80, 81]. Nutriose has found wide application in food and pharmaceutical industries, as components of fiber-enriched drinks [92], components of a fiber-enriched composition for enteral nutrition [88], granulation binders [93], in the preparation of low-calorie food [11], and sugar-free confectionery [91]. Very well tolerated, Nutriose may be 20-25% of a product's composition without causing discomfort or bloating. About 15% of Nutriose are absorbed in the small intestine, about 75% fermented in the large intestine, while the remainder (about 10%) is excreted in the faeces [72, 110]. Nutriose induced an increase of the colonic saccharolytic flora and decrease in potentially harmful *Clostridium perfringens* in human faeces [54, 72]. Nutriose induced a decrease in the faecal pH of human volunteers, increased production of short chain fatty acids (SCFAs) in rats [54], induced changes in faecal bacterial enzyme concentration [72, 110]. It was also shown that learning (respectively physical) performances are improved in rats 180 minutes (respectively 150 minutes) after the consumption of Nutriose, compared to dextrose. The glycaemic kinetics is not sufficient to predict this effect: even though the glycaemic peak was lower with the resistant dextrin than with dextrose, the glycaemia was the same between the two groups at 150 and 180 minutes after ingestion. These preliminary results are very encouraging [82].

Guérin-Deremaux et al. [37] found that the non-viscous soluble dietary fiber may influence satiety. The randomized, double-blind, placebo-controlled clinical study in 100 overweight healthy adults in China investigated the effect of different dosages of dietary supplementation with a dextrin, Nutriose, on short-term satiety over time. Subjects were randomized by body mass index and energy intake and then assigned to receive either placebo or 8, 14, 18, or 24 g/d of Nutriose mixed with orange juice (n = 20 volunteers per group). On days - 2, 0, 2, 5, 7, 14, and 21, short-term satiety was evaluated with a visual analog scale, and hunger feeling status was assessed with Likert scale. Nutriose exhibits a progressive and significant impact on short-term satiety, which is time and dosage correlated. Some statistical differences appear for the group 8 g/d from day 5, and from day 0 for the groups 14, 18, and 24 g/d. The hunger feeling status decreases significantly from day 5 to the end of the evaluation for the group 24 g and from day 7 for the groups 14 and 18 g. By day 5, the group 24 g showed significantly longer time to hunger between meals compared with placebo. These results suggest that dietary supplementation with a soluble fiber can decrease hunger feeling and increase short-term satiety over time when added to a beverage from 8 to 24 g/d with time- and dose-responses relationship [37].

Human clinical trials on healthy subjects have shown that Nutriose is well tolerated and able to stimulate the growth of acid-resistant bacteria. We particularly observed a beneficial shift in the bacterial microbiota profile to butyrogenic genera such as *Peptostreptococcus, Fusobacterium* and *Bifidobacterium* [37].

Resistant maltodextrins made from starch are commercially available. Fibersol-2 is well-known soluble, non-digestible, starch-derived resistant maltodextrin [67]. Fibersol-2 is produced from corn-starch by pyrolysis and subsequent enzymatic treatment (similar to the process to manufacture conventional maltodextrins) to convert a portion of the normal α-1,4 glucose linkages to random 1,2-, 1,3- α or β linkages [69]. Solubility of Fibersol-2 in water reaches 70% (w/w) at 20°C. It is readily dispersible in water and highly compatible with dry drink mix applications. At typical use levels, it yields clear, transparent solutions that are near water-like in performance. Fibersol-2 adds no flavor or odor. It has essentially no sweetness of its own. It shows stability to acid and heat/retort processing, including stability in high acid, hot filled, aseptic, or retorted products like juices, sauces, puddings, fluid milks, and sports. Fibersol-2 shows superior freeze-thaw stability. It shows precise and extremely low viscosity and very low hygroscopicity. It does not actively participate in non-enzymatic Maillard-type browning [69]. Fibersol-2 exhibited very important physiological properties. It was fermented slowly, producing less acid and gas than most soluble dietary fiber [27]. Studies indicated that Fibersol-2 could effectively reduce postprandial levels of blood glucose and insulin [109]. Fibersol-2 significantly reduced levels of blood triglycerides and serum cholesterol [125]. By adding stool volume, moisture, and reducing transit time, Fibersol-2 helped maintain good colon health, potentially reducing the incidence of various types of colon diseases and cancers [102]. Fibersol-2 effectively promoted the growth of a variety of beneficial bacteria (naturally occurring or ingested as probiotics) in the colon. In promoting the growth of beneficial bacteria, Fibersol-2 indirectly reduced the presence of undesirable bacterial species [41, 63].

Bodinhan et al. [8] found that a non-viscous resistant starch significantly lowered energy intake after intake of the supplement compared with placebo during both an ad libitum test meal (P = 0.033) and over 24 hours (P = 0.044). Cani et al [14] found that treatment with the fermentable dietary fiber oligofructose increased satiety after breakfast and dinner and reduced hunger and prospective food consumption after dinner, suggesting a role for the use oligofructose supplements in the management of food intake in overweight and obese patients [37].

Kapusniak et al. [47] and Śliżewska et al. [98] enzyme-resistant chemically modified dextrins resulting from heating of potato starch with hydrochloric acid as catalyst and additionally polycarboxylic acids (citric and tartaric acids) were tested as the source of carbon for probiotic bacteria (*Lactobacillus* and *Bifidobacterium*) and bacteria isolated from the human feces (*Escherichia coli, Enterococcus, Clostridium* and *Bacteroides*). It was shown that all of the tested bacteria, both probiotics and those isolated from human feces, were able to grow and utilize dextrin as a source of carbon, albeit to varying degrees. After 24 h the highest growth was recorded for the probiotic bacteria *Lactobacillus* and *Bifidobacterium*, the weakest for *Clostridium* and *Escherichia coli* bacteria. After prolonging culture time to 72–168 h, which corresponds to retarded or pathological passage of large intestine contents, the viability of intestinal bacteria in a medium with resistant dextrin was found to be lower by one or two orders of magnitude as compared to the viability of probiotic bacteria. The number of probiotics and bacteria isolated from fecal samples grown in media containing 1% glucose was lower by two or three orders of magnitude than that of corresponding bacteria grown in a medium containing dextrin. This may have been caused by lower pH values of the controls, in which the culture environment became unfavorable to preserving high viability by the studied bacteria. This may have also been caused by the protective effects of dextrin on the bacteria. After the completion of incubation, that is, at 168 h, lactobacilli and bifidobacteria were found to be highly viable. Their counts were higher by one or two orders of magnitude than those of the intestinal bacteria *E. coli, Enterococcus, Clostridium* and *Bacteroides* isolated from fecal samples. At 168 h of incubation, the probiotic bacteria amounted to over 44% of the whole population. The *Clostridium* strain showed the weakest growth, with a 9.5% share in the entire population, while *Enterococcus*, *Escherichia coli*, and *Bacteroides* amounted to from 15.3% to 15.8% of the population [48].

9. Conclusions

In today's world, life style is an important determinant of health in later life. Lack of physical activity, particularly if associated with over consumption, increases the risk of development of nutrition related chronic diseases, such as obesity, hypertension, cardiovascular diseases, osteoporosis, type II diabetes, and several cancers. Over the last decade, drastic changes have taken place in the image and assessment of the importance of the daily diet. Foods are no longer judged in terms of taste and immediate nutritional needs, but also in terms of their ability to improve the health and well-being of consumers. The role of diet in human health has led to the recent development of the so-called functional food concept. A functional food is dietary ingredient, that has cellular or physiological effects above the normal nutritional value. Functional food can contain probiotics and/or prebiotics [33, 40].

The studies suggests that soluble fibers help to regulate the digestive system, may increase micronutrient absorption, stabilize blood glucose and lower serum lipids, may prevent several gastrointestinal disorders, and have an accepted role in the prevention of cardiovascular disease. It is concluded that supplementation with soluble fibers (e.g. wheat dextrin) may be useful in individuals at risk of a lower than recommended dietary fiber intake.

A prebiotic is a nondigestible food ingredient that beneficially affects the host by selectively stimulating the growth and/or activity of one or, a limited number, of bacteria in the colon that can improve host health [19]. Some carbohydrates, such as fructooligosaccharides (FOS) [10, 34, 75, 101], inulin [75, 76, 77, 112] and galactooligosaccharides (GOS) [84, 108] are well-accepted prebiotics.

Promising sources of prebiotics are starch products, especially resistant starch (RS) [3, 19, 106] and products of partial degradation of starch [56].

The commercial degraded starches are known as converted starches and comprise the "thin-boiling" acid-converted starches, oxidized starches, and dextrins. There are four major groups of dextrins: maltodextrins produced by hydrolysis of dispersed starch by action of liquifying enzymes such as amylase, degradation products by acid hydrolysis of dispersed starch, cyclodextrins, and pyrodextrins produced by the action of heat alone or in a combination with acid on dry granular starch. On the market pyrodextrins are available in three major varieties: British gums, white dextrins and yellow dextrins [5, 105, 122].

In particular, almost every food oligosaccharide and polysaccharide has been claimed to have prebiotic activity, but not all dietary carbohydrates are prebiotics. Conventional fibers, like pectins, cellulose, etc. are not selectively metabolized by gut bacteria. Resistant maltodextrins, being a mixture of fractions of different molecular weight (different degree of polymerization), are dietary fibre, but there are not necessarily selective for desirable bacteria in the gut. Hence, research showing the effect of prebiotic should continue to be performed, both in the *in vito* and *in vivo.*

Author details

Katarzyna Śliżewska
Institute of Fermentation Technology and Microbiology, Faculty of Biotechnology and Food Sciences, Technical University of Lodz, Lodz, Poland

Janusz Kapuśniak, Renata Barczyńska and Kamila Jochym
Institute of Chemistry, Environmental Protection and Biotechnology, Jan Dlugosz University in Czestochowa, Czestochowa, Poland

10. References

[1] Abrams S.A., Griffin I.J., Hawthorne K.M., Ellis K.J. (2007) Effect of prebiotic supplementation and calcium intake on body mass index. J. Pediatr. 151: 293-298.
[2] Annison G., Illman R., Topping D. (2003) Acetylated, propionylated or butyrylated starches raise large bowel short-chain fatty acids preferentially when fed to rats. J. Nutr. 133: 3523-3528.
[3] Asp N.G., Van Amelsvoort J.M.M., Hautvast J.G.A.J. (1996) Nutritional implications of resistant starch. Nutr. Res. Rev. 9: 1-31.
[4] Behall K.M., Daniel J., Scholfield D.J., Hallfrisch J.G., Liljeberg-Elmståhl H.G.M. (2006) Consumption of both resistant starch and b-glucan improves postprandial plasma glucose and insulin in women. Diabetes Care 29, 976-981.
[5] BeMiller J.N. (1993) Starch-based gums. In: Industrial Gums. Polysaccharides and Their Derivatives. Whistler R.L. and BeMiller, J.N., editors. Academic Press, San Diego, USA, pp. 579-601.
[6] Berenal M.J., Periago M.J., Ros G. (2002) Effects of processing on dextrin, total starch, dietary fiber and starch digestibility in infant. J. Food Sci. 67: 1249-1254.
[7] Björck, L., Gunnarsson A., Østergård,K. (1989) A study of native and chemically modified potato starch. Part II. Digestibility in the rat intestinal tract. Starch/Stärke 41: 128-134.
[8] Bodinham B.L., Frost G.S., Robertson M.D. (2010) Acute ingestion of resistant starch reduces food intake in healthy adults. Br J. Nutr. 103: 917-22.
[9] Boehm G., Stahl B. Oligosaccharides. In: Mattila-Sandholm and Saarela M, editors. Functional Dairy Products, Woodhead Publishing, CRC Press, England, 2003, pp. 203–243.
[10] Bouhnik Y., Vahedi K., Achour L., Attar A., Salfati J., Pochart P., Marteau P., Flourie B., Bornet F., Rambaud J.C. (1999) Short-chain fructo-oligosaccharide administration dose-dependently increases fecal bifidobacteria in healthy humans. J. Nutr. 129: 113-116.

[11] Brendel R., Boursier B., Leroux P. (2002) U.S. Patent 2002/0192344 A1.
[12] Brigidi P., Swennen E., Vitali B., Rossi M., Matteuzzi D. (2003) PCR detection of *Bifidobacterium* strains and Streptococcus thermophilus in feces of human subjects after oral bacteriotherapy and yogurt consumption. Int. J. Food Microbiol. 81: 203–209.
[13] Brown I.L., (2004) Applications and uses of resistant starch. J. AOAC Int. 87: 727-732.
[14] Cani P.D., Joly E., Horsmans Y., Delzenne N.M. (2006) Oligofructose promotes satiety in healthy human: a pilot study. Eur. J. Clin. Nutr. 60: 567-72.
[15] Champ M., Langkilde A.M., Brouns F., Kettlitz B. (2003) Advances in dietary fiber characterization. 2. Consumption, chemistry, physiology and measurement of resistant starch; implications for health and food labeling. Nutr. Res. Rev. 16: 143–161.
[16] Coudray C., Bellanger J., Castiglia-Delavaud C., Rémésy C., Vermorel M.,Rayssignuier Y. (1997) Effect of soluble or partly soluble dietary fibres supplementation on absorption and balance of calcium, magnesium, iron and zinc in healthy young men. Eur. J. Clin. Nutr., 51: 375-380.
[17] Crittenden R.G., Playne M.J. Prebiotics. In: Lee Y.K. and Salminen S., edtitors. Handbook of probiotics and prebiotics, John Wiley & Sons, New Jersey, Canada, 2009, pp. 535-584.
[18] Crittenden R.G., Prebiotics. In: Tannock GW, edtitor. Probiotics: A Critical review. Horizon Scientific Press, Wymondham, United Kingdom, 1999, pp. 141–156.
[19] Cummings J.H., Beatty E.R., Kingman S.M., Bingham S.A., Englyst H.N. (1996) Digestion and physical properties of resistant starch in the human large bowel. Br. J. Nutr. 75: 733-747.
[20] Cummings J.H., Stephen A.M. (2007) Carbohydrate terminology and classification. Eur. J. Clin. Nutr. 61, 5-18.
[21] Douglas L.C., Sandres M.E. (2008) Probiotics and prebiotics in dietetics practice. J. Am. Diet. Associat. 108: 510-521.
[22] Duncan S.H., Holtrop G., Lobley G.E., Calder A.G., Stewart C.S., Flint H.J. (2004) Contribution of acetate to butyrate formation by human fecal bacteria. Br. J. Nutr. 91: 915-923.
[23] Englyst H.N., Kingman S.M., Cummings J.H. (1992) Classification and measurement of nutritionally important starch fractions. Eur. Clin. Nutr. 46: 33–50.
[24] FAO Technical Meeting on Prebiotics. (2007) Food Quality and Standards Service (AGNS), Food and Agriculture Organization of the United Nations (FAO). FAO Technical meeting Report, September 15-16
[25] Faraj A., Vasanthan T., Hoover R. (2004) The effect of extrusion cooking on resistant starch formation in waxy and regular barley flours. Food Res. Int. 37, 517-525.
[26] Figueroa-González I., Quijano G., Ramírez G., Cruz-Guerrero A. (2011) Probiotics and prebiotics – perspectives and challenges. J. Sci. Food Agric. 91, 1341-1348.
[27] Flickinger E.A., Wolf B.W., Garleb K.A., Chow J., Leyer G.J., Johns P.W., Fahey G.C. (2000) Glucose-based oligosaccharides exhibit different *in vitro* fermentation patterns and affect in vivo apparent nutrient digestibility and microbial populations in dogs. J. Nutr. 130: 1267-1273.
[28] Fouache C., Duflot P., Looten P. (2003) U.S. Patent 2003/6630586.

[29] Frohberg C., Quanz M. (2008) Use of Linear Poly-Alpha-1,4-Glucans as Resistant Starch, United States Patent Application 20080249297, 2008, Available on line at: http://www.wipo.int/pctdb/en/wo.jsp?WO=2005040223.
[30] Fuentes-Zaragoza E., Riquelme-Navarrete M.J., Sánchez-Zapata E., Pérez-Álvarez J.A. (2010) Resistant starch as functional ingredient: A review. Food Res. Int. 43: 931-942.
[31] Fuentes-Zaragoza E., Sánchez-Zapata E., Sendra E., Sayas E., Navarro C., Fernández-López J., Pérez-Alvarez J.A. (2011) Resistant starch as prebiotic: A review. Starch/Stärke 63: 406-415.
[32] Gibson G.R. (2004) Fibre and effects on probiotics (the prebiotic concept). Clin. Nutr. Suppl. 1: 25-31
[33] Gibson R., Roberfroid M. (1995) Dietary modulation of the human colonic microbiota: introducing the concept of prebiotics. J. Nutr. 125: 140-1412
[34] Gibson G. (1999) Dietary modulation of the human gut microflora using the prebiotics oligofructose and inulin. J. Nutr. 129 (7Suppl): 1438S-1441S.
[35] Goldring J.M. (2004) Resistant starch: Safe intakes and legal status. J. AOAC Int. 87, 733–739.
[36] González-Soto R.A., Sánchez-Hernández L., Solorza-Feria J. (2006) Resistant starch production from non-conventional starch sources by extrusion. Food Sci. Tech. Int. 12: 5-11.
[37] Guérin-Deremaux L., Pochat M., Reifer C., Wils D., Cho S., Miller L.E. (2011) The soluble fiber NUTRIOSE induces a dose-dependent beneficial impact on satiety over time in humans. Nutr. Res. 31: 665-672.
[38] Haralampu S.G. (2000) Resistant starch: A review of the physical properties and biological impact of RS3. Carbohydrate Polymers, 41, 285–292.
[39] Hautvast J.G.A.J., Meyer D. (2007) The science behind inulin and satiety. Sensus: 6 30.
[40] Holzapfel W.H., Schillinger U. (2002) Introduction to pre- and probiotics. Food Res. Int. 35: 109-116.
[41] Hopkins M.J., Cummings J.H., Macfarlane G.T. (1998) Inter-species differences in maximum specific growth rates and cell yields of bifidobacteria cultured on oligosaccharides and other simple carbohydrate sources. J. Appl. Microbiol. 85: 381-386.
[42] Huebner J., Wehling R.L., Parkhurst A., Hutkins R.W. (2008) Effect of processing conditions on the prebiotics activity of commercial prebiotics. Int. Dairy J. 18: 287-293.
[43] Imaizumi K., Nakatsu Y., Sato M., Sedamawati Y., Sugano M. (1991) Effects of xylooligosaccharides on blood glucose, serum and liver lipids and caecum short-chain fatty acids in diabetic rats. Agric. Biol. Biochem., 55: 199-205.
[44] ISAPP (2008) 6th Meeting of the International Scientific Association of Probiotics and Prebiotics, London, Ontario.
[45] Janzen G.J. (1969) Verdaulichkeit von Stärken und phosphatierten Stärken mittels Pankreatin. Stärke 21: 231-237.
[46] Jiang H., Jane J.L., Acevedo D., Green A. (2010) Variations in starch physicochemical properties from a generationmeans analysis study using amylomaize V and VII parents. J. Agric. Food Chem. 58, 5633-5639.

[47] Kapusniak J., Barczynska R. Slizewska K., Libudzisz Z. (2008a) Utilization of enzyme-resistant chemically modified dextrins from potato starch by Lactobacillus bacteria. Zeszyty Problemowe Postępów Nauk Rolniczych 530: 445-457. (in polish)
[48] Kapusniak J., Jochym K., Barczynska R. Slizewska, K., Libudzisz Z. (2008b) Preparation and characteristics of novel enzyme-resistant chemically modified dextrins from potato starch. Zeszyty Problemowe Postępów Nauk Rolniczych 530: 427-444. (in polish)
[49] Kayode O., Adebowale T., Adeniyi Afolabi B., Iromidayo Olu-Owolabi. (2006) Functional, physicochemical and retrogradation properties of sword bean (Canavalia gladiata) acetylated and oxidized starches. Carbohydr. Polym. 65: 93–101.
[50] Le Blay G.M., Michel C.D., Blottière, H.M., Cherbut, C.J. (2003) Raw potato starch and short-chain fructo-oligosaccharides affect the composition and metabolic activity of rat intestinal microbiota differently depending on the caeco-colonic segment involved. J. Appl. Microbiol. 94, 312-320.
[51] Leegwater D.C., Luten J.B (1971) A study on the *in vitro* digestibility of hydroxypropyl starches by pancreatin. Stärke 23: 430-432.
[52] Leeman A.M., Karlsson M.E., Eliasson A.Ch., Bjőrck I.M.E. (2006) Resistant starch formation in temperature treated potato starches varying in amylose/amylopectin ratio. Carbochydr. Polym. 65: 306-313.
[53] Lefranc-Millot C. (2008) NUTRIOSE 06: a useful soluble fibre for added nutritional value. Nutr. Bull. 33: 234-239.
[54] Lefranc-Millot C., Wils D., Neut C., Saniez-Degrave M.H. (2006) Effects of a soluble fiber, with excellent tolerance, NUTRIOSE®06, on the gut ecosystem: a review. In: Proceedings of The Dietary Fibre Conference. June, 2006. Helsinki, Finland.
[55] Lenoir-Wijnkoop I., Sanders M.E., Cabana M.D. (2007) Probiotic and prebiotic influence beyond the intestinal tract. Nutr. Rev. 65: 469-489.
[56] Leszczyński W. (2004) Resistant Starch – classification, structure, production. Pol. J. Food Nutr. Sci. 13/54: 37-50.
[57] Lim C.C., Ferguson L.R., Tannock G.W. (2005) Dietary fibres as "prebiotics": implications for colorectal cancer. Mol. Nutr. Food Res. 49: 609–619.
[58] Liu R., Xu G. (2008) Effects of resistant starch on colonic preneoplastic aberrant crypt foci in rats. Food Chem. Toxicol. 46: 2672-2679.
[59] Lopez H.W., Levrat-Verny M.A., Coudray C., Besson C., Krespine V., Messager A., Demigné C, Rémésy C. (2001) Class 2 resistant starches lower plasma and liver lipids and improve mineral retention in rats. J. Nutr. 131: 1283-1289.
[60] Lunn J., Buttriss J. L. (2007) Carbohydrates and dietary fibre. Nutr. Bull. 32: 21-64.
[61] Maccfarlane G.T., Steed H., Maccfarlane S. (2008) Bacterial metabolism and health-related effects of galacto-oligosaccharides and other prebiotics. J. Appl. Microbiol. 104: 305-344.
[62] Mana K.K. (2003) Enzyme resistant dextrins from high amylose corn mutant starches. Starch/Stärke. 53: 21–26.
[63] Matsuda I., Satouchi M. (1997) Agent for promoting the proliferation of *Bifidobacterium*. U.S. Patent 5698437.
[64] Mermelstein N.H. (2009) Analyzing for resistant starch. Food Technol. 4: 80-84.

[65] Morais M.B., Feste A., Miller R.G., Lifichitz C.H. (1996) Effect of resistant starch and digestible starch on intestinal absorption of calcium, iron and zinc in infant pigs. Paediatr. Res. 39(5): 872-876.

[66] Morell M.K., Konik- Rose Ch., Ahmed R., Li Z., Rahman S. (2004) Synthesis of resistant starches in plants. J. AOAC Int. 87: 740–748.

[67] Ohkuma K., Hanno Y., Inada K., Matsuda I., Katta Y. (1997) U.S. Patent 5620873.

[68] Ohkuma K., Matsuda I., Katta Y., Hanno Y. (1999) Pyrolysis of starch and its digestibility by enzymes – Characterization of indegestible dextrin. Denpun Kagaku. 37: 107-114.

[69] Ohkuma K., Wakabayashi S. (2001) Fibersol-2: A soluble, non-digestible, starch-derived dietary fibre. In: McCleary B.V., Prosky L., editors. Advanced Dietary Fibre Technology. Blackwell Science Ltd., Oxford, pp. 509-523.

[70] Onyango C., Bley T., Jacob A. (2006) Infuence of incubation temperature and time on resistant starch type III formation from autoclaved and acid-hydrolysed cassava starch. Carbohydr. Polym. 66: 497-499.

[71] Palframan R., Gibson G.R., Rastall R.A. (2003) Development of a quantitive tool for comparison of the prebiotic effect of dietary oligosaccharides. Lett. Appl. Microbiol., 37: 281-284.

[72] Pasman W.J., Wils D., Saniez M.H., Kardinaal A.F. (2006) Long term gastro-intestinal tolerance of NUTRIOSE® FB in healthy men. Eur. J. Clin. Nutr. 60(8): 1024-1034.

[73] Ring S.G., Gee J.M., Whittam M., Orford P., Johnson I.T. (1988) Resistant starch: its chemical from in foodstuffs and effect on digestibility in vitro. Food Chem. 28: 97-109.

[74] Roberfroid M., Gibson G.R., Hoyles L., McCartney A.L., Rastall R., Rowland I., Wolvers D., Watzl B, Szajewska H., Stahl B., Guarner F., Respondek F., Whelan K., Coxam V., Davicco M-J., Léotoing L., Wittrant Y., Delzenne N.M., Cani P.D., Neyrick A.M., Meheust A. (2010) Prebiotic effects: metabolic and health benefits. Br. J. Nutr. 104: S1-S61.

[75] Roberfroid M.B. (1998) Prebiotics and synbiotics: concepts and nutritional properties. Br. J. Nutr. 80: S197-S202.

[76] Roberfroid, M.B. (2005) Introducing inulin-type fructans. Br. J. Nutr. 93: S13-S25.

[77] Roberfroid, M.B. (2007) Inulin-type fructans: Functional food ingredients. J. Nutr. 137: 2493S-2502S.

[78] Rochfort S., Panozzo J. (2007) Phytochemicals for health, the role of pulses. J. Agric. Food Chem. 55, 7981-7994.

[79] Rodríguez-Cabezas M.E., Camuesco D., Arribas B., Garrido-Mesa, N. (2010) The combination of fructooligosaccharides and resistant starch shows prebiotic additive effects in rats. Clin. Nutr. 29, 832-839.

[80] Roturier J.M., Looten P.H., Osterman E. (2003) Dietary fiber measurement in food containing NUTRIOSE®FB by enzymatic-gravimetric-HPLC method. Proc. Dietary Fiber Conference, Helsinki, Finland. May, 2003.

[81] Roturier J.M., Looten P.H. (2006) Nutroise: Analytical Aspects. Proc. The Dietary Fibre Conference, Helsinki, Finland, June 2006.

[82] Rozan P., Deremaux L., Wils D., Nejdi A., Messaoudi M., Saniez M.H. (2008) Impact of sugar replacers on cognitive performance and function in rats. Brit. J. Nutr. 100: 1004-1010.

[83] Sajilata M.G., Singhal R.S., Kulkarni P.R. (2006) Resistant starch- a review, Comp. Rev. Food. Sci. Food Safety. 5: 1-17.

[84] Sako T., Matsumoto K., Tanaka, R. (1999) Recent progress on research and applications of non-digestible galacto-oligosaccharides. Int. Dairy J. 9: 69-80.

[85] Salminen S., Gorbach S., Lee Y.K., Benno Y. (2004) Human Studies on Probiotics: What is Scientifically Proven Today? In: Salminen S, von Wright A and Ouwerhand A, editors Lactic Acid Bacteria: Microbiological and Functional Aspects. Marcel Dekker, New York, 2004, pp. 515–530.

[86] Sang S., Chang J.L. (2007) Formation, characterization, and glucose response in mice to rice starch with low digestibility produced by citric acid treatment. J. Cereal Sci. 45: 24-33

[87] Sang Y., Seib P. (2006) Resistant starches from amylose mutants of corn by simultaneous heat-moisture treatment and phosphorylation. Carbohydr.Polym. 63: 167-175.

[88] Saniez M.H. (2004) U.S. Patent 2004/6737414 B2.

[89] Satokari R.M., Vaughan E.E., Akkermans A.D.L., Saarela M., de Vos W.M. (2001) Polymerase chain reaction and denaturing gradient gel electrophoresis monitoring of fecal *Bifidobacterium* populations in a prebiotic and probiotic feeding trial. Syst. Appl. Microbiol. 2001, 24: 227–231.

[90] Saulnier D.M., Spinler J.K., Gibson G.R., Versalovic J. (2009) Mechanisms of probiosis and prebiosis: considerations for enhanced functional foods. Curr. Opinion Biotechnol. 20: 135-141.

[91] Serpelloni M. (2002) U.S. Patent 2002/0192343 A1.

[92] Serpelloni M. (2003) U.S. Patent 2003/0077368 A1.

[93] Serpelloni M. (2006) U.S. Patent 2006/0112956 A1.

[94] Shin M., Song J., Seib P. (2004) In vitro digestibility of cross-linked starches – RS4. Starch/ Stärke. 56: 478-483.

[95] Shin S.I., Lee Ch.J. (2007) Formation, characterization, and glucose response in mice to rice starch with low digestibility produced by citric acid treatment. J. Cereal Sci. 45: 24-33

[96] Singha J., Kaurb L. (2007) Review factors influencing the physico-chemical, morphological, thermal and rheological properties of some chemically modified starches for food applications—A review. Food Hydr. 21: 1–22.

[97] Slavin J., Stewart M., Timm D., Hospattankar A. (2009) International Association for Cereal Science and Technology (ICC), 1–3 July 2009, van der Kamp J.W., Vienna, Austria, p. 35.

[98] Śliżewska K., Kapuśniak J., Barczyńska R., Jochym K. (2010) The preparation of prebiotic properties. The patent application RP P-392895

[99] Storey D., Lee A., Bornet F., Brouns F. (2007) Gastrointestinal responses following acute and medium term intake of retrograded resistant maltodextrins, classified as type 3 resistant starch. Eur. J. Clin. Nutr. 61, 1262–1270.

[100] Suzuki N., Aiba Y., Takeda H., Fukumori Y., Koga Y. Superiority of 1-kestose, the smallest fructo-oligosaccharide, to a synthetic mixture of fructo-oligosaccharides in the selective stimulating activity on Bifidobacteria. Biosci. Microflora. 25: 109–116.

[101] Swennen K., Courtin CH.M., Delcour, J.A., 2006. Non-digestible oligosaccharides with prebiotic properties. Crit. Rev. Food Sci. Nutr. 46: 459-471.

[102] Takagak K., Ikeguchi M., Artura Y., Fujinaga N., Ishibashi Y., Sugawa-Katayama Y. (2001) The effect of AOJIRU drink powder containing indigestible dextrin on defecation frequency and faecal characteristics. J. Nutr. Food 4(4): 29-35.

[103] Tannock G.W., Munro K., Bibiloni R., Simon M.A., Hargreaves P., Gopal P., Harmsen H., Welling G. (2004) Impact of consumption of oligosaccharide-containing biscuits on the fecal microbiota of humans. Appl. Environ. Microbiol. 70: 2129–2136.

[104] Themeier H., Hollmann J., Neese U., Lindhauer M.G. (2005) Structural and morphological factors influencing the quantification of resistant starch II in starches of different botanical origin. Carbohydr. Polym. 61: 72-79.

[105] Tomasik P., Wiejak S., Pałasiński M. (1989) The thermal decomposition of carbohydrates. Part II. The decomposition of starch. Adv. Carbohydr. Chem. Biochem. 47: 279-344.

[106] Topping D.L, Clifton, P.M. (2001) Short-chain fatty acids and human colonic function: roles of resistant starch and nonstarch polysaccharides. Physiol. Rev. 81(3): 1031-1064.

[107] Trinidad T.P., Wolever T.M.S., Thompson L.U. (1996) Effect of acetate and propionate on calcium absorption from the rectum and distal colon of humans. Am. J. Clin. Nutr. 63: 574-578.

[108] Tzortzis G., Goulas A.K., Gibson G.R. (2005) Synthesis of prebiotic galactooligosaccharides using whole cells of a novel strain, Bifidobacterium bifidum NCIMB 41171. Appl. Microbiol. Biotechnol. 68: 412-416.

[109] Unno T., Nagata K., Horiguchi T. (2002) Effects of green tea supplemented with indigestible dextrin on postprandial levels of blood glucose and insulin in human subjects. Journal of Nutritional Food 5(2): 31-39.

[110] van den Heuvel E.G., Wils S.D., Pasman W.J., Bakker M., Saniez M.H., Kardinaal A.F. (2004) Short-term digestive tolerance of different doses of NUTRIOSE®FB, a food dextrin, in adult men. Eur. J. Clin. Nutr. 58(7): 1046-1055.

[111] Van Loo J. (2006) Inulin-type fructans as prebiotics. In: Prebiotics: Development and Application. Gibson G.R. and Rastall RA, editors. John Wiley and Sons, Chichester, UK, pp. 57–100.

[112] Van Loo J., Cummings J., Delzenne N., Englyst H., Franck A., Hopkins M., Kok N., Macfarlane G., Newton D., Quigley M., Roberfroid M., van Vliet T., van den Heuvel, E. (1999) Functional food properties of non-digestible oligosaccharides: a consensus report from the ENDO project (DGXII AIRII-CT94-1095). Br. J. Nutr. 81(2): 121-132.

[113] Wang J., Jin Z., Yuan X. (2006) Preparation of resistant starch from starch-guar gum extrudates and their properties. Food Chem. 101: 20-25.

[114] Wang Y. (2009) Prebiotics: Present and future in food science and technology. Food Res. Int. 42: 8-12.

[115] Wang Y., Kozlowski R., Delgado G.A. (2001) Enzyme resistant dextrins from high amylose corn mutant starches. Starch-Starke. 53: 21-26.

[116] Weese J., Schrezenmeir J. (2008) Probiotics, prebiotics and synbiotics. Adv. Biochem. Eng. Biotechnol. 111: 1-6.

[117] Wepner B., Berghofer E., Miesenberger E., Tiefenbacher K. (1999) Citrate starch: Application as resistant starch in different food systems. Starch/Stärke 51: 354-361.

[118] Wolf B.W., Wolever T., Bolognesib C. (2001) Glycemic response to a rapidly digested starch is not affected by the addition of an indigestible dextrin in humans. Nutr. Res. 21: 1099–1106.

[119] Woo K. S., Seib P. A. (2002) Cross-linked resistant starch: preparation and properties. Am. Ass. Cereal Chem. 79: 819-826.

[120] Wootton M., Chaundry M.A. (1979) Enzymic digestibility of modified starches. Starch/Stärke 31: 224-228.

[121] Wootton M., Chaundry M.A. (1981) *In vitro* digestion of hydroxypropyl derivatives of wheat starch. I. Digestibility and action pattern using porcine pancreatic α-amylase. Starch /Stärke. 33: 135-137.

[122] Wurzburg O.B. (1986) Converted starches. In: Modified starches: properties and uses. Wurzburg O.B., editor. CRC Press, Boca Raton Florida, USA, pp. 17-40.

[123] Xie X., Liu Q. (2004) Development and physicochemical characterization of new resistant citrate starch from different corn. Starches/ Starch. 56: 364 – 370.

[124] Xie X., Liu Q., Cui S.W. (2006) Studies on the granular structure of resistant starches (type 4) from normal, high amylose and waxy corn starch citrates. Food Res. Int. 39: 332–341.

[125] Yamamoto T. (2007) Effect of indigestible dextrin on visceral fat accumulation. J. Jpn. Soc. Study. Obes. 13: 34-41.

[126] Yao N., Paez A.V., White P.J. (2009) Structure and function of starch and resistant starch from corn with different doses of mutant amylose-extender and floury-1 alleles. J. Agric. Food Chem. 57: 2040-2048.

[127] Younes H., Coudray C., Bellanger J., Demigné C., Rayssiquier Y., Rémésy C. (2001) Effects of two fermentable carbohydrates (inulin and resistant starch) and their combination on calcium and magnesium balance in rats. Brit. J. Nutr. 86, 479-485.

[128] Younes H., Levrat M.A., Demige C., Remesy C. (1995) Resistant starch is more effective than cholestyramine as a lipid-lowering agent in the rat. Lipids, 30: 847-853.

Chapter 13

Food Structure and Carbohydrate Digestibility

Suman Mishra, Allan Hardacre and John Monro

Additional information is available at the end of the chapter

http://dx.doi.org/10.5772/51969

1. Introduction

Carbohydrate is almost universally the major dietary source of metabolic energy. Nearly all of it is obtained from plants, and nearly all of it requires digesting before it is available for metabolism. While digestion is aimed at breaking down molecular structure within carbohydrate molecules, there is a raft of further plant structural impediments to be overcome before most plant carbohydrates are available for digestion.

Starch, for instance, represents energy stored, not for animals, but for the plant that made the starch. It is a reserve available to carry a plant between seasons, to sustain it during periods when photosynthesis is limited, to prepare it for times of intense energy use such as flowering, and to support its progeny in seeds before autonomous growth. But so accessible is free starch as a form of energy that plants have taken special measures, many of them structural, to protect it physically from all sorts of opportunist consumers – animals, fungi and bacteria - and from the effects of existing in a hydrating entropic environment. All these structural barriers have to be overcome before the carbohydrate becomes available for digestion, and can be used as a source of food-derived energy.

In the human diet, lack of available carbohydrate is associated with under-nutrition and its attendant problems, while at the same time a surfeit of available carbohydrate is associated with obesity, metabolic syndrome, and diabetes – scourges of the developed world. Therefore, as food structure can have a critical role in determining the proportion of carbohydrate that is made available by food processing and digestion (Bjorck et al. 1994), it is of fundamental importance to nutrition and health.

This chapter discusses the importance of carbohydrate digestibility to human health, various forms of plant and food structure that have an impact on carbohydrate digestibility, and how food processing methods of various types overcome them.

2. The nutritional importance of carbohydrate digestion

Digestibility, energy and the glycemic response

The nutritional importance of available carbohydrate currently extends far beyond its role as a major source of sustenance for humans. Thanks to modern agriculture, transport and food technology, and to the market-driven economy in which appetite-driven food wants, rather than nutritional needs and survival, have come to determine the types of foods available to consumers, energy intakes have far exceeded energy requirements. As a result, the "developed" world is now facing an obesity crisis. Carbohydrate digestibility has gained new importance, not only because of its contribution to obesity, but also because a secondary consequence of obesity is the metabolic syndrome for which a defining feature is glucose intolerance – an impaired ability to control blood glucose concentrations after a carbohydrate meal.

It is now evident that the adipose tissue of obesity is not a passive fat storage tissue, but is physiologically active and intimately involved in glucose homeostasis. It plays a key role in glucose intolerance and Type 2 diabetes by producing factors, including free fatty acids, that induce insulin resistance (Saltiel & Kahn 2001). Resistance to insulin leads to a reduced rate of clearance of glucose from the blood, and the resulting increased concentration of glucose in the blood leads to generalized damage throughout the body, from chemical bonding (glycation) of proteins, increased oxidative stress, and damage to numerous biochemical processes (Brownlee 2001). In response to increased blood glucose and to the rate of blood glucose loading, insulin production increases, with its own damaging effects (Guigliano et al. 2008). Ultimately, exhaustion of the capacity of the pancreas to produce adequate insulin means that the insulin resistance of Type 2 diabetes evolves into the insulin insufficiency of Type 1 diabetes. The generalized, cumulative, systemic damage of prolonged and/or repeated exposure to high blood glucose concentrations manifests itself as a raft of disorders associated with long-term diabetes – kidney failure, circulatory problems, neuropathy, heart disease, blindness and so on – that are imposing enormous costs in suffering and resources (Zimmet et al. 2001).

In the context of the pandemic of obesity and glucose intolerance in the modern world, new ways of manipulating the rate and extent of digestibility of carbohydrate are being sought. The rate of starch digestion is important because the degree to which blood glucose loading exceeds blood glucose clearance determines the acuteness of the net increase in blood glucose concentrations, and consequently, the intensity of the insulin response required to remove the glucose overload and restore normal blood glucose concentrations. The rate of digestion also determines how sustained will be the supply of glucose by continued digestion in the gut, and therefore, how prolonged its contribution to delaying the urge to eat again will be.

Carbohydrate digestibility and colonic health

The extent of digestion during transit through the foregut is important because it determines the proportion of starch that is available to the colon as polysaccharide for fermentation, which has a role in colonic health (Fuentes-Zaragoza et al. 2010) and probably also in appetite control through the colonic brake feedback mechanism (Brownlee 2011). Undigested food residues, including both food structures and the carbohydrates and other nutrients that they have protected from digestion, are now recognized as being not simply gastrointestinal refuse, but a valuable feedstock for the colonic ecosystem. Through both fermentation of the residues and through the ability of a proportion of them to survive colonic transit, they play an essential part in maintaining gut health and function, as well as good health in general (Buttriss & Stokes 2008).

It is increasingly recognized that events in the colon influence the body as a whole, through products of colonic fermentation, through effects on the immune system mediated by the colonic epithelium, and through neuronal and hormonal feedback from the colon to upstream regions of the digestive tract (Wikoff et al. 2009). Short chain fatty acid products of colonic fermentation, propionic acid in particular, may play a direct role in blood glucose control by suppressing the release of plasma triglycerides, which contribute to insulin resistance. Colonic fermentation also appears to have indirect effects on hormones from the pancreas and adipose tissue that are involved in the regulation of energy metabolism (Nilsson et al. 2008).

Recent research suggests that obesity is associated with a colonic microbiota that is more effective in scavenging energy from undigested food polysaccharides than the microbiota from lean individuals (Turnbaugh et al. 2006). Although the daily increments in energy gain may be small, over time they accumulate in expanding adipose tissue. Recovering undigested energy by colonic fermentation could make the important difference between starving and surviving in an energy-depleted environment where food is scarce and of poor quality, or under the precarious conditions in which we evolved. However, in the present developed world of plenty, it may contribute to the difference between remaining trim and being overtaken by creeping obesity.

3. Forms of food structure affecting carbohydrate digestion

Food structure can take a number of forms that can affect the availability of carbohydrate in a number of different ways and at a number of different levels – molecular, cellular, plant tissue and food.

3.1. Molecular level

In the case of short chain sugars, such as the disaccharides sucrose, maltose and lactose, the structural constraint on digestion to monosaccharides lies solely within the glycosidic linkage between monosaccharide units, and is easily overcome by disaccharidases of the gut brush border (Wright et al. 2006). But even then, the rate at which the monosaccharide units traverse the gut wall, and so the extent to which absorption is completed during small intestinal transit, depends on the ability of membrane-bound transporters to recognize the structure of monosaccharides. Glucose transporters (SGLT1, GLUT 2) achieve active ATP-driven facilitated transport against a gradient, whereas the transporters that recognize fructose as a structure carry out less effective absorption by facilitated transport, which may result in overflow of fructose into the terminal ileum and colon, leading to intestinal discomfort from the resulting osmotic and fermentative effects (Gibson et al. 2004). Similarly, the structural specificity of lactase means that decline in lactase activity leads to the severe gastrointestinal problems of lactose intolerance.

Starch

Starch presents a different challenge for digestion from that of the common food disaccharides. Although it consists solely of α-D-glucose units, it may have a degree of polymerization of thousands or millions, and the glucose units may be α(1-4) linked into long linear amylose chains, or shorter amylose chains may be connected at α(1-6)-linked branch points. Most starch (~70%) is branched (amylopectin) and has a molecular weight of 50-500 million, and a degree of polymerization in the millions, depending on the plant species (French 1984; James et al. 2003; Thomas & Atwell 1998). The long regular string of glucose units in both amylose and amylopectin provides the opportunity for interactions between starch chains, leading to the buildup of pseudo-crystalline regions, which may sterically inhibit amylase access.

a. Native starch and Starch granules (RS2)

Above the scale of amylose and amylopectin molecules, the starch is organized during growth in plants into granules that impose further restrictions on enzyme access (Ayoub et al. 2006; Gallant et al. 1997). Starch granules characteristically consist of concentric rings of alternating amorphous and pseudo-crystalline structures laid down during granule growth (**Figure 1**). The amorphous starch corresponds to regions that are rich in branches at (α(1-6) glycosidic bonds, while in the pseudo-crystalline regions the starch is highly organized as closely packed short branches, approximately 10-20 glucose subunits in length (Gallant et al. 1997; Ratnayake & Jackson 2007; Waigh et al. 2000). The high degree of organization of the pseudo-crystalline region is revealed by the typical Maltese cross birefringence pattern of

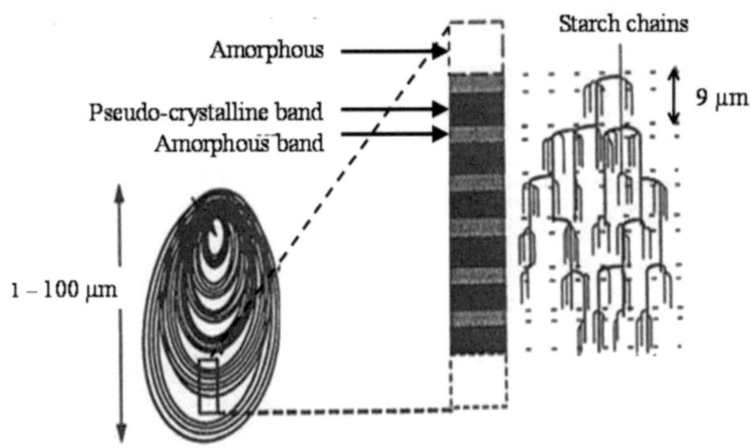

Figure 1. Schematic view of the organization of starch within a native starch granule

native starch when viewed in polarised light. The pseudo-crystalline regions are far more resistant to digestion by α-amylase than the amorphous regions (Donald 2004), and the highly organized starch granule as a whole may be relatively resistant to digestion, thanks to protein and lipid at the granule surface, which together form a coating resistant to water and digestive enzymes (Debet and Gidley, 2006).

Although covered with a resistant coating, almost all types of starch granules have been shown to bear surface pores that are entrances of channels that reach the near centre (hilum) of the granule (Huber & BeMiller 2000). The pores may be well developed in maize and nearly absent and much smaller in potato and tapioca (Juszczak et al. 2003). They may play an important role in digestion by allowing penetration of water and enzymes into the centre of the granules (Copeland et al. 2009) and leaching of glucose outwards, so the native starch granules often appear to be digested from the inside out (Gallant et al. 1997; Oates 1997; Planchot et al. 1995; Tester & Morrison 1990). However, digestion remains relatively slow while the starch is organized in its native (ungelatinized) state.

b. Gelatinized starch

Gelatinization is the loss of the pseudocrystalline structure of the starch granules and is characterised by a loss of the maltese cross pattern in polarised light and rapid water absorption and digestion in the presence of amylase. It involves a dramatic loss of structural organization of starch granules in response to temperatures above about 60°C in conjunction with excess moisture, or by processing at temperatures above 120°C at high shear, even at low moisture levels, such as during extrusion processing (**Figure 2**).

Various techniques used to study the gelatinization process suggest that the profound change in structure during gelatinization in moderate heat and in the presence of excess water is due principally to water invasion and swelling of the amorphous regions of the starch granule (Donald 2004). Because the molecules of the amorphous regions have connecting bonds with the semi-crystalline regions, as the amorphous regions swell they force the molecules of the pseudo-crystalline regions to dissociate. As the swelling and dispersion progresses, the starch becomes increasingly accessible to digestive enzymes, and the glycemic impact of the starch rises dramatically. Starch granule pores may assist by allowing water to invade deeply into the granule interior.

Starches differ in their susceptibility to gelatinization, and have been classified as those that swell rapidly, those that have restricted swelling associated with surface lipids and proteins (Debet & Gidley 2006), and a third group of granules that contain high amounts of amylose (high semi-crystalline content), which do not swell significantly at temperatures below 100°C.

c. Retrograded starch (RS3)

Retrogradation of starch is a form of structural change that has a large effect on digestibility. It occurs as the linear portions of starch molecules that have been dispersed during gelatinization randomly re-crystallize, without the organizing guidance of the living plant, when the gelatinized starch is cooled. Both amylose and amylopectin will retrograde.

However, amylose chains being less branched than amylopectin, will tend to re-crystallize almost irreversibly and again become nearly resistant to amylase digestion, while retrogradation of branched amylopectin is less complete and more reversible, and digestion by amylase is retarded less.

d. Modified starches (RS4)

As starch is a long digestible polymer covered in exposed hydroxyl groups, there are many ways that it may be modified. It may be partly depolymerized by enzymes or acid, substituent groups may be added (e.g. acetylated), it may be oxidized, cross-linked, pre-gelatinized and retrograded. Most modifications to starch are designed to change its functional properties as a food ingredient by altering its rheological characteristics (Taggart 2004; Whistler & BeMiller 1997). All the chemical/processing modifications involve structural change at the molecular level and many alter the digestion characteristics of the starch. Where chemical modification of starch causes resistance to digestion, type 4 resistant starch (RS4) is formed (Sajilata et al. 2006).

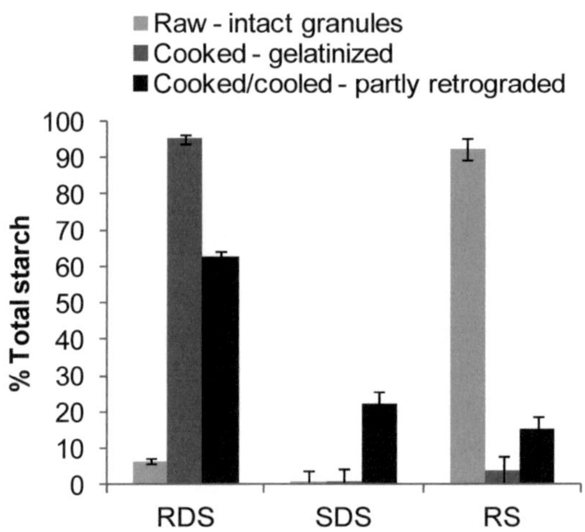

Figure 2. Effects of food structure at the molecular level - dependence of starch digestibility *in vitro* on its molecular form: rapidly digested (RDS), slowly digested (SDS) and digestion-resistant starch (RS) in potatoes digested raw (pseudo-crystalline, intact starch granules), freshly cooked (starch dispersed after gelatinizing), and cooked-cooled (starch partially recrystallized by retrogradation). (Mishra and Monro, unpublished)

e. Occluded starch (RS1)

In the mature endosperm of most cereals, the thin cell walls are largely obliterated and the endosperm becomes a protein matrix containing embedded starch granules (Eliasson & Wahlgren 2004; White & Johnson 2003). The density and occluding effect of the protein reduces water uptake during cooking by preventing the swelling of the starch granules and as a result reduces the rate of digestion. In species with hard endosperm, such as certain wheat and maize varieties and in rice, the protein matrix is almost continuous, whereas in wheat and maize cultivars with soft endosperm, and in cereals such as oats, rye and sorghum (Earp et al. 2004), there are many discontinuities that create pathways for water and enzyme penetration into the endosperm. As a result, soft endosperm variants hydrate more quickly and present a greater internal surface area of starch for water absorption and digestion.

In pastas based on high-protein durum wheat, a relatively slow rate of digestion and low glycemic impact has been attributed to protein coating the starch granules, inhibiting both gelatinization and amylase access to starch (Colonna et al. 1990; Jenkins et al. 1987). Microscopy has revealed that protein-starch conglomerates survive in cooked pasta (Kim et al. 2007). Because of the protein occlusion of starch, carbohydrate digestion in pasta may be enhanced by cooperative protease activity (Holm & Bjorck 1988). In fatty or oily tissues such as nuts, the hydrophobic nature of the fat may also be a factor protecting the starch from hydration, gelatinization and subsequent digestion.

f. Complexed starch

Complexing of starch with other macromolecules may involve a change in structure that is associated with reduced digestibility. Amylo-lipid complexes, formed when starch is gelatinized in the presence of lipid, are regarded as crystalline (Eliasson & Wahlgren 2004). The rate of digestion of amylose-lipid complexes is less than digestion of amylose, but greater than digestion of retrograded amylase (Holm et al. 1983).

3.2. Cell and tissue level

So far we have been discussing structural factors at the sub-cellular level that may affect carbohydrate digestibility. At the multicellular level, many sources of food carbohydrate are swallowed in the form of plant tissue fragments in which cell walls, and multiple overlying layers of cells, may act as partial barriers to both digestive enzyme penetration into, and carbohydrate diffusion out of the fragment or particle. In fruits, cereal kernels, nuts and pulses, tissue structure may influence the availability of carbohydrate and other nutrients (Mandalari et al. 2008; Palafox-Carlos et al. 2011; Tydeman et al. 2010a; Tydeman et al. 2010b).

Cereals

Seeds have evolved as dry, mechanically resistant structures that protect the embryo and the starchy endosperm from insect and animal attack until germination. In addition, in many mature grains such as rice, maize, the hard wheat varieties and some legumes, the molecular structural organization of starch and the protein that surrounds it results in a very hard endosperm that fragments into particles when crushed. Although the surface of such kernel particles is available for attack by digestive enzymes, penetration into the dense particles, especially when uncooked, is slow, and a high proportion of the starch may reach the colon. To obtain the digestible energy available in such grains, they must be subjected to processes such as grinding, flaking and cooking before being consumed. However, the dependence of digestibility on particle size provides a means by which the rate of starch availability, and the amount escaping foregut digestion to act as a substrate for colonic bacteria, may be influenced. The progressive decrease in rapidly digestible starch and increase in inaccessible (resistant) starch with increasing particle size is a very clear trend (**Figure 3**).

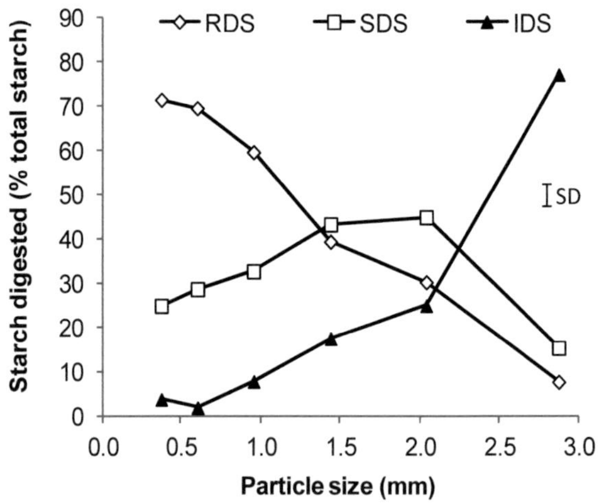

Figure 3. Effect of tissue structure on digestibility: Effect of particle size in chopped kernel fragments of the wheat cultivar 'Claire' on the *in vitro* digestibility of starch; RDS = rapidly digested (0-20 min), SDS = slowly digested (20-120 min), IDS = inaccessible digestible starch (undigested until residue homogenized) (Monro and Mishra, unpublished).

Pulses

In pulses, the starch-containing reserve tissues of the cotyledons differ in structure from the endosperm of cereals, in that the cells of the storage tissue are living and the walls retain an organized structure separating cells and contributing to tissue support in species in which the cotyledons become "seed leaves" after germination (Berg et al. 2012). In contrast, cereal endosperm cell walls are thin and usually disintegrated, and the structural integrity of the kernel is maintained by the starch/protein concretion of the endosperm, combined with the tough surrounding testa or seed coat (the bran in wheat). The differences in structure between the pulses and cereal products are reflected in the patterns of carbohydrate digestion from them (**Figure 4**).

The thick and resistant cell walls of pulses and may retard the gelatinization of starch by confining it within the cell lumen (Tovar et al. 1990; Tovar et al. 1992). When the starch is densely packed within resilient clusters of intact cells with robust cell walls, swelling is constrained. In addition, an encapsulating layer of gel from unconstrained starch in the outer cell layers of pulse fragments may create a barrier that impedes water penetration. However, partly because they are pectin rich compared with cereals, when processing is harsh or prolonged, the cell walls of pulses will degrade enough for the cells to separate. Then the starch becomes free to swell and disperse, and digestion is more rapid.

In domestic cooking, the robust cell walls of pulses are often able to survive moist heat enough to remain intact, so that cohesive plant cell clusters with encapsulated starch remain after cooking, making pulses some of the most slowly digested carbohydrate sources, with a typically low glycemic effect compared with other carbohydrate foods (Venn et al. 2006). Pulses such as kidney beans and chick peas are typical, and *in vitro* show a slow linear digestion that is usually incomplete (**Figure 4**). *In vivo* and particularly when cooking is incomplete or the food fragments are poorly comminuted before swallowing, they load the colon with fermentable starch (Type 1 resistant starch), which is largely responsible for the flatulence generated by pulses.

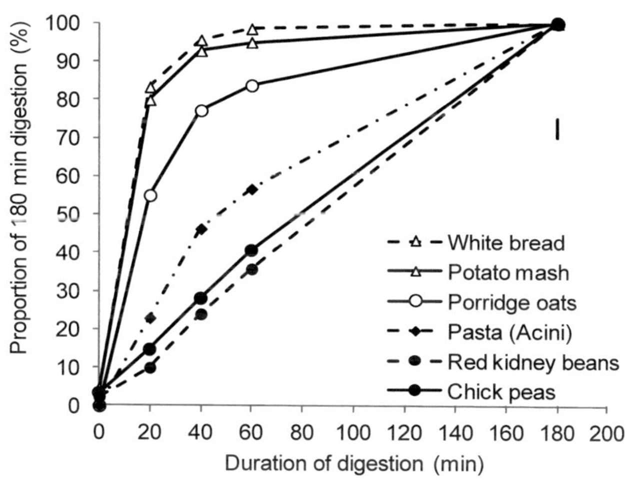

Figure 4. *In vitro* digestion patterns (%carbohydrate available after 180 min digestion) associated with different types of food structure: Porous, no intact cell walls (white bread), crushed and dispersed (mashed potato), crushed but partially intact native structure (porridge oats), dense non-porous structure (pasta; acini), robust and intact plant cell walls encapsulating starch (chick peas and red kidney beans). The bar represents the mean between duplicate ranges (5%). (Mishra and Monro, unpublished).

Fruits and vegetables

In most ripe fruits, available carbohydrates are in the form of soluble sugars – glucose, fructose and sucrose – which are highly soluble and mobile. Glucose and fructose are absorbed by specific transporters in the intestinal wall, while sucrose is hydrolyzed by a brush border invertase (Wright et al. 2006). Therefore, the only direct structural impediments to sugar availability from fruits are those that retard sugar diffusion. The parenchyma cell walls of fruits, even after mincing and digesting *in vitro*, can markedly retard sugar diffusion (**Figure 5**) and removing them from fruit puree increases its glycaemic impact (Haber et al. 1977). Such retardation can be regarded as a structural effect, as the presence of cell wall fragments with their enmeshed pectic polysaccharides increases the length of the diffusion pathways to a degree that would make a significant difference to blood glucose loading *in vivo*.

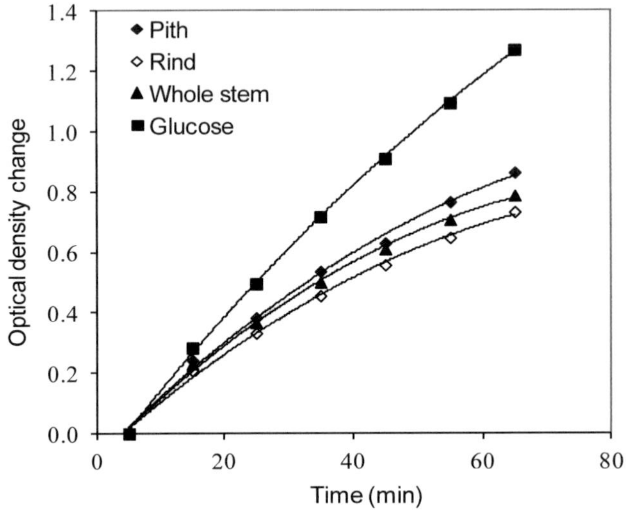

Figure 5. Effect of structure in digested plant tissue remnants on a process important to the digestive process - diffusion. Glucose diffusion was retarded about 40% by the presence of digestion-resistant remnants of broccoli tissue in an unstirred system. Pith – parenchyma cells. Rind (cortex) – parenchyma, fiber and vascular cells. The tissue remnants (cell walls) were at settled bed density, after they had been predigested *in vitro* and allowed to settle overnight by gravity. All the solutions contained 10% glucose (w/v) at the start of dialysis (Monro, unpublished). The mean between duplicate ranges was <0.1 OD units.

3.3. Food level - secondary structure established by processing

The characteristics of carbohydrate digestion in many carbohydrate foods are determined by the structure of the food matrix established during food processing involving cooking, with and without disruption of the original cellular structure of the plant source. Such secondary food structure exerts its influence largely by affecting the accessibility of the digestion medium to starch in the food. At the larger level, food particle geometry may create structures that have an enormous impact on digestibility through their influence on the surface area available for digestion (Monro et al. 2011).

a. Open porous structures

Foods with an open porous structure include leavened products such as breads and cakes, puffed products produced by steam expansion, including many snack foods, and breakfast cereals such as puffed rice. These structures have a high internal surface area that is almost immediately available for amylase attack, and as they are precooked to eliminate native starch granule structure, they typically digest very rapidly (**Figure 4**), causing an acute blood glucose response. Such foods are typified by a high rapidly digested starch (RDS) content, little slowly digested starch (SDS) and a small proportion of retrograded resistant starch (RS Type 3) (**Figure 6, Group A**).

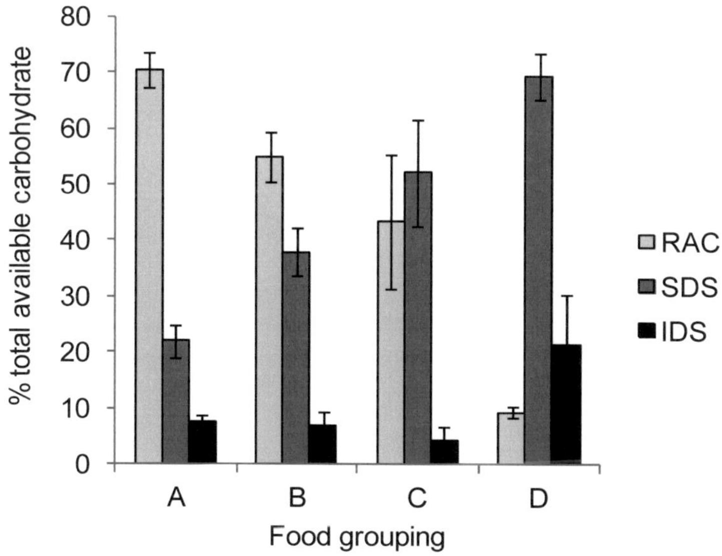

A: Little structure, starch gelatinized e.g. extruded, puffed and flaked precooked cereal products
B: Some structure, minimally processed, incomplete starch gelatinization e.g. rolled oats
C: Dense secondary structure, non-porous, surface digestion e.g. pasta
D: Intact, robust cell walls encapsulating starch in native tissue structure e.g. pulses.

Figure 6. Effects of food structure typical of various food groupings on the content of rapidly digested (RDS), slowly digested (SDS) and resistant (inaccessible; IDS) starch that they contain. Error bars are SDs of food means within groups. (Mishra and Monro, unpublished).

b. Dense low porosity structures

Dense low porosity structures include products such as pastas, in which hydrated flour is force-molded into a dense configuration and then cooked under conditions such as boiling, and in which porosity as a result of gas formation in the food matrix does not occur. Foods such as pastas allow digestion only as fast as digestive enzymes can erode superficial layers of the food, to expose underlying carbohydrate, so their rate of digestion depends strongly on their surface area, and is therefore dependent on particle geometry. The dependence of digestion rate on particle geometry has been examined in detail using pastas of different shapes and gelatinized sago as models. As surface area of a sphere depends on the square of the radius, small increases in particle size can have a large influence on digestion rate (Monro et al. 2011).

The role of the dense structure of unexpanded starch in retarding digestion becomes clear during *in vitro* digestion of solid foods such as pasta and tapioca with and without homogenizing to eliminate the occlusive effect of the starch. After homogenizing the pasta and tapioca, starch was immediately digested, in contrast to the starch in the intact pasta and tapioca, which was gradually released as the starch was digested by superficial erosion (**Figure 7**). In dense low-porosity structures such as pastas made from durum wheat, the occlusive effects of protein will also to retard starch digestion.

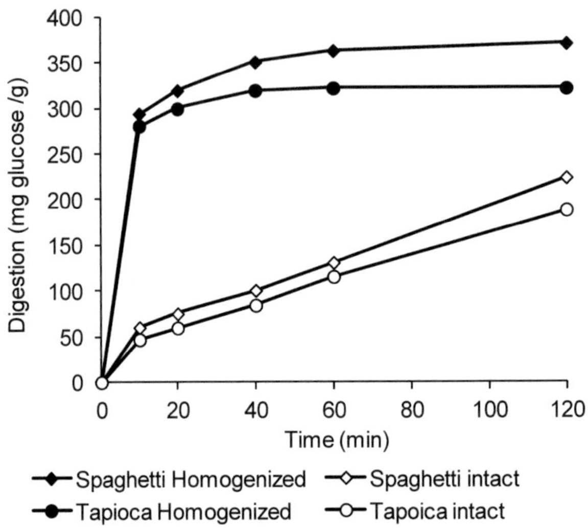

Figure 7. Effects of secondary structure on digestibility *in vitro*. Effect of occlusion by starch and dependence on surface area of food in dense non-porous foods revealed *in vitro* by the increased rate of digestion after homogenizing the foods to expose interior starch to digestive enzymes. The mean between duplicate range was < 5% (Mishra and Monro, unpublished).

4. Manipulating food structure to control carbohydrate digestibility

Many of the properties of foods that retard carbohydrate digestion discussed in the previous section are the result of the carbohydrate being stored by the plant during biosynthesis in a stable, organized, semi-crystalline and protected form, until it is required by the plant. For animals and humans to use plant carbohydrate as an energy source, it is necessary to overcome or reverse the steps the plant has taken to protect its energy reserves. Thus, before the process of enzymatic depolymerization of the individual starch molecules in the gut can provide the minimized molecular forms in which starch is absorbed by the gut border – as glucose, maltose and dextrins - there are several obstructions to be removed: firstly, the protective tissues of the plant; secondly, the barrier function of the starch granule surface; and thirdly, the obstructive molecular packing of starch within the starch granule must be overcome.

The inaccessibility of starch due to plant tissue structure has been overcome in four main ways, by external mechanical disruption, by cooking, by chewing, and in the intestine by the weak shearing and abrasion of intestinal contractions acting on digesta. In practice, with most starchy foods all four processes are used sequentially to obtain available carbohydrate from food for absorption, but the contributions that each may play in determining the amount of carbohydrate digested have been demonstrated individually.

4.1. By mechanical disruption during ingredient preparation

Mechanical disruption of food structure is one of the most effective ways of increasing carbohydrate energy availability from foods, and milling, crushing, pounding and such processes have been used for thousands of years to improve energy extraction from all types of plant tissue, but especially from the well protected form of seeds. On the other hand, reducing tissue disruption to lower carbohydrate digestibility of grain products has been found to be an effective strategy in reducing the glycemic impact of foods in populations with excessive energy intakes, obesity and diabetes (Venn & Mann 2004).

a. Milling

The most widely used procedures for milling of cereals such as wheat involve a combination of cutting and grinding, and applying strong shearing forces that break the seed coat and tear the endosperm tissue apart. More than any other process, milling can convert kernels of grain in which starch remains almost completely protected from digestion, even when cooked, to the finely divided form of flour, in which the same starch is rapidly digested because the inaccessibility of the starch to digestive enzymes caused by intact cell walls and protective plant structures has been eliminated. The effective protection of starch in cooked but intact plant tissue, and its susceptibility to digestion as soon as the tissue is disrupted by milling, is revealed by the changing distribution of starch between RDS (0-20 min), SDS (20-120 min), and RS (digestion-resistant) starch fractions measured by *in vitro* digestion (**Figure 2)**. With decreasing particle size the RDS fraction increases and the RS fraction decreases, while there is very little change in SDS; the cooked starch is either inaccessible ("resistant") or accessible, and if accessible, is rapidly digested - with the cell wall barrier removed starch is quickly degraded once gelatinized (Hallfrisch & Behall 2000).

The effects of milling do not result solely from the disruption of surrounding plant tissue, but damage to the starch granules, including cracking, fracturing, and internal changes to the granules occur that also increase their susceptibility to gelatinization and digestibility (Donald 2004).

b. Chopping/cutting

Cutting and chopping such as that occurring during the kibbling of grain does not cause the internal tissue disruption caused by the strong shearing force of grinding mills, and produce a more slowly digested product. Even simple crushing has an enormous influence on the availability of starch, because it involves forces strong enough to disrupt starch-protein conglomerates and cells containing encapsulated starch and other nutrients effectively, and creates pathways for ingress of digestive enzymes. Cutting and chopping without the shearing forces of grinding may increase digestibility much less because plant tissue damage is more restricted to the cut surfaces (**Figure 8**). Penetration of the effects of digestion through layers of intact cells underlying the cut surfaces may be a relatively slow process (Mandalari et al. 2008).

Recent detailed studies of the release of nutrients from nut fragments have shown that cell walls formed a very effective barrier against the intestinal environment (Mandalari et al. 2008; Palafox-Carlos et al. 2011; Tydeman et al. 2010a; Tydeman et al. 2010b). Mandalari et al. (2010) showed that after 3 h of simulated gastric plus duodenal digestion of almond fragments, the intracellular contents had been lost from only the first layer of the cells, at the fracture surface. After 12 h of digestion, the loss of nutrients had extended to only three to five cell layers deep. In the zone of digestion the cell walls appeared to change in structure, in that they swelled, but without any detectable change in composition of the cell wall polysaccharides. There was no evidence of cell wall fracture, so any enzyme penetration into the food particles could occur only by diffusion through the cell walls.

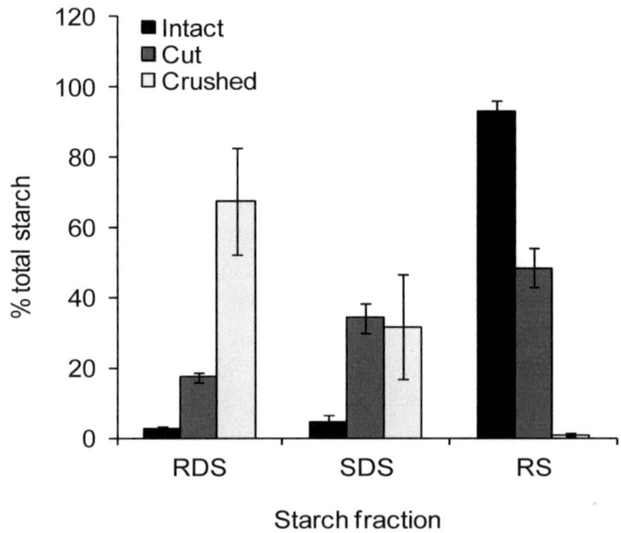

Figure 8. Influence of tissue structure on digestion revealed by the effect of cutting and crushing of cooked wheat kernels on *in vitro* digestion of starch. Individual whole hydrated kernels (n = 5/treatment) were cooked and digested either intact ("Intact"), after they had been cut transversely in half ("Cut") or crushed to 1 mm thickness ("Crushed"). RDS = rapidly digested, SDS = slowly digested, RS = resistant starch. Error bars are the standard deviations (Monro, unpublished).

4.2. By controlling gelatinization during cooking

Cooking starch under hydrating conditions brings about an often dramatic increase in starch digestibility as a result of gelatinization. In some plants, such as potato, the raw starch granules are virtually indigestible, but as soon as gelatinization occurs they, are rapidly and totally digested (**Figure 2**). Between the extremes of raw starch and total gelatinization, the degree of gelatinization may be controlled by limiting the amount of water available for hydration and by carefully controlling the cooking temperature. Because starch gelatinization requires a combination of heat and water, the availability of water during cooking can substantially modulate the effects of cooking on digestibility. In food products in which the water content is not adequate to gelatinize starch fully, the glycemic impact may be correspondingly reduced. Rolled oats, for instance, are prepared under conditions in which incomplete gelatinization of starch occurs. If consumed directly in the form of muesli, the starch digestibility is relatively low, but if further cooked and hydrated to form porridge, starch digestion is greatly increased along with the glycemic impact of the oats **(Figure 9).**

The sensitivity of starch digestion to the degree of hydration during cooking of a number of starches for 10 minutes at 95°C – maize (normal starch, 27% amylose), Hi maize (70% amylose), Mazaca (waxy maize, 2% amylose), pea, potato, rice, tapioca and wheat – increased steadily as moisture content was increased in 5 or 10% intervals from 0% moisture, reaching a maximum digestibility at about 60% moisture (**Figure 10**), which corresponds approximately to the degree of gelatinization that occurs in water-unlimited conditions at 100°C. The dependence of digestion on hydration during cooking was very similar for all the starches, even though the starches (controls) cooked with no added water differed considerably in their susceptibility to digestion in the raw (granular) state, as indicated by different Y intercepts in Figure 10. The differences in the digestibility of uncooked starch (Fig 10 A) probably reflect the importance of granule morphology, including pore size and surface chemistry, rather than any intrinsic differences in the starch molecules in determining the initial rate at which raw starch is digested. In food products containing hydrating components other than starch, such as non-starch polysaccharides in cell wall remnants, intrinsic and added gums, and sugars, competition for water may reduce the gelatinization of starch during cooking (Pomeranz et al. 1977), allowing the digestion-inhibiting effect of native starch structure to persist. Even in white bread, which contains about 60% moisture, partially intact birefringent starch granules remain in the cooked product.

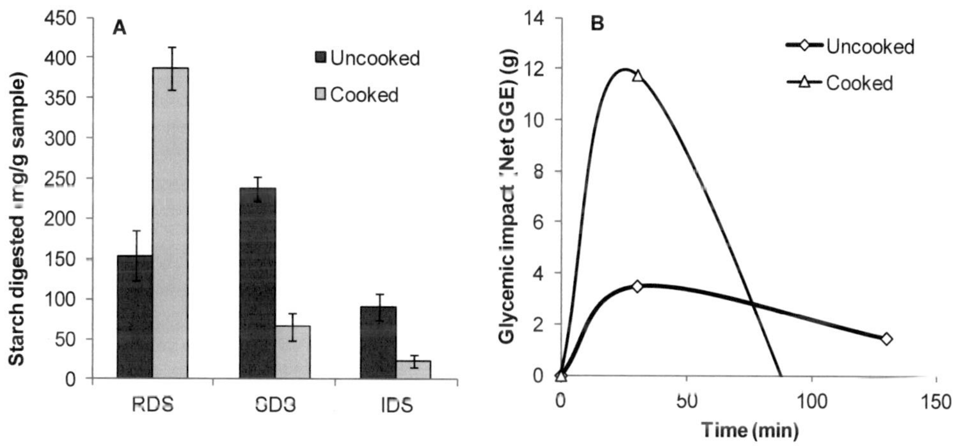

Figure 9. Effects on *in vitro* digestibility of further processing (boiling) of rolled oats (a minimally processed product) in preparing porridge, during which loss of starch structure (gelatinization) and extensive loss of oat grain structure occurs. **A** starch fractions of different digestibility defined in caption to Figure 3. **B** Theoretical blood glucose response curves based on RDS and SDS release with glucose disposal allowed for using the method of Monro et al. (2010). Error bars are standard deviations (Mishra and Monro, unpublished).

Two hydrothermal treatments of starch, annealing and heat-moisture treatment (HMT), cause changes in the structural and physicochemical properties and increase the digestibility of raw starch enough for it to qualify as slowly digested starch (digested between 20 and 120 min *in vitro*) (Lehmann & Robin 2007). In annealing, the starch is heated to below gelatinization temperature but long enough for some molecular rearrangement of the starch to occur. In HMT, higher temperatures are used but water is restricted so that full gelatinization does not occur. Although structural change is sufficient to increase digestibility, the granular structure and birefringence of the starch granules remain.

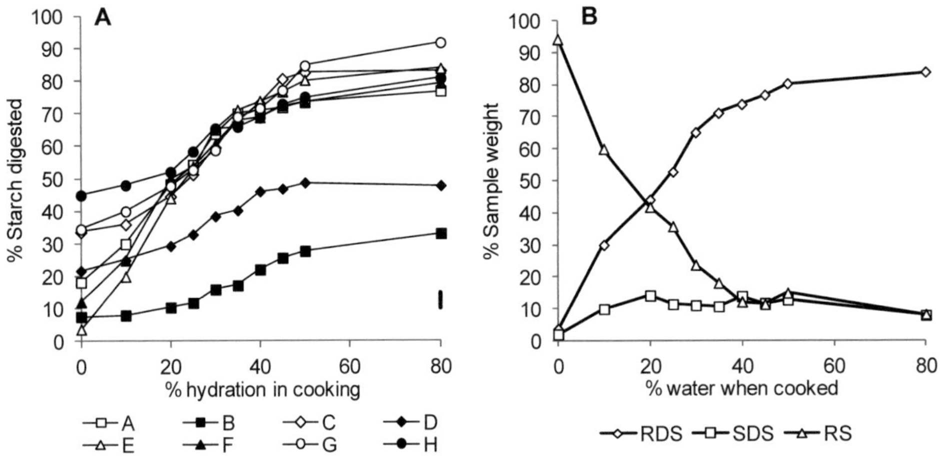

Figure 10. Effect of structural change in starch granules as a function of hydration during cooking, on *in vitro* digestion. Chart A. Rapidly digested starch as a % of total starch during cooking of seven starches at progressively increasing degrees of hydration: A. Maize; B. Hi Maize (high amylose); C. Mazaca (low amylose); D. Pea ('Sonata'); E. Potato; F. Rice; G. Tapioca; H. Wheaten cornflour. Chart B. Example of changes in starch fractions differing in digestibility: Potato starch showing that increased hydration caused the interconversion of resistant and rapidly digestible starch without substantially increasing slowly digested starch. The bar is the mean between-duplicate range.

4.3. By controlled retrogradation after cooking

During cooling of starch dispersed by gelatinization, the linear sections of amylopectin and amylose chains anneal to form hydrogen bonded alignments that limit access by digestive enzymes. The partial retrogradation of amylopectin in cooked potato to form slowly digested starch (SDS) is a good example of the effect of retrogradation (Figure 1; cooked-chilled), and has led to the suggestion that a way to reduce the glycemic impact of potatoes would be to consume them in the form of cold potato salad rather than as freshly cooked hot potato (Leeman et al. 2005; Monro & Mishra 2009).

Retrogradation is now used industrially to produce resistant starches (RS3). They have become widely used as bland low energy ingredients for which colonic benefits are claimed, in nutritionally enhanced bakery and other products.

4.4. By retaining tissue structure in whole foods – minimal processing

Tissue structure can be retained by consuming whole foods, in which the incomplete comminution achieved by chewing allows some survival and influence of food structure on carbohydrate availability. The diffusion of sugars from fruit pieces, for instance, is much slower than from fruits consumed as a puree (Haber et al. 1977).

The use of retained tissue structure in cooked food is most widely used in the baking industries, when fragments of intact kernels are included in grain breads. Chopping and cutting (kibbling) of grain kernels that would otherwise be ground to flour has found a place in food processing for populations with high rates of obesity and glucose intolerance, where there is need to reduce both the rate and extent of carbohydrate digestion. By substituting partially intact kibbled kernels for flour in bread products, the rate of digestion and the resulting glycemic impact of foods can be significantly reduced, as is revealed by the increases in digestion rates and *in vitro* glycemic index estimates when kernel-rich breads are homogenized to remove grain structure (**Figure 11**). Pumpernickel is an extreme example, as it consists largely of a conglomerate of rye grains, and in line with its low digestibility, it has a much lower glycemic index than most other breads (Jenkins et al. 1986).

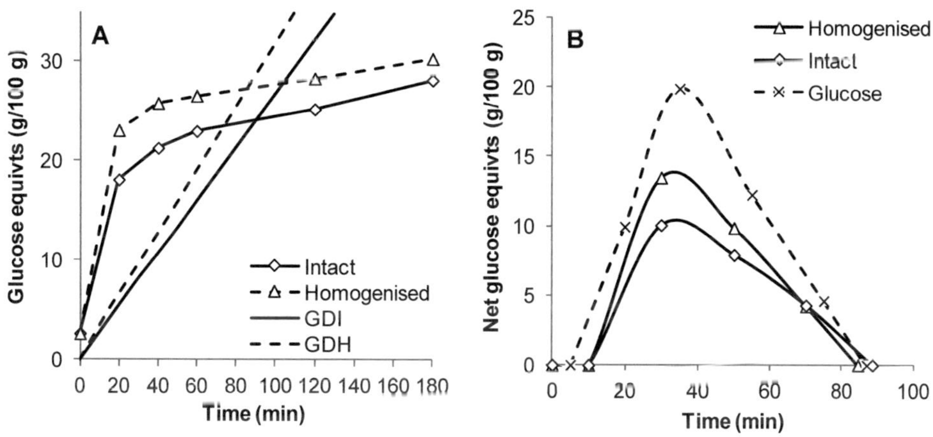

Figure 11. Effect of homogenizing to remove structure: A commercial wholegrain bread containing 25% kernel fragments > 2 mm in diameter was homogenized and the intact and homogenized breads digested *in vitro*. A. Digestion profiles (with markers) and lines of theoretical glucose disposal (GDI = glucose disposal for intact bread, GDH = glucose disposal for homogenized bread). Mean inter-duplicate range <5% at each time point. B. The curves after taking into account theoretical glucose disposal and response lag. The areas under the curves compared with the area under the glucose reference curve gave *in vitro* glycemic index values of 70 (high) for the homogenized bread compared with 55 (low) for the unhomogenized bread. Mean between-duplicate range was < 5% (Monro and Mishra, unpublished).

Minimal processing is not a term specific to any procedure, but it generally refers to processing that is the minimum required to make a product palatable, or saleable in a particular form for future cooking (Fellows 2000). The digestive advantages of minimal processing for nutrition are a reduced rate and extent of starch digestion, more resistant starch and non-starch polysaccharide for colonic function (as in bran), and greater nutrient retention than in a refined product from the same cereal source. The minimally processed category includes a range of cereal grains that have been steamed to partially precook and then rolled or flaked to eliminate hardness, such as rolled oats and barley. Depending on how thinly they are rolled they may have a relatively low glycemic impact until further processed into the form of hot porridge (Granfeldt et al. 2000) (**Figure 9**).

One of the main points supporting the argument in favor of consuming a greater proportion of dietary carbohydrate as whole grains, in which native structure is partially retained, is that "wholeness", in the sense of intactness, is associated with a reduced blood glucose loading, and increased colonic loading of resistant starch (Venn & Mann 2004). A significant association of increasing particle size with decreasing glycemic response (Fardet et al. 2006) and increased colonic fermentation (Bird et al. 2000) has been demonstrated and is consistent with results of *in vitro* analyses of grain starch digestibility.

4.5. By replacing native structure with secondary structure in food processing

a. Formation of open textures

Extrusion cooking under shear and pressure often at moisture contents of less that 15% is a means of producing highly digestible crisp, dry food products. The extrusion process involves high temperatures, extreme shearing forces, and release of hot product under high pressure, to yield expanded dry products that are gelatinized, porous and retain almost no native tissue or starch granule structure to resist amylase activity. Leavened products such as white bread are similar, except that tissue disruption is achieved by milling before cooking, and porosity is achieved by including a leavening agent such as yeast in the product formulation. The porosity of such products coupled with the absence of any integrated plant cell wall structures ensures rapid penetration of digestive enzymes and almost immediate collapse and digestion, so that the conversion of starch to sugar during digestion rapidly increases to a plateau where digestion is complete (**Figure 4**). Accordingly, such products are often of high glycemic impact (Foster-Powell et al. 2002).

b. Formation of dense low-porosity structures

Dense, low porosity food structures such as pasta are produced by extrusion at low temperatures with limited gelatinization and no puffing. In these foods, carbohydrate digestion is relatively slow and related to food geometry, because little enzyme penetration is possible, and any digestion is dependent on surface area (Monro et al. 2011). In such foods there is potential to use food shape to influence digestion rate as long as a proportion of the particles survive mastication. Dense foods that are soft enough to be swallowed partially intact, so that the influence of surface area on digestion rate is retained, may be useful in delivering available carbohydrate without an acute postprandial blood glucose response. Development of foods that use particle shape to determine digestion rate will therefore need also to consider the influence of food texture on the urge to chew.

4.6. By combinations of processes

Many food products are made using combinations of processes, each affecting food structure and carbohydrate digestibility in different ways. Some examples are given below (Table 1).

4.7. By changing the molecular structure of starch

Molecular modifications to starch that affect degree of branching or the ability of starch chains to interact or retrograde will affect digestibility, as already discussed. The feasibility of using starch modification at the point of biosynthesis in the plant is now being investigated, through genetic manipulation of the enzymes involved in establishing starch structure. Transgenic potato lines deficient in granule-bound starch synthase, and in two

Product type	Processes	Structure	Carbo-hydrate digestion	Relative Glycemic impact (GGE/g)[1]
Raw fruits	None	Plant tissue structure intact. Available carbohydrate as mono and disaccharides	High	Low
Pasta	Milling, cold low pressure, low shear extrusion, then time limited boiling	Dense, polymeric available carbohydrate (starch), incomplete starch gelatinization, protein occlusion of starch granules, superficial digestion	Slow but complete	Low-moderate
Biscuit	Milling then high temperature-low moisture cooking	Dense, friable, incomplete starch gelatinization. High fat levels coat starch	Moderate	Moderate
Bread, white	Milling then high temperature, high moisture cooking	Porous, more complete starch gelatinization than biscuits	High	High
Bread, kibbled grain	Milled flour plus cut grain, then high temperature, high moisture cooking	Porous gelatinized matrix containing intact kernel fragments with limited access of digestive enzymes	Moderate	Low-moderate[2]
Extrusion cooked/ puffed	Milling prior to high temperature, high shear, high pressure extrusion and puffing	All plant tissue structure and starch granule structure eliminated. Highly porous, readily accessible gelatinized starch	High	High
Muesli/ rolled oats	Crushing, steaming, limited cutting	Plant tissue structure partially intact, starch partially gelatinized	Moderate	Low[3]
Porridge	Crushing, steaming, then moist cooking	Plant tissue structure less intact than previous, starch gelatinized	High	Moderate - high

[1] Glycemic glucose equivalents/g sample

[2] Depends on inclusion rate of kibbled grain

[3] Depends on degree of crushing and chopping

Table 1. Examples of the relationship between processing combinations, food structure, carbohydrate digestibility, and relative glycemic impact per equal food weight in several food types.

starch-branching enzymes have been produced. Lines in which more linear starch was produced showed a lowered susceptibility to *in vitro* digestion than the parent lines (Karlsson et al. 2007).

Traditional plant breeding and selection of mutant types has already produced varieties with alterations in starch structure that affect starch digestibility. A high amylose maize cultivar, for instance, is used as a source of commercially available resistant starch (HiMaize®, Figure 10).

4.8. By structural breakdown during food consumption and digestion

a. Chewing

As chewing is the natural way to increase carbohydrate availability by reducing food structure, it has been seriously suggested that a way of using structure to reduce carbohydrate digestion rate, to reduce glycemic impact, is to swallow food without chewing it (Read et al. 1986). However, chewing of food is such an important part of enjoying it and converting it into a lubricated form that can be swallowed, that not chewing food is not a practical option for controlling glycemic impact.

Evolution has equipped humans with an effective grinding mechanism in the form of teeth, and the sensitive and dexterous combination of cheek and tongue to sort and position food particles accurately between the molar grinding surfaces. The effectiveness of crushing in bringing about the conversion of inaccessible digestible ("resistant"; RS) to rapidly digested (RDS) starch **(Figure 8)** underscores the importance of mastication to the successful exploitation of starch in an omnivorous diet.

Chewing has three very important functions, all related to food structure. Firstly, chewing crushes foods to release nutrients; secondly, it reduces the size of food fragments so they may be comfortably swallowed; and thirdly, and most importantly, it churns and mixes the food with saliva to convert it into the form of a well-lubricated semi-solid bolus that may be easily swallowed. Not until the food is swallowed can effective digestion commence.

The purpose of chewing has been revealed in studies of individual differences in chewing. Although individuals differ markedly in the mechanical details of how they go about their oral comminution of foods, they all arrive at a remarkably similar endpoint in terms of the particle size reduction in the mouth (**Figure 12**). It is apparent that the urge to stop chewing and swallow is determined more by the physical properties of the bolus that results from chewing, than by details of the mastication process that lead to the bolus. Dentition, the number of chews, the rate of chewing and so on are less important than the final result (Peyron et al. 2004).

In many processed foods produced nowadays, digestibility is not very dependent on structural degradation due to crushing by chewing, because the foods are based on ingredients, such as flour, that have been thoroughly comminuted by milling before cooking. Foods for which chewing makes a difference are usually those consisting of, or containing intact grains, such as rice and kibbled grains, and those consisting of dense, non-porous starch matrices. For many starchy processed foods the combination of mastication and salivary α-amylase activity reduces the adhesiveness of starch, and the structural cohesion of food in the form of a bolus, allowing the stomach to separate available carbohydrate from more fibrous components quite rapidly for transfer to the duodenum, which is the primary site of carbohydrate digestion in the gut.

The combination of salivary α-amylase with chewing, and the fact that starch is usually gelatinized and not intrinsically fibrous, means that most of the starch component is quickly reduced to a small particle-containing slurry that can be separated in the stomach and moved on to the ileum for digestion with little delay. The rapid dispersion of starch in most foods explains why, despite the tendency of the stomach to retain large particles, blood glucose responses to foods almost invariably commence after a lag of only about 10 min from ingestion and almost always reach a peak between 30 and 40 minutes from ingestion (Brand-Miller et al. 2009).

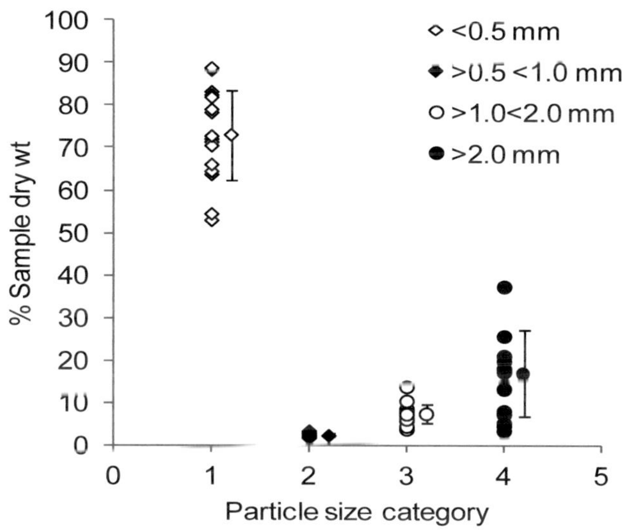

Figure 12. Effect of chewing on reduction of structure in a carbohydrate food – rice. Range of particle sizes in within size categories from cooked white rice chewed by 20 subjects. The subjects chewed intact whole rice grains in quantities they would normally consume and expectorated them when they felt the urge to swallow. For all subjects, most of the chewed sample was less than 0.5 mm in diameter. Means ± SD shown for each size category. Based on data used in Ranawana et. al. (2010).

b. Gastric and small intestinal digestion

The stomach and intestine are not passive reservoirs, but are motile reactors in which food, while undergoing enzymatic dismemberment, is also continually subjected to shearing and abrasion from circumferential, longitudinally migrating contractions of the gut wall (Lentle & Janssen 2008). Compared with the concentrated forces exerted at the molar surfaces by jaw muscles in chewing, the forces on food particles in the stomach and intestine due to peristalsis are very small (Lentle & Janssen 2008). However, in combination with digestive enzyme action they have the important role of reducing food structure by sloughing off the hydrated and digestively weakened surface layers of food particles, to improve access of digestive enzymes to the interior.

When the mechanical and processing steps of food preparation and mastication have reduced food structure to the extent that the food can be swallowed, and the stomach has then separated a starchy slurry from the large particles remaining, the starch is ready for digestion in the ileum by α-amylase from the pancreas working in concert with enzymes in the gut wall.

The rate of starch digestion depends partly on the rate of gastric emptying (Darwiche et al. 2001), but also the structural form in which it arrives in the ileum, as intact starch granules, as disorganized or dispersed gelatinized starch, or as once-disorganized starch that has re-aggregated to form retrograded starch. Digestion of intact food particles or starch granules is relatively slow in the gut, as it is *in vitro,* and the digestion pattern depends very much on the botanical origin of the granules (Donald 2004; Oates 1997).

a. Colonic digestion

In the colon, carbohydrate digestibility is governed by a completely different set of parameters than in the foregut, to which the discussion has referred so far. Carbohydrates entering the colon are those that were unable to be digested and/or absorbed during gastric/ileal transit. They include starch that has survived digestion for the reasons of structure and chemistry already discussed, including crystallinity in ungelatinized and retrograded starch, encapsulation by plant tissue cell walls, and occlusion by fat. However the main carbohydrate source entering the colon consists of the non-starch polysaccharides that constitute the plant cell wall and in the colonic ecosystem into which they pass they are exposed to a myriad of bacterial enzymes that are absent from the foregut. The colonic bacteria disassemble the cell walls to provide carbohydrate substrate for bacterial fermentation. The products of colonic fermentation are short chain fatty acids that are absorbed and enter intermediary metabolism, where they may provide as much as 10% of dietary energy requirements for humans.

The physical/structural constraints that modulate colonic fermentation of polysaccharide residues involve molecular structure, occlusion and particle size, which may all affect availability of substrate for bacterial enzymes and the ability of bacteria to colonize and invade fragments of plant tissue and cell walls. In the bacterial ecosystem that consists of thousands of species of bacteria that can adapt rapidly to changes in available substrates, there are few natural plant polysaccharides that on their own can resist the multipronged and coordinated attack of the diverse colonic microflora. However, some, such as psyllium gum, which is a complex and highly branched polysaccharide, are fermented so slowly that much of their molecular structure, and the hydration capacity that depends on it, remains intact after passage through the colon. Such polysaccharides make very effective faecal bulking agents but may lead to problems such as reduced mixing and fermentation in the colon when present at high concentrations. Cellulose, which exists as highly crystalline fibrils, is also slowly fermented, and is a major component of the dietary fiber that survives and contributes faecal bulk in plant-containing diets (Monro & Mishra 2010) .

Plant residues are often non-fermentable because of the occlusive effects of secondary thickening of the tissues, initially by cellulose but followed in many cases by the deposition of lignin, resulting from phenolic condensation within pre-existing cell walls, usually when they have already undergone secondary thickening by cellulose deposition (Esau 1967). The combined effects of increasing amounts of crystalline cellulose and lignin is clearly seen in the contrast between fermentation of parenchymatous pith cells and of secondarily thickened rind (cortex) cells of broccoli stem in the hind gut. Although derived from the same parenchymatous ground tissue, the pith remains as parenchyma, while the rind differentiates to contain a high proportion of secondarily thickened and lignified xylem tissue. The parenchyma cells are almost completely consumed by the bacterial flora in the hind gut, while the cells of the rind remain apparently intact and recognizable in the feces (Monro & Mishra 2010).

Even within tissues consisting entirely of parenchyma cells, the rate of fermentation can be modulated by structure. Paradoxically, multicellular clumps of carrot cells were found to be more rapidly fermented than cell wall fragments, probably because the intercellular spaces and angles between cells in multicellular clusters provided colonization sites for colonic bacteria (Day et al. 2012). There is, however, likely to be an optimal size at which increased colonization sites are counterbalanced by inaccessibility in large particles.

5. Conclusion

Food structure can affect carbohydrate digestibility in a range of ways. From the level of molecular conformation to plant anatomy, structure plays an important role in carbohydrate digestion. Attempts to manipulate structure as a means of controlling nutritional attributes of foods related to carbohydrate digestion have been in progress for thousands of years. They remain an important focus of modern food technology aimed at addressing the health problems associated with both inadequate energy intakes in developing countries, and excessive energy consumption in the developed world.

Author details

Suman Mishra and John Monro

Food Industry Science Centre, The New Zealand Institute for Plant & Food Research Limited, Palmerston North, New Zealand

Allan Hardacre

Institute of Food Nutrition and Human Health, Massey University, Palmerston North, New Zealand

6. References

Ayoub A, Ohtani T, Sugiyama S 2006. Atomic force microscopy investigation of disorder process on rice starch granule surface. Starch-Starke 58(9): 475-479.

Berg T, Singh J, Hardacre A, Boland MJ 2012. The role of cotyledon cell structure during in vitro digestion of starch in navy beans. Carbohydrate Polymers 87(2): 1678-1688.

Bird AR, Hayakawa T, Marsono Y, Gooden JM, Record IR, Carrol RL, Topping DL 2000. Coarse brown rice increases faecal and large bowel short chain fatty acids and starch but lowers calcium in the large bowel. J Nutr 130: 1780-1787.

Bjorck I, Granfeldt Y, Liljeberg H, Tovar J, Asp NG 1994. Food Properties Affecting the Digestion and Absorption of Carbohydrates. American Journal of Clinical Nutrition 59(3): S699-S705.

Brand-Miller JC, Stockmann K, Atkinson F, Petocz P, Denyer G 2009. Glycemic index, postprandial glycemia, and the shape of the curve in healthy subjects: analysis of a database of more than 1000 foods. American Journal of Clinical Nutrition 89(1): 97-105.

Brownlee IA 2011. The physiological roles of dietary fibre. Food Hydrocolloids 25(2): 238-250.

Brownlee M 2001. Biochemistry oand molecular biology of diabetic complications. Nature 414: 813-820.

Buttriss JL, Stokes CS 2008. Dietary fibre and health: an overview. Nutr Bull 33: 186-200.

Colonna P, Barry JL, Cloarec D, Bornet F, Gouilloud S, Galmiche JP 1990. Enzymic susceptibility of starch from pasta. Journal of Cereal Science 11(1): 59-70.

Copeland L, Blazek J, Salman H, Tang MCM 2009. Form and functionality of starch. Food Hydrocolloids 23(6): 1527-1534.

Darwiche G, Ostman EM, Liljeberg HGM, Kallinen N, Bjorgell O, Bjorck IME, Almer LO 2001. Measurements of the gastric emptying rate by use of ultrasonography: studies in humans using bread with added sodium propionate. American Journal of Clinical Nutrition 74(2): 254-258.

Day L, Gomez J, Oiseth SK, Gidley MJ, Williams BA 2012. Faster Fermentation of Cooked Carrot Cell Clusters Compared to Cell Wall Fragments in Vitro by Porcine Feces. Journal of Agricultural and Food Chemistry 60(12): 3282-3290.

Debet MR, Gidley MJ 2006. Three classes of starch granule swelling: Influence of surface proteins and lipids. Carbohydrate Polymers 64(3): 452-465.

Donald AM 2004. Understanding starch structure and functionality. In: A-C E ed. Starch in Food. Cambridge, UK, Woodhead Publishing Limited.

Earp CF, McDonough CM, Rooney LW 2004. Microscopy of pericarp development in the caryopsis of Sorghum bicolor (L.) Moench. Journal of Cereal Science 39(1): 21-27.

Eliasson AC, Wahlgren M 2004. Starch-lipid interactions and their relevance in food products. In: Eliasson AC ed. Starch in food. Cambridge, Woodhead Publishing Limited. Pp. 441-460.

Esau K 1967. Plant Anatomy. New York, John Wiley & Sons Inc.

Fardet A, Leenhardt F, Lioger D, Scalbert A, Remesy C 2006. Parameters controlling the glycaemic response to breads. Nutrition Research Reviews 19(1): 18-25.

Fellows P 2000. Food processing technology: principles and practice.

Foster-Powell K, Holt SHA, Brand-Miller JC 2002. International table of glycemic index and glycemic load values: 2002. American Journal of Clinical Nutrition 76(1): 5-56.

French D 1984. Organisation of starch granules. In: Whistler RL, BeMiller JN, Paschall EF eds. Starch chemistry and technology. New York, Academic Press. Pp. 184-242.

Fuentes-Zaragoza E, Riquelme-Navarrete MJ, Sanchez-Zapata E, Perez-Alvarez JA 2010. Resistant starch as functional ingredient: A review. Food Research International 43(4): 931-942.

Gallant DJ, Bouchet B, Baldwin PM 1997. Microscopy of starch: Evidence of a new level of granule organization. Carbohydrate Polymers 32(3-4): 177-191.

Gibson GR, Probert HM, Van Loo G, Rastall B, Roberfroid MB 2004. Dietary modulation of the human colonic microbiota: updating the concept of prebiotics. Nutrition Research Reviews 17: 259-75.

Granfeldt Y, Eliasson AC, Bjorck I 2000. An examination of the possibility of lowering the glycemic index of oat and barley flakes by minimal processing. Journal of Nutrition 130(9): 2207-2214.

Guigliano D, Ceriello A, Esposito K 2008. Glucose metabolism and hyperglycemia. Amer J Clin Nutr 87 (suppl): 217S-222S.

Haber GB, Heaton KW, Murphy D 1977. Depletion and disruption of dietary fibre. Effects on satiety plasma glucose and serum-insulin. Lancet ii: 679-682.

Hallfrisch J, Behall KM 2000. Mechanisms of the effects of grain on insulin and glucose response J Amer Coll Nutr 19: 320S-325S.

Holm J, Bjorck I 1988. Effects of Thermal-Processing of Wheat on Starch .2. Enzymic Availability. Journal of Cereal Science 8(3): 261-268.

Holm J, Bjorck I, Ostrowska S, Eliasson AC, Asp NG, Larsson K, Lundquist I 1983. Digestibility of Amylose-Lipid Complexes Invitro and Invivo. Starke 35(9): 294-297.

Huber KC, BeMiller JM 2000. Channels of maize and sorghum starch granules. Carbohydrate Polymers 41: 269-276.

James MG, K. , Denyer KaAM, Myers AM 2003. Starch synthesis in the cereal endosperm. Current Opinions in Plant Biology 6: 215-222.

Jenkins DJ, Wolever TM, Jenkins A, Giordano C, Giudici S, Thompson LU, Kalmusky J, Josse RG, Wong GS 1986. Low glycemic response to traditionally processed wheat and rye products: Bulgar and pumpernickel bread. American Journal of Clinical Nutrition 43: 516-520.

Jenkins DJA, Thorne MJ, Wolever TMS, Jenkins AL, Rao AV, Thompson LU 1987. The Effect of Starch-Protein Interaction in Wheat on the Glycemic Response and Rate of Invitro Digestion. American Journal of Clinical Nutrition 45(5): 946-951.

Juszczak L, Fortuna T, Krok F 2003. Non-contact Atomic Force Microscopy of Starch Granules Surface. Part I. Potato and Tapioca Starches. Starch/Staerke 55: 1-7.

Karlsson ME, Leeman AM, Bjorck IME, Eliasson A-C 2007. Some physical and nutritional characteristics of genetically modified potatoes varying in amylose/amylopectin ratios. Food Chemistry 100(1): 136-146.

Kim E, Petrie J, Motoi L, Sutton K, Morgenstern M, Mishra S, Simmons L 2007. Effect of structural and physicochemical characteristics of the protein matrix in past on in-vitro starch digestibility. Food Biophysics 3(2): 229-234.

Leeman M, Ostman E, Bjorck I 2005. Vinegar dressing and cold storage of potatoes lowers postprandial glycaemic and insulinaemic responses in healthy subjects. European Journal of Clinical Nutrition 59(11): 1266-1271.

Lehmann U, Robin F 2007. Slowly digestible starch - its structure and health implications: a review. Trends in Food Science & Technology 18(7): 346-355.

Lentle RG, Janssen PWM 2008. Physical characteristics of digesta and their influence on flow and mixing in the mammalian intestine: a review. Journal of Comparative Physiology B-Biochemical Systemic and Environmental Physiology 178(6): 673-690.

Mandalari G, Faulks RM, Rich GT, Lo Turco V, Picout DR, Lo Curto RB, Bisignano G, Dugo P, Dugo G, Waldron KW, Ellis PR, Wickham MSJ 2008. Release of protein, lipid, and vitamin E from almond seeds during digestion. Journal of Agricultural and Food Chemistry 56(9): 3409-3416.

Monro J, Mishra S 2009. Nutritional Value of Potatoes: Digestibility, Glycemic Index, and Glycemic Impact. In: Singh. J, Kaur. L eds. Advances in Potato Chemistry and Technology Elsevier Inc. Pp. 371-394.

Monro JA, Mishra S 2010. Digestion-resistant remnants of vegetable vascular and parenchyma tissues differ in their effects in the large bowel of rats. Food Digestion 1(1-2): 1 (1-2) 47-56.

Monro JA, Mishra SS, Hardacre A 2011. Glycaemic impact regulation based on progressive geometric changes in solid starch-based food particles during digestion. Food Digestion 2(1--3): 1-12.

Nilsson AC, Ostman EM, Hoist JJ, Bjorck IME 2008. Including indigestible carbohydrates in the evening meal of healthy subjects improves glucose tolerance, lowers inflammatory markers, and increases satiety after a subsequent standardized breakfast. Journal of Nutrition 138(4): 732-739.

Oates CG 1997. Towards understanding of starch granule structure and hydrolysis. Trends Food Sci Tech 8: 375-382.

Palafox-Carlos H, Ayala-Zavala JF, Gonzalez-Aguilar GA 2011. The Role of Dietary Fiber in the Bioaccessibility and Bioavailability of Fruit and Vegetable Antioxidants. Journal of Food Science 76(1): R6-R15.

Peyron MA, Mishellany A, Woda A 2004. Particle size distribution of food boluses after mastication of six natural foods. Journal of Dental Research 83(7): 578-582.

Planchot V, Colonna P, Gallant DJ, Bouchet B 1995. Extensive degradation of native starch granules by alpha-amylase from aspergillus fumigatus. Journal of Cereal Science 21(2): 163-171.

Pomeranz Y, Shogren MD, Finney KF 1977. Fiber in breadmaking – Effects on functional properties. Cereal Chem 54: 25-41.

Ranawana V, Monro JA, Mishra S, Henry CJK 2010. Degree of particle size breakdown during mastication may be a possible cause of interindividual glycemic variability. Nutrition Research 30(4): 246-254.

Ratnayake WS, Jackson DS 2007. A new insight into the gelatinization process of native starches. Carbohydrate Polymers 67(4): 511-529.

Read NW, Welch IM, Austen CJ, Barnish C, Bartlett CE, Baxter AJ, Brown G, Compton ME, Hume KE, Storie I, Worlding J 1986. Swallowing Food without Chewing - a Simple Way to Reduce Postprandial Glycemia. British Journal of Nutrition 55(1): 43-47.

Sajilata MG, Singhal RS, Kulkarni PR 2006. Resistant starch – a review. Comprehensive Review in Food Science Safety 5: 1-17.

Saltiel AR, Kahn CR 2001. Insulin signalling and the regulation of glucose and lipid metabolism. Nature 414(6865): 799-806.

Taggart P 2004. Starch as an ingredient: manufacture and applications. In: Eliasson AC ed. Starch in food. Cambridge, Woodhead Publishing Limited. Pp. 363-392.

Tester RF, Morrison WR 1990. Swelling and gelatinization of cereal starches .1. Effects of amylopectin, amylose, and lipids. Cereal Chemistry 67(6): 551-557.

Thomas DJ, Atwell WA eds. 1998. Starches. 2 ed. St Paul, Minnesota USA, Eagen Press. 100 p.

Tovar J, Bjorck IM, Asp NG 1990. Analytical and nutritional implications of limited enzymic availability of starch in cooked red kidney beans. Journal of Agricultural and Food Chemistry 38(2): 488-493.

Tovar J, Granfeldt Y, Bjorck IM 1992. Effect of processing on blood-glucose and insulin responses to starch in legumes. Journal of Agricultural and Food Chemistry 40(10): 1846-1851.

Turnbaugh PJ, Ley RE, Mahowald MA, Magrini V, Mardis ER, Gordon JI 2006. An obesity-associated gut microbiome with increased capacity for energy harvest. Nature 444(7122): 1027-1031.

Tydeman EA, Parker ML, Faulks RM, Cross KL, Fillery-Travis A, Gidley MJ, Rich GT, Waldron KW 2010a. Effect of Carrot (Daucus carota) Microstructure on Carotene Bioaccessibility in the Upper Gastrointestinal Tract. 2. In Vivo Digestions. Journal of Agricultural and Food Chemistry 58(17): 9855-9860.

Tydeman EA, Parker ML, Wickham MSJ, Rich GT, Faulks RM, Gidley MJ, Fillery-Travis A, Waldron KW 2010b. Effect of Carrot (Daucus carota) Microstructure on Carotene Bioaccessibilty in the Upper Gastrointestinal Tract. 1. In Vitro Simulations of Carrot Digestion. Journal of Agricultural and Food Chemistry 58(17): 9847-9854.

Venn BJ, Mann JI 2004. Cereal grains, legumes and diabetes. European Journal of Clinical Nutrition 58(11): 1443-1461.

Venn BJ, Wallace AJ, Monro JA, Perry T, Brown R, Frampton C, Green TJ 2006. The glycemic load estimated from the glycemic index does not differ greatly from that measured using a standard curve in healthy volunteers. Journal of Nutrition 136(5): 1377-1381.

Waigh TA, Kato KL, Donald AM, Gidley MJ, Clarke CJ, Riekel C 2000. Side-chain liquid-crystalline model for starch. Starch/Starke 52: 450-460.

Whistler RL, BeMiller JM 1997. Carbohydrate Chemistry for Food Scientists. Minnesota, USA, Eagan Press.

White PJ, Johnson LA 2003. Corn Chemistry and Technology, 2nd Edition, American Association of Cereal Chemists.

Wikoff WR, Anfora AT, Liu J, Schultz PG, Lesley SA, Peters EC, Siuzdak G 2009. Metabolomics analysis reveals large effects of gut microflora on mammalian blood metabolites. Proceedings of the National Academy of Sciences of the United States of America 106(10): 3698-3703.

Wright EM, Loo DDF, Hirayama BA, Turk E 2006. Sugar Absorption Physiology of the Gastrointestinal Tract. In: Johnson LR ed. Physiology of the gastrointestinal tract. 4th ed. New York, Elsevier Academic Press. Pp. 1653-1666.

Zimmet P, Alberti K, Shaw J 2001. Global and societal implications of the diabetes epidemic. Nature 414(6865): 782-787.

Chapter 14

Carbohydrate Metabolism in Drosophila: Reliance on the Disaccharide Trehalose

Alejandro Reyes-DelaTorre, María Teresa Peña-Rangel and Juan Rafael Riesgo-Escovar

Additional information is available at the end of the chapter

http://dx.doi.org/10.5772/50633

1. Introduction

Work on Drosophila has been so influential that it has even impinged on human metabolism and health. The reason for this is obvious: There is a strong evolutionary conservation between biological processes of flies and humans. Also, cloning and functional analyses of genes in Drosophila allowing study of many cellular processes have greatly aided in the identification of homologous genes and processes in humans and other organisms, as exemplified by the study of homeotic genes and mutations [1]. A great number of genes required for a myriad different functions (from transcription factors, structural proteins, ion channels and signaling molecules, to those required for behavior and sleep) is common between the two species [2]; in fact, [3] showed that from a pool of 287 genes known to be implicated in humans in cancer and malformations, and neurological, cardiovascular, hematological, immune, endocrine, renal and metabolic diseases, 178 (62%) had their counterparts in flies. These findings underscore the potential impact of Drosophila as a powerful system for gene discovery, but also for study of diseases' symptoms and complications, and the discovery of therapeutic treatments.

Drosophila melanogaster was originally introduced as a research organism and genetic model at the beginning of the twentieth century by T. H. Morgan and co-workers. From their pioneering studies on heredity and the chromosome theory of inheritance [4], work with this model organism has been the source of biological insights in many areas, including genetics and development [5]. Clearly, Drosophila offers an array of advantages: its cultures in the laboratory are easy and economical, its life cycle is short (approximately ten days at room temperature), its embryonic development lasts only twenty four hours, there is no meiotic recombination in males, it has clearly defined and easily identified structures in the cytoskeleton (subject to mutational change, and, correspondingly, easy to score), has a group of marked balancer chromosomes that impede meiotic recombination in females thus allowing the recovery and maintenance of other mutations of interest, has only four pairs of chromosomes, etc. These characteristics all greatly aid in genetic analyses [6], and since the sequencing of its genome (first released in 2000 [2, 7]), has arguably become the best characterized multicellular eukaryote model system. Critically, there are also large collections of mutant strains, with the one housed at the Bloomington Drosophila Stock Center at Indiana University [8] culturing over 49,000 different mutant stocks. Finally, since the publication of the fly genome, and its refinements with time, emphasis has also been placed on analyses and studies of physiology and signaling pathways involved, rather than just concentrating on 'gene hunts' or 'fishing expeditions' for candidate genes [9].

Since the discovery and characterization of abnormal phenotypes of insulin pathway mutants in flies [10], one of recent focuses of work in Drosophila have been metabolic studies, especially those of carbohydrates. One of the underlying reasons for this is, no doubt, the recent surge in obesity and diabetes mellitus type 2 among the world's population in both developing and developed countries [11, 12]. Just as it happens in vertebrates, *Drosophila melanogaster* regulates its levels of circulating carbohydrates, and stores excesses in the forms of glycogen and lipids, especially of triglycerides, in the fat body [13]. Another fruit of this recent surge in carbohydrate studies has been the elucidation of evolutionarily conserved signaling pathways, like the insulin/IGF/Target of Rapamycin (TOR) pathway [14].

2. The insulin pathway

Glucose homeostasis in mammals is maintained by feedback mechanisms balancing glucose cellular import and replenishment to the blood, so as to maintain a constant 5.5 mM level [15]. This level is the result of glucose cellular entry rates, glucose removal by the liver and new glucose coming from digestion of food glucose sources to the blood. Members of the Glut family of facilitated transporters, of which mammals possess several isoforms and genes, carry out glucose cellular import. It was recognized in the middle nineties that insulin stimulation in adipocytes in culture resulted in activation of a phosphatidyl-inositol-3-kinase (PI3K) isoform, and more glucose cellular import, resulting from Glut4 plasma membrane translocation [16]. Nowadays, Glut4 plasma membrane translocation from an internal vesicle pool is recognized as the rate-limiting step in glucose import by muscle (skeletal and cardiac) cells, and vertebrate adipocytes [17].

In contrast, the relation between insulin signaling and glucose cellular import in insect cells is not that clear. In Drosophila, there is a glucose transport system with similar kinetics to vertebrate Glut transporters [18]. A Drosophila Glut1 homolog cDNA has been cloned and described [19]. Nonetheless, experiments have shown that there is no increase in cellular import of 2-deoxyglucose labeled with ^{3}H in Drosophila Kc cells in culture after activation of the insulin pathway [20], and neither manipulation of activation levels of PI3K or protein kinase B (also called AKT or PKB), both molecules in the insulin pathway, in Drosophila S2 cells augments glucose cellular import [21].

These results could lead one to think that insulin pathway function in vertebrates and insects is not well conserved, and that insect insulin-like-peptides (ILPs) have no role in glucose and / or carbohydrate homeostasis, yet there is ample evidence to the contrary. The Drosophila genome possesses eight insulin-like–peptides (ILPs) [22-25] at last count, and these have been shown to interact with the Drosophila insulin receptor homolog (InR). Moreover, the Drosophila insulin receptor homolog can be functionally exchanged with the vertebrate insulin receptor [26], arguing that functional conservation must be significant.

Furthermore, ablation of the neurosecretory cells (NSC) that secrete DILPs 2, 3 and 5 in larvae gives rise to adults with developmental delays, reduced body size, and high levels of carbohydrates and lipids. Ablation of these same cells in adult flies lead to hypertrehalosemia without growth phenotypes. Knockdown of DILP2 alone by RNAi in NSC results in higher corporal trehalose levels [27, 28]. These effects, due to lower levels of circulating DILPs, causing lipid and carbohydrate accumulation, are analogous to effects observed in diabetic patients or diabetic mice when there is generalized insulin resistance, like in the insulin receptor knockout mice model [13].

Perhaps the clearest demonstration of the insulin pathway's role in carbohydrate and lipid metabolism comes from studies of Drosophila insulin pathway mutants: From the pioneering studies on Chico, a Drosophila insulin-receptor-substrate (IRS) homolog [10], it has now been demonstrated that practically all viable mutant combinations, that create partial loss-of-function or hypomorphic conditions for the insulin pathway, course with altered lipid and carbohydrate levels [29]. This is also akin to having congenital diabetes mellitus type 2, as these flies are born with a dysfunctional insulin pathway. Besides this metabolic disarray, these mutants have developmental delays, fewer and smaller cells, and concomitantly, smaller sizes, these last being cell independent [10]. In this regard, fly mutant growth phenotypes are similar to IRS1 loss-of-function mice models [30, 31] and other murine diabetic models [32] where growth phenotypes also exist; besides, in flies, the insulin and IGF (insulin growth factor) pathways are mediated by a common route. In addition, fly mutants also have nervous system abnormalities [29].

The insulin and insulin growth factor (IGF) pathways, as stated above, are united in Drosophila (see figure 1). Eight ILPs function as ligands for the sole InR in the fly genome [22, 23]. ILPs 1, 2, 3, 5 and 7 are secreted from NSC in the brain, whereas ILPs 4 and 6 are expressed in the medial intestine and the fat body. Binding of any of these ILPs to the InR causes receptor oligomerization and InR auto-phosphorylation, as the InR is a receptor tyrosine kinase. The InR itself is translated as a single polypeptide, but during its maturation within the cell and before insertion in the plasma membrane, it is proteolytically cleaved, both pieces held together by disulphur bridges. Once phosphorylated, the InRs recruit the IRS-like adaptor proteins, Chico and Lnk [10, 33]. Besides Chico and Lnk, the Drosophila InR has a long cytoplasmic tail that also acts as an IRS, in effect acting as a third IRS [26] (vertebrates have four IRS genes). InR phosphorylates the IRS proteins, like Chico, generating docking sites for other proteins in the pathway.

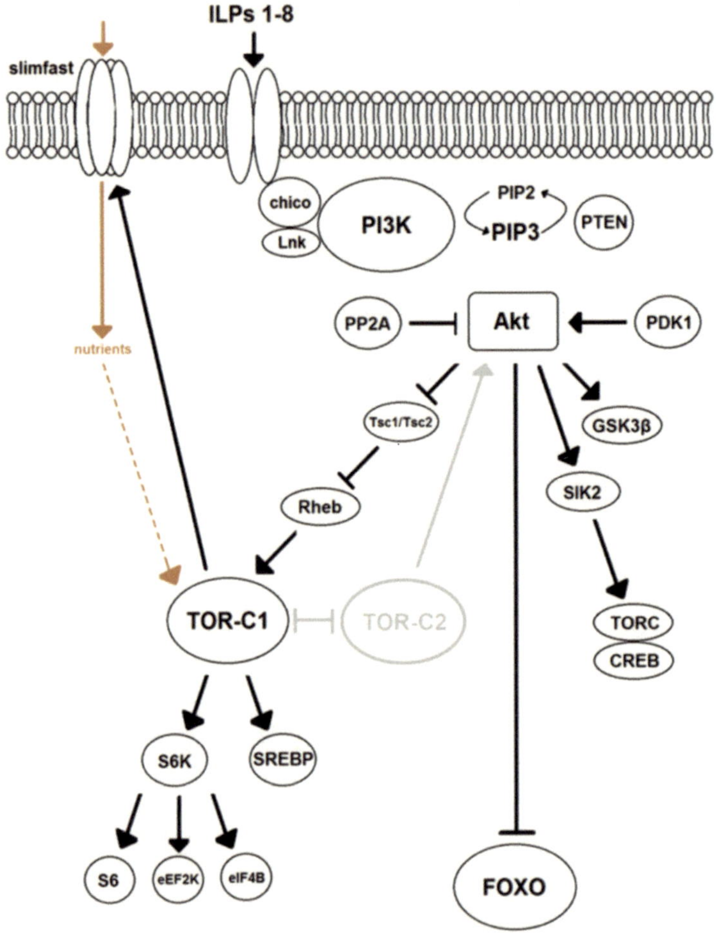

Figure 1. The insulin-signaling pathway in Drosophila. See text for details. Modified from [13].

The activated InR-IRS complex then recruits a PI3K complex (called Dp110 and Dp60, catalytic and regulatory subunits, respectively) to the membrane [14, 34]. Activated PI3K then generates phosphatidyl-inositol 3, 4, 5 trisphosphate (PIP3) from membrane phosphatidyl-inositol 4, 5 biphosphate (PIP2) and ATP. PTEN, a tumor suppressor gene, counteracts this enzymatic activity of PI3K, regulating PI3K output [35]. Plasma membrane accumulation of PIP3 then leads to membrane recruitment and activation by phosphorylation of two kinases, PDK1 and PKB (also called AKT). PKB in particular then phosphorylates and regulates several target proteins involved in metabolic control, including glycogen synthase kinase 3β (GSK-3β, or *shaggy* (*sgg*), as the gene is called in Drosophila), the transcription factor Foxo, the Tuberous sclerosis complex 2 protein, or Tsc2 (called *gigas* (*gig*) in Drosophila) and the salt-inducible kinase 2, or SIK2. PKB's activity is negatively regulated by a protein phosphatase, of class 2A (PP2A), which de-phosphorylates PKB and inactivates it. PKB can also be phosphorylated and activated by the Target of Rapamycin-complex 2 (TOR-C2) [13].

Activated *shaggy* then reduces glycogen synthesis by phosphorylation of glycogen synthase, besides participating in several other signaling networks, most notably the Wnt pathway [10, 13]. SIK2 phosphorylates and inactivates the transcriptional coactivator TORC (CRTC2 in mammals), precluding it from acting as a CREB co-activator. Foxo acts as a central catabolic regulatory point in cells; its phosphorylation by PKB enables its cytoplasmic retention, thereby inhibiting its nuclear localization, and transcriptional activity. Part of insulin's (or ILPs) anabolic effects relies on counteracting Foxo activity, which promotes energy conservation. Foxo also activates InR transcription, as part of a feedback loop.

Gigas' phosphorylation by PKB stimulates the activity of the Target of Rapamycin Complex 1 (TOR-C1) mediated by the Rheb GTPase. The Target of Rapamycin (TOR) kinase is a central anabolic kinase, regulating many aspects of general metabolism. Besides receiving input from the insulin pathway, TOR also receives input from nutrient sensors like Slimfast, an amino acid plasma membrane transporter. TOR regulates lipid, carbohydrate and autophagy levels. TOR itself forms part of two different complexes, TOR-C1 and TOR-C2. TOR-C1 is anabolic and regulates cellular growth and size, controlling cells' commitment to growth and energy utilization. Among its phosphorylation targets are the transcription factor SREBP (sterol-regulatory-element-binding-protein), and the ribosomal S6 kinase (S6K). TOR-C2 acts antagonistically to TOR-C1.

SREBP, a target of TOR-C1, regulates expression of genes involved in sterol biosynthesis. S6K, another TOR-C1 target, phosphorylates and regulates several proteins, chief of which is the ribosomal protein S6, besides the initiation and elongation factors elF4B and eEF2K. S6K promotes growth and protein synthesis [13]. In summary, Drosophila and vertebrates have great similarities in the insulin pathway, both at the structural and functional levels.

In vertebrates, glucose homeostasis also depends on another hormone that has opposite effects from those of insulin: Glucagon. Both are secreted from the pancreas. Insulin facilitates glucose translocation from the general circulation to the cell interior via plasma membrane translocation of Glut4 glucose transporters (acting as an anabolic hormone), besides activating growth. In contrast, glucagon activates glycogen hydrolysis (acting as a catabolic hormone), helping to maintain constant glucose levels in the blood. In insects, these basic homeostatic mechanisms are evolutionarily conserved, with a family of eight peptides akin to insulin acting as anabolic hormones in glucose homeostasis, as stated, and the adipokinetic peptides, a subfamily of which constitutes the so-called hypertrehalosemic hormones, playing the role of glucagon.

Despite the close homology throughout the insulin pathway and sugar metabolism, there is one clear and fundamental difference between carbohydrate metabolism of insects and vertebrates: Insects have trehalose (or α1-D-glucopyranosyl-α1-D-glucopyranoside; see figure 2). In fact, glucagon's functions are partially diverted to trehalose synthesis by means of the afore-mentioned hypertrehalosemic hormones. In having trehalose, insects are not alone: Many other species also use trehalose.

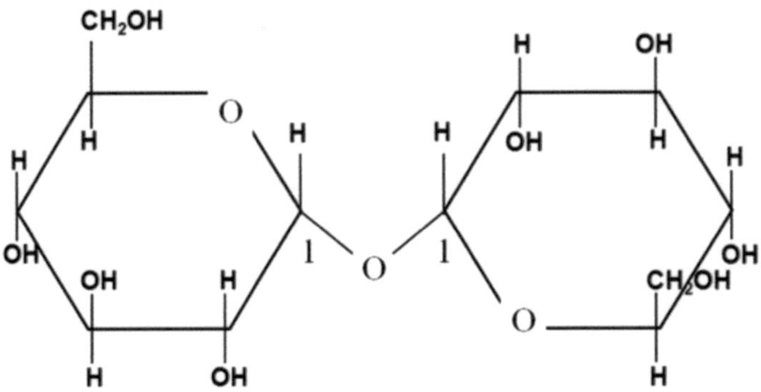

Figure 2. Structure of trehalose.

3. Why trehalose? Physicochemical properties of trehalose

Trehalose was discovered in the mid-nineteen century in rye ergot and then as a component of the cocoons of beetle species of the genus *Larinus* [36]. Many organisms use trehalose as a go-between between carbohydrate storage as glycogen and the cellular availability of glucose for energy needs (besides the other cellular roles trehalose plays in the cells' economy as an anti-stress factor): Why is this the case? Vertebrates have foregone altogether the use of trehalose, but still possess unique trehalose-degrading enzymes (trehalases; for example, human trehalase, used to digest trehalose from the diet and garner the glucose moieties within [37]). What advantages do trehalose imparts that has assured its selection in many diverse organisms (bacteria, algae, plants, fungi, invertebrates including insects), yet its loss in the vertebrate lineage? Part of the reason lies in the unique physicochemical properties of trehalose.

Trehalose is a disaccharide formed from two glucose moieties: It is synthesized by the union of two glucopyranose rings at the reducing end of the glycosyl residues. As a consequence, it is a non-reducing sugar. It demonstrates exceptional stability (with a high melting temperature, above 200°C, a high glass transition temperature, above 100°C, a slow rate of hydrolysis, and very high stability in extreme pHs; more than 99% of trehalose is still present in a pH range from 3.5 to 10 after 24 h at 100°C). It is not easily hydrolysed by acid (less so than sucrose, another non-reducing disaccharide found mainly in plant tissues), and its glycosidic bond is not cleaved by α-glycosidase, either. It also has low hygroscopicity, normally appearing as a dihydrated form. This form forms hydrogen bonds with the two water molecules, albeit at relatively short distances, making it an unusually strong interaction. In solution, all of the trehalose hydroxide groups make hydrogen bonds with water molecules. These bonds are easier to form than those with sucrose, leading to trehalose solutions with higher viscosity at equivalent concentrations, and less diffusion [36, 38].

The fact that trehalose is so resistant to acid hydrolysis may be key to understanding the use of trehalose in insect blood or hemolymph: Insect hemolymph has high amounts of peptides, free amino acids, and proteins, all of which have α amine and amino functional groups. A less resistant sugar might then react with these and be subject to condensation reactions resulting in imines and enamines, altering the structure of proteins and peptides. Also, it is less likely, from the foregoing, to cause 'browning', where the sugar's reactivity ultimately leads to the formation of pigments, and protein degradation, something sucrose is more prone to do. Sucrose circulates in plant sap without these problems mainly because these liquids have much lower levels of proteins, amino acids, and lipids. The slow rate of hydrolysis of trehalose also makes it a very stable compound, and one that is likely to aid in stabilizing other components of mixtures where trehalose is an ingredient [38].

Trehalose is highly soluble in water (up to about 50 g/100 g water at 20°C), and even if it is less water-soluble than sucrose (200 g/100g water), solubility is still high enough to allow it to accumulate to the 100 mM range in bodily fluids, like insect hemolymph. This allows trehalose to be readily transported and to accumulate in various extracellular and intracellular milieus. Also, it is a sweet compound, although, again, less so than sucrose. In solution, trehalose does not normally form intra-molecular hydrogen bonds [38].

On the other hand, trehalose also functions as a stabilizer in solutions for other organic molecules, like lipids and proteins. It seems to do so stemming from some of the physicochemical properties discussed: higher hydration capacity, higher viscosity and hydrogen bonding with proteins and other organic molecules, displacing water, than the disaccharide sucrose.

3.1. Hemolymph trehalose concentrations: Enantiostasis

Contrary to what happens with glucose in insect hemolymph (or for that matter, with glucose in vertebrates in their blood), where human glucose concentration is held constant at about 0.1% independently of nutritional status, ambient temperature, water availability, and / or oxidative stress, among other factors), trehalose levels vary normally within the hemolymph by an order of magnitude [15]. Since 1956 it was known that trehalose is a very abundant component of insect hemolymph [39], and also since then, various reports have shown that trehalose levels can vary greatly [36, 39]. As stated before, there are several reasons for this: Trehalose levels respond to a variety of environmental conditions including changes in ambient and concomitant bodily temperature, nutritional status, developmental stage, oxidative stress, salinity, etc., such that the term 'enantiostasis' was coined to describe this condition. Enantiostasis signifies that organisms' adaptations to various environmental variables require corresponding co-variance of some cellular and organismal parameters in order for the organisms to cope with the environmental challenge. In other words, trehalose variations are not only normal, but are actually adaptive, and a way of addressing changing environmental conditions. Trehalose levels can rise as much as up to 2% in hemolymph [36].

One condition that allows this compound to be used as enantiostatic is that it is non-toxic, even when in large amounts and / or high concentration. Again, the physical and chemical properties of trehalose are favorable: it is a very non-reactive sugar, much less likely than glucose or other reactive sugars to form adducts or derived metabolites. Also, a given concentration of trehalose exerts only half the osmotic effect than an equivalent concentration of glucose. Besides, as stated before, it is exceptionally stable within a large range of pH, salinity, temperature and osmotic values.

3.2. Trehalose functions

These properties lead to some of trehalose's functions in eukaryotes in general: a) as an energy source, b) as a cryoprotector, allowing hemolymph to remain liquid at normally freezing temperatures in insects, c) as a participant in a mechanism for anoxia tolerance, d) as a protein stabilizer during stress (osmotic, oxidative, thermal), and e) as a structural component for plant and bacterial cell wall polysaccharides as well as in chitin synthesis. In general, trehalose's roles can be viewed as double: on one side, as a source of energy and carbon, and on the other as a preservative and protectant against various forms of cellular and organismal stresses. These characteristics may explain in part why trehalose has been evolutionarily retained in so many organisms. Besides these roles, modern pharmacology and cosmetology makes use of trehalose in many formulations as a protein stabilizer, and for pill fabrication as a general preservative and excipient [38].

3.3. Trehalose's role as energy source

One of trehalose's most important functions, as stated before, is as a source of energy and carbon. This has been extensively reviewed in studies of metabolism in insect flight, since insect indirect flight muscles are the most energy-demanding in the animal kingdom, and are known to sustain the highest metabolic rates of all animal tissues (oxygen consumption when starting from the resting state to flight increases up to 70-fold, compared to a maximum of 20-fold increase in human muscles due to exercise). Besides, flight muscles account for about 20% of a flying insect's body mass (18% in the locust) [40, 41]. It becomes obvious that insects need "high octane" fuels to maintain these powerful motors running.

In flying insect species, flight muscles make use and oxidize different energy-rich molecules (glucose directly, glycogen, trehalose, proline, phosphoarginine and lipids) with a species-specific characteristic pattern of consumption and with time of intensive use, and according to its availability in hemolymph. The cockroach (*Periplaneta americana*) which uses glucose as its primary flying fuel [42], opts for trehalose as a high energy substrate for prolonged flights. Administration of an inhibitor of trehalase (the enzyme involved in trehalose breakdown) significantly decreases cockroach flight time [43]. Many other long distance fliers, like locusts, also use trehalose [36].

3.4. Trehalose anabolism

The ultimate source of trehalose comes from diet carbohydrates, but their use for trehalose synthesis is not necessarily direct. Under conditions where carbohydrates are abundant in the diet, hexoses are transported in the intestine by means of facilitated diffusion. A Drosophila Glut1 homolog was cloned and sequenced, with a 68% homology to human Glut1 [19]. Glut1 transporters are not specific for glucose, being able to transport other carbohydrates, like mannose, galactose and glucosamine. Insects with protein-rich diets utilize amino acids to synthesize *de novo* glucose, amino acids being also an indirect source for trehalose synthesis. When not required for energy, glucose from the diet can be stored as glycogen in the fat body. Glycogen is generally a more proximate source of glucose for trehalose synthesis.

Trehalose is synthesized in a condensation reaction of two glucose molecules, generally derived from glycogen (in effect, one of the constituent glucose substrates is in the form of UDP-glucose, derived from glucose-1P, whereas the other glucose substrate is glucose-6P). Since glucose is at the crux of trehalose anabolism, glucose regulation in the fat body impinges directly on trehalose regulation. Both vertebrates and invertebrates, in general, have very similar glucose regulatory mechanisms: insulin signaling and glucagon / hypertrehalosemic (HTH) signaling peptides, as stated above. Hypertrehalosemic hormones in particular are a family of adipokinetic (AKH) neuropeptides whose function is to derive glucose into trehalose synthesis in the fat body.

3.4.1. *Insect fat body*

Hemolymph circulating trehalose is synthesized in insects' fat body, a very important and extended organ analogous to both vertebrate liver and adipose tissue in insects [44]. Fat body is distributed throughout the insect body. It is also the principal organ for hemolymph circulating proteins and metabolites syntheses. It performs key roles in energy metabolism, as it both stores lipids and carbohydrates (these last in the form of glycogen), and responds to energy demands liberating both glucose and trehalose to the hemolymph. A *Drosophila melanogaster* fat body transcriptome analysis has shown that over 2200 genes are expressed in fat body cells, 290 of which attain high levels of expression. Among these are genes required for the insulin pathway, and lipid regulators. Many of the genes identified remain uncharacterized, and so, more work is required to define in greater detail the fat body transcriptome [45].

Several cell types constitute the fat body, the main one being the adipocyte (see figure 3). The adipocyte is a very dynamic cell type, with an active metabolism, regulating organismal carbohydrates and lipids, and also being an important endocrine cell [44].

The fat body is the main insect organ storing energy reserves. Most insect reserves are in the form of lipids, and these, mainly as triacylglycerols (TAGs). TAGs are stored in lipid droplets within adipocytes (figure 3). Lipid droplets have an outer coat of a monolayer of phospholipids and cholesterol molecules enclosing a core of TAGs and cholesterol molecules. In this outer coat some proteins are embedded, especially PATs (perilipin and adipocyte differentiation-related proteins). In insects two PAT proteins have been identified: Lsd1 and Lsd2 [46]. Together, they control synthesis and lipolisis of lipid stores.

Trehalose synthesis in response to energy needs normally courses through glycogen breakdown by glycogen phosphorylases in the fat body (see figure 4). Since the glucose thus obtained can also vie for glycolysis, this last path has to be inhibited. This is achieved in the fat body of cockroaches, where it has been studied, by reduction of fructose-2, 6-bisphosphate, an allosteric regulator of phosphofructokinase, the key regulatory enzyme in glycolysis [15, 47]. The adipokinetic hormones in turn, stimulate this decrease.

Figure 3. 2-D projection of a confocal stack of an abdominal adipocyte isolated from an Akt (PKB) *Drosophila melanogaster* mutant. The adipocyte was stained with Nile Red to image neutral lipids in order to evidence and quantify lipid droplets within the cell.

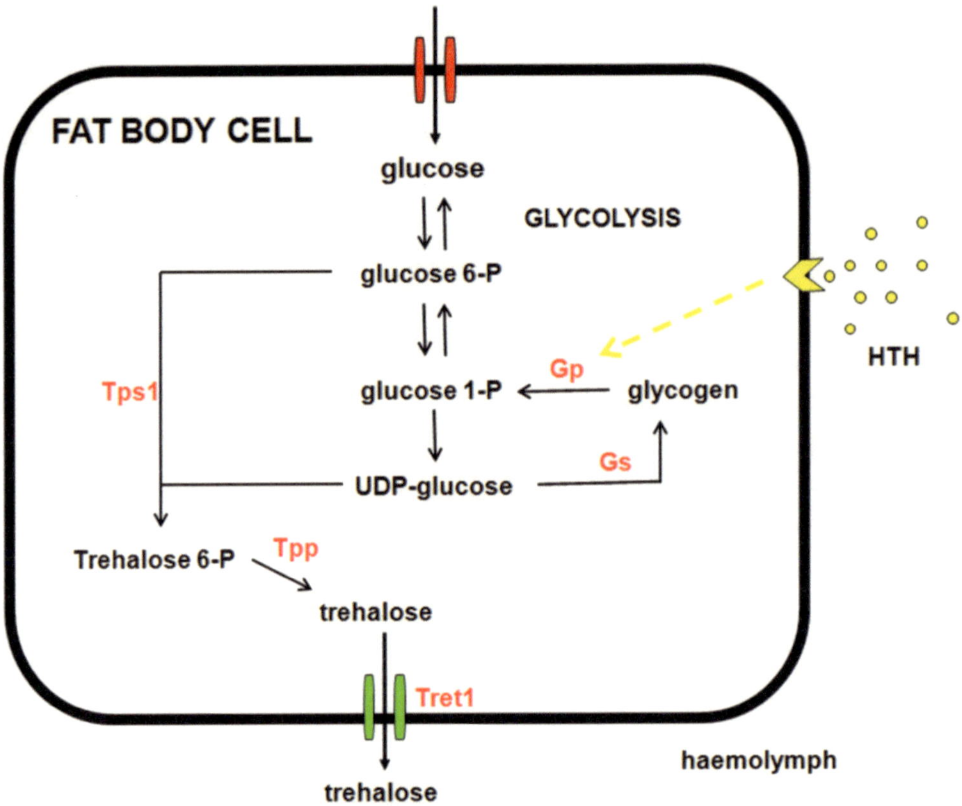

Figure 4. Trehalose synthesis in an insect fat body cell. See text for details. Abbreviations: HTH hypertrehalosemic hormones; Tret1, trehalose transporter 1; Tps1, trehalose phosphate synthase 1; Tpp, trehalose phosphate phosphatase; Gs, glycogen synthase; Gp, glycogen phosphorylase.

3.4.2. *Adipokinetic hormones (Hypertrehalosemic hormones)*

AKH were originally shown to induce lipid catabolism. In insects, as stated, one subgroup of these hormones produces an increase in circulating trehalose in the hemolymph [48]. About fifty HTH have been reported to date, and it is known that they participate in different functions. They are produced by neurosecretory cells of the *corpora cardiaca*, part of the ring gland of insects together with the prothoracic gland and the *corpus allatum*. The ring gland is located surrounding the anterior proventriculus in the thorax [49].

HTHs are synthesized from longer precursor peptides that are proteolitically processed to 8-10 amino acid long peptides. When the HTH neurosecretory cells are ablated, there is a significant drop in circulating trehalose levels in Drosophila larvae; over-expression of these HTHs leads to higher trehalose levels [48]. These hormones are secreted in response to various stimuli that would necessitate high circulating levels of trehalose, like lack of food, dehydration or thermal stress, or during flight. Some insects, like locusts, initiate flight utilizing trehalose, but switch to lipids later, whereas others, like some cockroaches, depend solely on trehalose supply. Still others use proline as an energy source, like some beetle species [44].

HTHs have membrane receptors. The Drosophila and homologous *Bombyx mori* HTH receptors have been cloned; they are G-protein-coupled-receptors (GPCRs) and are sequence-wise related to the gonadotrophin releasing hormone receptors of mammals [50]. Mutations in the Drosophila receptor show that it indeed mediates the HTH response: Mutants had glycogen and lipid accumulation in the fat body, irrespective of nutritional status, whereas rescue experiments returned glycogen and lipids to wild type levels [51].

There are scant data on the signaling pathways activated by HTHs. In cockroaches, HTH signaling increases inositol trisphosphate levels, leading to intracellular high Ca^{+2} concentrations and higher activity of protein kinase C (PKC). This produces an increase in the activity of glycogen phosphorylase, and trehalose synthesis. Inhibitors of PKC reduce trehalose production [52]. On the other hand, in locusts, AKH stimulation leads to cAMP production [53].

3.4.3. *Final reactions of trehalose synthesis*

Trehalose phosphate synthase catalyses condensation of glucose 6-P and UDP-glucose to generate trehalose 6-P in the fat body. Trehalose 6-P is then de-phosphorylated by the protein phosphatase coded by the *trehalose phosphate phosphatase (tpp)* gene to generate trehalose [54]. Both trehalose phosphate synthase and glycogen synthase use UDP-glucose as substrate, so besides competition for glucose between trehalose synthesis and glycolysis, the opposing trehalose phosphate synthase and glycogen synthase enzymes must be regulated too in order to shunt glucose to one or the other metabolic outcome [55]. Km values for trehalose phosphate synthase vary in different species where it has been studied (the moth *Hyalophora cecropia*, the cockroach *Periplaneta americana*, the blowfly *Phormia regina*, the sleeping chironomid *Polypedilum vanderplanki*, and the fruit fly *Drosophila melanogaster*). *trehalose phosphate synthase 1 (tps1)* overexpression in flies results in elevated trehalose levels, whereas mutations in this gene are lethal in the first larval instar. Trehalose phosphate synthases have also been studied in other organisms, like *Escherichia coli*, the yeast *Saccharomyces cerevisiae*, and plants like *Selaginella lepidophylla*, *Myrothamus spp.* and *Arabipdosis thaliana*. In *A. thaliana*, whose genome has been sequenced, there are eleven *tps* and ten *tpp* genes, suggesting that trehalose metabolism in plants is also very important [55, 56].

The aforementioned synthesis pathway is the one most commonly found; but there are, nevertheless, four other biosynthetic routes, three of them only present in prokaryotes. Some prokaryotes can use up to three different synthesis routes. The genes encoding trehalose synthase (*treS*), maltooligosyl-trehalose synthase (*treY*), maltooligosyl-trehalohydrolase (*treZ*) and trehalose glycosyltransfering synthase (*treT*) are prokaryotic, whereas trehalose phosphorylase (*treP*) is present in both [56] (see figure 5).

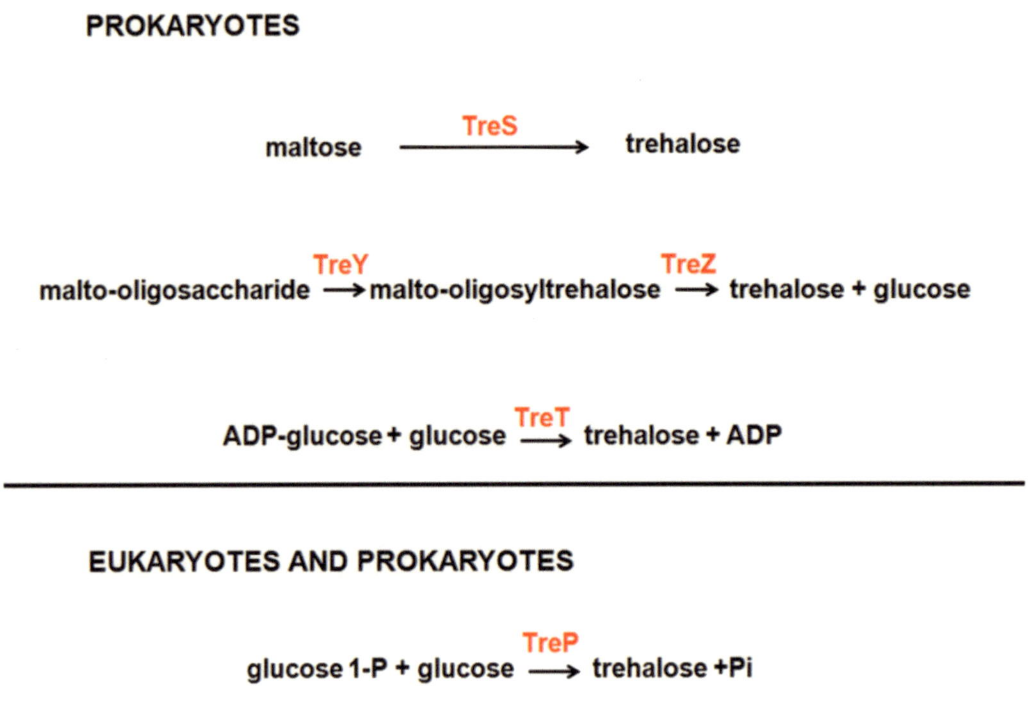

Figure 5. Alternative pathways of trehalose synthesis in prokaryotes and eukaryotes.

3.5. Trehalose in circulation

Once synthesized, trehalose is secreted to the hemolymph. A trehalose transporter cloned from several species, called Tret1, accomplishes this. It has been described in the mosquito *Anopheles gambiae*, the fruit fly *Drosophila melanogaster*, the moth *Bombyx mori* and the bee *Apis mellifera*, besides the sleeping chironomid *Polypedilum vanderplanki*. The *P. vanderplanki* trehalose transporter gets activated by dehydration conditions and that leads to trehalose accumulation such that up to 20% of total body weight can be composed of trehalose [57]. Besides Tardigrades, this last species is the pluricellular eukaryote that can withstand the harshest environmental conditions known, due in no small part to its trehalose accumulation strategy [58, 59].

The Tret1 transporter has twelve predicted membrane-spanning segments, similar to Glut family sugar transporters. Cloned Tret1 transporters from the aforementioned species were expressed in *Xenopus* oocytes. All cloned genes were functional, although kinetic parameters differed [60]. In *Drosophila melanogaster* there are two genes coding for trehalose transporters: Tret1-1 and Tret1-2, with Tret1-1 having several isoforms. Tret 1-2 shows no trehalose uptake and might be a pseudogene [60]. High expression has been noted in the fat body according to FlyAtlas for both genes [61]. Also, expression of both appears to be coincidental in many tissues, besides the fat body [62]. Transport of trehalose appears to be bi-directional for these transporters, such that not only secretion from the fat body cells goes through them, but also trehalose import to other cell types as well. In accordance with this, in *Bombyx mori* expression is also seen in muscle, testis and head, besides the fat body [60].

3.6. Trehalose catabolism

The enzyme responsible for trehalose breakdown is trehalase. This reaction produces two glucose molecules. The *treh* genes encode trehalase. *treh* has been cloned from several insect species, including the fruit fly *Drosophila melanogaster*. There are both membrane attached and soluble forms. In fruit flies it is expressed in many tissues, as expected. In many other insect species expression is seen in hemocytes (hemolymph cells).

There are trehalases in many other organisms, like yeast, thale cress and humans. Trehalase is regulated both at the transcriptional level (depending on nutritional status), and at the post-transcriptional level, by phosphorylation and by the presence of a trehalase inhibitor (this last in the cockroach *P. americana*) [36]. Altogether, trehalase appears to be finely regulated.

3.7. Trehalose and stress

Besides its role as an energy reserve, especially in response to high demands (flight, reproductive efforts like ovogenesis, etc.), trehalose functions as a molecule that aids organisms to cope with stresses from the environment. Organisms are frequently exposed to adverse environmental conditions like lack of water or food, lack of oxygen, oxidative stress and thermal variations, all of which conspire against the organism's survival. In response to these conditions living beings have evolved different strategies, involving the production of anti-freeze, chaperone and heat shock proteins, changes in membrane phospholipid composition, and cryoprotector metabolites such as glycerol, sorbitol, manitol and proline, besides trehalose. In extreme cases, organisms may even enter states of suspended animation, or cryptobiosis, where metabolism is stopped, or all but stopped. Depending on what causes it, cryptobiosis is called differently: anydrobiosis when induced by lack of water, cryobiosis when induced by cold temperatures, chemobiosis when induced by the presence of toxic chemicals, and anoxybiosis when induced by lack of oxygen.

Studies of trehalose have shown it to be a sort of jack-of-all-trades, employed in a variety of stress-provoking situations. This may explain its widespread occurrence and production in nature, despite its metabolic cost. Part of this is, of course, explained by the unusual physicochemical properties of trehalose.

3.7.1. Anhydrobiosis

Some organisms have evolved adaptations to cope with environments of transient very low humidity (mosses, famously the Tadigrades), where vital functions are just maintained or even stopped altogether. The chironomid *Polypedilum vanderplanki* can withstand up to 106°C while dessicated. Experiments with *P. vanderplanki* larvae show that gradual dehydration is accompanied by production and accumulation of trehalose, up to about 18% of total dry weight [59]. In another series of experiments, activity of the synthesis enzymes for trehalose, trehalose phosphate synthase and trehalose phosphate phosphatase was measured. These enzyme's activities increased during dehydration, whereas that of trehalase decreased. This was not only due to altered enzymatic activity, but also to transcriptional control, as the levels of the synthesis enzymes cDNAs increased at the beginning of dehydration, whereas that of trehalase increased only after forty eight hours since initation of dehydration, perhaps as preparation for subsequent rehydration [63]. Other species also use trehalose as a protectant during anhydrobiosis: the crustacean *Artemia salina* and the nematode *Avelenchus avenae*. It has been suggested that this increase in trehalose is driven by higher concentrations of ions in bodily fluids, rather than stemming from hormonal action.

Fruit flies can also withstand dehydration: 60% of fruit fly larvae subjected to conditions of less than 5% of relative humidity survived after subsequent rehydration. During dehydration and rehydration trehalose levels were measured, and went from 2 to 10 μg per larvae during the eight hours the gradual dehydration procedure lasted in the experiment. This correlated with a decrease in trehalase activity and an increase in trehalose phosphate synthase activity. Conditions reversed upon rehydration [64].

3.7.2. Cryobiosis

Bacteria like *Escherichia coli* use trehalose as a cryoprotectant. Changes in temperature from 37°C to 16°C drive an increase in trehalose levels of about an order of magnitude, and even though this does not affect growth, it does favorably affect survival when cells are subsequently transferred to a medium at 4°C. Bacteria mutant for *otsA* and *otsB*, the homologs of trehalose phosphate synthase and trehalose phosphate phosphatase die faster when transferred to 4°C; this lethality was rescued when transfected with a plasmid carrying these two genes, demonstrating the reliance on trehalose synthesis for survival for survival [65].

The lepidopteran *Cydia pomonella,* whose larvae are the common apple worm pest, also use trehalose as cryoprotectant. In a study that measured trehalose levels throughout the year, and correlated those measurements with temperature changes, they observed higher levels of trehalose (up to 18.4 mg/g of fresh weight) in January, where the temperature dipped to 0.4°C, compared to August (4.8 mg/g), where the temperature reached 23°C [66].

3.7.3. Thermal shock tolerance

Yeast cells (*Saccharomyces cerevisiae*) use trehalose as protection against heat shocks. In a microarray-based study, yeast cells subjected to a 30°C to 40°C heat shock survived better if trehalose accumulated within them. Especially successful were yeast cells with forced trehalose accumulation made by over-expressing *tps1,* the gene that codes for the biosynthetic enzyme for trehalose, together with mutations on the genes *nth1*, *nth2* and *ath1*, that code for trehalose degrading enzymes. The same set of heat shock-responsive genes were induced by heat treatment in both trehalose overexpressing and not overexpressing cells, showing that these heat shock genes are not responsible for increased survival (rather, trehalose is). Hexose transport genes were induced by heat shock, and heat shocked yeast cells transport more glucose into the cell, leading to subsequent increases in trehalose [67].

3.7.4. Oxidative stress

Metabolic activity generates, as a byproduct, reactive oxygen species (ROS) like hydrogen peroxide, superoxide ions, or hydroxyl radicals, among others, that can oxidize several types of biomolecules (like protein oxidation and crosslinking, nucleic acid modification and lipid peroxidation). Due to its prevalence, cells have evolved several different mechanisms to counter these toxic effects: Enzymes like superoxide dismutase or catalase, or molecules like thioredoxin and gluthatione, all counteract ROS toxic effects. Trehalose is also one of these protective mechanisms.

Yeast cells subjected first to a 28°C to 38°C heat shock to induce trehalose accumulation (see paragraph above), and then subjected to oxidative stress with a combination of hydrogen peroxide and ferric chloride incubation, fared much better (survival close to 100%) compared to other yeast cells not subjected to a heat shock first, whose measured trehalose levels were lower. Mutations in the trehalose synthesis genes (*tps1* and *tps2*, coding for trehalose phosphate synthase and trehalose phosphate phosphatase, respectively), did not fare better than control cells without heat shock even though they were subjected to a heat shock before incubation with hydrogen peroxide / ferric chloride. Under these conditions, mutant cells survived better if trehalose was also included in the medium, proving that trehalose accumulation is key for survival under these conditions. In these experiments *tps2* induction required YAP1 (stress response element binding protein), a protein required for catalase and thioredoxin expression, suggesting an integral response to oxidative stress [68].

3.7.5. Anoxybiosis

There is a very large body of literature concerning cellular and organismal responses to oxygen deprivation: cells are generally very sensitive to decreases of oxygen in the medium, especially nervous tissue. Yet some cells and organisms have trehalose-based anoxia tolerance mechanisms. Drosophila kept in a 100% nitrogen atmosphere for five minutes constitutively over-expressing *tps1* (trehalose phosphate synthase) compared to flies not over-expressing *tps1* fared better after the anoxic treatment. *tps1* over-expression lead, of course, to higher trehalose levels. Loss-of-function mutations in *tps1*, conversely, stimulated anoxia sensibility [55].

There are two basic ideas that try to explain in general why trehalose is so helpful in these stressful environments, and they both rely on the lack of reactivity of trehalose and its capacity to form hydrogen bonds with other biomolecules and / or its vitrification at high temperature. Trehalose hydrogen bonding with other biomolecules could substitute for water, preventing or reducing protein aggregration and denaturation, while not otherwise chemically reacting with them. Likewise, trehalose vitrification could lead to formation of trehalose "crystals" that would allow for stabilization of other biomolecules.

3.8. Trehalose in chitin synthesis

In insects, trehalose is also used as a chitin precursor. Chitin is an N-acetyl-β-D-glucosamine polymer. Chitin synthesis is not perfectly understood in detail but is known that it initiates with trehalose being metabolized to glucose in cells, and these glucose molecules, in turn, metabolized to glucosamine sugars, which are then used in chitin synthesis. The biosynthetic pathway is complex, involving at least eight different reactions [69].

4. Conclusion

Why has trehalose become an evolutionary 'winner' molecule? And this, despite the energetic costs associated with its reasonably convoluted metabolism, as a central go-between between the diet and other storage carbohydrates. The reasons seem to stem from the fact that trehalose is not only a high energy storage molecule, with low reactivity and high stability, but also, and precisely because of this stability and lack of reactivity, because it can stabilize and preserve other biomolecules, like proteins and lipids subjected to stressful environments, besides environments; besides, it can be transported it can be transported and accumulated to high concentrations without toxic effects.

Its prevalence in many different organisms of varied evolutionary origins and relatedness strongly suggest that trehalose has been retained in most lineages, rather than introduced, and that it is only in the vertebrate lineage where it is conspicuously missing. Why this is so in vertebrates, is difficult to explain.

Nowadays, trehalose has been subject to a revival in interest, since many of the same characteristics of non reactivity, stability and no toxicity that have propelled it to such a prominent role in many species in the natural world are now being newly appreciated by the pharmaceutical and cosmetic industries. Trehalose is being featured more and more in the excipients of drugs and pills, in creams, and as a sugar substitute in processed food formulas, as it is an energetic, non-toxic, stable meal ingredient, and has a sugary taste. Especially in Japan, where a procedure for industrial synthesis of trehalose was invented and is now being commercially applied, the allure of trehalose is proving to be irresistible, with promises in the near future to likewise take on the world as a whole [38].

Author details

Alejandro Reyes-DelaTorre, María Teresa Peña-Rangel and Juan Rafael Riesgo-Escovar*

Developmental Neurobiology and Neurophysiology Dept., Instituto de Neurobiología, Universidad Nacional Autónoma de México Querétaro, México

Acknowledgement

The corresponding author's laboratory is funded by PAPIIT # IN203110 and CONACYT # 81864. A. R. D. is funded by CONACYT scholarship # 369737.

5. References

[1] Gehring WJ (1998) Master control genes in development and evolution : the homeobox story. New Haven: Yale University Press. xv, 236 p.

[2] Kornberg TB, Krasnow MA (2000) The Drosophila genome sequence: implications for biology and medicine. Science. 287(5461): 2218-2220.

[3] Rubin GM, Yandell MD, Wortman JR, Gabor Miklos GL, Nelson CR, Hariharan IK, et al. (2000) Comparative genomics of the eukaryotes. Science. 287(5461): 2204-2215.

[4] Griffiths AJF (2008) Introduction to genetic analysis. 9th ed. New York: W.H. Freeman and Co. xxiii, 838 p.

[5] St Johnston D (2002) The art and design of genetic screens: Drosophila melanogaster. Nat Rev Genet. 3(3): 176-188.

[6] Ashburner M (1989) Drosophila. Cold Spring Harbor, N.Y.: Cold Spring Harbor Laboratory.

[7] Adams MD, Celniker SE, Holt RA, Evans CA, Gocayne JD, Amanatides PG, et al. (2000) The genome sequence of Drosophila melanogaster. Science. 287(5461): 2185-2195.

[8] Bloomington Drosophila Stock Center at Indiana University [database on the Internet] (2012) Available: http://flystocks.bio.indiana.edu/.

[9] Tickoo S, Russell S (2002) Drosophila melanogaster as a model system for drug discovery and pathway screening. Curr Opin Pharmacol. 2(5): 555-560.

* Corresponding Author

[10] Bohni R, Riesgo-Escovar J, Oldham S, Brogiolo W, Stocker H, Andruss BF, et al. (1999) Autonomous control of cell and organ size by CHICO, a Drosophila homolog of vertebrate IRS1-4. Cell. 97(7): 865-875.
[11] Baker KD, Thummel CS (2007) Diabetic larvae and obese flies-emerging studies of metabolism in Drosophila. Cell Metab. 6(4): 257-266.
[12] Musselman LP, Fink JL, Narzinski K, Ramachandran PV, Hathiramani SS, Cagan RL, et al. (2011) A high-sugar diet produces obesity and insulin resistance in wild-type Drosophila. Dis Model Mech. 4(6): 842-849.
[13] Teleman AA (2010) Molecular mechanisms of metabolic regulation by insulin in Drosophila. Biochem J. 425(1): 13-26.
[14] Oldham S, Hafen E (2003) Insulin/IGF and target of rapamycin signaling: a TOR de force in growth control. Trends Cell Biol. 13(2): 79-85.
[15] Becker A, Schloder P, Steele JE, Wegener G (1996) The regulation of trehalose metabolism in insects. Experientia. 52(5): 433-439.
[16] Cheatham B, Vlahos CJ, Cheatham L, Wang L, Blenis J, Kahn CR (1994) Phosphatidylinositol 3-kinase activation is required for insulin stimulation of pp70 S6 kinase, DNA synthesis, and glucose transporter translocation. Mol Cell Biol. 14(7): 4902-4911.
[17] Huang S, Czech MP (2007) The GLUT4 glucose transporter. Cell Metab. 5(4): 237-252.
[18] Wang M, Wang C (1993) Characterization of glucose transport system in Drosophila Kc cells. FEBS Lett. 317(3): 241-244.
[19] Escher SA, Rasmuson-Lestander A (1999) The Drosophila glucose transporter gene: cDNA sequence, phylogenetic comparisons, analysis of functional sites and secondary structures. Hereditas. 130(2): 95-103.
[20] Ceddia RB, Bikopoulos GJ, Hilliker AJ, Sweeney G (2003) Insulin stimulates glucose metabolism via the pentose phosphate pathway in Drosophila Kc cells. FEBS Lett. 555(2): 307-310.
[21] Hall DJ, Grewal SS, de la Cruz AF, Edgar BA (2007) Rheb-TOR signaling promotes protein synthesis, but not glucose or amino acid import, in Drosophila. BMC Biol. 5: 10.
[22] Brogiolo W, Stocker H, Ikeya T, Rintelen F, Fernandez R, Hafen E (2001) An evolutionarily conserved function of the Drosophila insulin receptor and insulin-like peptides in growth control. Curr Biol. 11(4): 213-221.
[23] Rajan A, Perrimon N (2012) An Interleukin-6 like cytokine regulates systemic insulin signaling by conveying the 'fed' state from the fat body to the brain. 53rd Annual Drosophila Conference; Chicago, IL.
[24] Claeys I, Simonet G, Poels J, Van Loy T, Vercammen L, De Loof A, et al. (2002) Insulin-related peptides and their conserved signal transduction pathway. Peptides. 23(4): 807-816.

[25] Garofalo RS (2002) Genetic analysis of insulin signaling in Drosophila. Trends Endocrinol Metab. 13(4): 156-162.

[26] Fernandez R, Tabarini D, Azpiazu N, Frasch M, Schlessinger J (1995) The Drosophila insulin receptor homolog: a gene essential for embryonic development encodes two receptor isoforms with different signaling potential. EMBO J. 14(14): 3373-3384.

[27] Rulifson EJ, Kim SK, Nusse R (2002) Ablation of insulin-producing neurons in flies: growth and diabetic phenotypes. Science. 296(5570): 1118-1120.

[28] Broughton S, Alic N, Slack C, Bass T, Ikeya T, Vinti G, et al. (2008) Reduction of DILP2 in Drosophila triages a metabolic phenotype from lifespan revealing redundancy and compensation among DILPs. PLoS One. 3(11): e3721.

[29] Murillo-Maldonado JM, Sanchez-Chavez G, Salgado LM, Salceda R, Riesgo-Escovar JR (2011) Drosophila insulin pathway mutants affect visual physiology and brain function besides growth, lipid, and carbohydrate metabolism. Diabetes. 60(5): 1632-1636.

[30] Araki E, Lipes MA, Patti ME, Bruning JC, Haag B, 3rd, Johnson RS, et al. (1994) Alternative pathway of insulin signalling in mice with targeted disruption of the IRS-1 gene. Nature. 372(6502): 186-190.

[31] Tamemoto H, Kadowaki T, Tobe K, Yagi T, Sakura H, Hayakawa T, et al. (1994) Insulin resistance and growth retardation in mice lacking insulin receptor substrate-1. Nature. 372(6502): 182-186.

[32] Sone H, Suzuki H, Takahashi A, Yamada N (2001) Disease model: hyperinsulinemia and insulin resistance: Part A–targeted disruption of insulin signaling or glucose transport. Trends in Molecular Medicine. 7(7): 320-322.

[33] Werz C, Kohler K, Hafen E, Stocker H (2009) The Drosophila SH2B family adaptor Lnk acts in parallel to chico in the insulin signaling pathway. PLoS Genet. 5(8): e1000596.

[34] Leevers SJ, Weinkove D, MacDougall LK, Hafen E, Waterfield MD (1996) The Drosophila phosphoinositide 3-kinase Dp110 promotes cell growth. EMBO J. 15(23): 6584-6594.

[35] Huang H, Potter CJ, Tao W, Li DM, Brogiolo W, Hafen E, et al. (1999) PTEN affects cell size, cell proliferation and apoptosis during Drosophila eye development. Development. 126(23): 5365-5372.

[36] Thompson SN (2003) Trehalose – The Insect 'Blood' Sugar. Advances in Insect Physiology: Academic Press. p. 205-285.

[37] Ishihara R, Taketani S, Sasai-Takedatsu M, Kino M, Tokunaga R, Kobayashi Y (1997) Molecular cloning, sequencing and expression of cDNA encoding human trehalase. Gene. 202(1-2): 69-74.

[38] Ohtake S, Wang YJ (2011) Trehalose: current use and future applications. J Pharm Sci. 100(6): 2020-2053.

[39] Wyatt GR, Kale GF (1957) The chemistry of insect hemolymph. II. Trehalose and other carbohydrates. J Gen Physiol. 40(6): 833-847.

[40] Wegener G (1996) Flying insects: model systems in exercise physiology. Experientia. 52(5): 404-412.

[41] Candy DJ, Becker A, Wegener G (1997) Coordination and Integration of Metabolism in Insect Flight*. Comparative Biochemistry and Physiology Part B: Biochemistry and Molecular Biology. 117(4): 497-512.

[42] Chino H, Lum PY, Nagao E, Hiraoka T (1992) The molecular and metabolic essentials for long-distance flight in insects. Journal of Comparative Physiology B: Biochemical, Systemic, and Environmental Physiology. 162(2): 101-106.

[43] Kono Y, Takahashi M, Matsushita K, Nishina M, Kameda Y, Hori E (1994) Inhibition of flight in Periplaneta americana (Linn.) by a trehalase inhibitor, validoxylamine A. Journal of Insect Physiology. 40(6): 455-461.

[44] Arrese EL, Soulages JL (2010) Insect fat body: energy, metabolism, and regulation. Annu Rev Entomol. 55: 207-225.

[45] Jiang Z, Wu XL, Michal JJ, McNamara JP (2005) Pattern profiling and mapping of the fat body transcriptome in Drosophila melanogaster. Obes Res. 13(11): 1898-1904.

[46] Gronke S, Beller M, Fellert S, Ramakrishnan H, Jackle H, Kuhnlein RP (2003) Control of fat storage by a Drosophila PAT domain protein. Curr Biol. 13(7): 603-606.

[47] Becker A, Liewald JF, Stypa H, Wegener G (2001) Antagonistic effects of hypertrehalosemic neuropeptide on the activities of 6-phosphofructo-1-kinase and fructose-1,6-bisphosphatase in cockroach fat body. Insect Biochem Mol Biol. 31(4-5): 381-392.

[48] Lee G, Park JH (2004) Hemolymph sugar homeostasis and starvation-induced hyperactivity affected by genetic manipulations of the adipokinetic hormone-encoding gene in Drosophila melanogaster. Genetics. 167(1): 311-323.

[49] Demerec M (1994) The biology of Drosophila. Facsim. ed. Plainview, N.Y.: Cold Spring Harbor Laboratory Press. x, 632 p.

[50] Staubli F, Jorgensen TJ, Cazzamali G, Williamson M, Lenz C, Sondergaard L, et al. (2002) Molecular identification of the insect adipokinetic hormone receptors. Proc Natl Acad Sci U S A. 99(6): 3446-3451.

[51] Bharucha KN, Tarr P, Zipursky SL (2008) A glucagon-like endocrine pathway in Drosophila modulates both lipid and carbohydrate homeostasis. J Exp Biol. 211(Pt 19): 3103-3110.

[52] Sun D, Garcha K, Steele JE (2002) Stimulation of trehalose efflux from cockroach (Periplaneta americana) fat body by hypertrehalosemic hormone is dependent on protein kinase C and calmodulin. Arch Insect Biochem Physiol. 50(1): 41-51.

[53] Zhiwei W, Hayakawa Y, Downer RGH (1990) Factors influencing cyclic AMP and diacylglycerol levels in fat body of Locusta migratoria. Insect Biochemistry. 20(4): 325-330.

[54] Candy DJ, Kilby BA (1961) The biosynthesis of trehalose in the locust fat body. Biochem J. 78: 531-536.

[55] Chen Q, Ma E, Behar KL, Xu T, Haddad GG (2002) Role of trehalose phosphate synthase in anoxia tolerance and development in Drosophila melanogaster. J Biol Chem. 277(5): 3274-3279.

[56] Paul MJ, Primavesi LF, Jhurreea D, Zhang Y (2008) Trehalose metabolism and signaling. Annu Rev Plant Biol. 59: 417-441.

[57] Kikawada T, Saito A, Kanamori Y, Nakahara Y, Iwata K, Tanaka D, et al. (2007) Trehalose transporter 1, a facilitated and high-capacity trehalose transporter, allows exogenous trehalose uptake into cells. Proc Natl Acad Sci U S A. 104(28): 11585-11590.

[58] Hinton HE (1960) Cryptobiosis in the larva of Polypedilum vanderplanki Hint. (Chironomidae). Journal of Insect Physiology. 5(3–4): 286-300.

[59] Watanabe M, Kikawada T, Minagawa N, Yukuhiro F, Okuda T (2002) Mechanism allowing an insect to survive complete dehydration and extreme temperatures. J Exp Biol. 205(Pt 18): 2799-2802.

[60] Kanamori Y, Saito A, Hagiwara-Komoda Y, Tanaka D, Mitsumasu K, Kikuta S, et al. (2010) The trehalose transporter 1 gene sequence is conserved in insects and encodes proteins with different kinetic properties involved in trehalose import into peripheral tissues. Insect Biochem Mol Biol. 40(1): 30-37.

[61] Chintapalli VR, Wang J, Dow JAT (2007) Using FlyAtlas to identify better Drosophila melanogaster models of human disease. Nature Genetics. 39(6): 715-720.

[62] McQuilton P, St. Pierre SE, Thurmond J, Consortium tF (2011) FlyBase 101 – the basics of navigating FlyBase. Nucleic Acids Research.

[63] Mitsumasu K, Kanamori Y, Fujita M, Iwata K, Tanaka D, Kikuta S, et al. (2010) Enzymatic control of anhydrobiosis-related accumulation of trehalose in the sleeping chironomid, Polypedilum vanderplanki. FEBS J. 277(20): 4215-4228.

[64] Thorat LJ, Gaikwad SM, Nath BB (2012) Trehalose as an indicator of desiccation stress in Drosophila melanogaster larvae: A potential marker of anhydrobiosis. Biochemical and Biophysical Research Communications. 419(4): 638-642.

[65] Kandror O, DeLeon A, Goldberg AL (2002) Trehalose synthesis is induced upon exposure of Escherichia coli to cold and is essential for viability at low temperatures. Proc Natl Acad Sci U S A. 99(15): 9727-9732.

[66] Khani A, Moharramipour S, Barzegar M (2007) Cold tolerance and trehalose accumulation in overwintering larvae of the codling moth, *Cydia pomonella* (Lepidoptera: Tortricidae). Eur J Entomol. 104(3): 385-392.

[67] Mahmud SA, Hirasawa T, Furusawa C, Yoshikawa K, Shimizu H (2012) Understanding the mechanism of heat stress tolerance caused by high trehalose accumulation in Saccharomyces cerevisiae using DNA microarray. J Biosci Bioeng. 113(4): 526-528.

[68] Benaroudj N, Lee DH, Goldberg AL (2001) Trehalose accumulation during cellular stress protects cells and cellular proteins from damage by oxygen radicals. J Biol Chem. 276(26): 24261-24267.

[69] Merzendorfer H, Zimoch L (2003) Chitin metabolism in insects: structure, function and regulation of chitin synthases and chitinases. J Exp Biol. 206(Pt 24): 4393-4412.

Chapter 15

Starvation Conditions Effects on Carbohydrate Metabolism of Marine Bacteria

Monia El Bour

Additional information is available at the end of the chapter

http://dx.doi.org/10.5772/50877

1. Introduction

In the coastal shorelines terrestrial organic materials transported by river runoff represent an important material source to the ocean and is estimated that 0.4×1015 g yr−1 organic carbon is discharged to ocean by land flows or rivers (Meybeck, 1982, He et al, 2010). This amount of riverine organic carbon is sufficient to support the entire organic carbon turnover in the ocean (Williams and Druffel, 1987) even though, little terrestrial signal by carbon isotopic ratio (δ13C) of the bulk DOC in ocean was shown, suggesting that terrestrial organic carbon undergo rapid removal and decomposition within estuarine mixing (Druffel et al, 1992; Hedges et al, 1997).

Carbohydrates are, among others, the major components of identified organic matter in ocean and account for 3% to 30% of the bulk Dissolved Organic Carbon (DOC) (Gueuen et al, 2006; Hung et al, 2003; Pakulski and Benner, 1994), whereas, in estuarine and marine surface waters they are considered the most labile fractions of bulk organic matter and may play key roles in the geochemical cycles as reported by several studies (Benner et al, 1992; Burdige and Zheng, 1998; Middelboe et al, 1995; Murrell and Hollibaugh, 2000).

In complex macro-aggregates mixtures, the carbohydrates are dominant with proteins and lipids, with evidence that in mucilage samples carbohydrates are relevant components of the organic carbon (17–45% on dry weight basis depending on the age of the aggregates) with heavy colonization by several heterotrophic bacteria and autotrophic organisms which embedded the organic matrix (Simon et al, 2002; Urbani and Sist, 2003). Thus high density of micro organisms enriches the matrix in organic and inorganic nutrients in comparison with the surrounding water and such macro flocks are important hot-spots for bacterial growth and carbon cycling (Kaltenbock and Herndl, 1992; Alldredge, 2000 and Azam and Long, 2001) and then significance of bacteria for the formation and decomposition of aggregates appears to be higher than previously estimated (Simon et al, 2002). .

2. Importance of carbohydrates availability for marine bacteria

Marine micro-organisms have important established roles which may reflect large-scale changes within inter tidal systems and heterotrophic bacteria, for example, are crucial to transformations and re-mineralization of organic carbon, nitrogen and other nutrients throughout oceans (Azam, 1998; Azam et al, 1993).

Bacterial utilization of carbohydrates depends heavily on their chemical composition (Aluwihare and Repeta, 1999; Arnosti, 2000, Zoppini et al, 2010) and slow hydrolysis may provide the pathway for the accumulation in aquatic environments (Cowie and Heges, 1984; Cowie et al, 1995; Benner et al, 1992).

Else, bacteria play an important role in regulation of the rate of organic matter mineralization, nutrient cycling, and energy transfer in aquatic environments and demineralization or the turnover of carbohydrates has been used to evaluate the efficiency of the microbial community and the liability of the organic carbon in coastal waters (Azam and Worden, 2004; Lennon, 2007).

Considering that total number of bacteria on Earth, the largest proportion of bacterial cells presumably reside in oceanic subsurface rather than in terrestrial areas (3.5 x ⊚10^{30} and 0.25–2.5 x ⊚10^{30} respectively) and therefore bacterial cells are estimated to contain, in total, 350–550 Pg of carbon, up to 60–100% of the total carbon found in plants, as well as large amounts of nitrogen and phosphorous (Whitman *et al* 1998). Thus, despite their modest size as individuals, as a group these organisms not only contribute to the flow of nutrients worldwide, but may also constitute a significant proportion of the nutrients in living biomass (Horner-Devine et al, 2009).

In estuarine areas, biodegradation by heterotrophic bacteria for Organic composition changed rapidly along the estuary, showing a selected removal of carbohydrates and amino acids within the DOC pool in the upper reach and mixing zone, and an autotrophic source of Particular Carbohydrates (PCHO) in the lower estuary, which gave an insight into the DOC estuarine process (He et al, 2010).

For hydrothermal marine compartment, heterotrophic thermophiles in culture far exceeds the number of their autotrophic counterparts, and many are known to metabolize carbohydrates *via* fermentation or respiration nevertheless, the energetic associated with these high-temperature microbial processes have been all but ignored (Amend et al, 2001).

In Coastal marine regions, sediments receive organic matter from a variety of sources, and its reactivity towards anaerobic fermentation and respiration is determined by the relative content of e.g. carbohydrates, lipids and proteins (Kristensen et al, 1995; Kristensen and Holmer, 2001). Proteins and labile carbohydrates derived from algae usually constitute the main organic sources for marine sediments, but coastal areas may also receive detritus rich in structural carbohydrates (e.g. cellulose) from vascular plants (Cowie and Hedges, 1992; Leeuw and Largeau, 1993; Kristensen, 1994). Furthermore, sediments of anthropogenic point sources, such as marine fish farms, may concentrate high loads of protein and lipid rich materials of animal origin (Ackefors and Enell, 1994). Proteins and simple carbohydrates are generally labile towards anaerobic degradation (Arnosti and Repeta, 1994; Holmer and Kristensen, 1994; Arnosti and Holmer, 1999), whereas anaerobic degradation of lipids and structural carbohydrates occur at much lower rates (Boetius and Lochte, 1996; Canuel and Martens, 1996). This difference is most likely caused by differential efficiency of extra-cellular enzymes produced by anaerobic fermenting bacteria (Valdemarsen et al, 2010)

2.1. Bio mineralization of carbohydrates by marine bacteria and production of carbohydrases

Bacteria are the most abundant and most important biological component involved in the transformation and mineralization of organic matter in the biosphere (Cho et al, 1988; , Pomeroy et al, 1991). Heterotrophic bacteria contribute to the cycles of nutrients and carbon in two major ways: by the production of new bacterial biomass (secondary production) and by the re mineralization of organic carbon and nutrients. Understanding this dual character of planktonic bacteria in aquatic ecosystems is a central paradigm of contemporary microbial ecology (Billen et al, 1984; Ducklow et al, 1992 and Del Giorgio and Cole, 1998).

Further, bacteria are capable of out competing, other organisms for organic compounds at low concentrations since they possess high substrate affinities, surface-to volume ratios, and metabolic rates. Natural DOM sources are heterogeneous mixtures of compounds, and bacteria can utilize several molecules simultaneously (Bott et Kaplan, 1985).

Extracellular enzymes, term generally utilised to define enzymes located outside the cytoplasmic membrane, are important catalyst in the decomposition of particulate organic matter (POM) and dissolved organic matter (DOM) including marine snow (Cho and Azam, 1988; Smith et al, 1992) and only small molecules, around 600 Da (Weiss et al, 1991), can be transported across the bacterial membrane and thus bacteria secrete enzymes to hydrolyze high-molecular-weight organic matter. Moreover extra cellular enzymes play an important role in nutrient cycling as they may be produced in order to acquire the limiting nutrient (Hoppe, 1983; Zoppini et al, 2010).

In the aggregates, the activity of the extracellular enzymes transforms non utilizable POM in DOM, thus bringing small nutrient molecules into solution (Smith et al, 1992; Simon et al, 2002). The metabolic activity of colonizing bacteria drives sinking particulate organic matter into non sinking DOM (Azam and Long, 2001; Smith et al, 1992) fueling free-living bacteria (Kiørboe and Jackson, 2001). This bun coupled solubilisation causes rapid POMYDOM transition with important biogeochemical implications for carbon flux in the oceans (Smith et al, 1992; Simon et al, 2002).

Further, almost heterotrophic bacteria have developed several mechanisms which allow the preferred utilization of the most efficiently metabolised carbohydrates when are exposed to a mixture of carbon sources. Interestingly, similar mechanisms are used by some pathogens to control various steps of their infection process. The efficient metabolism of a carbon source might serve as signal for proper fitness. Alternatively, the presence of a specific carbon source might indicate to bacterial cells that they thrive in infection-related organs, tissues or cells and that specific virulence genes should be turned on or switched off. Frequently, virulence gene regulators are affected by changes in carbon source availability. Thus, the activity of PrfA, the major virulence regulator in *Listeria monocytogenes,* seems to be controlled by the phosphorylation state of phosphotransferase system (PTS) components. In *Vibrio cholerae* synthesis of HapR, which regulates the expression of genes required for motility, is controlled via the Crp/cAMP CCR mechanism, whereas synthesis of *Salmonella enterica* HilE, which represses genes in a pathogenicity island, is regulated by the carbohydrate responsive, PTS-controlled Mlc (Poncet et al, 2009).

2.2. Use of carbohydrates mineralization for marine bacteria identification

The metabolic diversity of bacteria is perhaps as remarkable as their taxonomic and evolutionary diversity. Although culture-based studies are limited in their ability to estimate bacterial diversity, and several studies demonstrated varieties of modes of energy conversion of bacteria and wide ranges of substrate uses and metabolic pathways.). The flexibility of bacterial metabolism is similarly best illustrated by their abilities to degrade xenobiotic compounds such as malathion (an insecticide) and 2,4,5-trichlorophenoxyacetic acid (a herbicide), which are toxic to many other organisms (Brock et al, 1987 and Horner et al, 2003).

Metabolic assays mainly based on carbohydrates degradation on either aerobic or anaerobic conditions still effectively used to identify environmental bacteria besides molecular methods. Among these methods, the most widely used commercial identification systems are the API, the Biolog systems and more recently Microgen systems (Awong-Taylor et al, 2007).

The API 20 system included profiles of approximately 40 groups and species of bacteria in addition to the *Enterobacteriaceae* group in the API data set segregated mainly on assimilation of several carbohydrates with varied number depending on the type of API test kit used. (MacDonell et al, 1982; Bertone et al, 1996).

The Biolog identification system (Biolog, Inc., Hayward, CA, USA) is a bacterial identification method that establishes identifications based on the exchange of electrons generated during respiration, leading subsequently to tetrazolium-based color changes. This system tests the ability of micro organisms to oxidize a panel of 95 different carbon sources (Truu et al, 1999).

3. Effects of starvation conditions in carbohydrates metabolism

In their natural environment especially in marine ecosystems, bacteria are frequently exposed to major changes in growth conditions. These changes can include temperature, salt concentrations, essential nutrients, oxygen supply, pressure…etc. In order to sustain these changes, bacteria have developed mechanisms which allow them to adapt to drastic environmental alterations. Mainly, response to carbon source availability seems to be of special importance, because many bacteria dispose of more than one mechanism to adapt Carbon Metabolism and Virulence to changes in carbohydrate composition. These mechanisms include induction of specific carbohydrate transport and utilization system by the presence of the corresponding carbon source and their repression when a more efficiently utilizable carbohydrate is present in addition. The latter phenomenon is called carbon catabolite repression (CCR) and is often mediated by more than one signal transduction pathway, which can include regulation of transcriptional activators or repressors, anti terminators, carbohydrate transporters or metabolic enzymes (Poncet et al, 2009).

3.1. *In vivo* starvation effect on catabolic carbohydrates profiles of *E.coli*

Escherichia coli, a fecal coliform, was found to survive for longer periods of time in non sterile natural seawater when sediment material was present than in seawater without sediments , and growth was observed to occur by one occasion (Gerba et al, 1976). This enteric bacterium was found to increase rapidly in number in autoclaved natural seawater and autoclaved sediment taken from areas receiving domestic wastes, even when the seawater had salinities as high as 34 g/kg. Else, longer survival of *E. coli* in sediment is revealed and attributed to greater content of organic matter present in sediment than the seawater.

Experimental, mainly *in vitro,* studies have been performed on the survival of different enteric bacterial species in seawater and results showed solar radiation is the most adverse factor for enteric bacteria. Its effect was found to be wavelength-dependent (most inactivation caused by UVB solar spectrum with a range of dose from 0.98 to 4 kJ/m22) and thus restricted to shallow (45 to 90 cm) or clear waters (Rozen and Belkin, 2001; Sinton et al, 1994; Villarino et al, 2003; Whitman et al, 2004). Moreover, in cases of its release from wastewater into marine environment, enteric bacteria are subject to osmotic stress with effects depending on water salinity (a high loss of viability for *E. coli* occurred for salinities between 15 and 30 g/L) (Anderson et al, 1979; Pereira and Alcantara, 1993). Most coastal waters provide low concentrations of nutrients essential to the growth and survival of enteric bacteria (Barcina et al, 1997). These stresses induce different resistance mechanisms for at least part of the cellular population (Martin et al 1998; Rozen et al 2002). As a consequence, different cellular states occur in the stressed population, including the viable but nonculturable state (VBNC). The ability of such cells to recover their growth capacity is still under debate (Arana et al, 2007).

Thus, we investigated survival and virulence of *Escherichia coli* strains (*E.coli* O126:B16 and *E.coli* O55:B5) exposed to natural conditions in brackish water by incubation in water microcosms in the Bizerte lagoon in Northern Tunisia and exposed for 12 days to natural sunlight in June (231 to 386 W/m2, 26 6 1 uC, 30 g/L) and in April (227 to 330 W/m2, 17 6 1 uC, 27 g/L) or maintained in darkness for 21 days (17 6 1 uC, 27 g/L). The results revealed sunlight as the most significant inactivating factor (decrease of 3 Ulog within 48 hours for the two strains) compared to salinity and temperature (in darkness). Survival time of the strains was prolonged as they were maintained in darkness. Local strain (*E. coli* O55:B5) showed better survival capacity (T90 5 52 hours) than *E. coli* O126:B16 (T90 5 11 h).

For both, modifications were noted only for some metabolic activities of carbohydrates hydrolysis (Table 1). Thus, *E. coli* O126:B16 culturable cells lost the ability to assimilate amygdalin, and *E. coli* O55:B5 lost ability to assimilate melibiose, amygdalin, rhamnose, and saccharose.

Although, in previous data revealed exposure increase to sunlight, particularly UV radiation, alleviates some of the negative effects of altered organic matter quality through selective photolytic degradation. Both ultraviolet radiation (UV-B, 280–320 nm and UV-A, 320–400 nm) and visible light (photo synthetically active radiation, PAR, 400–700 nm) and induce major photolytic changes to complex organic molecules and generate large quantities of readily utilizable substrates for bacterial metabolism (Lindell et al 1995, Espeland & Wetzel 2001, Tietjen & Wetzel 2003). Thus, rates of carbon and nutrient cycling are often accelerated, stimulating ecosystem productivity. Interestingly, this enhanced microbial respiration may lead to increased CO2 production and evasion to the atmosphere, creating a potential positive feedback to atmospheric CO2 concentrations (Wetzel 2001; Lingo et al, 2007)

Strains	ONPG	ADH	LDC	ODC	CIT	H_2S	URE	TDA	IND	VP	GEL	GLU	MAN	INO	SOR	RHA	SAC	MEL	AMY	ARA	OX	CAT
Sunlight incubation (April and June) (n=2)																						
***E.coli* O126B16**[a]	+	+	+	+	-	-	-	+	+	-	-	+	+	-	+	-	-	-	+	+	-	+
after 3 days	+	+	+	+	-	-	-	+	+	-	-	+	+	-	+	-	-	-	+	+	-	+
***E.coli* O55B5**[a]	+	+	+	+	-	-	-	+	+	-	-	+	+	-	+	+	+	+	+	+	-	+
after 12 days	+	+	+	+	-	-	-	+	+	-	-	+	+	-	+	+	+	+	+	+	-	+
Darkness incubation (n=2)																						
***E.coli* O126B16**[a]	+	+	+	+	-	-	-	+	+	-	-	+	+	-	+	-	-	-	+	+	-	+
after 2 days	+	+	+	+	-	-	-	+	+	-	-	+	+	-	+	-	-	-	(-)	+	-	+
after 21 days	+	+	+	+	-	-	-	+	+	-	-	+	+	-	+	-	-	-	(-)	+	-	+
***E.coli* O55B5**[a]	+	+	+	+	-	-	-	+	+	-	-	+	+	-	+	+	+	+	+	+	-	+
after 2 days	+	+	+	+	-	-	-	+	+	-	-	+	+	-	+	+	+	(-)	(-)	+	-	+
after 16 days	+	+	+	+	-	-	-	+	+	-	-	+	+	-	+	(-)	(-)	(-)	(-)	+	-	+
after 21 days	+	+	+	+	-	-	-	+	+	-	-	+	+	-	+	(-)	(-)	(-)	(-)	+	-	+

(+) positive reaction; (-) : negative reaction; *n*: number of colony tested. ONPG: β-galactosidase, ADH: arginine dihydrolase, ODC: ornithine decarboxylase, CIT: use of citrate, URE: urease, TDA: tryptophane desaminase, IND: indole production, VP: acetoin production, GEL: gelatinase, GLU: glucose assimilation, MAN: mannitol, SOR: sorbitol, RHA: rhamnose, SAC: saccharose, MEL: melibiose, ARA: arabinose

Table 1. Characteristics of *E.coli* O55B5 and *E.coli* O126B16 before and during seawater incubation

3.2. *In vitro* starvation effect on catabolic carbohydrates profiles of *E.coli*

Depending on the nature of aquatic environmental receptor (seawater, wastewater or brackishwater), different physiological, metabolic and pathogenic modifications related to stress conditions have been reported by much studies (Gauthier *et al*, 1993; Dupray and Derrien, 1995; Baleux *et al*, 1998; Troussellier *et al*, 1998; Monfort and Baleux, 1999). And different studies showed that adaptation forms of enterobacteria to drastic conditions occurring gradually during discharging in wastewater (Dupray and Derrien, 1995), or during the transfer from a previous culture to a saline medium or to mixed wastewater/seawater medium (Munro *et al,* 1987), allowing better survival in seawater conditions. In this report, *E. coli* was chosen as a representative strain of Thermotolerant Coliform (TTC), used as fecal contamination indicator. Most of *E. coli* strains are normal intestinal flora components; however, certain pathogenic strains are important disease factors. Those pathogenic strains are classified into at least six distinct groups: enteropathogenic *E.coli* (EPEC), enterotoxigenic *E.coli* (ETEC), enteroinvasive *E.coli* (EIEC), diffuse-adhering *E.coli* (DAEC), enteroaggregative *E.coli* (EAEC), enterohemorrhagic *E.coli* (EHEC) (Schroeder *et al,* 2004). Thus, we studied the behavior of these 6 different types of *E.coli* when discharged in seawater with or without a previous incubation in wastewater microcosms. Else, we examined their metabolic, antibiotic and virulence resistance patterns after exposure to salinity gradient.

Therefore, the studies were carried out using membrane chambers with 120ml capacity, constituted by assembly of glass tubes allowing dissolved matter to penetrate, through 0.2 µm- pore- size, 25 mm-diameter filters, fixed between two Teflon joints. For each *E.coli* strains two types of microcosms were prepared: one type was inoculated by seawater and the second with wastewater.

For microcosm inoculation, the seawater used was brought from Bizerte lagoon (northern part in Tunisia of Tunis). Wastewater was brought from sewage treatment station (near Tunis town).

The results obtained (Table 2) showed most important modifications noted for the strains previously inoculated in wastewater and O126B16 *E.coli* strain lost ability of fermenting glucose and mannitol, producing indole and to use citrate, but it acquired urease and gelatinase activities. All the other pathogenic *E.coli* strains maintained their major metabolic characteristics except for the ability to use citrate and to produce indole.

In further study, we compared the behavior of *E.coli* (ATCC 14948), *Vibrio paraheamolyticus and Salmonella Typhymurium* (ATCC 17802) when discharged in seawater with or without previous incubation in wastewater microcosms and we examined mainly their metabolic, patterns after exposure to salinity gradient. The results obtained showed metabolic profiles changes both for *E.coli* and *Salmonella Typhymurium* and loss for production of several carbolases mainly glucose, mannose, rhamnase, sorbitol, sucrose, melibiose and arabinose hydrolases whereas for *Vibrio paraheamolyticus* no changes were observed.

Similar results were found previously by Ben Kahla –Nakbi et al (2007) for *V. alginolyticus* strains incubated in seawater microcosms for long period of starvation and for what the result of metabolic carbohydrates profiles tested by results of API system tests, showed no modification. These findings were explained by preparation of cells to enter into VBNC state characterized by low metabolic rate, reduction in size, morphological changes or synthesis of specific proteins. This adaptation strategy developed by bacteria in aquatic environments allows them to survive during long period of time. Modifications of cellular morphology should be related to modifications with carbohydrates metabolites contained in pariatal membranes for these species.

	API 20E*															
Strains	*ONPG*	*ADH*	*ODC*	*CIT*	*URE*	*TDA*	*IND*	*VP*	*GEL*	*GLU*	*MAN*	*SOR*	*RHA*	*SAC*	*MEL*	*ARA*
O126B16	+	+	-	+	-	+	+	-	-	+	+	+	+	+	+	+
S.W.	-	-	-	-	-	+	+	-	-	+	+	+	-	-	+	+
W.W.	-	-	-	+	+	+	-	+	+	-	-	-	-	+	-	-
ECEAgg	+	-	+	+	-	+	+	-	-	+	+	+	+	+	+	+
S.W.	+	-	+	-	-	+	-	-	-	+	+	+	+	+	+	+
W.W.	+	+	+	-	-	+	-	-	-	+	+	+	+	+	+	+
ECEI	+	-	+	+	-	+	+	-	-	+	+	+	+	-	+	+
S.W.	+	-	+	-	-	+	+	-	-	+	+	+	+	-	+	+
W.W.	+	+	+	+	-	-	-	-	-	+	+	+	-	-	-	-
ECEH	+	-	+	+	-	+	+	-	-	+	+	+	+	-	+	+
S.W.	+	-	-	-	-	+	+	+	-	+	+	+	+	-	-	+
W.W.	+	+	-	+	-	+	-	-	-	+	+	+	+	-	-	+
ECEP	+	-	+	+	-	+	+	-	-	+	+	+	+	-	+	+
S.W.	+	-	-	-	-	+	+	+	-	+	+	+	+	-	-	+
W.W.	+	-	-	+	-	+	-	-	-	+	+	+	+	-	-	+
ECET	+	-	+	+	-	+	+	-	-	+	+	+	+	+	+	+
S.W.	+	-	+	-	-	+	+	+	-	+	+	+	+	+	+	+
W.W.	+	+	+	+	-	+	-	-	-	+	+	+	-	+	-	+

*ONPG: β-galactosidase, **ADH:** arginine dihydrolase, **ODC:** ornithine decarboxylase, **CIT:** use of citrate, **URE:** urease, **TDA:** tryptophane desaminase, **IND:** indole production, **VP:** acetoin production **GEL:** gelatinase, **GLU:** glucose, **MAN:** mannitol, **SOR:** sorbitol, **RHA:** rhamnose, **SAC:** saccharose, **MEL:** melibiose, **ARA:** arabinose.

Table 2. Biochemical patterns. Modifications of some biochemical characters on APi 20E strips of different *E.coli* strains after survival in seawater with previous incubation in seawater (SW) or in wastewater (WW). O126B16 (reference *E.coli* strain), EaggEC (Enteroaggregative *E.coli* strain), EIEC (Enteroinvasive *E.coli* strain), EHEC (Enterohemorrhagic *E.coli* strain), EPEC (Enteropathogenic *E.coli* strain), ETEC (Entero toxigenic *E.coli* strain).

Keymer et al (2007) described similar results for *V.cholerae* and pointed that environmental parameters, measured *in situ* during sample collection, should correlated to the presence of specific dispensable genes and metabolic capabilities, including utilization of mannose, sialic acid, citrate, and chitosan oligosaccharides. Thus, gene content identified and metabolic pathways that are likely selected for in certain coastal environments and may influence *V. cholerae* population structure in aquatic environments.

Our previous data described *A.hydrophila* cultured for 30 days in marine water microcosms which mainained the ability to metabolize all the carbohydrates and continued to produce several enzymes with differences of its homologous cultured in waste water previous to incubation in marine water microcosm which failed to produce numerous enzymes and metabolize carbohydrates (El Mejri et al, 2008). Thus, for marine heterotrophic bacterial populations as *Vibrio sp* and *Aeromonas sp* halotolerant conserve major potential of their carbohydrate metabolism.

For high salinity or halotolerant bacteria, many strains that grew fermentatively on carbohydrates in the presence of air (determined by the acidification of the medium) grew especially well on glucose, totally by substrate- level phosphorylation in the absence of air. Tomlinson and Hochstein (1976) demonstrated that O2 consumption by *Halobacterium saccharovorum* in presence of glucose was 18% of the theoretical amount required for its complete oxidation it was for 83% in presence of galactose or fructose which like galactose, requires more oxygen for its oxidation and therefore was generally a poorer substrate for fermentation (Javor et al, 1984). The measurement of O2 consumption, growth rates, and fermentation products in the presence of glycerol, pyruvate, and acetate by strains that demonstrated relatively good anaerobic growth on these substrates would provide further evidence of the importance of fermentative metabolism in extreme halophiles.

In the water column, biofillm microenvironments in suspended flocks may form a stabilizing refugee that enhances the survival and propagation of pathogenic (i.e., disease-causing) bacteria entering coastal waters from terrestrial and freshwater sources. The EPS matrix offers microbial cells a tremendous potential for resiliency during periods of stress, and may enhance the overall physiological activities of bacteria as it's emphasized by Giller et al(1994) that influences small-scale microbial biofilms must be addressed in understanding larger-scale processes within inter tidal systems. Intertidal systems are a key interface of the ocean, atmosphere, and terrestrial environments, and as such, are characterized by frequent fluctuations in temperature, ion concentration, desiccation, UV-irradiation, and wave action. The relative frequency of these fluctuations poses both physical and biochemical challenges to micro organisms which inhabit this environments such estuarine areas. The characteristics and intensities of such stresses may vary substantially (Decho et al, 2000).

Therefore, in estuarine region with with salinity of 35 ppt it is been reported that the growth rate of lingo-cellulose-degrading populations (50%) is inhibited by NaCl (Liu and Boone, 1991). According to Park et al (2006) the high salinity seems to significantly reduce ecto enzyme activities. Thus, it should be possible that type of bacterial population found at particular station can be influenced by salinity changes. Also, it has been reported in most of the aquatic environments there is a significant relationship between extracellular enzyme activities, their corresponding substrates (polymers) and their hydrolysis products (monomers) (Münster et al, 1992). Thus, presence of cellulolytic and hemi-cellulolytic bacteria in one particular region and their absence in other region should indicated presence of wide-ranging organic matter in each region (Khanderparker et al, 2011). Previous research has also put forth a widely accepted concept that hydrolytic enzymes are induced by presence of polymeric substrates (Chrost, 1991; Vetter and Deming, 1999). It was observed that the genus *Bacillus* and *Vibrio* were the dominant hemicellulase and cellulase producers in both the estuaries. According to Ruger (1989) the genus *Bacillus* comprised phylogenetically and phenotypically diverse species, which are ubiquitous in terrestrial and fresh water habitats and are also widely distributed in seawater ecosystems. While the familyof *Vibrionaceae* represent the most important bacterial autochtonous groups in marine environments. Members of this family often predominate in the bacterial flora of seawater, plankton, and fish. In the West Pacific Ocean, , Vibrios accounted for nearly 80% of the bacterial population in surface seawater (Simidu et al, 1980) and Vibrio phylogenetic diversity of their culturable forms in Bulgarian hot springs as described along with their abilities to metabolize carbohydrates (Derekova et al,2008)..

For Moari et al, (2011), marine models tested showed, prokaryotic abundance in superficial marine sediments controlled by organic trophic resources, while in sub-surface sediments, prokaryotic activities and abundance were driven by environmental factors and predatory pressure, suggesting that the shift in prokaryotic community structure could be coupled to a change in life-style of microbial assemblages.

Previously, Bouvier et al, (2002), discussed the relationship between the compositional succession and changes in single-cell metabolic activitiesd in the Choptank River, and suggested that profound phylogenetic shifts were linked to cell stress, loss of activity, and death (Del Giorgio and Bouvier, 2002). Similarly both types of succession, i.e., activation/ inactivation and replacement, occured simultaneously, and understanding the relative importance of such processes in determining bacterial succession and community composition remain to be clarified considering environmental conditions that trigger bacterial succession observed within the mixing areas. .

4. Conclusion

Carbohydrates metabolism still basic to marine microbiota mainly for autochtonous communities as *Vibrio, Aeromonas* and *Bacillus* genus with high potential of degradation and considerable supply of big amounts of different carbolases and pathways which should take more attention in order to be clarified.

Increase in significant amounts of organic materials are being exchanged between the land and marine ecosystems with higher amounts of dissolved carbohydrates provided especially the polymeric substrates which represent the most abundant molecules used by marine bacteria as structural and /or storage compounds under extreme variations in environmental parametres and biological marine habitats.

The intrinseque ability of bacterial cells to enter in to viable but non culturable state (VBNC) characterized by low metabolic rate and morphological modifications (reduction in cell size and membrane alterations) should be related with carbohydrates metabolism and change observed experimentally. Therefore, investigations in carbohydrates metabolites which continue going on will certainly elucidate more molecular adaptations forms of bacteria (for mainly enteric species) in variable marine ecosystems .

Author details

Monia El Bour
National Institute of Sea Sciences and Technologies, Salammbô, Tunisia

5. References

Ackefors H. & Enell M. (1994) The release of nutrients and organic matter from aquaculture systems in Nordic countries. *Journal of Applied Ichthyology*. Vol.10, pp 225–241.Ackefors and Enell, 1994

Alldredge AL.(2000). Interstitial dissolved organic carbon (DOC) concentrations within sinking marine aggregates and their potential contribution to carbon flux. *Limnologie and Oceanography*. Vol45, pp1245–1253

Aluwihare, L.I., Repeta, D.J. & Chen, R.F., (1997). A major biopolymeric component to dissolved organic carbon in surface sea water. *Nature*. Vol387, pp166–169.

Amend J.P. & Plyasunov A.V. (2001). Carbohydrates in thermophile metabolism: Calculation of the standard molal thermodynamic properties of aqueous pentoses and hexoses at elevated temperatures and pressures. *Cosmochimica Acta*. Vol 65, pp. 3901-3917.

Anderson, I.C., Rhodes, M.W., Kator, H.I., (1979). Sublethal stress in *Escherichia coli*: a function of salinity. *Applied and Environmental Microbiology* Vol.38, pp1147-1152. Pereira and Alcantara, 1993

Arnosti C. and Repeta D. J. (1994). Oligosaccharide degradation by anaerobic marine bacteria : characterization of an experimental system to study polymer degradation in sediments. *Limnology and Oceanography*. Vol 39, pp1865–1877.

Arnosti C. and Holmer M. (1999). Carbohydrate dynamics and contributions to the carbon budget of an organic-rich coastal sediment. *Geochimistry and. Cosmochimestry*. Acta. Vol. 63, pp 393–403.

Arnosti C.(2000). Substrate specificity in polysaccharide hydrolysis: contrast between bottom water and sediments. *Limnologie and Oceanography*. Vol.45, pp1112– 9.

Azam F, & Long R. (2001). Sea snow microcosm. Nature. Vol. 8, pp414:495..

Azam, F. & Worden, A.Z., (2004). Microbes, molecules, and marine ecosystems. *Science*. Vol. 303, pp 1622-1624.

Baleux, B., Caro, A., Lesne, J., Got P., Binard, S., Delpeuch, B., 1998. Survie et maintien de la virulence de *Salmonella Typhimurium* VNC exposée simultanémènt à trois facteurs stressants expérimentaux. *Océanologica Acta* Vol.21, pp939-950.

Barcina I., LeBaron P., & Vives-Rego J. (1997). Survival of allochthonous bacteria in aquatic systems: A biological approach. *FEMS Microbiology and Ecology*. Vol.23, pp 1–9.

Ben Kahla-Nakbi A, Besbes A, Chaieb K, Rouabhia M, & Bakhrouf A. (2007). *Marine Environmental Research*, Vol. 64,pp 469

Benner, R., Pakulski, J.D., McCarty, M., Hedges, J.I. & Hatcher, P.G. (1992). Bulk chemical characteristics of dissolved organic matter in the ocean. *Science* Vol. 255, pp 1561-1564.

Billen G. (1984). Heterotrophic utilization and regeneration of nitrogen. In *Heterotrophic Activity in the Sea*, ed. JE Hobbie, PJleB Williams, pp. 313–55.New York: Plenum

Boetius, A., Ravenschlag, K., Schubert, C.J., Rickert, D., Widdel, F., Gieseke, A., Amann, R., Jørgensen, B.B., Witte, U. & Pfannkuche, O. (2000). A marine microbial consortium apparently mediating anaerobic oxidation of methane. *Nature*. Vol. 407, pp 623-626. .

Bott T.L.& Kaplan L.(1985). Bacterial biomass, metabolic state and activity in stream sediments: relation to environmental variables and multiple assay comparisons. *Applied and Environmental Microbiology*. Vol. 50, pp 508-522.

Bouvier T.C. & Del Giogio P.A.(2002). Compositional changes in free-living bacterial communities along a salinity gradient in two temperate estuaries. *Limnologie and Oceanography*. Vol. 2002, pp. 453-470.

Brock, T. D. (1987). The study of microorganisms *in situ*: progress and problems. *Symposium Of Society Of. General Microbiologyl* Vol.. 41, pp1–17.

Burdige, D.J. & Zheng, S.L.(1998). The biogeochemical cycling of dissolved organic nitrogen in estuarine sediments. *Limnology and Oceanography.* Vol.43, pp1796–1813.

Canuel E. A. and Martens C. S. (1996). Reactivity of recently deposited organic matter: degradation of lipid compounds near the sediment–water interface. *Geochimestry and Cosmochimestry* Acta 60, pp 1793–1806.

Cho BC & Azam F.(1988) Major role of bacteria in biogeochemical fluxes in the ocean's interior. *Nature*. Vol.332, pp441–3.

Chrost, R.J., 1991. Environmental control of the synthesis and activity of aquatic microbial ectoenzymes. In: *Microbial Enzymes in Aquatic Environments*. Springer, pp. 29-59.

Cowie G. L. and Hedges J. I. (1992) Sources and reactivities of amino-acids in a coastal marine environment. *Limnologie Oceanography.* Vol.37, pp703–724.

Cowie GL, Heges JI, Prahl FG & De Lange GJ.(1995) Elemental and major biochemical changes across an oxidation front in a relict turbidite: an oxygen effect. *Geochimestry Cosmochim Acta*. Vol.59. pp33– 46..

Del Giorgio P.A. & Cole J.J.(1998). Bacterial growth efficiency in natural aquatic systems. *Annual Review of Ecology*. Vol. 29, pp 503-541.

Decho A.W.,(2000). Microbial biofilms in intertidal systems: an overview. *Continental Shelf Research*. Vol.20, pp. 1257-1273.

Derekova, A., Mandeva, R., Kambourova, M., (2008). Phylogenetic diversity of thermophilic carbohydrate degrading Bacilli from Bulgarian hot springs. *World Journal of Microbiology and Biotechnology*. Vol. 24, pp 1697-1702..

Druffel, E.R.M., Williams, P.M., Bauer, J.E. & Ertel, J.R. (1992). Cycling of dissolved and particulate organic matter in the open ocean. *Journal of Geophysical Research.* Vol.97, pp15639–15659.

Ducklow HW & Carlson CA. (1992). Oceanic bacterial production *Advances in Microbiol Ecologie.* Vol.12, pp113–81

Dupray E., Derrien A. (1995). Influence du passage de *Salmonella* spp. Et *Escherichia coli* en eaux usées sur leur survie ulterieure en eau de mer. *Water Research.* Vol. 29, pp1005-1011.

Espeland, E. M. & Wetzel, R. G., (2001). Complexation, stabilization and UV photolysis of extracellular and surface-bound glucosidase and alkaline phosphatase: implications for biofilm microbiota. – *Microbiology and Ecology*. Vol. 42, pp 572–585.

Gauthier M.J., Munro P.M., Flatau G.N., Clément R.L. & Breittmayer V.A. (1993). Nouvelles perspectives sur l'adaptation des entérobactéries dans le milieu marin. *Marine Life,* Vol.3, pp1-18.

Gerba C.P. & McLeod J.S. (1976).Effect of Sediments on the Survival of *Escherichia coli* in Marine Waters. *Applied and Environmental Microbiology*. Vol.32, pp 114-120.

Giller, P.S., Hildrew, A.G., Ra!aelli, D.G. (Eds). (1994). Aquatic Ecology: Scale, Pattern and Processes. Blackwell Publishers, Oxford, 649 pp..

Gueuen, C., Guo, L., Wang, D., Tanaka, N., Hung, C.C. (2006). Chemical characteristics and origin of dissolved organic matter in the Yukon River. *Biochemistry* Vol. 77, pp139–155.

Holmer M. & Kristensen E. (1994) Organic-matter mineralization in an organic-rich sediment – experimental stimulation of sulfate reduction by fish food pellets. *FEMS Microbiology and Ecology*. Vol.14, pp 33–44.

Hedges, J.I., Keil, R.G., Benner, R., 1997. What happens to terrestrial organic matter in the ocean? *Organic Geochemistry*. Vol. 27, pp195–212.

Hoppe HG. (1983). Significance of exoenzymatic activities in the ecology of brackish water: measurements by means of methylumbelliferylsubstrates. Marine Ecological Progressive Series Vol.11, pp 299–308.

He B., Dai M., Zhai W., Wng L., Wang L., Chen J., Lin J., Han A. & Xu Y. (2010). Distribution, degradation and dynamics of dissolved organic carbon and its major compound classes in the Pearl River estuary, China. *Marine Chemistry*. Vol. 119, pp 52–64.

Horner-Devine M.C., Carney K.M. &. Bohannan .M.J. (2004). An ecological perspective on bacterial biodiversity. *Proceedings of the Royal Society of London*. Vol.271, 113-122

Hung, C.-C., Guo, L., Santschi, P.H., Alvarado-Quiroz, N. & Haye, J.M. (2003). Distributions of carbohydrate species in the Gulf of Mexico. Marine Chemistry. Vol. 81, pp119–135.

Javor B.J.,(1984).Growth Potential of Halophilic Bacteria Isolated from Solar Salt Environments: Carbon Sources and Salt Requirements. *Applied and Environmental Microbiology*. Vol.48, pp. 352-360.

Kaltenbock E & Herndl GJ.(1992). Ecology of amorphous aggregates (marine snow) in the northern Adriatic Sea IV Dissolved nutrients and autotrophic community associated with marine snow. *Marine Ecological Progressive Series* Vol.87, pp147–59.

Keymer D.P., Miller M.C., Schoolnik G.K, Boehm A.B. (2007). Genomic and phenotypic diversity of costal Vibrio cholerae strains is linked to environmental factors. *Applied and Environmental Microbiology*. Vol 73, pp 3705-3714.

Khanderparker R.,Verna P., Meena R.M. & Deobagkar D.D. (2011). Phylogenetic diversity of carbohydrate degrading culturable bacteria from Mandovi and Zuari estuaries, Goa, west coast of India. *Estuarine, Coastal and Shelf Science*. Vol. 95, pp 359-366.

Kiørboe T. & Jackson GA.(2001). Marine snow, organic solute plumes, and optimal chemosensory behaviour of bacteria. *Limnology Oceanography*. Vol.46, pp1309– 18.

Kristensen E. (1994). Decomposition of macroalgae, vascular plants and sediment detritus in seawater: use of step wise thermogravimetry. *Biogeochemistry* Vol.26, pp1–24.

Kristensen E., Ahmed S. I. and Devol A. H. (1995) Aerobic and anaerobic decomposition of organic matter in marine sediment: which is fastest? *Limnology and Oceanography*. Vol 40, pp1430–1437.

Kristensen E. and Holmer M. (2001) Decomposition of plant materials in marine sediment exposed to different electron acceptors (O2, NO3, and SO42), with emphasis on substrate origin, degradation kinetics, and the role of sulfate reduction. *Geochimestry and Cosmochimestry*. Acta Vol.65, pp419–433.

Lango Z., Leech D.M., Wetzel R.G., (2007). Indirect effects of elevated atmospheric CO2 and solar radiation on the growth of culturable bacteria. *Fundamental and Applied Limnology Archiv Für Hydrobiologie,* Vol. 168, pp. 327-333..

Lennon, J.T., 2007. Diversity and metabolism of marine bacteria cultivated on dissolved DNA. *Applied and Environmental Microbiology* Vol. 73, pp 2799-2805.

Leeuw J. W. D. and Largeau C. (1993). A review of macromolecular organic compounds that comprise living organisms and their role in kerogen, coal and petroleum formation. In Organic (eds. M. H. Engel and S. A. Macko). Plenum Press, New York.

Lindell, M. J., Granéli, H. W. & Tranvik, L. J., (1995). Enhanced bacterial growth in response to photochemical transformation of dissolved organic matter. *Limnologie and Oceanography*, Vol. 40, pp 195–199.

Liu, Y., Boone, D.R., 1991. Effects of salinity on methanogenic decomposition. *Bioresource Technology.* Vol. 35, pp 271-274.

Martin, Y., Troussellier, M., Bonnefont, J-L., 1998. Adaptative responses of *Escherichia coli* to marine environmental stresses: a modelling approach based on viability and dormancy concepts. *Oceanologica Acta* 21, 951-964.

Meybeck, M., (1982). Carbon, nitrogen, and phosphorus transport by world rivers. *American Journal of Science.* Vol.282, pp401–450.

Middelboe, M., Borch, N.H., Kirchman, D.L., 1995. Bacterial utilization of dissolved free amino acids, dissolved combined amino acids and ammonium in the Delaware Bay estuary: effects of carbon and nitrogen limitation. *Marine Ecology Progress Series*. Vol.28, pp109–120.

Molari M., Giovannelli D., D'Errico G. & Manini E. (2011). Factors influencing prokaryotic community structure composition in sub-surface coastal sediments. *Estuarine, Coastal and Shelf Science.* Vol. xxx, pp1-8.

Monfort P., Baleux B. (1999). Bactéries viables non cultivables: réalité et conséquences. *Bulletin de la Société Française de Microbiologie,* Vol.14, pp 201-207.

Murrell, M.C. & Hollibaugh, J.T. (2000). Distribution and composition of dissolved and particulate organic carbon in northern San Francisco Bay during low flow conditions. *Estuarine, Coastal and Shelf Science*. Vol. 51, pp75–90.

Munro P., Gauthier J., Laumond F. (1987). Changes in *Escherichia coli* cells starved in seawater or grown in seawater-wastewater mixtures. *Applied Environmental Microbiologie,* Vol.53, pp1476-1481.

Münster, U., Einiö, P., Nurminen, J. & Overbeck, J. (1992). Extracellular enzymes in a polyhumic lake: important regulators in detritus processing. *Hydrobiologia*. Vol. 229, pp. 225-238.

Pakulski, J.D. & Benner, R. (1994). Abundance and distribution of carbohydrates in the ocean. *Limnology and Oceanography*. Vol. 39, pp930–940..

Park, J.S., Choi, D.H., Hwang, G.J., & Park, B.C. (2006). Seasonal study on ectoenzyme activities, carbohydrate concentrations, prokaryotic abundance and production in a solar saltern in Korea. *Aquatic Microbial Ecology*. Vol. 43, pp 153-163.

Pereira, M.G., Alcantara, F., 1993. Culturability of *Escherichia coli* and *Streptococcus faecalis* in batch culture and "*in situ*" in estuarine water (Portugal). *Water Research* Vol.27, pp.1351-1360

Pomeroy LR, Wiebe WJ, Deibel D, Thompson RJ & Rowe GT,(1991). Bacterial responses to temperature and substrate concentration during the Newfoundland spring bloom. *Marine Ecology Progressive Series.*Vol 75, pp143–59

Poncet S., Milohanic E., Mazé A., Nait Abdallah J., Aké F., Larribe M., Deghmane A.E., Taha M.K., Dozot M., De Bolle X., Letesson. J.J & Deutscher J.(2009). Correlations between Carbon Metabolism and Virulence in Bacteria. In Contributions to Microbiology (Mattias Collins & Raymond Scush (Editors). Vol. 16, pp.88-102.

Rozen, Y. & Belkin, S., 2001. Survival of enteric bacteria in seawater. Microbiology Review 25, 513-529.

Rozen, Y., LaRossa, R.A., Templeton, L.J., Smulski, D.R. & Belkin, S. (2002). Gene expression analysis of the response by *Escherichia coli* to seawater. *Antonie van Leeuwenhoek.* Vol.81, pp15-25.

Ruger, H.J., 1989. Benthic studies of the northwest African upwelling region psychrophilic bacterial communities from areas with different upwelling intensities. *Marine Ecological Progress Series.* Vol. 57, pp. 45-52.

Simidu, U., Taga, N., Cohvell, R.R., Schwartz, J.R., (1980). Heterotrophic bacterial flora of the seawater from the Nansei Shoto (Ryukyu Retto) area. *Bulletine Japanese Society of Science and Fisheries.* Vol. 46, pp.505-510.

Simon M, Grossart HP, Schweitzer B & Ploug H.(2002).Microbial ecology of organic aggregates in aquatic ecosystems. Aquat Microb Ecol. Vol.28, pp175– 211.

Smith, D.C., Simon, M., Alldredge, A.L. & Azam, F., (1992). Intense hydrolytic enzyme activity on marine aggregates and implications for rapid particle dissolution. *Nature* Vol. 359, pp. 139-142.

Sinton, L.W., Davies-Colley, R.J., Bell, R., 1994. Inactivation of enterococci and fecal coliforms from sewage and meatworks effluents in seawater chambers, *Applied and Environmental Microbiology.* Vol60, pp2040-2048.

Tietjen, T. & Wetzel, R. G., (2003) Extracellular enzyme-clay mineral complexes: enzyme adsorption, alteration of enzym activity, and protection from photodegradation. *Aquatic Ecology*, Vol. 37, pp 331–339.

Tomlinson G. A., & Hochstein L.I. (1976). *Halobacterium saccharovorum sp.* nov., a carbohydrate-metabolizing, extremely halophilic bacterium. *Canadian Journal of Microbiologie.* Vol. 22, pp.587-591.

Troussellier, M., Got, P., Bouvy, M., M'Boup, M., Arfi, R., Lebihan, F., Monfort, P., Corbin, D. & Bernard, C., (2004). Water quality and health status of the Senegal River estuary. *Marine Pollution Bulletin.* Vol. 48, pp 852-862.

Truu J., Talpsep E., Heimrn E., Stottmeister U., Wand H., Hermarn A. (1999). Comparison system of Api 20NE and Biolog GN identification systems assessed by techniques of multivariantc analyses. *Journal of Microbiological Methods.* Vol.36, pp 193-201.

Urbani, R., Sist, P., 2003. Studio di caratterizzazione chimica delle componenti polisaccaridiche e del loro ruolo nel meccanismo di formazione degli aggregati gelatinosi. In: Istituto per la Ricerca Scientifica e Tecnologica Applicata al Mare, Ministero dell'Ambiente e della Tutela del Territorio, editors. Programma di monitoraggio e studio sui processi di formazione delle mucillagini nell'Adriatico e nel Tirreno — MAT. Final Report, June 2003. pp.135–70.

Valdemarsen T., & Kristensen E.(2010). Degradation of dissolved organic monomers and short-chain fatty acids in sandy marine sediment by fermentation and sulfate reduction. *Geochimica et Cosmochimica Acta*, Vol. 74, pp1593–1605

Vetter, Y.A. & Deming, J.W. (1999). Growth rates of marine bacterial isolates on particulate organic substrates solubilized by freely released extracellular enzymes. *Microbial Ecology*, Vol. 37, pp 86-94..

Weiss MS, Abele J, Weckesser W, Welte W, Schultz E, Schulz GE. (1991) Molecular architecture and electrostatic properties of a bacterial porin. *Science.* Vol.254, pp1627– 30.

Wetzel, R. G. & Tuchman, N. C. (2005) Effects of atmospheric CO2 enrichment and sunlight on degradation of plant particulate and dissolved organic matter and microbial utilization – *Archives of. Hydrobiology*, Vol. 162, pp 287–308.

Williams PJleB. (1981). Microbial contribution to overall marine plankton metabolism: direct measurements of respiration. *Oceanologica. Acta* Vol.4, pp359–64

Williams, P.M. & Druffel, E.R.M. (1987). Radiocarbon in dissolved organic matter in the Central North Pacific Ocean. *Nature.* Vol. 330, pp246–248.

Whitman, W. B., Coleman, D. C. & Wiebe, W. J. (1998). Prokaryotes: the unseen majority. *Proceedings of the. Natural Academic of Sciences USA,* Vol. 95, pp 6578–6583.

Zoppini A., Puddu A., Fagi S., Rosati M. & Sist P. (2005). Extracellular enzyme activity and dynamics of bacterial community in mucilaginous aggregates of the northern Adriatic Sea. *Science of the Total Environment*. Vol.353, pp 270– 286

Chapter 16

Adapting the Consumption of Carbohydrates for Diabetic Athletes

Anna Novials and Serafín Murillo

Additional information is available at the end of the chapter

http://dx.doi.org/10.5772/51570

1. Introduction

Nutritional recommendations associated with physical exercise performed by people with type 1 diabetes have changed in recent years, alongside the development of new drug therapies and discoveries in the field of nutrition. Before insulin was discovered in 1921, exercise was considered a dangerous activity, usually discouraged due to the high risk of metabolic disorder involved. Currently, exercise is not only considered safe but is prescribed as a basic treatment of the disease, essential for a healthy lifestyle and recommended for all patients with diabetes. The beneficial effects of regular physical exercise associated with diabetes are numerous, offering improved physical capacity, decreased cardiovascular risk and increased emotional and social wellbeing [1]. However, the specific role of exercise in improving glycemic control is currently under debate. The root of this debate lines in findings that show that the successful treatment of type 1 diabetes consists of creating and maintaining a balance among many different factors involved in glycemic control (kind of exercise, insulin dosage and diet). Therefore, the failure to adapt carbohydrate intake to insulin dosage or the characteristics of the physical activity to be performed may lead to significant glycemic imbalances.

In this sense, exercise-induced hypoglycemia is described as one of the main factors limiting physical exercise, as well as a cause of poor performance in athletes with type 1 diabetes. However, certain strategies, such as supplementing carbohydrate intake, can reduce the severity and duration of these episodes, thereby eliminating the fear of exercise-induced hypoglycemia [2].

The role of carbohydrates in the diet of athletes with diabetes is important as a prerequisite for achieving maximum athletic performance, not only in improving and accelerating the recovery of energy stores in the body, but also in regulating blood glucose levels during training sessions and competitions [3]. In general, the nutritional strategies used by athletes with and without diabetes should not differ. However, certain standards and guidelines must be established for including carbohydrate supplements within the set of specific treatments for diabetes.

2. Carbohydrates and glycemic control

Carbohydrates ingested in food are the main determinant of blood glucose levels following meals. Both the quantity and quality (or kind of carbohydrate) is particularly influential in the variability of blood glucose levels obtained after intake. One must bear in mind that foods rich in carbohydrates represent the basis of nutrition in most cultures, as they are important sources of vitamins, minerals and fiber [4].

Regarding the amount of carbohydrates, the minimum daily requirements are 130 g [5], with no studies available on the safety of nutritional guidelines calling for fewer carbohydrates during periods longer than one year. The quality or kind of carbohydrate ingested can cause significant variations in one's response to food, increasing or decreasing the rate at which the carbohydrates contained can be digested as well as their ability to raise blood sugar levels. Certain food-specific factors can increase the rate at which blood glucose levels rise, such as:

- *Thermal or mechanical processing*. Longer cooking times or certain mechanical processes, such as the milling of grain into flour, increase the food's absorption rate.
- *Degree of starch gelatinization*. Applying heat in the presence of water initiates the process of starch gelatinization, which facilitates the breakdown of food by intestinal enzymes during the digestive process, thereby increasing the absorption rate.
- *Amylose-amylopectin ratio of starches*. Starches are mainly made up of amylose and amylopectin chains. Amylose forms helical structures, unbranched, which are less accessible to digestive enzymes than amylopectin chains.
- *Kind of sugar*. Fructose has a slower absorption rate than glucose or sucrose. Furthermore, once absorbed by the intestine, it must undergo a series of processes by the liver in order to become glucose. Therefore, fructose-rich foods cause blood sugar to increase more slowly than those that contain other kinds of sugar, such as glucose or even starch, in their composition.
- *Other food components*. The presence of high amounts of protein, fiber or fat in food can alter the rate of absorption by slowing down the digestive process. Certain condiments such as vinegar are capable of acidifying food, which slows down the digestive process and therefore the absorption of food.

Furthermore, other factors specific to the individual must be taken into account that may also affect the rate of absorption. These factors include low blood glucose levels or fasting before eating. It should be noted that low carbohydrate intake during the hours or days prior to performing moderate or intense, mid- or long-duration physical activity can cause glycogen reserves in the liver and muscles to decrease, which then lowers the glycemic effect of any food eaten.

One way to know the rate at which carbohydrates contained in a certain food increase blood sugar levels is by looking at its glycemic index. This measurement is made by comparing the blood glucose curve obtained after a quantity of food is ingested that provides 50g of carbohydrates, as compared with the blood glucose curve obtained with a reference food, usually 50g of glucose. Foods are grouped into three different categories, low, moderate and high, according to their glycemic index, with high glycemic index foods producing abrupt spikes in glycemic levels. Recently the concept of glycemic load has also surfaced, which is determined by taking into account the amount of carbohydrates contained in the portion of food consumed.

It is important to remember that measurements are a theoretical concept and that there is a wide range of factors affecting the rate at which carbohydrates are absorbed and raise glycemic levels. These factors include the kind of food, brand or country of origin, cooking methods used, and degree of food processing.

At any rate, in diabetes management it is important to coordinate insulin doses with the carbohydrate content of meals, which can be estimated using a variety of methods including carbohydrate counting, the carbohydrate exchange system, or experience-based serving size approximations.

3. Effect of exercise on blood glucose balance and the risk of hypoglycemia

The effect of exercise on blood glucose levels is determined by the interaction that takes place between metabolic and hormonal effects, as well as changes produced in muscle glucose uptake.

3.1. Energy substrates during rest and exercise

Fat deposits of adipose tissue are the major source of energy in the human body, with a reserve of between 60,000 and 150,000 Kcal. This content is much greater than the energy that carbohydrates can provide, approximately 2,000 Kcal. Most of these, about 1500 kcal, are stored as glycogen in the muscles, and the rest comes from glycogen deposits accumulated in the liver and from glucose found in the blood and extracellular fluids.

During the early stages of exercise [6-8], muscle glycogen is the main energy source for muscle contraction. Subsequently, the depletion of muscle glycogen deposits activates the lipolysis of fatty acids stored in adipose tissue. Thus, the increase of free fatty acids and glycerol in plasma constitutes an additional energy source for muscle contraction. Glucose will then be provided by hepatic glycogenolysis followed by hepatic gluconeogenesis. The substrates used by the liver to synthesize new glucose are lactate, pyruvate and certain amino acids, mainly alanine, together with the glycerol derived from the metabolism of triglycerides. The contribution of different substrates depends on the intensity, duration and kind of exercise performed. The relative proportion of each seeks to maintain three fundamental physiological aspects: 1) to preserve blood glucose balance, 2) to maintain the efficient metabolism and storage of glucose, and 3) to preserve and maintain deposits of muscle glycogen to avoid and/or delay the onset of muscular fatigue.

3.2. Hormonal response to exercise

Metabolic adjustments to exercise are possible thanks to a highly efficient system that integrates nerve impulses and hormonal response. During rest, individuals without diabetes experience what is called a basal insulin secretion, which increases in response to the rise in blood glucose that occurs after ingesting food. It is well known that insulin stimulates glucose uptake by skeletal muscle and by the liver, to later facilitate its storage in the form of glycogen.

Exercise causes adrenergic nerve stimulation, acting on pancreatic beta-cells and leading to inhibited insulin secretion [9-10]. Fortunately, this decrease in insulin levels does not affect muscle glucose uptake, as exercise triggers other mechanisms that can improve muscle glucose uptake.

Stimulation of another pancreatic hormone, glucagon, and, in particular, its close and proper interaction with insulin, are phenomena taking place during the exercise that are essential to maintaining and regulating glucose production. Decreased insulin levels at baseline exercise promote increased glucagon secretion by alpha cells in pancreatic islets. This increase is critical and acts directly on the metabolic pathways of hepatic glucose production (glycogenolysis and gluconeogenesis). Also, during exercise, counter-regulatory hormones increase, such as catecholamines, cortisol and growth hormone, promoting the balance of the aforementioned metabolic pathways and increasing lipolysis in fat cells. High-intensity exercise can activate these counterregulatory hormones in an exaggerated way, resulting in a noticeable increase in hepatic glucose production and a moderate degree of hyperglycemia when exercise is over.

3.3. Metabolic responses to exercise in diabetes

As expected, athletes with type 1 diabetes are different from their non-diabetic counterparts, as they have an insulin secretion deficiency and, consequently, their counterregulatory hormones respond differently. Patients with diabetes treated with insulin should learn to mimic their own natural insulin secretory rhythm, in response to physiological changes induced by exercise.

Any patient with diabetes and, especially, diabetic athletes should prevent problems arising from poor insulin dosage. If an athlete with diabetes starts exercise with a significant insulin deficiency, his/her response to exercise may trigger a hyperglycemic overcompensation, even ketosis, since a lack of insulin induces: 1) increased hepatic glucose production; 2) decreased peripheral glucose utilization; and 3) excessive lipolysis with an increase in free fatty acid production.

If an athlete with diabetes, under any circumstances, experiences an excess of insulin when exercising, his/her response to it can trigger hypoglycaemia, due to: 1) decreased hepatic glucose production; 2) increased peripheral glucose utilization; and 3) reduced lipolysis [11-12].

It is also important to consider the changes that occur after physical activity. Muscle glucose uptake independent of insulin is still stimulated during the hours following activity, and muscles need more quantity of glucose to replenish glycogen stores. Exercise increases insulin sensitivity, which means that the effects of insulin last longer. If an athlete with diabetes not properly increase carbohydrate intake and/or reduce insulin dosage, a significant episode of hypoglycemia may occur hours after exercise has ended [13].

3.4. Effects of insulin and exercise on muscle glucose uptake

Both insulin and muscle contraction help glucose enter the muscles, where it is oxidized and subsequently turned into energy for muscle contraction. However, the mechanisms by which these two stimuli facilitate glucose transport are not entirely understood. There is evidence suggesting that they do not act in a similar manner. For example, exercise, unlike insulin, induces an increase in muscle blood flow and glucose transport, which persists for hours after exercise has ended. Experiments with laboratory animals have demonstrated that the induction of glucose transport produced by exercise is independent of insulin, as tyrosine kinase activity of the insulin receptor is not stimulated. Instead, it appears to act through other pathways, in particular, those involving the AMP kinase (AMPK).

Molecular characterization of glucose transporter GLUT 4, which is specifically expressed in muscle and fat cells, has shed new light on elucidating the mechanisms mentioned above. It appears that the translocation of GLUT 4 [14-15] from the cytosol to the cell membrane is one of the main mechanisms of glucose transport in muscles and can be stimulated both by insulin and by exercise. The existence of another type of transporter in muscle, GLUT 1, has also been demonstrated, although its function is not entirely known. The important role of certain enzymes such as hexokinase and glycogen synthase also highlights the complex system of glucose transport into muscles. AMPK [16] emerges as an important molecule regulating multiple metabolic processes that occur in skeletal muscle, in response to exercise. It appears that muscle contractions induced by exercise increase the activity of the enzyme, which, in turn, stimulates glucose transport.

4. Factors that determine carbohydrate needs during and after physical exercise

A prediction of one's consumption of carbohydrates during and in the hours after physical activity requires a great knowledge and analysis of factors specific to each type of activity. In this sense, the personal observations and experience of each athlete are highly important, assessing the effect of each type of activity on his/her blood sugar levels. It is recommended to monitor blood glucose levels before and after exercise, or, if possible, to use a continuous glucose sensor.

In general, the characteristics of the activity to be performed must be known, as these factors affect blood glucose levels to a high degree:

- *Type of exercise*: Forms of exercise dominated by the aerobic component, such as walking, running, swimming, skating or cycling, produce greater glucose consumption and, therefore, have a stronger hypoglycaemic effect. In contrast, forms of exercise with an elevated anaerobic component, like sprints, combat sports and working with heavy weights, can produce strong adrenergic stimulation (stimulating liver production of glucose) and thus usually have a lesser hypoglycaemic effect post-exercise. Competitive sports can also be associated with significant emotional stress (high adrenergic stimulation), which causes increased blood sugar after exercise, especially in children and adolescents.
- *Duration*: During the first 30-60 minutes of moderate- or high-intensity exercise, muscular and hepatic glycogen becomes the primary muscle fuel. Thereafter, the glycogen stores begin to decrease, and muscles increasingly obtain energy from fatty acids and plasmatic glucose. Following this is when the most significant changes in blood glucose are observed.
- *Intensity*: Glucose is the preferred muscle fuel for exercise performed at a moderate or high intensity, whereas low-intensity exercise uses fatty acids as energy source. Therefore, low-intensity activities such as walking may have a minimal effect on blood sugar, whereas intense activities such as running could cause a stronger and faster blood glucose lowering effect.
- *Frequency*: Hypoglycaemic effects, especially after exercise, increase after several consecutive days of physical exercise. In this situation, since it is virtually impossible to recover of glycogen stores from one day to another, the body is less able to use the mechanisms of hepatic glycogenolysis to regulate low blood glucose.
- *Schedule*: The timetable of insulin administration results in the presence of different blood insulin levels throughout the day, which brings about a greater tendency to develop hypoglycemia during physical activity. This effect is more likely when exercise is performed just within 2-3 hours after meals, when rapid-acting insulin acts the strongest.

It is important to keep in mind that an athlete's level of fitness also affects his/her glucose uptake. For this reason, athletes just starting a particular sport usually consume more glucose than those who are more advanced in the sport, as training helps the body to use fat as muscle fuel, preserving most glucose reserves as glycogen.

5. Recommendations for carbohydrate supplementation in diabetic athletes

Nutritional guidelines for athletes with diabetes are similar to those for all athletes, in accordance with the activity performed. Thus, the recommended total amounts of macro and micronutrients should be calculated taking into account the characteristics of each individual as well as the specific requirements of the different stages of training or competition. However, in the case of carbohydrates, in addition to overall quantitative control, the distribution of these carbohydrates throughout the day takes on even more importance, according to the exercise habits and insulin dosage followed by each athlete.

Carbohydrates represent the main nutrient for both ensuring optimal muscle work and for maintaining blood glucose levels as close to normal as possible. Since reserves of carbohydrates in the body are limited, nutritional strategies must aim at maintaining or quickly replenishing these deposits once consumed through physical exercise. In this sense, dietary guidelines that heavily restrict carbohydrates are not recommended, as they would decrease athletic performance and at the same time increase the risk of hypoglycemia.

The kind of carbohydrates recommended depends on the timing of when they are ingested in relation to when exercise is performed. Thus, it is possible to use slow-absorbing carbohydrates (low-glycemic index) in the hours prior to a workout or competition, but also fast-absorbing ones (high-glycemic index) just before or during the activity. Also, continous blood glucose measurements should be used to monitor the rate at which the body needs glucose, bearing in mind the type of food most appropriate for each situation.

On another front, studies in people without diabetes found a better recovery of glycogen stores after exercise when high glycemic index foods were taken [17]. In contrast, in athletes with type 1 diabetes, priority was given to the total amount of carbohydrates over the kind of carbohydrates ingested [18]. Further studies are needed to assess the importance of following nutritional strategies based on the use of high glycemic index foods in the post-exercise recovery of athletes with type 1 diabetes.

5.1. Dietary planning

Carbohydrate intake should be adapted to the needs of each given moment, in accordance with each phase of general conditioning or training for competitions.

5.1.1. Diet during the training phase

In daily training, an athlete's diet must maintain a high percentage of carbohydrates, in order to replenish the muscle and liver glycogen that has been spent. Conserving glycogen stores helps reduce the frequency of hypoglycemia during and in the hours following exercise. It should be noted that hepatic glycogenolysis is one of the main mechanisms involved in restoring blood glucose levels during the process of hypoglycemia and that a lack of liver glycogen stores decisively limits normal functioning.

In individuals without diabetes who train for less than 1 hour a day, the required carbohydrate intake is about 5-6g of carbohydrates / kg of body weight per day, reaching 8-10g/kg weight / day if the training exceeds 2 hours a day. Unfortunately, there are no studies evaluating the effectiveness of these nutritional strategies in athletes with type 1 diabetes, under the assumption that they should be any different for athletes with type 1 diabetes who maintain the disease under optimum control.

5.1.2. Diet during the pre-competition phase

Many athletes have adopted the strategy of carbohydrate loading during the week prior to a running or cycling competition. In general, carbohydrate loading is not recommended, especially when performed after 2-3 days of carbohydrate restriction. This strategy, used by some athletes to increase muscle glycogen stores, puts athletes with diabetes at risk, especially during the period of carbohydrate restriction, leading to even metabolic imbalance and ketosis. On the other hand, ingesting very high amounts of carbohydrates during the loading period can provoke blood sugar fluctuations that could even decrease the athlete's performance during competition. This kind of dietary requires increasing the frequency of capillary blood glucose measurements in order to maintain glycemia within normal range.

At any rate, on the days prior to a competition, a high-carbohydrate diet is recommended, near or, if possible, above the guideline of 8g/kg weight/day.

6. Nutritional composition of foods commonly used in sports

During exercise it is advisable to increase carbohydrate intake, of both low and high glycemic index values. However, glycemic levels must be considered before recommending a certain food or drink or another. Below is a description of the carbohydrate content of some of the foods and drinks most commonly used while performing physical exercise:

- *Isotonic drinks*: These contain a sugar content of 5 to 8% (between 5 and 8g for every 100cc), the concentration that is most easily digested, allowing for a good tolerance of the drink while practicing sports. Moreover, they provide sodium, potassium or chlorine, electrolytes that help replenish lost minerals, thus, they are especially appropriate for activities lasting 1 hour or longer.
- *Soft drinks:* This group includes cola and orange soft drinks, tonic water, and carbonated drinks containing sugar. Their sugar content is around 10%, which makes their digestion more difficult when exercise is high-intensity. Furthermore, cola soft drinks contain caffeine, a substance which increases dehydration during exercise.
- *Energy drinks:* These drinks have a high sugar content (above 10%) and, in addition, contain substances to which anti-fatigue properties are attributed, such as 2-aminoethanesulfonic acid or ginseng. Their high sugar content and stimulating effects do not make them recommendable as a supplement during physical exercise.
- *Fruit juices:* The distinction between natural and commercial fruit juices must be made. Natural juices have a relatively low carbohydrate content (4-6%), allowing them to be used in a way similar to isotonic drinks. This also applies to commercial juices with "no added sugar", as they contain only the sugar naturally found in fruit (fructose) without adding any other kind of sugar later during processing. In contrast, commercial fruit juices with added sugar during processing have a total sugar content of around 10% (similar to that of soft drinks), in which case consumers are advised to read nutrition labels, as manufacturers add different quantities of sugar.

Another difference between the two is that natural juices raise glycemic levels much more slowly than commercial juices, as they consist primarily of fructose.

- *Glucose tablets:* Pure glucose is the substance that raises glycemic levels most quickly. Occasionally it can produce some digestive discomfort, such as abdominal pain or diarrhea, produced when large quantities of glucose enter the intestine. Tablets are recommended to be taken little by little and always with water, in order to facilitate their absorption.
- *Glucose gels:* These consist of a mixture of glucose (or other forms of sugar) with water and fruit flavorings, forming a textured emulsion similar to that of honey or marmalade, more pleasing than that of glucose tablets or pills. However, they can also produce some digestive discomfort in certain individuals, so they are also recommended to be taken in small servings and always with abundant liquids.
- *Energy bars:* These are made from a grain or flour base, with a certain amount of added sugar or protein. Energy bars serve two different purposes during exercise: in addition to maintaining glycemic levels, they also help ward off hunger during long-duration exercise. They are very practical, as they are easy to transport and store and also easy to digest, rarely creating digestive discomfort.

In terms of their effects on glycemic levels, certain differences are found between energy bars, according to the ingredients used. Thus, grain bars increase glycemia faster than those with a higher protein content. There are also enery bars made with dried fruit as an alternative to traditional grain bars.

Food/drink	Serving	Carbohydrates (g) per serving	Carbohydrates (%)
Isotonic drinks			
Isostar®	200ml	14	7
Gatorade®	200ml	12	6
Aquarius®	200ml	12	6
Powerade®	200ml	13	6,5
Other drinks			
Soft drinks	200ml	20	10
Commercial fruit juices	200ml	24	12
Redbull	200ml	22	11,2
Foods			
Orange	Medium size, 130g	10	8
Apple	Medium size, 130g	12	9
Banana	Small size, 80g	16	20
Graham crackers	3 units, 21g	13	63
Bread	1 large slice, 30g	14	47
Glucose tablets	2 units, 10g	10	99,5
Energy bars	1 unit, 25g	15	60

Table 1. Carbohydrate content of certain foods and drinks used during exercise

7. Carbohydrates in preparing for a competition

7.1. Nutrition during the hours prior to competition

If exercise is performed right after fasting or an initial nutritional deficiency, the reserves of glycogen in the liver are reduced, which also leads to a premature decrease in glycemic levels, implicating a greater risk of hypoglycemia as well as a noticeable decrease in performance. Therefore, diabetic athletes must eat a meal rich in carbohydrates during the 3 hours prior to a training session or competition lasting 45-60 minutes or more. This meal must preferably include foods rich in low-glycemic index carbohydrates.

For short-duration exercise, pre-competition strategies are not as important, as the depletion of glycogen reserves is normally not as crucial to the overall performance of the athlete.

One strategy for controlling glycemic levels consists of measuring the levels of glucose in the blood 15-30 minutes before a competition and, depending on the levels obtained, taking a supplement of 15g of high glycemic index carbohydrates, in order to prevent hypoglycemia during exercise. This supplement is omitted if blood glucose levels are above 150-200mg/dl or higher, in accordance with the kind of exercise performed and the individual needs of each athlete. Including supplements of this nature has been found to reduce the frequency of hypoglycemia also during the hours following physical activity [19].

Before beginning any kind of exercise, the measurement of blood glucose levels is recommended. Thus, the intake of carbohydrates before physical activity begins is conditioned by the glycemic levels detected beforehand. On one hand, physical activity is not advised when levels are above 250mg/dl and ketone bodies are present, or should be performed with caution when glycemic levels are above 300mg/dl. In this case, performing exercise could represent a risk, as physical activity itself increases the production of ketone bodies, which could considerably worsen the metabolic state of the athlete.

On the other hand, glycemic levels before exercise found to be lower than 100mg/dl require the administration of high glycemic index carbohydrates in order to prevent hypoglycemia during the initial minutes of physical activity [21]. This extra supplement should provide 10 to 15 grams of carbohydrates.

7.2. Nutrition during competition

Before beginning mid- or long-duration exercise, reducing insulin doses prior to the activity may not be sufficient but also require an increased intake of carbohydrates during exercise. This is the case of mild- and long-duration exercise (more than 60-90 minutes) and, in particular, unplanned exercise sessions, reducing insulin doses beforehand is not possible.

These carbohydrate supplements should be individualized according to the previously described factors. Table 2 offers approximate target ranges that may prove useful for people with diabetes who are taking up a sport or physical activity.

Exercise intensity ($\%VO_{2max}$)	<20 minutes	20-60 minutes	>60 minutes
25 / light	0-10g	10-20g	15-30g/h
50 / moderate	10-20g	20-60g	20-100g/h
75 / intense	0-30g	30-100g	30-100g/h

Table 2. Carbohydrate supplements according to intensity and duration of exercise

These recommendations must be adapted and tailored to the individual needs of each athlete and the doses of carbohydrates confirmed for each of the training sessions or competitions performed. To do so, glycemic levels should be monitored both before and after physical activity, searching for patterns particular to each athlete.

Unplanned physical exercise deserves special attention. In this situation, adjusting insulin doses may not be possible, and hypoglycemia must be prevented exclusively by an increased intake of carbohydrates. The time of day at which the activity is performed must also be considered; when under the maximum effect of a fast-acting insulin analog or hypoglycemic drug, hypoglycemia must be prevented by administering a greater amount of carbohydrates. An initial supplement of 10-30g of carbohydrates should be taken, with continued carbohydrate intake depending on the intensity and duration of the exercise to be performed (see Table 2).

Diabetic athletes, in order to prevent hypoglycemia, often consume an excess of carbohydrates both before and after exercise. In other cases, some tend to overestimate the effects of physical activity on glucose consumption, with the belief that exercise alone will reduce any excesses in their diet. Special emphasis must be placed on adjusting carbohydrate intake to the expected duration and intensity of the exercise, as in some athletes who are very accustomed to a particular kind of exercise the reduced insulin doses and increased carbohydrate intake do not meet the recommendations in clinical guides [22].

7.3. Nutrition after competition

Once the activity is over, the replenishment of glycogen reserves and increased sensitivity to insulin continue for several hours, even until the day after the training session or competition.

Athletes with type 1 diabetes therefore must know that during this period they are more prone to suffer hypoglycemia. The optimal intake of carbohydrates consists of 1,5g per kg of body weight. These carbohydrates should preferably be high glycemic index foods, to be taken during the first 30 minutes immediately following physical activity. Ideally, another dose of 1,5g of carbohydrates per kg of body weight should be taken before 120 minutes have elapsed after the physical activity has ended.

Although this amount of carbohydrates is the recommendation for replenishing glycogen reserves in the muscles and liver in all athletes, the habits of athletes without diabetes make it even more difficult for athletes with type 1 diabetes to achieve proper glycemic control. The recommended quantities may need to be accompanied by insulin in order to prevent hypoglycemia, therefore, nutritional strategies are recommended only for those diabetic athletes with significant experience in handling the disease. At any rate, the amount of carbohydrates in post-exercise supplements should be adjusted to the duration and intensity of the exercise performed, beginning very gradually and in association with a greater number of capillary glycemia measurements.

During the hours following exercise, glucose needs are increased, even 12-16 hours after physical activity has ended. This phenomenon is due in part to an increased sensitivity of muscle cells to glucose and in part to the need for replenishing glycogen reserves depleted during exercise.

As a result, this increase in glucose consumption following exercise raises the risk of suffering episodes of hypoglycemia, which must be prevented by changes in diet and in insulin doses. For this reason, once physical activity has concluded, it is advisable to check glycemic levels and evaluate the need to take food supplements containing carbohydrates, in accordance with the obtained results:

- Glycemia less than 120 mg/dl: take a supplement containing 15-20g of high glycemic index carbohydrates.
- Glycemia between 120 and 200 mg/dl: supplements are usually not required.
- Glycemia greater than 200 mg/dl: carbohydrate supplements are usually not required and the possibility of including a fast-acting insulin analog should be considered[1].

Furthermore, in order to prevent post-exercise hypoglycemia, it may be necessary to compensate for the elevated consumption of glucose by the muscles through reducing insulin doses during the hours following exercise, both in fast-acting insulin as well as in slow-acting insulin for the basal needs of those using an insulin pump. The magnitude of this decrease in insulin depends mainly on the intensity and duration of the exercise performed, keeping in mind that the hypoglycemic effect of exercise is greater during the first 60-90 minutes afterwards [22].

8. Adapting nutrition to different kinds of physical exercise

The following are examples of nutritional strategies recommended for certain kinds of exercise:

- *Short-duration, low-intensity exercise:* Exercise such as walking, riding a bike or light swimming, for a period of time less than 30 minutes, is associated with low glucose consumption, thus, its hypoglycemic effect is small. Normally adjusting insulin treatment and increasing carbohydrate intake are not necessary.

---

[1] Athletes with diabetes must keep in mind that the effects of insulin following exercise are also much greater than normal.

- *Short-duration, high-intensity exercise:* This group includes swimming races or track and field events, such as sprints and hurling (for example, 100 meter dash, shot-putting, etc.), and also combat sports like judo or taekwondo. The brief duration of the exercise means that the complete depletion of glucose is not as significant. Moreover, the elevated intensity of the exercise produces adrenergic activation, which increases the production of glucose by the liver. For these reasons, adjusting insulin doses and supplementing with carbohydrates are usually not necessary.
- *Mid-duration aerobic exercise:* Training for cycling, swimming or track events, as well as spinning, aerobics or step aerobics sessions, lasting approximately one hour, usually produce a more pronounced hypoglycemic effect. Nevertheless, since the sessions are still relatively brief, the decrease in insulin should be in the range of 10-20% prior to the activity, accompanied by an intake of 20-30g of carbohydrates, preferably fast-absorbing.
- *Long-duration exercise:* Sporting events such as cycling competitions, triathlons or marathons involve performing exercise of a moderate or high intensity for prolonged periods of time. The consumption of glucose is quite high, depending on the intensity and duration of the activity. The risk of hypoglycemia is very high, both during the activity and up to 24 hours after the activity has ended.

In this case, adjustments in treatment are much more severe. A decrease in insulin of 25-50% is recommended prior to the onset of activity, which, in the case of long-duration competitions (more than 4 hours), could reach a 75% or greater reduction. The use of fast-acting insulin following the activity should also be reduced, by about 10 to 20%. In a similar fashion, the intake of carbohydrates should increase throughout the activity, according to the intensity of exercise involved.

- *Team sports:* In sports such as soccer, basketball or volleyball, the consumption of glucose can be quite variable, involving factors such as the position of the player on the team, the conditions of the game, and even the scoreboard.

Adjustments in treatment include reducing insulin doses prior to the activity by 10-20%, along with the intake of 10-20g of carbohydrates for every 30 minutes of activity.

9. Conclusions

A proper dietary education proves to be absolutely essential for athletes with type 1 diabetes to achieve optimal results. The objective of nutritional therapy is two-fold: on one hand, to maintain glycemic levels as normal as possible, and, on the other, to achieve the maximum sporting performance possible.

For this purpose, it is necessary to understand the functioning of the energy pathways of the body, as well as its hormonal response to exercise, with the goal of being able to foresee future glycemic response to the activity performed.

Nutritional recommendations regarding the intake of carbohydrates geared toward athletes with type 1 diabetes should not be any different from those geared toward other athletes. Nevertheless, diabetic athletes must alter the way in which they incorporate carbohydrates into their normal diet, in order to produce the smallest changes possible in their glycemic levels. Thus, it is important to properly distribute these carbohydrates among the various meals of the day, keeping in mind the dosage of insulin to be used, while planning different training sessions or competitions.

Finally, dietary changes should take into account the characteristics of the physical activity, such as the kind of exercise, intensity and duration, as well as time of day at which it is performed.

Author details

Anna Novials and Serafín Murillo
Department of Endocrinology and Nutrition, Hospital Clínic de Barcelona, Barcelona, Spain
Institut d'Investigacions Biomèdiques August Pi i Sunyer (IDIBAPS), Barcelona, Spain
Spanish Biomedical Research Centre in Diabetes and Associated Metabolic Disorders (CIBERDEM), Spain

Acknowledgement

We thank K. Katte [from Spanish Biomedical Research Centre in Diabetes and Associated Metabolic Disorders (CIBERDEM)] for language consultancy.

10. References

[1] Schneider SH, Ruderman NB (1990) Exercise and NIDDM (Technical Review). Diabetes Care.13:785–789.
[2] Vrazeau AS, Rabasa-Lhoret R, Strychar&Mircescu H (2008) Barriers to physical activity among patients with type 1 diabetes. Diabetes Care.31:2108-2109.
[3] Nagi D (2006) Exercise and Sport in Diabetes. Wiley & Sons Ltd.Chichester. England.
[4] American Diabetes Association (2008) Position Statement: Nutritional recommendations and principles for individuals with diabetes mellitus. Diabetes Care. 31 (Suppl 1): S61-S78.
[5] Institute of Medicine. (2002) Dietary Reference Intakes: Energy, Carbohydrate, Fiber, Fat, Fatty Acids, Cholesterol, Protein, and Amino Acids. Washington, DC, National Academies Press.
[6] Wasserman DH (1995) Control of glucose fluxes during exercise in the postabsortive state. Annual Review of Physiology. Palo Alto, CA, Annual Reviews, p 191-195.

[7] Zinker BA, Bracy D, Lacy DB, Jacobs J, Wasserman DH (1993) Regulation of glucose uptake and metabolism during exercise: an in vivo analysis. Diabetes.42: 956-965.

[8] Horowitz JF, Klein S (2000) Lipid metabolism during endurance exercise. Am J Clin Nutr. 72 (Suppl): 558S-563S.

[9] Hirsch IB, Marker JC, Smith LJ, Spina R, Parvin CA, Cryer PE (1991) Insulin and glucagon in the prevention of hypoglycemia during exercise in humans. Am J Physiol. 260: E695-704.

[10] Sigal RJ, Fisher SF, Halter JB, Vranic M, Marliss EB (1995) The roles of catecholamines in glucoregulation in intense exercise as defined by the islet cell technique. Diabetes.45: 148-156.

[11] Wasserman DH.,Zinman B (1994) Exercise in individuals with IDDM (Technical review) Diabetes Care. 17: 924-937.

[12] Davis SN, Galassetti, Wasserman DH, Tate DH (2000) Effects of antecedent hypoglycemia on subsequent counterregulatory responses to exercise. Diabetes. 49: 73-81.

[13] Hernandez JM, Moccia T, Fluckey JD, Ulbrecht JS, Farrell PA (2000) Fluid snacks to avoid late onset post exercise hypoglicemia. Med Sci Sports Exerc. 32: 904-910.

[14] Nakar VA, Downes M, Yu T, Embler E,Wang Y et al. (2008) AMPK and PPARδ agonists are exercise mimetics. Cell.134(3): 405–415.

[15] Goodyear LJ (2008) The Exercise Pill — Too Good to Be True? New Engl J Med.359(17):1842-1844.

[16] Fujii N, Aschenbach WG, Musi N, Hirshman MF, Goodyear LJ (2008) Regulation of glucose transport by the AMP-activated protein kinase. ProcNutr Soc. 63: 205-210.

[17] Burke LM, Collier GR, Hargreaves M (1993) Muscle glycogen storage after prolonged exercise: the effect of glycaemic index of carbohydrate feedings. J Appl Physiol.75:1019-1023.

[18] Diabetes UK (2003) The implementation of nutritional advice for people with diabetes. Diab Med.20:786-807.

[19] Nathan DN, Madnek S, Delahanty L (1985) Programming pre-exercise snacks to prevent post-exercise hypoglycaemia in intensively treated insulin-dependent diabetics. Ann Intern Med. 4:483-86

[20] American Diabetes Association. (2004) Diabetes and Exercise. Diabetes Care. 27(1):s58-s62.

[21] Grimm JJ, Ybarra J, Berné C, Muchnick S, Golay A (2004) A new table for prevention of hypoglucemia during physical activity in type 1 diabetic patients. Diabetes Metab.30:465-470.

[22] Murillo S, Brugnara L, Novials A (2010) One year follow-up in a group of half-marathon runners with type-1 diabetes treated with insulin analogues. J Sports Med Phys Fitness. 50(4):506-10.

[23] Mc Mahon SK, Ferreira LD, Ratnam N, Davey RJ, Youngs LM, Davis EA, et al. (2007) Glucose requirements to maintain euglycemia after moderate-intensity afternoon exercise in adolescents with type 1 diabetes are increased in a biphasic manner. J Clin Endocrinol Metab.92(3):963-8.

Section 4

Animal and Plant

Environmental and Genetic Variation for Water Soluble Carbohydrate Content in Cool Season Forage Grasses

Ali Ashraf Jafari

Additional information is available at the end of the chapter

http://dx.doi.org/10.5772/51575

1. Introduction

Forage dry matter has been divided into two fractions on the basis of nutritional availability. One fraction (non-structural) corresponds to the cell content and is composed of WSC (water soluble carbohydrates) (fructans and starches), lipids, most of the protein, nucleic acid and minerals. Fructan, a polymer of fructose, is accumulated as an energy reserve mainly in the leaf sheaths of temperate grasses which are made in the cytoplasm during periods of photosynthesis (Pollock and Cairns, 1991). The other dry matter fraction corresponds to the plant cell wall and consists mostly of structural carbohydrates (cellulase, hemicellulase and lignin) (Gill et al, 1989). Chemical composition of grasses changes with advancing maturity. As grass matures the proportion of the cell wall increases and the cell content fraction decreases. Together, non-structural and structural carbohydrates, depending on maturity stage, make up approximately 50 to 80% of dry matter of forages (Gill et al, 1989).

The most important traits that affecting the feeding value of herbage are digestibility (DMD), the ratios of crude protein, water-soluble carbohydrate, fiber, and the concentration of alkaloid toxins (Smith, et al., 1997, Wilkins and Humphreys, 2003). High WSC content is an important breeding goal for milk production and liveweight gain. Over a range of species, Grimes et al (1967) found strong positive correlation between WSC content and intake and liveweight gain of lambs. In the last decade, few improved ryegrass varieties had been bred for increase WSC concentration, in UK (Wilkins and Lovatt 2007; Wilkins et al. 2010), and New Zealand (Rasmussen et al., 2009). Most published data suggest that animals consuming grasses with high WSC concentrations were able to utilize the protein in their diet more efficiently, resulting in increased liveweight gain, milk production and lower loss of N (Lee et al., 2001; Evans et al., 2011; Miller et al., 2001). Similarly, Mayland et al (2001) and Smit (2006) suggested that forage with higher WSC concentration is related to preference. In contrast, few trials that animal grazing of pure swards with high versus low concentrations of WSC, showed low (Cosgrove et al. 2010) or non significant animal performance advantages than control (Allsop et al., 2009; Parsons et al. 2010).

WSC are completely digestible and have an important role in animal nutrition, as they are a primary source of the readily available energy necessary for efficient microbial fermentation in the rumen. Fermentation of structural carbohydrates also provides energy directly to the microbial population and indirectly to the animal. Increased microbial metabolism is a prerequisite to improving forage intake and nutrient utilisation but this only occurs if the interrelationships of carbohydrate and protein are optimized (Carlier, 1994; Evans et al., 2011). One of the major problems in grasses and legumes consumption is reduced efficiency of protein utilisation and unbalanced supplies of carbohydrate and protein to support microbial metabolism (Wilkins and Humphreys, 2003). A large proportion of the protein may be lost during rumen fermentation, mainly caused by a deficiency in available carbohydrates. To increase the carbohydrate/protein ratio in the diet, either grass composition should be changed or grass should be supplemented with feeds of readily available carbohydrate. Beever and Reynolds (1994) suggested that breeding for yield and digestibility alone is no longer adequate, and that forage breeding programs should concentrate on improved nutrient value of forages, such as improved carbohydrate availability and more controlled protein degradation. Beerepoot and Agnew (1997) advised against simple selection for increased carbohydrate levels alone because of possible negative effects upon rumen pH and cell wall degradation and suggested that selection for highly digestible cell walls is preferable.

Success in conserving legumes and grasses as silage often depends on the amount of readily available fermentable carbohydrate present in herbage. If the concentration is sufficiently high, conditions are more favorable for establishment and growth of lactic acid bacteria which leads to better fermentation and preservation (Humphreys, 1994). A minimum of 3.7% WSC in fresh grass (i.e. about 15% WSC in dry weight) is considered necessary to produce good quality silage without additives (Haigh, 1990).

2. Environmental and genetic effects on WSC content

2.1. Environmental effects

WSC in forage grasses appears to be associated with good tiller survival and sward persistency (Thomas and Norris, 1981). Stored WSC provides the energy and structural carbohydrate required for growth when demand cannot be met by contemporary photosynthesis e.g. regrowth after defoliation, recovery from drought and persistency during winter (Humphreys, 1989d, 1994).

The WSC level in grasses depend on a wide range of factors, including the plant part, ploidy level, plant maturity, diurnal and seasonal effects, temperature, light intensity, growth rate, endophyte, water status, as well as the inherent differences between cultivars (Stewart and Hayes, 2011). Accumulation of WSC in tiller bases of cocksfoot is highly correlated with summer survival and rapid recovery after drought (Volaire and Gandoin, 1996). The amount of WSC maintained through the winter appears critical. Factors which reduce carbohydrate reserves such as autumn disease, autumn growth forced by high nitrogen application, late cutting, or uncompensated respiratory loss through growth at low light intensity, can all seriously reduce winter hardiness. If carbohydrate levels can be maintained during periods of winter growth a decline in hardiness may not occur (Humphreys, 1989d).

Radojevic et al (1994) found significant variation for WSC and DMD among four European and New Zealand varieties of perennial ryegrass under Australian conditions. A variation of 20% for WSC concentration among six diploid ryegrasses was found to be consistent over several years (Smit, 2006), the largest variation among cultivars was found during summer, in July and August. Tetraploid cultivars generally had a higher WSC concentration than diploid cultivars (Smith et al. 2001; Gilliland et al. 2002). DMD concentration declined during summer due to the influence of higher temperatures, however, high WSC lines exhibited fewer declines in DMD than the lines that did not accumulate as much WSC.

Intake depends, to some extent, on digestibility but other factors such as palatability (animal preference) are important. Jones and Roberts (1991), in a survey comparing four varieties of perennial ryegrass with the same digestibility but different WSC contents, found that varieties with high WSC content were more palatable than varieties with moderate content of WSC.

WSC content is sensitive to temperature and light intensity. Low levels of WSC are found in Western Europe country as Ireland and UK due to cloudy weather conditions (i.e. low light intensity) (Jung et al, 1996). WSC generally decline during autumn. Fulkerson and Donaghy (2001) suggested that the lower WSC concentration in autumn has several causes. The reduced solar radiation will decrease photosynthetic activity, which results in a reduced primary production of non-structural carbohydrates and WSC concentrations are closely related with hours of sunlight. The higher night temperatures in autumn will induce respiration during night time (McGrath 1988). The altering source–sink relations deplete WSC reserves, and hence lower the WSC concentration in the herbage.

There is seasonal variation in WSC content. Dent and Aldrich (1963) reported WSC content was high in spring and fell to the lowest point in summer, rising again in autumn, to about half of spring level. McGrath (1988) reported that average concentration of WSC in perennial ryegrass was 20% (mean of three cuts per year) with maximum level in late April each year. WSC content was at least 50% higher in stem than in leaf. He also reported that, on average, fructans accounted for 70% of soluble carbohydrate, the remainder being fructose, glucose and sucrose. For these reasons (environmental effects) WSC level in forage grasses vary considerably ranging from less than 5% (Waite and Boyd, 1953) to up to 50% (Bugge, 1978).

Moisture stress retards growth of grasses and is accompanied by an increase in WSC (Brown and Blaser, 1970). WSC contents can be influenced by management practices such as defoliation and nitrogen application level. Application of N fertilizer allows more rapid growth, and thus tends to result in lower levels of WSC accumulation. WSC is greatest when swards are allowed to accumulate higher herbage mass, such as silage crops, with levels lower under very frequent grazing (Rasmussen et al. 2008). Cutting of herbage invariably causes a decrease in the amount of WSC in the remaining plant parts and this decrease reaches a minimum about one week after defoliation and is then followed by an increase associated with increasing photosynthetic activity. If cutting is repeated before the original level is restored, the successive herbage yields will be smaller and eventually such a treatment will lead to plant death. Based on this, Alberda (1966) concluded that WSC are of greater importance for the regrowth and recovery of sward after cutting. McGrath (1992) reported that application of nitrogen fertilizer increased DM yield and invariably increased CP content and decreased WSC content. However, he found no effect of nitrogen on total WSC yield.

Sowing density has a major effect on the development of individual plants. Smouter et al (1995) in a study on the effect of three tiller densities (100, 500, and 5000 plant m^{-2}) on WSC concentration of *Lolium rigidum*, found that the concentration of WSC varied in response to treatment. Yield of WSC in the low-density sward was 30 to 50% higher than other swards at anthesis. In addition, the digestibility of low density was higher than of the high-density sward.

Maturity and herbage age generally have a greater influence on forage quality than environmental factors. As plants advance in maturity, cell wall concentration within stems and leaves, increases and the proportion of cell soluble content decreases. The rate of decline in digestibility of herbage is greatest during reproductive growth (Buxton et al. 1996). Breese and Thomas (1967) showed that small differences in maturity of cocksfoot could greatly affect digestibility. Rezaeifard, et al (2010) in cocksfoot grown in Iran, found that both WSC and acid detergent fiber (ADF) values increased by plant growth, however, WSC increases were slower than for ADF by advancing maturity from vegetative to milky stage. Contents of WSC were higher in stem than in leaf, so by advance of plant maturity WSC contents increase. These results are expected because with plant maturity and reducing the ratio of leaf to stem, the WSC content will increase in stem. In general, DM yield, WSC and ADF values dramatically increased by advancing of plant maturity, in contrast, DMD, CP and ash values decreased (Rezaeifard, et al., 2010). Similarly, Jafari et al (2010) in tall wheatgrass *Agropyron elongatum* and Jafari and Rezaeifard (2010) in tall fescue found the higher values of WSC-yield in pollination and milky stage under conservation management, respectively.

Humphreys (1989c) found considerable genetic variation in WSC content of ryegrass during both vegetative and reproductive growth phases. The range of variation for WSC (13.9 to 28.8%) was greater than that found for CP or DMD and WSC showed generally high heritability. Frandsen (1986) found significant variation for WSC among 36 HS families (22.9 to 44.9%) in perennial ryegrass. Wilkins and Davies (1994) also reported significant differences for WSC among four varieties of perennial ryegrass under conservation management. But, the range was relatively small (30.5 to 33.6%). Dent and Aldrich (1963), using perennial ryegrass varieties under sward condition, reported a range of 14.1 to 19.3 and 17.6 to 26.3 WSC% for frequent and conservation management, respectively.

More recently, significant variation for WSC in five phenological stages (vegetative, heading, pollination, milky and dough seed stage), were obtained in by Jafari et al (2010) in tall wheatgrass (11.2 to 14.4%), Jafari and Rezaeifard (2010) in tall fescue (8.38 to 12.69) and by Rezaeifard, et al., (2010) in cocksfoot (7.76 to 11.37), respectively.

2.2. Gene action and heritability

By comparison to DM yield and DMD there is little information on genetic control of WSC and the extent of genetic variation in forage grasses. Estimates of three heritability estimates broad sense heritability (h^2_b), narrow sense heritability (h^2_n), and heritability, offspring/parent regression (h^2_{op}) are summarized in Table 1. Cooper (1962) reported h^2_{op} of 0.11 to 0.62 in perennial ryegrass (Table 1). Based on a diallel analysis, he estimated high h^2_n=0.84 and concluded that gene effects were additive (Cooper, 1973).

Contrasting results were reported by Humphreys (1989a, b). From an analysis of crosses between the early heading ryegrass, Aurora, and five late heading varieties, he found different degrees of directional dominance and over-dominance (heterosis) and concluded that WSC behaved as a complex polygenic trait, which was controlled by mainly non-additive gene effects. Genetic variation within crosses was relatively small and estimated F_2 family heritability was 0.08 to 0.14 and 0.20 to 0.38 under frequent cutting and conservation management, respectively (Table 1). In other experiments Humphreys (1995) found much higher estimates of h^2_b while the values for conservation management was again higher than that for frequent cutting (Table 1). While assessing winter hardiness of 86 accessions of perennial ryegrass under spaced plants condition, Humphreys (1989d) found significant genetic variation for WSC and estimated h^2_b from 0.38 to 0.60 for March and October harvests, respectively. In another survey he estimated h^2_b in the range of 0.27 to 0.70 for different cuts over two years in 81 accessions of perennial ryegrass (Table 1) (Humphreys, 1991).

Jafari (1998) using a polycross progeny test in ryegrass under spaced plants and swards, estimated relatively high, (h^2_b) and (h^2_n) for WSC (h^2_b=0.60-0.80), (h^2_n=0.50-0.54), respectively. The genetic analysis indicated additive genetic variances were more important for WSC and other quality traits. In another survey using full sib families of ryegrass he estimated relatively high h^2_b=0.60 and 0.80 for conservation and frequent cutting management, respectively (Table 1) (Jafari et al., 2003).

In other grass species some estimates of genetic control of WSC have been published (Table 1). Cooper (1962), in cocksfoot, estimated relatively high h^2_{op} which ranged from 0.78 to 0.56 for July and August cuts, respectively. Buckner et al (1981), using a polycross progeny test in *Lolium-festuca* hybrid derivatives, estimated h^2_b of 0.55 and h^2_{op} of 0.01 to 0.47 for WSC. Burner et al (1983) analyzed variance of tall fescue and obtained h^2_b =0.55. Grusea and Oprea (1994) found h^2_b estimates for WSC of 0.58 in cocksfoot. Jafari and Javarsineh (2005) in tall fescue and Jafari and Naseri (2007) in cocksfoot estimated relatively moderate to high values of h^2_b, while h^2_n estimates were low suggests that genetic variation in this trait is controlled by both additive and non additive gene action. Sanada, et al (2007) in cocksfoot found significant genetic variation and moderate h^2_n estimates, and suggested that forage quality of cocksfoot was influenced by an additive gene effect and could be improved genetically by recurrent and phenotypic-genotypic selection (Table 1).

Author (s)	Management/ Environment	Grass species	Basis of estimation	Heritability		
				h^2_b	h^2_n	h^2_{op}
Cooper (1962)	C/pots/ 2 cuts	*L.perenne*	FS-families			0.11 to 0.62
Cooper (1973)	C/SP/1st cut	*L.perenne*	Diallel		0.84	
Humphreys (1989d)	C/SP/2 cuts	*L.perenne*	Populations	0.38 to 0.60		
Humphreys (1991)	C/SP/2 years	*L.perenne*	Populations	0.27 to 0.70		
Humphreys (1989b)	C/Sward/2 yr.	*L.perenne*	FS-families	0.08 to 0.38		
Humphreys (1995)	C/Sward/2nd yr.	*L.perenne*	HS-families	0.34 to 0.71		
Jafari (1998)	C/SP/2 years	*L.perenne*	HS-families	0.75	0.50	0.26
Jafari (1998)	C/Sward/ 2 yr.	*L.perenne*	HS-families		0.54	0.17
Jafari (1998)	C/ Pots/3 yr.	*L.perenne*	HS-families	0.67	0.32	0.35
Jafari et al (2003)	C/Sward/ 2 yr.	*L.perenne*	FS-families	0.60		
Jafari et al (2003)	F/Sward/ 2 yr.	*L.perenne*	FS-families	0.80		
Jafari & Javarsineh (2005)	C/SP/ 2 yr.	*F.arundinacea*	HS-families	0.50	0.11	0.00
Burner et al (1983)	C/SP/ 2 cuts	*F.arundinacea*	Clones	0.55		
Jafari and Naseri (2007)	C/SP/2 yr.	*D.glomerata*	HS-families	0.50	0.23	0.62
Cooper (1962)	C/pots/ 2 cuts	*D.glomerata*	FS-families			0.56 to 0.78
Sanada et al (2007)	C/SP/	*D.glomerata*	HS-families	0.10	0.53	0.78
Sanada et al (2007)	C/Sward/	*D.glomerata*	Clones	0.50	0.59	
Grusea & Oprea (1994)	C/Sward/ 2 yr.	*D.glomerata*	FS-families	0.58		
Buckner et al (1981)	C/SP/ 2nd cut	*Festolulium*	Clones	0.55		
Buckner et al (1981)	C/Sward/2ndcut	*Festolulium*	HS-families		0.39 to 0.49	0.01 to 0.47

C, F, SP, HS, FS, Conservation management, frequent cutting management, spaced plants, Half-sib, Full-sib, respectively

Table 1. Heritability estimates for water soluble carbohydrates (WSC) in some cool-season grass species

2.3. Genotype × Environment interactions

There is little evidence for GE interaction of WSC. Burner et al (1983) reported that interaction between tall fescue genotypes and environments for WSC concentration were minor. Humphreys (1989b) reported that although WSC is a trait which is subject to large environmental fluctuations, genetic differences could remain fairly stable without large GE interaction. He also found relative WSC contents in parents and hybrids were consistent over generations under both spaced plants and sward conditions. In contrast, Buckner et al (1981), using a polycross progeny test in *Lolium-festuca* hybrid derivatives, reported when parent and HS families were evaluated in the same environment h^2_{op} were much higher than when they were grown in different environments. They suggested that a minimum of two locations with widely different environments was necessary to develop varieties for forage quality. Conaghan et al (2008) for DM yield came to same similar conclusions. They suggested at least three locations and 2 sowing years is necessary for evaluation of perennial ryegrass trials.

Sanada, et al (2007) in cocksfoot reported a significant genotype × year interaction for WSC and suggests that the evaluation of forage quality traits should be carried out by divergent selection under multiple environments, especially for the evaluation of parental clones. Jafari and Rezaeifard (2010) in tall fescue found non significant genotype × phenological stage interaction effect for WSC. Similarly, Rezaeifard et al (2010) in cocksfoot reported that genotypes by environments interaction for WSC concentration were minor. This finding were in agreement of Buxton and Casler (1993), that in a review, concluded that most environment stresses have a greater effect on DM yield than on quality traits and G x E interactions should be smaller for forage quality than for yield.

In conclusion, despite great fluctuations in WSC due to environmental and plant developmental factors, there is genetic variation within and between forage grasses populations for this character which could be exploited by selection.

3. Correlation between WSC and yield/quality traits

3.1. Correlation between WSC and Digestibility (DMD)

An understanding of the interrelationships between quality traits is important in the development of a selection program. Table 2 summarizes some of the results published for the relationships between DMD and WSC.

A broad pattern of positive phenotypic correlation between digestibility and WSC has been found for forage grasses under both conservation and frequent cutting management (Table 2). Since WSC is completely digestible a positive correlation between these two parameters is expected. The summary data for confirms that these characters show a moderate/strong positive correlation. Humphreys (1989c) found large positive correlation between DMD and WSC for spaced plants. However, under sward conditions with frequent cutting management only three of eight cuts showed significant correlation and over all cuts it was non significant (Table 2). Significant positive correlations between digestibility and WSC concentration in tall fescue forage have been reported (Burns and Smith, 1980). Bugge (1978), in Italian ryegrass, and Clements (1969), using canarygrass, found no significant correlation between DMD and WSC. Strong positive phenotypic and genotypic correlation between WSC and DMD were obtained by Jafari (1998), Jafari et al (2003) in ryegrass, Jafari and Naseri (2007) in cocksfoot, Jafari and Javarsineh (2005) in tall fescue (Table 2).

Source	Management / Nursery/ harvests	Grass species	WSC/ DMD	
			Phenotypic	Genotypic
Frandsen (1986)	C/SP/1st Cut	*Lolium perenne*	0.76**	
Humphreys (1989c)	C/SP/annual	*Lolium perenne*	0.70**	
Dent & Aldrich (1963)	C/Sward /2 Cuts	*Lolium perenne*	0.29 to 0.51**	
Jafari (1998)	C/SP/ 2 years	*Lolium perenne*	0.89**	0.94
Jafari (1998)	C/Sward 2years	*Lolium perenne*	0.69**	0.82
Jafari (1998)	C/ Pots/3 generation	*Lolium perenne*	0.55	0.67
Jafari et al (2003)	C/Sward/ 2 years	*Lolium perenne*	0.61**	0.58
Dent & Aldrich (1963)	F/Sward	*Lolium perenne*	0.02 to 0.56**	
Grimes et al (1967)	F/Sward/annual	*Lolium perenne*	0.87**	
Humphreys (1989c)	F/Sward/ 8Cuts	*Lolium perenne*	-0.20 to 0.90**	
Jafari et al (2003)	F/Sward/ 2 years	*Lolium perenne*	0.62**	0.67
Clements (1969)	C/ Pots/3 generation	*Phalaris tuberosa*	0.16	
Buckner et al (1981)	C/SP/ 2nd Cut	*Festololium*	0.17 to 0.57**	
Thomson & Rogers (1971)	C/Sward/2 years	*Phleum pratense*	0.12 to 0.49**	
Jafari and Naseri (2007)	C/SP/ 2 years	*Dactylis glomerata*	0.79**	0.98
Jafari and Javarsineh (2005)	C/SP/ 2 years	*Festuca arundinacea*	0.71**	0.77

*, **, Significant at 5%, 1%, respectively, respectively
SP, C, F, Spaced plant, Conservation and Frequent cutting management, respectively.

Table 2. Correlation coefficients (r) between dry matter digestibility (DMD) and water soluble carbohydrates (WSC) in some cool-season grass species.

3.2. Correlation between WSC and Crude protein (CP)

Phenotypic correlations between WSC and CP are consistent and negative across all species (Table 3). Cooper (1962) reported negative genetic correlation between WSC and CP and concluded that the limits are sufficiently wide to allow selection of a large range of protein/carbohydrate ratios. This negative relationship is present over a wide range of environment for all species. Apart from data in Table 3, there are other results which confirm this conclusion. Vose and Breese (1964), in a glasshouse study, reported high negative phenotypic relationship between WSC and CP in ryegrass. However, using covariance analysis, they suggested there is sufficient degree of genetic independence to allow simultaneous selection for both characters. Radojevic et al (1994) investigated the relationship between WSC and CP using a multiple regression analysis based on three variables (harvest date, genotype and WSC concentration). The analysis showed that the largest influences on CP concentration were time of year (72%) and genotype (5%). WSC did not explain a significant proportion of the variance of CP concentration and they concluded that, although WSC and CP were negatively correlated, this was due mainly to divergent seasonal variation in these components of herbage. Humphreys (1989c) suggested that as growth increases with rapid uptake of nitrogen fertilizer, increase of CP and decrease of WSC content are environmentally induced effects.

Strong negative phenotypic and genotypic correlation between WSC and CP were obtained by Jafari (1998), Jafari et al (2003) in ryegrass, Jafari and Naseri (2007) in cocksfoot, Jafari and Javarsineh (2005) in tall fescue. However, Sanada et al (2007) obtained weak negative phenotypic and genotypic correlation between two traits in cocksfoot (Table 3).

Source	Management / Nursery/ harvests	Species	WSC/CP	
			Phenotypic	Genotypic
Cooper (1962)	C/Pots/2 Cuts	*Lolium perenne*		-0.20 to -0.70**
Humphreys (1989c)	C/SP/annual	*Lolium perenne*	-0.20*	
Humphreys (1989d)	C/SP/Oct. Cut	*Lolium perenne*		-0.70**
Valentine & Charles (1979)	C/Sward/3rd Cut	*Lolium perenne*	-0.61**	
Bugge (1978)	C/SP/1st Cut	*Lolium perenne*	-0.50**	
Jafari (1998)	C/SP 2 years	*Lolium perenne*	-0.75**	*-0.84*
Jafari (1998)	C/Sward 2years	*Lolium perenne*	-0.69**	*-0.80*
Jafari et al (2003)	C/Sward/ 2 years	*Lolium perenne*	-0.65**	-0.60**
Dent & Aldrich (1963)	F/Sward	*Lolium perenne*	-0.44** to -0.69**	
Grimes et al (1967)	F/Sward/annual	*Lolium perenne*	-0.65**	
Humphreys (1989c)	F/Sward/ 8Cuts	*Lolium perenne*	0.30 to -0.90**	
Jafari et al (2003)	F/Sward/ 2 years	*Lolium perenne*	-0.51**	-0.66**
Marais et al (1993)	F/SP/annual	*Lolium multiflorum*	-0.17 to -0.26*	-0.56**
Cooper (1962)	C/Pots/2 Cuts	*Dactylis glomerata*		-0.48* to -0.57**
Sanada et al (2007)	C/Sward/	*Dactylis glomerata*	-0.34	-0.33
Sanada et al (2007)	C/SP/	*Dactylis glomerata*	-0.14	0.16
Jafari and Naseri (2007)	C/SP/ 2 years	*Dactylis glomerata*	–0.53*	–0.32
Clements (1969)	C/Pots/3generation	*Phalaris tuberosa*	-0.13	
Thomson & Rogers (1971)	C/Sward/2 years	*Phleum pratense*	-0.56** to -0.68**	
Jafari &Javarsineh (2005)	C/SP/ 2 years	*Festuca arundinacea*	-0.77**	–0.83

*, **, Significant at 5%, 1%, respectively, respectively
SP, C, F, Spaced plant, Conservation and Frequent cutting management, respectively.

Table 3. Correlation coefficients (r) between water soluble carbohydrates (WSC) and crude protein (CP) in some cool-season grass species.

In summary, data in Table 3 show a consistent negative correlation between WSC and CP at both phenotypic and genotypic levels and over various herbage species. It could be concluded that, due to genetic correlation, selection for one component alone is likely to have negative effect on the other component.

3.3. Correlation between WSC and DM yield

Genotypic and phenotypic correlations between DM yield and WSC in perennial ryegrass and in other forage species are inconsistent although significant correlations are generally positive (Humphreys, 1989c). In a glasshouse study, Vose and Breese (1964) found a low and non-significant correlation between DM yield and WSC. However, Valentine and Charles (1979) obtained a highly positive correlation between DM yield and WSC particularly at low nitrogen level (Table 4).

Jafari (1998) in ryegrass found, positive and significant correlation between WSC and DM yield under spaced plant. For sward experiments their relationships were positively non significant (Jafari, 1998; Jafari et al., 2003). In contrast, Jafari and Naseri (2007) in cocksfoot, found negative and significant relationships between two traits. Jafari and Javarsineh (2005) in tall fescue and Sanada et al (2007) in cocksfoot reported low and inconsistent values between traits. The overall pattern of results suggests that DM yield and WSC are independent or show a weak positive correlation (Table4).

Source	Management / Nursery/harvests	Grass Species	WSC vs. DM yield	
			Phenotypic	Genotypic
Valentine & Charles (1979)	C/Sward/3rd Cut	*Lolium perenne*	0.24 to 0.78**	
Humphreys (1989d)	C/SP/Oct. Cut	*Lolium perenne*		0.60**
Vose & Breese (1964)	C/Pots/4 Cuts	*Lolium perenne*		0.16
Jafari (1998)	C/SP 1 years	*Lolium perenne*	0.36*	0.61**
Jafari (1998)	C/Sward 2years	*Lolium perenne*	0.25	0.17
Jafari et al (2003)	C/Sward/ 2 years	*Lolium perenne*	0.39	0.04
Humphreys (1989c)	F/Sward/8 Cuts	*Lolium perenne*	-0.40 to 0.60**	
Jafari et al (2003)	F/Sward/ 2 years	*Lolium perenne*	0.36	0.08
Clements (1969)	C/Sward /3 years	*Phalaris Tuberosa*	0.20	
Brown & Blaser (1970)	C/Sward / 1st Cut	*Dactylis glomerata*	0.38 to 0.82**	
Sanada et al (2007)	C/Sward/	*Dactylis glomerata*	-0.11	-0.10
Jafari & Naseri (2007)	C/SP/ 2years	*Dactylis glomerata*	–0.45**	–0.51**
Jafari & Javarsineh (2005)	C/SP/ 2years	*Festuca arundinacea*	0.04	0.06

*, **, Significant at 5%, 1%, respectively, respectively.

SP, C, F, Spaced plant, Conservation and Frequent cutting management, respectively.

Table 4. Correlation coefficients (r) between DM yield and water soluble carbohydrates (WSC) in some cool-season grass species.

4. Conclusion and grass breeding strategy

Under rotational grazing and cutting there is frequently an excess of CP in most of grasses in terms of animal requirement, particularly in autumn cuts. This excess of CP leads to high levels of rumen degradable protein (RDP). In the absence of a readily available supply of fermentable carbohydrate this can result in elevated ruminal ammonia concentrations (Beever and Siddons, 1986; Van Vuuren et al, 1986). There is also evidence that this imbalance between CP and carbohydrate has a negative influence on reproductive behavior in grazing animals (Visek, 1984; Canfield et al, 1990). Given the relationship between WSC, CP, DM yield and DMD, it is tempting to suggest that selection for high WSC is a means to improve quality in general. Beerepoot and Agnew (1997) have argued that this simple approach may not result in improved herbage quality because of possible negative effects on rumen pH. There is, however, indirect evidence that higher WSC in ryegrass may result in improved animal performance (Lee et al., 2001; Evans et al., 2011; Miller et al., 2001). Tetraploid ryegrass varieties have higher levels of WSC compared to diploids (Jung et al, 1996; Smith et al. 2001; Gilliland et al. 2002). There are two large scale animal production trials in which tetraploid and diploid varieties were compared (Castle and Watson, 1971; Connolly et al, 1977). The results from both experiments indicated that tetraploids were superior to diploids of comparable DM yield. It is possible, that this superiority was due to the slightly higher WSC content of tetraploid varieties (Smith et al. 2001; Gilliland et al. 2002). Grimes et al (1967) also found strong positive correlation between WSC content and intake and liveweight gain of lambs. On the basis of published data, it is suggested that increased WSC, particularly in the vegetative leaf and in the summer/autumn period when CP is frequently in excess, would improve herbage quality. The data also indicate that response to combined selection for both WSC and DM yield should be possible.

Abbreviations

WSC	Water soluble carbohydrates	h^2_b	Broad sense heritability
DMD	Dry matter digestibility	h^2_n	Narrow sense heritability
DM yield	Dry matter yield	h^2_{op}	Heritability, offspring/parent regression
CP	Crude protein	r_p	Phenotypic correlation
HS families	Half -sib families	r_g	Genotypic correlation

Author details

Ali Ashraf Jafari
Plant Breeding, Gene Bank Division, Research Institute of Forests and Rangelands, Tehran, Iran

5. References

Alberda, T. 1966. The influence of reserve substances on dry matter production after defoliation. *Proceeding X International Grassland Congress* pages 140-147.

Allsop, A.R.C., Nicol, A.M. and Edwards, G.R. 2009. Effect of water soluble carbohydrate concentration of ryegrass on the partial preference of sheep for clover. Proceedings of the New Zealand Animal Production Society 69: 20–23.

Beerepoot, L. J. and Agnew, R. E. 1997. Breeding for improved herbage quality in perennial ryegrass. In. *"Seeds of Progress"* (ed. Weddell, J. R.) *Occasional Symposium of the British Grassland Society*. 32: 135-145.

Beever, D. E. and Reynolds, C. K. 1994. Forage quality, feeding value and animal performance. In: *"Grassland and society"* (eds. tMannetje and Frame), *Proceeding of the 15th EUCARPIA general Meeting of the European Grassland Federation,* Wageningen, Netherlands, pages 48-60.

Beever, D. E. and Siddons, R. C. 1986. Digestion and metabolism in the grazing ruminant. In control of digestion and metabolism in ruminants (eds. Milling, et al). Englewood Cliffs, NJ: Prentice-Hall, pages 479-497.

Breese, E. L. and Thomas, A. C. 1967. *In vitro* digestibility in cocksfoot. Report of the Welsh Plant Breeding Station for 1966, UK, pp. 35-41.

Brown, R. H. and Blaser, R. E. 1970. Soil moisture and temperature effects on growth and soluble carbohydrates of orchardgrass (*Dactylis glomerata*). *Crop Science* 10: 213-216.

Buckner, R. C., Burrus, P. B., Cornelius, L. P., Bush, L. P. and Leggett, J. E. 1981. Genetic variability and heritability of certain forage quality and mineral constituents in *Lolium-Festuca* hybrid derivatives. *Crop Science* 21: 419-423.

Bugge, G. 1978. Genetic variability in chemical composition of Italian ryegrass ecotypes. *Z. Pflanzenzuchtg* 81: 235-240.

Burner, D. M., Balasko, J. A. and Thayne, W. V. 1983. Genetic and environmental variance of water soluble carbohydrate concentration, yield and disease in tall fescue. *Crop Science* 23: 760-763.

Burns, J. C. and Smith, D. (1980). Non-structural carbohydrates residue, neutral detergent fiber, and *in vitro* dry matter disappearance of forages. *Agronomy Journal* 72: 276-281.

Buxton, D. R. and Casler, M. D. 1993. Environmental and genetic effects on cell wall composition and digestibility. In: *"Forage cell wall structure and digestibility"* (eds. Jung, et al), ASA, CSSA, and SSSA, Madison, USA, pages 685-714.

Buxton, D. R., Mertens, D. R. and Fisher, D. S. 1996. Forage quality and ruminant utilisation. In: *"Cool-season forage grasses"* (eds. Moser et al) ASA, CSSA, and SSSA, Madison, USA pages 229-266.

Canfield, R. W., Sniffen, C. J. and Butler, R. W. 1990. Effects of excess degradable protein on postpartum reproduction and energy blance in dairy cattle. *Journal of Dairy Science* 73: 2342-2349.

Carlier, l. 1994. Breeding, forage quality, feeding value and animal performance. *Proceeding of the 19th EUCARPIA Fodder Crops Section Meeting,* Brugge, Belgium, pages 25-27.

Castle, M. E. and Watson, J. N. 1971. A comparison between a diploid and a tetraploid perennial ryegrass for milk production. *Journal of agricultural Science Cambridge* 77: 69-76.

Clements, R. J. 1969. Selection for crude protein content in *Phalaris tuberosa* L. I. Response to selection and preliminary studies on correlated response. *Australian Journal of Agricultural Research* 20: 643-652.

Conaghan, P., M. D. Casler, D. A. McGilloway, P. O'Kiely and L. J. Dowley. 2008. Genotype x environment interactions for herbage yield of perennial ryegrass sward plots in Ireland *Grass and Forage Science* 63: 107–120.

Connolly, V., Ribeiro, M. do Valle and Crowley, J.G. 1977. Potential of grass and legume cultivars under Irish conditions. *Proceedings International Meeting on animal production from Temperate Grassland,* Dublin, pages 23-28.

Cooper, J. P. 1962. Selection for nutritive value. *Report of the Welsh Plant Breeding Station for 1961,* UK, pages 145-156.

Cooper, J. P. 1973. Genetic variation in herbage constituents. In: *"Chemistry and biochemistry of herbage"* (eds. Butler and Bailey), Vol. II. Academic press, London pages 379-417.

Cosgrove, G.P., Koolaard, J., Potter, F., Luo, D., Burke, J.L. and Pacheco, D. 2010. Milk solids yield from high water soluble carbohydrate ryegrasses: a combined analysis of multiyear data. In: "Proceedings of the 4th Australian Dairy Science Symposium", 31st Aug.2nd Sept. Lincoln University, New Zealand, pages 334–338.

Dent, J. W. and Aldrich, D. T. A. 1963. The inter relationship between heading date, yield chemical composition and digestibility in varieties of perennial ryegrass, timothy, cocksfoot and meadow fescue. *Journal of the National Institute of Agriculture Botany* 9: 261-281.

Evans, J. G., M. D. Fraser, I. Owen and D. A. Davies. 2011. An evaluation of two perennial ryegrass cultivars (AberDart and Fennema) for sheep production in the uplands. *Journal of Agricultural Science* 149: 235-248.

Frandsen, K. J. 1986. Variability and inheritance of digestibility in perennial ryegrass (*Lolium perenne*), meadow fescue (*Festuca pratensis*), and cocksfoot (*Dactylis glomerata*). II. F1 and F2 progeny. *Acta Agricuturae Scandinavica* 36: 241-263.

Fulkerson, W.J. and Donaghy, D.J., 2001. Plant-soluble carbohydrate reserves and senescence: key criteria for developing an effective grazing management system for ryegrass-based pastures: a review. *Australian Journal of Experimental Agriculture* 41 (2): 261-275.

Gill, M., Beever, D. E. and Osbourn, D. F. 1989. The feeding values of grass and grass products. In: *"Grass, its production and utilisation"* (ed. Holmes, W.), Blackwell Scientific Publications, London, pages 89- 129.

Gilliland, T.J., Barrett, P.D., Mann, R.L., et al., 2002. Canopy morphology and nutritional quality traits as potential grazing value indicators for *Lolium perenne* varieties. *Journal of Agricultural Science,* 139(3): 257-273.

Grimes, R. C., Watkin, B. R. and Gallagher, J. R. 1967. The growth of lambs on perennial ryegrass, tall fescue and cocksfoot, with and without white clover, as related to the botanical and chemical composition of the pasture and pattern of fermentation in rumen. *Journal of Agricultural Science Cambridge* 68: 11-21.

Grusea, A. and Oprea, G. 1994. Variation and inheritance of quality of *Dactylis glomerata* varieties which were obtained by different breeding methods. *Proceeding of the 19th EUCARPIA Fodder Crops Section Meeting* Brugge, Belgium, pages 145-149.

Haigh, P. M. 1990. Effect of herbage water-soluble carbohydrate content and weather conditions at ensilage on the fermentation of grass silage made on commercial farms. *Grass and Forage Science* 45: 263-271.

Humphreys, M. O. 1989a. Water-soluble carbohydrates in perennial ryegrass breeding. I. Genetic differences among cultivars and hybrid progeny grown as spaced plants. *Grass and Forage Science* 44: 231-236.

Humphreys, M. O. 1989b. Water-soluble carbohydrates in perennial ryegrass breeding. II. Cultivar and hybrid progeny performance in cut plot. *Grass and Forage Science* 44: 237-244.

Humphreys, M. O. 1989c. Water-soluble carbohydrates in perennial ryegrass breeding. III. Relationships with herbage production, digestibility and crude protein content. *Grass and Forage Science* 44: 423-430.

Humphreys, M. O. 1989d. Assessment of perennial ryegrass (*Lolium perenne* L.) for breeding. II. Components of winter hardiness. *Heredity* 41: 99-106.

Humphreys, M. O. 1991. A genetic approach to the multivariate differentiation of perennial ryegrass (*Lolium perenne* L.) populations. *Heredity* 66: 437-443.

Humphreys, M. O. 1994. Variation in the carbohydrate and protein content of ryegrass: Potential for genetic manipulation. *Proceeding of the 19th EUCARPIA Fodder Crops Section Meeting* Brugge, Belgium, pages 165-171.

Humphreys, M. O. 1995. Multitrait response to selection in *Lolium perenne* L. (perennial ryegrass) populations. *Heredity* 74: 510-517.

Jafari A. A. and Javarsineh S. 2005. Estimates of heritability and predicted of genetic gain for yield and quality traits in parents and half sib families of tall fescue (*festuca arundinacea* schreb). *Iranian Journal of Rangelands Forests Plant Breeding And Genetic Research* 13(1): 99-124. ISSN 1735-0891 (Abstract in English).

Jafari A. A. and Rezaeifard, M. 2010. Effects of maturity on yield and quality traits in tall fescue (*Festuca arundinace* Schreb) American-Eurasian J. Agric. & Environ. Sci. 9 (1): 98-104

Jafari, A. A. 1998. Genetic analysis of yield and quality in perennial ryegrass (*Lolium perenne L.*). Ph.D. Thesis, Department of Crop Science, Horticulture and Forestry. University College Dublin, Ireland.

Jafari, A. A. and Naseri, H. 2007. Genetic variation and correlation among yield and quality traits in cocksfoot (*Dactylis glomerata* L). *Journal of Agricultural Science, Cambridge.* 145: 599-610.

Jafari, A. A., V. Connolly and E.K. Walsh. 2003. Genetic analysis of yield and quality in full-sib families of perennial ryegrass *(Lolium perenne.* L) Under two cutting managements. *Irish Journal of Agricultural and Food Research* 42: 275-292.

Jafari, A.A., Anvari, H., Nakhjavan, S. and Rahmani, E. 2010. Effects of phenological stages on yield and quality traits in 22 populations of tall wheatgrass grown in Lorestan, Iran. *Journal of Rangeland Science* 1(2): 13-20

Jones, E. L. and Roberts, J. E. 1991. A note on the relationship between palatability and water-soluble carbohydrates content in perennial ryegrass. *Irish Journal of Agricultural and food Research* 30: 163-167.

Jung, G. A., Van Wijk, J. A. P., Hunt. W. F. and Watson, C. E. 1996. Ryegrass. In: "*Cool-season forage grasses*" (eds. Moser et al.), ASA, CSSA, and SSSA, Madison, USA, pages 605- 641.

Lee, M. R. F. , Jones, E. L. ,Moorby, J. M., Humphreys, M. O., Theodorou, M. K., MacRae, J. C., Scollan, N. D. 2001. Production responses from lambs grazed on *Lolium perenne* selected for an elevated water-soluble carbohydrate concentration Animal Research 50(6): 441-449.

Marais, J. P., de Figueiredo, M. and Goodenough, D. C. W. 1993. Dry mater and non-structural carbohydrate content as quality parameters in *Lolium multiflorum* breeding program. *African Journal Forage Science* 10: 118-123.

Mayland H. F., Shewmaker G.F., Harrison P. A. and Chatterton N. J. 2001. Nonstructural carbohydrate in tall fescue cultivars; relationship to animal preference. *Agronomy Journal*, 92: 1203-1206.

McGrath, D. 1988. Seasonal variation in the water-soluble carbohydrates of perennial and Italian ryegrass under cutting conditions. *Irish Journal of Agricultural and food Research* 27: 131-139.

McGrath, D. 1992. A note on the influence of nitrogen application and time of cutting on water soluble carbohydrate production by Italian ryegrass. *Irish Journal of Agricultural and food Research* 31: 189-192.

Miller, L. A., Moorby, J. M., Davies, D. R., Humphreys, M. O., Scollan, N. D., Macrae, J. C., Theodorou, M. K. 2001. Increased concentration of water-soluble carbohydrate in perennial ryegrass (*Lolium perenne* L.). Milk production from late-lactation dairy cows. *Grass and Forage Science* 56 (4): 383-394.

Parsons, A. J., Edwards, G.R., Newton, P. C. D., Chapman, D. F., Caradus, J. R., Rasmussen, S. and Rowarth, J.S. 2010. Past lessons and future prospects: plant breeding for cool temperate pastures. In: "Proceedings of the 4th Australasian Dairy Science Symposium", Lincoln University, New Zealand, pages 272–291.

Pollock C.J. and Cairns A.J. 1991. Fructan metabolism in grasses and cereals. Annual Review of Plant Physiology and Plant Molecular Biology 42:77-101.

Radojevic, I., Simpson, R. J., John, J. A. St. and Humphreys, M. O. 1994. Chemical composition and *in vitro* digestibility of lines of *Lolium perenne* selected for high concentrations of water-soluble carbohydrate. *Australian Journal of Agriculture Research* 45: 901-912.

Rasmussen, S., Parsons, A.J., Fraser, K., Xue, H. and Newman, J. 2008. Metabolic profiles of *Lolium perenne* are differentially affected by nitrogen supply, carbohydrate content, and fungal endophyte infection. Plant Physiology 146: 1440–1453.

Rasmussen, S., Parsons, A.J., Xue, H. and Newman, J.A. 2009. High sugar grasses harnessing the benefits of new cultivars through growth management. Proceedings of the New Zealand Grassland Association 71: 167–175.

Rezaeifard M., Jafari, A. A. and Assareh, M. H. 2010. Effects of phenological stages on forage yield quality traits in cocksfoot (*Dactylis glomerata*). *Journal of Food, Agriculture & Environment* 8(2): 365-369.

Sanada, Y., Takai, T. and Yamada, T. 2007. Inheritance of the concentration of water-soluble carbohydrates and its relationship with the concentrations of fiber and crude protein in herbage of cocksfoot (*Dactylis glomerata* L.). *Grass and Forage Science* 62 (3): 322–331.

Smit, H. J. 2006. Cultivar effects of perennial ryegrass on herbage intake by grazing dairy cows. In: A. Elgersma, J. Dijkstra and S. Tamminga (eds.), Fresh Herbage for Dairy Cattle, 45-62.

Smith K. F., Reed K.F.M. and Foot J.Z. 1997. An assessment of the relative importance of specific traits for the genetic improvement of nutritive value in dairy pasture. Grass and Forage Science, 52: 167-175.

Smith, K.F., Simpson, R.J., Culvenor, R.A., et al., 2001. The effects of ploidy and a phenotype conferring a high water-soluble carbohydrate concentration on carbohydrate accumulation, nutritive value and morphology of perennial ryegrass (*Lolium perenne* L.). *Journal of Agricultural Science*, 136 (1): 65-74.

Smouter, H. Simpson, R. J. and Pearce, G. R. 1995. Water-soluble carbohydrates and *in vitro* digestibility of annual ryegrass (*Lolium rigidum* Gaudin) sown at varying densities. *Australian Journal of Agricultural Research* 46: 611-625.

Stewart A. and R. Hayes. 2011. Ryegrass breeding-balancing trait priorities. *Irish Journal of Agricultural and Food Research* 50: 31–46.

Thomas, H. and Norris I. B. 1981. The effect of light and temperature during winter on growth and death in simulated swards of *Lolium perenne. Grass and Forage Science* 36: 107-116.

Thomson, A. J. and Rogers, H. H. 1971. The interrelationship of some components of forage quality. *Journal of Agricultural Science Cambridge* 76: 283-293.

Valentine, J. and Charles, A. H. 1979. The associations of dry matter yield with nitrogen and soluble carbohydrate concentration in perennial ryegrass (*Lolium perenne* L.). *Journal of Agricultural Science, Cambridge* 93: 657-667.

Van Vuuren, A. M., Van Der Koelen, C. J. and Vroons-De-Bruin, J. 1986. Influence of level and composition of concentrate supplements on rumen fermentation patterns of grazing dairy cows. *Netherlands Journal of Agricultural Science* 34: 457-467.

Visek, W. J. 1984. Ammonia: its effects on biological system, metabolic hormones, and reproduction. *Journal of Dairy Science* 67: 481-489.

Volaire, F. and Gandoin, J. M. 1996. The effect of age of the sward on the relationship between water-soluble carbohydrate accumulation and drought survival in two contrasted populations of cocksfoot (*Dactylis glomerata* L.). *Grass and Forage Science* 51: 190-198.

Vose, P. B. and Breese, E. L. 1964. Genetic variation in utilization of nitrogen by ryegrass species *Lolium perenne* and *Lolium multiflorum. Annals of botany* 110: 251-270.

Waite, R. and Boyd, J. 1953. The water-soluble carbohydrates of grasses. II. Grasses cut at grazing height several times during growing season. *Journal Science Food Agriculture* 4: 257-261.

Wilkins P. W. and Humphreys M. O. 2003. Progress in breeding perennial forage grasses for temperate agriculture. Journal of Agricultural Science, Cambridge, 140: 129-150.

Wilkins, P. W. and Davies, R. W. 1994. Progress in combining high dry matter yield with reduced flowering intensity and improved digestibility in perennial ryegrass. *Proceeding of the 19th EUCARPIA Fodder Crops Section Meeting* Brugge, Belgium, pages 247-250.

Wilkins, P.W. and Lovatt, J.A. 2007. "AberMagic (Ba13582)", IBERS Internal Reference 1340 4213. Institute of Biological, Environmental and Rural Sciences, University of Aberystwyth, UK

Wilkins, P.W., Lovatt, J. A., Hayes, R.C. and Thomas, G.L. 2010. "AberGreen (Ba13926)", IBERS Internal Reference. Institute of Biological, Environmental and Rural Sciences, University of Aberystwyth, UK.

Chapter 18

Alterations in Root Morphology of Rootstock Peach Trees Caused by Mycorrhizal Fungi

José Luis da Silva Nunes, Paulo Vitor Dutra de Souza, Gilmar Arduino Bettio Marodin, José Carlos Fachinello and Jorge Ernesto de Araújo Mariath

Additional information is available at the end of the chapter

http://dx.doi.org/10.5772/51586

1. Introduction

The morphology of plant roots have gained prominence in various branches of knowledge, especially in the Biological and Agricultural Sciences, according to the same being one of the main features of the plant body related to the supply and support of plant (Marschner, 1995). Agricultural practices of soil management require special attention in the relations of the roots of different plants with different managements employees, because the health of plants is dynamically linked to these delicate relatio (Silva et al, 2005). This is because the management practices linked monocultures allow the reproduction of micro-organisms that cause crop damage, and the common use of pesticides to alleviate this problem (Bressan & Vasconcelos, 2002).

In this sense, studies are being conducted with the objective of evaluating the possibility to reduce the use of these chemicals in the control of harmful micro-organisms, ranging from research on structural strength of the plant, past the front of the dynamic plant managements, to the use of microorganisms considered beneficial plants (Bressan & Vasconcelos, 2002). On this last point, the arbuscular mycorrhizal fungi (AMF) colonize the root system of most plants, and one of the most reported benefits has been a greater phosphorus absorption by the mycorrhized plants (Nunes et al, 2006), forming a mutualistic symbiosis type biotrophic (Dodd, 2000). This symbiosis is widely distributed in the plant kingdom, occurring in 83% of dicotyledonous plants, in 79% of monocots and in all Gymnosperms, without altering the external appearance of the root (Wilcox, 2002). Moreover, the occurrence of symbiosis is widespread in most habitats, both natural ecosystems and in ecosystems altered by human activities (Sylvia et al., 2001).

In this respect, mutualism is manifested in the bidirectional exchange of nutrients, where the plant comes from carbohydrates to the fungus, while it provides you with water and nutrients, especially for the case of phosphorus (Smith et al., 2003). Although the result of symbiosis be beneficial for the phytobionts, the effectiveness varies in function of the combination the vegetal species and fungus involved in the association (Smith et al., 2003).

By mechanisms promoted by the AMF, the external hypha and mycelia increase the root capacity to exploit the soil results in greater nutrient absorption (Siqueira et al., 2002). However, this absorption has also been related to alterations in the morphological properties of the root of the host plant (Moreira & Siqueira, 2002).

The root system morphology is determined genetically, and can vary among species and individuals in function of environmental factors, such as water availability, nutrients and temperature (Tokeshi, 2000) and the plasticity of the root system can also be influenced by AMF (Berta et al., 1995). The root morphology influences the fast development of the root system and is critical for the successful establishment of most horticultural and fruit plants (Bressan & Vasconcellos, 2002).

This fact is fundamental to the understanding of the effects of the AMF on root development, especially in the case of rootstock plants (Berta et al., 1995). However, the relationships involved in the formation of this symbiosis, since the signaling between the phytobionts, the early stages of the colonization process, as well as possible alterations in the morphological structure of the roots (Berta et al., 1995), in order to be considered a complete understanding relations between the symbionts

There is little information about such relationships, as well as morphological changes produced by mycorrhizal infection in plant tissues (Souza et al., 2000). Some authors report that the AMF does not cause major morphological changes in roots (Cooper, 1984), but studies showed that the AMF induces changes in the architecture (Berta et al., 1995; Norman et al., 1996), especially in the increase of the root ramification, in the morphology (Berta et al., 1995; Bressan & Vasconcelos, 2002; Kothari et al., 1990; Norman et al., 1996,) and the anatomy (Berta et al., 1995) the roots of different plant species.

Most infections of the root system of plants by soil microorganisms imply relations between the actors involved, these relations are based on compatibility between symbionts or the ability of the microorganism to overcome the defense mechanisms of plants (Paszkowski, 2006). The study of morphological relationships between the symbionts highlight the determinants of compatibility that allow the symbiosis occurs involving taxonomically distinct groups of plants and AMF infective (Panstruga, 2003).

The objective of this study was to relate the morphology and root system development of plants of the rootstock cultivars of the peach trees Aldrighi and Okinawa with root colonization by AMF species and the influence of this relationship on nitrogen, phosphorus and potassium absorption and the vegetative development of the plants.

2. Material and methods

2.1. Execution area

The study was carried out under shading (Okinawa cultivar) and a greenhouse (Aldrighi cultivar) at the UFRGS Agronomic Experimental Station, county of Eldorado do Sul, RS, located at latitude 30° 05′ South and longitude 51° 39′ West from 2004 to 2005.

2.2. Plant and fungal material

Seeds from the two rootstock cultivars were stratified in sterilized sand and placed in a refrigerator at 4°C for 45 days to break the seed dormancy.

Afterwards the seeds were sown on a bed of sterilized sand in a greenhouse. When they were about 5 cm long, the seedlings were replicated to 5 liters black plastic bags containing substrate consisting of clay soil, sand with medium particle size and decomposed black acacia bark residue (1:1:1, V:V.V). The substrate was previously disinfected with formaldehyde solution at 10%.

The AMF species tested were *Acaulospora sp.* (Trappe), *Glomus clarum* (Nicol. and Schenck) and *Glomus etunicatum* (Becker and Gerd) for the Okinawa cultivar and for the Aldrighi cultivar the same treatments were tested along with *Scutellospora heterogama* (Nicol. and Gerd.). The AMF species were inoculated by adding to each plastic bag 30 g of roots and rhyzospheric soil of oregano (*Origanum vulgare* Link) containing AMF structures. The inoculum was placed in a layer situated in the mid-part of the recipient.

A randomized block design was used, with 20 plants per plot and four replications, in a total of 320 plants for the Okinawa cultivar and 400 plants for the Aldrighi cultivar.

2.3. Determination of roots colonization and plant responses

When the plants had diameter for grafting (360 days for the Okinawa cultivar and 180 days for the Aldrighi cultivar) the height was assessed of the 20 plants in each plot, from the root-stem junction to the tip of the main stem, using a measuring tape, and the main stem diameter, at the root-stem junction and plant height using a pachymeter.

In addition, 5 plants were used from each replication of the treatments, for determination of leaf area, through the use of leaf area meter mark Li-Cor (LI - 3000). After, the shoot was dried and ground and where the fractions were removed for evaluation of plant tissue nitrogen, phosphorus and potassium content by digestion, distillation and spectrophotometry flames, following the methodology by Tedesco et al. (1995).

Five second order roots with similar length and diameter were collected from the root system to assess the root colonization rate (by the ratio number of infected segments/total analyzed). To determine the colonization rate the radicels were stained following methodology reported by Phillips and Hayman (1970).

2.4. Determination of reserve substances

Samples of the aerial part (leaves, stems and stem) and dried roots were ground in the mill, coupled with a sieve of 20 meshes per inch. Each sample was collected approximately one gram for determination of reserve substances.

A similar procedure was carried out with samples of roots. After each sample individually packaged in bags made of special screen for the filtration of food products and brought back to 65°C oven to constant weight, recording the weight of each bag, after, were digested in order to extract all components of plant tissue (carbohydrates, fats, fatty acids, etc.) that were not fibers (cellulose, hemicellulose and lignin), as conventionally known as reserve substances the method described by Priestley (1965).

The samples were placed in one liter Erlenmeyer flask containing an aqueous solution with 5% trichloroacetic acid (99%) and 35% methanol (99.8%) remained on heating gas burner, under a hood with hood, by eight hours. From the third hour to eight hours, distilled water was added to the solution, as it would evaporate in order to always maintain the same volume of liquid sufficient to maintain the samples immersed in the solution.

After the samples were rinsed with distilled water again and put in stove to dry at 65°C until constant weight. The difference in mass of the samples before and after digestion constisted substance content of the buffer that contained samples.

2.5. Histological studies

Secondary roots with similar diameter were used for the morphological studies, as shown in Tables 1 and 2. The histological studies followed the methods described by Johansen (1940), where 1 cm long samples were dehydrated and blocked in paraffin, and 10-15μm thick slices were made using a manual microtome.

The slices were placed on slides, removed the paraffin with xylol, rehydrated for later staining with aqueous Safranin (1%) and Toluidine Blue O (0.05%), and than dehydrated again and the preparations mounted in Canada balsam with a coverslip.

These sections were observed under a Leica DM microscope with 400X magnification. The images were captured with a Nikon CoolPix 990 digital camera (Photos of José Luis da Silva Nunes) and analyzed using the "WCIF Image J" software.

The morphometric parameters measured in the roots were area, diameter, number and perimeter of the tracheal element cells, regardless of the stage of ontogenetic development (primary or secondary) and, from the primary xylem, only the metaxylem was measured, because the protoxylem collapsed at the end of its differentiation (Figures 1 and 2).

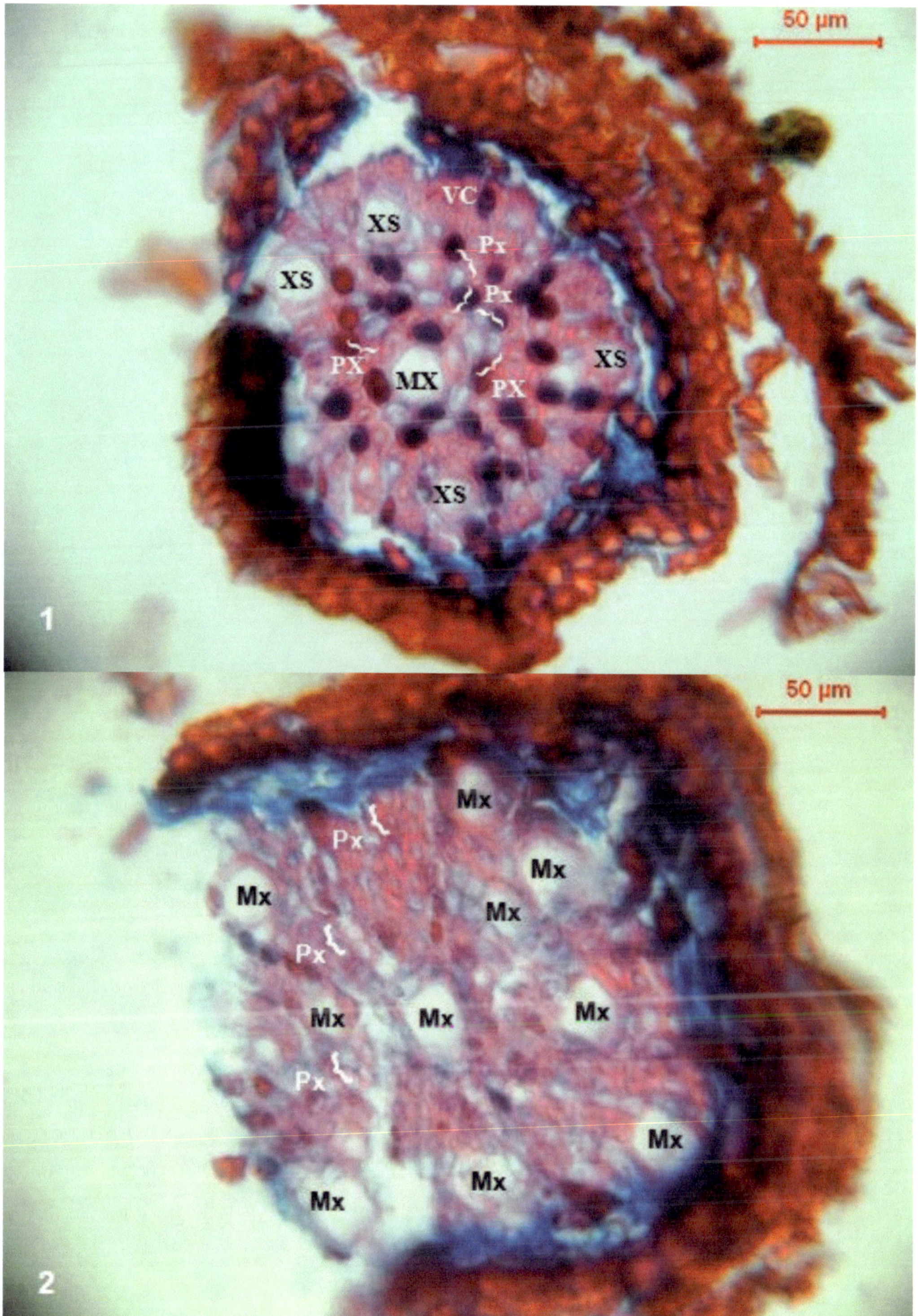
50 µm
VC
XS
Px
PX
MX
1
50 µm
Mx
Px
2

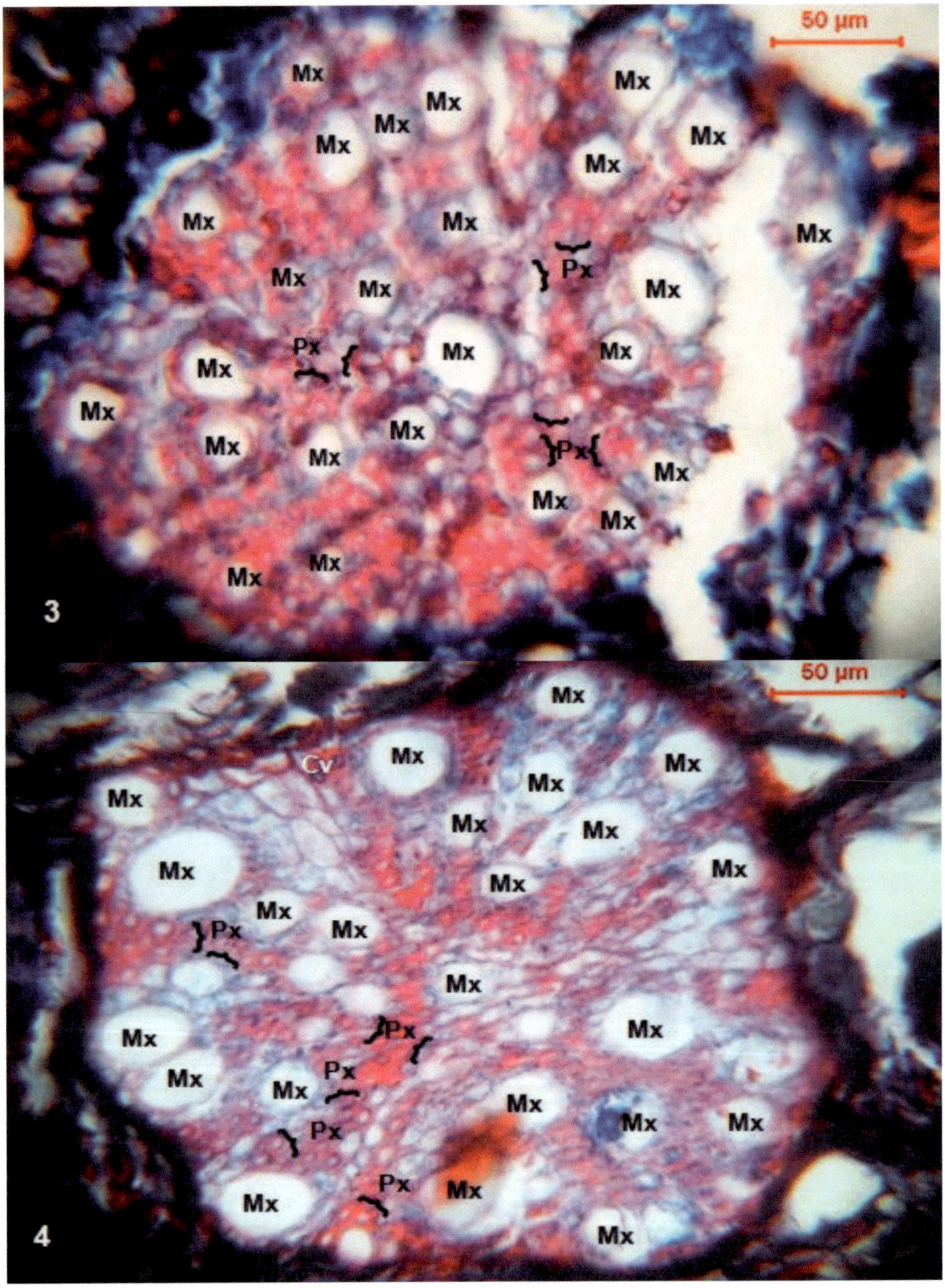

Figure 1. Cross sections of secondary roots of peach tree rootstock Okinawa without inoculation (1) and inoculated with AMF (2 – *G. clarum*, 3 – *G. etunicatum*, 4 – *Acaulospora sp.*). MX – metaxylem; PX – protoxylem; XS – secondary xylem. Scale 50 µm.

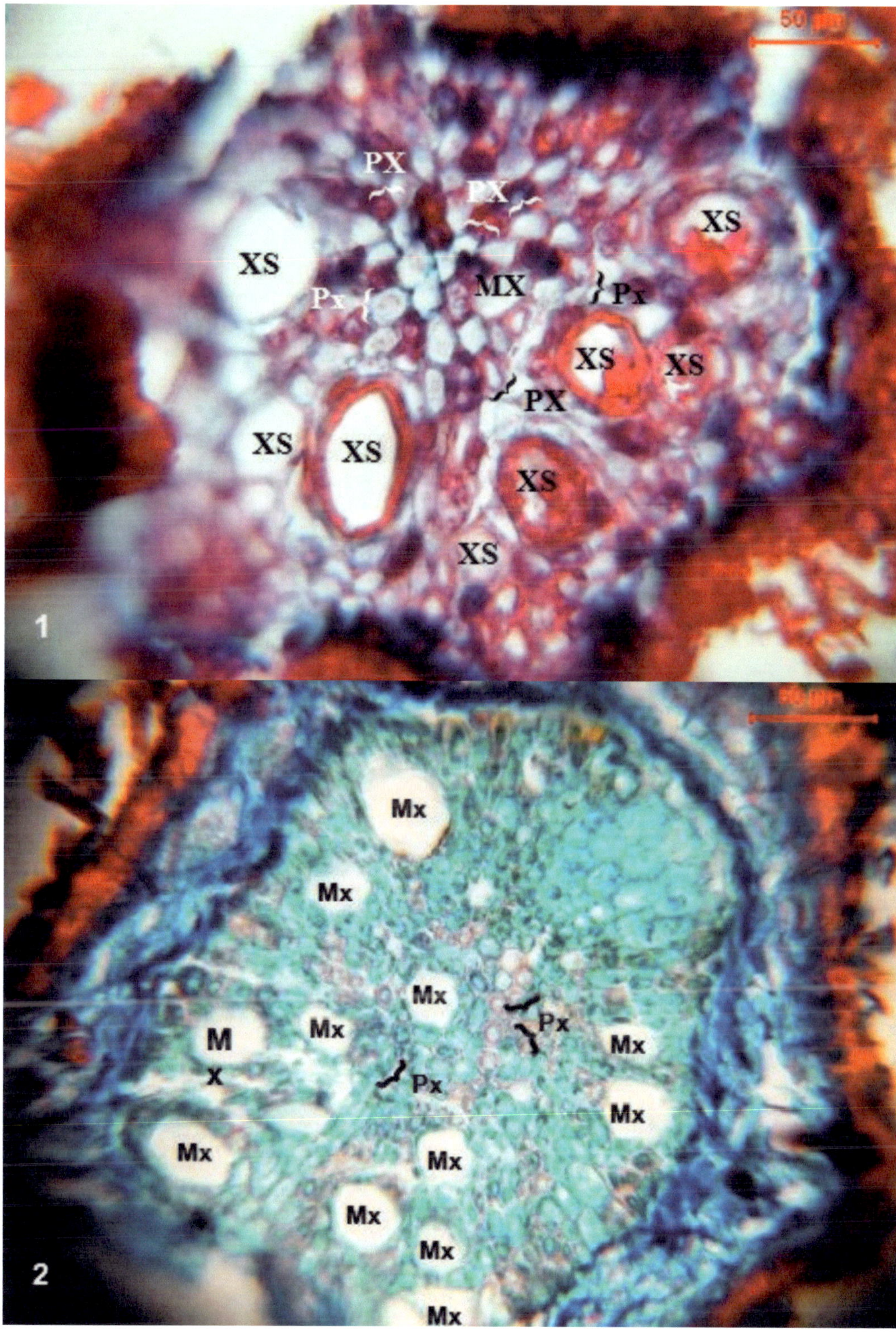
50 µm
PX
PX
XS
XS
MX
Px
Px
XS
XS
PX
XS
XS
XS
XS
1
Mx
Mx
Mx
Px
Mx
M
x
Mx
Px
Mx
Mx
Mx
Mx
Mx
Mx
2
Mx

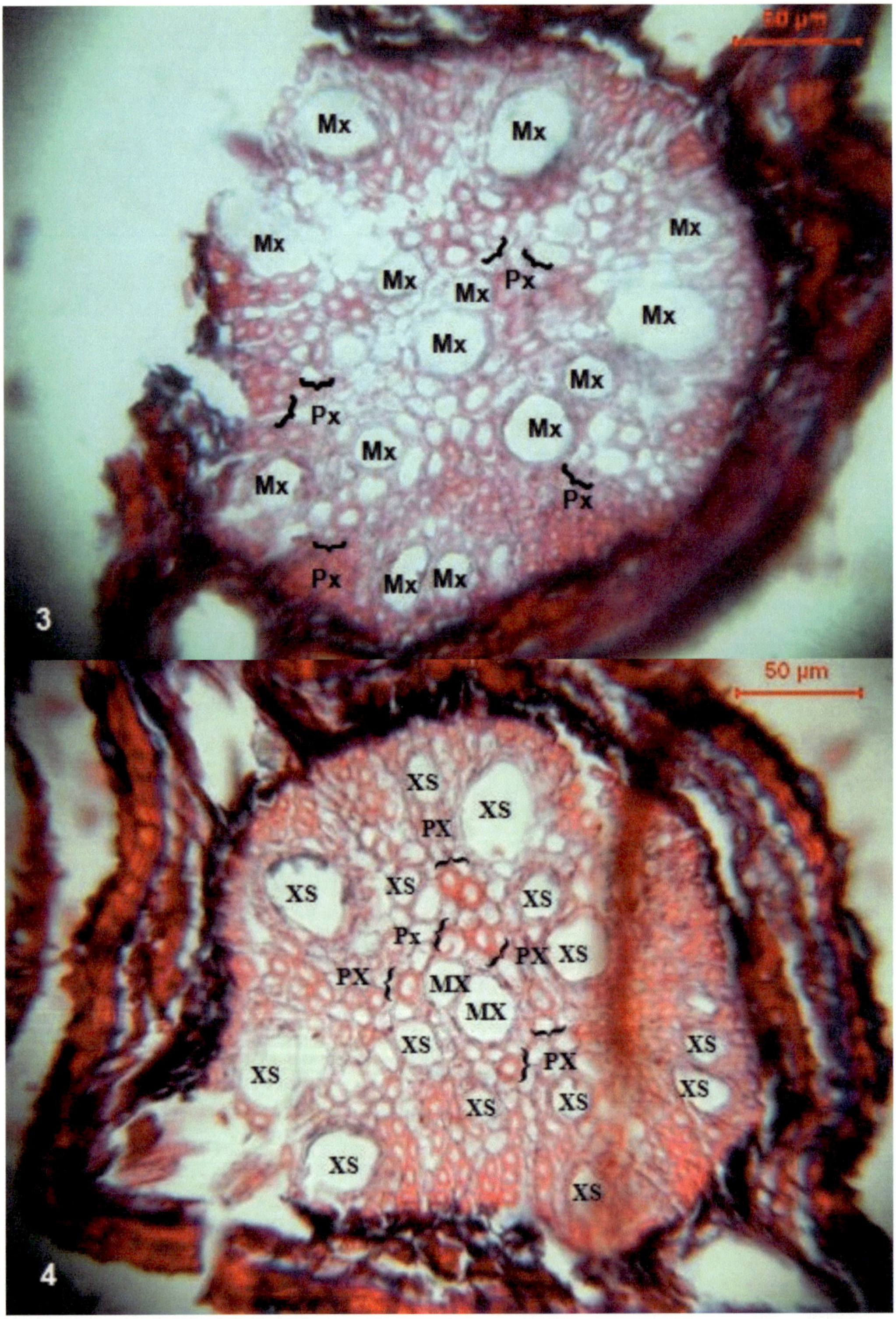
50 µm
Mx
Mx
Mx
Mx
Mx
Px
Mx
Mx
Mx
Mx
Px
Mx
Mx
Mx
Px
Px
Mx
Mx
3
50 µm
XS
XS
PX
XS
XS
XS
Px
PX
XS
PX
MX
MX
XS
PX
XS
XS
XS
XS
XS
XS
XS
4

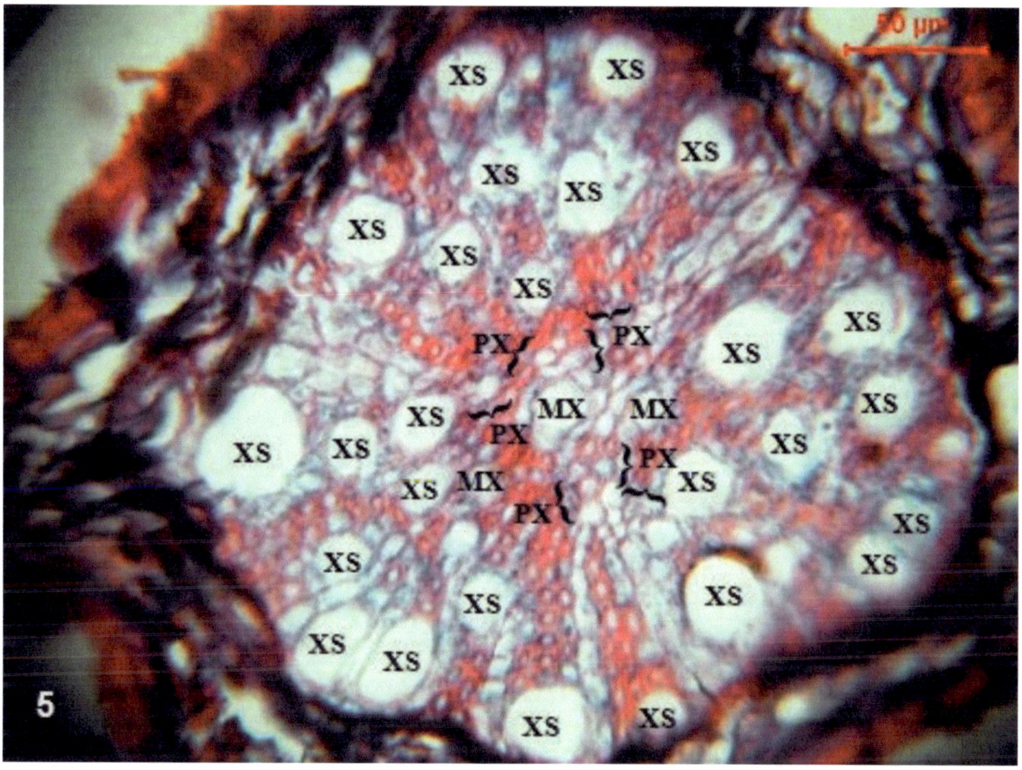

Figure 2. Cross sections of secondary roots of peach tree rootstock Aldrighi without inoculation (1) and inoculated with AMF (2 – *G. clarum*, 3 – *G. etunicatum*, 4 – *Acaulospora sp.*, 5 - *Scutellospora heterogama*). MX – metaxylem; PX – protoxylem; XS – secondary xylem. Scale 50 µm.

2.6. Statistics

The data were submitted to an analysis of variance by the SAS program and the measurements were compared by the Duncan test (Duncan, 1955) at the level of 5% significance.

3. Results

The results regarding the effect of the AMF on the conductor tissue of the roots of the Okinawa cultivar showed that the treatments with the *Acaulospora sp.* and *G. etunicatum* species performed similarly for the number and diameter of the metaxylem and secondary xylem cells, superior to the other treatments, while for cell area and perimeter, *Acaulospora sp.* was superior to the other treatments. The treatment with *G. etunicatum* presented cell area, perimeter and diameter similar to *G. clarum*, superior to the controls. The treatment with *G. clarum* presented the number of cells of these xylem classes similar to that of the controls. There was an inverse performance for cortical thickness, where the control plants presented the highest results, followed by the plants inoculated with *G. clarum* while those inoculated with *Acaulospora sp.* and *G. etunicatum* presented similar but lower results, compared to the control plants (Table 1).

Treatment	Diameter of root (μm)	Cortex thickness (μm)	Metaxylem e secondary xylem			
			Number of Cell	Cell diameter (μm)	Cell perimeter (μm)	Cell area (μm²)
Acaulospora sp.	957,14ns	118,12c	38,53a	5,50a	16,86a	58,55a
G. clarum	958,32ns	129,85b	26,00b	5,11b	15,61b	48,46b
G. etunicatum	961,35ns	119,31c	34,60a	5,26ab	15,70b	50,45b
Testemunha	952,61ns	139,39a	22,00b	4,04c	11,77c	41,14c
V. C. (%)	5,17	7,41	5,01	10,09	6,02	13,22

Table 1. Root morphology of secondary roots of the Okinawa cultivar rootstock inoculated with three AMF species (*Acaulospora sp., G. clarum* and *G. etunicatum*), collected 360 days after sowing. Eldorado do Sul, RS, 2005. Means followed by the same letter in the column do not differ by the Duncan test at 5% significance. nsNon-significative.

For the Aldrighi cultivar, the treatment with the *S. heterogama* species presented better results than the other treatments for the parameters number of cells, cell area and perimeter of the metaxylem and the secondary xylem, but performed similarly to *G. etunicatum* for cell diameter, and was superior to the other treatments for this parameter (Table 2).

Treatment	Diameter of root (μm)	Cortex thickness (μm)	Metaxylem e secondary xylem			
			Number of Cell	Cell diameter (μm)	Cell perimeter (μm)	Cell area (μm²)
Acaulospora sp.	945,01ns	118,08c	33,40b	4,87b	12,95c	52,95c
G. clarum	939,11ns	128,85b	24,93c	4,93b	13,10c	53,10c
G. etunicatum	955,23ns	107,31d	30,20b	6,17a	16,86b	56,86b
S. heterogama	960,12ns	105,22d	37,00a	7,06a	22,70a	62,70a
Testemunha	940,04ns	138,39a	16,33d	3,87c	11,57d	41,57d
V. C. (%)	4,76	6,11	5,54	7,84	6,84	11,32

Table 2. Root morphology of secondary roots of the Aldrighi cultivar rootstock inoculated with four AMF species (*Acaulospora sp., G. clarum, G. etunicatum* and *S. heterogama*) collected 180 days after sowing. Eldorado do Sul, RS, 2005. Means followed by the same letter in the column do not differ by the Duncan test at 5% significance. nsNon-significative.

G. etunicatum performed similarly to *Acaulospora sp.* regarding the number of cells that was better than *G. clarum* and the control. Furthermore, *G. etunicatum* performed better than *Acaulospora sp., G. clarum* and the controls for the other parameters. The plants inoculated with *Acaulospora sp.* performed better than *G. clarum* for the number of cells and similarly to the other parameters. The control plants presented lower results for all the parameters in the assessment of the xylem classes. However, regarding cortical thickness, the control plants performed better than the other treatments, followed by *G. clarum* that was better than *Acaulospora sp.* that was better than the treatments with *S. heterogama* and *G. etunicatum* that presented similar results.

The inoculation with AMF species accelerated the growth of the plants of the Okinawa cultivar rootstock, inducing greater height, diameter, leaf area and greater nitrogen, phosphorus and potassium content, compared with the control. All presented root colonization rates were over 90%. *Acaulospora sp.* was the species that was most efficient among the AMF tested giving greatest height, diameter and nutritional state compared to the other species. The plants inoculated with *G. clarum* and *G. etunicatum* presented to mediate growth and nutritional state, and were similar (Table 3).

Treatment	H (cm)	D (mm)	L.A. (cm^2/ plant)	Nutrients (%)			Colonization (%)
				N	P	K	
Acaulospora sp.	136,46a	8,42a	197,01a	2,35a	0,16a	2,07a	97,00a
G. clarum	126,65b	7,79b	163,00b	2,23b	0,15b	1,74b	91,76b
G. etunicatum	129,04b	7,87b	167,00b	2,22b	0,15b	1,82b	92,62b
Control	119,23c	7,24c	142,03c	2,05c	0,14c	1,60c	00,00c
V. C. (%)	3,88	2,17	2,54	2,56	2,61	4,75	2,42

Table 3. Height (H), root-stem junction diameter (D), leaf area (L.A.), percentage of nitrogen (N), phosphorus (P) and potassium (K) in the plant tissue and root colonization (%) of the Okinawa cultivar rootstock inoculated with three AMF species (*Acaulospora sp., G. clarum* and *G. etunicatum*), collected 360 days after sowing. Eldorado do Sul, RS, 2005. Means followed by the same letter in the column do not differ by the Duncan test at 5% significance.

For the plants of the Aldrighi cultivar, only the *S. heterogama* and *G. etunicatum* species were shown to be efficient for the height parameter (Table 4).

Treatment	H (cm)	D (mm)	L.A. (cm^2/ plant)	Nutrients (%)			Colonization (%)
				N	P	K	
Acaulospora sp.	130,94c	6,28c	125,00c	2,99c	0,16b	2,40b	30,00c
G. clarum	129,80c	6,24c	119,01c	2,96c	0,17b	2,44b	28,50c
G. etunicatum	138,36b	6,82b	137,00b	3,33b	0,20a	2,72a	91,50b
S. heterogama	143,97a	7,29a	173,10a	3,74a	0,22a	2,76a	97,75a
Control	129,70c	5,88d	105,00d	2,65d	0,16b	2,29c	00,00d
V. C. (%)	1,55	2,77	2,44	6,64	10,60	3,97	2,91

Table 4. Height (H), root-stem junction diameter (D), leaf area (L.A.), percentage of nitrogen (N), phosphorus (P) and potassium (K) in the plant tissue and root colonization (%) of the Aldrighi cultivar rootstock inoculated with four AMF species (*Acaulospora sp., G. clarum, G. etunicatum* and *S. heterogama*), collected 180 days after sowing. Eldorado do Sul, RS, 2005. Means followed by the same letter in the column do not differ by the Duncan test at 5% significance.

These species were the only ones to present root colonization rates of over 90%. All of the AMF species were efficacious for the root-stem junction diameter and leaf area parameters and only varied in the response intensity. In all the assessments of plant growth and nutritional states, invariably *S. heterogama* induced greater growth compared to the other AMF species, while *G. etunicatum* induced intermediate performance and *Acaulospora sp.* and *G. clarum* presented similar performance.

Inoculation with AMF increased content of reserve substances to plants of cv. Okinawa, especially when inoculated with *Acaulospora sp.* In the shoots of rootstock plants inoculated with *G. etunicatum* and *G. clarum* showed intermediate levels. In roots, the plants inoculated with *G. etunicatum* also showed intermediate values, while those inoculated with *G. clarum* not differ from the controls (Table 5).

Treatment	Reserve substances (% in the plant)	
	Shoot	Roots
Acaulospora sp.	39,81a	28,38a
G. clarum	35,05b	21,02c
G. etunicatum	35,53b	24,28b
Control	27,29c	19,41c
V. C. (%)	5,24	2,58

Table 5. Reserve substances of shoots (leaves and stems) and roots of plants of the Okinawa cultivar rootstock inoculated with three AMF species (*Acaulospora sp., G. clarum* and *G. etunicatum*), collected 360 days after sowing. Eldorado do Sul, RS, 2005. Means followed by the same letter in the column do not differ by the Duncan test at 5% significance.

In reviewing the data on the percentage of reserve substances from plants of cv. Aldrighi, present in the tissue of the shoot, it appears that the plants were inoculated with the AMF species had percentages higher than uninoculated plants (Table 6).

Treatment	Reserve substances (% in the plant)	
	Shoot	Roots
Acaulospora sp.	34,69b	22,03c
G. clarum	35,50a	22,99bc
G. etunicatum	36,61a	24,92b
S. heterogama	38,25a	28,27a
Control	28,57c	19,80c
V. C. (%)	2,76	3,26

Table 6. Reserve substances of shoots (leaves and stems) and roots of plants of the Okinawa cultivar rootstock inoculated with four AMF species (*Acaulospora sp., G. clarum, G. etunicatum* and *S. heterogama*), collected 180 days after sowing. Eldorado do Sul, RS, 2005. Means followed by the same letter in the column do not differ by the Duncan test at 5% significance.

For the shoot, plants inoculated with *G. clarum*, *G. etunicatum* and *S. heterogama* showed percentages of reserve substances statistically similar but superior to *Acaulospora sp.*, which in turn was higher than the control. In roots, the plants were inoculated with *S. heterogama* presented the greatest results, while those inoculated with *G. etunicatum* showed intermediate levels, statistically similar to *G. clarum* that, in turn, was similar to those inoculated with *Acaulospora sp.* and the control.

4. Discussion

4.1. Anatomy and morphology changes in roots

It was observed that inoculation with AMF reduced the cortex thickness of inoculated plants in both cultivars, associated to increase in most of the morphological parameters of the root xylem assessed for the Okinawa cultivar and for all of those of the Aldrighi cultivar (Tables 1 and 2).

The main effect of the AMF occurred on the metaxylem, that is, one of the categories of the primary xylem, whose conductor cells differentiate later and are larger in diameter (Costa et al., 2003) and also on the secondary xylem cells. On the other hand, the AMF not did not seem to exercise effect on the protoxylem, that are conductor cells of the primary xylem that differentiate first, that is, they acquire secondary lignin walls early (Apezzato-da-Glória & Hayashi, 2003) that reduce the possibility of the AMF acting on the growth of this category of cells of the primary xylem.

The decrease in the cortex area seems to be directly linked to the increase in the number of cells in the metaxylem and the secondary xylem of the plants inoculated with AMF. The control plants presented a smaller number of metaxylem and secondary xylem cells that were smaller in diameter compared to the cells of the inoculated plants, especially in the case of the treatments with the species *Acaulospora sp.* (cv. Okinawa) and *S. heterogama* (cv. Aldrighi).

The mycelia of endomycorrhizal fungi were extracted from roots of *Ophrys lutea* (Orchidaceae) and placed in culture medium kept and in the dark for three weeks (Barroso et al., 1986). Then the derivatives released in the culture medium were extracted, whose greatest concentrations consisted of indol-3-acetic acid and indol-3-ethanol acid, showing the ability of these fungi to synthesize hormones. The authors concluded that the identification of these compounds in the mycelia extract suggested the transference of these compounds from the fungi to the host plants in the phase when symbiosis was established.

Roots colonized by AMF presented an increase in auxin and cytokinin production that are involved in the increase or continuity of the growth of the conductor tissue cells, especially in the size and number of the cells of the metaxylem and the secondary xylem (Hirsch et al., 1997). According to the same authors, the establishment of symbiosis would lead to the production of biochemical signals that would activate genes involved in the production of these plant hormones, and thus the same signals would be responsible for the formation of the root nodes on legumes colonized by *Rhizobium sp*. These authors showed that in roots of the MN 1008 alfalfa mutant cultivar, that carries out the transcription of these signals in the absence of the symbiants, the responses of the root tissue were identical to those of roots of plants colonized both by AMF and *Rhizobium meliloti*.

Thus it can be inferred that the presence of AMF would favor the constant differentiation of the xylem tracheal elements, that coincides with the results obtained in this study for both the root stock cultivars.

There appear to be possible variable effects on root morphology, according to the AMF species and the plant species involved in the symbiosis that also influences the size and growth of the xylem cells, that was also observed in this study, because some species presented variable performance in increasing the size and number of cells, in function of the cultivar used, and in function of the AMF species used for the same cultivar (Atkinson et al., 1994). The species *G. fasciculatum* and *G. etunicatum* induced modification in the roots of the Elsanta and Cambridge Favorite strawberry cultivars, but did not cause any alteration in the morphology of the roots of the Rhapsody cultivar, shedding the variable effect of these AMF species on different cultivars of the same plant species (Norman et al., 1996).

Moreover, roots of plants were colonized by AMF may or may not show increases in longevity, depending on plant species and the fungi involved in symbiosis (Atkinson et al, 2003; Eissenstat et al., 2000; Hodge et al., 2000). However, the morphological attributes of the roots that may be affected by the AMF, such as roots and branches of the diameter of the conducting tissue, has a direct influence on increasing the longevity of roots (Wells et al., 2002). In addition to increasing longevity, root colonization by AMF provides a quick renewal of the root system, increasing the rate of substitution of roots that have collapsed.

4.2. Acquisition of nutrients and benefits

The increase in the absorption and transport volume of nutrients such as nitrogen, that is a constituent of proteins (Tedesco et al., 1995), phosphorus that is essential for a cell division and photosynthesis metabolism and potassium that acts on the electric equilibrium of the cells and on the stomata opening and closing (Tedesco et al., 1995), is vital for plant growth. This contributed to greater responses of the inoculated plants in terms of plant development that was observed in this study for both the cultivars, especially in the plants where there were the highest percentages of root colonization (Tables 3 and 4).

The AMF obtain carbohydrates from their host plants and provide nutrients, especially phosphate. In the case of phosphate, depending on the combination plant-fungus, the acquisition can be performed wholly or partly by the fungus (Smith et al., 2003). The metabolic pathway of nutrient acquisition starts with the uptake by hyphae-soil interface (Benedetto et al., 2005). In hyphae, the nutrient is transported to structures of the fungus in the roots (Ohtomo et al., 2005), where it is transferred to the plant via arbúsculo (Nagy et al.,

2005). The route of transfer of carbohydrates from the plant to the AMF follows the opposite direction (Nagy et al., 2005; Ohtomo et al., 2005).

The benefits given by the AMF on xylem development is associated to many action mechanisms of these fungi, that act directly or indirectly on the plants (Souza et al., 2000). One of the positive effects of the AMF is in function of the presence of the external mycelia, which play an important role in slow diffusion nutrient absorption, such as phosphorus and potassium (Minhoni & Auler, 2003; Souza et al., 2000; Tobar et al., 1994), increasing the nutritional content of the plants (An et al., 1993; Barea, 1991). Associated to this, the modifications caused by the AMF in the xylem structure, such as increase in the number and diameter of the metaxylem and secondary xylem cells, permitted a greater flow of nutrient absorption, such as nitrogen, phosphorus and potassium, translocated to the upper part of the plant, culminating in accelerated growth (Souza, 2000; Souza et al., 2000).

The fact that the AMF species induce major development parameters such as height, diameter and leaf area per plant provides greater photosynthesis and, consequently, a higher level of production of assimilates (Nunes et al., 2006). This report confirms the data obtained in this study with respect to the reserve substances of shoots of both cultivars, for all species used provided an increase in leaf area compared to control (Tables 3 and 4). There is also agreement with other authors, who found higher levels of reserve substances in the tissues of plants inoculated with AMF (Theodore et al., 2003; Sena et al., 2004, Souza et al., 2005).

Another fact to be noted is that only the AMF species that provided the greatest results for height, diameter and leaf area for both cultivars (*Acaulospora sp.* Okinawa cultivar, *S. heterogama* for Aldrighi cultivar and *G. etunicatum* for both cultivars) yielded significant differences in plant reserves, both to the tissues of shoots and roots to the tissues, compared to other treatments. Plants with greater height and leaf area, are more light-gathering capacity and production of assimilates, which allows a higher flow of carbohydrates into the root system where one part would be used by the AMF in its nutrition and accumulation in structures buffer (vesicles, where *Acaulospora sp.* and *G. etunicatum*), and the rest would be accumulated in the storage tissue of the plant in the form of reserve substances (Souza et al., 1999; Scatena & Scremin-Dias, 2003). Moreover, the larger diameter provided by the AMF, would increase the upward flow of water and nutrients, and sap formulated in the downward (Mazzoni-Viveiros & Trufem, 2004).

5. Conclusion

Plants inoculated with AMF have changes in the morphological structure of the roots, such as reduction of the cortex and increased the number and size of cells of the metaxylem, which provides greater volume of water and nutrients translocated to the top of the plant. This benefits plants, accelerating its vegetative growth, improving the content of macronutrients and allowing the production and accumulation of assimilates.

Author details

José Luis da Silva Nunes
BADESUL Desenvolvimento, Brasil

Paulo Vitor Dutra de Souza, Gilmar Arduino Bettio Marodin
and Jorge Ernesto de Araújo Mariath
Universidade Federal do Rio Grande do Sul, Brasil

José Carlos Fachinello
Universidade Federal de Pelotas, Brasil

Acknowledgement

To Ministério da Agricultura, Pecuária e Abastecimento (MAPA) and Conselho Nacional de Desenvolvimento Científico e Tecnológico (CNPq) for research support and grants of the authors.

6. References

An Z.Q., Shein T., Wang H.G. (1993). Mycorrhizal fungi in relation to growth and mineral nutrition of apple seedlings. *Scientia Horticulturae,* Vol.54, Nº 4 (July 1993), pp. 275 – 285, ISSN 0304-4238

Appezzato-da-Glória B., Hayashi A.H. (2003). Raiz. In: *Anatomia Vegetal,* Appezzato-da-Glória B., Carmello-Guerreiro S.M. (eds), pp. 267 – 287, UFV, ISBN 85-7269-240-1, Viçosa, Brasil

Atkinson D., Berta G., Hooker J.E. (1994). Impact of mycorrhizal colonization on root archicture, roots longevity and the formation of growth regulators. In: *Impact of Arbuscular Mycorrhizas on Sustainable Agriculture and Natural Ecosystems,* Gianinazzi S., Schüepp H. (eds), pp. 89 – 99, Birkhäuser Verlag, ISBN 3-7643-5000-8, Basel, Switzerland

Atkinson, D., Blanck, K.E., Forbes, P.J., Hooker, J.E., Baddeley, J.A., Watson, C.A. (2003). The influence of arbuscular mycorrhizal colonization and environment on root development in soil., Vol.54, Nº 4 (December 2003), pp. 751 – 757, ISSN 1365-2389

Barea J.M. (1991). Vesicular-arbuscular mycorrhizae as modifiers of soil fertility. In: *Advances in Soil Science,* STEWART, B.S. (ed), pp. 01 – 40, Springer-Verlag, ISBN 0-3879-7354-0, New York, USA

Barroso J., Neves H.C., Pais M.S. (1986). Production of indole-3-ethanol and indole-3-acetic acid by the mycorrhizal fungus of *Ophrys lutea* (Orchidaceae). *New Phytologist,* Vol.103, Nº 4 (December 1986), pp. 745 – 749, ISSN 1469-8137

Benedetto, A., Magurno, F., Bonfante, P., Lanfranco, L. (2005). Expression profiles of a phosphate transporter gene (GmosPT) from the endomycorrhizal fungus *Glomus mosseae. Mycorrhiza,* Vol.15, Nº 8 (December 2005), pp. 620 – 627, ISSN: 0940-6360

Berta G., Trotta A., Fusconi A., Hooker J.E., Munro M., Atkinson P., Giovannetti M., Morini S., Fortuna P., Tisseranti B., Gianinazzi-Pearson V., Gianinazzi S. (1995). Arbuscular mycorrhizal induced changes to plant growth and root system morphology in *Prunus cerasifera*. *Tree Physiology*, Vol.15, Nº 5 (May 1995), pp. 281 – 293, ISSN 0829-318X

Bressan W., Vasconcellos C.A . (2002). Alterações morfológicas no sistema radicular do milho induzidas por fungos micorrízicos e fósforo. *Pesquisa Agropecuária Brasileira*, Vol.37, Nº 4 (Abril 2002), pp. 509 – 517, ISSN 0100-204X

Cooper K.M. (1984). Physiology of VA Mycorrhizae associations. In: *VA Mycorrhiza*. Powel CL, Bagyaraj J (eds), pp. 155 – 186, CRC, ISBN 08-493-569-46, Boca Raton, USA

Costa C.G., Callado C.H., Coradin V.T.R., Carmello-Guerreiro S.M. (2003). Xilema. In: *Anatomia Vegetal*, Appezzato-da-Glória B., Carmello-Guerreiro S.M. (eds), pp. 129 – 154, UFV, ISBN 85-7269-240-1, Viçosa, Brasil

Dodd, J. C. (2000) The role of arbuscular mycorrizal fungi in agro – and natural ecosystems. *Outlook on Agriculture*, Vol.29, Nº 1 (March 2000), p. 55 – 62, ISSN 0030-7270

Duncan, D.B. (1955). Multiple range and multiple F tests. *Biometrics*, Vol.11, Nº 1 (March 1955), pp. 1- 42, ISSN 1947-2006

Eissenstat, D.M., Wells, C.E., Yanai, R.D., Whitbeck, V.L. (2000) Building roots in a changing environment: implications for root longevity. *New Phytologist*, Cambridge, Vol.147, Nº 1 (July 2000), pp. 33 – 42, ISSN 1469-8137

Hirsch A.M., Fang Y., Asad S., Kapulnik Y. (1997). The role of phytohormones in plant-microbe symbioses. *Plant and soil*, Vol.194, Nº 2 (January 1997), pp. 171 – 184, ISSN 0032-079X

Hodge, A., Robinson, D., Fitter, A.H. (2000). An arbuscular mycorrhizal inoculum enhances root proliferation in, but not nitrogen capture from, nutrient-rich patches in soil. *New Phytologist*, Cambridge, Vol.147, Nº 3 (September 2000), pp. 575 - 584, ISSN 1469-8137

Johansen D.A. (1940). *Plant microtechnique*. McGraw-Hill, ISBN 007592, New York, USA

Kothari B.K., Maschner, H., George, E. (1990) Effect of VA mycorrhizal fungi and rhizosphere microorganisms on root and shoot morphology, growth and water-relations in maize. *New Phytologist*, Vol.116, Nº 2 (October 1990), p. 303 – 311, ISSN 1469-8137

Marschner, H. (1995) *Mineral nutrition of higher plants*. Academic Press, ISBN 978-0124735415, San Diego, USA

Mazzoni-Viveiros, S.C., Trufem, S.F.B. (2004) Efeitos da poluição aérea e edáfica no sistema radicular de *Tibouchina pulchra* Cogn. (Melastomataceae) em área de mata Atlântica: associações micorrízicas e morfologia. *Revista Brasileira de Botânica*, Vol.27, Nº 2 (Abril/ Junho 2004), pp. 337 – 348, ISSN 0100-8404

Minhoni M.T.A., Auler P.A.M. (2003) Efeito do fósforo, fumigação do substrato e fungo micorrízico arbuscular sobre o crescimento de plantas de mamoeiro. *Revista Brasileira de Ciência do Solo*, Vol.27, Nº 5 (Outubro 2003), pp. 841 – 847, ISSN 0100-0683

Moreira F.M.S., Siqueira J.O. (2002) *Microbiologia e bioquímica do solo*. Editora UFLA, ISBN 85-8769-233-X, Lavras, Brasil

Nagy, R., Karandashov, V., Chague, V., Kalinkevich, K.; Tamasloukht, M., Xu, G., Jakobsen, I., Levy, A.A., Amrhein, N., Bucher, M. (2005). The characterization of novel mycorrhiza-specific phosphate transporters from *Lycopersicon esculentum* and *Solanum tuberosum* uncovers functional redundancy in symbiotic phosphate transport in solanaceous species. *Plant Journal*, Vol.42, Nº 2 (April 2005), pp. 236 – 250, ISSN 1365-313X

Norman J.R., Atkinson D., Hooker J.E. (1996). Arbuscular mycorrhizal-fungal-induced alteration to root architecture in strawberry and induced resistance to the pathogen *Phytophthora fragariae*. *Plant and Soil*, Vol.185, Nº 2 (September 1996), pp. 191 – 198, ISSN 0032-079X

Nunes M.S., Soares, A.C.F., Soares Filho, W.S., Lêdo, C.A.S. (2006). Colonização micorrízica natural de porta-enxertos de citros em campo. *Pesquisa Agropecuária Brasileira*, Vol.41, Nº 3 (Março 2006), pp. 525 – 528, ISSN 0100-204X

Ohtomo, R., Saito, M. (2005). Polyphosphate dynamics in mycorrhizal roots during colonization of an arbuscular mycorrhizal fungus. *New Phytologist*, Vol. 167, Nº 2 (August 2005), pp. 571 – 578, ISSN 1469-8137

Panstruga, R. (2003). Establishing compatibility between plants and obligate biotrophic pathogens. *Current Opinions in Plant Biology*, Vol.6, Nº 4 (August 2003), p. 320-326, ISSN: 1369-5266

Paszkowski, U. (2006). Mutualism and parasitism: the yin and yang of plant symbioses. *Current Opinions in Plant Biology*, Vol.9, Nº 4 (August 2006), p.364 – 370, ISSN: 1369-5266

Phillips J.M., Hayman D.S. (1970). Improved procedures for clearing roots and staining parasitic and vesicular-arbuscular mycorrhizal fungi for rapid assessment of infection. *Transactions of the British Mycological Society*, Vol.55, Nº 1 (January 1970), pp. 157-160, ISSN 0007- 1536

Priestley, G.A. (1965) New method for the estimation of the resources of apple tress. *Journal of the Science of Food and Agriculture*, Vol.16, Nº 12 (December 1965), pp. 717 – 721, ISSN 1097-0010

Scatena, V.L., Scremin-Dias, E. (2003). Parênquima, Colênquima e Esclerênquima. In: *Anatomia Vegetal*, Appezzato-da-Glória B., Carmello-Guerreiro S.M. (eds), pp. 109 – 127, UFV, ISBN 85-7269-240-1, Viçosa, Brasil

Sena, J.O.A., Labate, C.A., Cardoso, E.J.B.N. (2004) Caracterização fisiológica da redução de crescimento de mudas de citros micorrizadas em altas doses de fósforo. *Revista Brasileira da Ciência do Solo*, Vol.28, Nº 5 (Setembro/ Outubro 2004), pp. 827 – 832, ISSN 0100-0683

Siqueira J.O., Lambais M.R., Stürmer S.L. (2002) Fungos micorrízicos arbusculares. *Biotecnologia, Ciência & Desenvolvimento*, Vol.25 (Março/Abril 2002), pp. 12 – 21, ISSN 1414-4522

Silva, L.M.S.; Alquini, Y.; Cavallet, V.J. (2005) Inter-relações entre a anatomia vegetal e a produção vegetal. *Acta Botanica Brasilica*, Vol.19, Nº 1 (Janeiro/ Março 2005), p. 183 – 194, ISSN 0102-3306

Souza, F.A., Trufem, S.F.B., Almeida, D.L., Silva, E.M.R., Guerra, J.G.M. (1999) Efeito de pré-cultivos sobre o potencial de inoculo de Fungos Micorrízicos Arbusculares e produção de mandioca. *Pesquisa Agropecuária Brasileira*, Vol.34, Nº 10 (Outubro 1999), pp. 1913 – 1923, ISSN 0100-204X

Souza P.V.D., Agustí M., Abad M., Almela V. (2000). Desenvolvimento vegetativo e morfologia radicular de Citrange Carrizo afetado por ácido indolbutírico e micorrizas arbusculares. *Ciência Rural*, Vol.30, Nº 2 (Março/Abril 2000), pp. 249 – 255, ISSN 0103-8478

Souza P.V.D. (2000). Interação entre micorrizas arbusculares e ácido giberélico no desenvolvimento vegetativo de plantas de Citrange Carrizo. *Ciência Rural*, Vol.30, Nº 5 (Setembro /Outubro 2000), pp. 783 – 787, ISSN 0103-8478

Souza, P.V.D., Carniel, E., Schimitz, J.A.K., Silveira, S.V. (2005) Influência de substratos e fungos micorrízicos arbusculares no desenvolvimento do porta-enxerto Flying Dragon (*Poncirus trifoliata*, var. monstruosa Swing.). *Revista Brasileira de Fruticultura*, Vol.27, Nº 2 (Agosto 2005), pp. 285 – 287, ISSN 0100-2945

Smith, S.E., smith, F.A., Jakobsen, I. (2003) Mycorrhizal fungi can dominate phosphate supply to plants irrespective of growth responses. *Plant Physiology*, Vol.133, Nº 2 (October 2003), pp. 16–20, ISSN 0032-0889

Sylvia, D. M., Chellemi, D. O. (2001) Interations among root-inhabiting fungi and their implications for biological control of root pathogens. *Advances in Agronomy*, Vol.73, Nº 1 (April 2001), pp. 1 – 33, ISBN 978-0-12-000773-8

Tedesco M.J., Gianello C., Bissani C.A., Bohnen H., Volkweiss S.J. (1995). *Análises de solo, plantas e outros materiais (Boletim Técnico, 5)*, UFRGS/Departamento de solos, ISBN 000-148837, Porto Alegre, Brasil

Theodoro, V.C.A., Alvarenga, M.I.N., Guimarães, J., Mourão Junior, M. (2003) Carbono da biomassa microbiana e micorriza em solo sob mata nativa e agroecossistemas cafeeiros. *Acta Scientiarum: Agronomy*, Vol.25, Nº 1 (Maio 2003), pp. 147 – 153, ISSN 1679-9275

Tobar R., Azcón R., Barea J.M. (1994) Inproved nitrogen uptake and transport from ^{15}N-labelled nitrate by external hyphae of arbuscular mycorrhiza under water stressed condictions. *New Phytologist*, Vol.126, Nº 1 (January 1994), pp. 119 – 122, ISSN 1469-8137

Tokeshi, H. (2000) Doenças e pragas agrícolas geradas e multiplicadas pelos agrotóxicos. *Fitopatologia Brasileira*, Vol.25 (Janeiro 2000), pp. 264-270, ISSN 0100-4158

Wells, C.E., Glenn, D.M., Eissenstat, D.M. Changes in the risk of fine-root mortality with age: a case study in peach, *Prunus persica* (Rosaceae). *American Journal of Botany*, Vol.89, Nº 1 (January 2002), pp. 79 – 87, ISSN 0002-9122

Wilcox, H. E. (2002). Mycorrhizae. In: *Plants roots*, Waisel, Y., Eshel, A., Kafkafi, U. (eds), Marcel Dekker, ISBN 0-8247-0631-5, New York, USA

Chapter 19

Plant Responses to Sugar Starvation

Iwona Morkunas, Sławomir Borek, Magda Formela and Lech Ratajczak

Additional information is available at the end of the chapter

http://dx.doi.org/10.5772/51569

1. Introduction

The production of carbohydrates via photosynthesis is the most fundamental activity in plant life. Carbohydrate synthesis, transport, utilization, and storage are dynamic processes, strongly dependent on cell physiology, plant organ, environmental conditions, and developmental stage of the plant. The plant's ability to monitor and respond to the level of carbohydrates may act as a controlling mechanism, integrating the influence of environmental conditions (e.g. light, nutrients, biotic and abiotic stress factors) with internal developmental programs, controlled directly by hormones (Koch 2004, Rolland et al. 2006, Hammond and White 2008, Loreti et al. 2008, Ramon et al. 2008, Agulló-Antón et al. 2011). Studies conducted in recent years have provided an extensive body of information on the participation of carbohydrates in metabolic reactions of plant cells. In plants, sugars are essential as respiratory substrates for the generation of energy and metabolic intermediates that are then used for the synthesis of macromolecules. Binding to sugar is required for proper functioning of many proteins and lipids. Moreover, carbohydrates have important hormone-like functions as physiological signals, which cause activation or repression of many plant genes, and this in turn leads to specific metabolic effects. Sugar-signaling networks have the ability to regulate directly the expression of genes and to interact with other signaling pathways. The progress in research on molecular mechanisms of sugar sensing and signaling in plants shows that signal molecules include glucose, fructose, sucrose, and trehalose (Jang and Sheen 1994, Koch 1996, Müller et al. 1999, Rolland et al. 2002, Koch 2004, Gibson 2005, Gonzali et al. 2006, Rolland et al. 2006, Ramon et al. 2008, Rosa et al. 2009, Cho and Yoo 2011). Plants have developed effective mechanisms of perception and transduction of sugar signals. Perception of the sugar signal may take place already in the apoplast during transport across membranes or within the cell, in the cytosol. Sugar signal perception and transduction may involve cell wall invertases (CW-INV), sucrose and glucose transporters (and specific sugar receptors), and hexokinase (HXK) (Sheen et al. 1999, Smeekens 2000, Loreti et al. 2001, Rolland et al. 2002, Harrington and Bush 2003, Moore et al. 2003, Sherson et al. 2003, Koch 2004, Rolland and Sheen 2005,

Rolland et al. 2006, Ramon et al. 2008, Cho et al. 2009, Hanson and Smeekens 2009, Smeekens et al. 2010). Additionally, biochemical studies provide evidence for the involvement of a variety of protein kinases, i.e. Snf1-related kinases (SnRKs) (Rolland et al. 2006, Smeekens et al. 2010), calcium-dependent protein kinases (CDPKs), mitogen-activated protein kinases (MAPKs), and protein phosphatases, 14-3-3 proteins, Ca^{2+} ions as a second messenger, and G-proteins (Rolland et al. 2006) in sugar signal transduction (Rolland et al. 2002, Sinha et al. 2002). However, it must be stressed that HXK plays a significant role as a component of the sugar sensing machinery. Genetic analyses have revealed a central role for HXK as a conserved glucose sensor (Moore et al. 2003, Harrington and Bush 2003, Rolland and Sheen 2005, Rolland et al. 2006, Ramon et al. 2008, Cho et al. 2009, Hanson and Smeekens 2009, Smeekens et al. 2010). HXK sensing and signaling functions are probably dependent on HXK's subcellular localization, translocation, and/or interactions with downstream effectors. The HXK sugar sensor, as a cytosolic protein or associated with mitochondria or other organelles, then could activate a signaling cascade through HXK-interacting proteins (HIPs) or affect transcription directly after nuclear translocation (Rolland et al. 2002, Rolland et al. 2006, Hanson and Smeekens 2009). Various sugar signals activate many HXK-dependent and HXK-independent pathways and use different molecular mechanisms to control transcription, translation, protein stability, and enzymatic activity. It has been shown that in 7-day-old *Arabidopsis* seedlings, glucose (3%) during 6 h increased the expression of 983 genes and decreased the expression of 771 genes (Li et al. 2006). Sucrose (3%) in *Arabidopsis* cell suspension during 48 h modulated the expression of 243 genes in light and 193 genes in darkness (Nicolaï et al. 2006). Glucose-regulated transcription factors (TFs) account for 8.3% (82 factors among 978) of all glucose-regulated genes, where most of TFs showed repression (Price et al. 2004).

Genes encoding the enzymes of the phenylpropanoid biosynthesis pathway (Hara et al. 2003, Solfanelli et al. 2006, Morkunas et al. 2011) as well as sink-specific enzymes, such as sucrose synthase (Salanoubat and Belliard 1989), granule-bound starch synthase (Visser et al. 1991), and extracellular invertase (Roitsch et al. 1995), are induced by sucrose or glucose. Sugar-induced gene expression has also been detected for enzymes involved in pathogen and stress response, such as proteinase inhibitor II of potato (Johnson and Ryan 1990), chalcone synthase (Tsukaya et al. 1991), and flavonoid biosynthetic enzymes (Morkunas et al. 2011). In contrast, sugar repression of photosynthetic genes was observed. For example, genes encoding photosynthetic proteins, e.g. the small subunit of the Calvin cycle enzyme ribulose bisphosphate carboxylase/oxygenase (Rubisco) and the chlorophyll *a* binding protein, are repressed by carbohydrates (Sheen 1990, Krapp et al. 1993).

Due to their regulatory and signal function, sugars affect all phases of the plant life cycle by controlling the number of essential metabolic processes. In animals, the level of sugar in cells is strictly controlled, but in plants the level and composition of carbohydrates varies widely, depending on tissue type and environmental conditions. An excess or loss of carbohydrates or their derivatives triggers various reactions in plants and significantly affects the metabolism, growth, and development. Moreover, all abiotic and biotic stress responses are regulated, at least in part, by sugars (Koch 1996, Rolland et al. 2002). Plants are generally considered to be autotrophs but sometimes they can be heterotrophs, e.g. at some stages of development (such as seed germination, as long as the seedling grows in darkness, before it emerges above the soil surface) and in non-photosynthetic organs, such as most roots, stems, and flowers. In germinating seeds, in periods of unfavorable environmental conditions or after too deep sowing, mobilization of storage materials in cotyledons may be delayed or there may be some disturbances in distribution of carbohydrates. This may lead to a decrease in available carbohydrates in embryo axes (i.e. sugar starvation) and in a lower seed germination rate. Additionally, high carbohydrate losses can be observed in most plant species under the influence of environmental conditions, such as water or temperature, attack of pathogens or herbivores. These factors may lead to a remarkable decrease in photosynthetic rate in donor tissues (i.e. leaves that synthesize and export carbohydrates), and this reduces the supply of carbohydrates to acceptor tissues (i.e. non-photosynthetic tissues importing carbohydrates for respiration, growth, and development). Besides, in some conditions, e.g. during dormancy or leaf shedding, photosynthesis is switched off or slowed down. In such conditions, the stored carbohydrates must be used, so their reserves may be greatly diminished in non-photosynthetic tissues then. Knowledge of the response to sugar starvation and of adaptive mechanisms in plants is both fundamental and agronomically important.

2. Materials and methods of research on sugar starvation

Carbohydrate starvation has been studied in many of plant species, e.g. in common wheat (Wittenbach 1977, Wittenbach et al. 1982), maize (Saglio and Pradet 1980, Pace et al. 1990), barley (Farrar 1981), pearl millet (Baysdorfer et al. 1988), pea (Webster and van't Hof 1973, Sahulka and Lisa 1978, Webster and Henry 1987, Morkunas et al. 1999, 2000), soybean (Kerr et al. 1985, Walsh et al. 1987), sycamore (Journet et al. 1986, Dorne et al. 1987, Roby et al. 1987, Genix et al. 1990), tobacco (Moriyasu and Ohsumi 1996), lupine (Morkunas et al. 1999, Borek and Ratajczak 2002, Morkunas et. al. 2003, Borek et al. 2006, Borek and Nuc 2011, Borek et al. 2011, 2012a, 2012b), *Arabidopsis thaliana* (Rose et al. 2006). Metabolic changes caused by sugar starvation are usually analysed in cell cultures, callus tissue cultures, and in isolated plant organs cultured *in vitro*. In all types of the cultures it is easy to control nutrition of the plant material by manipulating the medium composition. Cell cultures clearly react by metabolic changes when sugar is omitted from the nutrient medium, e.g. in sycamore (*Acer pseudoplatanus*) cells (Journet et al. 1986, Gout et al. 2011), but similar sensitivity is observed also in excised root tips (Brouquisse et al. 1991, 1992). *In vitro* cultured embryo axes (isolated from seeds), cut off from the natural source of nutrients, i.e. cotyledons, are a good model for research on the effect of sugars on seedling development and metabolism (Borek et al. 2001, Borek and Ratajczak 2002, Borek et al. 2003, 2006, Borek and Nuc 2011, Borek et al. 2012a, 2012b), or even for research on the role of sugar in plant resistance to pathogens (Morkunas et al. 2005, Morkunas and Gmerek 2007, Morkunas et al. 2007, Morkunas and Bednarski 2008, Morkunas et al. 2008, 2011). The influence of sugar on leaf senescence was assessed on excised leaves and on leaves of intact plants. The effect of sugar starvation was achieved by keeping the plant material in darkness or by its spraying with photosynthetic inhibitors (Mohapatra et al. 2010). To examine the effects of sugar on senescence of various flower parts, cut flowers were treated with various sucrose concentrations in the vase solution (Azad et al. 2008, Arrom and Munné-Bosch 2012). Brouquisse et al. (1998) show that in roots (the main sink organs of maize plants) the consequences of carbon depletion induced by extended darkness are identical to those observed in the excised root tip, but some aspects of sugar starvation (e.g. effect on translocation of assimilates, links with the metabolism of nitrogen and phosphorus, interactions with plant hormones) were investigated in intact plants (Ciereszko et al. 2005). By spraying of sugar-starved plants with various sugars, it can be determined which of them are signal molecules and which of them are sugar sensors (Lothier et al. 2010).

Particularly valuable information is obtained as a result of research on various types of mutants, which sometimes supplies surprising information about the role of sugars as signal molecules (Gibson 2005, Ramon et al. 2008, Usadel et al. 2008, Hanson and Smeekens 2009).

Plant reactions to sugar deficits have been studied with the use of a wide range of research methods: electron microscopy (Borek and Ratajczak 2002, Borek et al. 2006, 2011, 2012a), confocal microscopy (Morkunas and Bednarski 2008, Morkunas et al. 2011, 2012), enzymatic activity assays and metabolite assays, e.g. by using HPLC (Morkunas et al. 2010), nuclear 1HNMR spectroscopy (Brouquisse et al. 2007, Kim et al. 2007), and ^{13}C and ^{31}P-NMR spectroscopy (Vauclare et al. 2010, Gout et al. 2011). Electron paramagnetic resonance (EPR) spectroscopy is applied to measure free radicals (Morkunas et al. 2004, Morkunas and Bednarski 2008, Morkunas et al. 2008, 2012). Novelties include various methods of molecular biology: assays of gene expression (Buchanan-Wollaston et al. 2005) and application of reporter genes (Lee et al. 2007). Moreover, sugar-starvation-induced promoters have been used to obtain useful recombinant proteins (Xu et al. 2011a, 2011b).

3. Morphological, anatomical, and ultrastructural changes under sugar starvation

Sugar starvation causes changes in growth, as shown in organ cultures *in vitro* and in cell suspensions. For example, significant differences are observed in morphological structure of embryo axes of narrowleaf lupine (*Lupinus angustifolius* L.), white lupine (*Lupinus albus* L.), yellow lupine (*Lupinus luteus* L.), Andean lupine (*Lupinus mutabilis* Sweet) and garden pea (*Pisum sativum* L.) cultured *in vitro* under sugar starvation and with 60 mM sucrose. Radicles of *Pisum* and *Lupinus* spp. cultured under sugar starvation were shorter and thicker than sugar-fed radicles. The length and fresh weight of starved isolated embryo axes was about 2-fold lower than those of sugar-fed axes (Morkunas 1997, Borek et al. 2012a). In contrast to lupine isolated embryo axes, lupine seedlings were much bigger when they were grown on medium without sucrose (Borek et al. 2012a).

Light micrographs of radicle cross sections revealed that starved radicles were thicker than sugar-fed radicles. Cortical cells of sugar-fed embryo axes were rounded, with large intercellular spaces. Many of the observed cells were dividing or have just divided. The cells lacked a central vacuole; instead, they contained very numerous small vacuoles. The major feature distinguishing cortical cells of starved lupine (as compared to sugar-fed lupine) was the presence of a large central vacuole with marginal cytoplasm.

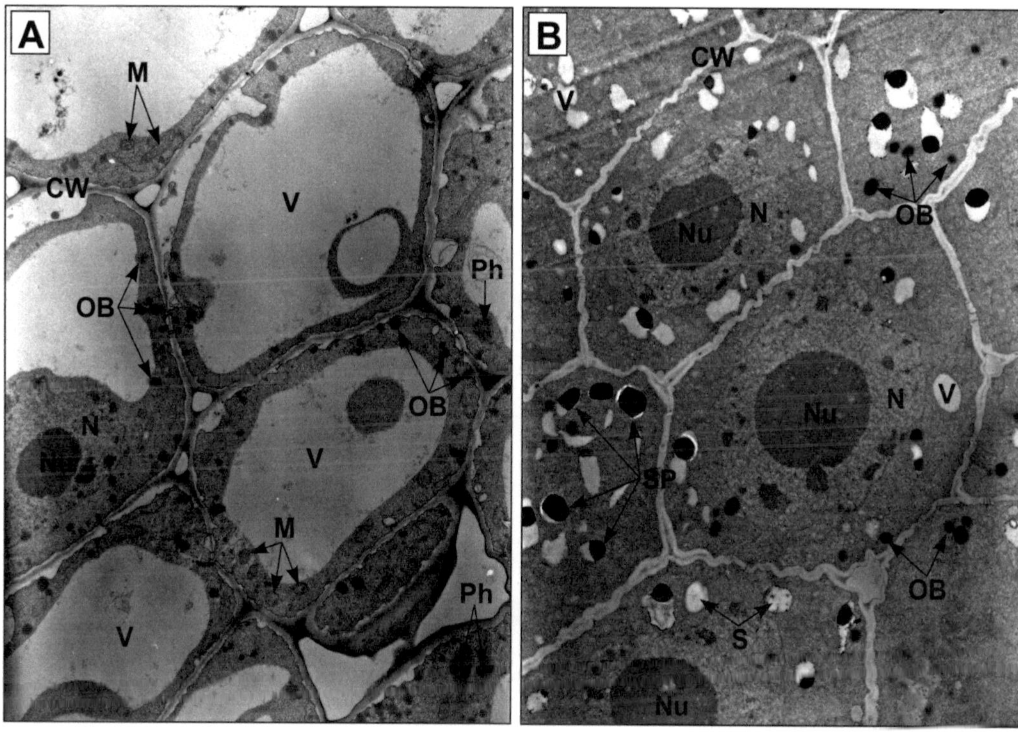

Figure 1. Ultrastructure of root meristematic zone cells of yellow lupine embryo axes grown for 96 h *in vitro* on medium without sucrose (A; sugar starvation) and on medium with 60 mM sucrose (B). CW cell wall, M mitochondrion, N nucleus, Nu nucleolus, OB oil body, Ph phytoferritin, S starch, SP storage protein, V vacuole.

Ultrastructural studies of the root meristematic zone of sucrose-starved lupine embryo axes showed that (in contrast to cells of embryo axes fed with sucrose) their cytoplasm, along with endoplasmic reticulum, mitochondria, plastids, ribosomes, and nuclei, were forced to the periphery by growing vacuoles (Fig. 1A, B). In spite of gradual autolysis of cytoplasmic proteins, in starved cells mitochondria were protected as organelles maintaining cellular respiration. Electron micrographs reveal that the inner mitochondrial membrane in starved pea cells is well developed, forming numerous cristae (Fig. 2A), in contrast to sucrose-fed cells, with less developed cristae (Fig. 2B). Oxygraphic studies of mitochondria isolated from starved embryos of *Pisum sativum* and *Lupinus angustifolius* have shown that they are active and exercise respiratory control (Morkunas et al. 2000, 2003). The phenomenon of mitochondria protection against autolysis under sugar starvation was observed also in other species and plant tissues (Baysdorfer et al. 1988, Brouquisse et al. 1991). However, Couée et al. (1992) reported that under conditions of carbohydrate starvation, mitochondria could be degraded. Those authors revealed heterogeneity of mitochondria isolated from root tips of maize (*Zea mays*) and fractionated on Percoll density gradients, i.e. higher- and lower-density mitochondria. The higher-density mitochondria from glucose-starved maize root tips retained the ultrastructure and most of the respiratory properties of non-starved mitochondria. By contrast, lower- and intermediate-density mitochondria were absent in the mitochondrial fractions from glucose-starved maize root tips and were not detected *in situ*. Interestingly, in plastids of starved pea cells, large amounts of phytoferritin were accumulated (Fig. 3A), while only starch grains and tubular structures were found in plastids of cells with a high level of sucrose (Fig. 3B) (Morkunas 1997). Lowering carbohydrate level in storage tissues of germinating seeds enhances the breakdown of reserves. In cotyledons of 4-day-old lupine seedlings and in excised lupine cotyledons grown on medium without sucrose, the deposits of storage protein were clearly smaller than in organs fed with sucrose. Starch granules and oil bodies were also smaller and less numerous (Fig. 4A, B). Similarly, the cell walls, containing hemicelluloses (important storage compounds in lupine seeds) were thinner (Borek et al. 2006, 2011, 2012a).

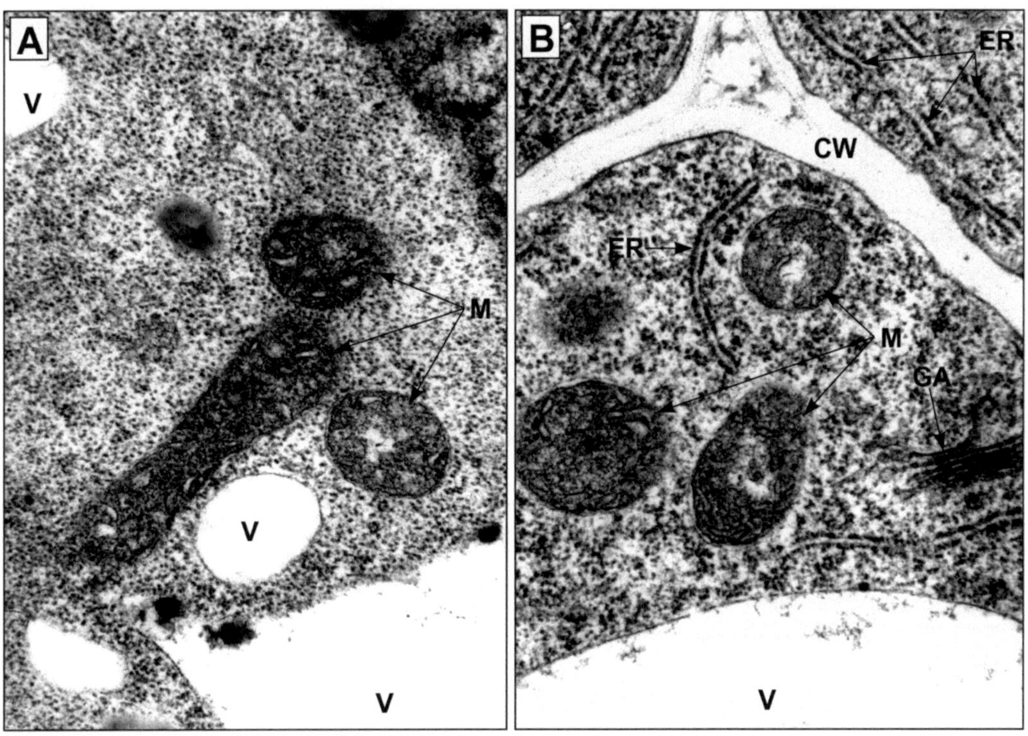

Figure 2. Mitochondria of pea embryo axes grown for 72 h *in vitro* on medium without sucrose (A; sugar starvation) and on medium with 60 mM sucrose (B). CW cell wall, ER endoplasmic reticulum, GA Golgi apparatus, M mitochondrion, V vacuole.

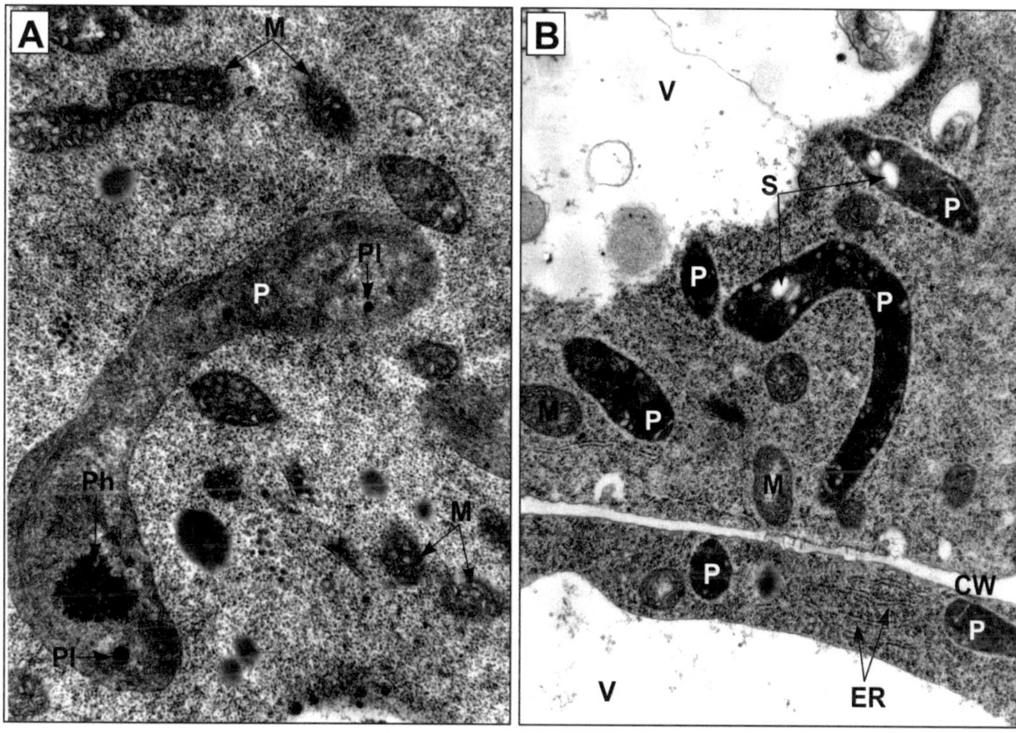

Figure 3. Plastids of pea embryo axes grown for 72 h *in vitro* on medium without sucrose (A; sugar starvation) and on medium with 60 mM sucrose (B). CW cell wall, ER endoplasmic reticulum, M mitochondrion, P plastid, Ph phytoferritin, Pl plastoglobule, S starch, V vacuole.

Research conducted on *Arabidopsis* cell suspension revealed that sugar starvation caused an immediate arrest of cell growth, together with a rapid degradation of cellular proteins. Cell divisions, as indicated by cell number and accumulation of biomass, were stopped immediately after the initiation of sucrose starvation. Observations of the morphology of starved cells using Nomarski interference microscopy and confocal laser-scanning microscopy revealed that in sucrose-free medium, the width of the cytoplasm was markedly decreased, while the vacuole was enlarged (Rose et al. 2006). Carbon starvation induced autophagy in plant cells, as confirmed by experiments on model systems, e.g. on cell suspensions of sycamore (Journet et al. 1986), rice (Chen et al. 1994), *Arabidopsis* (Rose et al. 2006) and in pea and lupine embryo axes grown *in vitro* (Morkunas 1997, Morkunas et al. 2003, Borek et al. 2011, 2012a, 2012b). Vacuolar autophagy starts as early as several hours after sugar starvation. Sequestration of portions of the cytoplasm with organelles (except for the nucleus) by endomembranes results in formation of autophagic vacuoles, followed by their fusion with other vacuoles. During periods of nutrient starvation, the autophagic process can be reinitiated in plant cells that are already vacuolated (Chen et al. 1994, Aubert et al. 1996, Moriyasu and Ohsumi 1996, Moriyasu and Klionsky 2003, Moriyasu et al. 2003). Carbohydrate starvation-induced autophagy is associated with an increase in intracellular proteolysis (James et al. 1996, Moriyasu and Ohsumi 1996, Thompson and Vierstra 2005) and a marked degradation of membrane polar lipids (Aubert et al. 1996, Inoue and

Moriyasu 2006). In addition, Rose et al. (2006) revealed that concomitantly, the number of transvacuolar strands is decreased dramatically. Starvation-induced acidic compartments were most frequent in the dense perinuclear cytoplasm, close to the large central vacuole. In vacuoles of starved cells, quinacrine was strongly accumulated. The role of autophagy in adaptation to sugar starvation has been investigated primarily because of its impact on basal metabolism and biomass production. The intracellular degradation of cytoplasmic components by means of autophagy produces amino acids, phospholipids, and other elements that are necessary for basic metabolism and essential biosynthetic pathways. In starved cells the initiation of autophagy is needed to obtain respiratory substrates (Journet et al. 1986, Chen et al. 1994, Aubert et al. 1996).

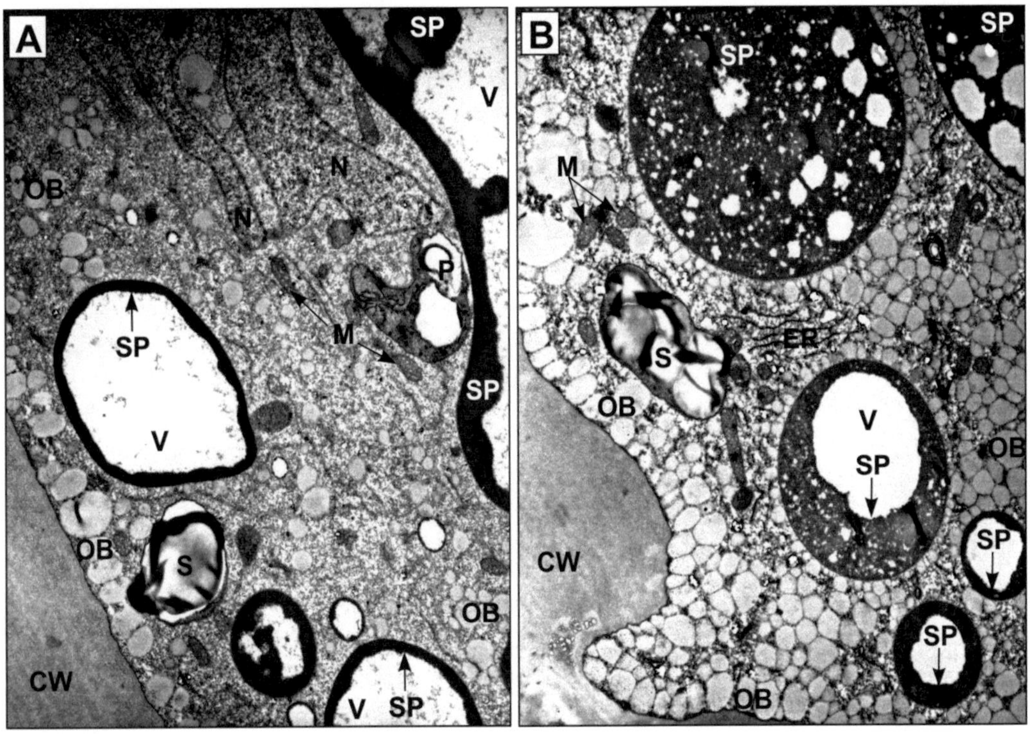

Figure 4. Ultrastructure of parenchyma cells of white lupine excised cotyledons grown for 96 h *in vitro* on medium without sucrose (A) and on medium with 60 mM sucrose (B). CW cell wall, ER endoplasmic reticulum, M mitochondrion, N nucleus, Nu nucleolus, OB oil body, Ph phytoferritin, S starch, SP storage protein, V vacuole.

4. Plant metabolism under sugar starvation

The studies have shown that in most cases, sugar starvation triggers a specific sequence of events in plant cells. The cells subjected to sugar starvation at first adapt to the lack of carbohydrates through gradual replacement of carbohydrate metabolism by protein and lipid metabolism. Such metabolic reorganization may result in autophagy (Saglio and Pradet 1980, Journet et al. 1986, Brouquisse et al. 1991, Rose et al. 2006). Brouquisse et al.

(1992) report that during sugar starvation of maize root meristems, 3 phases can be distinguished:

1. acclimation (from 0 to 30-35 h), when cellular carbohydrate levels and respiration rate decrease, while nitrogen is released from storage proteins through their degradation;
2. survival phase (from 30-35 to 90-100 h), involving intensive breakdown of proteins and lipids, release of P_i phosphorylcholine, and free amino acids; this phase can be reversed by sugar feeding;
3. cell disorganization (more than 100 h), when the level of all metabolites and enzymatic activity are significantly decreased and the changes are irreversible, leading to death.

During sugar starvation at the acclimation and survival phases, the total protein content decreases. This is associated with a temporary increase in free amino acids and an increase in proteolytic activity (Tassi et al. 1992, James et al. 1993, Moriyasu and Ohsumi 1996, Borek and Ratajczak 2002). *In vitro* studies of embryo axes of lupines and garden pea have revealed a dramatic decrease in protein concentration (especially the soluble fraction) in starved cells (Morkunas 1997, Borek et al. 2012a). The decrease in cytosolic proteins in starved cells was correlated with an increase in the activity of proteolytic enzymes, i.e. endo- and exopeptidases (Morkunas et al. 1999, Borek and Ratajczak 2002). The increased activity of proteases could result from activation of enzymatic proteins that were already present in cells but in a bound or inactive form. Their increased activity could also result from *de novo* synthesis of the enzymes, caused by release of their genes from catabolite repression. The increased activity of proteases does not confirm unambiguously that the lack of sugar causes an enhanced expression of genes of these enzymes, because enzymatic activity can be controlled in other ways, e.g. by allosteric regulation or by phosphorylation and dephosphorylation, which is the most common type of reversible covalent modification (Berg et al. 2002). However, a study of starved maize root meristems shows that the increased activity of proteolytic enzymes under sugar starvation is an effect of enhanced transcription and translation (James et al. 1993). It also has been evidenced that proteases are closely associated with some cell compartments. Optimal acidic pH for endo- and carboxypeptidase, which increased under starvation, suggested that proteases were located in vacuoles or in lysosomes (Feller 1986, Mikola and Mikola 1986, Huffaker 1990). Accumulation of autolysosomes was observed in tobacco suspension cells cultured under sucrose starvation conditions in the presence of a cysteine protease inhibitor E-64. Experiments with fluorescent dyes, green fluorescent protein (GFP), and endocytosis markers (FM4-64 and Lucifer Yellow CH) suggested that there is a membrane flow from the plasma membrane to autolysosomes. Using fluorescent dyes and markers of the central vacuole (GFP-AtVam3p, sporamin-GFP and gamma-VM23-GFP), the transport of components of the central vacuole to autolysosomes was displayed (Yano et al. 2004). When cells of tobacco were treated with E-64c cysteine protease inhibitor, then both protein degradation and protease activation were inhibited. Simultaneously, many spherical bodies accumulated in the cytosol, beside the nucleus. The bodies, with acidic pH inside, contained acid phosphatase, which is a marker enzyme of autolysosomes. The bodies accumulated because E-64c inhibitor blocked the degradation of the proteins contained in them.

Consequently, the normally short life span of these bodies was significantly prolonged, and their observation was possible. In all probability, the observed spherical bodies were autolysosomes participating in protein degradation during autophagy (Moriyasu and Ohsumi 1996). During advanced sugar starvation, autophagy involves cytosolic proteins and even whole organelles can be degraded. Plastids, ribosomes, and endoplasmic reticulum are consumed relatively early, whereas the plasma membrane, mitochondria, and peroxisomes persist longer (Baker and Graham 2002). The increased activity of proteolytic enzymes and autophagic processes observed under sugar starvation are certainly adaptive processes enabling the cells to maintain continuous energy supply under sugar starvation. Amino acids released from degraded cytosolic proteins, storage proteins, and organelles are used as respiratory substrates (Brouquisse et al. 1991, Ratajczak et al. 1996, Yu 1999, Borek et al. 2001, Gonzali et al. 2006).

A perfect example of enzymes induced under sugar starvation is glutamate dehydrogenase (GDH). This is a mitochondrial enzyme that catalyzes the synthesis and degradation of glutamate, which is one of the central amino acids of nitrogen metabolism in plants (Lehmann and Ratajczak 2008, Borek et al. 2011). The reversibility of the reaction catalyzed by GDH *in vivo* is considered as alternative and depends on carbon and ammonium status (Lehmann et al. 2011). In sucrose-starved pea and lupine embryo axes, a remarkable increase in GDH activity was observed (Morkunas et al. 2000, Lehmann et al. 2003, 2010, Borek and Nuc 2011, Borek et al. 2012a). Sugar starvation caused also a significant increase in a number of GDH isoenzymes (Morkunas et al. 2000, Lehmann et al. 2003, 2010). The increased GDH activity caused by sugar deficit was also recorded in callus of *Nicotiana plumbaginifolia* (Maestri et al. 1991) and in carrot cell suspension (Robinson et al. 1992). Under sugar starvation, plant mitochondria may adapt to the stress conditions by increasing their ability to use amino acids as respiratory substrates. Thus the increase in GDH activity under sugar starvation may be the reason for the higher oxidation of glutamate by mitochondria. GDH catalyzing the oxidative deamination of glutamate, in cooperation with the corresponding aminotransferases, is the main pathway of amino acid catabolism in plants (Lea 1993, Lehmann and Ratajczak 2008). During this process, a large quantity of ammonia is produced. Nitrogen in this form is toxic, so its utilization is necessary. One of the possibilities of ammonia detoxification is the synthesis of asparagine, whose content increases during sugar starvation. In sugar-starved suspension cells of sycamore, asparagine content increased steadily, whereas the cell protein content declined progressively (Genix et al. 1990).

The higher activity of the enzymes of protein and amino acid catabolism under sugar starvation probably results from the release of the genes encoding them from catabolite repression (caused by sugar). Catabolite repression is of fundamental importance since it allows cells to adapt their metabolism to carbon sources other than sugar when they are in a habitat where sugar is unavailable. If sugar is present in the medium, then the synthesis of enzymes of the catabolism of carbon sources other than carbohydrates is inhibited. This applies to proteolytic enzymes, amino acid dehydrogenases (mostly GDH), alcohol dehydrogenase, and isocitrate dehydrogenase. Other studies of the influence of sugar deficit on plant cell metabolism show that already in the initial phase of starvation (after 1.5 h), specific proteins appear, known as carbohydrate-responsive proteins (Baysdorfer et al. 1988). They may be the proteins associated with the phase of tissue acclimation to sugar starvation. Prolonged starvation results in synthesis of many specific proteins known as starvation-related proteins (STP) (Tassi et al. 1992). The genes encoding them (at least some of them) are normally under catabolite repression caused by carbohydrates. Their expression is initiated when the level of sugar in the cell falls below a critical level.

During sugar starvation, respiration rate declines rapidly in embryos of garden pea (Morkunas et al. 2000) and yellow lupine (Borek et al. 2011). The lower respiration rate is caused by the deficit of respiratory substrates rather than by mitochondrial degradation. In mitochondria isolated from starved pea embryos, respiratory activity (with glutamate as a substrate) is 60% higher than in those isolated from embryos fed with 60 mM sucrose (Morkunas et al. 2000). A similar relationship was observed in mitochondrial fractions from starved and control (sugar-fed) embryos of narrowleaf lupine (Morkunas et al. 2003). The concentration of mitochondrial proteins in starved cells was slightly lower than in control cells, and the activity of a mitochondrial marker enzyme, NAD^+-dependent isocitrate dehydrogenase, was reduced in starved embryos only slightly. Besides, mitochondria isolated from lupine embryo axes cultivated without sucrose, exhibited respiration coupled with oxidative phosphorylation. They exhibited a respiration control ratio of 2 for succinate as a respiratory substrate. Mitochondria isolated from starved embryo axes oxidized glutamate and malate more intensively than mitochondria from embryo axes fed with sucrose. Ultrastructural analysis of cells of starved pea embryo axes shows that the inner mitochondrial membrane is highly convoluted, forming many folds (cristae, Fig. 2A). A similar finding was reported by Journet et al. (1986) who studied sycamore cell suspension. Mitochondria were protected in sycamore cells even as late as after 60 h of starvation, and then an addition of sugar to the medium resulted in intensive respiration. In starved cultures, sugar reserves were used up very quickly. Respiration rate was maintained at a high level for up to 30 h of starvation. In that period, respiratory substrates were provided by degradation of starch, phosphate esters, and proteins. Morkunas et al. (2000) also showed that the transfer of pea embryo axes from sugar starvation conditions onto the medium with sucrose after 48 h recovered their respiration activity. Metabolic adaptations of starved embryos of lupine (Morkunas et al. 2003) and pea (Morkunas et al. 2000) were quite effective physiologically, enabling their survival, and their transfer to a medium with sucrose allowed restoration of normal growth. Similarly, Couée et al. (1992), who studied maize root tips cultured *in vitro*, found that after 48 h of sugar starvation they still had a significant pool of functional mitochondria. Carbohydrate stress caused degeneration of lower-density mitochondria, but not of higher-density mitochondria, which was proved by measurements of respiratory activity of mitochondria and of fumarase activity (mitochondrial marker enzyme).

Under sugar starvation, the increased utilization of cytosolic proteins and storage proteins, is accompanied by a more intensive catabolism of lipids, aimed to supply respiratory substrates. In turn, sugar feeding significantly retards the breakdown of storage lipid. In the research on the regulatory function of sugars in lipid metabolism, sucrose and glucose are predominantly taken under consideration. Sucrose is particularly important in seed tissues because storage lipid is synthesized from this sugar in developing seeds and sucrose is one of the main end products of storage lipid breakdown during seed germination. In seedlings of *Arabidopsis* (a model oil-seed plant), 1% sucrose significantly retards the breakdown of storage lipid (Eastmond et al. 2000), while 0.3 M glucose nearly eliminates the mobilization of this storage compound (To et al. 2002). Seedlings grown for 22 days on glucose solution contained about 80% of their seed storage lipid. In contrast, 22-day-old seedlings grown in equi-molar sorbitol retained only 4-5% of their seed storage lipid. This result additionally proves that the effect of glucose is not due to the osmotic potential of the media (To et al. 2002).

Transcriptome analysis of sucrose-starved rice suspension cells has revealed a decrease in the expression of transcripts involved in fatty acid synthesis but induced those involved in fatty acid degradation, such as 3-ketoacyl-CoA thiolase, fatty acid multifunctional proteins, and acyl-CoA oxidase (Wang et al. 2007). An increased expression of genes encoding lipase and enzymes linked to fatty acid β-oxidation was noted in sucrose-starved *Arabidopsis* suspension cells (Contento et al. 2004). Similarly, in dark-induced senescent *Arabidopsis* leaves an enhanced gene expression of lipase and enzymes involved in fatty acid β-oxidation was observed (Buchanan-Wollaston et al. 2005). Glucose starvation remarkably enhances the intensity of lipid degradation in isolated maize root tips. After 24 h of glucose starvation, the rate of oxidation of palmitic acid to CO_2 was increased 2.5-fold. The overall β-oxidation of fatty acids (measured as acetyl-CoA formation) was increased up to 5-fold in glucose-starved roots. An increase in catalase activity was observed as well (Dieuaide et al. 1992). Catalase is not directly involved in fatty acid β-oxidation but is involved in detoxification of H_2O_2 produced in glyoxysomes by the activity of acyl-CoA oxidase when acyl-CoA is converted into trans-2-enoyl-CoA. Sucrose starvation in *Arabidopsis* suspension cells caused a 5.7-fold increase in the expression of acyl-CoA oxidase-4, which is the first committed step of short-chain fatty acid β-oxidation. A significant increase in acyl-CoA oxidase-4 activity in suspension cells was observed after 12 h of sucrose starvation, whereas a similar effect in one-week-old seedlings was noted after 2 days of sucrose starvation. The significant increase in gene expression of catalase-3 (glyoxysomal isoform) was observed already after 6 h of the removal of sucrose from the medium of *Arabidopsis* suspension cells. An increase in catalase activity was observed 6 h later, i.e. after 12 h of sucrose starvation. In seedlings the increase in catalase activity was observed after 2 days of sucrose starvation. About half of the catalase activity observed during sucrose starvation was due to glyoxysomal isoform 3 (Contento and Bassham 2010). In cucumber suspension cells, the expression of genes encoding 2 marker enzymes of the glyoxylate cycle (isocitrate lyase and malate synthase) was induced during deficits of sucrose, mannose and fructose in the medium (Graham et al. 1994). The above-mentioned results suggest that the enhanced gene expression patterns and enzymatic activities involved in lipolysis, fatty acid β-oxidation, and glyoxylate cycle, facilitate the production of acetyl-CoA from the lipid, which may be directly utilized by respiration to sustain energy production and growth when carbohydrates are exhausted.

Another set of data concerns the regulation of lipid metabolism by sucrose in developing and germinating protein lupine seeds (storage protein content up to 45% of seed dry matter). Restriction in sucrose feeding of developing embryos of yellow lupine (lipid content about 6% of seed dry matter), white lupine (lipid content 7-14%), and Andean lupine (lipid content about 20%) caused a decrease in storage lipid accumulation during seed development (Borek et al. 2009). Ultrastructural investigations of cotyledons of 4-day-old yellow, white, and Andean lupine seedlings showed that oil bodies were smaller and less numerous when sugar level was decreased in tissues (Borek et al. 2006, 2011, 2012a). A peculiar and puzzling feature was observed in 4-day-old sucrose-starved isolated embryo axes of yellow, white, and Andean lupine grown *in vitro*. Their lipid level was by 43, 44, and 70% higher, respectively, than in sucrose-fed organs (Borek et al. 2012b). Lipase and catalase activity increased when sugar reserves were depleted in yellow lupine germinating seed organs (Borek et al. 2006). However, the enzymes involved in further steps of lipid breakdown (cytosolic aconitase, isocitrate lyase, $NADP^+$-dependent cytosolic isocitrate dehydrogenase) were more active in sucrose-fed tissues (Borek and Nuc 2011). Changes in enzymatic activity caused by sucrose starvation were accompanied by modification of gene expression. Lipase mRNA level was higher, while mRNA levels for cytosolic aconitase and $NADP^+$-dependent isocitrate dehydrogenase were lower in yellow lupine cotyledons and embryo axes whose sugar content was decreased (Borek and Nuc 2011). During yellow lupine seed germination, lipid-derived carbon skeletons are used for sugar synthesis but some of them are also used for amino acid formation (mainly asparagine, glutamine, and glutamate). In sugar-deficient conditions in tissues, this carbon flow from storage lipid to amino acids is significantly inhibited (Borek et al. 2003, Borek and Ratajczak 2010).

5. The role of sugar starvation in plant growth and development

As mentioned above, sugar in plants is not only an energetic and structural substrate but it also regulates the expression of many genes. Sugar may play a role of hormonal-like signal in a variety of eukaryotic cell types. The best known are the signaling pathways in yeasts (*Saccharomyces cerevisiae*), but many similar pathways have also been discovered in plants (Rolland et al. 2006). The time-course monitoring of gene expression profiles in suspension cells was investigated by Wang et al. (2007). After 12, 24, and 48 h of sucrose-starvation in rice cells, the expression of 867 genes was increased, while the expression of 855 genes was inhibited. Most of these genes encoded enzymes associated with metabolism. It is characteristic that genes associated with catabolism were up-regulated, while genes involved in biosynthetic pathways were down-regulated. Importantly, Wang et al. (2007) succeeded to identify the genes responsible for transcription and translation, which were the second and third major categories of genes in each group significantly affected by sucrose starvation. Response-to-stress genes were the fourth largest group of regulated genes, which suggests that most of the stress-associated genes were also significantly up-regulated by sucrose starvation. Admittedly, the model of cell suspension used by Wang and his research team is a rather "artificial" system, but identification of the genes affected by sugar starvation proved to be very useful for interpretation of results of research on the role of sugar starvation in the regulation of various processes of plant growth and development and in plant responses to biotic and abiotic stresses.

Most of the data on the regulatory role of sugar starvation in plant development was obtained as a result of research on senescence of plant organs, particularly of leaves and flowers. Controversies between groups of researchers on the role of sugar in leaf senescence may partly result from selection of various experimental models. Buchanan-Wollaston et al. (2005) identified over 800 genes of *Arabidopsis*, whose expression was up- or down-regulated during senescence. Those authors came to a conclusion that gene expression patterns during natural leaf senescence, compared with those identified when senescence is artificially induced in leaves by darkness or sucrose starvation in cell suspension cultures, showed not only similarities but also considerable differences. In this chapter, the role of sugar starvation in senescence is discussed in section 6.

There is also a rich literature on participation of the sugar signal in mobilization of storage materials during seed germination. Interaction between sugar and gibberellin in germinating barley grain is well documented (Thomas and Rodriquez 1994). Depending on the seedling's demand for carbohydrates, the rate of starch mobilization in barley endosperm is regulated by a combination of the sugar signal with gibberellin. This regulation is a textbook example of the role of feedback in metabolic adaptation to developmental needs of the plant (Bewley and Black 1994). Results of research on the role of sugar starvation in the expression of amylase genes have some practical applications. Promoter of the gene of rice α-amylase 3D (RAmy3D), which is induced by sugar starvation, has proved to be very useful in transgenic plant cell cultures producing recombinant proteins, including plant-made pharmaceutical and plant-made industrial proteins (Park et al. 2010, Huang and McDonald 2012). For example, this promoter has been used for achieving high-level expression in rice cell suspension cultures of many therapeutic proteins, including human growth hormone, human α1-antitrypsin, human granulocyte-macrophage colony stimulating factor, bryodin-1, lysozyme and human serum albumin (Xu et al. 2011a, 2011b). Virtually every recombinant protein expressed in this rice suspension cell system has produced significantly higher secreted protein levels than attained with any other plant cell expression system tested (Xu et al. 2011a).

The role of sugar starvation in the control of storage lipid mobilization in germinating seeds was mentioned above. Storage protein mobilization in germinating seeds is also affected by sugar starvation. A good model for research on this phenomenon is yellow lupine seed. Its main storage compounds are proteins, which account for up to 45% of seed dry weight (Duranti et al. 2008, Borek et al. 2012a, 2012b). In ripe lupine seeds, starch is absent (Duranti et al. 2008, Borek et al. 2006, 2011). Sugar starvation releases from catabolite repression not only the genes of proteolytic enzymes but also the genes of amino acid catabolism. This allows the use of amino acid carbon skeletons both as sources of energy and as structural components. Interestingly, transitional starch appears during germination of yellow lupine seeds both in cotyledons and in seedling embryo axes (Borek et al. 2006, 2011, 2012a). This phenomenon can be explained by analogy with the appearance of transitional starch in leaves during photosynthesis. Sugar signal inhibits the expression of genes of some photosynthetic proteins (Sheen 1990, Krapp et al. 1993), so carbohydrates must be quickly converted into the neutral starch. In cotyledons of germinating lupine, carbohydrates are synthesized through gluconeogenesis (Borek et al. 2003, Borek and Ratajczak 2010, Borek et al. 2011), so transitional starch synthesis prevents the inhibition by sugar of the continued mobilization of storage lipids and proteins. Transitional starch synthesis in seedling embryo axes may help to maintain the gradient necessary for sugar transport between the source (cotyledons) and the sink (embryo axes). Hypocotyl elongation of etiolated seedlings was suppressed by sugar. However, it is possible that maintenance of an appropriate level of carbohydrates in the embryo axis prevents the inhibition of hypocotyl elongation in etiolated seedlings (Jang et al. 1997, Rolland et al. 2002, Yang et al. 2004, Borek et al. 2012a). The major task of the hypocotyl during epigeal germination (e.g. in lupine) is to bring the cotyledons above the ground. Thus excessively deep sowing may lead to a loss of the ability to transform cotyledons from storage materials into photosynthetic organs (Elamrani et al. 1994).

Sugar signal in cooperation with hormone signaling pathways, may control a wide range of processes related to plant growth and development (Rolland et al. 2006). Sugar and nitrogen are the main factors that modify plant morphology, as they determine the growth rate and transition to the next developmental stages. Transcription analyses show that sugar and inorganic nitrogen act as both metabolites and signaling molecules (Price et al. 2004). There are many recent reports about the role of sugar starvation in processes of shoot growth reduction and root proliferation enhancement during phosphorus deficits (Ciereszko et al. 2005, Karthikeyan et al. 2007, Polit and Ciereszko 2009). The importance of sugar in plant responses to biotic and abiotic stresses is discussed in section 7.

6. Sugar starvation versus senescence

Shaded leaves age more quickly. Their sugar content declines as a result of limited photosynthesis, so their senescence may be induced by sugar starvation. Chung et al. (1997) discovered in *Arabidopsis* plants a gene called *sen 1*, which was expressed during leaf senescence. A particularly strong expression of the gene was observed in excised leaves kept in darkness, which simultaneously showed symptoms of accelerated senescence. When the excised leaves kept in darkness were treated with sugar, then the *sen 1* gene promoter was repressed. On the basis of the results, the cited authors postulated that sugar starvation induces leaf senescence. Further research on plant tissue senescence, e.g. with the use of a hexokinase mutant (as mentioned above, hexokinase is a sugar sensor), confirmed that a low level of sugar may induce leaf senescence (Moore et al. 2003). However, some other authors reported that leaf senescence is induced by a high level of sugar (Parrott et al. 2005, Pourtau et al. 2006). The dispute on which of these contradictory hypotheses is true was one of the most interesting scientific debates in the field of plant biology in the last decade. A detailed outline of the debate was presented by van Doorn (2008). He critically analyzed numerous works whose results support one or the other hypothesis. He did not question the results, but warned against their simplistic interpretation and unjustified generalization. Van Doorn did not support unambiguously any of the hypotheses, but concluded that "it is quite possible that neither of the two hypotheses is correct". On the basis of gene expression analyses during natural and induced leaf senescence, Buchanan-Wollaston et al. (2005) proposed a model of cooperation of the sugar signaling pathway with abscisic acid, jasmonate, ethylene, salicylic acid, and cytokinin signaling pathways.

As in the case of leaf senescence, there is some controversy about the role of sugar starvation in flower senescence. Sugar starvation may be involved in the process of flower senescence because application of sugars to cut flowers generally delays its visible symptoms (van Doorn 2004). However, petals showing symptoms of senescence have a relatively high sugar content. Arrom and Munné-Bosch (2012) assayed a large number of sugars and hormones in various organs of uncut and cut lily flowers. Addition of sucrose to the vase solution elevated the level of sugar in various organs of cut flowers, but primarily it altered the hormonal balance of floral tissues. Those results confirm the earlier findings that sucrose added to the vase solution accelerates the senescence of cut flowers with the participation of hormones. Hoeberichts et al. (2007) used cDNA microarrays to characterize senescence-associated gene expression in petals of cut carnation (*Dianthus caryophyllus*) flowers. The plants were treated with ethylene, silver thiosulphate (which blocks the ethylene receptor), or sucrose. Ethylene accelerated flower senescence and enhanced senescence-associated gene expression. Silver thiosulphate and sucrose had an opposite effect. The cited authors conclude that sucrose may slow down flower senescence by a negative regulation of the ethylene signal pathway. Cut flowers proved to be a good model for research on senescence regulation in plants, participation of hormones in this process (Tripathi and Tuteja 2007, and references cited there) and the role of sugar as a regulator of gene expression (van Doorn and Woltering 2008). Those studies enriched our knowledge but also provided some useful tips for horticultural practice.

At the cellular level, sugar starvation in senescent organs and in organs that have not initiated the senescence program causes similar or identical symptoms: degradation of cellular structures and degradation of many cellular components, such as proteins and lipids. Nevertheless, the goals of these transformations differ. In the case of senescence, valuable metabolites are removed from the cells destined to die, to feed other cells in other plant organs, e.g. in developing leaves or maturing seeds. During sugar starvation of non-senescent organs, protein and lipid degradation is aimed to supply substrates for respiratory processes, which provide the energy necessary for survival. Senescence is a stage of programmed death, during which a low level of sugar may be necessary for activation of some senescence-associated genes, whose expression is repressed by sugar. However, sugar starvation can be used in plant development in many other processes, not only in programmed cell death. As mentioned above, sugar signaling may, in cooperation with plant hormones, play a major role in plant response to environmental factors and in shaping of plant morphology. However, even at the level of cells, sugar starvation, if not associated with programmed cell death, may show other symptoms than during senescence. For example, generation of free radicals (which play an important role in programmed cell death) may be neutralized by initiation of scavenging free radicals whenever cells need to be protected against the negative effects of sugar starvation (Morkunas et al. 1999, 2003). The role of sugars in plant defense response against free radicals is discussed in section 7.

7. Effect of sugar starvation on plant response to biotic and abiotic stress factors

Depending on the duration of sugar starvation, considerable ultrastructural and metabolic changes were observed in plant cells (see above). Moreover, it was interesting whether sugar starvation (as a nutritional stress) is accompanied by changes in the redox status of the cell, like in the case of plant response to other stress factors (biotic and abiotic). In plant response to abiotic stress factors, the increased generation of free radicals, including reactive oxygen species (ROS), attests to a greater influence of the stress factor, whereas during the plant-pathogen interaction, particularly at an early stage of the infection, an increased generation of free radicals in plant cells is favorable (Morkunas and Bednarski 2008). Free radicals may be toxic to the pathogenic microorganism, hamper the penetration of host tissues (through initiation of lignification of host cell walls), activate the expression of host defense genes, and participate in signal transduction over short and long distances.

Under sugar starvation, synthesis of free radicals is intensified, particularly after 72 and 96 h of starvation of lupine embryo axes *in vitro* (Morkunas et al. 2003). Electron Paramagnetic Resonance (EPR) analyses of free radicals detected the presence of free radicals with a signal at $g = 2.0060$. Sugar starvation caused a 2-fold increase in the concentration of free radicals in embryo axes cultured *in vitro* for 72 h. Spectroscopic splitting factors (g-values) indicated that these radicals might be quinones involved in electron transport pathways. Thus, intensified synthesis of free radicals during nutritional stress and other stresses on plant cells may result in a transient oxidative stress due to enhanced ROS generation (Polle et al. 1990, Schraudner 1992, del Rio 1996, Polle 1996, Purvis 1997, Becana et al. 1998, Chamnongpol 1998, Prasad et al. 1999). Concentrations of free radicals increased also in aging plants (Leshem 1981, Thompson et al. 1987, Pastori and del Rio 1997). However, generation of free radicals during sugar starvation may be neutralized by a number of adaptive mechanisms protecting the cells from oxidative damage (Morkunas et al. 1999, 2003). An example of such a mechanism is the accumulation of large amounts of phytoferritin in plastids of starved lupine and pea cells. This fact was interpreted as a defense mechanism aimed at inactivation of iron ions (Fe^{2+}), as it prevents ROS generation. Results of malondialdehyde (MDA) assays supported this interpretation (Morkunas et al. 1999). Ferritin, which binds iron ions (Fe^{2+}) prevented ROS generation in Fenton reaction (Becana et al. 1998). Besides, in starved cells an antioxidant system was activated, i.e. superoxide dismutase (SOD), catalase (CAT) and peroxidase (POX) activity was higher than in sugar-fed tissues (Borek et al. 2006, Morkunas and Gmerek 2007, Morkunas and Bednarski 2008). Additionally, an increase was observed in the ability of starved embryo axes to take up Mn^{2+} from the growth medium, which seems to be an interesting adaptation, probably associated with increased SOD activity (Morkunas et al. 2003). All stresses generate ROS, potentially leading to oxidative damage. However, it must be emphasized that during nutritional stress, such as sugar starvation, specific defense mechanisms are induced, alleviating the results of oxidative stress.

Apart from the well-known classical antioxidant mechanisms, sugars and sugar-metabolizing enzymes are important players in the defense against oxidative stress (Bolouri-Moghaddam et al. 2010). Those authors hypothesized that the synergistic interaction of sugars (or sugar-like compounds) and phenolic compounds is a part of an integrated redox system. It quenches ROS and contributes to stress tolerance, especially in tissues or organelles with high soluble sugar concentrations.

Moreover, sugar-signaling pathways interact with stress pathways in a complex network, which modulates metabolic plant responses (Ho et al. 2001, Tran et al. 2007). Soluble sugars may either act directly as negative signals or as modulators of plant sensitivity, so they can also play important roles in cell responses to stress-induced remote signals (Rosa et al. 2009). Descriptive ecological and agronomic studies have revealed a strong correlation between soluble sugar concentrations and stress tolerance. However, since energy and resources are required for plants to cope with abiotic and biotic stress conditions, the source-sink partitioning between different organs is a crucial component of the mechanisms of stress tolerance (Ho 1988, Krapp and Stitt 1995). Recent studies for increasing tolerance to environmental stress, through metabolic engineering of compatible solutes have shown that increased concentrations of soluble sugars and/or other osmolytes increase plant tolerance to abiotic stresses, such as drought, salinity, and cold (Rathinasabapathi 2000).

Energy and resources are required for plants to cope with abiotic and biotic stress conditions, so in sugar starvation conditions, when metabolic processes are slowed down (e.g. respiration), the plants are more sensitive to stress factors. For example, the link between low sugar content and plant sensitivity to fungal diseases known as "low-sugar diseases", was first reported over 50 years ago by Horsfall and Dimond (1957). It must be emphasized that many environmental factors, such as insufficient light, high humidity, excessive nitrogen fertilization, and excessively deep sowing, may cause a decrease in the level of carbohydrates in host plant tissues, contributing to an increase in plant sensitivity to fungal infection (Vidhyasekaran 1974a, 1974b, Morkunas et al. 2004, Huber and Thompson 2007, Yoshida et al. 2008, Morkunas et al. 2010, Morkunas et al. 2012). By contrast, low temperature can induce plant resistance to specific pathogens, because it results in accumulation of soluble carbohydrates, affecting the cell water potential and other osmotically-active molecules, as well as accumulation of pathogenesis-related (PR) proteins (Tronsmo 1993, Thomashow 1998, Hiilovaara-Teijo et al. 1999, Płażek and Żur 2003, Yuanyuan et al. 2009). Sugar starvation of host plant cells facilitates the development of fungal pathogens because water potential is much higher there than when the concentration of soluble carbohydrates in cells is high. A lack of carbohydrates in the growth medium caused a decrease in endogenous levels of soluble carbohydrates in lupine embryo axes (Morkunas et al. 2005, Borek et al. 2006), and facilitated the development of *Fusarium* infection. This was reflected in an increased area of necrotic spots, more intensive post-infection changes in cell ultrastructure, and in limited elongation and fresh weight increments of embryo axes, as compared to inoculated axes with a high endogenous level of carbohydrates (Morkunas et al. 2005, Morkunas and Bednarski 2008). Besides, Morkunas and Bednarski (2008) observed on transmission electron micrographs the occurrence of hyphae of the pathogenic fungus *Fusarium oxysporum* in intercellular spaces of inoculated embryo axes cultured without sucrose as early as 72 h after inoculation, while no fungi were found in intercellular spaces of inoculated embryo axes cultured with sucrose. Cook and Papendick (1978) report that a high level of carbohydrates is a strong osmoticum in host plant cells, decreasing the water potential and inhibiting fungal growth. Moreover, in the absence of sugar, genes encoding cell wall-degrading extracellular enzymes secreted by fungal pathogens are released from catabolite repression (Akimitsu et al. 2004). These enzymes may contribute to pathogenesis by degrading wax, cuticle, and cell walls, which aids in tissue invasion and pathogen dissemination (Carapito et al. 2009, Kikot et al. 2009). Moreover, the induction of defense in host plant cells is cost-intensive (Swarbrick et al. 2006). Thus, under sugar starvation, where respiration rate is lowered, acquisition of the energy necessary to initiate defense responses, such as synthesis of PR proteins, phenylpropanoids, and papilla formation, is limited (Morkunas et al. 2005, Morkunas and Gmerek 2007, Morkunas et al. 2007, Morkunas and Bednarski 2008, Morkunas et al. 2011). It needs to be emphasized that a very important defense strategy against *Fusarium oxysporum* in inoculated lupine embryo axes cultured with sucrose involved a high accumulation of flavonoids, including isoflavonoids (Morkunas et al. 2005, 2007), fast stimulation of tissue lignification (Morkunas and Gmerek 2007), and intensive generation of superoxide anion radical (Morkunas and Bednarski 2008). This mechanism was not observed in inoculated embryo axes cultured without sucrose.

8. Conclusions

At some stages of plant development, sugar content may decrease considerably (so-called sugar starvation), usually under the influence of adverse environmental factors (biotic or abiotic). In this chapter, we have shown that sugar starvation in plant cells, depending on its duration, causes both ultrastructural and physiological-biochemical changes of varied severity. In non-senescent cells, the changes are aimed to maintain respiration and other basic metabolic processes at a specified level. Thus under conditions of sugar starvation, a priority is the acquisition of energy necessary for cell survival, even at the expense of organelles except the nucleus and the system supplying energy, i.e. mitochondria. The changes taking place during sugar starvation in the acclimation phase or even in the survival phase can be reversed, as long as the respiratory system functions properly. Carbon, nitrogen, and phosphorus are the major elements whose availability affects plant growth and development, so the sugar deficit signal in cooperation with nitrogen and phosphorus signaling pathways may influence plant morphology. Moreover, it is interesting that during sugar starvation, the observed processes resemble senescence. Many symptoms are identical, e.g. autophagy, decrease in protein content, and strong activation of proteolytic enzymes. However, senescence is controlled by programmed cell death. By contrast, if sugar starvation concerns non-senescent cells, then the initiated reactions cause cell acclimation to such conditions. At first these are reactions that protect cells against irreversible degradation of cell structures, especially those responsible for conservation and realization of genetic information as well as energy supply. At the second phase of sugar starvation, defense strategies are initiated, e.g. activation of the enzymatic antioxidant system, manganese ion uptake, phytoferritin accumulation, protecting cells against the destructive influence of an excess of free radicals (e.g. membrane damage). However, the mechanisms underlying the processes used by plant cells to survive sugar starvation require further research. Currently we should focus on explanation of the controlling mechanisms, which may involve signal molecules. Their synthesis may, to some extent, alleviate the influence of stress factors, and thus play an important role in plant acclimation to stress conditions. In plant cells, in response to environmental factors (either abiotic or biotic), many signaling pathways are initiated. Knowledge of the signaling pathways initiated under conditions of sugar starvation, and their cross-talk in the network, would certainly greatly help to explain the molecular mechanisms underlying plant reactions to sugar starvation.

Author details

Iwona Morkunas* and Magda Formela
Department of Plant Physiology, Faculty of Horticulture and Landscape Architecture, Poznań University of Life Sciences, Poznań, Poland

Sławomir Borek and Lech Ratajczak
Department of Plant Physiology, Faculty of Biology, Adam Mickiewicz University, Poznań, Poland

Acknowledgement

This study was supported by the Polish Committee for Scientific Research (KBN, grant no. N N303 414437).

9. References

Agulló-Antón MÁ, Sánchez-Bravo J, Acosta M, Druege U (2011) Auxins or sugars: what makes the difference in the adventitious rooting of stored carnation cuttings? J Plant Growth Reg 30: 100-113.

Akimitsu K, Isshikia A, Ohtanib K, Yamamotoa H, Eshelc D, Prusky D (2004) Sugars and pH: A clue to the regulation of fungal cell wall-degrading enzymes in plants. Physiol Mol Plant Pathol 65: 271-275.

Arrom L, Munné-Bosch S (2012) Sucrose accelerates flower opening and delays senescence through a hormonal effect in cut lily flowers. Plant Sci 188-189: 41-47.

Aubert S, Gout E, Bligny R, Marty-Mazars D, Barrieu F, Alabouvette J, Marty F, Douce R (1996) Ultrastructural and biochemical characterization of autophagy in higher plant cells subjected to carbon deprivation: control by the supply of mitochondria with respiratory substrates. J Cell Biol 133: 1251-1263.

* Corresponding Author

Azad AK, Ishikawa T, Ishikawa T, Sawa Y, Shibata H (2008) Intracellular energy depletion triggers programmed cell death during petal senescence in tulip. J Exp Bot 59: 2085-2095.

Baker A, Graham IA (2002) Plant peroxisomes: Biochemistry, cell biology and biotechnological applications. Kluwer Academic Publishers, Dordrecht, The Netherlands.

Baysdorfer C, Warmbrodt RD, Van Der Woude WJ (1988) Mechanisms of starvation tolerance in pearl millet. Plant Physiol 88: 1381-1387.

Becana M, Moran JF, Iturbe-Ormaetxe I (1998) Iron-dependent oxygen free radical generation in plants subjected to environmental stress: toxicity and antioxidant protection. Plant Soil 201: 137-147.

Berg JM, Tymoczko JL, Stryer L (2002) Biochemistry. Fifth edition, WH Freeman, New York.

Bewley JD, Black M (1994) Seeds. Physiology of development and germination. Second edition, Plenum Press, New York, London.

Bolouri-Moghaddam MR, Le Roy K, Xiang L, Rolland F, Van den Ende W (2010) Sugar signalling and antioxidant network connections in plant cells. FEBS J 277: 2022-2037.

Borek S, Kubala S, Kubala S (2012a) Regulation by sucrose of storage compounds breakdown in germinating seeds of yellow lupine (*Lupinus luteus* L.), white lupine (*Lupinus albus* L.) and Andean lupine (*Lupinus mutabilis* Sweet). I. Mobilization of storage protein. Acta Physiol Plant 34: 701-711.

Borek S, Kubala S, Kubala S, Ratajczak L (2011) Comparative study of storage compound breakdown in germinating seeds of three lupine species. Acta Physiol Plant 33: 1953-1968.

Borek S, Morkunas I, Ratajczak W, Ratajczak L (2001) Metabolism of amino acids in germinating yellow lupin seeds. III Breakdown of arginine in sugar-starved organs cultivated in vitro. Acta Physiol Plant 23: 141-148.

Borek S, Nuc K (2011) Sucrose controls storage lipid breakdown on gene expression level in germinating yellow lupine (*Lupinus luteus* L.) seeds. J Plant Physiol 168: 1795-1803.

Borek S, Pukacka S, Michalski K (2012b) Regulation by sucrose of storage compounds breakdown in germinating seeds of yellow lupine (*Lupinus luteus* L.), white lupine (*Lupinus albus* L.) and Andean lupine (*Lupinus mutabilis* Sweet). II. Mobilization of storage lipid. Acta Physiol Plant 34: 1199-1206.

Borek S, Pukacka S, Michalski K, Ratajczak L (2009) Lipid and protein accumulation in developing seeds of three lupine species: *Lupinus luteus* L., *Lupinus albus* L., and *Lupinus mutabilis* Sweet. J Exp Bot 60: 3453-3466.

Borek S, Ratajczak L (2010) Storage lipids as a source of carbon skeletons for asparagine synthesis in germinating seeds of yellow lupine (*Lupinus luteus* L.). J Plant Physiol 167: 717-724.

Borek S, Ratajczak W (2002) Sugars as a metabolic regulator of storage protein mobilization in germinating seeds of yellow lupine (*Lupinus luteus* L.). Acta Physiol Plant 24: 425-434.

Borek S, Ratajczak W, Ratajczak L (2003) A transfer of carbon atoms from fatty acids to sugars and amino acids in yellow lupine (*Lupinus luteus* L.) seedlings. J Plant Physiol 160: 539-545.

Borek S, Ratajczak W, Ratajczak L (2006) Ultrastructural and enzymatic research on the role of sucrose in mobilization of storage lipids in germinating yellow lupine seeds. Plant Sci 170: 441-452.

Brouquisse R, Gaudillère JP, Raymond P (1998) Induction of a carbon-starvation-related proteolysis in whole maize plants submitted to light/dark cycles and to extended darkness. Plant Physiol 117: 1281-1291.

Brouquisse R, James F, Pradet A, Raymond P (1992) Asparagine metabolism and nitrogen distribution during protein degradation in sugar-starved maize root tips. Planta 188: 384-395.

Brouquisse R, Rolin D, Cortès S, Gaudillère M, Evrard A, Roby C (2007) A metabolic study of the regulation of proteolysis by sugars in maize root tips: effects of glycerol and dihydroxyacetone. Planta 225: 693-709.

Brouquisse R. James F, Rajmond P, Pradet A (1991) Study of glucose starvation in excised maize root tips. Plant Physiol 96: 619-626.

Buchanan-Wollaston V, Page T, Harrison E, Breeze E, Lim PO, Nam HG, Lin JF, Wu SH, Swidzinski J, Ishizaki K, Leaveret CJ (2005) Comparative transcriptome analysis reveals significant differences in gene expression and signalling pathways between developmental and dark/starvation-induced senescence in Arabidopsis. Plant J 42: 567-585.

Carapito RI, Imberty A, Jeltsch JM, Byrns SC, Tam PH, Lowary TL, Varrot A, Phalip V (2009) Molecular basis of arabinobio-hydrolase activity in phytopathogenic fungi. Crystal structure and catalytic mechanism of *Fusarium graminearum* GH93 exo-α-L-arabinanase. J Biol Chem 284: 12285-12296.

Chamnongpol S (1998) Defense activation and enhanced pathogen tolerance induced by H_2O_2 in transgenic tobacco. Proc Natl Acad Sci USA 95: 5818-5823.

Chen MH, Liu LF, Chen YR, Wu HK, Yu SM (1994) Expression of α-amylases, carbohydrate metabolism, and autophagy in cultured rice cells is coordinately regulated by sugar nutrient. Plant J 6: 625-636.

Cho JI, Ryoo N, Eom JS, Lee DW, Kim HB, Jeong SW, Lee YH, Kwon YK, Cho MH, Bhoo SH, Hahn TR, Park YI, Hwang I, Sheen J, Jeon JS (2009) Role of the rice hexokinases OsHXK5 and OsHXK6 as glucose sensors. Plant Physiol 149: 745-759.

Cho YH, Yoo SD (2011) Signaling role of fructose mediated by FINS1/FBP in *Arabidopsis thaliana.* PLoS Genet 7: e1001263. doi:10.1371/journal.pgen.1001263.

Chung BC, Lee SY, Oh SA, Rhew TH, Nam HG, Lee CH (1997) The promoter activity of *sen1*, a senescence-associated gene of *Arabidopsis*, is repressed by sugars. J Plant Physiol 151: 339-345.

Ciereszko I, Johansson H, Kleczkowski LA (2005) Interactive effects of phosphate deficiency, sugar and light/dark conditions on gene expression of UDP-glucose pyrophosphorylase in Arabidopsis. J Plant Physiol 162: 343-353.

Contento AL, Bassham DC (2010) Increase in catalase-3 activity as a response to use of alternative catabolic substrates during sucrose starvation. Plant Physiol Biochem 48: 232-238.

Contento AL, Kim SJ, Bassham DC (2004) Transcriptome profiling of the response of Arabidopsis suspension culture cells to Suc starvation. Plant Physiol 135: 2330-2347.

Cook RJ, Papendick RI (1978) Role of water potential in microbial growth and development of plant disease with special reference to postharvest pathology. Hortsciences 13: 559-564.

Couée I, Murielle J, Carde JP, Brouquisse R, Raymond P, Pradet A (1992) Effects of glucose starvation on mitochondrial subpopulations in the meristematic and submeristematic regions of maize root. Plant Physiol 100: 1891-1900.

del Rio LA (1996) Peroxisomes as a source of superoxide and hydrogen peroxide in stressed plants. Biochem Soc Trans 24: 434-438.

Dieuaide M, Brouquisse R, Pradet A, Raymond P (1992) Increased fatty acid β-oxidation after glucose starvation in maize root tips. Plant Physiol 99: 595-600.

Dorne AJ, Bligny R, Rebeille F, Roby C, Douce R (1987) Fatty acid disappearance and phosphorylcholine accumulation in higher plant cells after a long period of sucrose deprivation. Plant Physiol Biochem 25: 589-595.

Duranti M, Consonni A, Magni C, Sessa F, Scarafoni A (2008) The major proteins of lupin seed: characterisation and molecular properties for use as functional and nutraceutical ingredients. Trends Food Sci Tech 19: 624-633.

Eastmond PJ, Germain V, Lange PR, Bryce JH, Smith SM, Graham IA (2000) Postgerminative growth and lipid catabolism in oilseeds lacking the glyoxylate cycle. Proc Natl Acad Sci USA 97: 5669-5674.

Elamrani A, Gaudillère JP, Raymond P (1994) Carbohydrate starvation is a major determinant of the loss of greening capacity in cotyledons of dark-grown sugar beet seedlings. Physiol Plant 91:56-64.

Farrar JF (1981) Respiration rate of barley roots: its relation to growth, substrate supply and the illumination of the shoot. Ann Bot 48: 53-63.

Feller U (1986) Proteolytic enzymes in relation to leaf senescence. In: Dalling MJ (ed.) Plant proteolytic enzymes. 2. CRC Press, Boca Raton, pp. 49-68.

Genix P, Bligny R, Martin JB, Douce R (1990) Transient accumulation of asparagine in sycamore cells after a long period of sucrose starvation. Plant Physiol 94: 717-722.

Gibson SI (2005) Control of plant development and gene expression by sugar signaling. Curr Opin Plant Biol 8: 93-102.

Gonzali S, Loreti E, Solfanelli C, Novi G, Alpi A, Perata P (2006) Identification of sugar-modulated genes and evidence for in vivo sugar sensing in Arabidopsis. J Plant Res 119: 115-123.

Gout E, Bligny R, Douce R, Boisson AM, Rivasseau C (2011) Early response of plant cell to carbon deprivation: *in vivo* ^{31}P-NMR spectroscopy shows a quasi-instantaneous disruption on cytosolic sugars, phosphorylated intermediates of energy metabolism, phosphate partitioning, and intracellular pHs. New Phytologist 189: 135-147.

Graham IA, Derby KJ, Leaver CJ (1994) Carbon catabolite repression regulates glyoxylate cycle gene expression in cucumber. Plant Cell 6: 761-772.

Hammond JP, White PJ (2008) Sucrose transport in the phloem: integrating root responses to phosphorus starvation. J Exp Bot 59: 93-109.

Hanson J, Smeekens S (2009) Sugar perception and signaling - an update. Curr Opin Plant Biol 12: 562-567.

Hara M, Oki K, Hoshino K, Kuboi T (2003) Enhancement of anthocyanin biosynthesis by sugar in radish (*Raphanus sativus*) hypocotyls. Plant Sci 164: 259-265.

Harrington GN, Bush DR (2003) The bifunctional role of hexokinase in metabolism and glucose signaling. Plant Cell 15: 2493-2496.

Hiilovaara-Teijo M, Hannukkala A, Griffith M, Yu XM, Pihakaski-Maunsbach K (1999) Snow-mold-induced apoplastic proteins in winter rye leaves lack antifreeze activity. Plant Physiol 121: 665-673.

Ho LC (1988) Metabolism and compartmentation of imported sugars in sink organs in relation to sink strength. Annu Rev Plant Physiol Plant Mol Biol 39: 355-378.

Ho SL, Chao YC, Tong WF, Yu SM (2001) Sugar coordinately and differentially regulates growth- and stress-regulated gene expression via a complex signal transduction network and multiple control mechanisms. Plant Physiol 125: 877-890.

Hoeberichts FA, van Doorn WG, Vorst O, Hall RD, van Wordragen MF (2007) Sucrose prevents up-regulation of senescence-associated genes in carnation petals. J Exp Bot 58: 2873-2885.

Horsfall JG, Dimond AE (1957) Interactions of tissue sugar, growth substances, and disease susceptibility. Z Pflanzenkr Pflanzenschutz 64: 415-421.

Huang TK, McDonald KA (2012) Bioreactor systems for *in vitro* production of foreign proteins using plant cell cultures. Biotechnol Adv 30: 398-409.

Huber DM, Thompson IA (2007) Nitrogen and plant disease. In: Datnoff LE, Elmer WH, Huber DM (eds.) Mineral nutrition and plant disease. APS, St Paul, MN, pp. 31-44.

Huffaker RC (1990) Proteolytic activity during senescence of plants. New Phytol 116: 199-231.

Inoue Y, Moriyasu Y (2006) Autophagy is not a main contributor to the degradation of phospholipids in tobacco cells cultured under sucrose starvation conditions. Plant Cell Physiol 47: 471-480.

James F, Brouquisse R, Pradet A, Raymond P (1993) Changes in proteolytic activities in glucose-starved maize root tips. Regulation by sugars. Plant Physiol Biochem 31: 845-856.

James F, Brouquisse R, Suire C, Pradet A, Raymond P (1996) Purification and biochemical characterization of a vacuolar serine endopeptidase induced by glucose starvation in maize roots. Biochem J 320: 283-292.

Jang JC, León P, Zhou L, Sheen J (1997) Hexokinase as a sugar sensor in higher plants. Plant Cell 9: 5-19.

Jang JC, Sheen J (1994) Sugar sensing in higher plants. Plant Cell 6: 1665-1679.

Johnson R, Ryan CA (1990) Wound-inducible potato inhibitor II genes: enhancement of expression by sucrose. Plant Mol Biol 14: 527-536.

Journet EP, Blingy R, Douce R (1986) Biochemical changes during sucrose deprivation in higher plant cells. J Biol Chem 261: 3193-3199.

Karthikeyan AS, Varadarajan DK, Jain A, Held MA, Carpita NC, Raghothama KG (2007) Phosphate starvation responses are mediated by sugar signaling in *Arabidopsis*. Planta 225: 907-918.

Kerr PS, Rufty TW Jr, Huber SC (1985) Changes in nonstructural carbohydrates in different parts of soybean (*Glycine max* [L.] Merr.) plants during a light/dark cycle and in extended darkness. Plant Physiol 78: 576-581.

Kikot GE, Hours RA, Alconada TM (2009) Contribution of cell wall degrading enzymes to pathogenesis of *Fusarium graminearum*: a review. J Basic Microbiol 49: 231-241.

Kim SW, Koo BC, Kim J, Liu JR (2007) Metabolic discrimination of sucrose starvation from *Arabidopsis* cell suspension by ^{1}H NMR based metabolomics. Biotech Bioprocess Eng 12: 653-661.

Koch K (2004) Sucrose metabolism: regulatory mechanisms and pivotal roles in sugar sensing and plant development. Curr Opin Plant Biol 7: 235-246.

Koch KE (1996) Carbohydrate-modulated gene expression in plants. Annu Rev Plant Physiol Plant Mol Biol 47: 509-540.

Krapp A, Hofmann B, Schäfer C, Stitt M (1993) Regulation of the expression of rbcSandother photosynthetic genes by carbohydrates: a mechanism for the "sink regulation" of photosynthesis? Plant J 3: 817-828.

Krapp A, Stitt M (1995) An evaluation of direct and indirect mechanisms for the "sink-regulation" of photosynthesis in spinach: changes in gas exchange, carbohydrates, metabolites, enzyme activities and steady state transcript levels after cold-girdling source leaves. Planta 195: 313-323.

Lea PJ (1993) Nitrogen metabolism. In: Plant Biochemistry and Molecular Biology (ed.) Lea PJ, Leegood RC, John Wiley and Sons, Chester, 5: 155-180.

Lee EJ, Matsumura Y, Soga K, Hoson T, Koizumi N (2007) Glycosyl hydrolases of cell wall are induced by sugar starvation in Arabidopsis. Plant Cell Physiol 48: 405-413.

Lehmann T, Dabert M, Nowak W (2011) Organ-specific expression of glutamate dehydrogenase (GDH) subunits in yellow lupine. J Plant Physiol 168: 1060-1066.

Lehmann T, Ratajczak L (2008) The pivotal role of glutamate dehydrogenase (GDH) in the mobilization of N and C from storage material to asparagine in germinating seeds of yellow lupine. J Plant Physiol 165: 149-158.

Lehmann T, Ratajczak L, Deckert J, Przybylska M (2003) The modifying effect of sucrose on glutamate dehydrogenase (GDH) activity in lupine embryos treated with inhibitors of RNA and protein synthesis. Acta Physiol Plant 25: 325-335.

Lehmann T, Skrok A, Dabert M (2010) Stress-induced changes in glutamate dehydrogenase activity imply its role in adaptation to C and N metabolism in lupine embryos. Physiol Plant 138: 35-47.

Leshem YY (1981) Oxy free radicals and plant senescence. In: What's new in Plant Physiol 12: 1-4.

Li Y, Lee KK, Walsh S, Smith C, Handingham S, Sorefan K, Cawley G, Bevan MW (2006) Estabilishing glucose- and ABA-regulated transcription networks in Arabidopsis by microarray analysis and promoter classification using a Relevance Vector Machine. Genome Res 16: 414-427.

Loreti E, De Bellis L, Alpi A, Perata P (2001) Why and how do plant cells sense sugars? Annals Bot 88: 803-812.

Loreti E, Povero G, Novi G, Solfanelli C, Alpi A, Perata P (2008) Gibberellins, jasmonate and abscisic acid modulate the sucrose induced expression of anthocyanin biosynthetic genes in Arabidopsis. New Phytologist 179: 1004-1016.

Lothier J, Lasseur B, Prudhomme MP, Morvan-Bertrand A (2010) Hexokinase-dependent sugar signaling represses fructan exohydrolase activity in *Lolium perenne*. Func Plant Biol 37: 1151-1160.

Maestri E, Restivo FM, Gulli M, Tassi I (1991) Glutamate dehydrogenase regulation in callus cultures of *Nicotiana plumbaginifolia*: effect of glucose feeding and carbon source starvation of the isoezymatic pattern. Plant Cell Environm 14: 613-618.

Mikola L, Mikola J (1986) Occurrence and properties of different types of peptidases in higher plants. In: Dalling MJ (ed.) Plant proteolytic enzymes. 1. Boca Raton, CRC Press, pp. 97-117.

Mohapatra PK, Patro L, Raval MK, Ramaswamy NK, Biswal UC, Biswal B (2010) Senescence-induced loss in photosynthesis enhances cell wall β-glucosidase activity. Physiol Plant 138: 346-355.

Moore B, Zhou L, Rolland F, Hall Q, Cheng WH, Liu YX, Hwang I, Jones T, Sheen J (2003) Role of the Arabidopsis glucose sensor HXK1 in nutrient, light, and hormonal signaling. Science 300: 332-336.

Moriyasu Y, Hattori M, Jauh GY, Rogers JC (2003) Alpha tonoplast intrinsic protein is specifically associated with vacuole membrane involved in an autophagic process. Plant Cell Physiol 44: 795-802.

Moriyasu Y, Klionsky DJ (2003) Autophagy in plants. In: Klionsky DJ (ed.) Autophagy. Landes Bioscience, Georgetown, TX, pp: 208-215.

Moriyasu Y, Ohsumi Y (1996) Autophagy in tobacco suspension-cultured cells in response to sucrose starvation. Plant Physiol 111: 1233-1241.

Morkunas I (1997) Regulation by carbohydrates of germinating legume seeds metabolism. PhD thesis, Department of Plant Physiology, Adam Mickiewicz University, Poznań, Poland.

Morkunas I, Bednarski W (2008) *Fusarium oxysporum* induced oxidative stress and antioxidative defenses of yellow lupine embryo axes with different level of sugars. J Plant Physiol 165: 262-277.

Morkunas I, Bednarski W, Kopyra M (2008) Defense strategies of pea embryo axes with different levels of sucrose to *Fusarium oxysporum* and *Ascochyta pisi*. Physiol Mol Plant Pathol 72: 167-178.

Morkunas I, Bednarski W, Kozłowska M (2004) Response of embryo axes of germinating seeds of yellow lupine to *Fusarium oxysporum*. Plant Physiol Biochem 42: 493-499.

Morkunas I, Borek S, Ratajczak W, Ratajczak L, Antonowicz J (1999) Accumulation of phytoferritin caused by carbohydrate stress in plastids of embryos in lupin and pea cultivated *in vitro*. In: Zesz Probl Post Nauk Rol, Wyd. ZFR PAN, Warszawa, 469: 247-256.

Morkunas I, Formela M, Marczak Ł, Stobiecki M, Bednarski W (2012) The mobilization of defense mechanisms in the early stages of pea seed germination against *Ascochyta pisi*. Protoplasma doi: 10.1007/s00709-012-0374-x.

Morkunas I, Garnczarska M, Bednarski W, Ratajczak W, Waplak S (2003) Metabolic and ultrastructural responses of lupine embryo axes to sugar starvation. J Plant Physiol 160: 311-319.

Morkunas I, Gmerek J (2007) The possible involvement of peroxidase in defense of yellow lupine embryo axes against *Fusarium oxysporum*. J Plant Physiol 164: 185-194.

Morkunas I, Kozłowska M, Ratajczak L, Marczak Ł (2007) Role of sucrose in the development of *Fusarium* wilt in lupine embryo axes. Physiol Mol Plant Pathol 70: 25-37.

Morkunas I, Lehmann T, Ratajczak W, Ratajczak L, Tomaszewska B (2000) The involvement of glutamate dehydrogenase in the adaptation of mitochondria to oxidize glutamate in sucrose starved pea embryos. Acta Physiol Plant 22: 389-394.

Morkunas I, Marczak J, Stachowiak M, Stobiecki (2005) Sucrose-induced lupine defense against *Fusarium oxysporum*: Sucrose-stimulated accumulation of isoflavonoids as a defense response of lupine to *Fusarium oxysporum*. Plant Physiol Biochem 43: 363-373.

Morkunas I, Narożna D, Nowak W, Samardakiewicz S, Remlein-Starosta D (2011) Cross-talk interactions of sucrose and *Fusarium oxysporum* in the phenylpropanoid pathway and the accumulation and localization of flavonoids in embryo axes of yellow lupine. J Plant Physiol 168: 424-433.

Morkunas I, Stobiecki M, Marczak Ł , Stachowiak J, Narożna D, Remlein-Starosta D (2010) Changes in carbohydrate and isoflavonoid metabolism in yellow lupine in response to infection by *Fusarium oxysporum* during the stages of seed germination and early seedling growth. Physiol Mol Plant Pathol 75: 46-55.

Müller J, Wiemken A, Aeschbacher RA (1999) Trehalose metabolism in sugar sensing and plant development. Plant Sci 147: 37-47.

Nicolaï M, Roncato MA, Canoy AS, Rouquié D, Sarda X, Freyssinet G, Robaglia C (2006) Large-scale analysis of mRNA translation states during sucrose starvation in Arabidopsis cells identifies cell proliferation and chromatin structure as targets of translational control. Plant Physiol 141: 663-673

Pace GGM, Volk RJ, Jackson WA (1990) Nitrate reduction in response to CO_2-limited photosynthesis. Plant Physiol 92: 286-292.

Park CI, Lee SJ, Kang SH, Jung HS, Kim DI, Lim SM (2010) Fed-batch cultivation of transgenic rice cells for the production of hCTLA4Ig using concentrated amino acids. Process Biochem 45: 67-74.

Parrott D, Yang L, Shama L, Fischer AM (2005) Senescence is accelerated, and several proteases are induced by carbon "feast" conditions in barley (*Hordeum vulgare* L.) leaves. Planta 222: 989-1000.

Pastori GM, del Rio LA (1997) Natural senescence of pea leaves. Plant Physiol 113: 411-418.

Płażek A, Żur I (2003) Cold-induced plant resistance to necrotrophic pathogens and antioxidant enzyme activities and cell membrane permeability. Plant Sci 164: 1019-1028.

Polit JT, Ciereszko I (2009) In situ activities of hexokinase and fructokinase in relation to phosphorylation status of root meristem cells of *Vicia faba* during reactivation from sugar starvation. Physiol Plant 135: 342-350.

Polle A (1996) Developmental changes of antioxidative systems in tobacco leaves as the affected by limited sucrose export in transgenic plants expressing yeast invertase in apoplastic space. Planta 198: 253-262.

Polle A, Chakrabarti K, Schürmann W, Rennenberg H (1990) Composition and properties of hydrogen peroxide decomposing systems in extracellular and total extracts from needles of Norway spruce (*Picea abies* L., Karst). Plant Physiol 94: 312-319.

Pourtau N, Jennings R, Pelzer E, Pallas J, Wingler A (2006) Effect of sugar-induced senescence on gene expression and implications for the regulation of senescence in *Arabidopsis*. Planta 224: 556-568.

Prasad KVSK, Saradhi PP, Sharmila P (1999) Concerted action of antioxidant enzymes and curtailed growth under zinc toxicity in *Brassica juncea*. Environ Exp Bot 42: 1-10.

Price J, Laxmi A, Martin SKST, Jang JC (2004) Global transcription profiling reveals multiple sugar signal transduction mechanisms in *Arabidopsis*. Plant Cell 2004 16: 2128-2150.

Purvis AC (1997) Role of the alternative oxidase in limiting superoxide production by plant mitochondria. Physiol Plant 100: 165-170.

Ramon M, Rolland F, Sheen J (2008) Sugar sensing and signaling. In: The Arabidopsis book. Rockville, MD: American Society of Plant Biologists; doi:10.1199/tab.0117.

Ratajczak W, Lehmann T, Polcyn W, Ratajczak L (1996) Metabolism of amino acids in germinating yellow lupin seeds. I. The decomposition of ^{14}C-aspartate and ^{14}C-glutamate during the imbibition. Acta Physiol Plant 18: 13-18.

Rathinasabapathi B (2000) Metabolic engineering for stress tolerance: Installing osmoprotectant synthesis pathways. Annals Bot 86: 709-716.

Robinson SA, Stewart GR, Phillips R (1992) Regulation of glutamate dehydrogenase activity in relation to carbon limitation and protein catabolism in carrot cell suspension cultures. Plant Physiol 98: 1190-1195.

Roby C, Martin JB, Bligny R, Douce R (1987) Biochemical changes during sucrose deprivation in higher plant cells. J Biol Chem 262: 5000-5007.

Roitsch T, Bittner M, Godt DE (1995) Induction of apoplastic invertase of *Chenopodium rubrum* by D-glucose and a glucose analog and tissue specific expression suggest a role in sink source regulation. Plant Physiol 108: 285-294.

Rolland F, Baena-Gonzalez E, Sheen J (2006) Sugar sensing and signaling in plants: conserved and novel mechanisms. Annu Rev Plant Biol 57: 675-709.

Rolland F, Moore B, Sheen J (2002) Sugar sensing and signaling in plants. Plant Cell 14 (suppl.): S185-S205.

Rolland F, Sheen J (2005) Sugar sensing and signalling networks in plants. Biochem Soc Trans 33: 269-270.

Rosa M, Prado C, Podazza G, Interdonato R, González JA, Hilal M, Prado FE (2009) Soluble sugars - metabolism, sensing and abiotic stress. Plant Signaling Behavior 4: 388-393.

Rose TL, Bonneau L, Der C, Marty-Mazars D, Marty F (2006) Starvation-induced expression of autophagy-related genes in Arabidopsis. Biol Cell 98: 53-67.

Saglio PH, Pradet A (1980) Soluble sugars, respiration, and energy charge during aging of excised maize root tips. Plant Physiol 66: 516-519.

Sahulka J, Lisa L (1978) The influence of exogenously supplied sucrose on glutamine synthetase and glutamate dehydrogenase levels in excised *Pisum sativum* roots. Biol Plant 20: 446-452.

Salanoubat M, Belliard G (1989) The steady-state level of potato sucrose synthase mRNA is dependent on wounding, anaerobiosis and sucrose concentration. Gene 84: 181-185.

Schraudner M (1992) Biochemical plant responses to ozone. III. Activation of the defense related proteins β-1,3-glucanase and chitinase in tabacco leaves. Plant Physiol 99: 1321-1328.

Sheen J (1990) Metabolic repression of transcription in higher plants. Plant Cell 2: 1027-1038.

Sheen J, Zhou L, Jang JC (1999) Sugars as signaling molecules. Curr Opin Plant Biol 2: 410-418.

Sherson SM, Alford HL, Forbes SM, Wallace G, Smith SM (2003) Roles of cell-wall invertases and monosaccharide transporters in the growth and development of *Arabidopsis*. J Exp Bot 54: 525-531.

Sinha AK, Romer U, Kockenberger W, Hofmann M, Elling L, Roitsch T (2002) Metabolism and non-metabolizable sugars activate different signal transduction pathways in tomato. Plant Physiol 128: 1480-1489.

Smeekens S (2000) Sugar induced signal transduction in plants. Annu Rev Plant Physiol Plant Mol Biol 51: 49-81.

Smeekens S, Ma J, Hanson J, Rolland F (2010) Sugar signals and molecular networks controlling plant growth. Curr Opin Plant Biol 13: 274-279.

Solfanelli C, Poggi A, Loreti E, Alpi A, Perata P (2006) Sucrose-specific induction of the anthocyanin biosynthetic pathway in Arabidopsis. Plant Physiol 140: 637-246.

Swarbrick PJ, Schulze-Lefert P, Scholes JD (2006) Metabolic consequences of susceptibility and resistance in barley leaves challenged with powdery mildew. Plant Cell Environ 29: 1061-1076.

Tassi F, Maestri E, Restivo FM, Marminoli N (1992) The effects of carbon starvation on cellular metabolism and protein and RNA synthesis in Gerbera callus cultures. Plant Sci 83: 127-136.

Thomas BR, Rodriquez R (1994) Metabolite signal regulate gene expression and source/sink relations in cereal seedlings. Plant Physiol 106: 1235-1239.

Thomashow MF (1998) Role of cold-responsive genes in plant freezing tolerance. Plant Physiol 118: 1-7.

Thompson AR, Vierstra RD (2005) Autophagic recycling: lessons from yeast help define the process in plants. Curr Opin Plant Biol 8: 165-173.

Thompson JE, Legge RL, Barber RF (1987). The role of free radicals in senescence and wounding. New Phytol 105: 317-344.

To JPC, Reiter WD, Gibson SI (2002) Mobilization of seed storage lipid by Arabidopsis seedlings is retarded in the presence of exogenous sugars. BMC Plant Biol 2: 4.

Tran LS, Nakashima K, Shinozaki K, Yamaguchi Shinozaki K (2007) Plant gene networks in osmotic stress response: from genes to regulatory networks. Methods Enzymol 428: 109-128.

Tripathi SK, Tuteja N (2007) Integrated signaling in flower senescence. Plant Signal Beh 2: 437-445.

Tronsmo AM (1993) Resistance to winter stress factors in half-sib families of *Dactylis glomerata*, tested in a controlled environment. Acta Agric Scand 43: 89-96.

Tsukaya H, Ohshima T, Naito S, Chino M, Komeda Y (1991) Sugar dependent expression of the CHS-A gene for chalcone synthase from petunia in transgenic *Arabidopsis*. Plant Physiol 97: 1414-1421.

Usadel B, Blasing OE, Gibon Y, Retzlaff K, Hohne M, Gunther M, Stitt M (2008) Global transcript levels respond to small changes of the carbon status during progressive exhaustion of carbohydrates in Arabidopsis rosettes. Plant Physiol 146: 1834-1861.

van Doorn WG (2004) Is petal senescence due to sugar starvation? Plant Physiol 134: 35-42.

van Doorn WG (2008) Is the onset of senescence in leaf cells of intact plants due to low or high sugar levels? J Exp Bot 59: 1963-1972

van Doorn WG, Woltering EJ (2008) Physiology and molecular biology of petal senescence. J Exp Bot 59: 453-480.

Vauclare P, Bligny R, Gout E, De Meuron V, Widmer F (2010) Metabolic and structural rearrangement during dark-induced autophagy in soybean (*Glycine max* L.) nodules: an electron microscopy and ^{31}P and ^{13}C nuclear magnetic resonance study. Planta 231: 1495-1504.

Vidhyasekaran P (1974a) Possible role of sugars in restriction of lesion development in finger millet leaves infected with *Helminthosporium tetramera*. Physiol Plant Pathol 4: 457-467.

Vidhyasekaran P (1974b) Finger millet helminthosporiose, a low sugar disease. Z Pflanzenkr Pflanzenschutz 81: 28-38.

Visser RG, Stolte A, Jacobsen E (1991) Expression of a chimaeric granule-bound starch synthase-GUS gene in transgenic potato plants. Plant Mol Biol 4: 691-699.

Walsh KB, Vessey JK, Layzell DB (1987) Carbohydrate supply and N_2 fixation in soybean. Plant Physiol 85: 137-144.

Wang HJ, Wan AR, Hsu CM, Lee KW, Yu SM, Jauh GY (2007) Transcriptomic adaptations in rice suspension cells under sucrose starvation. Plant Mol Biol 63: 441-463.

Webster PL, Henry M (1987) Sucrose regulation of protein synthesis in pea root meristem cells. Environ Exp Bot 27: 253-262.

Webster PL, van't Hof J (1973) Polyribosomes in proliferating and non-proliferating root meristem cells. Am J Bot 60: 117-121.

Wittenbach VA (1977) Induced senescence of intact wheat seedlings and its reversibility. Plant Physiol 59: 1039-1042.

Wittenbach VA, Lin W, Hebert RR (1982) Vacuolar localization of proteases and degradation of chloroplasts in mesophyll protoplasts from senescing primary wheat leaves. Plant Physiol 69: 98-102.

Xu J, Dolan MC, Medrano G, Cramer CL, Weathers PJ (2011a) Green factory: Plants as bioproduction platforms for recombinant proteins. Biotechnol Adv doi:10.1016/ j.biotechadv.2011.08.020 (in press).

Xu J, Ge X, Dolan MC (2011b) Towards high-yield production of pharmaceutical proteins with plant cell suspension cultures. Biotechnol Adv 29: 278-299.

Yang Z, Zhang L, Diao F, Huang M Wu N (2004) Sucrose regulates elongation of carrot somatic embryo radicles as a signal molecule. Plant Mol Biol 54: 441-459.

Yano K, Matsui S, Tsuchiya T, Maeshima M, Kutsuna N, Hasezawa S, Moriyasu Y (2004) Contribution of the plasma membrane and central vacuole in the formation of autolysosomes in cultured tobacco cells. Plant Cell Physiol 45: 951-957.

Yoshida M, Nakajima T, Tonooka T (2008) Effect of nitrogen application at anthesis on *Fusarium* head blight and mycotoxin accumulation in breadmaking wheat in the western part of Japan. J Gen Plant Pathol 74: 355-363.

Yu SM (1999) Cellular and genetic responses of plants to sugar starvation. Plant Physiol 121: 687-693.

Yuanyuan M, Yali Z, Jiang L, Hongbo S (2009) Roles of plant soluble sugars and their responses to plant cold stress. African J Biotech 8: 2004-2010.

Section 5

Biotechnology

Chapter 20

Carbohydrates from Biomass: Sources and Transformation by Microbial Enzymes

Luis Henrique Souza Guimarães

Additional information is available at the end of the chapter

http://dx.doi.org/10.5772/51576

1. Introduction

In the last decades the destination of biomass, especially agro-industrial residues, is an important world problem that has been target of many researches. Accumulation of agro-industrial residues in the environment can cause serious ecological problems. On the other hand, these kinds of rich carbohydrate materials can aggregate economical value to different biotechnological process as for example in the microbial fermentative processes. According to this, the proposal of this chapter is to describe and discuss the utilization of biomass from agro-industrial residues and products and its transformation by microbial enzymes to obtain products (saccharides) with industrial interest. This is a subject that has attracted the attention of many researches and industrial sectors. To organize the information concerning this subject in a chapter is very interesting to qualify the state of the art on the utilization and importance of carbohydrates from the agro-industrial residues and products. The importance of microorganism for the transformation of biomass is another important aspect that will be highlighted.

Microorganisms, as bacteria and fungi, are able to use a great variety of inorganic and organic compounds as nutrients, reflecting an interesting metabolic diversity. Among these nutrients, nitrogen and carbon sources are indispensible for a primary metabolism. Others nutrients are required at low concentration, as vitamins. According to the growing, microorganisms are able to produce many enzymes that can show interesting biochemical properties for biotechnological application (Guimarães et al., 2006). Among these enzymes, some are constitutive while others are inducible. The induction of enzyme production by microorganisms can be obtained by use of properly biomass as carbon sources. Microorganisms are able to produce a diversity of enzymes as, for instance, the carbohydrate-active enzymes (figure 1). The glycoside hydrolases are enzymes able to acts on disaccharides, oligosaccharides and polysaccharides where can be found important enzymes as cellulases, amylases, inulinases and invertases (Table 1). Carbohydrate esterase is involved in the removal of O-(ester) and N-Acetyl moieties from carbohydrates. The polysaccharide lyase catalyzes the β-elimination reaction on uronic acid glucosides while the glycosyltransferase acts forming glycosidic bonds using activated sugar donors.

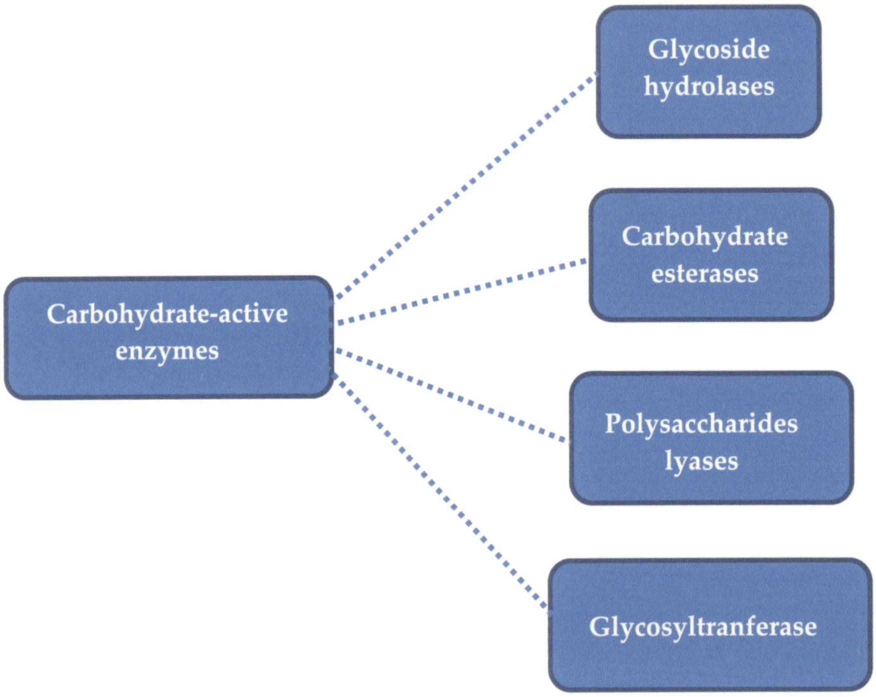

Figure 1. Classes of carbohydrate-active enzymes.

Microorganisms	Enzymes				
	Cellulase*	Xylanase	Invertase	Inulinase	Amylase
Bacteria					
Acremonium cellulolyticus	◆	◆			
Arthrobacter sp.				◆	
Bacillus amyloliquefaciens	◆				◆
Bacillus cellulyticus	◆				
Bacillus circulans		◆			◆
Bacillus licheniformis		◆			◆
Bacillus subtilis	◆	◆			◆
Bifidobacterium sp.			◆		
Cellulomonas sp.	◆	◆			
Clostridium cellulolyticum	◆				
Clostridium thermocellum	◆	◆			
Lactobacillus sp.			◆		◆
Pseudoalteromonas sp.	◆				
Streptomyces sp.		◆			

Microorganisms	Enzymes				
	Cellulase*	Xylanase	Invertase	Inulinase	Amylase
Fungi					
Aspergillus aculeatus	◆	◆			
Aspergillus caespitosus		◆	◆		
Aspergillus japonicus		◆	◆		
Aspergillus niger	◆	◆	◆	◆	◆
Aspergillus ochraceus		◆	◆		
Aspergillus oryzae	◆	◆	◆		◆
Aspergillus phoenicis		◆	◆		
Aspergillus terreus	◆	◆			
Chaetomium thermophilum		◆			◆
Emericella nidulans	◆	◆	◆		
Fusarium oxyporum		◆			
Humicola grisea	◆				◆
Humicola insolens	◆				◆
Neurospora crassa	◆	◆	◆		
Penicillium sp.	◆	◆		◆	◆
Trichoderma viride	◆	◆			

*Including endo-1,4- β-glucanases, exo-1,4- β-glucanases or cellobiohydrolases and 1,4- β-glucosidases. Data obtained from BRENDA (The Comprehensive Enzyme Information System).

Table 1. Some microbial (bacteria and fungi) sources of the enzymes involved in the utilization of carbohydrates found in the plant biomass.

2. Microbial cultivation using biomass

Different kinds of biomass have been used as carbon sources in the microbial cultivations under submerged and solid-state fermentations. Agro-industrial residues and products as, for example, rice straw, fruit peels, sugar cane bagasse and oat meal are important alternatives of carbon sources for both kinds of fermentation. The solid sate fermentation is characterized as a system constituted by solid material in absence of free water where microorganisms are able to grown. This condition is more similar than that found by microorganisms in the environment if compared to the submerged condition. In addition, some other advantages for use of solid-state fermentation has been mentioned as: i) higher yields of products; ii) similar or higher yield if compared to submerged fermentation; iii) uniform dispersion of spore suspension; iv) higher levels of aeration and v) reduction of problems with contamination by bacteria and yeast. It is also important to consider that the medium for solid-state fermentation is simple and low cost substrates as agro-industrial residues can be used.

According to the substrate nature, two processes can be used for solid-state fermentation. The solid substrate, in the first case, is used as both support and nutrient source. These substrates are obtained from agriculture activity or from by-products from food industry.

Generally, they are heterogeneous and water insoluble. When substrates with amylaceous or lignocellulosic nature will be used, a pre-treatment is required to convert raw substrate into a suitable substrate. After, the liquid medium containing nutrients necessaries to the microbial growth can be used to moisten the inert support. The microbial growth and the product synthesis in solid state fermentation is influenced by environmental factors, such as water activity, moisture content of the substrate, mass transfer processes, temperature and pH. The control of these factors is not easy, configuring a negative aspect from solid-state fermentation. However, under economic view, solid-sate fermentation can be applied in different sectors for biotransformation of crop residues, food additives, biofuels, bioactive products, production of organic acids, detoxification of agro-industrial wastes, bioremediation, biodegradation and enzyme production (Pérez-Guerra et al., 2003).

The use of agro-industrial residues and/or products as substrates/ carbon source for SSF should be considered under some aspects. According to the biomass characteristics a pre-treatment step is necessary as cited above. To transform the raw material to the available form for microbial utilization it is necessary, many times, to reduce the size of the material using for example grinding among others. Other possibility is to promote damages on the superficial substrate layers using cracking, grinding or pearling. The utilization of chemical or enzymatic pre-treatments, cooking or vapor treatment and elimination of contaminants can be also utilized. According to the nutritional exigency of the microorganism, supplementation with phosphorus and nitrogen sources and salts can improve the microbial growth and the product yield. On the other hand, the influence of environmental factors on the SSF system also deserves consideration. The microbial growth as well as the obtainment of the products in SSF is directly affected by the moisture content. Excessive or reduced moisture content is prejudicial to the microorganism and, consequently, to the product recovery. According to this, the moisture content should be adjusted for each microorganism used in process considering the nature of the matrix used as substrate, and it has been used water content of substrate from 30% to 70% (Pérez-Guerra et al., 2003).

In SSF the gases and nutrients diffusion are severally affected by the matrix structure and also by the liquid phase in the system. The aeration permits an effective supplement of oxygen that can be used for aerobic metabolism and, at the same time, it promotes the removing of CO_2 and water vapour as well as the heat and volatiles compounds produced by the microbial metabolism. The temperature in the SSF system is a consequence of the microbial metabolism if the heat is not removed. The acquirement of nutrients depends on both hydrolysis of the polymeric structure to obtain monomers and after diffusion through the cell membrane from outside to inside the cell. The pH is another factor that affects the SSF system but its control is difficult (for review, see Pérez-Guerra et al., 2007).

Advantages for enzyme and secondary metabolites productions have been reported for both fermentations. In addition, agro-industrial residues and products are excellent alternatives as substrates for solid-state fermentation. These substrates, and consequently their carbohydrate content, can be transformed by action of a set of enzymes instead the chemical conversion. Enzymatic technology is a clear and secure process minimizing the environment problems while chemical process can generate pollulents.

Sugar cane bagasse is one of the most important agro-industrial residues accumulating biomass in the environment (figure 2) that can be used for microbial transformation. In Brazil, around 80 million of ton of sugar cane bagasse is produced per year. This raw material is constituted by 26-47% cellulose, 19-33% hemicelluloses, 14-23% lignin and 1-5% ashes. Part of this biomass as used for electric energy generation, but the most part is accumulated without

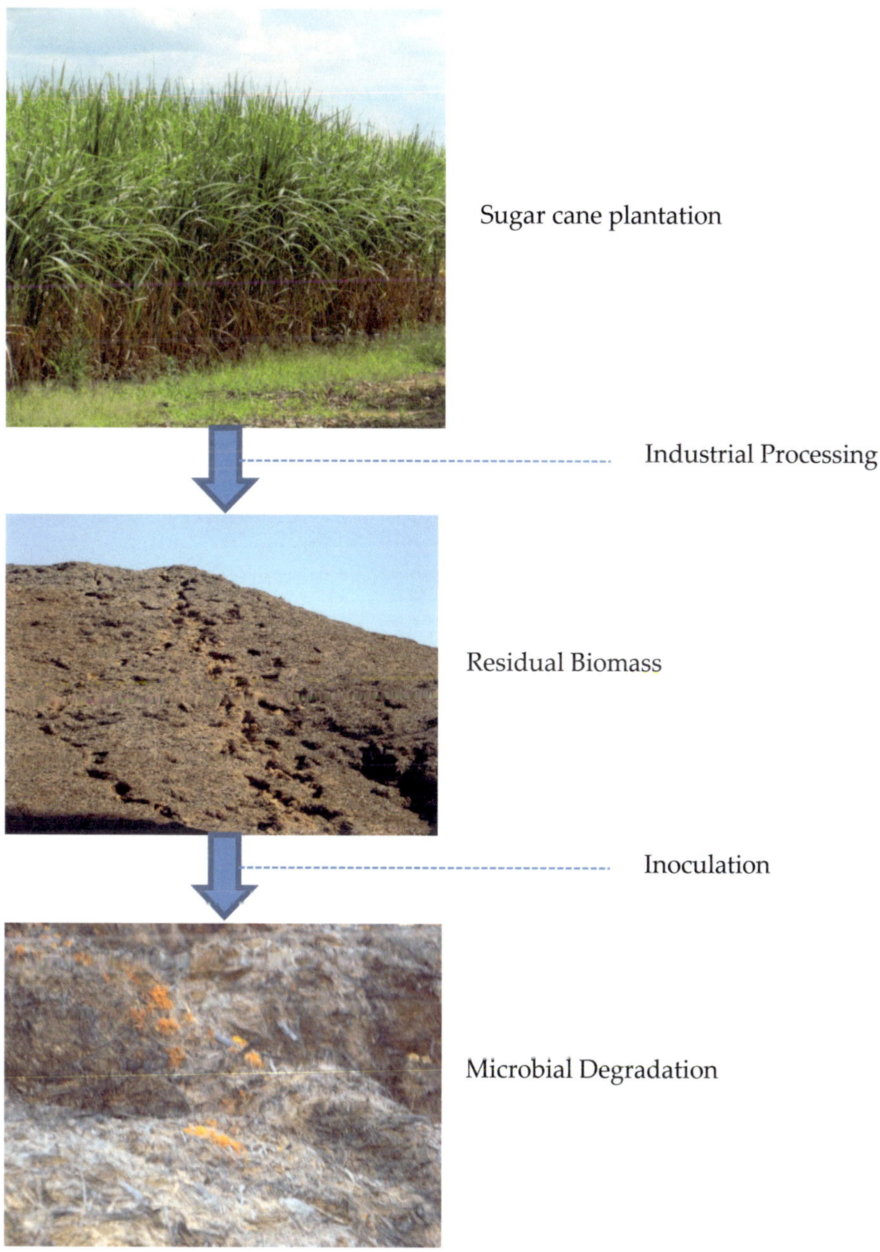

Figure 2. Utilization of sugar-cane bagasse as biomass source for microbial activity (photos by Guimarães L.H.S).

destination. So, sugar cane bagasse can be used for many other purposes as paper production, fertilizer, feeding for ruminant animals and ethanol production, among others. It is an interesting carbon source/substrate for microbial cultivation and enzyme production as verified for many filamentous fungi. High levels of β-fructofuranosidase were obtained using sugar cane bagasse as carbon source in submerged cultivation from *Aspergillus niveus* (Guimarães et al., 2009) and *Aspergillus ochraceus* (Guimarães et al., 2007).

Many others agro-industrial residues can be used for microbial cultivation such as wheat bran, which was used for invertase production by *Aspergillus caespitosus* under submerged and solid-state fermentation (Alegre et al., 2009).

Although the submerged fermentation and solid state fermentation as good option for microbial cultivation, a new kind of fermentation has been proposed. The biofilm fermentation (BF) is characterized by the fungal growing on the inert support as can be observed in the figure 3.

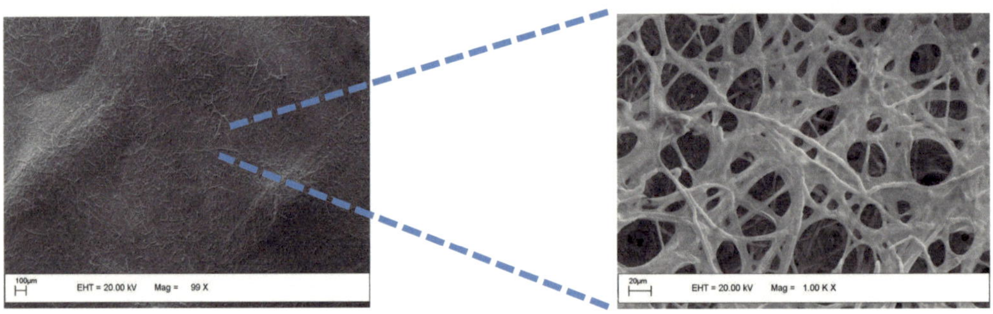

Figure 3. The *Aspergillus phoenicis* biofilm on the polyethylene as inert support.

The adhesion on surface is a natural process observed for the filamentous fungi in the environment. This complex process involves the production of adhesive compounds that fix the spore to the substrate, germ tube formation, hyphae elongation and, finally, surface colonization. These events can be observed in the *Aspergillus niger* biofilm formation on polyester cloth as reported by Villena and Gutiérrez-Correa (2007). These authors observed that the morphological pattern for *A. niger* growth attached to the surfaces is similar to that found in microbial biofilms with micro colonies development, extracellular matrix production and formation of pores and channels. The fungal morphology is an important factor to the enzyme production and, compared to the submerged fermentation, biofilms are more productive and more efficient if considered the metabolism associated with the specific enzymes which act on biomass, as for instance lignocellulosic enzymes. Recently, the use of *Aspergillus phoenicis* biofilms was reported for fructooligosaccharides production by one simple step (Aziani et al., 2012).

3. Biomass conversion by microbial enzymes

Residues and products of plant origin are recognized by their carbohydrate composition. According to this composition, the action of different enzymes on polysaccharides permits the obtainment of a variety of mono- and oligosaccharides which can be used by different sectors including biofuel, food, beverage and pharmaceutical among others.

Nowadays, the future of our energy sources and consequently the life in the planet is target of discussion around the world with participation of different sectors of the society as researches, politicians, undertakers and third sector. It has been noted that there is an increasing interest on biomass utilization as renewable energy source since that there is a conscience that the fossil fuels are restricted. In addition, this kind of fuel is a determinant factor of pollution to the atmosphere, where the CO_2 concentration has been increased. Hence, the utilization of biomass from plant residues for biofuel production is pointed-out as an important alternative for reduction of the energetic and environment problems. Brazil and USA are the main producer countries of ethanol to be used as fuel, the former using sugar cane and the later using the corn. For example, ton of sugar cane bagasse is generated as residue from the ethanol production in Brazil, which could be used for obtainment of fermentable sugars by enzymatic hydrolysis. In the next step, these sugars can be used for fermentation process to obtain ethanol.

3.1. Cellulases and lignocelullosic biomass

Considering the plant biomass, the main component from plant cell wall is the cellulose, the more abundant carbohydrate found in the planet. Structurally, this saccharide is constituted by glucopyranose monomers linked by β-1,4 glycosidic bonds with two distinct regions, the crystalline and amorphous regions. For the complete cellulose hydrolysis, an enzymatic complex (known as cellulases) constituted by endo-1,4- β-glucanases (EC 3.2.1.4), exo-1,4- β-glucanases or cellobiohydrolases (EC 3.2.1.91), and 1,4- β-glucosidases (EC 3.2.1.21), is necessary. Cellulases are modular enzymes included in the GH family (glycoside hydrolases). These enzymes have a complex structure with different modules as one or more catalytic domain and/or CBD module in the same protein. The CBD module is able to modify the catalytic domain and, consequently the cellulase properties, facilitating the interaction catalytic domain/crystalline cellulose. Cellulases can act using two main catalytic mechanisms, inversion or retention of the anomeric carbon. Two catalytic carboxylate residues are involved in both mechanisms and they are responsible for the acid-base catalysis in the reaction. Endoglucanases (EG; carboxymethiylcellulases, CMCase) catalyze random cleavage of cellulose internal bonds at amorphous region. Exoglucanases, also known as cellobiohydrolases (CBH) act at the chains ends (CBHI at the reduncing end and CBI III at non-reducing end), releasing cellobiose that can act as competitive inhibitor, while β-glucosidases (BGL) convert short cellooligosaccharides and cellobiose to glucose monomers. It is important to detach that the BGL activity is competitively inhibited by glucose. The GH1 family includes BGL obtained from bacteria, plant and mammalian and the GH3 family includes BGL from bacteria, fungi and plants. However, the full hydrolysis of cellulose depends on the previous hydrolysis of the other cell wall compounds, *i.e.* hemicelluloses and lignin (Dashtban et al., 2009; Bayer et al., 1998).

Hemicelluloses are polymeric molecules constituted by pentoses (as xylose and arabinose), hexoses (as mannose, glucose, galactose) and sugar acids. Because their heterogeneity, the hemicelluloses hydrolysis is only obtained by the action of different enzymes called hemicellulases. The most important hemicellulase is the enzyme that catalyzes the breakdown of β-1,4 linkages in the xylan, a polymer constituted by monomers of xylose. The oligomers obtained from this reaction are now substrates to the reaction catalyzed by β-xylosidase to obtain xylose (Dashtban et al., 2009).

Lignin is a heterogeneous aromatic polymer constituted by non-phenolic and phenolic structures that is able to link both cellulose and hemicelluloses making difficult the access of the enzymatic preparations to the cellulose and hemicelluloses. The enzyme able to catalyze the lignin hydrolysis are generically named as ligninases, which can be divided in two main families, phenol oxidase (laccase) and peroxidases that includes manganese peroxidase (MnP) and lignin peroxidase (LiP) (Dashtban et al., 2009).

The conversion of the lignocelluloses biomass from different sources to ethanol as can be observed in the figure 4 should take in account different steps as pre-treatment of the material, hydrolysis of the cellulose and hemicelluloses, fermentation and, finally distillation and evaporation. The pre-treatment will facilitate the asses of the hydrolases to the polyssacharides (they will be cleaved by cellulases and hemicellulases) through the lignin breakdown using physical-chemical or enzymatic process. The separated lignin can be used as matrix for energy production, as electricity. The hydrolysis and fermentation steps can be conducted separately or through of simultaneous saccharification and fermentation as shown in SSF square. Some considerations that will be done in next lines on the microorganism selection can be used for the others enzymes discussed in the next pages.

All enzymes from the cellulolitic complex, hemicellulases and lignin hydrolase can be obtained from microorganism as filamentous fungi. In the nature, filamentous fungi are able to produce and secrete different enzymes to the extracellular medium to hydrolyze polymeric compounds to obtain monomers that can be used as nutritional source.

3.2. Amylases, starch sources and structure

Another interesting carbohydrate found in the plant biomass is the starch, the main form carbohydrate reserve in these organisms. The starch is the result of interaction of two structure, one linear structure formed by glucose monomers linked by α-1,4 glycosidic bonds (amylose) with molar mass of 10^1 e 10^2 Kg/mol and another branched structure with α-1,6 bonds (amylopectin) with molar mass of 10^4 and 10^6 Kg/mol. The units of amylopectin can be classified in tree groups, A, B and C. The type A is simple and characterized by the α-1,6 linkage to the amylopectin structure. The type B is subdivided in B_1, B_2, B_3 e B_4 according to the size and the group formation. The type C is a mixture of A and B types. The starch granule is formed by 25% and 75% of amylase and amylopectin, respectively. The last one is responsible for the granule crystalinity. According to the crystallographic structure the native starch can be classified as cereal starch (type A), tuber starch (Type B) and leguminosea starch (type C), that corresponds to the amylopectin groups. The more stable structure for the α-1,4 chains, as observed for starch, is the helix with high degree of spiralization with intra chain hydrogen bridge. The helicoidal structure has six residues per loop where each glucose residue forming an angle of 60° with the next residue (Yoshimoto et al., 2000; Ritte et al., 2006).

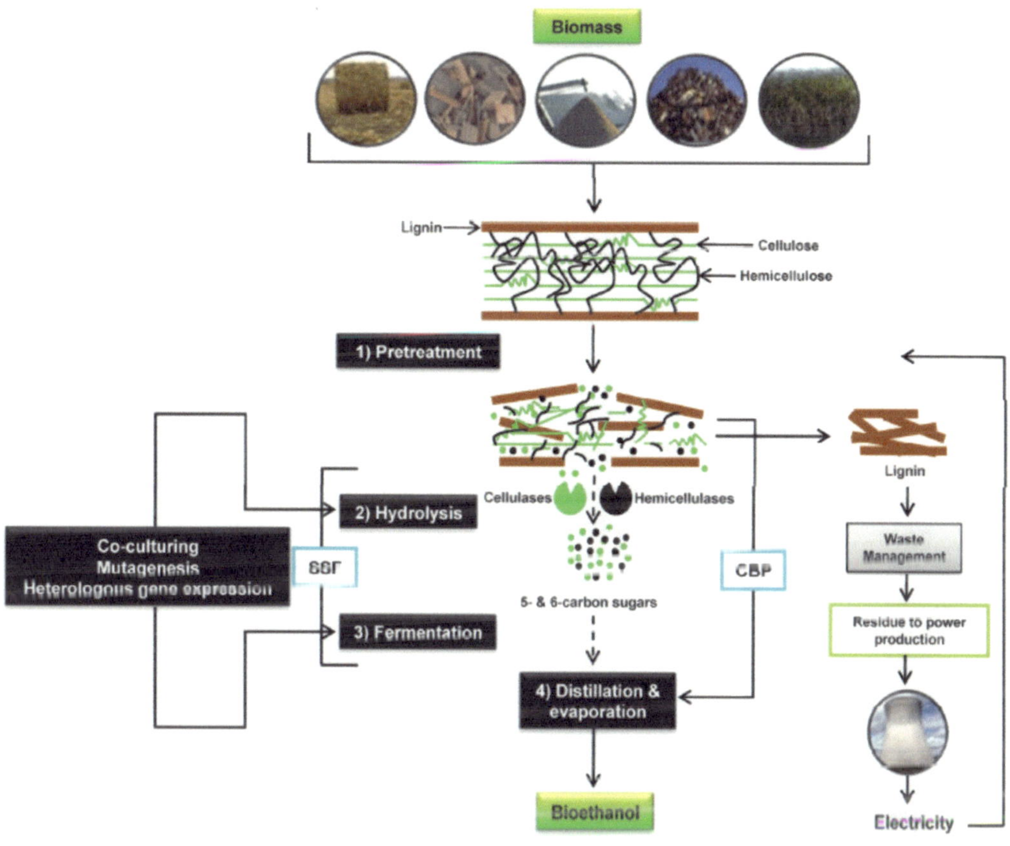

Figure 4. Schematic picture for the conversion of lignocellulosic biomass to ethanol, including the major steps (Original figure from Dashtban M., Schraft H., Qin W. Fungal Bioconversion of Lignocellulosic Residues; Opportunities & Perspectives. Int. J. Biol. Sci. 2009; 5(6):578-595. Available from http://www.biolsci.org/v05p0578.htm)

The starch synthesis is realized in the plastids and it is characterized, at the first step, by the conversion of glucose-1-phosphate and ATP to ADP-glucose and Pi by the ADP-glucose phosphorylase. After, the ADP-glucose can be used as a donor of glucose to different starch synthases that are able to elongate the glucan chain in the α-1,4 positions for both amylase and amylopectin. It is important to highlight the participation of the branching enzymes that are responsible to add α-1,6 linkages by re-organization of linear pre-existent chains. Some introduced branching can be removed by other enzymes to finalize the starch structure. The starch is highly hydrated because there are many hydroxyl groups permitting the interaction with the water. The starch is accumulated in the leaves during the day and it is used at night to maintain the respiration, the sucrose export and the grown. This transitory starch can be use through two ways, i) hydrolytic way to obtain maltose and ii) phosphorolytic way to provide carbon to the reactions in the chloroplast during the light phase.

Similar to the cellulose, the starch cleavage occurs under the action of an enzymatic complex (figure 5). The enzymes found in this complex are organized in four main groups: debranching enzymes; endoamylases; exoamylases; and transferases. The debraching enzymes are able to act exclusively on the α-1,6 glycosidic bonds and they are separated in isoamylase (EC 3.2.1.68) and pullulanase (EC 3.2.1.41). The former hydrolyze this kind of bond exclusively in amylopectin while the later hydrolyzes the α-1,6 bond from amylopectin and pullulan. Endoamylases acts on α-1,4 glycosidic bonds inside the amylose and amylopectin structure. In this group are located the α-amylases that release oligosaccharides with different length from their substrates. They can be found in many microorganisms as bacteria and fungi. Exoamylases are enzymes that hydrolyze the external bonds of amylase and amylopectin to release only glucose or maltose and β-limiting dextrin. Three main hydrolytic characteristics can be recognized for the exomylases: the specific breakdown of α-1,4 glycosidic bonds catalyzed by β-amylases (EC 3.2.1.2) and the breakdown of both α-1,4 and α-1,6 bonds catalyzed by amyloglucosidase (glucoamylase; EC 3.2.1.3) and α-glucosidase (EC 3.2.1.20). β-amylases as well as glucoamylase are able also to convert the anomeric configuration from α to β in the released maltose. In addition, glucoamylases act better on long-chain polysaccharide while α-glucosidases have preference on maltooligosaccharides. The last group of starch-converting enzymes, *i.e.* transferases, acts on α-1,4 glycosidic bonds of a donor molecule transferring part of this molecule to a glycosidic acceptor producing a new glycosidic bond. In this group are found the enzyme known as amylomaltase (EC 2.4.1.25), cyclodextrin glycosyltranferase (EC 2.4.1.19) and branching enzymes (EC 2.4.1.18). Amylomaltase and cyclodextrin glycosyltranferase have similar mechanism of reaction. Although the reduced hydrolytic activity from these enzymes, they are able to catalyze the transglycosylation reaction to obtain cyclodextrins by breakdown of α-1,4 glycosidic bonds and linkage of the reducing to the non-reducing end. However the product obtained by amylomaltase activity is linear while the cyclic product is obtained by cyclodextrin glycosyltranferase action (van der Maarel et al., 2002).

Many of these enzymes are involved in the complete hydrolysis of the starch. First, disbranching enzymes should act on the α-1,6 bonds to expose the linear structure that can be hydrolyzed by α-amylase and β-amylase (figure 5). Many authors have demonstrated the importance of obtainment of maltooligosaccharides from complex carbohydrate using amylolytic enzymes. Maltooligosaccharides can be used in different industrial sectors as food. Fungi are able to produce amylases. In addition, other important application of the starch is related with its use as source of renewable energy in the production of bioethanol as an alternative to the fossil fuels.

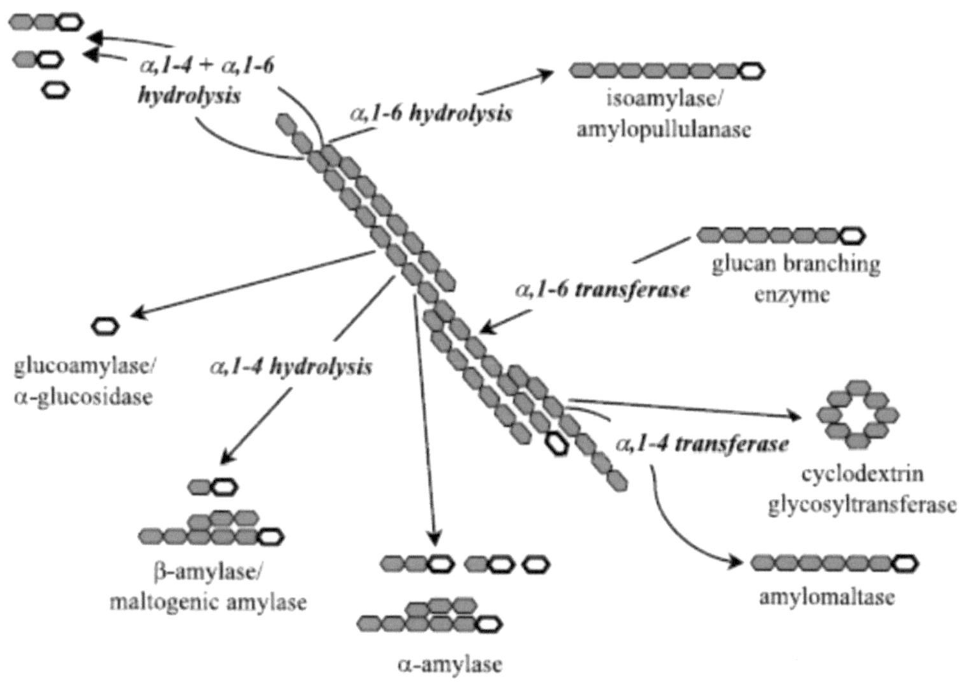

Figure 5. Different enzymes involve in the degradation of the starch. The open ring structure represents the reducing ends of a polyglucose molecule (Original figure from van der Maarel MJEC, van der Veen B, Uitdehaag JCM, Leemhuis H, Dijkhuizen L. Properties and applications of starch-converting enzymes of the α-amylase family. J. Biotechnol. 2002; 94: 137-155).

3.3. Fructosidases and other carbohydrates from biomass

Other saccharides can be also obtained from plant sources as inulin and sucrose, which are substrates for fructofuranosidase action, as β-fructofuranosidases and inulinases, which can be produced by microorganism as for example yeast and filamentous fungi. These enzymes have an important role in the microbial nutrition since monomers can be obtained and used in the metabolism. On the other hand, the enzymes involved in this process can be used with biotechnological goal.

3.3.1. The inulin and its utilization

Inulin, a polymer constituted by linear chain of β-2,1 fructufuranose residues terminated by a glucose residue, can be obtained from plant sources, especially from tubers and roots, as for example chicory, dahlia, yacon and Jerusalen artichoke. The plant sources of inulin have been considered as renewable raw material for many applications such as ethanol, obtainment of fructose syrup and fructooligosaccharides (FOS) production. This carbohydrate can be hydrolyzed by action of inulinases (2,1 β-D-fructan fructanohydrolase; EC 3.2.1.7), which can be classified into endoinulinase that hydrolyze the internal linkages from inulin to obtain inulotriose, inulotetraose and inulopentaose as end products, and exoinulinase that acts removing the terminal fructose from the non-reducing end from the inulin until the last linkage to release glucose (Ricca et al., 2007). It is important to observe that the type of enzymatic action depends on the microbial source. The most of fungal inulinases acts using the exo-mechanism. However, it was demonstrated that *Aspergillus ficuum* was able to produce endo- and exo-inulinases with different properties. The mixture of both enzymes can be considered as a good strategy to increase the conversion of the inulin to fructose (for review, see Ricca et al., 2007). Despite the action and affinity similarities for sucrose as substrate, with fructosidases as invertase, inulinases has been separated since invertase has reduced activity on high molecular mass substrates as inulin. The relation S/I has been used to separate inulinase from invertase. The S/I values depends on the inulin sources and also on the methodology used to determine the enzyme activity. However, kinetic studies are good methodologies that can be help the differentiation of these enzymes as for example considering the substrate affinity and catalytic efficiency. On the other hand, the enzymes recognized as true invertases have no activity on inulin.

The inulinases obtained from yeast are enzymes that can be linked to the cell membrane and partially secrete to the extracellular environment. In addition the synthesis of these enzymes is subject to the catabolic repression. In addition, inulinases are recognized as inducible enzymes and they are encoded by *INU* genes. The enzymes obtained from filamentous fungi has demonstrated optimum of pH activity from 4.5 to 6.0 differing than that observed for some bacterial inulinases with higher pH of activity. The optimum of temperature for activity for the most inulinases is from 30°C to 55°C but higher temperatures can be also found.

3.3.2. *The sucrose and its utilization*

Sucrose, a disaccharide constituted by D-glucose and D-fructose linked by α-1,2 glycosidic bond, is the main carbohydrate produced by plants using photosynthesis to generate ATP and NADPH, which will be used in the Calvin Cycle to fix CO_2 in the dark step. Two main enzymes are involved in the sucrose synthesis, the sucrose-phosphate syntase (EC 2.4.1.14) and the sucrose-phosphate phosphatase (EC 3.1.3.24) (Winter and Huber, 2000). After the synthesis, the sucrose produced in the photosynthetic leaves is distributed to the other plant organs and tissues. The ethanol production in Brazil is performed using rich-sucrose sugar cane juice. However, after the sucrose extraction, the residual sugar cane bagasse also has residual sucrose that can be used to obtain monosacharides by microbial action. These monomers can be used together with other monosaccharides obtained from the hydrolysis of polysaccharides present in the sugar cane bagasse, as cellulose, to fermentation process. The sucrose hydrolysis (figure 6) is catalyzed by the β-fructofuranosidases (invertases; EC 3.2.1.26), which are found in many microorganisms. The product obtained is an equimolar (1:1) mixture of the monosaccharides and residual sucrose known as invert sugar with wide application in the food and beverage industries. The fructose is much more attractive for application since it can has liquid and non-crystalizable constitution. The β-fructofuranosidases are located in the GH32 family of the glycosil hydrolases and grouped in different isoforms according to their pH of actuation as acid, alkaline and neutral enzymes (Vargas et al., 2003).

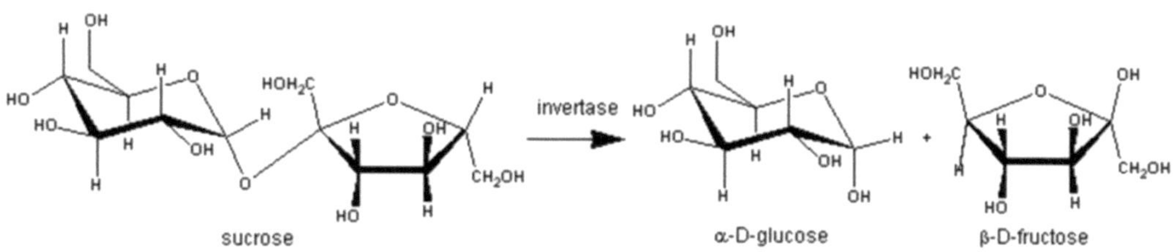

Figure 6. Hydrolysis of sucrose by invertase.

The production of β-fructofuranosidases (FFases) by microorganisms has been characterized, especially for the yeast *Saccharomyces cerevisiae*. In this microorganism it was observed the synthesis of two isoforms of FFases where one is gylcosylated and another non-glycosylated. Both enzymes are result from the two mRNA (1.8 and 1.9 kb) encoded by the same gene *SUC2*. The glycosylated enzyme is found in the periplasmic space while the non-glycosylated is found in the citosol (Belcarz *et al.*, 2002). In *S. cerevisiae* the sucrose metabolism occurs throughout two main ways: i) the sucrose is hydrolyzed in the extracellular environment by extracellular invertases to liberate glucose and fructose, which can be transported to inside of the cell by hexose transporters and ii) the sucrose is actively transported to inside of the cell by proton-symport mechanism and after hydrolyzed by intracellular invertase. The expression of the *SUC2* gene that encodes both enzymes is severally regulated by glucose (Basso et al., 2011).

The production of FFases by other microorganism, especially filamentous fungi as *Aspergillus* genera, among others, has been studied. In this situation, different fermentation processes are used, which many times the residual biomass are used as carbon sources (submerged fermentation) and/or substrates (solid-sate fermentation). Some microorganisms are able to produce multiple β-fructofuranosidases as observed for *Aureobasidium pullulans* (Yoshikawa et al., 2006). In this situation the authors observed the presence of five FFases (I, II, III, IV and V) with high FFase I activity at the initial times of culture and reduced FFase II-V activities. After initial times the FFase II-V activities are increased. In addition, the multiple FFases produced by *A. pullulans* have distinct properties as suggested by authors. FFase IV has high hydrolytic activity acting as FOS-degrading enzyme at the FOS degrading period while the participation of FFases II, III and V is uncertain, since they have significant transfructosylating activity and they are present in the FOS-degrading periods (Yoshikawa et al., 2006).

The most of β-fructofuranosidases are dimmers but monomers also can be found. The optimum of temperature and pH of reaction considering all microbial sources are variable. This carbohydrate has a negative influence on the invertase synthesis by *A. niger* as well as fructose. Only β-fructofuranoside saccharides were able to induce the invertase synthesis

(Rubio & Navarro, 2006). It was observed that the *A. niger* is able to produce two β-fructosidases known as SUC1 and SUC2. Both enzymes catalyzed the sucrose hydrolysis but only SUC 2 was able to act on inulin. Other fungal strains have been used for invertase production as *Aspergillus ochraceus* (Guimarães et al., 2007), *Aspergillus niveus* (Guimarães et al., 2009), *Aspergillus caespitosus* (Alegre et al., 2009), *Aspergillus phoenicis* (Rustiguel et al., 2011) and *Paecylomyces variotii* (Giraldo et al., 2012) using both submerged and solid-state fermentation with agro-industrial residues as carbon source/substrate. Thermostable FFases has been obtained by cultivation of *A. ochraceus* and *A. niveus* using sugar cane bagasse as carbon source.

At high sucrose concentration, some β-fructofuranosidases are able to catalyze transfructosylation reaction to obtain fructooligosaccharides (FOS) as 1-kestose (GF2), 1-nystose (GF3) and fructofuranosyl nystose (GF4). The molecular structure of these FOS can be observed in the figure 7. The GF2 is constituted by two molecules of fructose binding to the D-glycosyl unit at the non reducing end while the GF3 and GF4 by three and four fructose residues, respectively. These oligosaccharides have functional properties that have attracted the attention of different sectors. FOS are no caloric sugars that can be used by diabetic peoples with security since they are not metabolized by the organism. In addition, FOS can also stimulate the bifidobacteria development in the intestine and minimize the colon tumor. It has been demonstrated that some components of plant sources (biomass) used in pet foods exhibit FOS concentration of GF2, GF3 and GF4, as for example wheat bran, peanut hulls and barley, among others. The hydrolysis of FOS by microbial sources as bacteria using enzymes that act on these saccharides was demonstrated in some reports. Hence, enzymes that are able to act on FOS can be used to obtain saccharides as glucose from the biomass containing GF2, GF3 and/or GF4. On the other hand, different approaches have been used to obtain FOS as the utilization of immobilized enzymes on lignocellulosic materials and by substrate and enzyme engineering. Recently, the one-step FOS production was obtained using *Aspergillus phoenicis* biofilms in rich-sucrose medium as demonstrated in our laboratory (Aziani et al., 2012).

In the same way, the fuctosyltransferases as levansucrases (sucrose:2,6-β-fructans: 6-β-D-fructosyltransferase; EC 2.4.1.10), inulosucrases (sucrose:2,1-β-D-fructan: 1-β-D-fructosyltransferase; EC 2.4.1.9) and fructosyltransferase (sucrose:2,6-β-fructan:6-β-D-fructosyltranferase; EC 2.4.1.10) should be considered. The former is responsible by the synthesis of microbial levans using glucose or levan as acceptor to the β-D-fructosyl residues while the inulosucrases are able to catalyze the transference of the β-D-fructosyl residues to the sucrose or inulin as acceptors. The later are involved in the levan synthesis but it is not able to catalyze hydrolysis or exchange reactions as observed for the levansucrase (Velázques-Hernandez et al., 2009). These enzymes that catalyze the fructans synthesis are inserted in the GH68 family of glycoside hydrolases. In general, five main domains are recognized in the fructosyltransferases of microorganisms: a signal peptide; an N-terminal domain with variable length; a catalytic domain with around of 500 amino acids; a cell wall binding domain; and a C-terminal domain with variable length (Van Hijum et al., 2006).

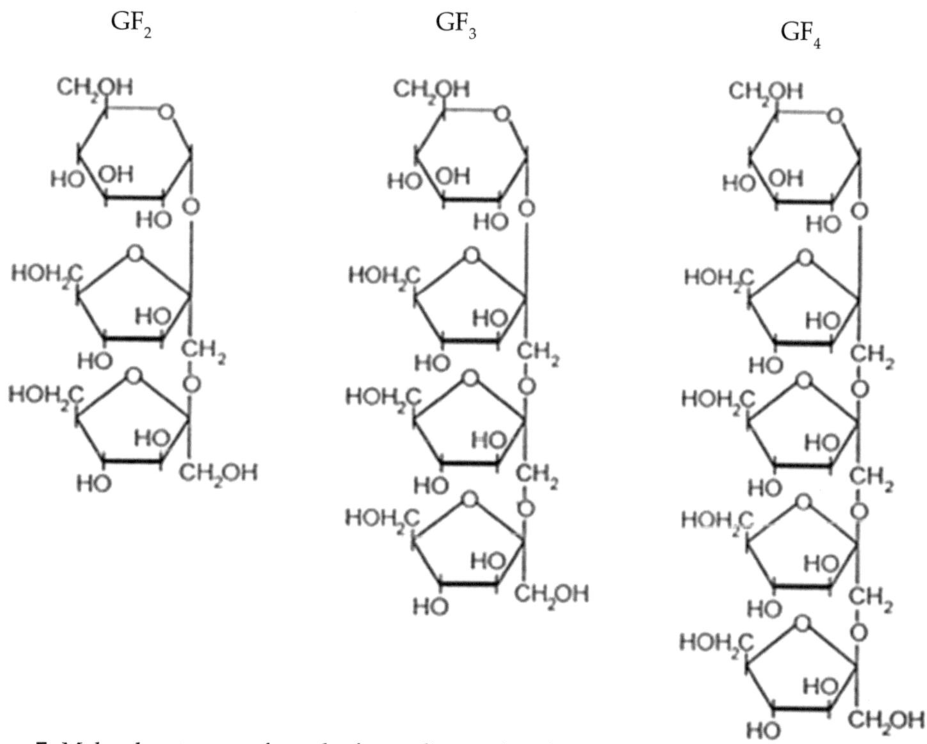

Figure 7. Molecular structure from the fructooligosaccharides nystose (GF2), 1-kestose (GF3) and fructosyl nystose (GF4).

4. Conclusion

In conclusion, the biomass that has been accumulated around the world as residue can be widely used for different applications considering its carbohydrate composition which can be accessed by microbial activity according to the enzymatic potential of each one. Microorganisms show metabolic versatility permitting the carbohydrate utilization and transformation from biomass since they are important sources of enzymes with biotechnological potential. According to this, different products can be obtained from biomass and applied in different industrial sectors. The view of the biomass as an important renewable energy source is very important to the future of the life in our planet, especially if considered the agro-industrial residues. In addition, the environment problems of bioaccumulation of residues can be reduced. Future studies to improve the biomass utilization are important as well as on the carbohydrate-active enzymes produced by microorganisms to optimize this process.

Author details

Luis Henrique Souza Guimarães

Faculdade de Filosofia, Ciências e Letras de Ribeirão Preto, University of São Paulo, Brasil

5. References

Alegre, A.C.P., Polizeli, M.L.T.M., Terenzi, H.F., Jorge, J.A. & Guimarães, L.H.S. (2009) Production of thermostable invertases by *Aspergillus caespitosus* under submerged or solid state fermentation using agro-industrial residues as carbon source. *Brazilian Journal of Microbiology*, Vol. 40, pp. 612-622.

Aranda, C., Robledo, A., Loera, O., Contreras-Esquiavel, J., Rodríguez, R. & Aguilar, C.N. (2006). Fungal invertase expression in solid-state fermentation. *Food Technology and Biotechnology*, Vol. 44, pp. 229-233.

Aziani, G., Terenzi, H.F., Jorge, J.A., Guimarães, L.H.S. (2012) Production of fructooligosaccharides by *Aspergillus phoenicis* biofilm on polyethylene as inert support. *Food Technology and Biotechnology*, Vol. 50, pp. 40-45.

Basso, T.O., de Kok, S., Dario, M., Espirito-Santo, J.C., Müller, G., Schlölg, P.S., Silva, C.P., Tonso, A., Daran, J.M., Gombert, A.K., van Maris, A.J., Pronk, J.T. & Stambuk, B.U. (2011). Engineering topology and kinetics of sucrose metabolism in *Saccharomyces cerevisiae* for improved ethanol yield. *Metabolism Engineering*, Vol. 13, pp. 694-703.

Bayer, E.A., Chanzy, H., Lamed, R. & Shoham, Y. (1998). Cellulose, cellulases and cellulosomes. *Current Opinion in Structural Biology* Vol. 8, pp. 548-557.

Belcarz, A., Ginalska, G., Lobarzewski, J. & Penel, C. (2002). The novel non-glycosilated invertase from *Candida utilis* (the properties and the conditions of production and putification). *Biochemistry and Biophysics Acta*, Vol. 1594, pp. 40-53.

Chi, Z., Chi, Z., Zhang, T., Liu, G. & Yue, L. (2009). Inulinase-expressing microorganisms and applications of inulinases. *Applied Microbiology and Biotechnology*, Vol. 82, pp. 211-220.

Dashtban, M., Schraft, H. & Qin, W. (2009). Fungal bioconversion of lignocellulosic residues: opportunities & perspectives. *International Journal of Biological Sciences*, Vol. 5, pp. 578-595.

Davies, G.J., Gloster, T.M. & Henrissat, B. (2005). Recent structural insights into the expanding world of carbohydrate-active enzymes. *Current Opinion in Structural Biology*, Vol. 15, pp. 637-645.

Fettke, J., Hejazi, M., Smirnova, J., Höchel, E., Stage, M. & Steup, M. (2009). Eukaryotic starch degradation: integration of plastidial and cytosolic pathways. *Journal of Experimental Botany*, Vol. 60, pp. 2907-2922.

Geigenberger, P. (2011). Regulation of starch biosynthesis in response to a fluctuating environment. *Plant Physiology*, Vol.155, pp. 1566–1577.

Gilbert, R.G. (2011). Size-separation characterization of starch and glycogen for biosynthesis–structure–property relationships. *Analytical and Bioanalytical Chemistry*, Vol.399, pp.1425–1438.

Giraldo, M.A., Silva, T.M., Salvato, F., Terenzi, H.F., Jorge, J.A. & Guimarães, L.H.S. (2012). Thermostable invertases from *Paecylomyces variotii* produced under submerged and solid-state fermentation using agroindustrail residues. *World Journal of Microbiology and Biotechnology*, Vol 28, pp. 463-472.

Guimarães, L.H.S., Somera, A.F., Terenzi, H.F., Polizeli, M.L.T.M. & Jorge, J.A. (2009). Production of β-fructofuranosidases by *Aspergillus niveus* using agroindustrail residues as carbon sources: characterization of an intracellular enzyme accumulated in the presence of glucose. *Process Biochemistry*, Vol. 44, pp. 237-241.

Guimarães, L.H.S., Terenzi, H.F., Polizeli, M.L.T.M. & Jorge, J.A. (2007). Production and characterization of a thermostable β-D-fructofuranosidase produced by *Aspergillus ochraceus* with agro-industrial residues as carbon source. *Enzyme and Microbial Technology*, Vol. 42, pp. 52-57.

Guzmán-Maldonado, H. & Paredes-López, O. (1995). Amylolytic enzymes and products derived from starch: A review. *Critical Review in Food Science and Nutrition*, Vol. 35, pp. 373-403.

Hizukuri, S. (1996). Starch analytical aspects. In: *Carbohydrates in Foods*. pp. 347–429. Eliasson, A.C., Ed., Marcel Dekker Inc., New York.

Hoover, R. (2010). The impact of heat-moisture treatment on molecular structures and properties of starches isolated from different botanical sources. *Critical Reviews in Food Science and Nutrition*, Vol. 50, pp. 835–847.

Jobling, S. (2004). Improving starch for food and industrial applications. *Current Opinion in Plant Biology*, Vol. 7, pp. 210-218.

Mischnick, P. & Momcilovic, D. (2005). Chemical structure analysis of starch and cellulose derivatives. *Advances in Carbohydrate Chemistry and Biochemistry*, Vol. 64, pp. 118-210.

Morrison,W.R. & Karkalas, J. (1990). Starch. In: Methods in Plant Biochemistry, Vol. 2. pp. 323–352. Academic Press, Inc., New York.

Parada, J. & Aguilera, J.M. (2011). Review: starch matrices and the glycemic response. *Food Science Technology International*, Vol. 17, pp. 187-204.

Pérez-Guerra, N., Torrado-Agrasar, A., López-Macias, C. & Pastrana, L. (2003). Main characteristics and applications of solid substrate fermentation. *Electronic Journal of Environmental, Agricultural and Food Chemistry*, Vol. 2, pp. 343-350.

Ricca, E., Calabrò, V., Curcio, S. & Iorio, G. (2007). The state of the art in the production of fructose from inulin enzymatic hydrolysis. Critical Reviews in Biotechnology, Vol. 27, pp. 129-145.

Ritte, G., Heydenreich, M., Mahlow, S., Haebel, S., Kötting, O. & Steup, M. (2006). Phosphorylation of C6 and C3-positions of glucosyl residues in starch is catalysed by distinct dikinases. *FEBS Letters*, Vol. 580, N° 20, pp. 4872-4876.

Rubio, M.C. & Navarro, A.R. (2006). Regulation of invertase synthesis in *Aspergillus niger*. *Enzyme and Microbial Technology*, Vol. 39, pp. 601-606.

Rustiguel, C.B., Oliveira, A.H.C., Terenzi, H.F., Jorge, J.A. & Guimarães, L.H.S. (2011). Biochemical properties of an extracellular β-D-fructofuranosidase II produced by *Aspergillus phoenicis* under solid-state fermentation using soy bran as substrate. *Electronic Journal of Biotechnology*, Vol. 14, N° 2 (doi: 10.2225/vol14-issue2-fulltext-1).

Szydlowski, N., Ragel, P., Hennen-Bierwagen, T.A., Planchot, V., Myers, A.M., Mérida, A., d'Hulst, C. & Wattebled, F. (2011). Integrated functions among multiple starch synthases determine both amylopectin chain and branch linkage location in *Arabidopsis* leaf starch. *Journal of Experimental Botany*, Vol. 62, N° 13, pp. 4547-4559.

Szydlowski, N., Ragel, P., Raynaud, S., Lucas, M., Roldan, I., Montero, M., Muñoz, F.J., Ovecka, M., Bahaji, A., Planchot, V., et al. (2009). Starch granule initiation in *Arabidopsis* requires the presence of either class IV or class III starch synthases. *Plant Cell*, Vol. 21, pp. 2443–2457.

Teixeira, F.A., Pires, A.V. & Nascimento, P.V.N. (2007). Sugarcane pulp in the feeding of bovine. *Revista Eletrônica de Veterinária*, Vol. 8, pp. 1-9.

van der Maarel, M.J.E.C., van der Veen, B., Uitdehaag, J.C.M., Leemhuis, H. & Dijkhuisen, L. (2002). Properties and applications of starch-converting enzymes of the α-amylase family. *Journal of Biotechnology*, Vol. 94, pp. 137-155.

Van Hijum, S.A.F.T., Kralj, S., Ozimek, L.K., Dijkhuizen, L. & van Geel-Schutten, I.G.H. (2006). Structure-function relationships og glucasucrase and fructansucrase enzymes from lactic acid bacteria. *Microbiology and Molecular Biology Review*, Vol. 70, pp. 157-176.

Vargas, W., Cumino, A. & Salerno, G.L. (2003). Cyanobacterial alkaline/neutral invertases. Origin of sucrose hydrolysis in the plant cytosol? *Planta*, Vol. 216, 951-960.

Velázquez-Hernández, M.L.; Baizabal-Aguirre, V.M., Bravo-Patiño, A., Cajero-Juárez, M., Chávez-Moctezuma, M.P. & Valdez-Alarcón, J.J. (2009). Microbial fructosyltransferases and the role of fructans. *Journal of Applied Microbiology*, Vol. 106, N°6, pp. 1763-1768.

Vijayaraghavan, K., Yamini, D., Ambika, V. & Sowdamini, N.S. (2009). Trends in inulinase production – a review. *Critical Reviews in Biotechnology*, Vol. 29, pp. 67–77.

Villena, G.K. & Gutiérrez-Correa, M. (2007). Morphological patterns of *Aspergillus niger* biofilms and pellets related to lignocellulosic enzyme productivities. *Letters in Applied Microbiology*, Vol. 45, pp. 231-237.

Weise, S.E., van Wijk, K.J. & Sharkey, T.D. (2011). The role of transitory starch in C(3), CAM, and C(4) metabolism and opportunities for engineering leaf starch accumulation. *Journal of Experimental Botany*, Vol. 62, pp. 3109-3118.

Winter, H. & Huber, S.C. (2000). Regulation of sucrose metabolism in higher plants: localization and regulation of activity of key enzymes. *Critical Review in Plant Science*, Vol. 19, pp. 31-67.

Yoshimoto, Y., Tashiron, J., Takenouchi, T. & Takeda, Y. (2000). Molecular structure and some physicochemical properties of high amylose barley starch. *Cereal Chemistry*, Vol. 77, pp. 279–285.

Chapter 21

Starch and Microbial α-Amylases: From Concepts to Biotechnological Applications

Amira El-Fallal, Mohammed Abou Dobara, Ahmed El-Sayed and Noha Omar

Additional information is available at the end of the chapter

http://dx.doi.org/10.5772/51571

1. Introduction

Starch is a polymer of glucose linked to one another through the C1 oxygen, known as the glycosidic bond. Amylases are capable of digesting these glycosidic linkages found in starch. Amylases have been isolated from diversified sources including plants, animals, and microbes, where they play a dominant role in carbohydrate metabolism. In spite of the wide distribution of α-amylase, microbial sources are used for the industrial production. This is due to their advantages such as cost effectiveness, consistency, less time and space required for production as well as ease of process modification and optimization.

In the present day scenario, α-amylases have applications in all the industrial processes such as in food, detergents, textiles and paper industry, for the hydrolysis of starch. They can also be of potential use in the pharmaceutical and fine chemical industries. In this light, microbial α-amylases have completely replaced chemical hydrolysis in the starch processing industry. Despite this, interest in new and improved α-amylase is growing and consequently, the research is intensified as well to meet requirements set by specific applications.

2. Starch

Starch and starch-containing substrates are wide spread in nature and also in industrial praxis. They can predominantly find their application in many industrial processes.

2.1. Sources and utilization

Starch occurs mainly in the seeds, roots and tubers of higher plants. Some algae produce a similar reserve polysaccharide called phytoglycogen. Plants synthesize starch as a result of photosynthesis. It is synthesized in plastids as a storage compound for respiration during dark periods. It is also synthesized in amyloplasts found in tubers, seeds, and roots as a long-term storage compound. In these latter organelles, large amounts of starch accumulate as water-insoluble granules. The shape and diameter of these granules depend on the botanical origin. Regarding to commercial starch sources, the granule sizes range from 2–30 μm (maize starch) to 5–100 μm (potato starch) (Robyt, 1998). A variety of different enzymes are involved in the synthesis of starch. Sucrose is the starting point of starch synthesis. It is converted into the nucleotide sugar ADP-glucose that forms the actual starter molecule for starch formation. Subsequently, enzymes such as soluble starch synthase and branching enzyme synthesize the amylopectin and amylose molecules (Smith, 1999).

Starch-containing crops form an important constituent of the human diet. Besides the direct use of starch-containing plant parts as a food source, starch is harvested and chemically or enzymatically processed into a variety of different products such as starch hydrolysates, glucose syrups, fructose, starch or maltodextrin derivatives, or cyclodextrins. In spite of the large number of plants able to produce starch, only a few plants are important for industrial starch processing. The major industrial sources are maize, tapioca, potato, and wheat.

2.2. Structure and properties

Starch is a polymer of glucose linked to one another through the C1 oxygen by a glycosidic bond. This glycosidic bond is stable at high pH but hydrolyzes at low pH. At the end of the polymeric chain, a latent aldehyde group is present. This group is known as the reducing end. Two types of glucose polymers are present in starch: (i) amylose and (ii) amylopectin. While amylopectin is soluble in water, amylose and the starch granule itself are insoluble in cold water.

Amylose is a linear polymer consisting of up to 6000 glucose units with α, 1-4 glycosidic bonds (Fig. 1a.). The number of glucose residues, also indicated with the term DP (degree of polymerization), varies with the origin. The relative content of amylose and amylopectin varies with the source of starch. The average amylose content in most common starches, e.g. in barley, corn and potato, is 20-30% (Marc et al., 2002).

Amylopectin consists of short α, 1-4 linked linear chains of 10–60 glucose units and α,1-6 linked side chains with 15–45 glucose units (Fig. 1b.). The average number of branching points in amylopectin is 5% (Thompson, 2000), but varies with the botanical origin. The complete amylopectin molecule contains about 2 000,000 glucose units, thereby being one of the largest molecules in nature (Marc et al., 2002). The most commonly accepted model of the structure of amylopectin is the cluster model, in which the side chains are ordered in clusters on the longer backbone chains (Bertoft, 2007; Thompson, 2000). In general, Zhu et al. (2011) suggested that the internal part of amylopectin is critical to the physical behavior of granular starch.

The diameter of starch granules ranges from 2 to 100μm (Whistler & Daniel, 1985) depending on its source. The orientation of the starch chains is thought to be perpendicular to the granule surface (French, 1984). Native starch is partly crystalline. The crystallinity of native starch varies between 15 and 45% depending on the origin and pretreatment (French, 1984). According to the currently accepted concept, amylopectin forms the crystalline component whereas amylose exists mainly in the amorphous form (Hanashiro et al., 1996; Marc et al., 2002; Zobel, 1992). Structural studies have shown that native starch has crystalline polymorphism. In x-ray diffraction, cereal starch typically gives A-type patterns of monoclinic symmetry, and tuber starch gives B-type patterns of hexagonal symmetry (Gerard et al., 2000; Imberty et al., 1991). The crystal lattice of B-type starch contains more water molecules than the A- structures, which is proposed to be the reason for higher stability of the A- structure. Both structures' molecular conformations are practically identical. They have left-handed double helices with parallel strands. Double helices contain six glucose units per turn in each chain and the glucose units are in a chair conformation. With in the double helix, there are inter-chain but no intra-chain hydrogen bonds. In additional, parallelly packed double helices are connected through a hydrogen bonding network.

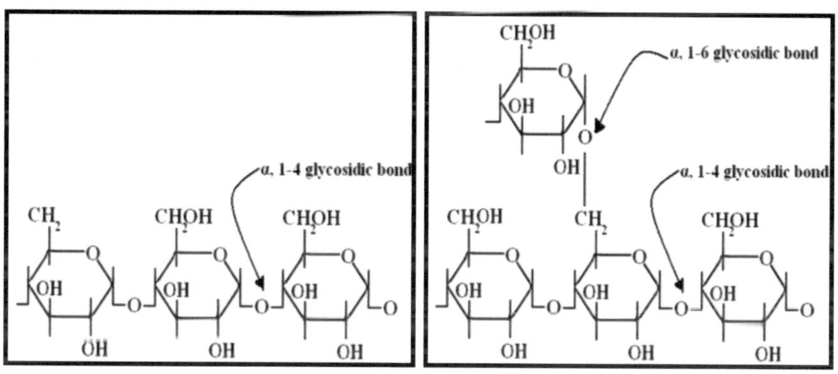

(a) amylose (b) amylopectin

Figure 1. Amylose and amylopectin chain structure

3. Amylases

Amylases are a class of enzymes that are capable of digesting these glycosidic linkages found in starches. Amylases can be derived from a variety of sources. They are present in all living organisms, but the enzymes vary in activity, specificity and requirements from species to species and even from tissue to tissue in the same organism. Raw-starch digesting amylases are produced by a variety of living organisms, ranging from microorganisms including fungi, yeast, and bacteria to plants and humans.

3.1. Microbial sources of amylases

Several amylase-producing bacteria, fungi and other microrganisms have been isolated and characterized over many decades. Bacteria and fungi secrete amylases outside their cells to carry out extra-cellular digestion.

Among mold species producing high levels of amylase, those of *Aspergillus niger, Aspergillus oryzae* (Aunstrup, 1979), *Thermomyces lanuginosus* (Arnesen et al., 1998) and *Penicillium expansum* (Doyle et al., 1989) in addition to many species of the genus *Mucor* (Domsch et al., 1995; Petruccioli & Federici, 1992; Zare-Maivan & Shearer, 1988). It was reported that four species of *Ganoderma* mushrooms could produce relatively weak amylase in sawdust medium (Y.W. Wang & Y. Wang, 1990). Amylolytic yeasts differ strongly with regard to amylase secretion and the extent of starch hydrolysis (De Mot et al., 1984a, 1984b). Strains of *Filobasidiuim capsuligenum* are capable of extensive starch hydrolysis (De Mot et al., 1984c; McCann& Barnett, 1984).

Regarding to bacteria, *Bacillus* spp and the related genera produce a large variety of extracellular enzymes, of which amylases are of particular significance to the industry e.g., *B. cereus* (Rhodes et al., 1987), *B. circulans* (Siggens, 1987), *B. subtilis* (El-Banna et al., 2007), *B. licheniformis* (El-Banna et al., 2008) and *Clostridium thermosulfurogenes* (Hyun & Zeikus, 1985a). Bacteria belonging mainly to the genus *Bacillus* have been widely used for the commercial production of thermostable α-amylase (Tonkova, 2006). However, most of the *Bacillus* liquefying amylases, such as the enzymes from *B. amyloliquefaciens* and *B. stearothermophilus* have pH optima of between 5 and 7.5 (Yamamoto, 1988). Many alkaline amylases have been found in cultures of *Bacillus* sp. (Hayashi et al., 1988; Kim et al., 1995). This alkaline amylases are all of the saccharifying type, except for the enzymes from *Bacillus* sp. strain 707 (Kimura et al., 1988) and *B. licheniformis* TCRDC-B13 (P. Bajpai and P.K. Bajpai, 1989). Thermostable β-amylases have been isolated from *Bacillus* species (Shinke et al., 1974; Takasaki, 1976). Also, *Lactobacillus plantarum* strain A6 was selected for its ability to synthesize large amounts of extracellular α-amylase (Giraud et al., 1991). Furthermore, a variety of ruminal bacteria exhibit the ability to utilize starch as a growth substrate and are present in the rumen in sufficient numbers to be of quantitative significance in the fermentation of this substrate. These species include *Bacteroides ruminicola, Ruminobacter amylophilus, Butyrivibrio fibrisolvens, Selenomonas ruminantium,* and *Streptococcus bovis* (Russell, 1984).

Genes encoding intracellular α-amylases have been reported for *Escherichia coli* and *Streptococcus bovis* (Satoh et al., 1997; Whitehead & Cotta, 1995). Although there has been some characterization of these activities, no clear physiological role for intracellular α-amylase has been established for either *E. coli* or *Streptococcus bovis*. However, it is postulated that it plays an important role in rapid cell growth in *Streptococcus bovis* (Brooker & McCarthy, 1997).

Many hyperthermophilic microorganisms possess starch-hydrolyzing enzymes in their genomes even though they live in environments where starch is rare (Sambrook et al., 1989). Among the polysaccharide-degrading enzymes of *Thermotoga maritime* described so far are two α-amylases, one is an extracellular putative lipoprotein (AmyA) (Liebl et al., 1997) and one is located in the cytoplasm (AmyB) (Lim et al., 2003). *Geobacillus thermoleovorans* has been found to produce hyperthermostable, high maltose-forming and Ca^{2+} independent α-amylase (Malhotra et al. 2000; Narang & Satyanarayana 2001). Numerous hyperthermophilic Archaea, especially deep-sea *Thermococcale* and *Sulfolobus* species have been reported to produce α-amylases (Leuschner and Antranikian, 1995; Sunna et al., 1997).

The industrial potential of high-maltose forming α-amylases from *Thermomonospora curvata* (Collins et al., 1993) is limited by their moderate thermostability and Ca^{2+} requirement.

α-Amylases are secreted by several species of *Streptomyces*, for example *S. albus* (Andrews & Ward, 1987), *S. griseus* IMRU3570 (Vigal et al., 1991), *S. thermocyaneoviolaceus* (Hang et al., 1996). Gene encoding extracellular α-amylase has been cloned from many *Streptomyces* species (Bahri & Ward, 1990; Virolle et al., 1988). In addition, α-amylase activity of *Thermoactinomyces* species was first reported by Kuo & Hartman (1966). After that, several α-amylases with different characters were found in other studies (Obi & Odibo, 1984; Omar et al., 2011; Shimizu et al., 1978; Uguru et al., 1997). Within actinomycetes, available reports on β-amylase production are scanty and refer mainly to nonthermostable enzyme (Shinke et al., 1974).

3.2. Amylases types

Enzymes belonging to amylases, endoamylases and exoamylases, are able to hydrolyse starch. These enzymes are classified according to the manner in which the glycosidic bond is attacked. The starch degrading enzymes are found in the numerous glycoside hydrolase (GH) families (13, 14 and 15), mainly in GH family 13 (Coutinho & Henrissat, 1999; Henrissat, 1991).

Endoamylases are able to cleave α,1-4 glycosidic bonds present in the inner part (endo-) of the amylose or amylopectin chain. α-amylase (EC 3.2.1.1) is a well-known endoamylase. It is found in a wide variety of microorganisms, belonging to the Archaea as well as the Bacteria (Pandey et al., 2000). The end products of α-amylase action are oligosaccharides with varying length with α-configuration and α-limit dextrins, which constitute branched oligosaccharides. α-amylases are often divided into two categories according to the degree of hydrolysis of the substrate (Fukumoto & Okada, 1963). Saccharifying α-amylases hydrolyze 50 to 60% and liquefying α-amylases cleave about 30 to 40% of the glycosidic linkages of starch.

Enzymes belonging to the second group, the exoamylases, either exclusively cleave α,1-4 glycosidic bonds such as β-amylase (EC 3.2.1.2) or cleave both α,1-4 and α,1-6 glycosidic bonds like amyloglucosidase or glucoamylase (EC 3.2.1.3) and α-glucosidase (EC 3.2.1.20). Exoamylases act on the external glucose residues of amylose or amylopectin and thus produce only glucose (glucoamylase and α-glucosidase), or maltose and β-limit dextrin. β-amylase and glucoamylase also convert the anomeric configuration of the liberated maltose from α to β. Glucoamylase and α-glucosidase differ in their substrate preference: α-glucosidase acts best on short maltooligosaccharides and liberates glucose with α-configuration while glucoamylase hydrolyzes long-chain polysaccharides best. β-amylases and glucoamylases have also been found in a large variety of microorganisms (Pandey et al., 2000).

3.3. α-amylases actions and structure

3.3.1. *Mode of action*

In general, it is believed that α-amylases are endo-acting amylases which hydrolyze α-(1-4) glycosidic bonds of the starch polymers internally. Several models for amylase action pattern have been proposed, such as the random action and the multiple attack action. Random action has also been referred to as a single attack or multi-chain attack action (Azhari & Lotan, 1991). In the former, the polymer molecule is successively hydrolysed completely before dissociation of the enzyme-substrate complex. While, in the latter, only one bond is hydrolysed per effective encounter. The multiple attack action is an intermediate between the single-chain and the multi-chain action (Bijttebier et al., 2008) where the enzyme cleaves several glycosidic bonds successively after the first (random) hydrolytic attack before dissociating from the substrate.

In short, it can clearly be seen that the multiple attack action is generally an accepted concept to explain the differences in action pattern of amylases (Kramhøft et al. 2005; Svensson et al. 2002). However, most of the endoamylases have a low to very low level of multiple attack action (Bijttebier et al., 2008). Although only few reports deal with the influence of pH and temperature on the action pattern of amylases, this influence was confirmed. Bijttebier et al. (2007) showed that the level of multiple attack of several endoamylases increased with temperature to a degree depending on the amylase itself.

3.3.2. *Molecular weight*

Despite wide difference of microbial α-amylases characters, their molecular weights are usually in the same range 40-70 kDa (Gupta et al., 2003). Ratanakhanokchai et al. (1992) reported the highest molecular weight of α-amylases, 210 kDa, for *Chloroflexus aurantiacus*. Whereas, 10 kDa of *Bacillus caldolyticus* α-amylase was reported to be the lowest value (Gupta et al., 2003).

This molecular weight may be raised due to glycosylation as in the case of *T. vulgaris* α-amylase that reach 140 kDa (Omar et al., 2011). In contrast, proteolysis may lead to decrease in the molecular weight. For example, α-amylase of *T. vulganis* 94-2A (AmyTV1) is a protein of 53 kDa and smaller peptides of 33 and 18 kDa that have been shown to be products of limited AmyTV1 proteolysis (Hofemeister et al., 1994).

3.3.3. *Modular structure*

α-amylases from different organisms share about 30% amino acid sequence identity and all belong to the same glycosyl hydrolase family 13 (Henrissat & Bairoch, 1993). The three dimensional (3D) structures of α-amylases have revealed monomeric, calcium-containing enzymes, with a single polypeptide chain folded into three domains (A-C).

The most conserved domain in α-amylase family enzymes, the A-domain, consists of a highly symmetrical fold of eight parallel β-strands arranged in a barrel encircled by eight α-helices.

The highly conserved amino acid residues of the α-amylase family involved in catalysis and substrate binding are located in loops at the C-termini of β-strands in this domain. This is typical to all enzymes belonging to the α/β –barrel protein family (Farber & Petsko, 1990).

α-amylases have a B-domain that protrudes between β-sheet no 3 and α-helix no. 3. It ranges from 44 to 133 amino acid residues and plays a role in substrate or Ca^{2+} binding (Marc et al., 2002). The sequence of this domain varies most; in *Bacillus* α-amylases it is relatively long and folds into a more complex structure of β-strands (Machius et al., 1995), whereas in barley α-amylase there is an irregularly structured domain of 64 residues (Kadziola et al., 1994).

All known α-amylases, with a few exceptions, contain a conserved Ca^{2+} binding site which is located at the interface between domains A and B (Linden et al., 2003; Prakash & Jaiswal, 2010). In addition, α-amylase produced by *Bacillus thermooleovorans* was found to contain a chloride ion binding site in their active site (Malhotra et al., 2000), which has been shown to enhance the catalytic efficiency of the enzyme, presumably by elevating the pKa of the hydrogen-donating residue in the active site (Prakash & Jaiswal, 2010).

α-amylases have a domain C which is relatively conserved and folds into an antiparallel β-barrel. The orientation of domain C relative to domain A varies depending on the type and source of amylase (Bayer et al., 1995). The function of this domain is unknown.

Structural studies have confirmed that the active sites of glycosyl hydrolases are composed of multiple binding sites, or subsites, for the sugar units of polymeric substrates. The open active site cleft is formed between domains A and B, so that residues from domain B participate in substrate binding. The substrate binding sites are commonly lined with aromatic residues which make hydrophobic stacking interactions with the sugar rings. In addition, the active sites contain many residues which form hydrogen bonds to the substrate either directly or via water molecules (Aleshin et al., 1994; Svensson & Sogaard, 1993).

In Taka-amylase A, the first examined protein α-amylase by X-ray crystallography, three acidic residues, i.e., one glutamic and two aspartic acids were found at the centre of the active site (Matsuura et al., 1984), and subsequent mutational studies have shown that these residues are essential for catalysis (Janecek, 1997; Svensson, 1994) The glutamic acid residue is now believed to be the proton donor, while the first of the two conserved aspartic acids appearing in the amino acid sequence of an α-amylase family member is thought to act as the nucleophile. The role of the second aspartic acid is less certain, but it has been suggested to be involved in stabilising the oxocarbenium ion-like transition state and also in maintaining the glutamic acid in the correct state of protonation for activity (Uitdehaag et al., 1999). These residues occur near the ends of strands 3, 4, 5 and 7 of the α/β -barrel and are found in four short sequences, long-recognised as being conserved in α-amylase family enzymes.

3.3.4. *Glycosylation*

Glycosylation is one of the major post-translation modifications that affect a variety of enzyme functions including secretion, stability, and folding (Barros et al., 2009; Shental-

Bechor & Levy, 2009). Oligosaccharides are usually linked to asparagine side chains (N-linked glycosylation) or to serine and threonine hydroxyl side chains (O-linked glycosylation) (Shental-Bechor & Levy, 2009).

Glycoproteins have been detected in α-amylases of *A. oryzae* (Eriksen et al., 1998), *B. stearothermophilus* (Srivastava, 1984) and *B. subtilis* strains (Matsuzaki et al., 1974; Yamane et al., 1973). Generally, this is about 10 % for most α-amylases (Vihinen & Mantsala, 1989). These carbohydrate moieties are thought to be responsible for high molecular weight of some α-amylases. A carbohydrate content as high as 56 % has been reported in *S. castelii* (Sills et al., 1984). Also, the high molecular weight α-amylase of 140 kDa produced by *T. vulgaris* (Abou Dobara et al., 2011) is a good example of highly glycosylated α-amylase (Omar et al., 2011). Using SDS-PAGE, glycoproteins can be detected by initial oxidation of carbohydrates by periodic acid and subsequent staining with cationic dyes such as alcian blue (Wardi & Michos, 1972).

3.4. Production of microbial α-amylases

The major advantage of using microorganisms for the production of amylases is the economical bulk production capacity and easy manipulation of microbes to obtain enzymes of desired characteristics (Lonsane & Ramesh, 1990). Screening for the α-amylase producers is a key step for production. Starch hydrolysis is usually detected directly on plates as clear zones surrounding the colonies. The diameter of the area of hydrolysis, with in limits, was always related to the potency of the amylase (Dhawale et al., 1982).

3.4.1. Factors affecting production

The production and stability of α-amylase in the medium is affected by a variety of physicochemical factors. In spite of expression's possibility under a wide range of culturing conditions, α-amylase could be denatured under some conditions. Many proteins easily aggregate into so-called inclusion bodies during expression in bacterial systems (Espargaro et al., 2008). Inhibition of protein aggregation during fermentation/expression can be achieved by adjusting the production conditions (Bahrami et al., 2009; Hao et al., 2007).

Regarding to the incubation period, many investigators have found that extracellular α-amylase production is growth associated (Abou Dobara et al., 2011; Asoodeh et al., 2010; Murthy et al., 2009). The changes in productivity of extracellular enzymes can be attributed to the differences in the timing of induction of separate components of the enzyme system, the inhibition by products of substrate hydrolysis and differential inactivation by proteases and/or variation in the pH during cultivation conditions (Tuohy & Coughlan, 1992; J.P. Wang et al., 1993). The accumulation of sugars over a critical concentration in the medium is well documented to inhibit the enzyme production (Dona et al., 2010; J.P. Wang et al., 2006).

Among the physical parameters, the temperature and pH of the medium play an important role in α-amylase production and stability. Generally, the influence of temperature on amylase production is related to the growth of the organism. Hence, the optimum temperature depends on whether the culture is mesophilic, thermophilic or psycrophilic. Among the fungi and actinomycetes, most amylase production studies achieved the optimum yields within the range 25°C- 40°C (Gupta et al., 2003). However, thermophilic fungi ,such as *Thermomyces lanuginosus* (Mishra & Maheshwari, 1996), and actinomycetes, namely; *Thermomonospora fusca* (Busch & Stutzenberger, 1997) and *Thermoactinomyces vulgaris* (Abou Dobara et al., 2011) have been reported to produce α-amylase optimally at 50 °C, 55 °C and 55 °C, respectively. On the other hand, it has been produced at a wider range of optimal temperature by bacteria reaching to 90 °C in *Thermococcale* and *Sulfolobus* species (Leuschner & Antranikian, 1995; Sunna et al., 1997). Also, the pH values were reported to serve as an indicator of the initiation and end of enzyme synthesis (Friedrich et al., 1989) because the change in pH affects α-amylase stability in the medium (Calamai et al., 2005). It is worth noting that the α-amylase active site consists of a large number of charged groups (Lawson et al., 1994; Strokopytov et al., 1996; Uitdehaag et al., 1999) which explain the fact that most α-amylases had optimum pH in the acidic to neutral range (Bozic et al., 2011; Pandey et al., 2000; Sun et al., 2010).

In general, amylase activity is connected with the substrate utilization. The inducibility nature of α-amylase has been assured in different microorganisms (Abou Dobara et al., 2011; Aiyer, 2004; Asoodeh et al., 2010; Ryan et al., 2006). α-amylase production is also appeared to be subjected to catabolite repression by maltose and glucose, like most other inducible enzymes that are affected by substrate hydrolytic products (Bhella & Altosaar, 1988; Morkeberg et al., 1995). However, α-amylase synthesis by *Bacillus* strains was reported to not subject to catabolite repression by monosaccharides (Kalishwaralal et al., 2010). Gupta et al. (2003) have classified xylose and fructose as strongly repressive to α-amylase synthesis. Addition of starch to the medium has normally been employed for the production of α-amylase from various microorganisms as reported in the literature.

Nitrogen source as a basal component of the medium is a major factor affecting α-amylase production. Its effect was not only as a nitrogen source but also as a metal ion source and a pH controller as well. Many investigators had recorded that organic nitrogen sources supported maximum α-amylase production by various bacteria (Abou Dobara et al., 2011; Aqeel & Umar, 2010; Mrudula & Kokila, 2010; Saxena et al., 2007). The increased α-amylase production by organic nitrogen sources could be attributed to the high nutritional amino acids and vitamins content. However, various inorganic salts have been reported to support better production in fungi (Gupta et al., 2003). As a metal ion source, ammonium chloride was found to enhance the production of the α-amylase by *T. vulgaris*, where chloride is a stabilizer, over that of other ammonium salts (Abou Dobara et al., 2011). In addition, the same authors also reported different productivity of α-amylase by using sodium nitrate from potassium nitrate.

3.4.2. Activity measurement of enzyme

The diversity and heterogeneity of natural substrates coupled with the mixed specificities of individual enzymes presents a problem in the characterization of amylases. Furthermore, the enzymatic degradation of native insoluble substrates involves steps and mechanisms which are not yet understood at the molecular level. Therefore biochemical studies always use starch in some modified form to simplify analyses. There are basically four different types of substrates used for activity measurements: purified insoluble substrates approximated to a native substrate, modified insoluble substrates, soluble modified polysaccharides and soluble oligosaccharides. Catalytic activity is usually measured by quantifying formed soluble saccharides or chromophoric aglycon. The action of enzyme on insoluble substrates can also be assayed by other means. For example, a viscosimetric method has been used to measure α-amylase activity on starch pastes (Marciniak & Kula, 1982).

The measurement of soluble products from insoluble or soluble polymeric substrates often means assaying the formed reducing sugars. One of the simplest and most widely used is the 3, 5-dinitrosalisylic acid (DNS) method (Miller, 1959). However, the colour development in the reaction is not strictly proportional to the number of reducing sugars present, but also to the length of the oligosaccharides, leading to higher apparent reducing values with longer sugars (Robyt & Whelan, 1972). DNS itself also breaks down the substrate. Several other reducing sugar determination methods have also been developed. In some cases dye groups have been attached to the polymeric substrate, e.g. dyed amylose-and amylopectin (Klein et al., 1970) and dyed and cross-linked starch (Cesk et al., 1969). The enzymatic assay is based on colour released from the substrate.

Starch forms a deep blue complex with iodine and with progressive hydrolysis of the starch, it changes to red brown. Several procedures have been described for the quantitative determination of amylase based on the reduction in blue colour intensity resulting from enzyme hydrolysis of starch (Swain et al., 2006). This method determines the dextrinising activity of α-amylase in terms of decrease in the iodine colour reaction. Also, the coupled assay methods have been used for amylases, in which the concentration of released glucose is determined either by glucose oxidase/peroxidase (Kunst et al., 1984) or by hexokinase/glucose-6-phosphate dehydrogenase method (Rauscher, 1984).

Generally, various available methods for the determination of α-amylase activity are based on decrease in starch–iodine colour intensity, increase in reducing sugars, degradation of colour-complexed substrate and decrease in viscosity of the starch suspension.

3.4.3. *Purification of enzyme*

Purification is a key step in the enzymes production where residual cell proteins and other contaminants are removed. Different techniques have been developed for purification of enzymes based on their properties, prior to their characterization or use in biotechnological and industrial processes. The commercial use of α-amylase generally does not require purification of the enzyme but enzyme applications in food industries, pharmaceutical and clinical sectors require high purity amylases. The enzyme in purified form is also a prerequisite in studies of structure-function relationships and biochemical properties.

The used methods to purify amylases can vary considerably, but most purification protocols involve a series of steps (Sun et al., 2010). The choice of purification protocol naturally depends on the intended use, the highest purity usually being required for basic purposes in which even separation of isozymes may be important. The purity and the yield attained depend on the number of steps and separation techniques employed.

The purification of α-amylases from microbial sources in most cases has involved classical purification methods. These methods involve separation of the culture from the fermentation broth, selective concentration by precipitation using ammonium sulphate or organic solvents. The crude enzyme is then subjected to chromatography. The most commonly used techniques are usually affinity chromatography, ion exchange, and/or gel filtration. Cross-linked starch or starch derivatives are useful affinity adsorbents for the isolation of bacterial α-amylases (Somers et al., 1995). Primarini & Ohta (2000) isolated and separated two pure α-amylases from *Streptomyces* sp. using starch adsorption, α-CD Sepharose 6B and DEAE-Toyopearl 650M. Adsorption of α-amylase, from *Streptomyces* sp. E-2248; on starch followed by separation on DEAE-Toyopearl and Toyopearl-HW55S gave the highest purification (2130-fold) (Kaneko et al., 2005). Also, α-amylase from *Bacillus licheniformis* has been purified 6-fold with a yield of 38% using by two gel filtration chromatography steps on Sephadex G-100 and Superose 12 column (Bozic et al., 2011).

In addition to the classical chromatographic techniques, immunoaffinity chromatography has been applied for the preparation of highly purified amylases (Jang et al., 1994). Recent advances in the understanding of the physical and functional properties of amylases, and of the selectivity and capacity of the adsorbents, have led to greater rationality in the design of separation methods. However, the potential of the methods for the separation of amylases has not been fully exploited.

3.4.4. *Industrial desirable aspects*

The stability of biocatalysts is often a limiting factor in the selection of enzymes for industrial applications due to the elevated temperature or extreme pH of many biotechnological processes. Therefore, there is a continuing demand to improve the stability of the enzymes and thus meet the requirements set by specific applications.

As an example, the problem with traditional detergent enzymes is that they have to function in a washing machine under conditions that are very unfavorable for the stability of the enzyme. The pH is highly alkaline in washing conditions. The high temperature (55–60°C) in a dishwasher requires thermostable enzymes. In addition, it is preferred to be resistant to various detergent ingredients, such as surfactants, chelating and oxidative agents (bleach).

In general, temperature has a complex effect on protein either directly or indirectly for both physical and chemical induced aggregation processes (Y.W. Wang et al., 2010). Therefore, it is the most critical environmental factor for consideration when proteins are handled during the entire development and commercialization processes. The advantages for using thermostable α-amylases in industrial processes include the decreased risk of contamination, the increased diffusion rate and the decreased cost of external cooling. In short, almost all industries need thermostable enzymes. Besides thermostability and other factors such as activity with high concentrations of starch, i.e. more than 30% dry solids, or the protein yields of the industrial fermentation are important criteria for commercialization (Schäfer et al., 2000). Also, α-amylases with wide pH range is desired to satisfy all applications either acidic as glucose syrup production or alkali as detergents industries.

However, there is a recent trend to use intermediate temperature stable (ITS) α-amylases (Ahuja et al., 1998, as cited in Gupta et al., 2003). Olesen (1991) found that this feature render the enzyme to be useful for baking industry through avoiding stickiness in bread. Also, a modern trend among consumers is to use colder temperatures for doing the laundry or dishwashing. At these lower temperatures, detergents with α-amylases optimally working at moderate temperatures and alkaline pH would be favourable (Marc et al., 2002). Although a wide variety of microbial α-amylases is known, α-amylase with 'ITS' property has been reported from only a few microorganisms (Gigras et al., 2002).

Another important desirable feature is calcium independency. Most known α-amylases, with a few exceptions, contain a conserved Ca^{2+} binding site (Linden et al., 2003; Prakash & Jaiswal, 2010) which make calcium be important to the enzyme activity. In manufacture of fructose syrup, the Ca^{2+} ions inhibit the glucose isomerase enzyme used in the final step of the process (Tonkova, 2006) and may lead to the formation of inorganic precipitates which have deleterious effects on fermentation and downstream processing (Kelly et al., 2009). Because the removal of these metal ions is both cost and time consuming to the overall industrial process (Kelly et al., 2009), the use of stable and functional α-amylases in the absence of Ca^{2+} ions at high temperatures would be highly favored.

4. Biodegradation of starch

The degradation of starch occurs mainly through the action of microorganisms in plant litter and soil. Since the native substrate is water-insoluble and cannot penetrate into cells, the biodegradation of starch occurs extracellularly. Amylases are mainly secreted into the medium or are found membrane-bound. Some microbial strains are known to produce intracellular amylases; the reason for this is unknown (Vihinen & Mantsala, 1989).

4.1. Enzymatic degradation of starch

The effective hydrolysis of starch demands the action of many enzymes due to its complexity, although a prolonged incubation with one particular enzyme can lead to (almost) complete hydrolysis. Few microorganisms produce a complete set of enzymes capable of degrading starch efficiently. There are basically four groups of starch-converting enzymes: (i) endoamylases; (ii) exoamylases; (iii) debranching enzymes; and (iv) transferases.

Endoamylases are able to cleave α,1-4 glycosidic bonds present in the inner part (endo-) of the amylose or amylopectin chain. Exoamylases act on the external glucose residues of amylose or amylopectin and thus produce only glucose (glucoamylase and α-glucosidase), or maltose and β-limit dextrin (β-amylase).

The third group of starch-converting enzymes is the debranching enzymes that exclusively hydrolyze α,1-6 glycosidic bonds: isoamylase (EC 3.2.1.68) and pullulanase type I (EC 3.2.1.41). These enzymes exclusively degrade amylopectin, thus leaving long linear polysaccharides. There are also a number of pullulanase type enzymes that hydrolyze both α, 1-4 and α,1-6 glycosidic bonds. These belong to the group II pullulanase and are referred to as α-amylase–pullulanase or amylopullulanase. The main degradation products are maltose and maltotriose.

The fourth group of starch-converting enzymes are transferases that cleave an α,1-4 glycosidic bond of the donor molecule and transfer part of the donor to a glycosidic acceptor with the formation of a new glycosidic bond. Enzymes such as amylomaltase (EC 2.4.1.25) and cyclodextrin glycosyltransferase (EC 2.4.1.19) form a new α, 1-4 glycosidic bond while branching enzyme (EC 2.4.1.18) forms a new α,1-6 glycosidic bond. Cyclodextrin glycosyltransferases have a very low hydrolytic activity and make cyclic oligosaccharides with 6, 7, or 8 glucose residues and highly branched high molecular weight dextrins, the cyclodextrin glycosyltransferase limit dextrins. Amylomaltases are very similar to cyclodextrin glycosyltransferases with respect to the type of enzymatic reaction. The major difference is that amylomaltase performs a transglycosylation reaction resulting in a linear product while cyclodextrin glycosyltransferase gives a cyclic product.

Depending on the relative location of the bond under attack as counted from the end of the chain, the products of this digestive process are dextrin, maltotriose, maltose, and glucose, etc. Dextrins are shorter, broken starch segments that form as the result of the random hydrolysis of internal glucosidic bonds. A molecule of maltotriose is formed if the third bond from the end of a starch molecule is cleaved; a molecule of maltose is formed if the point of attack is the second bond; a molecule of glucose results if the bond being cleaved is the terminal one; and so on.

Most of the enzymes that convert starch belong to one family based on the amino acid sequence homology. the α-amylase family or family 13 glycosyl hydrolases according to the classification of Henrissat (1991). Other little enzymes that convert starch don't belong to family 13 glycosyl hydrolases like β-amylases that belong to family 14 glycosyl hydrolases (Henrissat & Bairoch, 1993); and glucoamylases which belong to family 15 glycosyl hydrolases (Aleshin et al., 1992).

4.2. Catalytic mechanism and substrate binding

The α-glycosidic bond is very stable having a spontaneous rate of hydrolysis of approximately 2×10^{-15} s^{-1} at room temperature (Wolfenden et al., 1998). Members of the α-amylase family enhance this rate so enormously that they can be considered to belong to the most efficient enzymes known.

The α-amylase family enzymes always carry strictly conserved three essential catalytic residues (Matsuura et al., 1984). Of these three residues, the roles of Glu230 and Asp206 have been generally accepted as working for acid (proton donor) and base (nucleophile) catalyst, respectively (Janecek, 1997; Svensson, 1994). The catalytic mechanism has been discussed mostly on the basis of these two residues. However, the critical role of the third residue Asp297 seems to be still undefined and under dispute, except the facts that it plays an important role in the distortion of the substrate (Uitdehaag et al., 1999).

The generally accepted catalytic mechanism of the α-amylase family is that of the α-retaining double displacement (Koshland, 1953). The mechanism involves two catalytic residues in the active site; a glutamic acid as acid/base catalyst and an aspartate as the nucleophile. It involves five steps: (i) after the substrate has bound in the active site, the glutamic acid in the acid form donates a proton to the glycosidic bond oxygen, i.e. the oxygen between two glucose molecules at the subsites −1 and +1 and the nucleophilic aspartate attacks the C1 of glucose at subsite −1; (ii) an oxocarbonium ion-like transition state is formed followed by the formation of a covalent intermediate; (iii) the protonated glucose molecule at subsite +1 leaves the active site while a water molecule or a new glucose molecule moves into the active site and attacks the covalent bond between the glucose molecule at subsite −1 and the aspartate; (iv) an oxocarbonium ion-like transition state is formed again; (v) the base catalyst glutamate accepts a hydrogen from an incoming water or the newly entered glucose molecule at subsite +1, the oxygen of the incoming water or the newly entered glucose molecule at subsite +1 replaces the oxocarbonium bond between the glucose molecule at subsite −1 and the aspartate forming a new hydroxyl group at the C1 position of the glucose at subsite −1 (hydrolysis) or a new glycosidic bond between the glucose at subsite −1 and +1 (transglycosylation). Studies with cyclodextrin glycosyltransferase have shown that the intermediate indeed has a covalently linked bond with the enzyme (Uitdehaag et al., 1999).

Other conserved amino acid residues e.g. histidine, arginine, and tyrosine play a role in positioning the substrate into the correct orientation into the active site, proper orientation of the nucleophile, transition state stabilization, and polarization of the electronic structure of the substrate (Lawson et al., 1994; Strokopytov et al., 1996; Uitdehaag et al., 1999).

5. Biotechnological application

Nowadays, α-amylases represent one of the most important enzyme groups within the field of biotechnology. These enzymes are present in numerous biotechnological and industrial applications such as in food, detergents and textiles as well as in paper industry, for the hydrolysis of starch. They can also be of potential use in the pharmaceutical and fine chemical industries.

5.1. Industrial production of glucose and fructose from starch

The acid hydrolysis method for glucose production has been replaced recently by enzymatic treatment, with three or four different enzymes, in which α-amylase is the first (Crabb & Shetty, 1999).

For the complete conversion into high glucose syrup, the first step is the liquefaction into soluble, short-chain dextrins. Dry solids starch slurry (30–35%) of pH 6 is mixed with α-amylase and passed through a jet cooker after which the temperature is kept at 95–105 °C for 90 min to assure the removal of lipid–starch complexes. The dextrose equivalent value of starch hydrolysate syrup depends on the time of incubation and the amount of added enzyme. The drawback of the currently used α-amylases is that they are not active at a pH below 5.9 at the high used temperatures. Therefore, the pH has to be adjusted from the natural pH 4.5 of the starch slurry to pH 6. Also Ca^{2+} needs to be added because of the Ca^{2+}-dependency of these enzymes (Tonkova, 2006). The next step is the saccharification of the starch hydrolysate syrup to high concentration glucose syrup, with more than 95% glucose. This is done by using an exo-acting glucoamylase. The final step is the conversion of high glucose syrup into high fructose syrup by using glucose isomerase (Bhosale et al., 1996).

5.2. Bakery and anti-staling

The baking industry is a large consumer of starch and starch-modifying enzymes. Amylases can be added to degrade the damaged starch in the flour into smaller dextrins, which are subsequently fermented by the yeast. Upon storage, all undesirable changes together are called staling. Retrogradation of the starch fraction in bread is considered very important in staling (Kulp & Ponte, 1981). Staling is of considerable economic importance for the baking industry since it limits the shelf life of baked products. Several additives may be used in bread baking (Spendler & Jórgensen, 1997) to delay staling and improve texture, volume and flavor of bakery products. Enzymes active on starch have been suggested to act as anti-staling agents, especially α-amylases (Sahlstrom & Brathen, 1997).

α-amylase supplementation in flour not only enhances the rate of fermentation and reduces the viscosity of dough resulting in improvements in the volume and texture of the product (De Stefanis & Turner, 1981) but also it generates additional sugar in the dough, which improves the taste, crust colour and toasting qualities of the bread (Van Dam & Hille, 1992). A recent trend is to use intermediate temperature stable (ITS) α-amylases (Ahuja et al., 1998, as cited in Gupta et al., 2003) since they become inactive much before the completion of the baking process which avoid sickliness in bread.

5.3. Cyclodextrin formation

Cyclodextrins are cyclic α,1-4 linked oligosaccharides mainly consisting of 6, 7, or 8 glucose residues. The glucose residues in the rings are arranged in such a manner that the inside is hydrophobic while the outside is hydrophilic. This enables cyclodextrins to form inclusion complexes with a variety of hydrophobic guest molecules. The formation of inclusion complexes leads to changes in the chemical and physical properties of the guest molecules. These altered characteristics of the encapsulated compounds have led to various applications of cyclodextrins in analytical chemistry, agriculture, pharmacy, food and cosmetics. For the industrial production of cyclodextrins, starch is first liquefied by a heat-stable α-amylase and then the cyclization occurs with a cyclodextrin glycosyltransferase (Riisgaard, 1990).

5.4. Detergent industries

A growing area of application of α-amylases is in the fields of laundry, dish-washing detergents and spot removers (Borchet et al., 1995; Kennedy et al., 1988). Amylases have the function of facilitating the removal of starchy stains by means of catalytic hydrolysis of the starch polysaccharide, and have been used for this purpose for a fairly long time in dishwashing detergents and textile laundering (Speckmann et al., 2001).

Traditional detergent enzymes are functional under alkaline conditions, thermostable and resistant to various detergent ingredients, such as surfactants, chelating and oxidative agents. On the other hand, a modern trend among consumers is to use colder temperatures for doing the laundry or dishwashing. At these lower temperatures, the removal of starch from cloth and porcelain becomes more problematic. Detergents with α-amylases optimally working at moderate temperatures and alkaline pH can help solve this problem (Marc et al., 2002).

5.5. Ethanol production

For large-scale processing, the bioconversion of starchy materials to ethanol is very useful because it can be used as a biofuel and as the starting material for various chemicals. The production of ethanol from starchy biomass commonly involves three-step processes: liquefaction of starch by an endoamylase such as α-amylase to reduce the viscosity of the gelatinized starch produced after the cooking of the grains, enzymatic saccharification of the low-molecular-weight liquefaction products to produce glucose, and fermentation of glucose.

However, the present process for ethanol production from starchy materials via fermentation requires improvement of cost production. Although noncooking and low-temperature-cooking fermentation systems (Matsumoto et al., 1982) have succeeded in reducing energy consumption by approximately 50% (Matsumoto et al., 1982), it is still necessary to add large amounts of amylolytic enzymes to hydrolyze the starchy materials. Many researchers have reported attempts to resolve this problem by using recombinant glucoamylase-expressing yeasts with the ability to ferment starch to ethanol directly (Kondo et al., 2002). Also, a noncooking fermentation system using a cell surface-engineered yeast strain promises to be very effective in reducing the production costs of ethanol (Shigechi et al., 2004). On the other hand, fermentation of starch to ethanol in one step using co-cultures of two different strains has been suggested and has potential application for the direct bioconversion of starch into ethanol (Zeikus, 1979).

5.6. Miscellaneous applications

Besides amylases' use in the saccharification or liquefaction of starch, these are also used for the clarification of formed haze in fruit juices, the pretreatment of animal feed to improve the digestibility (Marc et al., 2002). α-amylase is used for the production of low viscosity, high molecular weight starch for coating of paper (Bruinenberg et al., 2004). Starch is a good sizing agent for the finishing of paper. It is added to the paper in the size press and paper picks up the starch by passing through two rollers that transfer the starch slurry. The temperature of this process lies in the range of 45–60 °C. A constant viscosity of the starch is required for reproducible results at this stage. The mill also has the flexibility of varying the starch viscosity for different paper grades. The viscosity of the natural starch is too high for paper sizing and is adjusted by partially degrading the polymer with α-amylases in a batch or continuous processes. Also, good desizing of starch sized textiles is achieved by the application of α-amylases, which selectively remove the size and do not attack the fibers. It also randomly cleaves the starch into dextrins that are water soluble and can be removed by washing.

Furthermore, high molecular weights amylases were found in culture supernatants of an environmentally derived microbial mixed culture selected for its ability to utilize starch-containing plastic films as sole carbon sources (Burgess-Cassler et al., 1991). This suggests a new application for amylases in biodegradation. With the advent of new frontiers in biotechnology, the spectrum of amylase applications has expanded into many other fields, such as clinical, medicinal and analytical chemistry (Becks et al., 1995).

A modern trend is to use starch for production of a more efficient and specific degradation products through a particular combination of activities. Amylase from *Aspergillus niger*, a saccharifying enzyme which produces maltose, maltotriose and some glucose, is capable of alcoholysis for the synthesis of methyl-glucosides from starch in the presence of methanol. As these products are a series of methyloligosaccharides, from methyl-glucoside to methyl hexomaltoside, the biotechnological applications of using starch as substrate for the production of alkyl-glucosides is analyzed (Santamaria et al., 1999).

Moreover, it becomes possible to produce lactic acid directly from starch by an efficient simultaneous saccharification and fermentation from soluble starch by recombinant *Lactobacillus* strains (Okano et al., 2009). Finally, α-amylase is suggested as an enzyme that contributes to the reduction of $AuCl_4^-$ to gold nanoparticles (Au-NPs) which makes it ideal for the production of Au-NPs (Kalishwaralal et al., 2010).

6. Conclusion

Despite the fact that several different α-amylase preparations are available with various enzyme manufacturers for specific use in varied industries, amylase biotechnology demands extension in terms of both quality and quantity. Qualitative improvements in amylase gene and its protein can be achieved by recombinant DNA technology and protein engineering. Quantitative enhancement needs strain improvement through site directed mutagenesis and/or standardizing the nutrient medium for the overproduction of active α-amylases. Another approach is to screen for novel microbial strains from extreme environments.

All the above-mentioned approaches are aimed to increase stability, improve product specificity, alter pH optimum, improve thermostability, achieve free Ca^{+2} requirement; by using the currently available insights into the structure–function relationships of the amylase family enzymes.

Author details

Amira El-Fallal, Mohammed Abou Dobara, Ahmed El-Sayed and Noha Omar
Faculty of Science, Damietta Branch, Egypt

Acknowledgement

The authors are greatly acknowledging Damietta Branch, Mansoura University for supporting this work.

7. References

Abou Dobara, M.I., El-Sayed, A.K., El-Fallal, A.A & Omar, N.F. (2011). Production and partial characterization of high molecular weight extracellular α-amylase from *Thermoactinomyces vulgaris* isolated from Egyptian soil. *Polish Journal of Microbiology*, Vol. 60, No. 1, (January 2011), pp. 65- 71, ISSN 1733-1331

Aiyer, P. (2004). Effect of C: N ratio on alpha amylase production by *Bacillus licheniformis* SPT 27. *African Journal of Biotechnology*, Vol. 3, No. 10, (October 2004), pp. 519-522, ISSN 1684-5315

Aleshin, A., Firsov, L. & Honzatko, R. (1994). Refined structure for the complex of acarbose with glucoamylase from *Aspergillus awamori* Var. X100 to 2.4-A resolution. *J. Biol. Chem.*, Vol. 269, No. 22, (June 1994), pp. 15631-15639, ISSN 1083-351X

Aleshin, A., Firsov, L. & Honzatko, R. (1994). Refined structure for the complex of acarbose with glucoamylase from *Aspergillus awamori* Var. X100 to 2.4-A resolution. *J. Biol. Chem.*, Vol. 269, No. 22, (June 1994), pp. 15631-15639, ISSN 1083-351X

Andrews, L. & Ward, J. (1987). Extracellular amylases from *Streptomyces albus*. *J. Biochem. Soc. Trans.*, Vol. 15, No.3, (June 1987), pp. 522-523, ISSN 1470-8752

Aqeel, B. & Umar, D. (2010). Effect of alternative carbon and nitrogen sources on production of alpha-amylase by Bacillus megaterium. *World Applied Sciences Journal*, Vol. 8, Special Issue of Biotechnology & Genetic Engineering, pp. 85-90, ISSN 0264-8725

Arnesen, S., Eriksen, S.H., Olsen, J. & Jensen, B. (1998). Increased production of alpha amylase from *Thermomyces lanuginosus* by the addition of Tween-80. *Enzyme Microb. Technol.*, Vol. 23, No. 3-4, (November 1998), pp. 249–252, ISSN 0141-0229

Asoodeh, A., Chamanic, J. & Lagzian, M. (2010). A novel thermostable, acidophilic α-amylase from a new thermophilic "*Bacillus* sp. Ferdowsicous" isolated from Ferdows hot mineral spring in Iran: Purification and biochemical characterization. *Int. J. Biol. Macromol.*, Vol. 46, No. 3, (April 2010), pp. 289-297, ISSN 0141-8130

Aunstrup, K. (1979). Production, isolation and economics of extracellular enzymes, In: *Applied Biochemistry and Bioengineering*, J. Wingard, L. Katchalski – Katzir & L. Golstein (Ed.), pp. (27-69), Academic Press, ISBN 0120411024, New York, USA

Azhari, R. & Lotan, N. (1991). Enzymic hydrolysis of biopolymers via single-scission attack pathways: a unified kinetic model. *J. Mater. Sci. Mater. Med.*, Vol. 2, No. 1, (January 1991), pp. 9–18, ISSN 1573-4838

Bahrami, A., Shojaosadati, S., Khalilzadeh, R., Mohammadian, J., Farahani, E. & Masoumian, M. (2009). Prevention of human granulocyte colony-stimulating factor protein aggregation in recombinant Pichia pastoris fed-batch fermentation using additives. *Biotechnol. Appl. Biochem.*, Vol. 52, No. 2, (February 2009), pp. 141–148, ISSN 1470-8744

Bahri, S. & Ward, J. (1990). Cloning and expression of an α-amylase gene from *Streptomyces thermoviolaceus* CUB74 in *Escherichia coli* JM107 and *S. lividans* TK24. *J. Gen. Microbiol.*, Vol. 136, No. 5, (May 1990), pp. 811-818, ISSN 1465-2080

Bajpai, P. & Bajpai, P.K. (1989). High-temperature alkaline α-amylase from *Bacillus licheniformis* TCRDC-B13. *Biotechnol. Bioeng.*, Vol. 33, No. 1, (January 1989), pp. 72–78, ISSN 1097-0290

Barros, M., Silva, R., Ramada, M., Galdino, A., Moraes, L., Torres, F. & Ulhoa, C. (2009). The influence of N-glycosylation on biochemical properties of Amy1, an α-amylase from the yeast *Cryptococcus flavus*. *Carbohydrate Research*, Vol. 344, No. 13, (September 2009), pp. 1682–1686, ISSN 0008-6215

Bayer, E., Morag, E., Wilcheck, M., Lamed, R., Yaron, S. & Shoham, Y. (1995). Cellulosome domains for novel biotechnological application. In: *Carbohydrate bioengineering*, B. Petersen, B. Srensson & S. Pedersen (Ed.), pp. (251-259), Progress in biotechnology, Elsevier, ISBN 0-444-82223-2, The Netherlands.

Becks, S., Bielawaski, C., Henton, D., Padala, R., Burrows, K. & Slaby, R. (1995). Application of a liquid stable amylase reagent on the ciba corning express clinical chemistry *system*. *Clin. Chem.*, Vol. 41, No. 2, (February 1995), pp. 186, ISSN 1530-8561

Bertoft, E. (2007). Composition of clusters and their arrangement in potato amylopectin. *Carbohydrate Polymers*, Vol. 68, No. 3, (April 2007), pp. 433-446, ISSN 0144-8617

Bhella, R.S. & Altosaar, I. (1988). Role of CAMP in the mediation of glucose catabolite repression of glucoamylase synthesis in *Aspergillus awamori*. *Curr. Genet*, Vol. 14, No. 3, (September 2004), pp. 247-252, ISSN 1432-0983

Bhosale, S., Rao, M. & Deshpande, V. (1996). Molecular and industrial aspects of glucose isomerase. *Microbiol. Rev.*, Vol. 60, No. 2, (June 1996), pp. 280–300, ISSN 1098-5557

Bijttebier A., Goesaert H. & Delcour J.A. (2007). Temperature impacts the multiple attack action of amylases. *Biomacromolecules*, Vol. 8, No. 3, (March 2007), pp. 765–772, ISSN 1526-4602

Bijttebier, A., Goesaert, H. & Delcour, J. (2008). Amylase action pattern on starch polymers. *Biologia*, Vol. 63, No. 6, (December 2008), pp. 989-999, ISSN 1336-9563

Borchet, T., Lassen, S., Svendsen, A. & Frantzen, H. (1995). Oxidation stable amylases for detergents. In: *Carbohydrate bioengineering: Progress in biotechnology*. Vol. 10, S. Petersen,

B. Svensson & S. Pedersen (Ed.), pp. (175-179), Elsevier, ISBN 0444822232, The Netherlands.

Bozic, N., Ruizb, J., López-Santínb, J. & Vujci´c, Z. (2011). Production and properties of the highly efficient raw starch digesting α-amylase from a *Bacillus licheniformis* ATCC 9945a. *Biochemical Engineering Journal*, Vol. 53, No. 2, (January 2011), pp. 203–209, ISSN 1369-703X

Brooker, J.D. & McCarthy, J.M. (1997). Gene knockout of the intracellular amylase gene by homologous recombination in *Streptococcus bovis*. *Curr. Microbiol.*, Vol. 35, No. 3, (September 1997), pp. 133–138, ISSN 1432-0991

Bruinenberg, P.M., Hulst, A.C., Faber, A. & Voogd, RH. (2004). A process for surface sizing or coating of paper, In: *European Patent Application* no. EP0690170, April 2012, Available from :< http://www.freepatentsonline.com/EP0690170.html>

Burgess-Cassler, A., Imam, S. & Gould, J. (1991). High-Molecular-Weight Amylase Activities from Bacteria Degrading Starch-Plastic Films. *Appl. Environ. Microbiol.*, Vol. 57, No. 2, (February 1991), pp. 612–614, ISSN 098-5336

Busch, J.E. & Stutzenberger, F.J. (1997). Amylolytic activity of *Thermomonospora fusca*. *World J. Microbiol. Biotechnol.* Vol. 13, No. 6, (November 1997), pp. 637-642, ISSN 1573-0972

Calamai, M., Canale, C., Relini, A., Stefani, M., Chiti, F. & Dobson, C.M. (2005). Reversal of protein aggregation provides evidence for multiple aggregated states. *J. Mol. Biol.* Vol. 346, No. 2, (February 2005), pp. 603–616, ISSN 0022-2836

Cesk, M., Brown, B. & Birath, K. (1969). Ranges of α-amylase activities in human serum and urine and correlations with some other α-amylase methods. *Clin. Chem. Acta.*, Vol. 26, No. 3, (December 1969), pp. 445-453, ISSN 0009-8981

Collins, B., Kelly, C., Fogarty, W. & Doyle, E. (1993). The high maltose-producing alpha-amylase of the thermophilic actinomycete *Thermomonospora curvata*. *Appl. Microbiol. Biotechnol.*, Vol. 39, No.1 , (April 1993), pp. 31–35, ISSN 1432-0614

Coutinho, P. & Henrissat, B. (1999). Carbohydrate active enzymes. In: *Recent Advances in Carbohydrate bioengineering*, H. Gilbert, G. Davies, B. Svensson & B. Henrissat (Ed.), pp. (3-12), Royal Society of Chemistry, ISBN 0854047743, Cambridge

Crabb, W. & Shetty, J. (1999). Commodity scale production of sugars from starches. *Curr. Opin. Microbiol.*, Vol. 2, No. 3, (June 1999), pp. 252– 256, ISSN 1369-5274

De Mot, R., Andries, K. & Verachtert, H. (1984a). Comparative study of starch degradation and amylase production by ascomycetous yeast species. *Syst. Appl. Microbiol.* Vol. 5, No. 1, (April 1984), pp. 106-118, ISSN 0723-2020

De Mot, R., Demeersman, M. & Verachtert. H. (1984b). Comparative study of starch degradation and amylase production by non-ascomycetous yeast species. *Syst. Appl. Microbiol.* Vol. 5, No. 3, (October 1984), pp. 421-432, ISSN 0723-2020

De Mot, R., Van Oudendijck, E., Hougaerts, S. & Verachtert. H. (1984c). Effect of medium composition on amylase production by some starch-degrading yeasts. *FEMS Microbiol. Lett.*, Vol. 25, No. 2, (December 1984), pp. 169-173, ISSN 1574-6968

De Stefanis, V. & Turner, E. (1981). Modified enzyme system to inhibit bread firming method for preparing same and use of same in bread and other bakery products, In:

United states Patent Application no. US4299848, April 2012, Available from: <http://patft.uspto.gov/netacgi/nphParser?Sect2=PTO1&Sect2=HITOFF&p=1&u=/netahtml/PTO/searchbool.html&r=1&f=G&l=50&d=PALL&RefSrch=yes&Query=PN/4299848 >

Dhawale, M., Wilson, J., Khachatourians, G. & Mike, W. (1982). Improved method for detection of starch hydrolysis. *Appl. Eenviron. Microbiol.*, Vol. 43, No. 4, (April 1982), pp. 747-750, ISSN 1098-5336

Domsch, K.H., Gams, W. & Anderson, T.H. (1995). *Compendium of soil fungi*, p. 859, IHW-Verlag, ISBN 3980308383, Alemanha

Dona, A., Pages, G., Gilbert, R. & Kuchel, P. (2010). Digestion of starch: In vivo and in vitro kinetic models used to characterise oligosaccharide or glucose release. *Carbohydrate Polymers.*, Vol. 80, No. 3, (May 2010), pp. 599-617 , ISSN 0144-8617

Doyle, E.M., Noone, A.M., Kelly, C.T., Quigley, T.A. & Fogarty, W.M. (1998). Mechanisms of action of the maltogenic α-amylase of *Byssochlamys fulva*. *Enzyme Microb. Technol.* Vol. 22, No. 7, (May 1998), pp. 612–616, ISSN 0141-0229

EL-Banna, T.E., Abd-Aziz, A.A., Abou-Dobara, M.I. & Ibrahim, R.I. (2007). Production and immobilization of α-amylase from *Bacillus subtilis*. *Pakistan J. of Biological Sciences*, Vol. 10, No. 12, (June 2007), pp. 2039-2047, ISSN 1812-5735

EL-Banna, T.E., Abd-Aziz, A.A., Abou-Dobara, M.I. & Ibrahim, R.I. (2008). Optimization and immobilization of α-amylase from *Bacillus licheniformis*, *Proceedings of the second international conference on the role of genetics and biotechnology in conservation of natural resources*, Ismailia, Egypt, July 2007.

Eriksen, S., Jensen, B. & Olsen, J. (1998). Effect of N-linked glycosylation on secretion, activity and stability of α-amylase from *Aspergillus oryzae*. *Curr. Microbiol.*, Vol. 37, No. 2, (August 1998), pp. 117-122, ISSN 1432-0991

Espargaro, A., Castillo, V., de Groot, N. & Ventura, S. (2008). The in vivo and in vitro aggregation properties of globular proteins correlate with their conformational stability: the SH3 case. *J. Mol. Biol.*, Vol. 378, No. 5, (May 2008), pp. 1116–1131, ISSN 0022-2836

Farber, G. & Petsko, G. (1990). The evolution of α/β–barrel enzymes. *TIBS*, Vol. 15, No. 6, (June 1990), pp. 228-234, ISSN 0968-0004

French, D. (1984). Organization of starch granules, In: *starch: chemistry and technology*, R.L. Whistler, J.N. Bemiller & E.F. Paschall (Ed.), pp. (183-207), Academic press, ISBN 78-0-12-746275-2, London.

Friedrich, J., Cimerman, A. & Steiner, W. (1989). Submerged production of pectinolytic enzymes by *Aspergillus niger*: effect of different aeration/agitation regimes. *Appl. Microbiol. Biotechnol.*, Vol. 31, No. 5-6, (Octobar 1989), pp. 490-494, ISSN 1432-0614

Fukumoto, J. & Okada, S. (1963). Studies on bacterial amylase, Amylase types of *Bacillus subtilis* species *J. Ferment. Technol.*, Vol. 41, No. 1, pp. 427-434, ISSN 0385-6380

Gerard, C., Planchot, V., Colonna, P. & Bertoft, E. (2000). Relationship between branching density and crystalline structure of A- and B-type maize mutant starches. *Carbohydr. Res.* Vol. 326, No. 2, (June 2000), pp. 130–144, ISSN 0008-6215

Gigras, P., Sahai, V. & Gupta, R. (2002). Statistical media optimization and production of its alpha amylase from *Aspergillus oryzae* in a bioreactor. *Curr. Microbiol.*, Vol. 45, No. 3, (September 2002), pp. 203–208, ISSN 1432-0991

Giraud, E., Brauman, A., Keleke, S., Lelong, B. & Raimbault, M. (1991). Isolation and physiological study of an amylolytic strain of *Lactobacillus plantarum*. *Appl. Microbiol. Biotechnol.*, Vol. 36, No. 3, (December 1991), pp. 379-383, ISSN 1432-0614

Gupta R., Paresh, G., Mohapatra, H., Goswami, V., & Chauhan, B. (2003) Microbial α-amylases: a biotechnological perspective. *Process Biochem.*, Vol. 38, No. 12, (July 2003), pp. 1599–1616, ISSN 1359-5113

Hanashiro, I., Abe, J. & Hizukuri, S. (1996). A periodic distribution of the chain length of amylopectin as revealed by high-performance anion-exchange chromatography. *Carbohydr. Res.* Vol. 283, No. 1, (March 1996), pp. 151-159, ISSN 0008-6215

Hang, M., Furuyoshi, S., Yagi, T. & Yamamoto, S. (1996). Purification and characterization of raw starch digesting α-amylase from *Streptomyces thermocyaneoviolaceus* IFO14271. *J. Appl. Glycosci.*, Vol. 43, pp. 487-497, ISSN 1880-7291

Hao, Y., Chu, J., Wang, Y., Zhuang, Y. & Zhang, S. (2007). The inhibition of aggregation of recombinant human consensus interferon-alpha mutant during *Pichia pastoris* fermentation. *Appl. Microbiol. and Biotechnol.*, Vol. 74, No. 3, (March 2007), pp. 578–584, ISSN 1432-0614

Hayashi, T., Akiba, T. & Horikoshi, K. (1988). Production and purification of new maltohexaose-forming amylases from alkalophilic *Bacillus* sp. H-167. *Agric. Biol. Chem.*, Vol. 52, No. 2, (February 1988). pp. 443–448, ISSN 1881-1280

Henrissat, B. & Bairoch, A. (1993). New families in the classification of glycosyl hydrolases based on amino acid sequence similarities. *Biochem. J.*, Vol. 293, No. 3, (August 1993), pp. 781-788, ISSN 0264-6021

Henrissat, B. (1991). A classification of glycosyl hydrolases based on amino acid sequence similarities. *Biochem.J.*, Vol. 280, No. 2, (December 1991), pp. 309–316, ISSN 0264-6021

Hofemeister, B., Konig, S., Hoang, V., Engel, J., Mayer, G., Hansen, G. & Hofemeister, J. (1994). The gene *amyE* (TV1) codes for a nonglucogenic α-amylase from *Thermoactinomyces vulgaris* 94-2A in *Bacillus subtilis*. *Appl. Environ. Microbiol.*, Vol. 60, No. 9, (September 1994), pp. 3381-3389, ISSN 1098-5336

Hyun, H.H. & Zeikus, J.G. (1985a). General biochemical characterization of thermostable extracellular β-amylase from *Clostridium thermosulfurogenes*. *Appl. Environ. Microbiol.*, Vol. 49, No. 5, (May 1985), pp. 1162-1167, ISSN 1098-5336

Imberty, A., Buleon, A., Tran, V. & Perez, S. (1991). Recent advances in knowledge of starch structure. *Starch/Starke*, Vol. 43, No. 10, (October 1991), pp. 375-384, ISSN 1521-379X

Janecek, S. (1997). α-Amylase family: molecular biology and evolution. *Prog. Biophys. Mol. Biol.* Vol. 67, No. 1, (January 1997), pp. 67-97, ISSN 0079-6107

Jang, S., Cheong, T., Shim, W., Wan Kim, J. & Park, K. (1994). Purification of *Bacillus licheniformis* thermostable α-amylase by immunoaffinity chromatography. *Korean Biochem. J.*, Vol. 27, No. 1, (January 1994), pp. 38-41, ISSN 0368-4881

Kadziola, A., Abe, J., Svensson, B. & Haser, R. (1994). Crystal and molecular structure of barley α-amylase. *J.Mol.Biol.*, Vol. 239, No. 1, (may 1994), pp. 104-121, ISSN 0022-2836

Kalishwaralal, K., Gopalram, S., Vaidyanathan, R., Deepak, V., Pandian, S. & Gurunathan, S. (2010). Optimization of α-amylase production for the green synthesis of gold nanoparticles. *Colloids and Surfaces B: Biointerfaces*, Vol. 77, No. 2, (June 2010), pp. 174–180, ISSN 0927-7765

Kaneko, T., Ohno, T. & Ohisa, N. (2005). Purification and characterization of a thermostable raw starch digesting amylase from a *Streptomyces* sp. isolated in a milling factory. *Biosci. Biotechnol. Biochem.*, Vol. 69, No. 6, (June 2005), pp. 1073-1081, ISSN 1347-6947

Kelly, R., Dijkhuizen, L. & Leemhuis, H. (2009). Starch and glucan acting enzymes, modulating their properties by directed evolution. *Biotechnology*, Vol. 140, No. 3-4, (March 2009), pp. 184–193, ISSN 0168-1656

Kennedy, J., Cabalda, V. & White, C. (1988). Enzymic starch utilization and genetic engineering. *Tibtech.*, Vol. 6, pp. 184-189, ISSN 0167-7799

Kim, T.U., Gu, B.G., Jeong, J.Y., Byun, S.M. & Shin, Y.C. (1995). Purification and characterization of maltotetraose-forming alkaline α-amylase from an alkalophilic *Bacillus* strain, GM8901. *Appl. Environ. Microbiol.*, Vol. 61, No. 8, (August 1995), pp. 3105–3112, ISSN 1098-5336

Kimura, K., Tsukamoto, A., Ishii, Y., Takano, T. & Yamane, K. (1988). Cloning of a gene for maltohexaose producing amylase of an alkalophilic *Bacillus* and hyper-production of the enzyme in *Bacillus subtilis* cells. *Appl. Microbiol. Biotechnol.* Vol. 27, No. 4, (January 1988), pp. 372–377, ISSN 1432-0614

Klein, B., Foreman, J. & Searcy, R. (1970). New chromogenic substrate for determination of serum amylase activity. *Clin. Chem.*, Vol. 16, No. 1, (January 1970), pp. 32-38, ISSN 1530-8561

Kondo, A., Shigechi, H., Abe, M., Uyama, K., Matsumoto, T., Takahashi, S., Ueda, M., Tanaka, A., Kishimoto, M. & Fukuda, H. (2002). High-level ethanol production from starch by a flocculent *Saccharomyces cerevisiae* strain displaying cell-surface glucoamylase. *Appl. Microbiol. Biotechnol.*, Vol. 58, No. 3, (March 2002), pp. 291–296, ISSN 1432-0614

Koshland, D.E. (1953). Stereochemistry and the mechanism of enzymatic reactions. *Biol. Rev.*, Vol. 28, No. 4, (November 1953), pp. 416-436, ISSN 1469-185X

Kramhøft B., Bak-Jensen K., Mori H., Juge N., Nohr J. & Svensson B. (2005). Involvement of individual subsites and secondary substrate binding sites in multiple attack on amylase by barley α-amylase. *Biochemistry*, Vol. 44, No. 6, (February 2005), pp. 1824–1832, ISSN 1520-4995

Kulp, K. & Ponte, J.G. (1981). Staling white pan bread: fundamental causes. *Crit. Rev. Food Sci. Nutr.*, Vol. 15, No. 1, (January 1981), pp. 1–48, ISSN 1549-7852

Kunst A., Drager, B. & Ziegenhorn, J. (1984). Colorimetric methods with glucose oxidase and peroxidase. In: *Methods of enzymatic analysis*, H. Bermeyer (Ed.), pp. (178-185), Weiheim, ISBN 3527255982, Verleg Chemie

Kuo, M. & Hartman, P. (1966). Isolation of amylolytic strains of *Thermoactinomyces vulgaris* and production of thermophilic actinomycete amylases. *J. Bacteriol.*, Vol. 92, No. 3, (September 1966), pp. 723-726, ISSN 1098-5530

Lawson, C.L., van Montfort, R., Strokopytov, B., Rozeboom, H.J., Kalk, K.H., de Vries, G.E., Penninga, D., Dijkhuizen, L. & Dijkstra, B.W. (1994). Nucleotide sequence and X-ray structure of cyclodextrin glycosyltransferase from *Bacillus circulans* strain 251 in a maltose-dependent crystal form. *J. Mol. Biol.* Vol. 236, No. 2, (February 1994), pp. 590–600, ISSN 0022-2836

Leuschner, C. & Antranikian, G. (1995). Heat-stable enzymes from extremely thermophilic and hyperthermophilic microorganisms. *World J Microb Biot.*, Vol. 11, No. 1, (January 1995), pp. 95–114, ISSN 1573-0972

Liebl, W., Stemplinger, I. & Ruile, P. (1997). Properties and gene structure of the *Thermotoga maritima* alpha-amylase AmyA, a putative lipoprotein of a hyperthermophilic bacterium. *J. Bacteriol.* Vol. 179, No.3, (February 1997), pp. 941-948, ISSN 1098-5530

Lim, W., Park, S., An, C., Lee, J., Hong, S., Shin, E., Kim, E., Kim, J., Kim, H.& Yun, H. (2003). Cloning and characterization of a thermostable intracellular alpha-amylase gene from the hyperthermophilic bacterium *Thermotoga maritima* MSB8. Res. *Microbiol.* Vol. 154, No.10, (December 2003), pp. 681–687, ISSN 0923-2508

Linden, A., Mayans, O., Meyer-Claucke, W., Antranikian, G. & Wilmanns, M. (2003). Differential regulation of a hyperthermophilic α-amylase with a novel (Ca, Zn) two-metal center by zinc. *J. Biological Chemistry*, Vol. 278, No. 11, (March 2003), pp. 9875-9884, ISSN 1083-351X

Lonsane, B. & Ramesh M. (1990). Production of bacterial thermostable [alpha]-Amylase by solid-state fermentation: a potential tool for achieving economy in enzyme production and starch hydrolysis. In: *Advances in Applied Microbiol.*, Vol. 35, S. Neidleman & A. Laskin (Ed.), pp. (1-56), Elsevier, ISBN 978-0-12-002635-7, The Netherlands.

Machius, M., Wiegand, G. & Huber, R. (1995). Crystal structure of calcium depleted *Bacillus licheniformis* α-amylase at 2.2 Å resolution. *J.Mol.Biol.*, Vol. 246, No. 4, (March 1995), pp. 545-559, ISSN 0022-2836

Malhotra, R., Noorvez, S. & Satyanarayana, T. (2000). Production and partial characterization of thermostable and calcium-independent α-amylase of an extreme thermophile *Bacillus thermooleovorans* NP54. *Letters in Applied Microbiology*, Vol. 31, No. 5, (November 2000), pp. 378-384, ISSN 1472-765X

Marc, J.E., van der Maarel, C., Joost, B.V., Uitdehaag, C.M., Leemhuis, H. & Dijkhuizen, L. (2002). Properties and applications of starch-converting enzymes of the α-amylase family. *J. Biotechnology*, Vol. 94, No. 2, (March 2002), pp. 137-155, ISSN 0168-1656

Marciniak, G. & Kula, M. (1982). Vergleichende untersuchung der methoden zur bestimmung der aktivität bakterieller alpha amylasen. *Starch/starke*, Vol. 34, No. 12, (December 1982), pp. 422-430, ISSN 1521-379X

Matsumoto, N., Fukushi, O., Miyanaga, M., Kakihara, K., Nakajima, E. & Yoshizumi, H. (1982). Industrialization of a noncooking system for alcoholic fermentation from grains. *Agric. Biol. Chem.*, Vol. 46, No. 6, (June 1982), pp. 1549–1558, ISSN 1881-1280

Matsuura, Y., Kusunoki, M., Harada, W. & Kakudo, M. (1984). Structure and possible catalytic residues of Taka amylase A. J. Biochem., Vol. 95, No. 3, (March 1984), pp. 697-702, ISSN 1756-2651

Matsuzaki, H., Yamane, K., Yamaguchi, K., Nagata, Y. & Maruo, B. (1974). Hybrid α-amylase produced by transformants of *Bacillus subtilis* I. Immunological and chemical properties of α-amylases produced by the parental strain and the transformants. *Biochim Biophys Acta.*, Vol. 367, No. 2, (October 1974), pp. 235-247, ISSN 0005-2736

McCann, A.K. & Barnett. J.A. (1984). Starch utilization by yeasts: mutants resistant of carbon catabolite repression. *Curr. Genet.* Vol. 8, No. 7, (September 1984), pp. 525-530, ISSN 1432-0983

Miller, G.L. (1959). Use of dinitrosalisylic acid reagent for determination of reducing sugar. *Anal. Chem.*, Vol. 31, No. 3, (March 1959), pp. 426-428, ISSN 1520-6882

Mishra, R.S. & Maheshwari, R. (1996). Amylases of the thermophilic fungus *Thermomyces lanuginosus*: their purification, properties, action on starch and response to heat. *J Biosci.*, Vol. 21, No. 5, (September 1996), pp. 653-672, ISSN 0973-7138

Morkeberg, R., Carlsen, M. & Nielsen, J. (1995). Induction and repression of α-amylase production in batch and continuous cultures of *Aspergillus oryzae*. *Microbiol.*, Vol. 141, No. 10, (October 1995), pp. 2449-2454, ISSN 1465-2080

Murthy, P., Naidu, M. & Srinivas, P. (2009). Production of α-amylase under solid-state fermentation utilizing coffee waste. *J. Chem. Technol. Biotechnol.*, Vol. 84, No. 8, (August 2009), pp. 1246–1249, ISSN 1097-4660

Narang, S. & Satyanarayana, T. (2001). Thermostable a-amylase production by an extreme thermophile *Bacillus thermooleovorans*. *Letters in Applied Microbiology*, Vol. 32, No. 1, (January 2001), pp. 31–35, ISSN 1472-765X

Obi, S. & Odibo, F. (1984). Some properties of a highly thermostable α-Amylase from a *Thermoactinomyces* sp. *Can. J. Microbiol.*, Vol. 30, No. 6, (June 1984), pp. 780-785, ISSN 1480-3275

Okano, K., Zhang, O., Shinkawa, S., Yoshida, S., Tanaka, T., Fukuda, H. & Kondo, A. (2009). Efficient Production of Optically Pure D-Lactic Acid from Raw Corn Starch by Using a Genetically Modified L-Lactate Dehydrogenase Gene-Deficient and ά-Amylase-Secreting *Lactobacillus plantarum* Strain. *Appl. Environ. Microbiol.*, Vol. 75, No. 2, (January 2009), pp. 462-467, ISSN 098-5336

Olesen, T. (1991). Antistaling process and agent. Patent application, In: *European Patent Application* no. EP19900915100, April 2012, Available from: <http://www.freepatentsonline.com/EP0494233.html >

Omar, N., Abou-Dobara, M. & El-Sayed, A. (2011). *Studies on amylase produced by some actinomycetes*, Lambert, Academic publishing ISBN 9783847334361, Germany

Pandey, A., Nigam, P., Soccol, C., Soccol, V., Singh, D. & Mohan, R. (2000). Advances in microbial amylases. *Biotechnol. Appl. Biochem.*, Vol. 31, No. 2, (April 2000), pp. 135–152, ISSN 1470-8744

Petruccioli, M. & Federici, R.G. (1992). A note on the production of extracellular hydrolytic enzymes by yeast-like fungi and related microorgnisms. *Ann. Microbiol. Enzimol.*, Vol. 42, No.1 , pp. 81-86, ISSN 0003-4649

Prakash, O. & Jaiswal, N. (2010). α-Amylase: An Ideal Representative of Thermostable Enzymes. *Appl Biochem Biotechnol.*, Vol. 160, No. 8, (April 2010), pp. 2401-2414, ISSN 1559-0291

Primarini, D. & Ohta, Y. (2000). Some enzyme properties of raw starch digesting amylases from *Streptomyces* sp. no. 4. *Starch/Starke*, Vol. 52, No. 1, (January 2000), pp. 28–32, ISSN 1521-379X

Ratanakhanokchai, K., Kaneko, J., Kamio, Y. & Izaki, K. (1992). Purification and properties of a maltotetraose and maltotriose producing amylase from *Chloroflexus aurantiacus*. *Appl. Environ. Microbiol.* Vol. 58, No. 8, (August 1992), pp. 2490-2494, ISSN 1098-5336

Rauscher, E. (1984). Determination of the degradation products maltose and glucose. In: *Methods of enzymatic analysis*, H. Bermeyer (Ed.), pp. (890-894), Weiheim, ISBN 3527255982, Verleg Chemie

Rhodes, C., Strasser, J. & Friedberg, F. (1987). Sequence of an active fragment of *Bacillus polymyxa* β-amylase. *Nucleic Acids Res.*, Vol. 15, No. 9, (May 1987), pp. 3934, ISSN 1362-4962

Riisgaard, S. (1990). The enzyme industry and modern biotechnology. In: *Proceedings of the Fifth European Congress on Biotechnology*, Vol. 1, C. Christiansen, L. Munck, J. Villadsen (Ed.), pp. (31-40), ISBN 8716106172, Munksgaard, Copenhagen.

Robyt, J.F. & Whelan, W.J. (1972). Reducing value methods for maltodextrins: chain length dependence of alkaline 3, 5-dinitrosalisylate and chain length independence of alkaline copper. *Anal. Biochem.*, Vol. 45, No. 2, (February 1972), pp. 510-516, ISSN 0003-2697

Robyt, J.F. (1998). *Essentials of Carbohydrate Chemistry*, Springer, ISBN 0-387-94951-8, New York, USA.

Russell, J.B. (1984). Factors influencing competition and composition of the rumen bacterial flora. In: *Herbivore nutrition in the subtropics and tropics*. F.M.C. Gilchrist and R.I. Mackie (Ed.), pp. (313-345), The Science Press, ISBN 0907997031, Craighall, South Africa

Ryan, S.M., Fitzgerald, G.F. & Van Sinderen, D. (2006). Screening for and identification of starch-, amylopectin-, and pullulan-degrading activities in *Bifidobacterial* strains. *Appl. Environ. Microbiol.*, Vol. 72, No. 8, (August 2006), pp. 5289-5296, ISSN 1098-5336

Sahlstrom, S. & Brathen, E. (1997). Effects of enzyme preparations for baking, mixing time and resting time on bread quality and bread staling. *Food Chem.*, Vol. 58, No. 1, (January 1997), pp. 75–80, ISSN 0308-8146

Sambrook, J., Fritsch, E.F. & Maniatis, T. (1989). *Molecular cloning: a laboratory manual*, 2nd ed. N. Harbor (Ed.), Cold Spring Harbor Laboratory Press, ISBN 0879693096, New York

Santamaria, R.I., Del Rio, G., Saab, G., Rodriguez, M.E., Soberön, X. and Loöpez-Munguia, A. (1999). Alcoholysis reactions from starch with α-amylases. *FEBS Letters*, Vol. 452, No. 3, (June 1999), pp. 346-350, ISSN 0014-5793

Satoh, E., Uchimura, T., Kudo, T. & Komagata, K. (1997). Purification, characterization, and nucleotide sequence of an intracellular maltotrioseproducing α-amylase from

Streptococcus bovis 148. *Appl. Environ. Microbiol.*, Vol. 63, No. 12, (December 1997), pp. 4941-4944, ISSN 1098-5336

Saxena, R.K., Dutt, K., Agarwal, L. & Nayyar, P. (2007). A highly thermostable and alkaline amylase from a *Bacillus* sp. PNS. *Biores. Technol.*, Vol. 98, No. 2, (January 2007), pp. 260-265, ISSN 0960-8524

Schäfer, T., Duffner, F. and Borchert, T.V. (2000). Extremophilic enzymes in industry: screening, protein engineering and application, *Proceedings of Third International Congress on Extremophiles*, Hamburg, Germany, 3-7 September 2000.

Shental-Bechor, D. & Levy, Y. (2009). Folding of glycoproteins: toward understanding the biophysics of the glycosylation code. *Current Opinion in Structural Biology*, Vol. 19, No. 5, (October 2009), pp. 524–533, ISSN 0959-440X

Shigechi, H., Koh, J., Fujita, Y., Matsumoto, T., Bito, Y., Ueda, M., Satoh, E., Fukuda, H. & Kondo, A. (2004). Direct Production of Ethanol from Raw Corn Starch via Fermentation by Use of a Novel Surface-Engineered Yeast Strain Codisplaying Glucoamylase and α-Amylase. *Appl. Environ. Microbiol.*, Vol. 70, No. 8, (August 2004), pp. 5037-5040, ISSN 098-5336

Shimizu, M., Kanno, M., Tamura, M. & Suckane, M. (1978). purification and some properties of a novel α-Amylase produced by a strain of *Thermoactinomyces vulgaris*. *Agric. Biol. Chem.*, Vol. 42, No. 9, (September 1978), pp. 1681-1688, ISSN 1881-1280

Shinke, R., Nishira, H. & Mugibayashi, N. (1974). Isolation of, Β-amylase producing microorganisms. *Agric. Biol. Chem.*, Vol. 38, No. 3, (March 1974), pp. 665- 666, ISSN 1881-1280

Siggens, K. (1987). Molecular cloning and characterization of the Β-amylase gene from *Bacillus circulans*. *Mol. Microbiol.* Vol. 1, No.3, (November 1987), pp. 86-91, ISSN 1365-2958

Sills, A., Sauder, M. & Stewart, G. (1984). Isolation and characterization of the amylolytic system of *Schwanniomyces castelli* . *J Inst Brew.*, Vol. 90,pp. 311-314, ISSN 0046-9750

Smith, A.M. (1999). Making starch. *Curr. Opin.* Plant Biol., Vol. 2, No. 6, (December 1999), pp. 223–229, ISSN 1369-5266

Somers, W., Lojenja, A.K., Bonte, A., Rozie, H., Visser, J., Rombouts, F. & Riet, V. (1995). Isolation of α-amylase on crosslinked starch. *Enzyme and Microb. Technol.*, Vol. 17, No. 1, (January 1995), pp. 56-62, ISSN 0141-0229

Speckmann, H., Kottwitz, B., Nitsch, C. & Maurer, K.H. (2001). Detergents containing amylase and protease, In: *Patent Application* no. US6380147B1, April 2012, Available from: <http://www.google.com.eg/patents?hl=ar&lr=&vid=USPAT6380147&id=jd8KAAAAEBAJ&oi=fnd&dq=Detergents+containing+amylase+and+protease&printsec=abstract#v=onepage&q=Detergents%20containing%20amylase%20and%20protease&f=false>

Spendler, T. & Jørgensen, O. (1997). Use of a branching enzyme in baking, In: *Patent Application* no. WO97/41736, April 2012, Available from <http://www.wipo.int/patentscope/search/en/detail.jsf?docId=WO1997041736&recNum=1&maxRec=&office=&prevFilter=&sortOption=&queryString=&tab=PCT+Biblio>

Srivastava, R. (1984). Studies on extracellular and intracellular purified amylases from a thermophilic *Bacillus stearothermophilus*. *Enzyme Microb Technol.*, Vol. 6, No. 9, (September 1984), pp. 426-422, ISSN 0141-0229

Strokopytov, B., Knegtel, R.M.A., Penninga, D., Roozeboom, H.J., Kalk, K.H., Dijkhuizen, L. and Dijkstra, B.W. (1996). Structure of cyclodextrin glycosyltransferase complexed with a maltononaose inhibitor at 2.6 A° resolution. Implications for product specificity. *Biochemistry*. Vol. 35, No. 13, (April 1996), pp. 4241–4249, ISSN 1520-4995

Sun H., Zhao, P., Ge, X., Xia, Y., Hao, Z., Liu, J. & Peng, M. (2010). Recent Advances in Microbial Raw Starch Degrading Enzymes. *Appl. Biochem. Biotechnol.*, Vol. 160, No. 4, (February 2010), pp. 988–1003, ISSN 1559-0291

Sunna, A., Moracci, M., Rossi, M. & Antranikian, G. (1997). Glycosyl hydrolases from hyperthermophiles. *Extremophiles*, Vol. 1, No. 1, (February 1997), pp. 2–13, ISSN 1433-4909

Svensson B., Jensen M., Mori H., Bak-Jensen K., Bonsager B., Nielsen P., Kramhøft B., Praetorius-Ibba M., Nohr J., Juge N., Greffe L., Williamson G. & Driguez H. (2002). Fascinating facets of function and structure of amylolytic enzymes of glycoside hydrolase family 13. *Biologia Bratislava*, Vol. 57, Suppl. 11,pp. 5–19, ISSN 0006-3088

Svensson, B. & Sogaard, M. (1993). Mutational analysis of glycosylase function. *J. Biotechnol.* Vol. 29, No. 1, (May 1994), pp. 1-37, ISSN 0168-1656

Svensson, B. (1994). Protein engineering in the α-amylase family: catalytic mechanism, substrate specificity, and stability. *Plant Mol. Biol.*, Vol. 25, No. 2, (May 1994), pp. 141-157, ISSN 1573-5028

Takasaki, Y. (1976). Purification and enzymatic properties of β- amylase and pullulanase from *Bacillus cereus* var. mycoides. *Agric. Biol. Chem.*, Vol. 40, No. 8, (August 1974), pp. 1523-1530, ISSN 1881-1280

Thompson, D.B. (2000). On the non-random nature of amylopectin branching. *Carbohydrate Polymers*, Vol. 43, No. 3, (November 2000), pp. 223–239, ISSN 0144-8617

Tonkova, A., (2006). Microbial starch converting enzymes of the α-Amylase family. In: *Microbial Biotechnology in Horticulture*, R.C. Ray & O.P. wards (Ed.), pp. (421-472), , Science Publishers, Enfield, ISBN 9781578084173, New Hampshire, USA

Tuohy, M. & Coughlan, M. (1992). Production of thermostable xylan degrading enzymes by *Talaromyces emersonii*. *Bioresource Technol.*, Vol. 39, No. 2, (February 1992), pp. 131-137, ISSN 0960-8524

Uguru, G., Akinyanju, J. & Sani. A. (1997). The use of sorghum for thermostable amylase production from *Thermoactinomyces thalpophilus*. *Letters in Applied Microbiology.*, Vol. 25, No. 1, (July 1997), pp. 13–16, ISSN 1472-765X

Uitdehaag, J., Mosi, R., Kalk, K., van der Veen, B., Dijkhuizen, L., Withers, S. & Dijkstra, B. (1999). X-ray structures along the reaction pathway of cyclodextrin glycosyltransferase elucidate catalysis in the α-amylase family. *Nat. Struct. Biol.*, Vol. 6, No. 5, (May 1999), pp. 432-436, ISSN 1545-9985

Van Dam, H. & Hille, J. (1992). Yeast and enzymes in bread making. *Cereal Foods World*, Vol. 37, No. 5, (September 1992), pp. 245–252, ISSN 0146-6283

Vigal, T., Gil, J., Daza, A., Garcia-Gonzales, M. & Martin, J.F. (1991). Cloning, characterization and of an α-amylase gene from *Streptomyces griseus* IMRU3570. *Mol. Gen. Genet.*, Vol. 225, No. 2, (February 1991), pp. 278-288, ISSN 1432-1874

Vihinen, M. & Mantsala, P. (1989). Microbial amylolytic enzymes. *Crit Rev Biochem Mol Biol.*, Vol. 24, No. 4, (April 1989), pp. 329-418, ISSN 1549-7798

Virolle, M., Long, C., Chang, S. & Bibb, M. (1988). Cloning, characterization and regulation of an α-amylase gene from *Streptomyces venezuelae*. *Gene*, Vol. 74, No. 2, (December 1988), pp. 321-334, ISSN 0378-1119

Wang, J.P., Mason, J.C. & Broda, P. (1993). Xylanases from *Streptomyces cyaneus*: their production, purification and characterization. *J. Gen. Micrbiol.*, Vol. 139, No. 9, (September 1993), pp. 1987-1993, ISSN 0022-1287

Wang, J.P., Zeng, A.W., Liu, Z. & Yuan, X.G. (2006). Kinetics of glucoamylase hydrolysis of corn starch. *J. Chem. Techno. and Biotechno.*, Vol. 81, No. 5, (May 2006), pp. 727–729, ISSN 1097-4660

Wang, Y.W. & Wang, Y. (1990) Study on nutrient physiology of some species of *Ganoderma*. *Edible Fungi of China*, Vol. 9, No. 5, (September 1990), pp. 7–10, ISSN 1003-8310

Wang, Y.W., Nema, S. & Teagarden, D. (2010). Protein aggregation Pathways and influencing factors. *Int. J. Pharm.*, Vol. 390, No. 2, (May 2010), pp. 89-99, ISSN 0378-5173

Wardi, A. & Michos, G. (1972). Alcian blue staining of glycoproteins in acrylamide disc electrophoresis. *Anal. Biochem.*, Vol. 49, No. 2, (October 1972), pp. 607-609, ISSN 0003-2697

Whistler, R.L. & Daniel, J.R. (1985). Carbohydrates, In: *Food chemistry*, O.R. Fennema, (Ed.), pp. (69-137), Marcel Dekker, ISBN 0-87055-504-9, New York, USA

Whitehead, T.R. & Cotta, M.A. (1995). Identification of intracellular amylase activity in *Streptococcus bovis* and *Streptococcus salivarius*. *Curr. Microbiol.*, Vol. 30, No. 3, (March 1995), pp. 143–148, ISSN 1432-0991

Wolfenden, R., Lu, X. & Young, G. (1998). Spontaneous hydrolysis of glycosides. *J. Am. Chem. Soc.*, Vol. 120, No. 27, (July 1998), pp. 6814–6815, ISSN 1520-5126

Yamamoto, T. (1988). Bacterial a-amylase (liquefying- and saccharifying types) of *Bacillus subtilis* and related bacteria. In: *Handbook of amylases and related enzymes*. The amylase research society of Japan (Ed.), pp. (40 45), Pergamon Press, ISBN 0080361412, Oxford

Yamane, K., Yamaguchi, K & Maruo, B. (1973). Purification and properties of a cross-reacting material related to α-amylase and biochemical comparison with parental α-amylase. *Biochim Biophys Acta.*, Vol. 298, No. 2, (March 1973), pp. 295-323, ISSN 0005-2736

Zare-Maivan, H. & Shearer, C.A. (1988). Extracellular enzyme production and cell wall degradation by freshwater lignicolous fungi. *Mycology*, Vol. 80, No. 3, (May 1988), pp. 365-375, ISSN 00275514

Zeikus, J. (1979). Thermophilic bacteria: ecology, physiology, and technology. *Enzyme Microb. Technol.*, Vol. 1, No. 4, (October 1979), pp. 243-251, ISSN 0141-0229

Zhu, F., Corke, H. & Bertoft, E. (2011).Amylopectin internal molecular structure in relation to physical properties of sweetpotato starch. *Carbohydrate Polymers,* Vol. 84, No. 3, (March 2011), pp. 907-918, ISSN 0144-8617

Zobel, H.F. (1992). Starch granule structure, In: *Developments in Carbohydrate chemistry,* R. J. Alexander & H.F. Zobel (Ed.), pp. (1-36), American association of cereal chemistry, ISBN 0913250767, Minnesota, USA.

Chapter 22

Algal Polysaccharides, Novel Applications and Outlook

Stefan Kraan

Additional information is available at the end of the chapter

http://dx.doi.org/10.5772/51572

1. Introduction

Marine algae contain large amounts of polysaccharides, notably cell wall structural, but also mycopolysaccharides and storage polysaccharides (Kumar et al. 2008b; Murata and Nakazoe 2001). Polysaccharides are polymers of simple sugars (monosaccharides) linked together by glycosidic bonds, and they have numerous commercial applications in products such as stabilisers, thickeners, emulsifiers, food, feed, beverages etc. (McHugh 1987; Tseng 2001; Bixler and Porse, 2010). The total polysaccharide concentrations in the seaweed species of interest range from 4-76 % of the dry weight (Table 1). The highest contents are found in species such as *Ascophyllum*, *Porphyra* and *Palmaria*, however, green seaweed species such as *Ulva* also have a high content, up to 65 % of dry weight.

Seaweeds are low in calories from a nutritional perspective. The lipid content is low and even though the carbohydrate content is high, most of this is dietary fibres and not taken up by the human body. However, dietary fibres are good for human health as they make an excellent intestinal environment (Holt and Kraan, 2011).

The cell-wall polysaccharides mainly consist of cellulose and hemicelluloses, neutral polysaccharides, and they are thought to physically support the thallus in water. The building blocks needed to support the thalli of seaweed in water are less rigid/strong compared to terrestial plants and trees. The cellulose and hemicellulose content of the seaweed species of interest in this review is 2 10 % and 9 % dry weight respectively. Lignin is only found in *Ulva* sp. at concentrations of 3 % dry weight (Table 2).

The cell wall and storage polysaccharides are species-specific and examples of concentrations are given in Table 3. Green algae contain sulphuric acid polysaccharides, sulphated galactans and xylans, brown algae alginic acid, fucoidan (sulphated fucose), laminarin (β-1, 3 glucan) and sargassan and red algae agars, carrageenans, xylans, floridean starch (amylopectin like glucan), water-soluble sulphated galactan , as well as porphyran as

		Brown Algae					Green Algae	Red Algae			
		Laminaria & Saccharina	Fucus	Ascophyllum	Undaria	Sargassum	Ulva	Chondrus	Porphyra	Gracilaria	Palmaria
Polysaccha-rides	Total	38 % [b] 48 % [s] 58 % [a] 61 % [s]	62 % [s] 66 % [a]	42-64 % [s] 44 % [c] 70 % [a]	35-45 % [d, e]	4 % [f] 68 % [f]	15 % -65 % [g-i, q] 18 % [p] 42-46 % [s]	55-66 % [s]	40 % [e] 41 % [r] 50-76 % [s] 54 % [b]	36 % [f] 62 % [b] 63 % [f]	38-74 % [j] 50 % [s] 66 % [k]
Structural and dietary fibres	Total	36 % [l]			35-46 % [l, m]	49-62 % [l, m]	38 % [l, m]		35-49 % [l, m]		
	Soluble			38 % [c]	30 % [l, m]	33 % [l, m]	21 % [l, m]		18 % [l, m]		
	Lignin						3 % [g]				
	Cellulose	10 % in stipe [n] 4.5-9 % [o]	2-4.5 % [o]	2 % [o] 3.5-4.6 % [n]			9 % [g]				
	Hemicelluloses						9 % [g]				

a= (Rioux et al. 2007), b= (Wen et al. 2006), c= (Tseng 2001), d= (Je et al. 2009), e= (Murata and Nakazoe 2001), f= (Marinho-Soriano et al. 2006), g= (Ventura and Castañón 1998), h= (Ortiz et al. 2006), i= (Sathivel et al. 2008), j= (Heo and Jeon 2008), k= (Mishra et al. 1993), l= (Dawczynski et al. 2007), m= (Lahaye 1991), n= (Horn 2000), o= (Black 1950), p= (Foster and Hodgson 1998), q= (Wong and Cheung 2000), r= (Arasaki and Arasaki 1983), s= (Morrissey et al. 2001)

Table 1. Content of total polysaccharides (% of dry weight) and structural and dietary fibres (% of dry weight) in seaweed species of interest in North-west Europe.

Class	Genus	Uses
Chlorophyta		
	Monostroma	edible, human food
	Enteromorpha	edible, human food
Phaeophyta		
	Laminaria	alginates, edible, human food
	Undaria	edible, human food
	Cladosiphon	edible, human food
Rhodophyta		
	Asparagopsis	medical applications
	Gelidiella	agar, food and medical
	Gelidiopsis	agar, food and medical
	Gelidium	agar, food and medical
	Graciluria	agar, food and medical
	Pterocladia	agar, food and medical
	Chondrus	carrageenan, human food
	Eucheuma	carrageenan, human food
	Kappaphycus	carrageenan, human food
	Gigartina	carrageenan, human food
	Hypnea	carrageenan, human food
	Iridaea	carrageenan, human food
	Palmaria	human and animal feed
	Porphyra	human food

Table 2. Most common genera and uses of seaweeds produced in aquaculture.

Product	Global Production	Retail Price	Approximate Gross Market Value	Amount Used	
	(ton/year)	(US$/kg)	(US$million/year)	Food (%)	Pharmacy(%)
Agars	9,600	18	173	80	ca 15
Alginates	26,500	12	318	30	5
Carrageenans	50,000	10.5	525	80	10

Table 3. Phycocolloids and their global production, retail price, gross value and percentage used in food and pharmacy(Data; McHugh, 2003; Bixler and Porse, 2010)

mucopolysaccharides located in the intercellular spaces (Table 3; Kumar et al. 2008b; Murata and Nakazoe 2001). Contents of both total and species-specific polysaccharides show seasonal variations. The mannitol content varied markedly in the fronds of *Saccharina* and *Laminaria* species with maximum amounts found during summer and autumn, from June to November. The laminaran showed extreme variations during the year with very small amounts or none at all in February to June and maximum in September to November (Haug and Jensen 1954; Jensen and Haug 1956). The maximum content of alginic acid in the fronds of *Saccharina* and *Laminaria* species was generally found from March to June and the minimum from September to October (Haug and Jensen 1954). However, highest contents of alginic acid were found during winter in other seasonal studies on *Laminaria* species from the same areas in Norway (Jensen and Haug 1956).

Further investigations on the hydrolysates of some brown algae showed complex mixtures of monosaccharides. The components of galactose, glucose, mannose, fructose, xylose, fucose and arabinose were found in the total sugars in the hydrolysates. The glucose content was 65 %, 30 % and 20 % of the total sugars in an autumn sample of 50 individual plants of *Saccharina, Fucus* (*serratus and spiralis*) and *Ascophyllum,* respectively (Jensen 1956).

Several other polysaccharides are present in and utilised from seaweed e.g. furcellaran, funoran, ascophyllan and sargassan, however these are not described in this chapter.

Seaweed polysaccharides are separated into dietary fibres, hydrocolloids etc. in the following paragraphs, even though the polysaccharides belong to more than just one of the functional groups.

2. Seaweed production and extraction

Harvesting or aquaculture of marine algae or seaweeds is an extensive global industry.

This seaweed industry provides a wide variety of products that have an estimated total annual production value of about US$ 6 billion. Food products for human consumption contribute about US$ 5 billion to this figure. Substances that are extracted from seaweeds – hydrocolloids – account for a large part of the remaining billion dollars, while smaller, miscellaneous uses, such as fertilizers and animal feed additives, make up the rest. The industry uses almost 20 million ton of wet seaweed annually (FAO, 2012 http://www.fao.org/fishery/statistics/global-aquaculture-production/en), harvested either from naturally growing (wild) seaweed or from cultivated (farmed) crops. The farming of seaweed has expanded rapidly as demand has outstripped the supply available from natural resources. Commercial harvesting occurs in about 35 countries, spread between the northern and southern hemispheres, in waters ranging from cold, through temperate, to tropical (Mc Hugh, 2003).

In Asia, seaweed cultivation is by far more important in terms of output and value than any other form of aquaculture. Looking at a global scale cultivated, seaweeds account for 87% of all seaweed harvested and processed of which the bulk is derived from aquaculture in Asia. The most valued of the cultivated seaweeds is the red alga Porphyra, or Nori. It is a major source of food for humans throughout the world, although it is almost exclusively cultivated in Japan, South Korea and China. Worldwide production has an annual value of over € 1.5 billion. In addition to Porphyra, other edible seaweeds include Gracilaria, Undaria, Laminaria and Caulerpa with their collective value exceeding € 3.0 billion. New applications of seaweeds and specific seaweed compounds in different sectors, such as food supplements, cosmetics, biomedicine and biotechnology are developed (Chritchley et al., 2006)

Seaweeds are also the industrial sources of polysaccharides such as carrageenans (Chondrus, Eucheuma and Kappaphycus), alginates (Ascophyllum, Laminaria, and

Macrocystis) and agars (Gelidium and Gracilaria; Table 2) and have a global value of approximately $ US 1 billion (Table 3). These important polysaccharides are used in the food, textile, paint, biotechnological and biomedical industries and have recently come under the spotlight as functional food ingredients. (Critchley et al., 2006).The majority of these species are used in some form for food or, in a few cases, for chemical extracts. The costs of production of the biomass tend to exceed the value of the biomass as a raw material for phycocolloid extraction, although it is known that some Chinese kelps produced by aquaculture are used for the production of salts of alginic acid, and applying low-technology extensive forms of aquaculture are used to produce Gracilaria for agar extraction (table 2). The value of red seaweed produced by aquaculture showed a declining trend to US$1.3 - 1.4 billion over the 1997 - 2000 period. This is probably due to the high volumes of carrageenophytes Kappaphycus and Eucheuma produced by cultivation in south-east Asia and, to some extent, Gracilaria cultivated in South America, and as such, showing trends towards commoditization.

3. Phycolloid industry

The term phycocolloid is used to refer to three main products (alginate, carrageenan and agar) which are extracted from Brown and red seaweeds respectively. The estimated world market value for phycocolloids is $ US 1 Billion (Bixler and Porse, 2010). European output of phycocolloids is estimated to have an annual wholesale value of around €130 million which is 97.5% of the total for all algal products in Europe (Earons, 1994).

3.1. Brown seaweeds and alginates

Global alginate production is ca. 26,500 tons and valued at US$318 million annually, and is extracted from brown seaweeds, most of which are harvested from the wild. The more useful brown seaweeds grow in cold waters, thriving best in waters up to about 20 °C. The main commercial sources of phaeophytes for alginates are *Ascophyllum, Laminaria,* and *Mycrocystis*. Other minor sources include *Sargassum, Durvillea, Eklonia, Lessonia,* and *Turbinaria* (Bixler and Porse, 2010)..

Brown seaweeds are also found in warmer waters, but these are less suitable for alginate production and are rarely used as food due to non-desirable chemical compounds such as terpenes. A wide variety of species are used, harvested in both the northern and southern hemispheres. Countries producing alginate include Argentina, Australia, Canada, Chile, Japan, China, Mexico, Norway, South Africa, the United Kingdom (mainly Scotland) and the United States. Most species are harvested from natural resources; cultivated raw material is normally too expensive for alginate production with the exception of China. While much of the *Laminaria* cultivated in China is used for food, when there is surplus production this can also be used in the alginate industry.

Seaweed hydrocolloids, such as alginate, agar and carrageenan, compete with plant gums (such as guar and locust bean) and cellulose derivatives (such as carboxy methyl cellulose (CMC) and methyl cellulose) that are often cheaper.

Alginates are used in the manufacture of pharmaceutical, cosmetic creams, paper and cardboard, and processed foods (Chapman, 1970). Alginate represents the most important seaweed colloid in term of volume. Grades of alginate are available for specific applications. Sodium alginate, pharmaceutical grade (US$ 13-15.5 per kg), food grade (US$ 6.5-11.0 per kg). In Japan and Korea, high demands for *Laminaria japonica* (trade name kombu) have resulted in high prices and necessitated the import of supplies for alginate extraction (Critchley, 1998). Annual growth rate for alginates is ca. 2-3 percent, with textile printing applications accounting for about half of the global market. Pharmaceutical and medical uses are about 20 percent by value of the market and have stayed buoyant, with 2-4 percent annual growth rates, driven by ongoing developments in controlled release technologies and the use of alginates in wound care applications. Food applications are worth about 20 percent of the market. That sector has been growing only slowly, and recently has grown at only 1-2 percent annually. The paper industry takes about 5 percent and the sector is very competitive, not increasing but just holding its own. The alginate industry faces strong competition from Chinese producers, whose prices do not reflect the real expense of cultivating Laminaria japonica, even in China, yet they do not appear to import sufficient wild seaweeds to offset those costs. The result is low profitability for most of the industry, with the best opportunities lying in the high end of the market, such as pharmaceutical and medical applications.

3.2. Red seaweeds and Carrageenans

Carrageenans are a group of biomolecules composed of linear polysaccharide chains with sulphate half-esters attached to the sugar unit. These properties allow carrageenans to dissolve in water, form highly viscous solutions, and remain stable over a wide pH range. There are three general forms (kappa, lambda and iota), each with their own gelling property. Kappa carrageenan today is almost exclusively obtained from farmed *Kappaphycus alvarezii* and iota from farmed *Eucheuma denticulatum* (Rasmussen and Morrissey 2007).

The Carrageenan market is worth US$ 527 m with most Carrageenan used as human food-grade semi-refined carrageenan (90%) and the rest going into pet food as semi-refined carrageenan. *Chondrus crispus* and *Kappaphycus* sp. are species containing up to 71 % and 88 % of carrageenan, respectively (Chopin et al. 1999; Rodrigueza and Montaño 2007). Food applications for carrageenans (E 407) are many, including canned foods, desert mousses, salad dressings, bakery fillings, ice cream, instant deserts and canned pet-foods. In the 1970s, an energy efficient process was developed in the Philippines to make a lower cost, strong-gelling kappa carrageenan and a weakly gelling iota. These semi-refined products gradually replaced the use of refined carrageenan as the gelling agent in canned meat pet foods. The process required lower capital investment than standard carrageenan refineries and the semi-refined extracts could be profitably sold for about two-thirds the price of conventionally refined carrageenan (Bixler and Porse, 2010). Industrial applications for purified extracts of carrageenans are equally diverse. They are used in the brewing industry for clarifying beer, wines and honeys, although less commonly than previously. There has been a fairly significant increase in production capacity for gel-press-refined carrageenan in recent years, particularly in Asia-Pacific.

3.3. Red seaweeds and agar

Agar is a mixture of polysaccharides, which can be composed of agarose and agropectin, with similar structural and functional properties as carrageenans. It is extracted from red seaweed such as *Gelidium* spp. and *Gracilaria* spp. (FAO 2008; Rasmussen and Morrissey 2007; Jeon et al. 2005). A total of 9600 mt were sold in 2009 with a value of US$ 173 m. Agar–agar is a typical and traditional food material in Japan and it is used as a material for cooking and Japanese-style confectionary. In addition, agar-agar is used in the manufacture of capsules for medical applications and as a medium for cell cultures, etc. Food applications continue to grow as shown by an increase of 2100 mt in over the last decade, and have been driven by the growth of processed foods in developing countries (Bixler and Porse, 2010).

Agar melts and gels at higher temperatures than carrageenan so it finds uses in pastry fillings and glazes that can be applied before the pastry is baked without melting in the pastry oven. In processed meats, carrageenan is the favored water binder or texturing agent, but agar hold on to the gelatine replacement market in canned meats and aspics. The texture of agar in fruit jellies also compete with kappa carrageenan jellies, but the agar texture is preferred in parts of Asia, particularly in Japan. Although significantly smaller, the markets for the specialty grades are quite attractive because of better profit margins. Agarose, a sulfate-free very pure form of agar finds widespread use today in gel electrophoretic analysis of the molecules of biotechnology.

4. Bioactive algal polysaccharides and functional properties

While food has long been used to improve health, our knowledge of the relationship between food components and health is now being used to improve food. Strictly speaking, all food is functional, in that it provides energy and nutrients necessary for survival. But the term "functional food" in use today conveys health benefits that extend far beyond mere survival. Food and nutrition science has moved from identifying and correcting nutritional deficiencies to designing foods that promote optimal health and reduce the risk of disease.

The combination of consumer desires, advances in food technology, and new evidence-based science linking diet to disease and disease prevention has created an unprecedented opportunity to address public health issues through diet and lifestyle. Widespread interest in select foods that might promote health has resulted in the use of the term "functional foods." Although most foods can be considered "functional," in the context of this Chapter the term is reserved for algal polysaccharides that have been demonstrated to provide specific health benefits beyond basic nutrition, such as, alginates, agars, carrageenans, fucoidan, mannitol, laminarin, and ulvan.

4.1. Dietary fibres

The dietary fibres are very diverse in composition and chemical structure as well as in their physicochemical properties, their ability to be fermented by the colonic flora, and their biological effects on animal and human cells (Lahaye and Kaeffer 1997). Edible seaweed contain 33-50 % total fibres on a dry weight basis, which is higher than the levels found in higher plants, and these fibres are rich in soluble fractions (Lahaye 1991). The dietary fibres included in marine algae are classified into two types, i.e. insoluble such as cellulose, mannans and xylan, and water soluble dietary fibres such as agars, alginic acid, furonan, laminaran and porphyran. The total content of dietary fibres is 58 % dry weight for *Undaria*, 50 % for *Fucus*, 30 % for *Porphyra* and 29 % for *Saccharina* (Murata and Nakazoe 2001). *Fucus* and *Laminaria* have the highest content of insoluble dietary fibres (40 % and 27 % respectively) and *Undaria pinnatifida* (wakame), *Chondrus* and *Porphyra* have the highest content of soluble dietary fibres (15 % – 22 %; Fleury and Lahaye 1991).

The undigested polysaccharides of seaweed can form important sources of dietary fibres, although they might modify digestibility of dietary protein and minerals. Apparent digestibility and retention coefficients of Ca, Mg, Fe, Na and K were lower in seaweed-fed rats (Urbano and Goñi 2002). The seaweed dietary fibres contain some valuable nutrients and substances, and there has been a deal of interest in seaweed meal, functional foods, and nutraceuticals for human consumption (McHugh 2003), because among other things polysaccharides show anti-tumour and anti-herpetitic bioactivity, they are potent as an anti-coagulant and decrease low density lipid (LDL)-cholesterols in rats (hypercholesterolemia), they prevent obesity, large intestine cancer and diabetes, and they have anti-viral activities (Lee et al. 2004; Murata and Nakazoe 2001; Amano et al. 2005; Athukorala et al. 2007; Ghosh et al. 2009; Murata and Nakazoe 2001; Ye et al. 2008). Moreover, glucose availability and absorption are delayed in the proximal small intestine after the addition of soluble fibres, thus reducing postprandial glucose levels (Jenkins et al. 1978). Water insoluble polysaccharides (celluloses) are mainly associated with a decrease in digestive tract transit time (see also digestibility of polysaccharides; Mabeau and Fleurence 1993).

4.2. Alginates

Alginates were discovered in the 1880s by a British pharmacist, E.C.C. Stanford; industrial production began in California in 1929. Algins/alginates are extracted from brown seaweed and are available in both acid and salt form. The acid form is a linear polyuronic acid and referred to as alginic acid, whereas the salt form is an important cell wall component in all brown seaweed, constituting up to 40 % - 47 % of the dry weight of algal biomass (Arasaki and Arasaki 1983; Rasmussen and Morrissey 2007). Alginates are anionic polysaccharides. They contain linear blocks of covalently (1–4)-linked β-D-mannuronate with the C5 epimer α-L-guluronate. The blocks may contain one or both of the monomers and the ratio of monomers A and B, as well as a number of units (m and n) in a block is species dependent.

It has been reported that alginic acid leads to a decrease in the concentration of cholesterol, it exerts an anti-hypertension effect, it can prevent absorption of toxic chemical substances, and it plays a major role as dietary fibre for the maintenance of animal and human health (Kim and Lee 2008; Murata and Nakazoe 2001; Nishide and Uchida 2003). These dietary polysaccharides are not found in any land plants. They help protect against potential carcinogens and they clear the digestive system and protect surface membranes of the stomach and intestine. They absorb substances like cholesterol, which are then eliminated from the digestive system (Burtin 2003; Ito and Tsuchida 1972) and result in hypocholesterolemic and hypolipidemic responses (Kiriyama et al. 1969; Lamela et al. 1989; Panlasigui et al. 2003). This is often coupled with an increase in the faecal cholesterol content and a hypoglycaemic response (Dumelod et al. 1999; Ito and Tsuchida 1972; Nishide et al. 1993).

Alginates, fucoidans and laminarin extracts were tested against nine bacteria, including *Escerichia coli, Staphycococcus, Salmonella* and *Listeria*. They appeared to be effective against *E. coli* and *Staphylocococcus*. Sodium alginate seemed to demonstrate a strong anti-bacterial element. It not only binds but also kills the bacteria. Studies conducted on seaweed extracts found that fucoidan appeared to function as a good prebiotic (a substance that encourages the growth of beneficial bacteria in the intestines). An anti-inflammatory effect from some of the extracts has also been found, and so far no toxic effects have emerged in use for human health (Hennequart 2007). Furthermore, alginates with molecular weights greater than or equal to 50 kDa could prevent obesity, hypocholesterolemia and diabetes (Kimura et al. 1996). Clinical observations of volunteers who were 25 % - 30 % overweight showed that alginate, a drug containing alginic acid, significantly decreased body weight (Zee et al. 1991). In type II diabetes treatment, taking 5 g of sodium alginate every morning was found to prevent a postprandial increase of glucose, insulin, and C-peptide levels and slowed down gastric transit (Torsdottir et al. 1991). Meal supplemented with 5 % kelp alginates decreased glucose absorption balance over 8 hours in pigs. Similar studies have been done on rats and humans (Vaugelade et al. 2000). Another health effect is that the binding property of alginic acid to divalent metallic ions is correlated to the degree of the gelation or precipitation in the range of Ba<Pb<Cu<Sr<Cd<Ca<Zn<Ni<Co<Mn<Fe<Mg. No intestinal enzymes can digest alginic acid. This means that heavy metals taken into the human body are gelated or rendered insoluble by alginic acid in the intestines and cannot be absorbed into the body tissue (Arasaki and Arasaki 1983).In several countries such as the USA, Germany, Japan, Belgium and Canada, the use of alginic acid and its derivatives for the treatment of gastritis and gastroduodenal ulcers, as well as the use of alginates as anti-ulcer remedies, is protected by patents (Bogentoff 1981; Borgo 1984; Reckitt and Colman Products Ltd 1987; Sheth 1967). Several products of alginate containing drugs have been shown to effectively suppress postprandial (after eating) and acidic refluxes, binding of bile acids and duodenal ulcers in humans. Examples are "Gaviscon" (sodium alginate, sodium bicarbonate, and calcium carbonate), "Algitec" (sodium alginate and cimetidine, an $H2$ antagonist) and "Gastralgin" (alginic acid, sodium alginate, aluminium hydroxide, magnesium hydroxide and calcium carbonate) (Khotimchenko et al. 2001; Washington and Denton 1995). Clinical trials showed that sodium alginate promotes regeneration of the mucous membrane in the stomach, suppresses inflammation, eradicates colonies of *Helicobacter pylori* in the mucous membrane and normalizes non-specific resistance of the latter in 4 to 15-year-old children. It also promotes restoration of the intestinal biocenosis (Miroshnichenko et al. 1998). Other studies show positive dietry effects of alginates on faecal microbial fauna, changes in concentrations of compounds and acids, and prebiotic properties that can promote health (Terada et al. 1995; Wang et al. 2006).

Sodium alginate is often used as a powder, either pure or mixed with other drugs, on septic wounds. The polysaccharide base stimulates reparative processes, it prepares the wound for scarring, and it displays protective and coating effects, shielding mucous membranes and damaged skin against irritation from unfavourable environments. Calcium alginate promotes the proliferation of fibroblasts and inhibits the proliferation of microvascular endothelial cells and keratinocytes (Doyle et al. 1996; Glyantsev et al. 1993; Swinyard and Pathak 1985). Profound wound healing effects have also been reported for a gelatine-alginate sponge impregnated with anti-septics and anti-biotics (Choi et al. 1999).

Another use of alginates is the absorbing hemostatic effect exploited in surgery. Gauze dressings, cotton, swabs, and special materials impregnated with a solution of sodium alginate are produced and used for external use and for application onto bleeding points during abdominal operations on parenchymatous organs (Khotimchenko et al. 2001; Savitskaya 1986). Studies on the effect of alginate on prothrombotic blood coagulation and platelet activation have shown that the degree of these effects depends on the ratio between the mannuronic and guluronic chains in the molecule, as well as on the concentration of calcium. However, a zinc ion containing alginate was shown to have the most profound hemostatic effects (Segal et al. 1998). A "poraprezinc-sodium alginate suspension" has been suggested as a high performance mixture for the treatment of severe gingivostomatitis (cold sores) complicated by hemorrhagic erosions and ulcers (Katayama et al. 1998). When applied to the tooth surface, alginate fibres swell to form a gel like substance, a matrix for coagulation. Alginate dressings are used to pack sinuses, fistulas, and tooth cavities (Reynolds et al. 1982).

Furthermore, the algins have anti-cancer properties (Murata and Nakazoe 2001) and a bioactive food additive "Detoxal", containing calcium alginate, has anti-toxic effects on hepatitis. This drug decreases the content of lipid peroxidation products and normalizes the concentrations of lipids and glycogen in the liver (Khotimchenko et al. 2001).

4.3. Carrageenans

From a human health perspective it has been reported that carrageenans have anti-tumour and anti-viral properties (Skoler-Karpoff et al. 2008; Vlieghe et al. 2002; Yan et al. 2004; Zhou et al. 2006b). Furthermore, Irish Moss or Carrageen (*C. crispus* and *Mastocarpus stellatus*) has a large number of medical applications, some of which date from the 1830s. It is still used in Ireland to make traditional medicinal teas and cough medicines to combat colds, bronchitis, and chronic coughs. It is said to be particularly useful for dislodging mucus and has anti-viral properties. Carrageenans are also used as suspension agents and stabilisers in other drugs, lotions and medicinal creams. Other medical applications are as an anti-coagulant in blood products and for the treatment of bowel problems such as diarrhoea, constipation and dysentery. They are also used to make internal poultices to control stomach ulcers (Morrissey et al. 2001).

New research on the biocide properties shows that applications of carrageenan gels from *C. crispus* may block the transmission of the HIV virus as well as other STD viruses such as gonorrhoea, genital warts and the herpes simplex virus (HSV) (Caceres et al. 2000; Carlucci et al. 1997; Luescher-Mattli 2003; Shanmugam and Mody 2000; Witvrouw and DeClercq 1997). In addition, carrageenans are good candidates for use as vaginal microbicides because they do not exhibit significant levels of cytotoxicity or anti-coagulant activity (Buck et al. 2006; Zeitlin et al. 1997). Results of sexual lubricant gels raised the possibility that use of such lubricant products, or condoms lubricated with carrageenan-based gels, could block the sexual transmission of HPV (human papillomavirus) types that can cause cervical cancer and genital warts. However, carrageenan inhibition of herpes simplex virus and HIV-1 infectivity were demonstrated as about a thousand-fold higher than the IC50s observed for genital HPVs *in vitro* (Witvrouw and de Clerck 1997; Luescher-Mattli et al. 2003). A carrageenan-based vaginal microbicide called Carraguard has been shown to block HIV and other sexually transmitted diseases *in vitro*. Massive clinical trials by the Population Council Centre began in two severely affected African countries; Botswana and South Africa in 2002. Carraguard entered phase III clinical trials involving 6000 non-pregnant, HIV-negative women in South Africa and Botswana in 2003 (Spieler 2002).

Many reports exist of anti-coagulant activity and inhibited platelet aggregation of carrageenan (Hawkins et al. 1962; Hawkins and Leonard 1963; Kindness et al. 1979). Among the carrageenan types, λ carrageenan (primarely from *C. crispus*) has approximately twice the activity of unfractioned carrageenan and four times the activity of κ-carrageenan (*Eucheuma cottoni* and *E. spinosum*). The most active carrageenan has approximately one-fifteenth the activity of heparin (Hawkins et al. 1962), but the sulphated galactan from *Grateloupa indica* collected from Indian waters, exhibited anti-coagulant activity as potent as heparin (Sen et al. 1994). The principal basis of the anti-coagulant activity of carrageenan appeared to be an anti-thrombotic property. λ-carrageenan showed greater anti-thrombotic activity than κ-carrageenan, probably due to its higher sulphate content, whereas the activity of the unfractionated material remained between the two. It was found that toxicity of carrageenans depended on the molecular weight and not the sulphate content. Similar results were obtained with λ-carrageenan of *Phyllophora brodiaei* which gave the highest blood anti-coagulant activity (Sen et al. 1994). In addition, the hypoglycaemic effect of carrageenan may prove useful in the prevention and management of metabolic conditions such as diabetes (Dumelod et al. 1999).The use of carrageenan for food applications started almost 600 years ago. Due to its long and safe history of use, carrageenan is generally recognised as safe (GRAS) by experts from the US Food and Drug Administration (21 CFR 182.7255) and is approved as a food additive (21 CFR 172.620). The World Health Organisation (WHO) Joint Expert Committee of Food Additives has concluded that it is not necessary to specify an acceptable daily intake limit for carrageenans (van de Velde et al. 2002). Although carrageenans are used widely as a food ingredient, they are also used in experimental research in animals where they induce pleurisy and ulceration of the colon (Noa et al. 2000). Furthermore, carrageenans can cause reproducible inflammatory reaction and they remain a standard irritant for examining acute inflammation and anti-inflammatory drugs. Two test systems are used widely for the evaluation of non-steroidal anti-inflammatory drugs and cyclooxygenase activity: (1) The carrageenan air pouch model and (2) The carrageenan-induced rat paw edema assay (Dannhardt and Kiefer 2001). The role of carrageenans in promotion of colorectal ulceration formation is controversial and much seems to depend on the molecular weight of the carrageenan used. The international agency for research on cancer classified degraded carrageenan as a possible human carcinogen but native carrageenan remains unclassified in relation to a causative agent of human colon cancer and as mentioned it has GRAS status (Carthew 2002; Tobacman 2001).

4.4. Agar

The agar content in *Gracilaria* species can reach 31 % . It has been reported that agar-agar leads to decreases in the concentration of blood glucose and exerts an anti-aggregation effect on red blood cells. It has also been reported to affect absorption of ultraviolet rays (Murata and Nakazoe 2001). Anti-tumour activity was also found in an agar-type polysaccharide from cold water extraction of another *Gracilaria* species and hydrolysates of agar resulted in agaro-oligosaccharides with activity against α-glucosidase and antioxidant ability (Chen et al. 2005; Fernandez et al. 1989). Agarose can be separated from the agar with a yield of 42 %, and the agar content varied seasonally from 26 % - 42 % in *Gelidium* spp. in another experiment (Mouradi-Givernaud et al. 1992; Jeon et al. 2005). Agaro-oligosaccharides have also been shown to suppress the production of a pro-inflammatory cytokine and an enzyme associated with the production of nitric oxide (Enoki et al. 2003).

4.5. Fucoidan/fucans/fucanoids

Fucoidans are a group of polysaccharides (fucans) primarily composed of sulphated L-fucose with less than 10 % of other monosaccharides. They are widely found in the cell walls of brown seaweed, but not in other algae or higher plants (Berteau and Mulloy 2003).

Fucoidan is considered as a cell wall reinforcing molecule and seems to be associated with protection against the effects of desiccation when the seaweed is exposed at low-tide. Fucoidans were first isolated by Kylin almost one century ago and have interesting bioactivities (Kylin 1913). According to Table 4b the species *Fucus vesiculosus* contains the highest concentration of fucoidans (up to 20 % on a dry weight basis). Fucanoids can make up more than 40 % of dry weight of isolated algal cell walls and can easily be extracted using either hot water or an acid solution (Berteau and Mulloy 2003). Fucoidan is viscous in very low concentrations and susceptible to breakdown by diluted acids and bases. Fucoidans are produced as complex, heterogeneous polysaccharides, which contribute to intercellular mucilage. Their structural complexity varies in the degree of branching, substituents,sulfatation and type of linkages, the fine structure depending on the source of the polysaccharide. Although their composition varies with species and geographical origin, fucoidans always contain fucose and sulfate with small proportions of uronic acids, galactose, xylose, arabinose and mannose. Algal fucoidans have one of two types of homofucose backbone chains, with either repeated (1→3)-linked α-L-fucopyranosyl residues or alternating (1→3)- and (1→4)-linked α-L-fucopyranosyl residues Cumashi et al 2007). Sulfonato- and acetyl-groups as well as α-L-fucopyranosyl, α-D-glucuronopyranosyl and some other sugar residues may occur at O-2 and/or at O-4 of the α-L-fucose units of the backbone. Fucoidans with backbones of first type have been isolated from seaweeds *Saccharina latissima, Laminaria digitata, Chorda filum,* and *Cladosiphon okamuranus*. The second type of backbone was found in fucoidans from *Fucus evanescense, Fucus distichus,* and *Ascophyllum nodosum*.

A Tasmanian company, Marinova Pty Ltd, is able to supply commercial volumes of fucoidan extract and their derivates, formulated to purity levels of up to 95 %. All fucoidans of the species *Undaria* sp., *Lessonia* sp., *Macrocystis* sp., *Cladosiphon* sp., *Durvillea* sp., *Laminaria* sp., *Ecklonia* sp., *Fucus* sp., *Sargassum* sp., *Ascophyllum* sp., and *Alaria* sp. are Halal and Kosher certified. Marinova has isolated fucoidans from a range of species (species-specific), and can provide characterised fractions for either investigational research or as ingredients for nutraceutical and cosmetic applications. Different therapeutic profiles are primarily due to the molecular structure. The company has developed the Maritech™ coldwater extraction process, which maintains the integrity of fucoidans, and produces nature-equivalent high molecular weight molecules with optimal bioactivity. Solvent-based extraction, which is commonly used, causes degradation of fucoidans, and limits the activity of these molecules in biological assays.

Sector	% Demand	Sector Growth	Ton	Value US$
Textile printing and technical grades	41.5	Flat	11,000	140.0m
Food and Pharma	49	60%	13,000	150.0m
Animal feed	3.8	-75%	1,000	18.0m
PGA	5.7	-25%	1,500	10.0m
Total	100		26,500	$318m

Table 4. Alginate markets (2009) – Tonnage, value and sector growth over last decade (Data; Bixler and Porse, 2010)

Although the major physiological purposes of fucans in the algae are not thoroughly understood, they are known to possess numerous biological properties with potential human health applications (Berteau and Mulloy 2003). The list of bioactivity of fucoidan for human health is long. Fucoidan found in seaweed such as *Undaria* and *Laminaria* shows anti-coagulant, anti-viral and anti-cancer properties; Chevolot et al. 1999; Zhuang et al. 1995).

Fucoidan preparations have been proposed as an alternative to the injectable anti-coagulant heparin, because fucoidan originates from plant matter and is less likely to contain infectious agents such as viruses (Berteau and Mulloy 2003). No toxicological changes were observed when 300 mg/kg body weight per day fucoidan was administered to rats, however, significantly prolonged blood-clotting times were observed when concentrations were increased three-fold (Li et al. 2005). The biological activity (e.g. antioxidant and anti-coagulant) of sulphated polysaccharides is not only related to molecular weight and sulphated ester content (role in the charge of the molecule), as previously determined, but also to glucuronic acid and fucose content, together with the position of the sulphate groups on the sugar resides (Berteau and Mulloy 2003; Li et al. 2005; Zhao et al. 2008). A large molecular weight is required to achieve anti-coagulant activity, as fucoidan needs a long sugar chain in order to be able to bind the thrombin (coagulation protein in the blood stream). Some researchers have measured fucoidan's molecular weight at approximately 100 kDa. while others have observed a molecular weight of 1600 kDa (Rioux et al. 2007). The native fucoidan from *Lessonia vadosa* with a molecular weight of 320 kDa showed good anti-coagulant activity compared to a smaller depolymerised fraction with a molecular weight of 32 kDa, which presented weaker anti-coagulant activity (Li et al. 2008a). Some structural features of fucoidan are most likely required for certain specific activities.

Fucoidan stimulates the immuno system in several ways, and the numerous important biological effects of fucoidans are related to their ability to modify cell surface properties (Usov et al. 2001). Oral intake of the fucoidans present in dietary brown seaweed might take the protective effects through direct inhibition of viral replication and stimulation of the immune system (innate and adaptive) functions (Hayashi et al. 2008). Fucoidan has been found to restore the immune functions of immune-suppressed mice, act as an immunomodulator directly on macrophage, T lymphocyte, B cell, natural killer cells (NK cell; Wang et al. 1994), promote the recovery of immunologic function in irradiated rats (Wu et al. 2003), induce the production of interleukin (IL-1) and interferon-γ (IFN-γ) *in vitro*, and promote the primary antibody response in sheep red blood cells *in vivo* (Yang et al. 1995).

The mechanism of anti-viral activities of fucoidan is to inhibit viral sorption so as to inhibit viral-induced syncytium formation (Mandal et al. 2007). Sulphate is necessary for the anti-viral activity and sulphate located at C-4 of (1-3)-linked fucopyranosyl units appears to be very important for the anti-herpetic activity of fucoidan (Mandal et al. 2007). Some anti-viral properties of sulphated fucans have also been characterised, for example inhibition of infection of human immunodeficiency virus (HIV), Herpes Simplex Virus (HSV) (Iqbal et al. 2000; Mandal et al. 2007; Witvrouw and DeClercq 1997), poliovirus III, adenovirus III, ECH06 virus, coxsackie B3 virus, coxsackie A16 (Li et al. 1995), cytomegalovirus and bovine viral diarrhea virus (Iqbal et al. 2000).

Fucoidan is known to have anti-tumour effects but its mode of action is not fully understood. A study done by Alekseyenko et al. 2007 demonstrated that when 10 mg/kg of fucoidan was administered in mice with transplanted Lewis lung adenocarcinoma, it produced moderate anti-tumour and anti-metastatic effects (Li et al. 2008b). These polyanionic polysaccharides have anti-angiogenesis, antiproliferation for tumour cells, they inhibit tumour growth and reduce tumour size (Ellouali et al. 1993; Li et al. 2008a), inhibit tumour cell adhesion to various substrata (Liu et al. 2005), and have direct anti-cancer effects on human HS-Sultan cells through caspase and ERK pathways (Aisa et al. 2005).

Besides directly inhibiting the growth of tumour cells, fucoidans can also restrain the development and diffusion of tumour cells through enhancing the body's immuno-modulatory activities, because fucoidan mediates tumour destruction through type 1 T-helper (Th1) cell and NK cell responses (Maruyama et al. 2007). In addition, at a dose of 25 mg/kg, fucoidan potentiated the toxic effect of cyclophosphamide used to treat various types of cancer and some auto-immune disorders (Alekseyenko et al. 2007).

Many studies suggest that fucoidan has potential for use as an anti-inflammatory agent. A study showed that fucoidan treatment led to less severe symptoms in the early stages of *Staphylococcus aureus*-triggered arthritis in mice, but delayed phagocyte recruitment and decreased clearance of the bacterium (Verdrengh et al. 2000). Additionally, injection of fucoidan into sensitized mice before hapten challenge can reduce contact hypersensitivity reactions (Nasu et al. 1997). Furthermore, recruitment of leukocytes into cerebrospinal fluid in a meningitis model is reduced by fucoidan (Granert et al. 1999) as is IL-1 (interleukin-1) production in a similar model (Ostergaard et al. 2000).

Fucoidan can act as a ligand for either L- or P- selectins, both of which interact with the sulphated oligosaccharides and this interaction has physiological consequences that could be therapeutically beneficial (Omata et al. 1997). Selectins are a group of lectins (sugar-binding proteins) that interact with oligosaccharides clustered on cell surfaces during the margination and rolling of leukocytes prior to firm adhesion, extravasation and migration to sites of infection (Lasky 1995).

In addition, fucoidan is an excellent natural antioxidant and presents significant antioxidant activity in experiments *in vitro*. In recent years, sulphated polysaccharides from the marine algae *Porphyra haitanesis* (Zhang et al. 2003), *Ulva pertusa* (Qi et al. 2005b; Qi et al. 2005a), *Fucus vesiculosus* (Ruperez et al. 2002), *Laminaria japonica* (Xue et al. 2004) and *Ecklonia kurome* (Hu et al. 2001) have been demonstrated to possess antioxidant activity. There are few reports however detailing the relationship between structure and antioxidant activity of sulphated polysaccharides from marine algae. Fucan showed low antioxidant activity relative to fucoidan (Rocha de Souza et al. 2007) and as mentioned previously, the ratio of sulphate content/fucose and the molecular weight were effective indicators to antioxidant activity of the samples (Wang et al. 2008). Fucoidan may have potential for preventing free radical mediated diseases such as Alzheimer's and the aging process. Previously, fucoidan was extracted from *L. japonica*, a commercially important algae species in China. Three sulphated polysaccharide fractions were successfully isolated through anion-exchange column chromatography and had their antioxidant activities investigated employing various established *in vitro* systems, including superoxide and hydroxyl radical scavenging activity, chelating ability, and reducing power (Wang et al. 2008). All fractions were more effective than the unprocessed fucoidan. Two galactose-rich fractions had the most potent scavenging activity against superoxide (generated in the PMS-NADH system) and hydroxyl radicals with EC_{50} values of 1.7 μg mL^{-1} and 1.42 mg mL^{-1}, respectively. One of these fractions also showed the strongest ferrous ion chelating ability at 0.76 mg mL^{-1} (Wang et al. 2008). Additionally, fucoidan (homofucan) from F. vesiculosus and fucans (heterofucans) from Padina gymnospora had an inhibitory effect on the formation of hydroxyl radical and superoxide radical (Rocha de Souza, et al. 2007). Healing of dermal wounds with macromolecular agents such as natural polymers is a growing area of research interest in pharmaceutical biotechnology. Fucoidan has been shown to modulate the effects of a variety of growth factors through mechanisms thought to be similar to the action of heparin. Fucoidan-chitosan films can promote re-epithelization and contraction of the wound area. Moreover, fucoidan-chitosan films may be suitable for use in hydrogel formulations for the treatment of dermal burns (Sezer, et al. 2007).

4.6. Mannitol

Mannitol is an important sugar alcohol which is present in many species of brown algae, especially in *Laminaria* and *Ecklonia*. The mannitol content is subject to wide seasonal fluctuations and varies with environment. Mannitol is the sugar alcohol corresponding to mannose. It usually constitutes less than 10 % of the dry weight in both *Ascophyllum nodosum* and *L. hyperborea* stipe. In autumn fronds of *L. hyperborea*, however, the content may be as high as 25 % of the dry weight .

Applications of mannitol are extremely diverse. It is used in pharmaceuticals, in making chewing gum, in the paint and varnish industry, in leather and paper manufacture, in the plastics industry and in the production of explosives. The US, the UK, France and Japan are the main centres of production. Mannitol can be used in a variety of foods, candies and chocolate-flavoured compound coatings because it can replace sucrose to make sugar-free compound coatings. Sugar-free chocolates are especially popular for people with diabetes, a growing problem in modern society. It is used as a flavour enhancer because of its sweet and pleasantly cool taste. Mannitol can be used to maintain the proper moisture level in foods so as to increase shelf-life and stability, because it is non-hygroscopic and chemically inert. Mannitol is the preferred excipient for chewable tablets due to its favourable feel in the mouth. It is non-carcinogenic and can be used in pediatric and geriatric food products, as it will not contribute to tooth decay (Nabors 2004).

4.7. Laminarin

Laminaran is a glucan, built up from (1→3)- and (1→6)-β-glucose residues. It is a linear polysaccharide, with a β(1→3):β(1→6) ratio around 3:1. Laminarin is found in the fronds of Laminaria/Saccharina and, to a lesser extent, in Ascophyllum and Fucus species and Undaria. The content varies seasonally and with habitat and can reach up to 32 % of the dry weight. Laminaran does not gel nor form any viscous solution and its main potential appears to lie in medical and pharmaceutical uses.

Commercial applications of the extract have so far been limited, although some progress has been made in France as an anti-viral in agricultural applications (Goemar 2010) or as dietary fiber (Deville et al. 2004). Especially the use of laminarin as substratum for prebiotic bacteria seems to have a good commercial application (Deville et al. 2004). Laminarin does not gel or form any viscous solution, and its main potential appears to lie in medical and pharmaceutical uses. It has been shown to be a safe surgical dusting powder, and may have value as a tumour-inhibiting agent and, in the form of a sulphate ester, as an anti-coagulant (Miao et al. 1999). The presence of anti-coagulant activity in brown algae was first reported in 1941, when *Laminaria* showed anti-coagulant properties with its active compound being located in the holdfasts (Shanmugam and Mody 2000). There are about 60 brown algal species identified to have blood anti-coagulant properties. Laminarin only shows anti-coagulant activity after structural modifications such as sulphation, reduction or oxidation. The anti-coagulant activity is improved chemically by increasing the degree of sulphation (Shanmugam and Mody 2000).

Preparations containing 1→3:1→6-β-D-glucans, laminarin, and fucoidan are manufactured by the health industry and marketed for their beneficial properties on the immune system. The producers of these tablets cite numerous papers discussing the biological activity of these glucans.

Laminarin provides protection against infection by bacterial pathogens, and protection against severe irradiation, it boosts the immune system by increasing the B cells and helper T cells, reduces cholesterol levels in serum and lowers systolic blood pressure, among other effects (Table 4c; Hoffman et al. 1995) lower the levels of total cholesterol, free cholesterol, triglyceride and phospholipid in the liver (Miao et al. 1999; Renn et al. 1994a; 1994b). The hypocholesterolemic and hypolipidemic responses are noted to be due to reduced cholesterol absorption in the gut (Kiriyama et al. 1969; Lamela et al. 1989; Panlasigui et al. 2003). This is often coupled with an increase in the faecal cholesterol content and a hypoglycaemic response (Dumelod et al. 1999; Ito and Tsuchida 1972; Nishide et al. 1993). A high level of low density lipid (LDL) cholesterol can lead to plaque forming and clog arteries and lead to cardiovascular diseases and heart attacks or strokes, a major cause of disease in the US. Laminarin as a potential cancer therapeutic is under intensive investigation (Miao et al. 1999).

4.8. Ulvan

The name ulvan is derived from the original terms ulvin and ulvacin introduced by Kylin in reference to different fractions of *Ulva lactuca* water-soluble sulphated polysaccharides. It is now being used to refer to polysaccharides from members of the Ulvales, mainly, *Ulva* sp. Ulvans are highly charged sulphated polyelectrolytes, composed mainly of rhamnose, uronic acid and xylose as the main monomer sugars, and containing a common constituting disaccharide; the aldobiuronic acid, [→4)-D-glucuronic acid-(1→4)- L-rhamnose3-sulfate-(1→]. Iduronic acid is also a constituent sugar. The average molecular weight of ulvans ranges from 189 to 8,200 kDa (Lahaye 1998). The cell-wall polysaccharides of ulvales represent 38 % to 54 % of the dry algal matter (Lahaye, 1998). Two major kinds have been identified: water soluble ulvan and insoluble cellulose-like material.

The mechanism of gel formation is unique among polysaccharide hydrogels. It is very complex and not yet fully understood. The viscosity of ulvan solutions as isolated polysaccharides has been compared to that of arabic gum. Whether ulvans present other functional properties of this gum remains to be established. The gelling properties of ulvans are affected by boric acid, divalent cations and pH. They are thermoreversible without thermal hysteresis. The gelling properties can be of interest for applications where gel formation needs to be precisely controlled (by pH or temperature), like the release of trapped molecules or particles under specific conditions (Percival and McDowell 1990; Lahaye et al. 1998). As already mentioned, highly absorbent, biodegradable hydrocolloid wound dressings limit wound secretions and minimise bacterial contamination. Polysaccharide fibres trapped in a wound are readily biodegraded. In the context of BSE (mad cow disease) and other prion contamination diseases, macromolecular materials from algal biomasses such as ulvans can constitute an effective and low-cost alternative to meat-derived products, because their rheological and gelling properties make them suitable as a substitute for gelatin and related compounds (Choi et al. 1999).

Ulvans are a source of sugars for the synthesis of fine chemicals. In particular, they are a potential source of iduronic acid, the only occurrence to date of this rare uronic acid in plants (Lahaye and Ray 1996). Iduronic acid is used in the synthesis of heparin fragment analogues with anti-thrombotic activities, and obtaining it requires a lengthy synthetic procedure that could be more cost-effectively replaced by a natural source (Lahaye 1998). Oligosaccharides from *Ulva* could be used as reference compounds for analyzing biologically active domains of glycosaminoglycans (GAG) like heparin. The use of oligo-ulvans as anti-coagulant agents could be expected since other, rarer, sulphated polysaccharides, like dermatan sulphate or fucan in brown algae, have shown this anti-thrombinic activity. Regular oligomers from ulvans may provide better-tolerated anti-thrombinic drugs (Paradossi et al. 2002).

Rhamnose, a major component of ulvans, is a rare sugar, used as a precursor for the synthesis of aroma compounds. Combinatorial libraries in glycopeptide mimetics are another example of the use of L-rhamnose in the pharmaceutical industry. The production of rhamnose from *Monostroma*, a Japanese species of Codiales, has been patented. Rare sulphated sugars such as rhamnose 3-sulphate and xylose 2-sulphate are also of interest (Lahaye and Robic 2007).

Other potential applications of ulvan oligomers and polymers are related to their biological properties. Recent studies have demonstrated that ulvans and their oligosaccharides were able to modify the adhesion and proliferation of normal and tumoural human colonic cells as well as the expression of transforming growth factors (TGF) and surface glycosyl markers related to cellular differentiation (Lahaye and Robic 2007) . Earlier work demonstrated strain-specific anti-influenza activities of ulvan from *U. lactuca* and the use of rhamnan, rhamnose and oligomers from desulphated *Monostroma* ulvans has been patented for the treatment of gastric ulcers (Fujiwara-Arasaki et al. 1984; Nagaoka et al. 2003).

Oligomers from seaweed species such as *Laminaria* sp. or *Fucus serratus* have also been studied as plant elicitors. These are natural compounds which stimulate the natural defences of plants. Several products derived from brown algae are already marketed worldwide. The success is because of their size and availability rather than their chemical composition. *Ulva* cell walls bind heavy metals and ulvans are the main contributors with 2.8 to 3.77 meq g^{-1} polysaccharide. The ion-exchange property of ulvans explains why they have been chosen as bioindicators for monitoring heavy metal pollution in coastal waters (Nagaoka et al. 1999).

4.9. Porphyran and Xylans

In red algae, the fibrillar network is made of low crystalline cellulose, mannan or xylan and represents only about 10% of the cell wall weight. It can also contain minor amounts of sulphated glucans, mannoglycans and complex galactans. Most of our current knowledge of red algal cell wall polysaccharides is on the gelling and thickening water soluble galactans, agars and carrageenans, used in various applications. Unlike most red seaweed generally studied, *Palmaria palmata* does not produce matrix galactans, but instead (10/4)- and (10/3)-linked b-D-xylan together with a minor amount of fibrillar cellulose and b-(10/4)-Dxylan. Xylans can be 35 % of the dry weight of *Palmaria* (Lahaye et al. 2003). Xylans have not yet been of economic interest and only few applications are known. Species of *Porphyra* contain a sulphated polysaccharide called porphyran; a complex galactan. Porphyran is dietary fiber of good quality, and chemically resembles agar. A powder consisting of 20 % nori mixed with a basic diet given orally to rats prevented 1,2-dimethylhydrazine-induced intestinal carcinogenesis. Porphyran showed appreciable anti-tumour activity against Meth-A fibrosarcoma. In addition, it can significantly lower the artificially enhanced level of hypertension and blood-cholesterol in rats (Noda 1993).

4.10. Digestibility of polysaccharides

The majority of edible seaweed fibres are soluble anionic polysaccharides which are little-degraded or not fermented by the human colonic micro flora (Lahaye and Kaeffer 1997). The amount of dietary fibres in marine algae not digested by the human digestive tract is higher than that of other food materials (Murata and Nakazoe 2001). Most of the total algal fibres disappeared after 24 h (range 60 % - 76 %) in *in vitro* fermentation of e.g. *L. digitata* and *U. pinnatifida* using human faecal flora. However, unlike the reference substrate (sugar beet fibres), the algal fibres were not completely metabolized to short chain fatty acids (SCFA; range 47 % - 62 %). Among the purified algal fibres, disappearance of laminarins was approximately 90 % and metabolism to SCFA was approximately 85 %, in close agreement with the fermentation pattern of reference fibres. Sulphated fucans were not degraded. Sodium alginates (Na-alginates) exhibited a fermentation pattern quite similar to that of the whole algal fibres, with a more pronounced discrepancy between disappearance and production of SCFA: disappearance was approximately 83 % but metabolism was only approximately 57 %. Laminarin seemed to be a modulator of the intestinal metabolism by its effects on mucus composition, intestinal pH and short chain fatty acid production, especially butyrate. The characteristic fermentation pattern of the total fibres from the brown algae investigated was attributed to the peculiar fermentation of alginates (Michel et al. 1996; Deville et al. 2007).

Phycolloids are more or less degraded following adaptation of the human micro flora, but none of the seaweed polysaccharides have been shown to be metabolized, although some may be partly absorbed. Nothing is known about the fate of other algal polysaccharides in the human digestive tract, except that they cannot be digested by human endogenous enzymes. Results of fermentation *in vitro* with human faecal bacteria, indicate that brown seaweed fibres exhibit an original fermentation pathway (Mabeau and Fleurence 1993). Carrageenan is a good source of soluble fibre (Burtin 2003). Rats excrete carrageenan quantitatively in the faeces, if it is administered in the diet at levels of 2 % - 20 % and it therefore has no direct nutritive value (Hawkins and Yaphe 1965). Weight gain was significantly reduced, especially at higher levels. Furthermore, food efficiency showed interference with utilization of other nutrients in the diet. Only 10 % - 15 % appeared digestible from faecal examination (Hawkins and Yaphe 1965). An experimental feeding with *L digitata* seaweed extract in pig resulted in a higher production of butyric acid in the caecum and colon compared to the control group (Reilly et al. 2008). Butyrate is a beneficial metabolite for intestinal bacteria, because it is quickly metabolised by colonoyctes and accounts for about 70 % of total energy consumption of the colon (Reilly et al. 2008). Therefore, it is desirable to promote butyrate production in the colon by laminarin fermentation.

The particular chemical structure of ulvan (and of *Ulva*) is responsible for the resistance of this polysaccharide to colonic bacterial fermentation. Consumption of dietary fibres from *Ulva* sp. could be expected to act mainly as bullring agents with little effect on nutrient metabolism due to colonic bacterial fermentation products (short-chain fatty acids; Bobin-Dubigeon et al. 1997). All soluble fibre fractions of *P palmata* consisted of linear beta-1,4/beta1,3 mixed linked xylans containing similar amounts of 1,4 linkages (70.5 % - 80.2 %). The insoluble fibres contained essentially 1,4 linked xylans with some 1,3 linked xylose and a small amount of 1,4-linked glucose (cellulose). Soluble fibres were fermented within 6 hours by human faecal bacteria into short chain fatty acids (Lahaye et al. 1993).

4.11. Commercial products, patents and applications

Due to their plethora of bioactive molecules, marine macroalgae have great potential for further development as products in the nutraceutical, functional food, and pharmaceutical markets. Patent activity in this area has increased and several novel products based on macroalgae have entered the market in recent years. In respect of carbohydrates for example, the Kabushiki Kaisha Yakult Honsha company, Japan has patented a polysaccharide derivative (which contains fucoidan and rhamnan or rhamnan sulphate polysaccharides), extracted from the marine brown macroalgae, such as, *Cladosiphon okamuranus, Chordaricles nemacystus, Hydrilla* sp., *Fucus* sp., and a green alga *Monostroma nitidum*. The purpose of this compound is as a therapeutic agent for the prevention and treatment of gastric ulcers (specifically inhibiting the adhesion of *Helicobacter pylori* and administered as tablets, granules, powders or capsules (Nagaoka et al., 2003). Furthermore, Takara Shuzo Company, Kyoto, Japan has developed a medicinal composition exemplified by viscous polysaccharides isolated from red algae (specifically: *Gelidium amansil, G. japonicum, G. pacificum, G. subcostatum, Pteocladia tennis* and *Hypneaceae* species consisting of at least one 3,6- anhydrogalactopyranose. This compound is proposed for the treatment or prevention of diabetes, rheumatism, cancer and contains various inhibitory factors (Enoki et al., 2003). Sulphated fucans from *Fuscus vesiculosis* and *Ascophyllum nodosum* have been patented as anticoagulant substances (Smit, et al., 2004). In practical gastroenterology, mixtures of alginic acid and alginates with antacids are used to prevent gastro-esophageal reflux and to cure epigastric burning (Klinkenberg- Knol et al., 1995; Zeitoun et al., 1998). Indeed, in several countries such as the US, Germany, Japan, Belgium and Canada the use of alginic acid and its derivatives for the treatment of gastritis and gastroduodenal ulcers, as well as the use of alginates as atiulcer remedies, is protected by patents (Bogentoff et al, 1981; Borgo et al., 1984; Reckitt et al., 1987; Sheth et al., 1967). The medical application of pure fucoidan fraction has also been patented by the French research institute IFREMER-CNRS (PATENT WO/32099, however the extract of brown seaweeds (containing fucoidan fractions) can still be applied in cosmetology as fibroblast proliferation activators in the context of treatments aimed at aesthetics, for example of antiwrinkle treatments or of prevention of skin ageing without patent infringement.

Company	Compound	Activity / disorder	Development
Various	Heparin and derivatives	Anti-coagulants	Since 1940's
Astellas	Auranofin (Ridaura)	Anti-rheumatic	1983
GSK	Zanamivir (Relenza)	Anti-influenza	1992
Johnson & Johnson	Topiramate (Topamax)	Anti-epileptic	1987
Bayer	Acarbose (Glucobay) (Pseudo-oligosaccharide)	Type II diabetes (a-glucosidase, a-amylase inhibitor)	1990
Ortho-McNeil Janssen Pharmaceutical	Elmiron (Pentosan polysulfate)	Cystitis (for CJD)	1996
Alfa Wassermann	Sulodexide (Vessel™)	cardiovascular indications	Marketed since 1980's
Hunter Fleming (now Newron)	HF0420 – low MWt oligosaccharide	neuroprotective	Phase I
Progen (Australia)	PI-88 (Phosphomannopentaose sulphate, Heparan sulfate mimetics)	Anti-angiogenic/anti-metastatic. (hepatocellular carcinoma)	Phase III
	PG500 series (Heparan sulfate mimetics)	Anti-angiogenic/anti-metastatic.	Preclinical
Endotis Pharma	EP42675 (org)	Anticoagulant	Phase I
	EP224283 (org)	Neutralizable antithrombotic	Preclinical
	EP37	venous and arterial thromboses	Preclinical
	EP8000 programme	Anti-angiogenesis, anti-tumour growth /metastasis	Preclinical
Biotec Pharmacon (Norway)	SBG (Soluble beta glucan – beta-1,3/1,6 glucan)	Immune stimulation Anti-cancer	GRAS nutraceutical

Table 5. Examples of carbohydrate-based drugs in use or development

The macroalgae polysaccharides described in this chapter show that these can be an interesting natural source of potential functional ingredients. As the content of proteins, carbohydrates, lipids and fiber can be influenced by the growing parameters (water, temperature, salinity, light and nutrients), macroalgae can be considered bioreactors that may be able to provide different polysaccharides at different quantities (Rui et. Al., 1990). From the descriptions above it is clear that macroalgae polysaccharides possess a multitude of bioactivities that might have antioxidant, antibacterial, antiviral, anticarcinogenic, anticoagulant and other bioactive properties for use as functional foods, pharmaceuticals and cosmeceuticals. The use of carbohydrate-based drugs is in its infancy, although there are several well-known examples (Table 5). Heparin is the key example of a major medicinally used carbohydrate based molecule. Low molecular weight heparins (eg. Certoparin, Dalteparin) and various derivatives (Fondaparinux – fully synthetic) have been developed to improve efficacy and half-life, and some are now being trialed for non-thrombotic/vascular applications (eg. Certoparin for inflammatory aspects of Alzheimers disease). It is only comparatively recently that the anti-inflammatory properties of heparin have been discovered. The presence of abundant 3-O-sulfated glucosamines in Heparan sulfates from human follicular fluid has suggested that some biological activities could be mimicked by other sulfated polysaccharides or derivatives (de Agostini *et al.*, 2008). Low molecular weight fucoidan (LMWF) has been used to demonstrate this (Colliec *et al.*, 1991, Millet *et al.*, 1999 and Durand *et al.*, 2008).

The drugs listed in Table 5 have overcome some of the perceived limitations of sugar-based molecules in terms of delivery, synthesis and immunogenicity. Although many still require intravenous delivery, several are available orally (e.g. Pentosan polysulfate), and further research is targeting to the improvement of oral availability by reducing compounds to their smallest active components, or by combining with other molecules (e.g. sulodexide). Improved synthesis has meant that some compounds can be synthesised (e.g. Fondaparinux), and small active components can be selected or modified to improve efficacy. No major problems have been reported with immunogenicity either in animal trials or as approved drugs. Sulphated polysaccharides may also be good anticoagulants through non-antithrombin mediated mechanisms, such as inhibition of thrombin mediated by heparin cofactor II (HC2). Like heparin, sulphated polysaccharides modulate cell growth-related activities, with inhibitory or promoting effects, depending on their carbohydrate backbones, degrees of sulphation, and distribution of sulphate groups. These activities can be usefully modulated through modification of sulphation patterns (Casu *et al.*, 2002; Naggi *et al.*, 2005). Several biological activities of sulphated polysaccharides are also molecular weight-dependent and the average length of their chains should be carefully controlled in order to maximize the desired activity. In some systems, opposite activities can be achieved with chains of different lengths of the same sulphated polysaccharide. Typically, relatively small oligosaccharides of heparin and heparinoids bind to growth factors and act as inhibitors of angiogenesis. On the other hand, the corresponding highermolecular weight species induce growth factors activation through oligomerization and binding with their receptors, thus promoting cell growth signalling (Casu and Lindahl, 2002; Goodner *et al.*, 2008). Also inhibiton of heparanase and heparanase-related biological activities such as angiogenesis and metastasis by heparin species (Naggi *et al.*, 2005) and sulfated hyaluronates (Naggi *et al.*, 2005) are dependent on molecular weights and sulfation patterns. Similarly, sulphated polysaccharides are expected to be exploitable as drugs either in their natural form or with depolymerization and/or chemical derivatisation. Macroalgal suphated polysaccharides need to be fully characterised and the biological activities associated with specific structures determined so that their development can progress.

5. Brown algae – Growth, cultivation and productivity

Due to the relatively high carbohydrate content in percentage of the dryweight (Table 1) brown macroalgae in particular species of the Laminariaceae have come under the spotlight for mass-cultivation. Specifically in respect of carbohydrate production for fermentation purposes for ethanol and or biodegradable plastics production or other medical and food applications. Brown macroalgae exploitation in Europe is currently restricted to manual and mechanised harvesting of natural stocks, although several EU projects explore mass cultivation in European waters. The majority of Asian seaweed resources are cultivated. The most common system in Europe to obtain seaweed biomass is by harvesting natural stocks in coastal areas with rocky shores and a tidal system. The natural population of seaweed is a significant resource. Depending on water temperature, some groups will dominate, like brown seaweeds in cold waters and reds in warmer waters. In Europe the main harvesting countries are Norway and France. Between 120,000 and 130,000 tonnes of *Laminaria* are harvested annually in Norway. The standing stock is estimated to be 10 million tonnes (Vae and Ask, 2011). France harvests about 30,000 – 50,000 tonnes annually, mainly *Laminaria* species for hydrocolloid production. The green alga *Ulva*, is commonly encountered in estuaries and inshore coastal areas. When mass proliferation occurs they are known as 'green tides.' They tend to develop at more and more locations along European coasts, due to eutrophication, and are used to a small extent as fertilisers but not yet used for industrial applications. Nevertheless they may form an interesting source and feedstock for carbohydrates (Kraan & Guiry, 2006).

In Europe, knowledge of seaweed cultivation is scattered across several R&D groups and a few industrial groups. The amount of cultivated seaweed is very low, mainly very small companies with local facilities for cultivating high value species. Existing industries having large scale cultivation are located in Asian countries (China, Philippines, Korea, Indonesia, and Japan) and in Chile. The main obstacle in European countries will be labour cost. Development of mechanized seaweed cultivation will be required in Europe to achieve cost objectives. In Ireland also the aquaculture sector is gradually building know-how and basic infrastructure for *Laminaria* cultivation. Technologies to cultivate *Laminaria* are well known. For instance, the FAO publish a guide to *Laminaria* culture which is very detailed (Chen, 2005). The main producers of *Laminaria* are located in China, Korea and Japan, where preservation of natural stocks is not always sustainably managed. The main reason for an increased harvest is increased productivity due to selection of the best performing strains, improved crop-care, less variability, fertilizing techniques and faster harvesting. This strategy will also be required to achieve the low material cost with a high carbohydrate content needed for biofuel applications.

Brown seaweeds from the *Laminariaceae* or *Fucaceae* families have a generative cycle including *microstages,* i.e. zoospores and gametophytes, before fertilization results into new sporophytes. Gametophytes are independently living organisms which can be grown and propagated vegetatively. The interesting thing is that the gametophytes can be grown in biofermentors in a controlled way like microalgae or bacteria. This procedure is used for example for production of fucoxanthin from *Undaria pinnatifida* cultivated under microscopic gametophyte form. Because cloning of these gametophyte structures is possible, genetically uniform strains can be obtained quickly. These clones can be used also for breeding purposes towards varieties of the seaweed in their mature stage. Selection and breeding is an essential step to reach a uniform crop and to optimize yield of the targeted compounds. Little is known about the composition of these microstages, but based on their survival rate in non favorable (a)biotic conditions interesting levels of bioactive polysaccharides may be obtained. Cultivation is the most efficient solution to guarantee consistent contents in bioactive compounds, whereas seaweeds harvested from wild resources undergo uncontrolled composition variations due to changes in growing environment. If highly valuables molecules such as fucoidans are aimed as products, seaweed aquaculture is therefore a very interesting option. The current cultivation methods are still based on Asian techniques although currently programs are initiated in the EU to develop open sea based seaweed cultivation technology. Although seaweeds are known for their richness in bioactive substances like polysaccharides, proteins, lipids, vitamins and polyphenols and have been shown to have a wide range of potential cosmetic, pharmaceutical and medical applications, their economic potential is still insufficiently developed.

5.1. Eutrophication and suitable carbohydrate feed stocks

Inputs of biodegradable organic matter and inputs deriving from fertilizer run-off together with run off or dilution from finfish and shellfish rearing in near-shore waters and land based activities have many effects on the quality of coastal inshore waters and are a primary cause of eutrophication due to increased availability of nutrients (EPA, 2003). In estuaries and shallow coastal bays, this can lead to the proliferation of vast green algal mats, known as 'green tides'(Fletcher 1996). Kelp farms (inshore and near-shore) are able to act as bio-filters and are able to remove nitrates and phosphates from the surrounding eutrophic inshore waters. This allows for increased production of farmed seaweed as demonstrated by Chopin et al.,2001, 2008b Eutrophe waters are high in ammonia and phosphorous which can be stripped from the water by seaweed at rates varying from 60% up to 90% of the nutrient input. Macroalgae are able to take up nitrogen from seawater with rates to allow for a biomass increase of ca. 10% day-1. It is well documented that in tank systems the green alga Ulva is able to remove 90% of the nitrogenous compounds and the red alga Gracilaria up to 95% of dissolved ammonium from fish effluent (Neori et al., 2000). By cultivating and harvesting macroalgae as biofilters integrated with other shellfish or fish production systems, nutrient polution from these aquaculture systems could be alleviated through a process called IMTA (Integrated Multitrophic Aquaculture) while increasing production and carrying capacity (Chopin et al.,2001, 2008; Troell et al., 2009). Production of macroalgae in near-shore sea cultivation can be harvested for the bioethanol market to produce a value added marketable product acting as both an economic incentive and environmental incentive.

5.2. Effect of enrichment on carbohydrates

High ammonia and nitrate concentrations will alter the proximate composition in the macroalgae and cause a shift with higher protein and generally lower carbohydrate levels such as starch or dietary fiber as demonstrated by Rosenberg and Ramus, (1982); Pinchetti et al., (1998). How exactly it effects the carbohydrate composition is not known but might be an interesting way to manipulate carbohydrate content and composition in algae. In the red carrageenophyte Kappaphycus alvarezii the effect of ammonium addition in an otherwise nitrogen starved environment had a profound impact on the carrageenan content and showed increased gel strength of the carrageenan with an increase of ammonia (Rui et. al., 1990).

6. From a hydrocarbon society to carbohydrate society

Global demand for bio-fuels continues unabated. Rising concerns over environmental pollution and global warming, has encouraged the movement to alternate fuels, the world ethanol market is projected to reach 100 billion litres this year. Bioethanol is currently produced from carbohydrates from and-based crops such as corn and sugar cane. A continued use of these crops drives the food versus fuel debate. An alternate feed-stock which is abundant and carbohydrate-rich is necessary. The production of such crop needs to be sustainable and reduce competition with production of food, feed and industrial crops on agricultural inputs (pesticides, fertiliser, land, water). Macroalgae, in specific brown seaweeds could meet these challenges, being an abundant and carbon neutral renewable resource with potential to reduce green-house gas (GHG) emissions and the man-made impact on climate change.

Macroalgae are fast growing marine plants that can grow to considerable size (up to tens of meters in length in the case of Pacific kelp species), although Atlantic species would be smaller at ~ 3 m length (Lüning 1990). Growth rates of marine macroalgae far exceed those of terrestrial biomass. The large brown algae of kelp forests found on rocky shores inhabit an environment of vigorous water movement and turbulent diffusion. This allows very high levels of nutrient uptake, photosynthesis and growth. Highest productivity of kelp forests is found along the North American Pacific coast, which out-performs that of the most productive terrestrial systems (Velimirov et al. 1977). *Laminaria*-dominated communities of the European coasts have an annual productivity of approximately 2 kg carbon per m^2, which is still higher than, for example, temperate tree plantations or grasslands with a productivity of generally less than 1 kg carbon per m^2 (Thomas 2002). Production figures have been reported in the range of 3.3 – 11.3 kg dry weight m^{-2} yr^{-1} for non-cultured and up to 13.1 kg dry weight m^{-2} over 7 month for cultured brown algae compared with 6.1 – 9.5 kg fresh weight m^{-2} yr^{-1} for sugar cane, a most productive land plant. In addition marine biomass does not require fertilisation as currents and water exchange provide a continuous flow of a base level of nitrates and phosphates and large scale cultures may be useful in alleviating increased nitrogen levels in inshore waters. Due to the absence of lignin and a low content of cellulose, brown macroalgae carbohydrates may be easily convertible in biological processes compared to land plants.

Seaweeds are already farmed on a massive scale in Asia and substantial quantities are also harvested from natural populations. Recent research has shown the potential for large scale culture of macroalgae in Atlantic waters (Germany; Buck and Buchholz 2004: Canada; Chopin et al. 2008: France; Kaas 2006: Ireland; Kraan et al. 2000: Isle of Man, UK; Kain et al. 1990, Spain; Peteiro et. al., 2006). The challenges now lie in further developing cost-effective methodologies to grow, harvest transport and process large quantities of macroalgae.

A large body of research on fermentation of seaweeds into methane exists starting in the early 1980s and is extensively reviewed in Kelly and Dworjanyn (2008).

The only commercially available biofuels today are first generation biofuels, mainly bioethanol and biodiesel, produced from e.g. sugar cane and corn, and rapeseed, respectively. However, continued use of these crops will drive the food versus fuels debate even more as demand for ethanol increases. Not only does the large-scale production of corn and sugar cane damage the environment by the use of pesticides, it uses two other valuable resources: arable land and enormous quantities of water. For instance, the production of corn in the USA uses over 3 trillion litres of water a year in 2007 (Chiu et al. 2009).

Increased demand and the competition with food production has called for the development of second generation biofuels, based on utilization of lignocellulosic biomass, such as wood and agricultural waste. Second generation biofuels do not compete with food as a feedstock, but they compete for land and fresh water resources. Therefore the challenge is to find a feedstock which is abundant and carbohydrate-rich. This crop must be sustainable, use no agricultural inputs (pesticides, fertiliser, land, water), and must not be part of the human or animal food chain. Such a feedstock and an alternative to terrestrial biomass are marine macroalgae or seaweeds. Macroalgae and aquatic biomass are emerging as one of the most promising potential sources for biofuels production.

6.1. Suitable species and production

Several species of macroalgae accumulate high levels of carbohydrates, which are suitable as substrate for microbial conversion processes, e.g. for production of bioethanol, biobutanol as biofuels or other desirable chemicals with an attractive high product value. Green algae species such as the *Ulva* sp Linnaeus with high levels of the polysaccharide Ulvan (Lahaye and Ray 1996; Lahaye 1998) have been used in ethanol and methane production production (Morand et al. 1991, Adams et al. 2009). Brown macroalgae in particular kelp contain 50-60% carbohydrates of the dryweight and cultivation techniques have been firmly established for the last 50 years. Moreover, kelp is cultivated in large quantities up to 15.5 million wet tonnes in the Far East (FAO 2010).

Five Atlantic kelp species are suitable for cultivation and have a high carbohydrate level, i.e., *Saccorhiza polyschides* (Lightfoot) Batters; *Alaria esculenta* (L.) Greville; *Laminaria hyperborea* (Gunnerus) Foslie; *Laminaria digitata* (Hudson) J.V. Lamouroux; and *Saccharina latissima* (Linnaeus) C.E. Lane, C. Mayes, Druehl & G.W. Saunders (Werner and Kraan 2004). They differ in various aspects, such as morphology, ecophysiology and longevity. *Laminaria digitata* and *Laminaria hyperborea* are the only species that form extended monospecific kelp beds.

The biomass productivity of macroalgae ranges converted to carbon is about 1 to 3.4 kg carbon m^{-2} $year^{-1}$ (Gao and McKinley 1994; Mann 1982; Mohammed and Fredriksen 2004). Seaweed communities of the North Atlantic coasts have an annual productivity of approximately 2 kg C per m^2, which is far higher than, for example, temperate tree plantations or grasslands with a productivity of generally less than 1 kg C m^{-2}, $year^{-1}$ (Mann 1982; Chapman 1987; Thomas 2002; Lüning and Pang 2003; Mohammed and Fredrikson 2004), and 2.8 times higher than for sugar cane (Gao and McKinley 1994). Macroalgae can be cultivated in the open sea (Zhang et al. 2008; Bartsch et al. 2008; Kelly and Dworjanyn 2008). Ocean farming of seaweed does not depend on fresh water and does not occupy land areas (Yarish and Pereira 2008). Sustainable utilization of algal biomass - a largely unexplored feed stock resource can be a complement to terrestrial biomass for the future global energy and carbon security and thereby also strengthen the maritime economies.

Ocean farming of seaweed has the potential to produce in the order of 40 ton dry weight biomass per hectare per annum. An area of 2500 km2, the size of Luxembourg, would be able to provide 10 million ton dry biomass, representing 5.6-5.8 million ton carbohydrates. With current 90% enzymatic conversion into ethanol (Wargacki et al., 2012) this would yield close to 2 billion litres of bioethanol. This is about 2-3 % of the world's global bioethanol production (F.O. Licht 2009); however, it would cover about 50% of the EU's ethanol demand (Annon 2008b). The use of algal biomass has the potential to not only replace fossil resources, and thereby mitigate climate change, but also aid in the recycling potential of nitrates and phosphates in near and inshore waters.

6.2. Processing and fermentation of macroalgal biomass

The water content in macroalgae is higher than in terrestrial biomass (80-85 %), making seaweeds more suited for microbial conversion than for direct combustion or thermo-chemical conversion processes, which is an alternative for land-based biomass (Horn et al. 2000a; Ross et al. 2007). Seaweed carbohydrates may be used as substrates for microbial production of a wide range of fuels and chemicals. Ethanol production from hexose sugars such as glucose, sucrose, laminarin etc. derived from e.g. corn stover or sugar cane, is a well-known process. However, hexose-based polysaccharides constitute only about 30-40% of the carbohydrates in kelp. The remaining fraction is composed of C-5 sugars that until now have not been applied as substrates for industrial microbial production processes. However, recent breakthroughs have been made in C5 sugar fermentation technology allowing up to 90% of the available carbohydrates to be fermented (Wargacki, et al., 2012).

6.3. Pre-treatment of the seaweed biomass and hydrolysis of the polysaccharides of brown seaweed biomass

Fresh harvested brown seaweed contains about 15-20 % carbohydrates of the total wet weight, which equals about 200 g carbohydrates per kg wet weight, which is an appropriate substrate concentration for microbial conversion processes (Horn et al. 2000a). Lack of lignin in seaweeds implies that the harsh pre-treatment applied for release of fermentable sugars from lignocellulosic biomass is not required. Laminaran and mannitol can easily be extracted by water (Horn et al. 2000b). Alginates (consisting of polymer blocks of Uronic and guluronic acid) is present in the macroalgae biomass at 30-40% (Honya et al. 1993; McHugh 2003). Sodium alginate can be removed from the initial extraction solution by adding a calcium salt. This causes calcium alginate to form with a fibrous texture; it does not dissolve in water and can be separated from it. The separated calcium alginate is suspended in water and acid is added to convert it into alginic acid. This fibrous alginic acid is easily separated, placed in a planetary type mixer with alcohol, and sodium carbonate is gradually added to the paste until all the alginic acid is converted to sodium alginate. The paste of sodium alginate is sometimes extruded into pellets that are then dried and milled (McHugh 2003) . Due to the high viscosity, dilution with large amounts of water is required. The process is operating at alginate concentrations in the order of ~2 %. Such a dilution cannot be applied to processes aimed at use of alginate as fermentation substrate. Preferably, no water should be added, as it will increase the downstream processing costs (McHugh 2003).

Bioethanol production from cellulosic materials can be achieved by running simultaneously enzymatic hydrolysis and fermentation. For the wet seaweed biomass it is not as easy due to the alginates which are harder to release from the biomass causing enzymatic degradation of un-treated biomass and will be rate-limiting step if combined with fermentation. Hydrolysis should therefore be a part of the biomass pre-treatment. This hydrolysis can be carried out mechanically through grinding and emulsifying equipment, chemically using acid or alkali, or enzymatic. Several methods for partial or complete degradation of alginate as an integrated part of the mechanical pre-treatment of the biomass are known. Chemical hydrolysis should follow existing technologies , e.g. by modification and adaptation of methods used for acid and alkali pre-treatment of wood biomass (e.g. Ballesteros 2001; Klinke et al. 2001) and by combination of acid- or alkali with steam treatment. Other studies with macroalgae demonstrated the need for pre-treatment at 65°C, pH 2 for 1 h prior to fermentation (Horn, 2000a; 200b, Percival and McDowell, 1967). This in contrast with Adams et al. (2009) who found that these pre-treatments are not required for the fermentations with *Saccharina latissima* conducted, with higher ethanol yields being achieved in untreated fermentations than in those with altered pH or temperature pre-reatments. This result was seen in fresh and defrosted macroalgae samples using *Saccharomyces cerevisiae* and 1 unit of the enzyme laminarinase per kg of defrosted macroalgae. Nevertheless, the easiest and environmental friendliest way of pre-treatment of algal biomass is through a combination of mechanical and enzymatic hydrolysis (Doubet and Quatrano 1982).

6.4. Ethanol and butanol from brown seaweeds

Ethanol production from fermentation is the most obvious one as it has a direct application in the replacement of fossil fuels. However other products such as butanol and itaconic acid can be produced as well which can substitute and/or replace similar products produced from fossil resources. There are many microorganisms in the marine environment that can degrade and utilize algal carbohydrates as a carbon source for energy. Often these microorganisms are associated with the seaweeds being present on the blade surface or in tissue as many kelp species produce exo-polysaccharides as mucus layer or shed entire skin . This would imply that these organisms possess the necessary enzymes for cleavage of the algal polysaccharides. However, compounds such as ethanol and butanol are produced by anaerobic fermentation that require the presence of specific metabolic pathways generating these compounds as end products, e.g. yeast for ethanol production and *Clostridia* for butanol production. Limited information is available on the efficiency of these processes with seaweed carbohydrates (Horn et al. 2000a; 2000b), although several brake troughs have recently been made in respect of ethanol production from brown seaweeds (Wargacki, et al., 2012).

6.4.1. Ethanol

The potential of ethanol production from seaweeds can be calculated and is based on the following assumptions: A carbohydrate content of 60 % of dry weight and a 90 % conversion ratio to ethanol. Through fermentation one gram of sugar can yield 0.4 g ethanol. This will yield 0.22 kg or 0.27 l ethanol from 1 kg dry weight seaweed biomass, corresponding to approximately 0.05 l ethanol per kg wet weight.

Bacteria can be metabolize uronic acids to pyruvate and glyceraldehyde-3-P, which may then be fermented to ethanol by the glycolytic pathway (van Maris et al. 2006). In anaerobic fermentation processes, as ethanol and butanol production, oxygen is not available for removal of excess hydrogen generated. This implies that the conversion reaction from substrate to product must be red-ox balanced. Ethanol-production from hexose sugars is red-ox balanced, while production from pentoses or mannitol generates excess hydrogen. In many bacteria but not in yeast the enzyme transhydrogenases, solves this problem. Yeasts can avoid the problem by receiving a small, controlled supply of oxygen. However, oxygen leads to complete oxidation of the substrate to CO_2 and water, and reduced ethanol yields. Another strategy is introduction of transhydrogenase into strains that lack this, through genetic engineering (Fortman et al. 2008; Lee et al. 2008). Prospecting macroalgae (seaweeds) as feedstocks for bioconversion into biofuels and commodity chemical compounds is limited primarily by the availability of tractable microorganisms that can metabolize alginate polysaccharides. Wargacki, et al., (2012) present the discovery of a 36–kilo–base pair DNA fragment from Vibrio splendidus encoding enzymes for alginate transport and metabolism. The genomic integration of this ensemble, together with an engineered system for extracellular alginate depolymerization, generated a microbial platform that can simultaneously degrade,uptake, and metabolize alginate. They further engineered for ethanol synthesis, this platform enables bioethanol production directly from macroalgae via a consolidated process.

6.4.2. *Butanol*

Butanol is an alternative to ethanol with a higher energy content (butanol 29.2 MJ/l, ethanol 19.6 MJ/l), compared to gasoline (32 MJ/l). It can be used to supplement both gasoline and diesel fuels and can be handled by existing infrastructures (Fortman et al. 2008). Butanol is an important industrial chemical and is currently produced via petrochemical processes. In the last century butanol was produced through bacterial fermentation of starch rich compounds using *Clostridia* strains (Zverlov et al. 2006), which can use hexose as well as pentose sugers.

7. Outlook

Macroalgae are an interesting source for a myriad of different bioactive polysaccharides ranging from industrial applications to novel food applications. They possess many different interesting and often exotic polysaccharides that are currently explored for their functional properties in food and biomedicine. However, a far larger application would be the use of carbohydrates from cultivated seaweeds for alternative fuel sources. It is the exploitation of nature's energy cycle, photosynthesis and the resulting plant biomass that can accelerate this application. Society has to make a transition from a hydrocarbon to a carbohydrate economy, with the accrued benefits of carbon neutral biofuel, plastics and medecine. Macroalgae are efficient solar energy converters, and can create large amounts of biomass in a short-term, however, marine biomass is often an overlooked source, and potentially represents a significant source of carbohydrates as a renewable energy source.

Author details

Stefan Kraan
Ocean Harvest Technology Ltd., N17 Business Park, Milltown, Co. Galway, Ireland

8. References

Adams, J.M., Gallagher, J.A., Donnison, I.S. (2009). Fermentation study on *Saccharina latissima* for bioethanol production considering variable pre-treatments. *J. Appl. Phycol.*, 21(5), 569-574

Aisa Y, Miyakawa Y, Nakazato T, Shibata H, Saito K, Ikeda Y, Kizaki M (2005) Fucoidan induces apoptosis of human HS-Sultan cells accompanied by activation of caspase-3 and down-regulation of ERK pathways. *Am J Hematol* 78:7–14

Alekseyenko TV, Zhanayeva SY, Venediktova AA, Zvyagintseva TN, Kuznetsova TA, Besednova NN et al. (2007).Antitumor and antimetastatic activity of fucoidan, a sulfated polysaccharide isolated from the Okhotsk sea Fucus evanescens brown alga. *Bulletin of Experimental Biology and Medicine* 143(6), 730-732.

Amano, H., Kakinuma, M., Coury, D. A., Ohno, H., Hara, T. (2005). Effect of a seaweed mixture on serum lipid level and platelet aggregation in rats. *Fisheries Science*, 71(1160-1166.

Annon (2008b). Http://www.euractiv.com/29/images/UEPA%20Balanced%20Approach_tcm29-172883.pdf. Accessed 15 April 2012

Arasaki, S. & Arasaki, T. (1983). Low calorie, high nutrition vegetables from the sea. To help you look and feel better. Japan Publications, Tokyo, 196 pp.

Athukorala, Y., Lee, K. W., Kim, S. K., Jeon, Y. J. (2007). Anticoagulant activity of marine green and brown algae collected from Jeju Island in Korea. *Bioresource Technology*, 98(9), 1711-1716.

Ballesteros, I. Oliva, J.M., Negro, M.J., Manzanares, P., Ballesteros, M. (2002). Enzymic hydrolysis of steam exploded herbaceous agricultural waste (*Brassica carinata*) at different particle sizes. *Process Biochem*. 38(2), 187–92

Bartsch, I., Wiencke, C., Bischof, K., Buchholz, C.M., Buck, B.H., Eggert. A., Feuerpfeil, P., Hanelt, D., Jacobsen, S., Karez, R., Karsten. U., Molis, M., Roleda, M.Y., Schubert, H., Schumann, R., Valentin, K., Weinberger, F., Wiese, J. (2008). The genus *Laminaria sensu lato*: Recent insights and developments. *Eur. J. Phycol.*, 43, 1-86

Berteau, O. & Mulloy, B. (2003). Sulfated fucans, fresh perspectives: structures, functions, and biological properties of sulfated fucans and an overview of enzymes active toward this class of polysaccharide. *Glycobiology*, 13(6), 29R-40R.

Bixler, H.J. and Porse, H. (2010). A decade of change in the seaweed hydrocolloids industry. *Journal of Applied Phycology*, 23, 321-335

Bobin-Dubigeon, C., Lahaye, M., Barry, J. L. (1997). Human colonic bacterial degradability of dietary fibres from sea-lettuce (*Ulva* sp). *Journal of the Science of Food and Agriculture*, 73(2), 149-159.

Bogentoff, C. B. (1981). 2722484).

Borgo, E. (1984). 1176984).

Buck, B. H. and Buchholz, C.M. (2004). The offshore-ring: A new system design for the open ocean aquaculture of macroalgae. *J. Appl. Phycol*. 16, 355–368

Buck, C. B., Thompson, C. D., Roberts, J. N., Muller, M., Lowy, D. R., Schiller, J. T. (2006). Carrageenan is a potent inhibitor of papillomavirus infection. *Plos Pathogens*, 2(7), 671-680.

Burtin, P. (2003). Nutritional value of seaweeds. *Electronic Journal of Environmental, Agricultural and Food Chemistry*, 2(4), 498-503.

Caceres, P. J., Carlucci, M. J., Damonte, E. B., Matsuhiro, B., Zuniga, E. A. (2000). Carrageenans from chilean samples of *Stenogramme interrupta* (Phyllophoraceae): structural analysis and biological activity. *Phytochemistry*, 53(1), 81-86.

Carlucci, M. J., Pujol, C. A., Ciancia, M., Noseda, M. D., Matulewicz, M. C., Damonte, E. B. et al. (1997). Antiherpetic and anticoagulant properties of carrageenans from the red seaweed *Gigartina skottsbergii* and their cyclized derivatives: Correlation between structure and biological activity. *International Journal of Biological Macromolecules*, 20(2), 97-105.

Carthew, P. (2002). Safety of carrageenan in foods. *Environmental Health Perspectives*, 110(4), A176-A176.

Casu B, Lindahl U (2001). Structure and biological interactions of heparin and heparan sulfate. Adv. Carbohydr. Chem. Biochem. 57:159-206.

Casu B, Naggi A, Torri G (2002). Selective chemical modification as a strategy to study structure-activity relationships of glycosaminoglycans. Sem. Thromb. Hemost 28:335-342.

Cérantola, S., Breton, F., Ar Gall, E., Deslandes, E. (2006). Co-occurrence and antioxidant activities of fucol and fucophlorethol classes of polymeric phenols in *Fucus spiralis*. *Botanica Marina*, 49(4), 347-351.

Cha, S. H., Ahn, G. N., Heo, S. J., Kim, K. N., Lee, K. W., Song, C. B. et al. (2006). Screening of extracts from marine green and brown algae in Jeju for potential marine angiotensin-I converting enzyme (ACE) inhibitory activity. *Journal of the Korean Society of Food Science and Nutrition*, 35(307-314.

Chapman, A. R. O. (1987). The wild harvest and culture of *Laminaria longicruris* de la Pylaie in Eastern

Canada. In M.S. Doty, T.F. Caddy & B. Santelices (Eds.). *Case studies of seven commercial seaweed resources*. FAO Fisheries Technical Paper, 181: 193-238

Chapman, V. J. (1970). Seaweeds and their uses, 2nd edn. Methuen,London, 304 pp

Chen, H.-M., Zheng, L., Yan, X.-J. (2005). The preparation and bioactivity research of agaro-oligosaccharides. *Food Technology and Biotechnology*, 43(1), 29-36.

Chevolot, L., Foucault, A., Chaubet, F., Kervarec, N., Sinquin, C., Fisher, A. M. et al. (1999). Further data on the structure of brown seaweed fucans: relationships with anticoagulant activity. *Carbohydrate Research*, 319(1-4), 154-165.

Chiu, Y. W., Walsh, B. Suh, S. (2009). Water Embodied in Bioethanol in the United States. *Environ. Sci.Technol.*, 43, 2688–2692

Choi, Y. S., Hong, S. R., Lee, Y. M., Song, K. W., Park, M. H., Nam, Y. S. (1999). Study on gelatin-containing artificial skin: I. Preparation and characteristics of novel gelatin-alginate sponge. *Biomaterials*, 20(5), 409-417.

Chopin, T. et al. (2008). Mussels and kelps help salmon farmers reduce pollution. SeaWeb Vol 1, No2, 3 July Département d'Ecologie Côtière, B.P. 70, 29280 Plouzané http://www.ifremer.fr/envlit/documentation/documents.htm. Accessed 10 May 2010

Chopin, T., Buschmann, A.H., Halling, C., Troell, M., Kautsky, N., Neori, A., Kraemer, G.P., Zertuche-Gonzalez, J.A., Yarish, C., Neefus, C. (2001). Integrating seaweeds into marine aquaculture systems: a key towards sustainability. *Journal of Phycology* 37:975-986

Chopin, T., Robinson, S.M.C., Troell, M., Neori, A., Buschmann, A.H., and Fang, J., (2008b). Multitrophic integration for sustainable marine aquaculture: 2463-2475. *In: The Encyclopedia of Ecology, Ecological Engineering (Vol. 3)*. S.E. Jørgensen and B.D. Fath (Eds.). Elsevier, Oxford

Chopin, T., Sharp, G., Belyea, E., Semple, R., Jones, D. (1999). Open-water aquaculture of the red alga *Chondrus crispus* in Prince Edward Island, Canada. *Hydrobiologia*, 398/399, 417-425.

Colliec S, Fischer AM, Tapon-Bretaudière J, Boisson C, Durand P, Jozefonvicz J. (1991). Anticoagulant properties of a fucoidan fraction. Thromb Res 64(2),143-154

Critchley, A. T., R. D. Gillespie & K. W. G. Rotmann, 1998. The seaweed resources of South Africa. In Ohno, M., A. T. Critchley, D. Largo & R. D. Gillespie (eds), Seaweed Resources of the World. JICA, Japan: 413–427.

Critchley, A., Ohno, M. Largo, D. (2006). *The seaweed resources of the world*. A CD-rom project Eds. Alan Critchley , Masao Ohno and Danilo Largo, ETI, Amsterdam The Netherlands

Cumashi A, Ushakova NA, Preobrazhenskaya ME, D'Incecco A, Piccoli A, Totani L, Tinari N, Morozevich GE,

Berman AE, Bilan MI, Usov AI, Ustyuzhanina NE, Grachev AA, Sanderson CJ, Kelly M, Rabinovich GA, Iacobelli S, Nifantiev NE (2007). A comparative study of the anti-inflammatory, anticoagulant, antiangiogenic, and antiadhesive activities of nine different fucoidans from brown seaweeds. Glycobiology 17:541-552.

Dannhardt, G. & Kiefer, W. (2001). Cyclooxygenase inhibitors - current status and future prospects. *European Journal of Medicinal Chemistry*, 36(2), 109-126.

de Agostini AI, Dong JC, de Vantéry Arrighi C, Ramus MA, Dentand-Quadri I, Thalmann S, Ventura P, Ibecheole V, Monge F, Fischer AM, HajMohammadi S, Shworak NW, Zhang L, Zhang Z, Linhardt RJJ (2008). Human follicular fluid heparan sulfate contains abundant 3-O-sulfated chains with anticoagulant activity. Biol Chem., 283(42):28115-24.

Deville, C., Damas, J., Forget, P., Dandrifosse, G., Peulen, O. (2004). Laminarin in the dietary fibre concept. *Journal of the Science of Food and Agriculture*, 84(9), 1030-1038.

Deville, C., Gharbi, M., Dandrifosse, G., Peulen, O. (2007). Study on the effects of laminarin, a polysaccharide from seaweed, on gut characteristics. *Journal of the Science of Food and Agriculture*, 87(9), 1717-1725.

Doubet, R.S., Quatrano, R.S. (1982). Isolation of marine bacteria capable of producing specific lyases for alginate degradation. *Applied and Environmental Microbiology*, 44, 3754-756

Doyle, J. W., Roth, T. P., Smith, R. M., Li, Y. Q., Dunn, R. M. (1996). Effect of calcium alginate on cellular wound healing processes modeled in vitro. *Journal of Biomedical Materials Research*, 32(4), 561-568.

Dumelod, B. D., Ramirez, R. P. B., Tiangson, C. L. P., Barrios, E. B., Panlasigui, L. N. (1999). Carbohydrate availability of arroz caldo with lambda-carrageenan. *International Journal of Food Sciences and Nutrition*, 50(4), 283-289.

Durand E, Helley D, Al Haj Zen A, Dujols C, Bruneval P, Colliec Jouault S, Fischer A-M, Lafont A (2008). Effect of Low molecular weight Fucoidan and Low molecular weight Heparin in a Rabbit Model of Arterial Thrombosis.J. Vascular Res 45:529-537.

Earons, G. (1994) Littoral Seaweed Resource Management. A Report Prepared for the Minch Project By Environment & Resource Technology Ltd Prepared for the Web by Gavin Earons, IT Unit, Comhairle nan Eilean Siar, 37p.

Ellouali, M., Boissonvidal, C., Durand, P., Jozefonvicz, J. (1993). Antitumor-activity of low-molecular-weight fucans extracted from brown seaweed *Ascophyllum nodosum. Anticancer Research*, 13(6A), 2011-2019.

Enoki, T., Sagawa, H., Tominaga, T., Nishiyama, E., Komyama, N., Sakai, T. et al. (2003). Drugs, foods or drinks with the use of algae-derived physiologically active substances. US Patent 0105029 A1).

EPA (2003). National Environmental Monitoring Programme for Transitional, Coastal and Marine Waters, Environmental Protection Agency.

F.O. Licht (2009). World Ethanol and Biofuels Report , vol. 7, no. 18, p. 365.

FAO (2012). FAO-statistics. Fishery Statistical Collections Global Aquaculture Production. http://www.fao.org/fishery/statistics/global-aquaculture-production/en. Accessed 15 April 2012.

Fernandez, L. E., Valiente, O. G., Mainardi, V., Bello, J. L., Velez, H., Rosado, A. (1989). Isolation and characterization of an antitumor active agar-type polysaccharide of *Gracilaria dominguensis*. *Carbohydrate Research*, 190(1), 77-83.

Fletcher, R. L. (1996). The occurrence of "green tides" - a review. *Marine benthic vegetation: recent changes and the effects of eutrophication*. W. S. A. P. H. Nienhuis. Berlin, Springer: 7-43.

Fleury, N. & Lahaye, M. (1991). Chemical and physicochemical characterization of fibers from *Laminaria digitata* (Kombu Breton) - A physiological approach. *Journal of the Science of Food and Agriculture*, 55(3), 389-400.

Fortman, J.L., Chhabra, S., Mukhopadhyay, A., Chou, H., Lee, T.S., Steen, E., Keasling, J.D. (2008). Biofuels alternatives to ethanol: pumping the microbial well. *Trends Biotechnol* , 26:375-381.

Fujiwara-Arasaki, T., Mino, N., Kuroda, M. (1984). The protein value in human nutrition of edible marine algae in Japan. *Hydrobiologia*, 116/117(513-516.

Gao, K., McKinley, K.R. (1994). Use of macroalgae for marine biomass production and CO2 remediation: a review. *J. Appl.Phycol*. 6, 45–60.

Ghosh, T., Chattopadhyay, K., Marschall, M., Karmakar, P., Mandal, P., Ray, B. (2009). Focus on antivirally active sulfated polysaccharides: From structure-activity analysis to clinical evaluation. *Glycobiology*, 19(1), 2-15.

Glyantsev, S., Annaev, A., Savvina, T., V (1993). Morphological basis for selecting composition and structure of biologically active compounds based on sodium alginate for wound treatment. *Byulleten' Eksperimental'noi Biologii i Meditsiny*, 115(1), 65-67.

Goemar (2010). Goemar. *http://www goemar fr*.

Granert, C., Raud, J., Waage, A., Lindquist, L. (1999). Effects of polysaccharide fucoidin on cerebrospinal fluid interleukin-1 and tumor necrosis factor alpha in pneumococcal meningitis in the rabbit. *Infection and Immunity*, 67(5), 2071-2074.

Haug, A. & Jensen, A. (1954). Seasonal variation in the chemical composition of *Alaria esculenta, Laminaria saccharina, Laminaria hyperborea* and *Laminaria digitata* from Northern Norway. 4,1-14.

Hawkins, W. W. & Leonard, V. G. (1963). Antithrombic Activity of Carrageenin in Human Blood. *Canadian Journal of Biochemistry and Physiology*, 41(5), 1325-&.

Hawkins, W. W. & Yaphe, W. (1965). Carrageenan as a dietary constituent for rat - faecal excretion nitrogen absorption and growth. *Canadian Journal of Biochemistry*, 43(4), 479-&.

Hawkins, W. W., Leonard, V. G., Maxwell, J. E., Rastogi, K. S. (1962). A study of the prolonged intake of small amounts of EDTA on the utilization of low dietary levels of calcium and iron by the rat. *Canadian Journal of Biochemistry and Physiology*, 40(391-.

Hayashi, L., Yokoya, N. S., Ostini, S., Pereira, R. T. L., Braga, E. S., Oliveira, E. C. (2008). Nutrients removed by *Kappaphycus alvarezii* (Rhodophyta, Solieriaceae) in integrated cultivation with fishes in re-circulating water. *Aquaculture*, 277, 185-191

Hennequart, F. (2007). Seaweed applications in human health/nutrition: The example of algal extracts as functional ingredients in novel beverages. 4th International Symposium Health and Sea, Granville, France. http://www.sante-mer.com/pdf_dwnld/O17_Presentation_H_HENNEQUART.pdf, accessed 14 April 2012

Hoffman, R., Paper, D. H., Donaldson, J., Alban, S., Franz, G. (1995). Characterization of a laminarin sulfate which inhibits basic fibroblast growth-factor binding and endothelial-cell proliferation. *Journal of Cell Science*, 108(3591-3598.

Holdt, S and Kraan, S. (2011). Bioactive compounds in seaweed; functional food applications and legislation. Review Article. *Journal of Applied Phycology*, 23, 543-597

Honya, M., Kinoshita, T., Ishikawa, M., Mori, H., Nisizawa, K. (1993). Monthly determination of alginate M/G ratio, mannitol and minerals in cultivated *Laminaria japonica*. *Nippon Suisan Gakkaishi* 59(2), 295-299.

Horn, S.J., Aasen, I.M., Østgaard, K. (2000a). Production of ethanol from mannitol by *Zymobacter palmae*. *J. Ind. Microbiol. Biotechnol.*, 24, 51-57.

Horn, S.J., Aasen, I.M., Østgaard, K. (2000b). Ethanol production from seaweed extract. *J. Ind. Microbiol. Biotechnol.*, 25, 249-254

Hu JF, Geng MY, Zhang JT, Jiang HD (2001) An in vitro study of the structure–activity relationships of sulfated polysaccharide from brown algae to its antioxidant effect. J Asian Nat Prod Res 3:353–358

Iqbal, M., Flick-Smith, H., McCauley, J. W. (2000) Interactions of bovine viral diarrhoea virus glycoprotein E-rns with cell surface glycosaminoglycans. *Journal of General Virology*, 81(451-459.

Ito, K. & Tsuchida, Y. (1972). The effect of algal polysaccharides on depressing of plasma cholesterol level in rats. *Proceeding of The Seventh International Seaweed Symposium*, 451-455.

Jenkins, D. J. A., Wolever, T. M. S., Leeds, A. R., Gassull, M. A., Haisman, P., Dilawari, J. et al. (1978). Dietary fibers, fiber analogs, and glucose-tolerance - importance of viscosity. *British Medical Journal*, 1(6124), 1392-1394.

Jensen, A. & Haug, A. (1956). Geographical and seasonal variation in the chemical composition of *Laminaria hyperborea* and *Laminaria digitata* from the Norwegian coast.

Norwegian Institute of Seaweed Research, Akademisk Trykningssentral, Blindern, Oslo. Report 14, pp 1–8

Jensen, A. (1956). Component sugars of some common brown algae. Norwegian Institute of Seaweed Research, Akademisk Trykningssentral, Blindern, Oslo. Report 9, pp 1–8

Jeon Y-J, Athukorala Y, Lee J (2005) Characterization of agarose product from agar using DMSO. Algae 20:61–67

Kaas, R. (2006). The seaweed resources of France. In: Critchly, A.T. Ohno, M. Largo D.B. *The seaweed resources of the world*. Interactive CD ROM, ETI, The Netherlands.

Kain, J.M., Holt T.J., Dawes C.P. (1990). European Laminariales and their cultivation. In Yarish C., Penniman C.A., Van Patten P. (eds), Economically Important Marine Plants of the Atlantic: Their Biology and Cultivation. Connecticut Sea Grant College Program, Groton: 95-111.

Katayama, S., Ohshita, J., Sugaya, K., Hirano, M., Momose, Y., Yamamura, S. (1998). New medicinal treatment for severe gingivostomatitis. *International Journal of Molecular Medicine*, 2(6), 675-679.

Kelly M, Dworjanyn, S. (2008). The potential of marine biomass for anaerobic biogas production. The Crown Estate, 103 pages, London.

Khotimchenko, Y. S., Kovalev, V. V., Savchenko, O. V., Ziganshina, O. A. (2001). Physical-chemical properties, physiological activity, and usage of alginates, the polysaccharides of brown algae. *Russian Journal of Marine Biology*, 27(1), 53-64.

Kim, I. H. & Lee, J. H. (2008). Antimicrobial activities against methicillin-resistant *Staphylococcus aureus* from macroalgae. *Journal of Industrial and Engineering Chemistry*, 14(5), 568-572.

Kimura Y, Watanabe K, Okuda H (1996) Effects of soluble sodium alginate on cholesterol excretion and glucose tolerance in rats. *J. Ethnopharmacol* 54:47–54

Kindness, G., Williamson, F. B., Long, W. F. (1979). Effect of polyanetholesulphonic acid and xylan sulphate on antithrombin III activity. *Biochemical and Biophysical Research Communications*, 13(88), 1062-1068.

Kiriyama, S., Okazaki, Y., Yoshida, A. (1969). Hypocholesterolemic effect of polysaccharides and polysaccharide-rich foodstuffs in cholesterol-fed rats. *Journal of Nutrition*, 97(3), 382-&.

Klinke, H.B., Thomsen, A., Ahring, B.K. (2001) Potential inhibitors from wet oxidation of wheat straw and their effect on growth and ethanol production by *Thermoanaerobacter mathranii*. *Appl. Microbiol. Biotechnol.* 57,631–638.

Klinkenberg-Knol EC, Festen HP, Meuwissen SG (1995) Pharmacological management of gastro-oesophageal reflux disease. Drugs 49:495–710

Kraan, S., Guiry, M.D. (2006). The seaweed resources of Ireland. In: *Seaweed resources of the world*, A CD-rom project Eds. Alan Critchley , Masao Ohno and Danilo Largo, ETI, Amsterdam The Netherlands.

Kraan, S., Verges Tramullas, A., Guiry, M. D. (2000). The edible brown seaweed *Alaria esculenta* (Phaeophyceae, Laminariales): hybridisation, growth and genetic comparisons of six Irish populations. *Journal of Applied Phycology*, 12, 577-583

Kumar, C. S., Ganesan, P., Bhaskar, N. (2008a). In vitro antioxidant activities of three selected brown seaweeds of India. *Food Chemistry*, 107(2), 707-713.

Kylin, H. (1913). Zur Biochemie der Meeresalgen. *Hoppe-Seyler's Zeitschrift für physiologische Chemie*, 83(171-197.

Lahaye, M. & Kaeffer, B. (1997). Seaweed dietary fibres: structure, physico-chemical and biological properties relevant to intestinal physiology. *Sciences des Aliments*, 17(6), 563-584.

Lahaye, M. & Ray, B. (1996). Cell-wall polysaccharides from the marine green alga *Ulva rigida* (Ulvales, Chlorophyta) - NMR analysis of ulvan oligosaccharides. *Carbohydrate Research*, 283,161-173.

Lahaye, M. & Robic, A. (2007). Structure and functional properties of Ulvan, a polysaccharide from green seaweeds. *Biomacromolecules*, 8(6), 1765-1774.

Lahaye, M. (1991). Marine-algae as sources of fibers - determination of soluble and insoluble dietary fiber contents in some sea vegetables. *Journal of the Science of Food and Agriculture*, 54(4), 587-594.

Lahaye, M. (1998). NMR spectroscopic characterisation of oligosaccharides from two *Ulva rigida* ulvan samples (Ulvales, Chlorophyta) degraded by a lyase. *Carbohydrate Research*, 314(1-2), 1-12.

Lahaye, M., Michel, C., Barry, J. L. (1993). Chemical, physicochemical and in vitro fermentation characteristics of dietary-fibers from *Palmaria palmata* (L) Kuntze. *Food Chemistry*, 47(1), 29-36.

Lamela, M., Anca, J., Villar, R., Otero, J., Calleja, J. M. (1989). Hypoglycemic activity of several seaweed extracts. *Journal of Ethnopharmacology*, 27(1-2), 35-43.

Lasky, L. A. (1995). Selectin-carbohydrate interactions and the initiation of the inflammatory response. *Annual Review of Biochemistry*, 64(113-139.

Lee, J. B., Hayashi, K., Hashimoto, M., Nakano, T., Hayashi, T. (2004). Novel antiviral fucoidan from sporophyll of *Undaria pinnatifida* (Mekabu). *Chemical & Pharmaceutical Bulletin*, 52(9), 1091-1094.

Lee, S.K., Chou, H., Ham, T.S., Lee, T.S., Keasling, J.D. (2008). Metabolic engineering of microorganisms for biofuels production: from bugs to synthetic biology to fuels. *Current Opinion in Biotechnology*, 19,556–563.

Li F, Tian TC, Shi YC (1995) Study on anti virus effect of fucoidan in vitro. J Norman Bethune Univ Med Sci 21:255–257

Li, B., Lu, F., Wei, X. J., Zhao, R. X. (2008a). Fucoidan: Structure and bioactivity. *Molecules*, 13(8), 1671-1695.

Li, N., Zhang, Q. B., Song, J. M. (2005). Toxicological evaluation of fucoidan extracted from *Laminaria japonica* in Wistar rats. *Food and Chemical Toxicology*, 43(3), 421-426.

Li, Y., Qian, Z. J., Le, Q. T., Kim, M. M., Kim, S. K. (2008b). Bioactive phloroglucinol derivatives isolated from an edible marine brown alga, Ecklonia cava. *Journal of Biotechnology*, 136(Supplement 1), S578-S578.

Liu, R. M., Bignon, J., Haroun-Bouhedja, F., Bittoun, P., Vassy, J., Fermandjian, S. et al. (2005). Inhibitory effect of fucoidan on the adhesion of adenocarcinoma cells to fibronectin. *Anticancer Research*, 25(3B), 2129-2133.

Luescher-Mattli, M. (2003). Algae, a possible source for new drugs in the treatment of HIV and other viral diseases. *Current Medicinal Chemistry*, 2(219-225.

Lüning, K. (1990). *Seaweeds, Their environment, Biogeography and Ecophysiology*. Eds. C. Yarish and H. Kirkman, John Wiley & Sons, Inc., New York, 527 pp.

Lüning, K., Pang, S. (2003). Mass cultivation of seaweeds: current aspects and approaches. *J. Appl. Phycol.*, 15, 115–119.

Mabeau, S. & Fleurence, J. (1993). Seaweed in food products: biochemical and nutritional aspects. *Trends in Food Science & Technology*, 4(4), 103-107.

Mandal, P., Mateu, C. G., Chattopadhyay, K., Pujol, C. A., Damonte, E. B., Ray, B. (2007). Structural features and antiviral activity of sulphated fucans from the brown seaweed *Cystoseira indica*. *Antiviral Chemistry and Chemotherapy*, 18(153-162.

Mann, K.L. (1982). *Ecology of coastal waters: a systems approach.* University of California Press: Berkeley, CA.

Maruyama, H., Tanaka, M., Hashimoto, M., Inoue, M., Sasahara, T. (2007). The suppressive effect of Mekabu fucoidan on an attachment of *Cryptosporidium parvum* oocysts to the intestinal epithelial cells in neonatal mice. *Life Sciences*, 80(8), 775-781.

McHugh, D. J. (1987). Production and utilization of products from commercial seaweeds. 288, 1-189. FAO Fisheries Technical Paper No. 288, pp 1–189

Miao, H. Q., Elkin, M., Aingorn, E., Ishai-Michaeli, R., Stein, C. A., Vlodavsky, I. (1999). Inhibition of heparanase activity and tumor metastasis by laminarin sulfate and synthetic phosphorothioate oligodeoxynucleotides. *International Journal of Cancer*, 83(3), 424-431.

Michel, C., Lahaye, M., Bonnet, C., Mabeau, S., Barry, J. L. (1996). *In vitro* fermentation by human faecal bacteria of total and purified dietary fibres from brown seaweeds. *British Journal of Nutrition*, 75(2), 263-280.

Millet J, Colliec-Jouault S, Mauray S, Theveniaux J, Sternberg C, Boisson Vidal C, Fischer AM (1999). Antithrombotic and Anticoagulant Activities of a Low Molecular Weight Fucoidan by Subcutaneous Route. Thromb Haemostasis 81:391-395.

Mohammed, A. I., Fredriksen, S. (2004). Production, respiration and exudation of dissolved organic matter by the kelp *Laminaria hyperborea* along the west coast of Norway. *J. Mar. Biol. Ass. U.K.*, 84, 887-894.

Morand, P., Carpentier, B., Charlier, R.H., Mazé, J., Orlandini, M., Plunkett, B.A., de Waart, J. (1991). Bioconversion of seaweeds. In: Seaweed resources in Europe, uses and potential. Eds. Guiry, M.D. and Blunden, G. John Wiley &Sons, Chichester 432 pp.

Morrissey, J., Kraan, S., Guiry, M. D. (2001). A guide to commercially important seaweeds on the Irish coast. 1-66.

Mouradi-Givernaud A, Givernaud T, Morvan H, Cosson J (1992) Agar from Gelidium latifolium (Rhodophyceae, Gelidiales)—biochemical composition and seasonal variations. Bot Mar 35:153–159

Murata, M. & Nakazoe, J. (2001). Production and use of marine algae in Japan. *Jarq-Japan Agricultural Research Quarterly*, 35(4), 281-290.

Nabors, L. O. B. (2004). Alternative sweeteners. *Agro Food Industry Hi-Tech*, 15(4), 39-41.

Nagaoka, M., Shibata, H., Kimura, I., Hashimoto, S. (2003). Oligosaccharide derivatives and process for producing the same. United States Patent 6,645,940).

Nagaoka, M., Shibata, H., Kimura-Takagi, I., Hashimoto, S., Kimura, K., Makino, T. et al. (1999). Structural study of fucoidan from *Cladosiphon okamuranus* Tokida. *Glycoconjugate Journal*, 16(1), 19-26.

Naggi A, Casu B, Perez M, Torri, G, Cassinelli, G, Penco, S, Pisano, C, Giannini, G, Ishai-Michaeli R, Vlodavsky I (2005). Modulation of heparanase-inhibiting activity of heparin through selective desulfation, graded N-acetylation, and glycol-splitting. J. Biol. Chem 280:12103-12113.

Nasu, T., Fukuda, Y., Nagahira, K., Kawashima, H., Noguchi, C., Nakanishi, T. (1997). Fucoidin, a potent inhibitor of L-selectin function, reduces contact hypersensitivity reaction in mice. *Immunology Letters*, 59(1), 47-51.

Neori, A., Shpigel, M., Ben-Ezra, D. (2000). A sustainable integrated system for culture of fish, seaweed and abalone. *Aquaculture* 186, 279–291.

Nishide, E., Anzai, H., Uchida, N. (1993). Effects of alginates on the ingestion and excretion of cholesterol in the rat. *Journal of Applied Phycology*, 5(2), 207-211.

Noa, M., Mas, R., Carbajal, D., Valdes, S. (2000). Effect of D-002 on acetic acid-induced colitis in rats at single and repeated doses. *Pharmacological Research*, 41(4), 391-395.

Noda, H. (1993). Health benefits and nutritional properties of nori. *Journal of Applied Phycology*, 5(2), 255-258.

Omata, M., Matsui, N., Inomata, N., Ohno, T. (1997). Protective effects of polysaccharide fucoidin on myocardial ischemia-reperfusion injury in rats. *Journal of Cardiovascular Pharmacology*, 30(6), 717-724.

Ostergaard, C., Yieng-Kow, R. V., Benfield, T., Frimodt-Moller, N., Espersen, F., Lundgren, J. D. (2000). Inhibition of leukocyte entry into the brain by the selectin blocker fucoidin decreases interleukin-1 (IL-1) levels but increases IL-8 levels in cerebrospinal fluid during experimental pneumococcal meningitis in rabbits. *Infection and Immunity*, 68(6), 3153-3157.

Panlasigui, L. N., Baello, O. Q., Dimatangal, J. M., Dumelod, B. D. (2003). Blood cholesterol and lipid-lowering effects of carrageenan on human volunteers. *Asia Pacific Journal of Clinical Nutrition*, 12(2), 209-214.

Paradossi G, Cavalieri F, Chiessi E (2002) A conformational study on the algal polysaccharide ulvan. Macromolecules 35:6404–6411

Peteiro, C. Salinas, j. M & Fuertes, C. (2006). Cultivation of the autoctonous seaweed *Laminaria saccharina* off the galician coast (nw spain): Production and features of the sporophytes for An annual and biennial harvest. Thalassas, 22 (1): 45-53

Percival E, McDowell RH (1990) Algal polysaccharides. In: Dey PM (ed) Methods in plant biochemistry, volume 2: carbohydrates. Academic, London, pp 523–547

Pinchetti, J.L.G., del Campo Fernández, E., Moreno Díez, P., and and García Reina, G. (1998). Nitrogen availability influences the biochemical composition and photosynthesis of tank-cultivated Ulva rigida (Chlorophyta). Journal of Applied Phycology, 10, 383-389.

Qi, H. M., Zhang, Q. B., Zhao, T. T., Chen, R., Zhang, H., Niu, X. Z. et al. (2005a). Antioxidant activity of different sulfate content derivatives of polysaccharide extracted from *Ulva pertusa* (Chlorophyta) in vitro. *International Journal of Biological Macromolecules*, 37(4), 195-199.

Qi, H. M., Zhao, T. T., Zhang, Q. B., Li, Z., Zhao, Z. Q., Xing, R. (2005b). Antioxidant activity of different molecular weight sulfated polysaccharides from *Ulva pertusa* Kjellm (Chlorophyta). *Journal of Applied Phycology*, 17(6), 527-534.

Rasmussen, R. S. & Morrissey, M. T. (2007). Marine biotechnology for production of food ingredients. 237-292.

Reckitt and Colman Products Ltd (1987). 858 003.

Reilly, P., O'Doherty, J. V., Pierce, K. M., Callan, J. J., O'Sullivan, J. T., Sweeney, T. (2008). The effects of seaweed extract inclusion on gut morphology, selected intestinal microbiota, nutrient digestibility, volatile fatty acid concentrations and the immune status of the weaned pig. *Animal*, 2(10), 1465-1473.

Renn, D. W., Noda, H., Amano, H., Nishino, T., Nishizana, K. (1994a). Antihypertensive and antihyperlipidemic effects of funoran. *Fischeries Science*, 60(423-427.

Renn, D. W., Noda, H., Amano, H., Nishino, T., Nishizana, K. (1994b). Study on hypertensive and antihyperlipidemic effect of marine algae. *Fischeries Science*, 60(83-88.

Reynolds JEF, Prasad AB (1982) Martindale the extra pharmacopoeia. 28th Pharmaceutical Press, London, 735 pp

Rioux, L. E., Turgeon, S. L., Beaulieu, M. (2007). Characterization of polysaccharides extracted from brown seaweeds. *Carbohydrate Polymers*, 69(3), 530-537.

Rocha de Souza M, Marques C, Guerra Dore C, Ferreira da Silva F, Oliveira Rocha H, Leite E (2007) Antioxidant activities of sulfated polysaccharides from brown and red seaweeds. J Appl Phycol 19:153–160

Rodrigueza, M. R. C. & Montaño, M. N. E. (2007). Bioremediation potential of three carrageenophytes cultivated in tanks with seawater from fish farms. *Journal of Applied Phycology*, 19(6), 755-762.

Rosenberg, C. and Ramus, J. (1982). Ecological growth strategies in the seaweeds *Gracilaria foliifera* (Rhodophyceae) and *Ulva* sp. (Chlorophyceae): Soluble nitrogen and reserve carbohydrates. Marine Biology, 251-259

Ross, A., Jones, J.M., Kubacki, M.L., Bridgeman, T.G. (2008). Classification of macroalgae as fuel and its thermochemical behaviour. *Bioresource Technology*, 6494-6504.

Rui, L., Jiajun, L. and Chaoyuan, W. (1990). Effect of ammonium on growth and carrageenan content in *Kappaphycus alvarezii* (Gigartinales, Rhodophyta). Hydrobiologia 204-205, 499-503.

Rupérez, P. & Saura-Calixto, F. (2001). Dietary fibre and physicochemical properties of edible Spanish seaweeds. *European Food Research and Technology*, 212(3), 349-354.

Ruperez, P., Ahrazem, O., Leal, J. A. (2002). Potential antioxidant capacity of sulfated polysaccharides from the edible marine brown seaweed *Fucus vesiculosus*. *Journal of Agricultural and Food Chemistry*, 50(4), 840-845.

Savitskaya, I. M. (1986). Trial of a local hemostatic with alginate base. *Klin Khirurgiya*, 3(39-40.

Segal, H. C., Hunt, B. J., Gilding, K. (1998). The effects of alginate and non-alginate wound dressings on blood coagulation and platelet activation. *Journal of Biomaterials Applications*, 12(3), 249-257.

Sen, A. K., Das, A. K., Banerji, N., Siddhanta, A. K., Mody, K. H., Ramavat, B. K. et al. (1994). A new sulfated polysaccharide with potent blood anti-coagulant activity from the red seaweed *Grateloupia Indica*. *International Journal of Biological Macromolecules*, 16(5), 279-280.

Sezer, A.D.; Hatipoğlu, F.; Cevher, E.; Oğurtan, Z.; Baş, A.L. & Akbuğa, J. (2007). Chitosan films containing fucoidan as a wound dressing for dermal burn healing: Preparation and in vitro/in vivo evaluation, AAPS PharmSciTech, 8, 1-8.

Shanmugam, M. & Mody, K. H. (2000). Heparinoid-active sulphated polysaccharides from marine algae as potential blood anticoagulant agents. *Current Science*, 79(12), 1672-1683.

Sheth, B. B. (1967). 3 326755.

Skoler-Karpoff, S., Ramjee, G., Ahmed, K., Altini, L., Plagianos, M. G., Friedland, B. et al. (2008). Efficacy of Carraguard for prevention of HIV infection in women in South Africa: a randomised, double-blind, placebo-controlled trial. *The Lancet*, 372(9654), 1977-1987.

Smit, A. J. (2004). Medicinal and pharmaceutical uses of seaweed natural products: A review. *Journal of Applied Phycology*, 16(4), 245-262.

Spieler, R. (2002). Seaweed compound's anti-HIV efficacy will be tested in southern Africa. *Lancet*, 359(9318), 1675-1675.

Terada, A., Hara, H., Mitsuoka, T. (1995). Effect of dietary alginate on the faecal microbiota and faecal metabolic activity in humans. *Microbial Ecology in Health and Disease*, 8(6), 259-266.

Thomas, D.N. (2002). *Seaweeds*, Natural History Museum, London

Tobacman, J. K. (2001). Review of harmful gastrointestinal effects of carrageenan in animal experiments. *Environmental Health Perspectives*, 109(10), 983-994.

Torsdottir, I., Alpsten, M., Holm, G., Sandberg, A. S., Tolli, J. (1991). A small dose of soluble alginate-fiber affects postprandial glycemia and gastric-emptying in humans with diabetes. *Journal of Nutrition*, 121(6), 795-799.

Troell, M., Joyce, A., Chopin, T., Neori, A., Buschmann, A.H., Fang, J-G (2009). Ecological engineering in aquaculture — Potential for integrated multi-trophic aquaculture (IMTA) in marine offshore systems. *Aquaculture* 297, 1–9.

Tseng, C. K. (2001). Algal biotechnology industries and research activities in China. *Journal of Applied Phycology*, 13(4), 375-380.

Urbano, M. G. & Goñi, I. (2002). Bioavailability of nutrients in rats fed on edible seaweeds, Nori (*Porphyra tenera*) and Wakame (*Undaria pinnatifida*), as a source of dietary fibre. *Food Chemistry*, 76(3), 281-286.

Usov, A. I., Smirnova, G. P., Klochkova, N. G. (2001). Polysaccharides of algae: 55. Polysaccharide composition of several brown algae from Kamchatka. *Russian Journal of Bioorganic Chemistry*, 27(6), 395-399.

van de Velde, F., Lourenco, N. D., Pinheiro, H. M., Bakker, M. (2002). Carrageenan: A food-grade and biocompatible support for immobilisation techniques. *Advanced Synthesis & Catalysis*, 344(8), 815-835.

van Maris, A.J.A., Abbot, D.A., Bellisimi, E., van den Brink, J., Kuyper, M., Luttik, M.A.H., Wisselink, H.W., Scheffers, W.A., van Dijken, J.P., Pronk, J.T. (2006). Alcoholic fermentation of carbon sources in biomass hydrolysates by *Saccharomyces cerevisiae*: current status. *Antonie van Leeuwenhoek*, 90, 391-418.

Vaugelade, P., Hoebler, C., Bernard, F., Guillon, F., Lahaye, M., Duee, P. H. et al. (2000). Non-starch polysaccharides extracted from seaweed can modulate intestinal absorption of glucose and insulin response in the pig. *Reproduction Nutrition Development*, 40(1), 33-47.

Vea, J. and Ask, E.,(2011). Creating a sustainable commercial harvest of Laminaria hyperborea, in Norway. Journal of Applied Phycology 23, 489-494

Velimirov, B, Field, J.G., Griffiths, C.L., Zoutendyk, P. (1977). The ecology of kelp bed communities in the Benguela upwelling system. Analysis of biomass and spatial distribution. *Helgoländer wissenschaftliche Meeresuntersuchungen* 30,495-518.

Verdrengh, M., Erlandsson-Harris, H., Tarkowski, A. (2000). Role of selectins in experimental *Staphylococcus aureus*-induced arthritis. *European Journal of Immunology*, 30(6), 1606-1613.

Vlieghe, P., Clerc, T., Pannecouque, C., Witvrouw, M., De Clercq, E., Salles, J. P. et al. (2002). Synthesis of new covalently bound kappa-carrageenan-AZT conjugates with improved anti-HIV activities. *Journal of Medicinal Chemistry*, 45(6), 1275-1283.

Wang, J., Zhang, Q. B., Zhang, Z. S., Li, Z. (2008). Antioxidant activity of sulfated polysaccharide fractions extracted from *Laminaria japonica*. *International Journal of Biological Macromolecules*, 42(2), 127-132.

Wang, W. T., Zhou, J. H., Xing, S. T. (1994). Immunomodulating action of marine algae sulfated polysaccharides on normal and immunosuppressed mice. *Chinese Journal of Pharm Toxicol*, 8(199-202.

Wang, Y., Han, F., Hu, B., Li, J. B., Yu, W. G. (2006). In vivo prebiotic properties of alginate oligosaccharides prepared through enzymatic hydrolysis of alginate. *Nutrition Research*, 26(11), 597-603.

Wargacki,A.J., Leonard, E., Nyan Win, M., Regitsky,D.D., Santos, C.N.S., Kim,P.B., Cooper, S.R., Raisner, R.M., Herman, A., Sivitz, A.B., Lakshmanaswamy, A., Kashiyama,Y.,

Baker, D., Yoshikuni, Y. (2012). An Engineered Microbial Platform for Direct Biofuel Production from Brown Macroalgae. *Science* 335, 308.

Washington, N. & Denton, G. (1995). Effect of alginate and alginate-cimetidine combination therapy on stimulated postprandial gastro-oesophageal reflux. *Journal of Pharmacy and Pharmacology*, 47(11), 879-882.

Werner, A., Kraan, S. (2004). Review of the potential mechanisation of kelp harvesting in Ireland. Marine Environment & Health Series, No.17 pp 1-52, Marine Institute, Galway.

Witvrouw, M. & DeClercq, E. (1997). Sulfated polysaccharides extracted from sea algae as potential antiviral drugs. *General Pharmacology-the Vascular System*, 29(4), 497-511.

Wu, X. W., Yang, M. L., Huang, X. L., Yan, J., Luo, Q. (2003). Effect of fucoidan on splenic lymphocyte apoptosis induced by radiation. *Chin J radiol Med Prot*, 23(430-.

Xue, C. H., Chen, L., Li, Z. J., Cai, Y. P., Lin, H., Fang, Y. (2004). Antioxidative activities of low molecular fucoidans from kelp *Laminaria japonica*. 139-145.

Yan, X. J., Zheng, L., Chen, H. M., Lin, W., Zhang, W. W. (2004). Enriched accumulation and biotransformation of selenium in the edible seaweed *Laminaria japonica*. *Journal of Agricultural and Food Chemistry*, 52(21), 6460-6464.

Yang XL, Sun JY, Xu HN (1995). An experimental study on immunoregulatory effect of fucoidan. Chin J Mar Drugs

Yarish, C., Pereira, R. (2008). Mass production of marine macroalgae. In Sven Erik Jørgensen and Brian D. Fath (Editor-in-Chief), *Ecological Engineering*. Vol. [3] of Encyclopedia of Ecology, 5 vols. pp. [2236-2247] Oxford, Elsevier.

Ye, H., Wang, K., Zhou, C., Liu, J., Zeng, X. (2008). Purification, antitumor and antioxidant activities in vitro of polysaccharides from the brown seaweed Sargassum pallidum. *Food Chemistry*, 111(2), 428-432.

Zee S (1991) Body weight loss with the aid of alginic acid. Arc Intern Med 3-4:113–114

Zeitlin, L., Whaley, K. J., Hegarty, T. A., Moench, T. R., Cone, R. A. (1997). Tests of vaginal microbicides in the mouse genital herpes model. *Contraception*, 56(5), 329-335.

Zeitoun P, Salmon L, Bouche O, Jolly, D, Thiefin G (1998) Outcome of erosive/ulcerative reflux oesophagitis in 181 consecutive patients 5 years after diagnosis. Ital J Gastroenterol Hepatol 30:470–474

Zhang, Q. B., Li, N., Zhou, G. F., Lu, X. L., Xu, Z. H., Li, Z. (2003). In vivo antioxidant activity of polysaccharide fraction from *Porphyra haitanesis* (Rhodephyta) in aging mice. *Pharmacological Research*, 48(2), 151-155.

Zhang, Q.S., Qu, S.C., Cong, Y.Z., Luo, S.J., Tang, X.-X. (2008). High throughput culture and gametogenesis induction of *Laminaria japonica* gametophyte clones. *J. Appl. Phycol.* 20, 205-211.

Zhao, X., Xue, C. H., Li, B. F. (2008). Study of antioxidant activities of sulfated polysaccharides from Laminaria japonica. *Journal of Applied Phycology*, 20(4), 431-436.

Zhou, Y., Yang, H. S., Hu, H. Y., Liu, Y., Mao, Y. Z., Zhou, H. et al. (2006b). Bioremediation potential of the macroalga *Gracilaria lemaneiformis* (Rhodophyta) integrated into fed fish culture in coastal waters of north China. *Aquaculture*, 252(2-4), 264-276.

Zhuang, C., Itoh, H., Mizuno, T., Ito, H. (1995). Antitumor active fucoidan from the brown seaweed, Umitoranoo (*Sargassum-thunbergii*). *Bioscience Biotechnology and Biochemistry*, 59(4), 563-567.

Zverlov, V.V., Berezina, O.V., Velikodvirskaya, G.A., Schwarz, W. H. (2006). Bacterial acetone and butanol production by industrial fermentation in the Soviet Union: use of hydrolyzed agricultural waste for biorefinery. *Applied Microbiology and Biotechnology* 71, 587-597.

Chapter 23

Biological Applications of Plants and Algae Lectins: An Overview

Edson Holanda Teixeira, Francisco Vassiliepe Sousa Arruda,
Kyria Santiago do Nascimento, Victor Alves Carneiro, Celso Shiniti Nagano,
Bruno Rocha da Silva, Alexandre Holanda Sampaio and Benildo Sousa Cavada

Additional information is available at the end of the chapter

http://dx.doi.org/10.5772/50632

1. Introduction

More than 120 years ago, Peter Hermann Stillmark in his doctoral thesis presented in 1888 to the University of Dorpat, gave the earliest step in the study of proteins that have a very interesting feature: the ability to agglutinate erythrocytes. These proteins were initially referred as to hemagglutinins or phytoagglutinins, since they were originally isolated from extracts of plants [1]. The first hemagglutinin isolated by Stillmark was extracted from seeds of the castor tree (*Riccinus communis*) and was named ricin [2]. This hemagglutinin was strongly used by Paul Ehrlich as model antigens for immunological studies [2,3].

Thirty-one years after Stillmark, James B. Sumner, isolated from jack bean (*Canavalia ensiformis*) a protein that he called concanavalin A (ConA). For the first time a pure hemagglutinin had been obtained [4]. However, the report that ConA agglutinates cells such as erythrocytes and yeasts and also precipitates glycogen from solution was given by Summer and Howell nearly two decades after its isolation. In addition, the findings of these researchers showed that the hemagglutination by ConA was inhibited by sucrose, demonstrating for the first time the sugar specificity of lectins. Thus, they suggested that the hemagglutination induced by ConA might be due to the reaction of the plant protein with carbohydrates expressed on the surface of the erythrocytes [1,4].

In 1907, Landsteiner and Raubitschek analyzed the hemagglutination of red blood cells from different animals by various seeds extracts. They found that the relative hemagglutinating activity was quite different for each extract tested [1]. However, it was only in the 1940s that Willian Boyd and Karl Renkonen, working independently, made the important discovery of human blood groups specificity for hemagglutinins. They found that crude extracts of two leguminous plants, *Phaseolus limensis* and *Vicia cracca*, agglutinated blood type A erythrocytes but not blood type B or O cells, whereas the extract of *Lotus tetragonolobus* agglutinated only blood type O erythrocytes [1,5].

The specific interaction between lectins and carbohydrates of erythrocytes played a crucial role in the investigations of the antigens associated with the ABO blood group system. In the subsequent decade, Morgan and Watkins found that the agglutination of type A erythrocytes by extracts of *Phaseolus limensis* was best inhibited by α-linked N-acetyl-D-galactosamine, while the agglutination of O cells by the extract of *L. tetragonolobus* was best inhibited by α-linked L-fucose [6].

Around thirty years after Boyd, the research on lectins reached the molecular level studies. It was clear the need to a better understanding on the structural aspects of lectins. Then, in 1972 Edelman and colleagues established the primary sequence of ConA [6]. In the same year, Edelman's group and independently Karl Hardman with Clinton Ainsworth, solved the 3D structure of ConA by X-ray crystallography [7,8].

2. What exactly is a lectin?

In 1954 Boyd and Shapleigh proposed the term lectin, from the Latin verb *legere* (which means "to select"). This term was based on the fact that these proteins have the ability to distinguish between erythrocytes of different blood types [9].

Lectins were early defined as carbohydrate-binding proteins of nonimmune origin that agglutinate cells or as carbohydrate-binding proteins other than antibodies or enzymes. However, these definitions were updated, since some plant enzymes are fusion proteins composed of a carbohydrate-binding and a catalytic domain, for instance, type 2 RIPs, such as ricin and abrin, are fusion products of a catalytically active A-chain (which has the N-glycosidase activity) and a carbohydrate-binding B-chain, both linked by a disulfide bond [10]. Furthermore, there is in nature carbohydrate-binding proteins possessing only one binding site and, therefore, are not capable of precipitating glycoconjugates or agglutinating cells [11].

Thus, in 1995 Peumans and Van Damme proposed the most suitable definition for lectins. According to the "new" definition, all plant proteins that possess at least one noncatalytic domain that binds reversibly to a specific mono- or oligosaccharide are considered as lectins [12,13].

2.1. Plant lectins

Lectins are proteins widely distributed in nature such in microorganisms, plants, animals and humans, acting as mediators of a wide range of biological events that involve the crucial step of protein-carbohydrate recognition, such as cell communication, host defense, fertilization, cell development, parasitic infection, tumor metastasis, inflammation, etc [14-15].

Peumans and Van Damme classified the plant lectins according to their overall structure. *Merolectins* consist exclusively of a single carbohydrate-binding domain (e.g. hevein, a chitin-binding latex protein isolated from the rubber tree *Hevea brasiliensis*). Since merolectins have a unique carbohydrate-binding site, they are incapable of precipitating glycoconjugates or agglutinating cells. *Hololectins* are also built of carbohydrate-binding domains. However, they contain at least two such domains that are identical or very similar. Because these lectins are di- or multivalent they can agglutinate cells and/or precipitate glycoconjugates. Most plant lectins are hololectins. *Superlectins* are built of at least two carbohydrate-binding domains. Unlike hololectins, these domains are not identical or similar. Thus, superlectins recognize structurally different sugar (e.g. TxLCI, a tulip bulb lectin that recognizes mannose and N-acetyl-galactosamine). *Chimerolectins* are fusion proteins that consist of two different chains, one of them with a remarkable catalytic activity (or another biological activity). RIPs type 2 are examples of chimerolectins [11-12].

The most thoroughly investigated lectins have been isolated from plants, particularly that extracted from members of the Leguminosae family. Legume lectins are a large group of proteins that share a high degree of structural similarity with distinct carbohydrate specificities. The subtribe Diocleinae (Leguminosae) comprises 13 genera, mostly of them from the New World. However, only 3 of these genera (i.e. *Canavalia*, *Cratylia* and *Dioclea*) are considered as the main sources for protein purification [16].

Concerning the biological activity, legume lectins are considered as enigmatic proteins. Despite the philogenetic proximity as well the high degree of similarity shared between them, they possess different biological activities such as histamine release from rat peritoneal mast cells, lymphocyte proliferation and interferon-γ production, peritoneal macrophage stimulation and inflammatory reaction as well as induction of paw edema and peritoneal cell immigration in rats [16].

2.2. Plant lectins as biotechnological tools

Significant progress has been reached in last years in understanding the crucial roles of lectins in several biological processes [17]. The importance of lectins as biotechnological tools has been established early in the studies involving its biological application. In 1960 a major step in immunology was given in order to determine the role of these proteins on the lymphocytes cell division. It was found that the lectin of the red kidney bean (*Phaseolus vulgaris*), known as phytohemagglutinin (PHA), possesses the ability to stimulate lymphocytes to undergo mitosis [18]. After these findings, many studies have been performed to evaluate the role of lectins on different models involving the immune response and its products, for instance the stimulation of citokine secretion [19], functional activation of monocytic and macrophage-like cells [20] and ROS production by spleen cells [21].

In addition to immunological studies, recent works have been investigated the influence of lectins in the field of microbiology, since lectins can be considered as valuable tools to verify the role of interaction between the pathogen and carbohydrates present in host cells and its importance to disease development. For instance, it has been proposed that the pathogen *Helicobcter pylori* infects human cells through an interaction involving a lectin [22].

3. Carbohydrates and the neoplasic process

Currently, malignancies are considered a major problem in the public health, especially given their increasing incidence and prevalence rates observed in recent decades. In this context, the malignant tumors, or cancers, account for approximately 7.8 million deaths per year, thus becoming the second greatest cause of death worldwide, only behind the cardiovascular disease [23].

The cancer can be defined as a set of more than 100 diseases that have in common the uncontrolled growth of cells, which invade tissues and organs and can spread to other body regions [23]. Thus, both the processes of cellular mutation that affects the neoplastic cell and metastasis, involve a series of genetic changes that culminate in modifications in the pattern of several receptors and signaling molecules present on the cell surface [24].

Carbohydrates are biomolecules that have enormous potential for encoding biological information. These combined-molecules (Glycoproteins and Glycolipids) are responsible for different biological interactions between the cell and the extracellular environment [26]. Regarding the neoplastic cells, the glycosylation of these proteins and lipids is changed, which generates membrane signaling molecules capable of inducing several processes directly related to tumor progression such as cell adhesion, angiogenesis, cellular mitosis and metastasis, in addition, in some cases, be responsible for inhibition of apoptosis induction triggered by the cells of the immune system [26].

Certain changes in glycans occur frequently in neoplastic cells and may be considered "tumor-specific", establishing a correlation between the stage of disease progression and prognosis of the same [25]. Some classic examples of these changes are the antigens of the ABH and Lewis system. ABH antigens are not expressed in cells of healthy human colon but significantly expressed on tumor cells [27]. Since the antigen Lewis^y can be observed in several carcinomas and has been correlated with poor prognosis in breast tumors [28]. Another example is the glycosylated antigens sialyl-Lewis^A and sialyl-Lewis^x, which are significantly up-regulated in carcinomas of the colon and appear to be related to tumor progression [29].

Thus, due to the intrinsic role of carbohydrates in the tumorigenesis, the glycosylation process as well as the identification of glycosylated antigens have been intensively focused, given the fact that glycosylation of antigens can vary extensively depending on the stage of the disease, which can provide, when properly identified, a better possibility of correct diagnosis and treatment.

4. Application of plant lectins in the diagnosis of malignant tumors

Because the peculiar characteristic of specific binding to carbohydrates, lectins have been used as tools to identify aberrant glycans expressed by neoplastic cells. Such methods have been essential to obtain a more precise diagnosis that allows a more accurate prognosis.

Several methods regarding the use of lectins as tools for recognizing aberrant glycans have been proposed in recent days [30,31,32]. The technique most common and widespread is the use of lectins in immunohistochemical assays.

In this context, the study conducted by [30], showed that leguminous lectins from *Canavalia ensiformis* (ConA) and *Ulex europaeus* (UEA-1) were used as histochemical markers of parotid gland mucoepidermoid carcinoma with low, intermediate, and high grade dysplasia. The authors stated that ConA localization in the cytoplasm and/or plasma membrane was significantly associated with neoplastic cells from the three grades of severity, whereas UEA-1 was associated with low and intermediate grade dysplasia. The authors obtained similar results after the analysis of other cell regions.

Another recent methodology was addressed by [33]. This methodology exploits the fact that glycoproteins produced by cancer cells have altered glycan structures, although the proteins themselves are common, ubiquitous, abundant, and familiar. However, as cancer tissue at the early stage probably constitutes less than 1% of the normal tissue in the relevant organ, only 1% of the relevant glycoproteins in the serum should have altered glycan structures [34]. With that in mind, the strategy to approach the detection of these low-level glycoproteins is based in: (a) a quantitative real-time PCR array for glycogenes to predict the glycan structures of secreted glycoproteins; (b) analysis by lectin microarray to select lectins that distinguish cancer-related glycan structures on secreted glycoproteins; and (c) an isotope-coded glycosylation site-specific tagging high-throughput method to identify carrier proteins with the specific lectin epitope [33].

Therefore, further analyses of lectins as biomarkers have been undertaken to improve our understanding of the processes involved in malignant tumor formation. As well as enable us to acquire new methods of identification of neoplastic cells at an early stage, enabling a better prognosis with appropriate treatment and low cost.

5. Application of plant lectins in the treatment of malignant tumors

Apoptosis is a mechanism by which cells undergo death to control cell proliferation or in response to irreparable DNA damage. It is featured by unique morphological and biochemical changes, such as nucleus condensation and margination, membrane blebbing, and internucleosomal DNA cleavage [35]. As the type I programmed cell death (PCD), apoptosis occurs through two major pathways, the extrinsic pathway triggered by the Fas death receptors, and the mitochondria-dependent pathway that brings about the release of Cytocrome *c* (Cyto *c*) and activation of the death signals under stimulus. In both ways the caspases, which belong to a family of cysteine proteases, have been well established as major players in apoptosis-causing mechanisms [36].

Several studies have demonstrated that lectins can induce apoptotic cell death. In the mitochondrial-dependent pathway, ConA treatment results in a decrease of mitochondrial membrane potential, and thus collapsing mitochondrial transmembrane potential. Cyto *c* is subsequently released, making up apoptosome with Apaf-1 and procaspase-9. After conjugating apoptosome, procaspase-3 turns into active caspase-3 that eventually triggers apoptosis [37].

In [38], it was evaluated the pro-apoptotic activity of a lectin isolated from *Artocarpus incisa* (frutalin) on HeLa cells derived from human cervical cancer. In this study, frutalin possessed a remarkable antiproliferative effect on HeLa cells. This effect was irreversible as well as time and dose dependent. When the lectins were added, serious visible cellular morphology changes were observed, possibly as a result of cell stress.

An interesting study conducted by [39] showed the pro-apoptotic caspase-dependent activity of the lectin isolated from *Astragalus membranaceus* in leukemia cell line (CML K562). These results showed a close relationship with the low expression of BCL-2 (anti-apoptotic protein) indicating that the lectin is active through the mitochondrial apoptotic pathway [40]. Furthermore, structural changes in cell membrane and different levels of Caspase-3 contributed to support their hypothesis.

Despite the apoptotic activity of several lectins have been demonstrated in different studies, the precise mechanism of how this process is triggered as well the mode of internalization is unknown until the present date.

6. Carbohydrates and the inflammatory process

The inflammation is a nonspecific event of immune response that occurs in reaction to any type of tissue injury. This process is capable of triggering a series of physiological changes such as increased blood flow, elevated cellular metabolism, vasodilation, release of soluble mediators, extravasation of fluids and cellular influx [41].

The continuity of cell recruitment and tissue damage in addition to chemical mediators released by the injured tissue as well as resident cells on site activate various mechanisms, in turn, induce the migration of more immune cells as well as increasing local tissue perfusion [42].

Both, acute and chronic inflammations have specific characteristics and the innate immune system plays a central role, since it mediates the initial response. Infiltration of innate immune cells, specifically neutrophils and macrophages, characterizes the acute inflammation, while infiltration of T lymphocytes and plasma cells are features of chronic inflammation [41,42].

As discussed previously, carbohydrates can act as the intermediates of communication in biological processes such as differentiation, proliferation and certain cell–cell interactions that are crucially important in both physiological and pathological phenomena [43,44]. The information contained in the enormous variety of oligosaccharide structures normally conjugated to proteins or lipids on cell surfaces (glycocodes) is recognized and deciphered by a specialized group of structurally diverse proteins, the lectins [44].

Galectins (formerly "S-type lectins"), an evolutionarily conserved family of endogenous animal lectins, share unique features, including their highly conserved structure, exquisite carbohydrate specificity, and ability to differentially regulate a myriad of biological responses [45].

Although galectins have been implicated in many biological activities, most of the functional studies reported to date link galectins to early developmental processes, such as neovascularization, regulation of immune cell homeostasis and inflammation [44,46,47]. Through deciphering glycan-containing information about host immune cells or microbial structures, galectins can modulate a diversity of signaling events that lead to cellular proliferation, survival, chemotaxis, trafficking, cytokine secretion and cell–cell communication [46,47].

These findings are extremely important because they demonstrate the importance of the glicocodes in the process of cell recognition and inflammation. In this context, plant lectins have been widely used to understand the pro-inflammatory mechanisms, as well as the design of new compounds with pro-healing effect.

7. Plant lectins and its immunomodulatory activity

An immune system is a system of biological structures and processes within an organism that protects against disease. In order to function properly, an immune system must detect a wide variety of agents, from viruses to cancer cells, and distinguish them from the organism's own healthy tissue [41]. The immune system is composed of many cells and molecules that act in a complex and harmonious way with the ultimate goal of annihilating the aggressive factor [42].

In the immune system, two phases of activity can be clearly established: the innate immune response and the adaptive immune response. In the innate immune response, there is the activity of cells and cytokines in a nonspecific way with the main purpose to annihilate quickly the local damaging agents. At this stage, we highlight the neutrophils, eosinophils, basophils and macrophages, cells with well-established activities but with the common function of production and release of cytokines. These cytokines are molecules with various functions in the inflammatory process, such as chemotaxis, activation of certain cell groups and increased tissue perfusion [48].

On the other hand, the adaptive immune response is composed by another set of cells that acts in a more specific way, the lymphocytes. Such cells are responsible for producing anti-bodies specific for certain invading microorganisms and the activation of mechanisms of apoptosis in abnormal cells [49].

Thus, the use of molecules capable of inducing cell recruitment as well as cytokine production and lymphocytes proliferation is of special scientific interest.

Korean mistletoe (*Viscum album* L. var. *coloratum*) is traditionally used as a sedative, analgesic, anti-spasmolytic, cardiotonic and anticancer agent, in Korea. An important lectin has been isolated from this plant and its immunomodulatory activity was analyzed [50]. It was shown that KML differentially modulated macrophage-mediated immune responses. It also enhanced the expression of various cytokines (IL-3, IL-23 and TNF-a), ROS generation, phagocytic uptake and surface levels of some glycoproteins (co-stimulatory molecules, PRRs and adhesion molecules). Nevertheless, the functional activation of adhesion molecules assessed by cell–cell or fibronectin adhesion events was up-regulated by KML treatment.

A recent study [51] demonstrated the immunomodulatory activity of ConBr, a lectin isolated from *Canavalia brasiliensis* seeds. The assays showed that ConBr was able to induce *in vitro* proliferation of splenocytes with minimal damage to the cellular structure. Furthermore, ConBr increased in the production of cytokines such as IL-2, IL-6 and IFN-γ production and decreased IL-10. These findings indicate the potential immunomodulatory effect of this lectin in conjunction with the intrinsic role of carbohydrates in intercellular communication related to the inflammatory process.

Regarding the activity of lectins on lymphocytes, a recent study [52] evaluated the effect of lectin extracted from seeds of *Cratylia mollis* Mart. (Cramoll 1,4) on experimental cultures of mice lymphocytes. In this study, aspects directly related to inflammation as cytokine production, cytotoxicity and cellular production of nitric oxide (NO) were evaluated. Cramoll 1,4 did not show cytotoxicity at the concentrations tested, in addition, was able to induce IFN-γ and showed an anti-inflammatory activity through the supression of NO production.

The biological function of carbohydrates in inflammation events is well-defined. In this context, proteins that bind specifically to such glycans are of great interest because of their possible functions and applications in biotechnological studies.

8. Plant lectins and its pro-healing activity

Recently, researches have undertaken efforts at the possible pro-healing activity of some lectins [53,54]. This goal is supported by the fact that such molecules may interfere with the inflammatory process. This effect is not yet fully elucidated, however peculiar and interesting results can be observed.

Experiment conducted in a murine model, employing the lectin isolated from *Bauhinia variegata* seeds (BVL) topically on surgically induced wounds, revealed the pro-healing potential of such molecule. Although not yet elucidated, it is suggested that this lectin appears to stimulate the mitogenic activity of resident cells, turning them into potent chemotactic agents for the recruitment of neutrophils through the release of cytokines [54]. Furthermore, it is suggested that the BVL stimulates the differentiation of fibroblasts into myofibroblast, which is an extremely important event in the remodeling of connective tissue [55].

Although promising, this issue requires further studies to better characterize the mechanisms involved in pro-healing role played by lectins.

9. Marine algae lectins

To date, there are fewer than 100 publications describing the presence of lectins in marine red, green and brown macroalgae. Moreover, and in marked contrast to higher land plant lectins, marine algal lectins have been isolated and characterized at a much lower pace since the first report of haemagglutinating activity in these organisms appeared more than 46 years ago [56]. Thereafter, other studies describing the presence and/or purification of algal lectins were reported by groups from England [57], Japan [58], Spain [59], United States [60] and Brazil [61].

Currently, the presence of lectins was analyzed at about 800 algae species. However, this number is still small, considering that there are thousands of species of marine algae. Together, the research shows that approximately 60% of the analyzed species show hemagglutinating activity. The number of positive species could be higher since in the first screenings the authors used a limited number of red blood cells and without enzyme treated erythrocytes.

The improvement in the methodologies of both, extraction and hemagglutination assays could increase the number of positive species. In fact, there appears to be coincidence that the rabbit erythrocytes treated with papain are most suited for the hemagglutinating activity detection in marine macroalgae [62,63].

Although marine algal lectins show proteinaceous content similar to lectins from terrestrial plants, they differ in some aspects. Early publications on this issue, reported that in general, lectins from algae have low molecular masses, no affinity for monosaccharides, strong specificity for complex oligosaccharides and/or glycoproteins. Moreover, they appear to have no requirement for metal ions, showing high content of acidic residues and even in high concentrations tend to stay in the monomeric form [64,65,66]. However there are a few reports showing that some of these molecules may be inhibited by simple sugars and are cation dependent as showed for the lectins from the green marine alga genus, *Codium* [67] and red marine alga genus, *Ptilota* [68,69,70].

The classical methods used to purify marine algae lectins include methods such as protein precipitation (using salt or ethanol), liquid chromatography (especially affinity) and electrophoresis [69,71]. Ion exchange chromatography has been effectively used in the isolation of lectins from seaweed, mainly in initial stages in purification. In this step, the lectins were separated from pigments present in the extracts [66,72,73]. In the protein extracts, phycobilins are co-extracted with lectins, becoming an undesirable contaminant in the purification process [72].

Lectins from marine algae *Cystoclonium purpureum* [74], *Gracilaria verrucosa* [75], *Palmaria palmata* [76], *Solieria robusta* [62], *Gracilaria tikvahiae* [60], *Bryothamnion seaforthii* and *B.triquetrum* [66], *Solieria filiformis* [77], *Enantiocladia duperreyi* [78], *Amansia multifida* [73], *Hypnea musciformis* [79], *Gracilaria ornata* [80], *Hypnea cervicornis* [81] and *Georgiela confluens* [82] were isolated by exchange chromatography, usually on DEAE cellulose. In contrast, due usually algal lectins have binding specificity of complex sugars, affinity chromatography has been used a few times, such as the lectins of green algae of the genus *Codium* [83,84,85], lectins from *Ulva lactuca* [86], *Caulerpa cupressoides* [87], *Enteromorpha prolifera* [88], *Ulva pertusa* [89], *Bryopsis plumosa* [90,91,92], *Bryopsis hypnoides* [93] and lectins from red marine algae of the genus *Ptilota* [68,69,70].

To date, only 31 lectins from Rhodophyceae and 17 lectins from Chlorophyceae were isolated and characterized. The virtual absence of lectins isolated from brown algae (Phaeophyta) is mainly due to the amount of polyphenols present in plants. It is known that polyphenols are released in extraction and that these compounds and their oxidation products, quinones, bind tightly to proteins [94] causing a false hemagglutination [83,95].

Even with the increase in the publications related to marine algae lectins, biochemical and structural information on algal lectins is scarce and from only a few species, and hence the functional and phylogenetic classification of these lectins remains unclear. The available structural information indicates the existence of different carbohydrate-binding proteins in the marine algae investigated. Moreover, the complete amino acid sequences of only 14 algal lectins have been determined. In red marine algae, *Bryothamnion triquetrum* lectin (BTL) was the first lectin to be determined its primary structure [96].

In same year, [97] reported the primary structure of *Hypnea japonica* agglutinin (HJA) that shares sequence similarity with BTL and with lectin from *Bryothamnion seaforthii* [98] and these lectins constitute the first marine red alga lectin family. Based on identity between HML and HCA and in the differences in amino acid sequences compared with BTL/HJA, [99] suggest that HCA/HML constitute another algal lectin family.

On the other hand, the lectins isolated from *Eucheuma serra, E. amakusaensis, E. cottonii* [100] have masses around 28 kDa, presents a monomer, share N-terminal sequence similarity with the complete amino acid sequence of isolectin 2 from *Eucheuma serra* (ESA-2) [101]. Also, the primary structure of ESA-2 contains repeated domains in their primary structure. These data suggest that lectins from genus *Eucheuma* can be grouped in a thirty family of red marine alga haemagglutinins. The amino acid sequence of lectin from red marine alga *Griffthisia sp.* [102] displays sequence similarity with lectin from jack fruit (*Artocarpus integrifolia*). The common methodology employed to determine the primary structures from red marine alga lectins was a combination of Edman degradation of sets of overlapping peptides and mass spectrometry. From green marine algae, the first primary structure determined was the lectin from *Ulva pertusa* (UPL-1) [89]. [103] described a 19 kDa protein expressed in strictly freshwater conditions in species of *Ulva limnetica Ichihara et Shimada* (freshwater alga), and this protein (named ULL) was identified and sequenced by cDNA cloning. The protein encoded by the cDNA showed 30% identity to UPL-1. However, the ULL should be considered a lectin-like since its haemaglutination activity was not yet characterized. UPL-1 and ULL did not show amino acid sequence similarity with known plant and animal lectins.

Recently [90] reported that the aggregation of cell organelles of *Bryopsis plumosa* in seawater was mediated by a lectin–carbohydrate complementary system and the purified lectin (named bryohealin) is involved in protoplast regeneration. The primary structure of bryohealin and of lectin from *Bryopsis hypnoides* [104] had few similarity with any known plant lectin, but rather resembled animal lectins with fucolectin domains. In addition, *Bryopsis plumosa*, has other two lectins described.

The lectin BPL-2 is a 17 kDa protein specific to D-mannose (ref). The authors found no similarity with others proteins in specific databases. The BPL-3 [92] possesses specificity to *N*-acetyl-D-galactosamine/*N*-acetyl-D-glucosamine and share the same sugar specificity with bryohealin. However, the primary structures of the two lectins were completely different. The homology sequence analysis of BPL-3 showed that it might belong to H lectin group from Roman snails (*Helix pomatia*). The latest primary structure published was of lectin from *Boodlea coacta* (BCA) [105]. BCA consisted of 3 internal tandem-repeated domains, each one containing the sequence motif similar to the carbohydrate-binding site of land plant lectin from *Galanthus nivalis*. It should be noted that the primary structures from green marine algal lectins were mainly determined with combination of Edman degradation and cDNA cloning.

Another observation is that a large number of sequenced algal lectins have the presence of cysteinyl residues or the duplication of internal domain. Still a lot of work needs to be done on the structure of algal lectins, since a few amount of primary structures has been determined. Further structural studies will contribute to understanding the differences in their biochemical characteristics as well as to the evolutionary aspects upon lectin presence in land plants and marine algae.

10. Marine algae lectins and its biotechnological role

There are few studies in the literature about the biotechnological applicability of lectins from marine algae. Probably due to low rentability of lectins obtained through the purification processes. It is noteworthy that the majority of lectins isolated so far were extracted from red algae, which are rich in carbohydrates and not in proteins. Moreover, during the extraction of algae proteins, phycobiliproteins are extracted simultaneously, and therefore the addition of steps to remove these phycobiliproteins causes losses of other proteins, among these are the lectins.

However, several studies on biological applications of algae lectins demonstrate that these molecules have an additional benefit; they are molecules with low molecular weight and may be less antigenic when used in biological models.

In cancerology, it was demonstrated that a lectin from the red marine algal *Eucheuma serra* (ESA) induced cell death in human cancer cells through the induction caspase-3 activity and DNA fragmentation in human colon adenocarcinoma (Colo201) cells [106]. ESA also induced cell death in a dose-dependent manner via apoptosis pathway in *in vitro* studies with Colon adenocarcinoma (Colon26) Cells derived from BALB/c mice [107].

In current studies, Span 80 vesicles (a potential type of nonionic vesicular drug delivery system) with ESA and PEGylated (EPV) lipids immobilized, showed hemagluttinating activity similar to free ESA and decreased the viability of Colo201 cancer cells *in vitro* and not affected the growth of normal cells. EPV caused the anti-tumor effect *in vivo* by inducing apoptosis in tumor cells [108].

Bryothamnion seaforthii lectin (BSL) and *Bryothamnion triquetrum* lectin (BTL) were able to differentiate human colon carcinoma cell variants. Differentiation was, probably, in function to cell membrane glycoreceptors and could be exploited to investigate structural modification of cell membrane glycoconjugates in cancer cell systems. In addition, it has been shown that the binding of both lectins to the carcinoma cells results in their internalization, which is a very interesting property that could be used in future applications, such as drug delivery [109].

The lectins isolated from the red marine algae *Hypnea cervicornis* (HCA), *Pterocladiela capillacea* (PcL) and *Caulerpa cupressoides* (CcL) have been tested as anti-inflamatory and antinociceptive agents. The data indicated that HCA, PcL and CcL have actions anti-inflammatory and antinociceptive (in formalin and acetic acid models). However, these lectins did not present significant antinociceptive effects in the hot plate test [110,111,112].

Concerning the mitogenic activity, the lectin from the red marine alga *Carpopeltis flabellate* (Carnin) was the first lectin that showed mitogenic activity for T lymphocytes from mouse spleen at a concentration of 10,5 μg/ml. Carnin inhibited the normal embryonic development of the sea urchin *Hemicentrotus pulcherrimus* at the stage of blastula and also inhibited the gasturulation induced in starfish *Asterina pectinifera,* at concentrations of 10 and 5 μg/mL, respectively. In addition, it was showed an interaction between a lectin of macroalgae with a microorganism of the marine ecosystem [113].

The antibacterial effect was too evaluated. The lectins from the red marine algal *Eucheuma serra* (ESA) and *Galaxaura marginata* (GMA) have an antibiotic activity. ESA and GMA strongly inhibited *Vibrio vulnificus*, a fish pathogen, but not were able to inactive *V. neresis* and *V. pelagius* [114]. BSL and BTL also were able to avoid the bacterial adhesion of streptococci strains in enamel pellicles. BTL was more efficient in avoiding the adherence of *Streptoccocus sobrinus* and *S. mitis* and BSL was able to reduce adherence of *S. mutans*. This fact can contribute with preventing caries at early stages [115].

Red marine algae lectins from *Amansia multifida* (AML), *Bryothamnion seaforthii* (BSL), *Bryothamnion triquetrum* (BTL) and *Gracilaria caudata* (GCL) induced neutrophil migration *in vitro* and *in vivo* in the peritoneal cavity or dorsal air pouch of rats or mice. The results showed that BT had the most potent effect in neutrophil migration when tested on rats. In mice, BS required four times higher than the dose of BT to induce neutrophil migration. However, when the algae lectins were assayed in mice, AML was the most potent [116].

Concerning the antiviral effects, a lectin named KAA-2 from the red marine algal *Kappaphycus alvarezii* (specific to high mannose type *N*-glycans) showed antiviral action against H1N1 virus [117]. The red alga lectin from *Griffithsia* sp. called griffithsin (GRFT) is a small mannose specific lectin that binds carbohydrates on HIV envelope glycoproteins and block HIV entry into target cells. GRFT is a candidate for development of anti-HIV microbicides [118,119]. GRFT was also able to inhibit the action of the pathogenic agent responsible for respiratory syndrome (SARS), the SARS coronavirus (SARS-CoV) [120].

11. Lectins as biotechnological tools to study the microbial biofilms

Biofilms are microbial complex communities established in a wide variety of surfaces and are generally associated with an extracellular matrix composed by several types of polymers [121]. This type of microbial association can develop on biotic and abiotic surfaces, including living tissues, medical devices and/or industrial water piping systems and marine environments [122,123,124].

Bacterial infections involving biofilm formation are usually chronic and often present a arduous treatment [125,126,127]. The growth and proliferation of microorganisms inside the biofilm provides reduction or prevents the penetration of various antimicrobial agents [128,129] and thus become extremely difficult or impossible to eradicate them [130,131,132]. For some antibiotics, the concentration required to eliminate the biofilm can be up to a thousand times higher than required to planktonic form of the same specie [127].

The ability of microorganisms to form pathogenic cell aggregates is a worldwide concern. In attempt to remedy this problem, pharmaceutical companies associated to research groups work avidly to the development of new options for the treatment of infections caused by bacteria organized in biofilms.

Biofilm formation is a process in which bacteria has a change in lifestyle, it goes from a state unicellular in suspension to a multicellular sessile, where the growth and cell differentiation results in structured communities. The biofilm begins with the setting free of microorganisms in a given area. The first microorganisms adhere initially by weak interactions, mainly of the van der Waals forces [133]. If the colonies are not immediately removed from the surface, they can anchor by cell adhesion molecules existing in the pili and / or flagella [134]. The first colonies facilitate the arrival of other cells through adhesion sites and begin to build the matrix that will form the biofilm. Only a few species are able to adhere to a surface *per se*. Others may anchor to the matrix or directly on existing colonies. Since the colonization has started, the biofilm grows through cell division and combination of the recruitment of other cells [135].

According to [136], the biofilm development occurs through three events. At first, there is a distribution of fixed cells in surface through cell motility. Then, occurs the proliferation of fixed cells by division, expanding to upward and sides forming agglomerated of cells, similar to the formation of colonies on agar plates [137]. Finally, clusters of cells attached to the biofilm are recruited by the action of the environment itself to the development of other biofilms, reaching a climax community [136]. These general stages provide guidance to the study of biofilms by bacteria furniture, although many details of regulation of this process may vary between species.

One of the biofilm-forming components of great importance in the maintenance of cell clusters is the matrix polymeric called EPS (*Extracellular Polymeric Substances*) [138].

Consisting of proteins, polysaccharides and environmental waste results in a solid structure highly hydrated with small channels between the microcolonies [139]. This matrix holds the biofilm together, one of those factors responsible for providing an increased resistance to antibiotics, disinfectants, and ultraviolet radiation [140]. In most biofilms, microorganisms that make up these agglomerates constitute less than 10% dry weight, while the extracellular matrix may contribute up to 90% [141].

The first step to biofilm formation, the early adhesion, is considered essential for colonization and infection by pathogenic bacteria. Macromolecules surfaces are directly involved in this stage [142]. Proteins known as adhesins are able to recognize specific polysaccharide substrate present on the surface to be colonized, for example, the presence of carbohydrates existing in the film of saliva that covers the teeth in the oral environment [143]. Glycidic epitopes present on surfaces of microorganisms (early colonizers) can also be recognized by these proteins to mediate an event known as coaggregation, which will start the formation of a multi-species community [144].

One etiological factor for the development of teeth biofilms is the adhesion of pathogenic bacteria in the dental enamel [145]. However, the microorganisms are not deposited directly on the tooth surface, but bind to a thin acellular layer composed of salivary proteins and other macromolecules that cover the tooth surfaces called acquired pellicle [146,147].

The acquired pellicle is formed by glycoproteins and carbohydrates that serve as receptors for bacteria containing proteins with glucan-binding domains [148]. Bacteria interact with the film by several specific mechanisms, including the interaction lectin-like involving the bacterial adhesins and receptors existing in the pellicle [148,149]. Next, other bacteria can adhere to the film as well as the bacteria pre-existing in the biofilm [150].

The event coaggregation is a phenomenon widely observed in diverse microbial communities [151,152,153]. Cells can interact in suspension, forming cell aggregates, as well as connect directly on biofilm in the process of formation. In the first case, plancktonic cells recognize specifically species genetically distinct creating the coaggregates. In other situations, coaggregates in the form of secondary colonizers can adhere on biofilm in development, a process known as coadhesion. Both cases have an important role in the integration and establishment of a mature biofilm [154].

Thus, carbohydrate residues have an important role in formation and maintainability of microbial biofilms. They act as mediators of the binding between bacteria and the surface that will serve as substrate for biofilm formation [155], as well as site of interaction between microorganisms to form cell aggregates [156,157]. Furthermore, through EPS matrix, maintains the biofilm attached in the surface, conferring a greater resistance to antimicrobial agents in general [158,159].

Molecules able to bind specifically and selectively to carbohydrates have a key importance in the development of research related to microbial biofilms. Thus, lectins have been shown as powerful tools for analyze the glycidic structures of those aggregates from microbial origin [160,161].

Studies of microbial biofilms through the interaction with lectins have two main objectives: visualization and characterization of polymeric matrix (EPS) formed by different species of microorganisms [162,163] and inhibition of oral biofilm formation by blocking the bacterial binding sites present in the pellicle of saliva in the form of glycoproteins and / or carbohydrates [164].

The application of lectins in the characterization of EPS is already widely exploited by many research groups. Lectin of wheat germ (WGA) was used as a marker of *Staphylococcus epidermidis* microcolonies, mainly in the study of the mechanisms involved in bacterial organization during the formation of the cell aggregates [165]. In a study published by the same group, WGA was used to quantify the production of GlcNAcβ-1,4n, a sugar component of the extracellular matrix involved in biofilm formation [166].

In 1980s was developed a system called ELLA (*Enzyme-linked Lectin Assay*) that uses enzyme-lectin conjugates to detect specific carbohydrate units on the surface of cells. This assay allows better detection and quantitation of the sugars by standard immunofluorescence with fluorescein-conjugated lectins [167]. The lectins concanavalin A (ConA) and WGA peroxidase-labeled were used to detect D-glucose or D-mannose and N-acetyl-D-glucosamine or N-acetyl-neuraminic acid, respectively, on the biofilm of several species [168].

Recent studies demonstrated the characteristics and distribution of glycoconjugates in cyanobacteria biofilm using lectins with different specificities. In this study the authors stated that the distribution of carbohydrates in the matrix is very variable. Based on lectin specificity, glycoconjugates produced by cyanobacteria biofilm contained mainly fucose, N-acetyl-glucosamine or -galactosamine and sialic acid [169].

Lectins may be a suitable antiadhesion agent for *streptococci*, since it has been reported that a lectin dependent mechanism is involved in its adhesion [170]. However, the application of lectins as tools to interfere with biofilm formation is still poorly explored [115,171]. As shown by [172], plant lectins appear as a new strategy to reduce the development of dental caries by inhibiting the early adherence and subsequent formation of of *Streptococcus mutans* biofilm.

Author details

Edson Holanda Teixeira, Francisco Vassiliepe Sousa Arruda and Bruno Rocha da Silva
LIBS, Integrated Laboratory of Biomolecules, Faculty of Medicine of Sobral, Federal University of Ceará, Fortaleza, CE, Brazil

Kyria Santiago do Nascimento, Victor Alves Carneiro and Benildo Sousa Cavada
BioMol, Laboratory of Biologically Active Molecules, Department of Biochemistry and Molecular Biology, Federal University of Ceará, Fortaleza, CE, Brazil

Celso Shiniti Nagano and Alexandre Holanda Sampaio
BioMar, Laboratory of Marine Biotechnology, Department of Fishing Engineering, Federal University of Ceará, Fortaleza, CE, Brazil

Acknowledgement

We would like to thank CNPq, CAPES, FUNCAP and UFC for financial support. AHS, BSC, CSN, EHT, KSN are senior investigators of CNPq.

12. References

[1] Sharon N, Lis H (2004) History of lectins: from hemagglutinins to biological recognition molecules. Glycobiology. 14: 53R-62R.

[2] Franz H (1988) The ricin story. Adv Lectin Res. 1: 10-25.

[3] Olsnes S (2004) The history of ricin, abrin and related toxins. Toxicon. 44: 361-70.

[4] Sumner JB, Howell SF (1936) The identification of the hemagglutinin of the jack bean with concanavalin A. J Bacteriol. 32: 227-237.

[5] Boyd WC, Shapleigh E (1954) Specific precipitation activity of plant agglutinins (lectins). Science. 119: 419.

[6] Morgan WT, Watkins WM (2000) Unraveling the biochemical basis of blood group ABO and Lewis antigenic specificity. Glycoconj J. 17: 501-530.

[7] Edelman GM, Cunningham BA, Reeke GN, Becker JW, Waxdal MJ, Wang JL (1972) The covalent and threedimensional structure of concanavalin A. Proc Natl Acad Sci USA. 69: 2580-2584.

[8] Hardman KD, Ainsworth CF (1972) Structure of concanavalin A at 2.4-A° resolution. Biochemistry. 11: 4910-4919.

[9] Hou FJ, Xu H, Liu WY (2003) Simultaneous existence of cinnamomin (a type II RIP) and small amount of its free A- and B-chain in mature seeds of camphor tree. Int J Biochem Cell Biol. 35: 455-64.

[10] Van Damme EJM, Balzarini J, Smeets K, Van Leuven F, Peuinans WJ (1994) The monomeric and dimeric mannose binding proteins from the Orchidaceae species *Listeru ovata* and *Epipactis helleborine:* sequence homologies and differences in biological activities. Glycoconjugate J. 11: 321-332.

[11] Peumans WJ, Van Damme EJ (1995) Lectins as plant defense proteins. Plant Physiol. 109: 347-52.

[12] Van Damme EJ, Peumans WJ, Barre A, Rougé P (1998) Plant lectins: a composite of several distinct families of structurally and evolutionary related proteins with diverse biological roles. Crit Rev Plant Sci. 17: 645-662.

[13] Sharon N, Lis H (1989) Lectins as cell recognition molecules. Science. 246: 227-234.

[14] Sharon N (2007) Carbohydrate-specific reagents and biological recognition molecules. J Biol Chem. 282: 2753-2764.

[15] Cavada BS, Barbosa T, Arruda S, Grangeiro TB, Barral-Netto M (2001) Revisiting *proteus*: do minor changes in lectin structure matter in biological activity? Lessons from and potential biotechnological uses of the Diocleinae subtribe lectins. Curr Protein Pept Sci. 2: 123-135.

[16] Wu AM, Lisowska E, Duk M, Yang Z (2009) Lectins as tools in glycoconjugate research. Glycoconj J. 26: 899-913.

[17] Nowell PC (1960) Phytohemagglutinin: an initiator of mitosis in culture of animal and human leukocytes. Cancer Res. 20: 462-466.

[18] de Oliveira Silva F, das Neves Santos P, de Melo CM, Teixeira EH, de Sousa Cavada B, Arruda FV, Cajazeiras JB, Almeida AC, Pereira VA, Porto AL (2011) Immunostimulatory activity of ConBr: a focus on splenocyte proliferation and proliferative cytokine secretion. Cell Tissue Res. 346: 237-44.

[19] Lee JY, Kim JY, Lee YG, Byeon SE, Kim BH, Rhee MH, Lee A, Kwon M, Hong S, Cho JY (2007) *In vitro* immunoregulatory effects of Korean Mistletoe lectin on functional activation of monocytic and macrophage-like cells. Biol Pharm Bull 30: 2043-2051.

[20] Melo CML, Paim BA, Zecchin KG, Morari J, Chiarrati MR, CorreiaMTS, Coelho LCBB, Paiva PMG (2010) Cramoll 1,4 lectin increases ROS production, calcium levels and cytokine expression in treated spleen cells of rats. Mol Cell Biochem. 342: 163-169.

[21] Ringnér M, Valkonen KH, Wadström T (1994) Binding of vitronectin and plasminogen to *Helicobacter pylori*. FEMS Immunol Med Microbiol. 9: 29-34.

[22] Bennett HJ, Roberts IS (2005) Identification of a new sialic acid-binding protein in *Helicobacter pylori*. FEMS Immunol Med Microbiol. 44: 163-9.

[23] Curado M, Edwards B, Shin H, Storm HH, Ferlay J, Heanue M, Boyle P (2008) Cancer incidence in five continents. Lyon: International Agency for Research on Cancer. 837 p.

[24] Saeland E, van Kooyk Y (2011) Highly glycosylated tumour antigens: interactions with the immune system. Biochem Soc Trans. 39: 388-92.

[25] Nangia-Makker P, Conklin J, Hogan V, Raz A (2002) Carbohydrate-binding proteins in cancer, and their ligands as therapeutic agents. Trends Mol Med. 8: 187-92.

[26] Taniguchi N, Korekane H (2011) Branched N-glycans and their implications for cell adhesion, signaling and clinical applications for cancer biomarkers and in therapeutics. BMB Rep. 44: 772-81.

[27] Cooper HS, Haesler WE Jr (1978) Blood group substances as tumor antigens in the distal colon. Am J Clin Pathol. 69: 594-8.

[28] Madjd Z, Parsons T, Watson NF, Spendlove I, Ellis I, Durrant LG (2005) High expression of Lewis y/b antigens is associated with decreased survival in lymph node negative breast carcinomas. Breast Cancer Res. 7: 80-7.

[29] St Hill CA, Farooqui M, Mitcheltree G, Gulbahce HE, Jessurun J, Cao Q, Walcheck B (2009) The high affinity selectin glycan ligand C2-O-sLex and mRNA transcripts of the core 2 beta-1,6-N-acetylglucosaminyltransferase (C2GnT1) gene are highly expressed in human colorectal adenocarcinomas. BMC Cancer. 9: 79.

[30] Sobral AP, Rego MJ, Cavalacanti CL, Carvalho LB Jr, Beltrão EI (2010) ConA and UEA-I lectin histochemistry of parotid gland mucoepidermoid carcinoma. J Oral Sci. 52: 49-54.

[31] de Lima AL, Cavalcanti CC, Silva MC, Paiva PM, Coelho LC, Beltrão EI, dos S Correia MT (2010) Histochemical evaluation of human prostatic tissues with *Cratylia mollis* seed lectin. J Biomed Biotechnol. 2010: 179817

[32] Beltrão EI, Medeiros PL, Rodrigues OG, Figueredo-Silva J, Valença MM, Coelho LC, Carvalho LB Jr (2003) *Parkia pendula* lectin as histochemistry marker for meningothelial tumour. Eur J Histochem. 47: 139-42.

[33] Narimatsu H, Sawaki H, Kuno A, Kaji H, Ito H, Ikehara Y (2010) A strategy for discovery of cancer glyco-biomarkers in serum using newly developed technologies for glycoproteomics. FEBS J. 277: 95-105.
[34] Rüegg C (2006) Leukocytes, inflammation, and angiogenesis in cancer: fatal attractions. J Leukoc Biol. 80: 682–684.
[35] Ghobrial IM, Witzig TE, Adjei AA (2005) Targeting apoptosis pathways in cancer therapy. CA Cancer J Clin. 55: 178-94.
[36] Michael OH (2000) The biochemistry of apoptosis. Nature. 407: 770–7.
[37] Li CY, Xu HL, Liu B, Bao JK (2010) Concanavalin A, from an old protein to novel candidate anti-neoplastic drug. Curr Mol Pharmacol. 3: 123-8.
[38] Oliveira C, Nicolau A, Teixeira JA, Domingues L (2011) Cytotoxic effects of native and recombinant frutalin, a plant galactose-binding lectin, on HeLa cervical cancer cells. J Biomed Biotechnol. 2011: 568932.
[39] Huang LH, Yan QJ, Kopparapu NK, Jiang ZQ, Sun Y (2011) *Astragalus membranaceus* lectin (AML) induces caspase-dependent apoptosis in human leukemia cells. Cell Prolif. 45:15-21.
[40] Lavrik IN, Golks A, Krammer PH (2005) Caspases pharmacological manipulation. J Clin Invest. 115: 2665–72.
[41] Ferrero-Miliani L, Nielsen OH, Andersen PS, Girardin SE (2007) Chronic inflammation: importance of NOD2 and NALP3 in interleukin-1beta generation. Clin Exp Immunol. 147: 227-35.
[42] Margetic S (2012) Inflammation and haemostasis. Biochem Med (Zagreb). 22: 49-62.
[43] Hurwitz ZM, Ignotz R, Lalikos JF, Galili U (2012) Accelerated porcine wound healing after treatment with α-gal nanoparticles. Plast Reconstr Surg. 129: 242e-251e.
[44] Di Lella S, Sundblad V, Cerliani JP, Guardia CM, Estrin DA, Vasta GR, Rabinovich GA (2011) When galectins recognize glycans: from biochemistry to physiology and back again. Biochemistry. 50: 7842-57.
[45] Rabinovich GA, Toscano M, Jackson DA, Vasta G (2007) Functions of cell surface galectin-glycoprotein lattices. Curr Opin Struct Biol. 17: 513–520.
[46] Rabinovich GA, Ilarregui JM (2009) Conveying glycan information into T-cell homeostatic programs: A challenging role for galectin-1 in inflammatory and tumor microenvironments. Immunol Rev. 230: 144–159.
[47] Rabinovich GA, Toscano M (2009) Turning "sweet" on immunity: Galectin-glycan interactions in immune tolerance and inflammation. Nat Rev Immunol. 9: 338–352.
[48] Rodríguez RM, López-Vázquez A, López-Larrea C (2012) Immune systems evolution. Adv Exp Med Biol. 739: 237-51.
[49] Fiocchi C (2011) Early versus late immune mediated inflammatory diseases. Acta Gastroenterol Belg. 74: 548-52.
[50] Lee JY, Kim JY, Lee YG, Byeon SE, Kim BH, Rhee MH, Lee A, Kwon M, Hong S, Cho JY (2007) *In vitro* immunoregulatory effects of Korean mistletoe lectin on functional activation of monocytic and macrophage-like cells. Biol Pharm Bull. 30: 2043-51.
[51] de Oliveira Silva F, das Neves Santos P, de Melo CM, Teixeira EH, de Sousa Cavada B, Arruda FV, Cajazeiras JB, Almeida AC, Pereira VA, Porto AL (2011)

Immunostimulatory activity of ConBr: a focus on splenocyte proliferation and proliferative cytokine secretion. Cell Tissue Res. 346: 237-44.

[52] de Melo CM, de Castro MC, de Oliveira AP, Gomes FO, Pereira VR, Correia MT, Coelho LC, Paiva PM (2010) Immunomodulatory response of Cramoll 1,4 lectin on experimental lymphocytes. Phytother Res. 24: 1631-6.

[53] Brustein VP, Souza-Araújo FV, Vaz AF, Araújo RV, Paiva PM, Coelho LC, Carneiro-Leão AM, Teixeira JA, Carneiro-da-Cunha MG, Correia MT (2012) A novel antimicrobial lectin from *Eugenia malaccensis* that stimulates cutaneous healing in mice model. Inflammopharmacology. 20.

[54] Neto LG, Pinto Lda S, Bastos RM, Evaristo FF, Vasconcelos MA, Carneiro VA, Arruda FV, Porto AL, Leal RB, Júnior VA, Cavada BS, Teixeira EH (2011) Effect of the lectin of *Bauhinia variegata* and its recombinant isoform on surgically induced skin wounds in a murine model. Molecules. 16: 9298-315.

[55] Li B, Wang JH (2011) Fibroblasts and myofibroblasts in wound healing: force generation and measurement. J Tissue Viability. 20: 108-20.

[56] Boyd WC, Almodovar LR, Boyd LG (1966) Agglutinins in marine algae for human erythrocytes. Transfusion 6: 82-83.

[57] Blunden G, Rogers DJ, Farnham WF (1975) Survey of British seaweeds for hemagglutinins. Lloydia 38: 162-168.

[58] Kamiya H, Ogata K, Hori K (1982) Isolation and characterization of a new lectin in the red alga *Palmaria palmata* (L.) O. Kuntze. Bot Marina. 15: 537-540.

[59] Fabregas J, Munoz A, Llovo J, Abalde J (1984) Agglutinins in marine red algae. IRCS Medical Science 12: 298-299.

[60] Chiles TC, Bird KT (1989) A comparative study of animal erythrocyte agglutinins from marine algae. Comp Biochem Physiol 94: 107-111.

[61] Ainouz IL, Sampaio AH (1991) Screening of Brazilian marine algae for hemaglutinins. Bot Marina 34: 211-214.

[62] Hori K, Ikegami S, Miyazawa K, Ito K (1988) Mitogenic and antineoplastic isoagglutinins from red alga *Solieria robusta.* Phytochemistry. 27: 2063-2067.

[63] Ainouz IL, Sampaio AH, Benevides NMB, Freitas ALP, Costa FHF, Carvalho MR, Pinheiro-Joventino F (1992) Agglutination of enzyme treated erythrocytes by Brazilian marine algal extracts. Bot Marina 35: 475-479.

[64] Rogers DJ, Hori K (1993) Marine algal lectins: new developments. Hydrobiologia, 260/261: 589-593.

[65] Hori K, Miyazawa K, Ito K (1990) Some common properties of lectins from marine algae. Hydrobiologia. 205: 561-566.

[66] Ainouz IL, Sampaio AH, Freitas ALP, Benevides NMB, Mapurunga S (1995) Comparative study on hemagglutinins from the red marine algae *Bryothamnion triquetrum* and *B. Seaforthii.* Rev Bras Fisiol Vegetal. 7: 15-19.

[67] Rogers DJ, Swain L, Carpenter BG, Critchley AT (1994) Binding of N-acetyl-D-galactosamine by lectins from species of green marine alga genus, *Codium.* Clin Biochem. 10: 162-165.

[68] Sampaio AH, Rogers DJ, Barwell CJ (1998) A galactose specific lectin from the red marine alga *Ptilota filicina*. Phytochemistry 48: 765-769.

[69] Sampaio AH, Rogers DJ, Barwell CJ, Saker-Sampaio S, Costa FHF, Ramos MV (1999) A new isolation and further characterization of the lectin from the red marine alga *Ptilota serrata*. J Appl Phycol 10: 539-546.

[70] Sampaio AH, Rogers DJ, Barwell CJ, Saker-Sampaio S, Nascimento KS, Nagano CS, Farias WRL (2002) New affinity procedure for the isolation and further characterization of the blood group B specific lectin from the red marine alga *Ptilota plumosa*. J Appl Phycol. 14: 489-496.

[71] Sharon N, Lis H (1990) Legume lectins – a large family of homologous proteins. FASEB J. 4: 3198-3208.

[72] Rogers DJ, Fish B, Barwell CJ, Loveless RW (1988) Lectins from marine algae associated with photosynthetic accessory proteins. In.INTERNATIONAL LECTIN MEETING, Berlin/New York. Proceedings of the 9th Lectin Meeting. Berlin/New York: Walter de Gruyter. 6: 373-376.

[73] Costa FHF, Sampaio AH, Neves SA, Rocha MLA, Benevides NMB, Freitas ALP (1999) Purification and partial characterization of a lectin from the red marine alga *Amansia multifida*. Physiol Mol Biol Plants. 5:53-61.

[74] Kamiya H, Shiomi K, Shimizu Y (1980) Marine biopolymers with cell specificity III agglutinins in the red alga *Cystoclonium purpureum*: isolation and characterization. J Natur Products. 43: 136-139.

[75] Shiomi K, Yamanaka H, Kikuchi T (1981) Purification and physicochemical properties of a hemagglutinin (GVA-1) in the red alga *Gracilaria verrucosa*. Bull Jap Soc Sci Fisheries. 47: 1079-1084.

[76] Kamiya H, Ogata K, Hori K (1982) lsolation and characterization of a: new agglutinin in the red alga *Palmaria palmata* (L.) O. Kuntze. Bot Marina. 25: 537-540.

[77] Benevides NMB, Leite AM, Freitas ALP (1996) Atividade hemaglutinante na alga vermelha *Solieria filiformis*. R Bras Fisiol Vegetal 8: 117-122.

[78] Benevides NMB, Holanda ML, Melo FR, Freitas ALP, Sampaio AH (1998) Purification and partial characterization of the lectin from the red marine alga *Enantiocladia duperreyi* (C. Agardh) Faalkenberg. Bot Marina 41: 521-525.

[79] Nagano CS, Moreno FB, Bloch Jr C, Prates MV, Calvete JJ, Saker-Sampaio S, Farias WR, Tavares TD, Nascimento KS, Grangeiro TB, Cavada BS, Sampaio AH (2002) Purification and characterization of a new lectin from the red marine alga *Hypnea musciformis*. Protein Pept Lett. 9: 159-66.

[80] Leite YFMM, Silva LMCM, Amorim RCN, Freire EA, Jorge DMM, Grangeiro TB, Benevides NMB (2005) Purification of a lectin from the marine red alga *Gracilaria ornata* and its effect on the development of the cowpea weevil *Callosobruchus maculatus* (Coleoptera: Bruchidae). Bioch et Bioph Acta. 1724: 137-145.

[81] Nascimento KS, Nagano CS, Nunes EV, Rodrigues RF, Goersch GV, Cavada BS, Calvete JJ, Saker-Sampaio S, Farias WRL, Sampaio AH (2006) Isolation and characterization of a new agglutinin from the red marine alga *Hypnea cervicornis* J. Agardh. Biochem Cel Biol. 84: 49-54.

[82] Souza BWS, Andrade FK, Teixeira DIA, Mansilla A, Freitas ALP (2010) Haemagglutinin of the antarctic seaweed *Georgiella confluens* (Reinsch) Kylin: isolation and partial characterization. Polar Biol. 1: 1-8.

[83] Rogers DJ, Loveless RW, Balding P (1986) Isolation and characterization of the lectins from sub-species of *Codium fragile*. In. INTERNATIONAL LECTIN MEETING, Berlin/New York. Proceedings of the 7th Lectin Meeting. Berlin/New York: Walter de Gruyter. 5: 155-160.

[84] Fabregas J, Muñoz A, Llovo J, Carracedo A (1988) Purification and partial purification of tomentine. An N-acetylglucosamine-specific lectin from green alga *Codium tomentosum* (huds) Stackh. J Exp Mar Biol Ecology. 124: 21-30.

[85] Rogers DJ, Flangu H (1991) Lectins from *Codium* species. Br Phycol J. 26: 95-96.

[86] Sampaio AH, Rogers DJ, Barwell CJ (1998) Isolation and characterization of the lectin from the green marine alga *Ulva lactuca*. Bot Marina 41: 765-769.

[87] Benevides NMB, Holanda ML, Melo FR, Pereira MG, Monteiro ACO, Freitas ALP (2001) Purification and partial characterization of the lectin from the marine green alga Caulerpa cupressoides (Vahl) C. Agardh. Bot Marina 44: 17-22.

[88] Ambrosio A, Sanz L, Sanchez EI, Wolfenstein-Todel C, Calvete JJ (2003) Isolation of two novel mannan- and L-fucose-binding lectins from the green alga *Enteromorpha prolifera*: biochemical characterization of EPL-2. Archives of Biochemistry and Biophysics 415: 245-250.

[89] Wang S, Zhong FD, Zhang YJ, Wu ZJ, Lin QY, Xie LH (2004) Molecular characterization of a new lectin from the marine alga *Ulva pertusa*. Acta Biochim Biophys Sin. 36: 111-117.

[90] Kim GH, Klochkova TA (2005) Purification and Characterization of a Lectin, Bryohealin, Involved in the Protoplast Formation of a Marine Green Alga *Bryopsis Plumosa* (Chlorophyta). J Phycol. 42: 86-95.

[91] Han JW, Jung MG, Kim MJ, Yoon KS, Lee KP, Kim GH (2010) Purification and characterization of a D-mannose specific lectin from the green marine alga, *Bryopsis plumose*. Phycol Res. 58: 143-150.

[92] Han JW, Yoon KS; Klochkova TA, Hwang M-S, Kim GH (2011) Purification and characterization of a lectin, BPL-3, from the marine green alga *Bryopsis plumosa* J Appl Phycol. 23: 745-753.

[93] Niu J, Wang G, Lü F, Zhou B, Peng G (2009) Characterization of a new lectin involved in the protoplast regeneration of *Bryopsis hypnoides*. Chin J Oceanol Limnol. 27: 502-512.

[94] Loomis WD (1974) Overcoming problems of phenolics and quinines in the isolation of plant enzymes and organelles. Methods Enzymol. 16: 528-544.

[95] Blunden G, Roger DJ, Loveless RW, Patel A.V (1986) Haemagglutinins in marine algae: Lectins or Phenols? In: "Lectins: Biology, Biochemistry, Clin Biochem. 5: 139-145.

[96] Calvete JJ, Costa FHF, Saker-Sampaio S, Moreno Murciano MP, Nagano CS, Cavada BS, Grangeiro TB, Ramos MV, Bloch Jr C, Silveira SB, Freitas BT, Sampaio AH (2000) The amino acid sequence of the agglutinin isolated from the red marine alga *Bryothamnion triquetrum* defines a novel lectin structure. Cell Mol Lif Sci 57: 343-350.

[97] Hori K, Matsubara K, Miyazawa K (2000) Primary structures of two hemagglutinins from the red alga *Hypnea japonica*. Bioch Biophys Acta. 1474: 226-236.

[98] Medina-Ramirez G, Gibbs RV, Calvete JJ (2006) Micro-heterogeneity and molecular assembly of the haemagglutinins from the red algae *Bryothamnion seaforthii* and *B. triquetrum* from the Caribbean Sea. Eur J Phycol. 42: 105-112.

[99] Nagano CS, Debray H, Nascimento KS, Pinto VPT, Cavada BS, Saker-Sampaio S, Farias WRL, Sampaio AH, Calvete JJ (2005) HCA and HML isolated from the red marine algae *Hypnea cervicornis* and *Hypnea musciformis* define a novel lectin family. Protein Sci. 14: 2167-2176.

[100] Kawabuko A, Makino H, Ohnishi J, Hirohara H, Hori K (1999) Occurrence of highly yielded lectins homologous within the genus *Eucheuma*. J Appl Phycol. 11: 149-156.

[101] Hori K, Sato Y, Ito K, Fujiwara Y, Iwamoto Y, Makino H, Kawakubo A (2007) Strict specificity for high-mannose type N-glycans and primary structure of a red alga *Eucheuma serra* lectin. Glycobiology, 17: 479-9.

[102] Mori T, O'Keefe BR, Sowder II RC, Bringans S, Gardella R, Berg S, Cochran P, Turpin JA, Buckheit Jr RW, MacMahon JB, Boyd MR (2005) Isolation and characterization of griffithsin, a novel HIV-inactivating protein, from the red alga *Griffithsia* sp. J Biol Chem, 280: 9345-9353.

[103] Ishihara K, Arai S, Shimada S (2009) cDNA cloning of a lectin-like gene preferentially expressed in freshwater from macroalga *Ulva limnetica* (ulcales, Chlorophyta). Phycol Res 57: 104-110.

[104] Yoon KS, Lee KP, Klochkova TA, Kim GH (2008) Molecular characterization of the lectin, bryohealin, involved in protoplast regeneration of the marine alga *Bryopsis plumosa* (chlorophyta). J Phycol 44: 103-112.

[105] Sato Y, Hirayama M, Morimoto K, Yamamoto N, Okuyama S, Hori K (2011) High Mannose-binding Lectin with preference for the cluster of α1-2 Mannose from green Alga *Boodlea coacta* is potent inhibitor of HIV-1 and Influenza Viruses. J Biol Chem. 286: 19446-19458.

[106] Sugahara T, Ohama Y, Fukuda A, Hayashi M, Kawakubo A, Kato K (2001) The cytotoxic effect of *Eucheuma serra* agglutinin (ESA) on cancer cells and its application to molecular probe for drug delivery system using lipid vesicles. Cytotechnology. 36: 93-99.

[107] Fukuda Y, Sugahara T, Ueno M, Fukuta Y, Ochi Y, Akiyama K, Miyazaki T, Masuda S, Kawakubo A, Kato K (2006) The anti-tumor effect of Euchema serra agglutinin on colon cancer cells *in vitro* and *in vivo*. Anticancer Drugs. 17: 943-7.

[108] Omokawa Y, Miyazaki T, Walde P, Akiyama K, Sugahara T, Masuda S, Inada A, Ohnishi Y, Saeki T, Kato K (2010) In vitro and in vivo anti-tumor effects of novel Span 80 vesicles containing immobilized *Eucheuma serra* agglutinin. Int J Pharm. 15: 157-167.

[109] Pinto VP, Debray H, Dus D, Teixeira EH, de Oliveira TM, Carneiro VA, Teixeira AH, Filho GC, Nagano CS, Nascimento KS, Sampaio AH, Cavada BS (2009) Lectins from the Red Marine Algal Species *Bryothamnion seaforthii* and *Bryothamnion triquetrum* as Tools to Differentiate Human Colon Carcinoma Cells. Adv Pharmacol Sci. 2009: 862162.

[110] Bitencourt FS, Figueiredo JG, Mota MR, Bezerra CC, Silvestre PP, Vale MR, Nascimento KS, Sampaio AH, Nagano CS, Saker-Sampaio S, Farias WR, Cavada BS, Assreuy AM, de Alencar NM (2008) Antinociceptive and anti-inflammatory effects of a mucin-binding agglutinin isolated from the red marine alga *Hypnea cervicornis*. Naunyn Schmiedebergs Arch Pharmacol. 377: 139-48.

[111] Silva LM, Lima V, Holanda ML, Pinheiro PG, Rodrigues JA, Lima ME, Benevides NM (2010) Antinociceptive and anti-inflammatory activities of lectin from marine red alga *Pterocladiella capillacea*. Biol Pharm Bull. 33: 830-5.

[112] Vanderlei ES, Patoilo KK, Lima NA, Lima AP, Rodrigues JA, Silva LM, Lima ME, Lima V, Benevides NM (2010) Antinociceptive and anti-inflammatory activities of lectin from the marine green alga *Caulerpa cupressoides*. Int Immunopharmacol. 10: 1113-8.

[113] Hori K, Matsuda H, Miyazawa K, Ito K (1987) A mitogenic agglutinin from the red alga *Carpopeltis flabellate*. Phytochemistry. 26: 1335-1338.

[114] Liao WR, Lin JY, Shieh WY, Jeng WL, Huang R (2003) Antibiotic activity of lectins from marine algae against marine vibrios. J Ind Microbiol Biotechnol. 30: 433-439.

[115] Teixeira EH, Napimoga MH, Carneiro VA, de Oliveira TM, Nascimento KS, Nagano CS, Souza JB, Havt A, Pinto VP, Gonçalves RB, Farias WR, Saker-Sampaio S, Sampaio AH, Cavada BS (2007) *In vitro* inhibition of oral streptococci binding to the acquired pellicle by algal lectins. J Appl Microbiol. 103: 1001-6.

[116] Neves SA, Dias-Baruff M, Freitas AL, Roque-Barreira MC (2001) Neutrophil migration induced in vivo and in vitro by marine algal lectins. Inflamm Res. 50: 486-90.

[117] Sato Y, Morimoto K, Hirayama M, Hori K (2011) High mannose-specific lectin (KAA-2) from the red alga Kappaphycus alvarezii potently inhibits influenza virus infection in a strain-independent manner. Biochem Biophys Res Commun. 405: 291-6.

[118] Mori T, O'Keefe BR, Sowder RC 2nd, Bringans S, Gardella R, Berg S, Cochran P, Turpin JA, Buckheit RW Jr, McMahon JB, Boyd MR. Isolation and characterization of griffithsin, a novel HIV-inactivating protein, from the red alga *Griffithsia sp*. J Biol Chem. 280: 9345-53.

[119] Emau P, Tian B, O'keefe BR, Mori T, McMahon JB, Palmer KE, Jiang Y, Bekele G, Tsai CC (2007) Griffithsin, a potent HIV entry inhibitor, is an excellent candidate for anti-HIV microbicide. J Med Primatol. 36: 244-53.

[120] O'Keefe BR, Giomarelli B, Barnard DL, Shenoy SR, Chan PK, McMahon JB, Palmer KE, Barnett BW, Meyerholz DK, Wohlford-Lenane CL, McCray PB Jr (2010) Broad-spectrum *in vitro* activity and *in vivo* efficacy of the antiviral protein griffithsin against emerging viruses of the family *Coronaviridae*. J Virol. 84. 2511-21.

[121] Abee T, Kovács AT, Kuipers OP, Van Der Veen S (2010) Biofilm formation and dispersal in Gram-positive bacteria. Curr Opin Biotechnol. 22: 1-8.

[122] Merritt J, Anderson MH, Park NH, Shi W (2001) Bacterial biofilm and dentistry. J Calif Dent Assoc. 29: 355-60.

[123] Donlan RM, Costerton JW (2002) Biofilms: survival mechanisms of clinically relevant microorganisms. Clin Microbiol Rev. 15: 167-193.

[124] Onurdağ FK, Ozkan S, Ozgen S, Olmuş H, Abbasoğlu U (2010) *Candida albicans* and *Pseudomonas aeruginosa* adhesion on soft contact lenses. Graefes Arch Clin Exp Ophthalmol. Dez. 2010.

[125] Prosser BT, Taylor PA, Cix PA, Cluland R (1987) Method of evaluating effects of antibiotics on bacterial biofilms. Antimicrob. Agents Chemother. 31: 1502-1506.

[126] Costerton JW, Lewandowski Z, Caldwell DE, Korber DR, Lappin-Scott HM (1995) Microbial biofilms. Ann. Rev. Microbiol. 49: 711-745.

[127] Mah TF, O'Toole GA (2001) Mechanisms of biofilm resistance to antimicrobial agents. Trends Microbiol. 9(1): 34-39.

[128] Costerton JW, Stewart PS, Greenberg EP (1999) Bacterial biofilms: A common cause of persistent infections. Scien. 284: 1318-1322.

[129] Costerton W, Veeh R, Shirtliff M, Pasmore M, Post C, Ehrlich G (2003) The application of biofilm science to the study and control of chronic bacterial infections. J. Clin. Invest. 112: 146-1477.

[130] Fux CA, Stoodley P, Hall-Stoodley, Costerton JW (2003) Bacterial biofilms: a diagnostic and therapeutic challenge. Exp Rev Anti-infect Ther. 1(4): 667-683.

[131] Alhede M, Bjarnsholt T, Jensen PO (2009) *Pseudomonas aeruginosa* recognizes and responds aggressively to the presence of polymorphonuclear leukocytes. Microbiol. 155: 3500-3508.

[132] Van Gennip M, Christensen LD, Alhede M (2009) Inactivation of the rhlA gene in *Pseudomonas aeruginosa* prevents rhamnolipid production, disabling the protection against polymorphonuclear leukocytes. APMIS. 117: 537-546.

[133] Busscher HJ, Bos R, van der Mei HC (1995) Initial microbial adhesion is a determinant for the strength of biofilm adhesion. FEMS Microbiol. Lett. 128: 229–234.

[134] Wu C, Mishra A, Reardon ME, Huang IH, Counts SC, Das A, Ton-That H (2012) Structural determinants of *Actinomyces sortase* SrtC2 required for membrane localization and assembly of type 2 fimbriae for interbacterial coaggregation and oral biofilm formation. J Bacteriol. Mar 23.

[135] Periasamy S, Kolenbrander PE (2010) Central role of the early colonizer *Veillonella sp.* in establishing multispecies biofilm communities with initial, middle, and late colonizers of enamel. J Bacteriol. 192(12): 2965-72.

[136] Stoodley P, Sauer K, Davies DG, Costerton JW (2002) Biofilms as complex differentiated communities. Annu Rev Microbiol. 56: 187-209.

[137] Heydorn A, Ersboll BK, Hentzer M, Parsek MR, Givskov M, Molin S (2000) Experimental reproducibility in flow-chamber biofilms. Microbiol. 146: 2395-2407.

[138] Xiao J, Klein MI, Falsetta ML, Lu B, Delahunty CM, Yates JR 3rd, Heydorn A, Koo H (2012) The Exopolysaccharide matrix modulates the interaction between 3D architecture and virulence of a mixed-species oral biofilm. PLoS Pathog. 8(4): e1002623.

[139] López D, Vlamakis H, Kolter R (2010) Biofilms. Cold Spr Harb Perspect Biol. 2(7): a000398.

[140] Flemming HC, Neu TR, Wozniak DJ (2007) The EPS matrix: the "house of biofilm cells". J Bacteriol. 189(22): 7945-7.

[141] Flemming HC, Wingender J (2010) The biofilm matrix. Nat Rev Microbiol. 8(9): 623-33.

[142] Fronzes R, Remaut H, Waksman G (2008) Architectures and biogenesis of non-flagellar protein appendages in Gram-negative bacteria. EMBO J. 27: 2271-80.

[143] Dorkhan M, Chávez de Paz LE, Skepö M, Svensäter G, Davies J (2012) Effects of saliva or serum coating on adherence of *Streptococcus oralis* strains to titanium. Microbiol. 158(2): 390-7.

[144] Mukherjee J, Karunakaran E, Biggs CA (2012) Using a multi-faceted approach to determine the changes in bacterial cell surface properties influenced by a biofilm lifestyle. Biofouling. 28(1): 1-14.

[145] Marsh PD (1994) Microbial ecology of dental plaque and its significance in health and disease. Adv. Dent. Res. 8: 263-271.

[146] Nyvad B (1993) Microbial colonization of human tooth surfaces. APMIS Suppl. 32: 1-45.

[147] Thylstrup A, Fejerskov O (1995) Cariologia clínica 2 ed. São Paulo: Santos, Cap. 3, p. 45-49.

[148] Ruhl S, Sandberg AL, Cisar JO (2004) Salivary receptors for the proline-rich protein-binding and lectin-like adhesins of oral actinomyces and streptococci. J Dent Res. 83(6): 505-10.

[149] Jenkinson HF, Lamont RJ (1997) Streptococcal adhesion and colonization. Crit Rev Oral Biol Med. 8(2): 175-200.

[150] Korea GK, Ghigo JM, Beloin C (2011) The sweet connection: Solving the riddle of multiple sugar-binding fimbrial adhesins in *Escherichia coli*: Multiple *E. coli* fimbriae form a versatile arsenal of sugar-binding lectins potentially involved in surface-colonisation and tissue tropism. Bioessays 33: 300-311.

[151] Kolenbrander PE, Andersen RN, Blehert DS, Egland PG, Foster JS, Palmer Jr RJ (2002) Communication among oral bacteria. Microbiol Mol Biol Rev. 66: 486–505.

[152] Ferreira CL, Grześkowiak L, Collado MC, Salminen S (2011) In vitro evaluation of *Lactobacillus gasseri* strains of infant origin on adhesion and aggregation of specific pathogens. J Food Prot. 74(9): 1482-7.

[153] Okuda T, Okuda K, Kokubu E, Kawana T, Saito A, Ishihara K (2012) Synergistic effect on biofilm formation between *Fusobacterium nucleatum* and *Capnocytophaga ochracea*. Anaerobe. 18(1): 157-61.

[154] Handley PS, Rickard AH, High NJ, Leach SA (2001) Coaggregation - is it a universal phenomenon? In: Biofilm community interactions: Chance or Necessity (Gilbert, P., Allison, D., Verran, J., Brading, M. and Walker, J., Eds.), pp. 1–10.

[155] Rodrigues DF, Elimelech M (2009) Role of type 1 fimbriae and mannose in the development of *Escherichia coli* K12 biofilm: from initial cell adhesion to biofilm formation. Biofoul. 25(5): 401-11.

[156] Holmes AR, Gopal PK, Jenkinson HF (1995) Adherence of *Candida albicans* to a cell surface polysaccharide receptor on *Streptococcus gordonii*. Infect Immun. 63(5): 1827-34.

[157] Rosen G, Sela MN (2006) Coaggregation of *Porphyromonas gingivalis* and *Fusobacterium nucleatum* PK 1594 is mediated by capsular polysaccharide and lipopolysaccharide. FEMS Microbiol Lett. 256(2): 304-10.

[158] Ryu JH, Beuchat LR (2005) Biofilm formation by *Escherichia coli* O157:H7 on stainless steel: effect of exopolysaccharide and curli production on its resistance to chlorine. Appl Environ Microbiol. 71(1): 247-54.

[159] Simões LC, Lemos M, Pereira AM, Abreu AC, Saavedra MJ, Simões M (2011) Persister cells in a biofilm treated with a biocide. Biofoul. 27(4): 403-11.

[160] Becer CR (2012) The glycopolymer code: Synthesis of glycopolymers and multivalent carbohydrate-lectin interactions. Macromol Rapid Commun. Apr 16.

[161] Kurz K, Garimorth K, Joannidis M, Fuchs D, Petzer A, Weiss G (2012) Altered immune responses during septicaemia in patients suffering from haematological malignancies. Int J Immunopathol Pharmacol. 25(1): 147-56.

[162] Strathmann, M, Wingender J, Flemming H (2002) Application of fluorescently labelled lectins for the visualization and biochemical characterization of polysaccharides in biofilms of *Pseudomonas aeruginosa*. J. Microbiol. Methods 50: 237-248.

[163] Wawrzynczyk J, Szewczyk E, Norrlöw O, Dey E (2007) Application of enzymes, sodium tripolyphosphate and cation exchange resin for the release of extracellular polymeric substances from sewage sludge. Characterization of the extracted polysaccharides/glycoconjugates by a panel of lectins. J. Biotechnol. 130: 274-281.

[164] Cavalcante TT, Rocha BAM, Carneiro VA, Arruda FVS, AS FN, Sá NC, Nascimento KS, Cavada BS, Teixeira EH (2011) Effect of lectins from Diocleinae subtribe against oral Streptococci. Molecul. 16(5): 3530-43.

[165] Sanford BA; De Feijter AW, Wade MH, Thomas VL (1996) A dual fluorescence technique for visualization of *Staphylococcus epidermidis* biofilm using scanning confocal laser microscopy. J Ind Microbiol. 16(1): 48-56.

[166] Thomas VL, Sanford BA, Moreno R, Ramsay MA (1997) Enzyme-linked lectin sorbent assay measures N-acetyl-D-glucosamine in matrix of biofilm produced by *Staphylococcus epidermidis*. Curr Microbiol. 35(4): 249-54.

[167] Mccoy Jr JP, Varani J, Goldstein IJ (1984) Enzyme-linked lectin assay (ELLA): Detection of carbohydrate groups on the surface of unfixed cells. Exper Cell Res. 1(51): 96-103.

[168] Leriche V, Sibille P, Carpentier B (2000) Use of an enzyme-linked lectin sorbent assay to monitor the shift in polysaccharide composition in bacterial biofilms. Appl Environ Microbiol. 66(5): 1851-6.

[169] Zippel B, Neu TR (2011) Characterization of glycoconjugates of extracellular polymeric substances in tufa-associated biofilms by using fluorescence lectin-binding analysis. Appl Environm Microbiol. 77: 505-516.

[170] Rebiere-Huët J, Di Martino P, Hulen C. (2004) Inhibition of *Pseudomonas aeruginosa* adhesion to fibronectin by PA-IL and monosaccharides: involvement of a lectin-like process. Can J Microbiol. 50(5): 303-12.

[171] Teixeira EH, Napimoga MH, Carneiro VA, de Oliveira TM, Cunha RM, Havt A, Martins JL, Pinto VP, Gonçalves RB, Cavada BS (2006) In vitro inhibition of Streptococci binding to enamel acquired pellicle by plant lectins. J Appl Microbiol. 101(1): 111-6.

[172] Islam B, Khan SN, Naeem A, Sharma V, Khan AU (2009) Novel effect of plant lectins on the inhibition of *Streptococcus mutans* biofilm formation on saliva-coated surface. J Appl Microbiol. 106(5): 1682-9.